Gas Purification
Fourth Edition

Gas Purification

Fourth Edition

Arthur L. Kohl
Fred C. Riesenfeld

AIR PRODUCTS AND CHEMICALS, INC.
ALLENTOWN, PA 18195
ATTN: INFO. SERVICES (R&D #1)

Gulf Publishing Company
Book Division
Houston, London, Paris, Tokyo

Gas Purification
Fourth Edition

Library of Congress Cataloging in Publication Data

Kohl, Arthur L.
    Gas purification.

    Includes bibliographies and index.
    1. Gases—Purification. I. Riesenfeld, Fred C.
II. Title.
TP754.K6   1985   665.7   85-4148
ISBN 0-87201-314-6

Contents

Preface ix

1. Introduction 1

Absorption 2; Adsorption, 21; Chemical Conversion, 24; References, 25

2. Alkanolamines for Hydrogen Sulfide
and Carbon Dioxide Removal 29

Basic Chemistry of the Process, 31; Selection of Process Solutions, 35; Flow Systems, 45; Design of Absorption Column, 49; Design of Stripping Column, 91; Commercial Data, 99; Removal of Organic Sulfur Compounds, 102; The Sulfiban Process, 104; References, 105

3. Mechanical Design and Operation
of Alkanolamine Plants 110

Corrosion, 110; Chemical Losses, 127; Nonacidic-Gas Entrainment in Solution, 140; References, 145

4. Hydrogen Sulfide and Carbon Dioxide
Removal with Ammonia 148

Basic Data, 150; Selective Hydrogen Sulfide Removal, 159; Carbon Dioxide Removal, 177; References, 181

5. Alkaline Salt Solutions for Hydrogen
Sulfide and Carbon Dioxide Absorption 182

Absorption at Ambient Temperature, 183; Absorption at Elevated Temperature, 211; References, 244

6. Water as an Absorbent for Gas Impurities 247

Carbon Dioxide Absorption in Water, 249; Hydrogen Sulfide Removal by Absorption in Water, 265; Absorption of Fluorides, 268; Hydrogen Chloride Absorption, 288; Chlorine Absorption in Water, 291; References, 296

7. Sulfur Dioxide Removal 299

Requirements for SO_2 Removal, 299; Sulfur Trioxide Formation, 302; Process Categories and Economics, 303; Alkaline Earth Processes, 307; Alkali Metal Processes, 340; Sulfur Dioxide—Recovery Processes Employing Ammonia, 359; Aqueous Aluminum Sulfate Processes, 371; Dilute Acid Processes, 374; Sulfur Dioxide—Recovery Processes Employing Aromatic Amines, 378; Sulfur Dioxide Removal by Absorption in Molten Salts, 386; Spray Dryer Processes, 388; Sulfur Dioxide Removal by Dry Sorption with Reaction, 396; Sulfur Dioxide Recovery by Adsorption, 403; Sulfur Dioxide Recovery by Catalytic Oxidation, 409; References, 412

8. Dry Oxidation Processes for Removal of Sulfur Compounds 420

Oxidation to Sulfur, 420; Oxidation to Oxides of Sulfur, 465; References, 479

9. Liquid Phase Oxidation Processes for Hydrogen Sulfide Removal 483

Polythionate Solutions, 484; Iron Oxide Suspensions, 486; Thioarsenate Solutions, 497; Iron Cyanide Solutions and Suspensions, 509; Iron Complex Solutions, 515; Organic Compounds, 519; Sulfur Dioxide, 539; Other Oxidation Processes, 545; References, 548

10. Removal of Basic Nitrogen Compounds from Gas Streams 551

Removal of Ammonia, 554; Removal of Pyridine Bases, 576; References, 580

11. Absorption of Water Vapor by Dehydrating Solutions 582

Glycol Dehydration Processes, 587; Dehydration with Saline Brines, 617; References, 627

12. Gas Dehydration and Purification by Adsorption 630

Water Vapor Adsorption, 632; Use of Molecular Sieves for Gas Purification, 669; Air and Gas Purification with Silica Gel, 682; Organic-Vapor Adsorption on Active Carbon, 684; Adsorption of Hydrocarbons from Gas Streams with Active Carbon, 712; References, 713

13. Catalytic Conversion of Gas Impurities 717

Conversion of Organic Sulfur Compounds to Hydrogen Sulfide, 721; Complete Desulfurization of Hydrocarbon and Synthesis Gas Steams, 753; Catalytic Removal of Carbon Oxides from Synthesis Gases, 754; Catalytic Removal of Acetylenic Compounds from Olefins by Selective Hydrogenation, 762; Purification of Gas Streams by Catalytic Oxidation and Reduction, 767; References, 783

14. Miscellaneous Gas-Purification Techniques 787

Absorption with Complex Formation, 787; Cosorb Process, 808; Low-Temperature Gas-Purification Processes, 809; Removal of Gas Impurities by Condensation and Absorption in Liquid Nitrogen; 810; Removal of Hydrocarbon Vapors by Oil Absorption, 828; Acid-Gas Removal by Physical Absorption in Organic Solvents, 840; References, 875

Appendix: Units and Conversion Factors 880

Index 883

Preface

This fourth edition of *Gas Purification* represents a substantial revision aimed at incorporating all significant advances in this field since publication of the previous edition in 1979. Emphasis has been placed on expanding the coverage for widely used technologies, such as the alkanolamine processes for H_2S and CO_2 removal and lime/limestone-based processes for flue gas desulfurization, while adding descriptions of new processes that have attained commercial status.

Among the new processes included are the use of sterically hindered amines for H_2S and CO_2 absorption; the Chiyoda Thoroughbred 121, Dowa basic aluminum sulfate, and lime slurry spray dryer processes for SO_2 removal from stack gas; the Sulfint chelated iron process, Unisulf and Fumaks catalytic oxidation processes, and EIC copper sulfate process for absorbing and reacting H_2S; the Selectox and MCRC processes for gas-phase conversion of sulfur compounds to elemental sulfur; the Cosorb process for removing and recovering CO; the Purasiv HR fluidized bed activated carbon system for adsorbing solvent vapors; the CNG process which uses liquefied CO_2 as the solvent for H_2S removal; the Sepasolv MPE physical solvent process for acid gas removal; and the membrane permeation process for a variety of gas separation operations.

In some areas, the field of gas purification has expanded to the extent that detailed coverage in a single book is impractical. Fortunately, other publications have appeared to fill the gap and references to these are provided throughout. Examples are the application of the theory of gas absorption with chemical reaction to the design of gas purification systems, and the use of computers for process design, equipment sizing, and cost estimation.

It is the aim of this book to provide a practical engineering description of techniques and processes in widespread use. Subject matter is limited to the removal from gas streams of gas-phase impurities which are present in relatively minor proportions. The removal of discrete solid or liquid particles is not discussed, nor are processes which would more appropriately be classified as separation rather than purification.

No attempt has been made to define the patent status of the processes described. Many of the basic patents have expired on well-known processes; however, patented improvements may be critical to commercial application. In fact, there appears to be a growing trend for older "generic" processes to be offered as proprietary systems based on relatively minor changes in process configuration and reagent composition.

In general, data are presented in the units in the original publication. Practically, this means that most U.S. data on commercial operations are given in English units while foreign data and U.S. scientific data are presented in metric or, occasionally, SI units. To aid in the conversion between systems, a table of units and conversion factors covering those most frequently used in gas purification plant design calculations is included as an appendix to this edition.

The assistance of many individuals who contributed material and suggested improvements is gratefully acknowledged. Thanks are also due to the organizations who have graciously provided data and given permission for reproducing charts and figures and to Rockwell International and the Ralph M. Parsons Company for their generous help during preparation of the manuscript. Finally, we wish to express our gratitude to our wives, Evelyn Kohl and Elsie Riesenfeld, for their valuable support and remarkable patience during the preparation of all four editions of this book.

Arthur L. Kohl
Fred C. Riesenfeld

1

Introduction

Gas purification, as discussed in this text, involves the removal of vapor-phase impurities from gas streams. The processes which have been developed to accomplish gas purification vary from simple once-through wash operations to complex multiple-step recycle systems. In many cases, the process complexities arise from the need for recovery of the impurity or reuse of the material employed to remove it. The primary operation of the gas-purification processes generally falls into one of the following three categories:

1. Absorption into liquid
2. Adsorption on a solid
3. Chemical conversion to another compound

The designer of gas-purification process units must, therefore, be particularly familiar with the principles of design of absorbers, adsorption units, and reactors. In addition, such unit operations as distillation, crystallization, and filtration may be encountered in the overall process scheme. The basic principles of design of equipment to handle the various unit operations are well covered in standard engineering texts, but the data required to apply these principles to specific cases are not always available. Furthermore, commercial applications of the various processes are frequently plagued by unpredictable problems of corrosion, side reactions, foaming, catalyst deactivation, and the like, so that actual plant (or pilot-plant) operating data are invaluable adjuncts to a theoretical design. Operating data and empirical design correlations are, therefore, presented in addition to basic design information wherever possible in the descriptions of processes given in subsequent chapters.

A valuable compilation of shortcut design techniques, with sample calculations, for processes commonly used in the purification of hydrocarbon gases and liquids has been published by Maddox (60). This reference also includes discussions of operating problems, and provides guidelines for the selection of appropriate processes for specific purification requirements.

In the absence of simple correlations and shortcut techniques based on experience, it is necessary to fall back upon chemical engineering fundamentals for the design of plants to use the gas purification processes which are described. The remaining sections of this chapter are, therefore, intended to provide a brief review of the three principal operations involved in gas purification and, more importantly, to provide references which offer more complete technical discussions of these subjects.

Absorption

Absorption is probably the most important gas-purification technique and is common to a great number of processes. It involves the transfer of a substance from the gaseous to the liquid phase through the phase boundary. The absorbed material may dissolve physically in the liquid or react chemically with it. Desorption (or stripping) represents a special case of the same operation in which the material moves from the liquid to the gaseous phase.

The great majority of absorbers used for gas-purification operations are packed, plate, or spray towers. These absorber types are interchangeable to a considerable extent although certain specific conditions may favor one over the other.

In general, packed towers are preferred for small installations, corrosive service, liquids with a tendency to foam, very high liquid/gas ratios, and applications where a low pressure drop is desired. Although many packing shapes are available, the most commonly used are rings, saddles, and grids. The formerly popular raschig rings and Berl saddles are being replaced to a large extent by proprietary shapes such as the Pall ring developed by B.A.S.F., Germany, and Intalox saddles developed by the United States Stoneware Co.

Rings and saddles are randomly packed. Systematically packed elements (also called ordered packing) are also used for absorption. An example is Flexipack offered by Koch Engineering Company. Each Flexipack element is a circular pad that occupies the entire cross section of the column and is composed of vertical corrugated sheets arranged so that corrugations of adjacent sheets are inclined in opposite directions. This type of packing is particularly recommended as a substitute for poorly performing dumped packing in existing towers to improve performance and reduce pressure drop.

Plate columns are frequently more economical because a higher gas velocity can usually be tolerated, and therefore a column of smaller diameter is required. They are particularly suitable for large installations; clean, noncorrosive, nonfoaming liquids; and low liquid-flow rate applications. Perforated trays (also called sieve trays) are probably the most widely used because of their simplicity and low cost. The formerly popular bubble-cap design is now used primarily for columns requiring a very low liquid-flow rate. In order to overcome some of the limitations of simple bubble-cap and perforated-tray designs, a number of special tray designs have been developed including the Turbogrid (1), Uniflux (2), valve (3), Flexitray (4), Kittel plate (5) (see also Chapter 4), and shed or shower trays (see Chapter 6). The latter are specifically applicable to very high liquid-flow rates such as those encountered in the absorption of carbon dioxide from ammonia-synthesis gas with water.

Spray contactors are of importance primarily where pressure drop is a major consideration and where solid particles are present in the gas, e.g., in the scrubbing of exhaust gas at atmospheric pressure. Several types of spray contactors, including the venturi scrubber and ejector, are utilized for the removal of hydrofluoric acid, silicon tetrafluoride, and sulfur dioxide from stack gases (see Chapters 6 and 7).

In addition to conventional packed, plate, and spray towers, specialized contactors have been developed to meet specific process requirements. These include impingement-plate scrubbers, which have been studied for SO_2 removal (6,7), and turbulent contact scrubbers (8,9,61), which use a mobile packing and are particularly suitable for application where particulate removal and mass transfer are required.

A categorization and simplified guide to the selection of gas-liquid contactors is presented in Table 1-1 which is based on the work of Frank. (62) For purposes of the selection guide, the turbulent contact absorber (TCA) design can be considered a special case of downcomerless trays and would generally be applicable to the same type of service (i.e., solids present, dirty or polymerizing solution, high liquid rates, and low cost). Venturi scrubbers represent a single-contact stage and are generally followed by a packed, perforated plate, or turbulent contact absorber if effective gas absorption is required.

As a further generalized guide to the selection of absorbers, the relative costs of six types of tray columns and of ten types of column packing are presented in Table 1-2 (63). It should be noted that the relative costs are for equal column volumes, not for equivalent performance.

Detailed design procedures for absorption operations are adequately described in such texts as *The Chemical Engineer's Handbook* (10), Sherwood and Pigford (11), and Treybal (12), and only a brief review of the principal design equations is presented here.

Column Height. The concept of absorption coefficient, which is the most convenient approach for packed-tower design, is based upon a two-film theory originally proposed by Whitman (13). It is assumed that the gas and liquid are in equilibrium at the interface and that thin films separate the inter-

Table 1-1. Selection Guide for Gas-Liquid Contactors (based on data of Frank) (62)

Conditions in Column	Rating of Column Internals					
	Staged Columns		Differential Columns		Pseudo-Equilibrium	
	Perforated, or Valve Trays	Bubble Cap or Tunnel Trays	Randomly Packed	Systematically Packed	Downcomer-less	Disc and Doughnut
Low pressure (<100 mm Hg)	2	1	2	3	0	1
Moderate pressure	3	2	2	1	1	1
High pressure (>50% of critical)	3	2	2	0	2	0
High turndown ratio	2	3	1	2	0	1
Low liquid rates	1	3	1	2	0	0
Foaming systems	2	1	3	0	2	1
Internal tower cooling	2	3	1	0	1	0
Solids present	2	1	1	0	3	1
Dirty or polymerized solution	2	1	1	0	3	2
Multiple feeds and sidestreams	3	3	1	0	2	1
High liquid rates (scrubbing)	2	1	3	0	3	2
Small diameter columns	1	1	3	2	1	1
Columns with diameter 3-10 ft	3	2	2	2	2	1
Large diameter columns	3	1	2	1	2	1

Corrosive fluids	2	2	3	1	2	2
Viscous fluids (at boiling pt)	2	1	3	0	1	0
Low ΔP (efficiency no concern)	1	0	2	2	0	3
Expanded column capacity	2	0	2	3	2	0
Low cost (performance no concern)	2	1	2	1	3	3
Available design procedures	3	2	2	1	1	1

Notes:

Rating key: 0 - Do not use
            1 - Evaluate carefully
            2 - Usually applicable
            3 - Best selection

Staged columns: Tray columns with separate liquid and vapor flow paths.
  Common types: Bubble cap, sieve, valve.
  Proprietary types: Angle, Uniflux, Montz, Linde, Thorman, Jet

Differential columns: True countercurrent flow of gas and liquid.
  Randomly packed: Raschig rings, saddles, slotted rings, Tellerettes, Maspac.
  Systematically packed: Flexipac, Goodloe, Hyperfil, Sulzer, Glitch Grid.

Pseudo-equilibrium stages: Countercurrent flow of gas and liquid with discrete trays.
  Downcomerless trays: Perforated, Turbogrid, Ripple.
  Low pressure drop trays: Disc and doughnut, shower deck.

Special devices (not rated in table):
  Venturi scrubber, turbulent contact absorber, marble bed absorber, horizontal spray chamber.

Table 1-2. Relative Costs of Columns (Data of Blecker and Nichols (63) based on 1972 Technology)

Relative Costs of Tray Columns (for equal diameter and height)

Bubble Cap	1.00
Koch Kascade	1.243
Plate Tray	0.842
Sieve Tray	0.874
Turbogrid	0.855
Valve Tray	0.911

Relative Costs of Column Packings (installed cost for equal volumes of packing)

	1 in. Dia.	2 in. Dia.
Berl saddles, stoneware	13.10	—
Berl saddles, steel	26.30	—
Berl saddles, stainless steel	32.80	—
Intalox saddles, ceramic	13.30	10.40
Pall rings, polypropylene	36.90	26.30
Pall rings, stainless steel	13.60	9.80
Raschig rings, stoneware	6.30	4.38
Raschig rings, stainless steel	15.70	10.90
Raschig rings, steel	12.60	8.79
Tellerettes, HD polyethylene	26.30	

face from the main bodies of the two phases. Two absorption coefficients are then defined as k_L, the quantity of material transferred through the liquid film per unit time, per unit area, per unit of driving force in terms of liquid concentration; and k_G, the quantity transferred through the gas film per unit time, per unit area, per unit of driving force in terms of pressure. Since the quantity of material transferred from the body of the gas to the interface must equal the quantity transferred from the interface to the body of the liquid, the following relationship holds:

$$N_A = k_G(p - p_i) = k_L(c_i - c) \tag{1-1}$$

Where N_A = quantity of component A transferred per unit time, per unit area

p = partial pressure of A in main body of gas

p_i = partial pressure of A in gas at interface

c = concentration of A in main body of liquid

c_i = concentration of A in liquid at interface

Any consistent set of units may be used; however, it is convenient to express p in atmospheres and c in pound moles per cubic foot, in which case k_G

is expressed as lb moles/(hr)(sq ft)(atm) and k_L as lb moles/(hr)(sq ft)(lb moles/cu ft).

The use of Equation (1-1) for design requires a knowledge of both k_G and k_L as well as the equilibrium relationship and the interfacial area per unit volume of absorber. Although these factors can be estimated for special design cases, it is more practical to use overall coefficients which are based on the total driving force from the main body of the gas to the main body of the liquid and which relate directly to the contactor volume rather than to the interfacial area. These overall coefficients, $K_G a$ and $K_L a$, are defined as follows:

$$N_A a \, dV = K_G a(p - p_e) \, dV = K_L a(c_e - c) \, dV \qquad (1\text{-}2)$$

where a = interfacial area per unit volume of absorber

p_e = partial pressure of A in equilibrium with a solution having the composition of main body of liquid

c_e = concentration of A in a solution in equilibrium with main body of gas

V = volume of packing

The overall coefficients are related to the individual film coefficients as follows:

$$\frac{1}{K_G a} = \frac{1}{k_G a} + \frac{H}{k_L a} \qquad (1\text{-}3)$$

$$\frac{1}{K_L a} = \frac{1}{k_L a} + \frac{1}{H k_G a} \qquad (1\text{-}4)$$

where H is Henry's law constant, p_i/c_i, or, in cases where Henry's law does not hold, $(p_i - p_e)/(c_i - c_e)$.

The use of overall coefficients is strictly valid only where the equilibrium line is straight over the operating region. However, because of their convenience, they are widely used for reporting test data, particularly on commercial equipment, and are therefore very useful for design.

In order to apply absorption coefficient data to the design of commercial columns, it is necessary to consider the changes in liquid and gas compositions which occur over the length of the column. This involves equating the quantity of material transferred (as indicated by gas- or liquid-composition change) to the quantity indicated to be transferred on the basis of the absorption coefficient and driving forces and then integrating this equation over the length of the column. For the individual film coefficients, this results in the following expression for column height (11):

$$h = G'_M \int_{p_2}^{p_1} \frac{p_{BM}\, dp}{k'_G a (P - p)^2 (p - p_e)}$$

$$= \frac{L}{\rho_L} \int_{c_2}^{c_1} \frac{dc}{k_L a (c_e - c)} \tag{1-5}$$

where h = height of packed zone, ft

$\quad G'_M$ = superficial molar mass velocity of inert gas, lb moles/(hr)(sq ft)

$\quad p_1$ = partial pressure of solute in entering gas, atm

$\quad p_2$ = partial pressure of solute in leaving gas, atm

$\quad P$ = total pressure of system, atm

$\quad p$ = partial pressure of solute in main gas stream, atm

$\quad p_e$ = partial pressure of solute in equilibrium with main body of solution, atm

$\quad p_{BM}$ = log mean of inert gas pressures

$\quad k'_G = k_G\,(p_{BM}/p)$ = special mass-transfer coefficient which is independent of gas composition, lb moles/(hr)(sq ft)(atm)

$\quad L$ = liquid-flow rate, lb/(hr)(sq ft)

$\quad \rho_L$ = liquid density, lb/cu ft (assumed constant)

$\quad c_1$ = solute concentration in liquid leaving bottom of column, lb moles/cu ft

$\quad c_2$ = solute concentration in liquid fed to top of column, lb moles/cu ft

$\quad c$ = solute concentration in main body of liquid, lb moles/cu ft

$\quad c_e$ = concentration of solute in liquid phase in equilibrium with main body of gas, lb moles/cu ft

The above equations may be integrated graphically by a method developed by Walker, Lewis, McAdams, and Gilliland (14) and which is also described by Sherwood and Pigford (11). Simplified forms of the equations have been developed which are much more readily used and which are sufficiently accurate for most engineering-design calculations. Two of these forms which are particularly adapted to low gas and liquid concentrations, such as are frequently encountered in gas purification, are presented below. These assume that the following conditions hold:

1. The equilibrium curve is linear over the range of concentrations encountered (and therefore overall coefficients can be used).
2. The partial pressure of the inert gas is essentially constant over the length of the column.
3. The solute contents of gaseous and liquid phases are sufficiently low that the partial pressure and liquid concentration values may be assumed proportional to the corresponding values when expressed in terms of moles of solute per mole of inert gas (or of solvent).

In terms of the overall gas coefficient and gas-phase compositions, the tower height can be estimated by Equation 1-6

$$h = \frac{G_M}{K_G a P} \int_{y_2}^{y_1} \frac{dy}{y - y_e} \tag{1-6}$$

or, where the overall liquid absorption coefficient is available, the column height may be calculated in terms of liquid-phase compositions.

$$h = \frac{L_M}{\rho_M K_L a} \int_{x_2}^{x_1} \frac{dx}{x_e - x} \tag{1-7}$$

In Equations (1-6) and (1-7) y and x refer to the mole fractions of solute in the gas and liquid streams, respectively; L_M and ρ_M represent the molar values of liquid-flow rate and density, i.e., lb moles/(hr)(sq ft) and lb moles/cu ft. The subscript 1 refers to the bottom of the column, subscript 2 to the top of the column, and subscript e to the equilibrium composition with respect to the main body of the other phase. The other symbols have the same significance as in the previous equations. In general, it is preferable to employ the overall gas-film coefficient for cases where the gas-film resistance is predominant and the overall liquid coefficient for cases where the principal resistance to absorption is in the liquid phase.

The above equations may be solved by relatively simple graphical integration. However, a further simplification, which can frequently be employed, is the use of a logarithmic mean driving force in the rate equation rather than graphical integration. This can be shown to be theoretically correct where the equilibrium curve and operating line are linear over the composition range of the column. The equations then reduce to

$$h = \frac{G_M(y_1 - y_2)}{K_G a P(y - y_e)_{LM}} \tag{1-8}$$

or

$$h = \frac{L_M(x_1 - x_2)}{\rho_M K_L a(x_e - x)_{LM}} \tag{1-9}$$

where $(y - y_e)_{LM}$ and $(x_e - x)_{LM}$ are equal to the logarithmic mean of the driving forces at the top and bottom of the column. Although not theoretically correct, the logarithmic mean driving force is often used to correlate $K_G a$ values for systems where the equilibrium curve is not a straight line and even for cases of absorption with chemical reaction. This greatly simplifies data reduction but can lead to serious errors. In general, the procedure is useful for comparing similar systems within narrow ranges of liquid composition and gas partial pressure.

A considerable amount of data on absorption-column performance is presented in terms of the "height of the transfer unit" (HTU), and design procedures based on this concept are preferred by many because of their simplicity and similarity to plate-column calculation methods. The basic concept which was originally introduced by Chilton and Colburn (15) is that the calculation of column height invariably requires the integration of a relationship such as (from Equation 1-6):

$$\int_{y_2}^{y_1} \frac{dy}{y - y_e}$$

The dimensionless value obtained from the integration is a measure of the difficulty of the gas-absorption operation. In the above case, it is called the number of transfer units based on an overall gas driving force, N_{OG}, and Equation 1-6 can be reduced to

$$h = \frac{G_M}{K_G aP} N_{OG} \qquad (1\text{-}10)$$

The HTU for this case (based on an overall gas-phase driving force) is then defined as

$$H_{OG} = \frac{h}{N_{OG}} = \frac{G_M}{K_G aP} \qquad (1\text{-}11)$$

Since N_{OG} is dimensionless, H_{OG} will have the same units as h. Similarly, for the overall liquid case

$$H_{OL} = \frac{h}{N_{OL}} - \frac{L_M}{\rho_M K_L a} \qquad (1\text{-}12)$$

As in the calculation of column height from $K_G a$ or $K_L a$ data, it is theoretically correct to use a logarithmic mean driving force when both the equilibrium and operating lines are straight. For this case, the number of transfer units (overall gas) may be calculated from the simple expression

$$N_{OG} = \frac{y_1 - y_2}{(y - y_e)_{LM}} \qquad (1\text{-}13)$$

This equation may be combined with the equilibrium relation

$$y_e = mx$$

and the material-balance expression

$$L_M(x_1 - x) = G_M(y_1 - y)$$

to eliminate the need for values of y_e. The resulting equation which was proposed by Colburn (16) is given below:

$$N_{OG} = \frac{\ln\left[\left(1 - \dfrac{mG_M}{L_M}\right)\left(\dfrac{y_1 - mx_2}{y_2 - mx_2}\right) + \dfrac{mG_M}{L_M}\right]}{1 - (mG_M/L_M)} \tag{1-14}$$

where N_{OG} = number of overall transfer units

   m = slope of equilibrium curve dy_e/dx

   x_2 = mole fraction solute in liquid fed to top of column

   y_1 = mole fraction solute in gas fed to bottom of column

   y_2 = mole fraction in gas leaving top of column

   G_M = superficial molar mass velocity of gas stream, lb moles/(hr)(sq ft)

   L_M = superficial molar mass velocity of liquid stream, lb moles/(hr)(sq ft)

It will be noted that the parameter mG_M/L_M appears several times in Equation 1-14. This parameter can be considered to be the ratio of m, the slope of the equilibrium curve, to L_M/G_M, the slope of a line representing the material-balance equation (the operating line).

A considerable number of alternate equations and graphical techniques have been developed to calculate N_{OG} from other data and for other design conditions (17, 18). A summary of useful design equations for transfer-unit calculations is presented by Sherwood et al. (19).

The HTU concept can also be employed for analysis of the contributions of the individual film resistances although, in general, the individual absorption coefficients are preferred for basic studies. Values of N_{OG} are particularly useful for expressing the performance of equipment in which the volume is not of fundamental importance. In spray chambers, for example (see Chapter 6), the effectiveness of the equipment is more a function of liquid-flow rate and spray-nozzle pressure than of tower volume. The use of a volume coefficient of absorption for such units is quite meaningless.

The calculation of packed column height by the techniques discussed above requires a knowledge of either the overall coefficient (e.g., K_Ga) or the height of a transfer unit (e.g., H_{OG}), and estimation of these values is usually the most difficult column design task. Although some success has been achieved in predicting packed tower mass transfer coefficients from a purely theoretical basis (20), the use of empirical correlations and experimental data represents the usual design practice. Absorption coefficient data are therefore presented whenever possible for processes described in subsequent chapters. Examples of K_Ga values for a variety of gas absorption operations are presented in Table 1-3 which is based on the data of Eckert et al. (21).

Generalized correlations, based on the individual mass transfer units, have been proposed by Onda et al. (70), Bolles and Fair (71), and Bravo and Fair (72). The correlations cover commonly used packing such as Raschig rings, Berl saddles, Pall rings, and related configurations. The key element in the development of generalized correlations has been the definition of a, the effective interfacial area.

The basic design concept for plate columns is the "theoretical plate." This concept is based on the assumption that, with a theoretically perfect contact plate, the gas and liquid leaving will be in equilibrium. Although this assumption does not exactly represent the operation of any actual plate (where much of the gas will not even come in contact with the leaving liquid), it greatly simplifies the design procedure, and the departure of actual plates from this ideal situation can conveniently be accounted for by an expression known as the "plate efficiency."

Table 1-3. Typical K_ga Values for Packed Absorption Columns

Absorbed Gas	Absorbent	K_Ga* lb moles/hr ft³ atm
Cl_2	$H_2O \cdot NaOH$	20.0
HCl	H_2O	16.0
NH_3	H_2O	13.0
H_2S	$H_2O \cdot MEA$	8.0
SO_2	$H_2O \cdot NaOH$	7.0
H_2S	$H_2O \cdot DEA$	5.0
CO_2	$H_2O \cdot KOH$	3.10
CO_2	$H_2O \cdot MEA$	2.50
CO_2	$H_2O \cdot NaOH$	2.25
H_2S	H_2O	0.400
SO_2	H_2O	0.317
Cl_2	H_2O	0.138
CO_2	H_2O	0.072
O_2	H_2O	0.0072

Source: Eckert et al. (21)

*K_Ga values for 1½ inch Intalox saddles at 25 percent completion of absorption reaction.

The simplest design procedure for estimating the number of theoretical plates for a required absorption job is frequently a graphical analysis such as shown in Chapter 11 for the removal of water vapor from gas with triethylene glycol solution. Analytical procedures have also been developed for special cases which closely resemble those employed for calculating the number of transfer units. A particularly useful equation suggested by Colburn (16) for the case of low solute concentrations and a straight equilibrium line is given below:

$$N_P = \frac{\ln\left[\left(1 - \frac{mG_M}{L_M}\right)\left(\frac{y_1 - mx_2}{y_2 - mx_2}\right) + \frac{mG_M}{L_M}\right]}{\ln\left(L_M/mG_M\right)} \qquad (1\text{-}15)$$

where N_P = number of theoretical plates and the other symbols have the same meaning as in Equation 1-14.

The parameter mG_M/L_M represents the ratio of the slope of the equilibrium curve to the slope of the operating line. When it is greater than unity, complete removal of the solute cannot be effected in the absorber. As can be seen, this equation closely resembles the one proposed for calculating the number of overall gas-transfer units based on the same assumptions.

As mentioned above, the number of actual plates in an absorber is related to the number of theoretical plates by a factor known as the "plate efficiency." In its simplest definition, the "overall plate efficiency" is defined as "the ratio of theoretical to actual plates required for a given separation." For individual plates, the Murphree vapor efficiency (22) more closely relates actual performance to the theoretical-plate standard. It is defined by the equation:

$$E_{MV} = \frac{y_p - y_{p+1}}{y_{pe} - y_{p+1}} \qquad (1\text{-}16)$$

where y_p = average mole fraction of solute in gas leaving plate

y_{p+1} = average mole fraction of solute in gas entering plate (leaving plate below)

y_{pe} = mole fraction of solute in gas in equilibrium with liquid leaving plate

Murphree plate efficiency values can be used to correct the individual steps in graphical analyses of the number of plates required. The overall efficiency, on the other hand, can only be used after the total number of theoretical plates has been calculated by a graphical or analytical technique. When operating and equilibrium lines are nearly parallel, the two efficiencies can be considered to be equivalent. Under other conditions they may vary widely.

Principal factors affecting plate efficiencies are gas solubility and liquid viscosity, and a correlation based on these two variables has been developed by O'Connell (23). His correlation for absorbers is reproduced in Figure 1-1. Unfortunately, other factors such as the absorption mechanism, liquid depth, gas velocity, bubble-plate design, and liquid velocity also influence the plate efficiency so that no simple correlation can adequately cover all cases. A comprehensive study of bubble-tray efficiency has been made by the Distillation Subcommittee of the American Institute of Chemical Engineers. The Design Manual (24) resulting from this work provides a standardized procedure for estimating efficiency which takes into account

1. The rate of mass transfer in the gas phase
2. The rate of mass transfer in the liquid phase
3. The degree of liquid mixing on the tray
4. The magnitude of liquid entrainment between the trays

The simplified design equations presented in the preceding discussion, specifically 1-8 and 1-9 for column height using overall film coefficients, Equations 1-13 and 1-14 for calculating the number of overall transfer units, and Equation 1-15 for calculating the number of theoretical plates, involve the assumption of a linear equilibrium line. In actual practice, particularly in gas-purification operations, the equilibrium line may be far from linear; and in many cases, the equilibrium line may have little significance in defining the driving force for mass transfer because of the occurrence of chemical reactions in solution. Such reactions can range from hydration involving only a portion of the dissolved molecules (e.g., chlorine absorption in water; see Chapter 6) to irreversible neutralization in which the compound formed exerts essentially zero vapor pressure (e.g., carbon dioxide in strong alkali; see Chapter 5).

Properly designed sieve trays are generally somewhat more efficient than bubble cap trays. A simplified approach for predicting the efficiency of sieve trays is given by Zuiderweg (73) who presents a series of correlations defining their overall performance. The Zuiderweg study relies heavily on data released by Fractionation Research, Inc. (FRI) on the performance of two types of sieve trays (74).

The effect of a chemical reaction in the liquid phase is generally to increase the liquid-film absorption coefficient over that which would be observed with simple physical absorption. With very slow reactions, however (such as that between carbon dioxide and water), the dissolved molecules apparently migrate well into the body of the liquid before reaction occurs so that the overall mass transfer rate is not appreciably increased by the occurrence of the chemical reaction. In this case, the liquid-film resistance is the controlling factor, the liquid at the interface can be assumed to be in

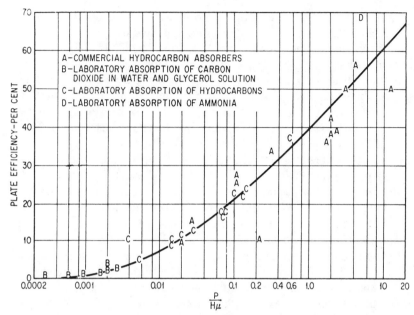

Figure 1-1. Correlation of overall plate efficiencies for commercial and laboratory absorbers; H = Henry's law constant in atm/(lb mole/cu ft), P = total pressure in atmospheres, and μ = liquid viscosity in centipoises. *From O'Connell (23).*

equilibrium with the gas, and the rate of mass transfer is governed by the molecular CO_2 concentration-gradient between the interface and the body of the liquid. At the other extreme are very rapid reactions (such as those of ammonia with strong acids) where the dissolved molecules migrate only a very short distance before reaction occurs. The location of the reaction zone (and the value of the absorption coefficient) will depend primarily upon the diffusion rate of reactants and reaction products to and from the reaction zone, the concentration of solute at the interface, and the concentration of the reactant in the body of the liquid. However, since the distance which the solute must diffuse into the liquid is extremely small compared to the distance which it would have to travel for simple physical absorption, a high liquid-film coefficient is observed, and, in many cases, the gas-film resistance becomes the controlling factor.

Since the effect of chemical reaction is to increase the liquid film coefficient, k_L, over the value it would have in the absence of chemical reaction, k_L^o, a common approach is to utilize the ratio, k_L/k_L^o, in correlations. This ratio is called the enhancement factor. Both k_L and k_L^o are affected by the fluid mechanics, but fortunately their ratio, E, has been found to be relatively independent of these factors. It is primarily a function of concentrations, reaction rates, and diffusivities in the liquid phase.

The theoretical evaluation of absorption followed by liquid-phase chemical reaction has received a great deal of attention although the results are not yet routinely useful for design purposes. Early studies of several reaction types were made by Hatta (25, 26) and Van Krevelen and Hoftijzer (27). This work has been expanded by more recent investigators to cover reversible and irreversible reactions, various reaction orders, and reaction rates from very slow to instantaneous. Important contributions have been made by Perry and Pigford (28), Brian et al. (29), Gilliland et al. (30), Brian (31), Danckwerts and Gillham (32), Decoursey (75), Matheron and Sandall (76), and Olander (77). The application of the theory to specific gas purification cases has been described by Joshi et al. (78) (absorption of CO_2 in hot potassium carbonate solution), and by Ouwerkerk (79) (selective absorption of H_2S in the presence of CO_2 into amine solutions).

Stripping in the presence of chemical reaction has been considered by Astarita and Savage (80), Savage, Astarita, and Joshi (81), and Weiland et al. (82). In general, it is concluded that the same mathematical procedures may be used for stripping as for absorption; however, the results may be quite different because of the different ranges of parameters involved. It is always necessary to consider reaction reversibility in the calculation of stripping with chemical reaction.

It is beyond the scope of this introductory discussion to present even a listing of the numerous mathematical equations developed to correlate the effects of chemical reactions on mass transfer. Detailed equations and examples of their application are presented in comprehensive books on the subject by Astarita (83), Danckwerts (84), and Astarita, Savage, and Bisio (85).

Column Diameter. The diameter of packed columns is usually established on the basis of flooding correlations such as those developed by Sherwood, Shipley, and Holloway (33), Elgin and Weiss (34), Lobo et al. (35), and Eckert (36, 37).

The Sherwood, Shipley, and Holloway study presented the basic form of the correlation; the other investigators developed improved modifications. Eckert's correlation appears to provide more predictable performance in designs because it is based on a packing factor (F) which is derived from an operating range of bed performance rather than flooding conditions. This correlation is reproduced in Figure 1-2, and recommended packing factor values are presented in Table 1-4. In using the correlation of Figure 1-2, it is normally considered good practice to design for a gas rate not over about 75 percent of that indicated to cause flooding in order to allow for the accumulation of sediment, foaming, and other operating difficulties. With special in-

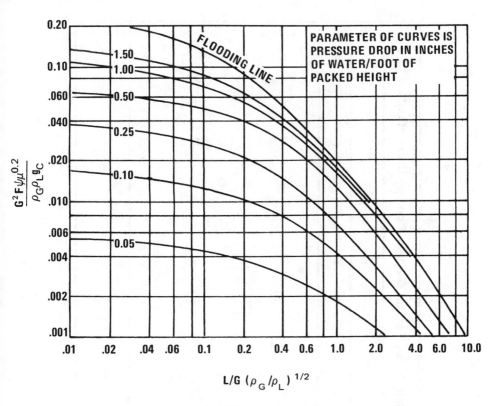

$$\frac{G^2 F \psi \mu^{0.2}}{\rho_G \rho_L g_c}$$

L/G $(\rho_G / \rho_L)^{1/2}$

L = LIQUID RATE, LB/SEC, SQ FT
G = GAS RATE, LB/SEC, SQ FT
ρ_L = LIQUID DENSITY, LB/CU FT
ρ_G = GAS DENSITY, LB/CU FT
F = PACKING FACTOR
μ = VISCOSITY OF LIQUID, CENTIPOISE
ψ = RATIO, DENSITY OF WATER/DENSITY OF LIQUID
g_c = GRAVITATIONAL CONSTANT = 32.2

Figure 1-2. Generalized pressure drop and flooding correlation for packed towers. *From Eckert (36).*

strumentation and a clean system, it is possible to operate up to as high as 90 percent of flooding. On the other hand, vacuum systems are ordinarily designed for very low-pressure drop, and many gas-purification operations, which operate near atmospheric pressure, are designed well below the flooding point in order to maintain a relatively low-pressure drop (e.g., 0.2 to 0.4 in. water per foot).

Table 1-4. Values of the Packing Factor, F, for Packed Column Flooding Correlation*

Type packing	Material	Nominal packing size (in.)										
		1/4	3/8	1/2	5/8	3/4	1	1¼	1½	2	3	3½
Super Intalox	Ceramic	—	—	—	—	—	60	—	—	30	—	—
Intalox saddles	Ceramic	725	330	200	—	145	98	—	52	40	22	—
Intalox saddles	Plastic	—	—	—	—	—	33	—	—	21	16	—
Raschig rings	Ceramic	1,600	1,000	580	380	255	155	125	95	65	37	—
(wall thickness, in.)		(1/16)	(1/16)	(3/32)	(3/32)	(3/32)	(1/8)	(3/16)	(3/16)	(1/4)	(3/8)	
Berl saddles	Ceramic	900	—	240	—	170	110	—	65	45	—	—
Pall rings	Plastic	—	—	—	97	—	52	—	32	25	—	—
Pall rings	Metal	—	—	—	70	—	48	—	28	20	—	16
Raschig rings, 1/32-in. wall	Metal	700	390	300	170	155	115	110	—	57	32	—
Raschig rings, 1/16-in. wall	Metal	—	—	410	290	220	137	—	83	—	—	—
Tellerettes	Plastic					No nominal size				45		
Maspac	Plastic					No nominal size				30(FN-200)		20(FN-90)

*Data of Eckert (37). Values given are for wet and dump packed packings in 16- and 30-in. ID towers. Some values are extrapolated.

The design of tray columns is somewhat more complicated than packed columns due to the internal mechanical features of the trays and the complex hydraulic requirements imposed by the gas and liquid flow patterns. In addition to column diameter, it is necessary to consider such factors as tray spacing, active tray area, hole size and number (for sieve trays), tray pressure drop, turndown limits, froth height, and the design of weirs and downcomers. Detailed discussions of tray column design procedures are given in a number of chemical engineering texts (10, 12, 38) and in several excellent review articles which have appeared in the literature (39, 62). Design data and procedures have also been published for specific tray column types, including sieve trays (64), valve trays (65), and Linde slotted sieve trays (66).

Most tray column design procedures are based on limiting the gas velocity through the available column cross section (A_n) to a value which will not cause flooding or excessive entrainment. The exact column diameter, tray spacing, and design of column internals are then established on the basis of the liquid and gas properties and flow rates and special requirements of the system. Typical conditions and dimensions for tray towers are listed in Table 1-5 as recommended by Treybal (12). The vapor-velocity limitation is usually established from a correlation of the general form.

Table 1-5. Recommended Conditions and Dimensions for Tray Tower (from Treybal (12))

Tray Spacing:

Tower Diameter, ft	Tray Spacing, in.
4 or less	18-20
4-10	24
10-12	30
12-24	36

Liquid Flow:
    a) Not over 0.165 cu ft/(sec)(ft diam) for single-pass cross-flow trays
    b) Not over 0.35 cu ft/(sec)(ft weir length) for others

Downspout Holdup: 8 sec minimum (superficial)

Downspout Seal: 0.5 in. minimum at no liquid flow

Weir Length: Straight rectangular weirs of cross-flow trays, 0.6-0.8 times the column diameter

Liquid Gradient: 0.5 in. (1.25 in. maximum)

Pressure Drop per Tray:

Operating Pressure	Pressure Drop
35 mm Hg abs	3 mm Hg or less
Atmospheric	0.07-0.12 psi
300 psi	0.15 psi

$$U = K_v \sqrt{\frac{\rho_L - \rho_G}{\rho_G}} \qquad (1\text{-}17)$$

where U = allowable superficial vapor velocity, ft/sec
ρ_L = liquid density, lb/cu ft
ρ_G = gas density, lb/cu ft
K_v = empirical constant

The equation was originally proposed by Souders and Brown (40), on the basis of an analysis of the frictional upward drag of the moving gas stream upon suspended liquid droplets. A number of other factors such as plate design and spacing have been found to affect entrainment; however, Equation 1-17 is still widely used as an empirical expression by adjusting K_v to the tray conditions. Typical values of K_v based on a correlation proposed by Fair (38) are given in Table 1-6. The Flow Parameter F_v is defined by the equation:

$$F_v = \frac{L}{G} \sqrt{\frac{\rho_G}{\rho_L}} \qquad (1\text{-}18)$$

where:

L = Liquid flow rate, lb/sec
G = Gas flow rate, lb/sec

The values given are for a liquid with a surface tension, σ, of 20 dynes/cm. The calculated gas velocity, U, may be corrected for other surface tension values by multiplying by the correction factor $(\sigma/20)^{0.2}$. The correlation provides a means for estimating the maximum allowable gas velocity for all types of plate columns subject to the following restrictions:

1. The system is low or nonfoaming
2. Weir height is less than 15% of the tray spacing
3. Sieve plate perforations are ¼-in. or less in diameter
4. The ratio of bubble-cap slot, sieve tray hole, or valve tray full opening area, A_h, to the active tray area, A_a, is 0.1 or greater

The key column areas involved in the correlation are:
A_a = Active area, the area on the tray actively involved in gas/liquid contact, typically the column cross section less two downcomers, sq ft

Table 1-6. Typical Design Values of K_v for Sieve, Bubble Cap, and Valve Plates			
Plate Spacing in.	K_v - When Flow Parameter, F_v is:		
	0.01	0.1	1.0
6	0.15	0.14	0.065
9	0.18	0.17	0.070
12	0.22	0.20	0.079
18	0.30	0.25	0.095
24	0.39	0.33	0.13
36	0.50	0.42	0.15
Based on correlation of Fair (38)			

A_h = Total slot, perforated, or open valve area on plate, sq. ft

A_n = Net area for vapor flow, typically the column cross section less one downcomer (used for calculating U), sq. ft

Adsorption

In adsorption, as employed for gas purification, the impurities are removed from the gas stream by concentration on the surface of a solid material. The principles of this operation and the nature of typical adsorbents used are discussed in some detail in Chapter 12. In general, the commercial adsorbents are granular solids which have been prepared to have a very large surface area per unit weight. Fixed beds of these materials are most frequently used for gas purification and dehydration, and these are commonly designed by rule-of-thumb techniques.

For such installations the adsorption operation can be likened to a packed absorption column in which the absorbent is stationary on the packing. At any instant a concentration gradient exists from the gas inlet to the gas outlet with respect to solute concentration in both the gas and solid phases. As in conventional liquid absorbers, the steepness of the concentration-gradient curve is an indication of the rate of mass transfer, and, as would be expected, this rate is a function of such factors as the gas velocity and the packing size—the packing in this case being the adsorbent itself. Unlike the counter-current liquid-absorption process, where the concentration gradient is maintained at the same level in the column by continuous addition of lean solvent at one end and removal of enriched solvent at the other, the concentration gradient in the adsorbent bed shifts toward the outlet end as the adsorbate is picked up and held by the adsorbent. This unsteady-state condition greatly complicates the mathematical analysis required for a logical design of fixed-bed adsorbers.

Equations have been developed by Hougen and Marshall (41) for calculating the fraction of the original adsorbate present in the fluid stream leaving the bed at any time and the fraction of adsorbent saturation at any time and position in the bed. Since these equations are too cumbersome for direct application to design, Hougen and Marshall represented them in the form of charts which are applicable for the specific condition of isothermal operation, linear equilibrium curves, negligible adsorbate gas in bed voids, and gas film, the controlling factor. These charts have been reproduced in a number of standard texts (42, 43, 44). Particularly accurate versions of the charts based on numerical solutions to the Hougen and Marshall equations are given by Ufrecht et al. (86). A modified chart based on logarithmic-probability coordinates is presented by Vermuelen (45).

In most actual fixed-bed adsorption systems the adsorption process is exothermic, and the heat release raises the bed temperature so that the assumption of isothermal operation is not valid. Adiabatic adsorption models have been described by Acrivos (46), Leavitt (47), Carter (48), Meyer and Weber (49), and Chi and Wasan (50). Unfortunately, the adiabatic adsorption models are quite cumbersome to úse and require property data, such as pore and solid diffusivites, which are seldom available. To overcome these problems, attempts have been made to correlate nonisothermal experimental data with isothermal models to obtain useful design equations. The procedure is described by Lee and Cummings (51) for the case of air drying with silica gel and by Lee and Chi (52) for CO_2 removal from air by molecular sieves. The latter authors present step-by-step calculation procedures for predicting break time and estimating the required bed height.

The application of any of the techniques mentioned for predicting overall adsorbent bed performance requires information on the rate of adsorption at all points in the bed. As in absorption, the rate data can be expressed in terms of an adsorption coefficient, height of a transfer unit, or, for a given bed, the number of transfer units. Adsorption rates can be controlled by external diffusion in the gas-phase, pore diffusion, surface reaction rate, or solid-phase internal diffusion. Generally, only the first two are of concern in gas-purification operations. The gas-phase external diffusion rates follow general mass transfer correlations such as that proposed by Gamson et al (53) for the range of Reynolds numbers usually encountered in commercial operations.

$$H_G = \frac{0.55}{a_v} \left(\frac{D_p G}{\mu}\right)^{0.51} \left(\frac{\mu}{\rho D_v}\right)^{0.67} \tag{1-19}$$

where H_G = height of gas-film transfer unit for adsorption, ft
  a_v = external area of adsorbent particles, sq ft/cu ft
  D_p = diameter of a sphere having the same surface area as the particle, ft
  G = mass velocity of gas, lb/(hr)(sq ft)
  μ = gas viscosity, lb/(ft)(hr)
  ρ = gas density, lb/cu ft
  D_v = diffusivity of vapor in the carrier gas, sq ft/hr

Pore-diffusion rates are discussed in detail by Wheeler (54). The application of Wheeler's equations to a specific case is described by Lee and Chi (52). A model based on combining the gas phase and particle resistance into an overall transfer coefficient, thereby eliminating the need for solving a diffusion equation, has been proposed by Chi & Wasan (50). This approach was further extended by Collins & Chao (67) to cover the case of multicomponent fixed-bed adsorption.

Inspection of Equation 1-19 reveals that the height of a transfer unit can be expected to decrease with decreased particle size and to increase with increased gas-flow rate. This explains why (for a given bed size) maximum adsorbate removal and adsorbent capacity are obtainable with relatively small particles of adsorbent and low gas velocity. It should be noted, however, that H_G increases with only the 0.51 power of the gas-flow rate. If a gas-flow rate increase is accompanied by a corresponding increase in the bed depth to maintain approximately the same cycle time, the net effect will therefore be an increase in the number of transfer units. As a result, it is more efficient to use a deep bed with a high gas velocity than a shallow bed with a correspondingly low gas velocity. In practical design, the velocity is normally limited by considerations of attrition and bed movement, and the bed volume is extablished by consideration of the desired cycle time and expected operating capacity of the adsorbent. This basic design is then modified as required by pressure-drop considerations. Capacity and pressure-drop data for typical adsorption operations are presented in Chapter 12.

An alternative to the use of mass transfer coefficients for transfer unit height values is the Mass-Transfer-Zone concept first proposed by Michaels for fixed-bed ion exchange columns (68). This approach provides a simple method for correlating data and designing fixed-bed adsorption systems. A detailed review of design by the Mass-Transfer-Zone concept has been presented by Lukchis (69). The approach is based on the assumption that when a gas containing an adsorbable component passes through a bed of regenerated adsorbent, adsorption occurs only in a portion of the bed called the Mass-Transfer-Zone.

The Mass-Transfer-Zone passes through the bed as a stable wave with saturated bed behind the front and clean bed ahead of it. The shape of the wave is determined by the mass transfer rate conditions of the system, and its velocity through the bed can be predicted by a simple material balance equation. Since the depth of the Mass-Transfer-Zone is relatively independent of bed size, the concept is quite valuable for use in designing large installations based on data from small units. A more detailed discussion of the Mass-Transfer-Zone concept is presented in Chapter 12 under the subheading "Drying-Tower Dimensions."

Chemical Conversion

With few exceptions, chemical conversion of gas-phase impurities for gas-purification purposes is accomplished by heterogeneous catalysis using solid catalysts. Fixed-bed catalytic reactors are by far the most common for this type of operation, and the construction of these units is very similar to that of adsorption beds. A general discussion of catalytic operations of this type with details of several typical processes is presented in Chapter 13, and a related operation (iron oxide dry boxes) where the solid may either act catalytically or participate in the reaction is discussed in detail in Chapter 8.

The rigorous design of fixed-bed catalytic reactors involves most of the problems of adsorbers plus some additional complications occasioned by the occurrence of chemical reactions. It is generally accepted that gas-phase reactions which are catalyzed by solids actually proceed on the surface of the solids, and it has been postulated that the reactions occur at specific "active centers." The activity of a catalyst is thus related to the number of active centers per unit area and the total area available. In view of the area effect, it is not surprising that many adsorbents (such as silica gel and activated carbons) also exhibit marked catalytic activity.

Catalytic conversion of gas-phase components by surface catalysis requires the transfer of the reactants from the gas stream to the solid surface, activated adsorption onto the surface, chemical reaction on the surface to form an adsorbed product, desorption of the product, and transfer of the product back into the gas stream. Hougen and Watson (55) have expressed the relative importance of the various steps with regard to the overall rate of conversion in a catalytic reactor as follows:

1. Mass transfer of reactants and products between the exterior of solid particles and the main gas stream is important in the case of very rapid reactions or unfavorable flow conditions. It becomes more important with increased temperature as this normally increases the rate of reaction without greatly affecting the rate of mass transfer.

2. Mass transfer of reactants and products within the pore structure of the catalyst is important in the case of relatively large catalyst particles having a large active internal surface with restricted pore size.

3. Chemical phenomena including activated adsorption and the various chemical reaction steps which result in the overall chemical conversion are highly sensitive to temperature. Since rates of individual reactions

may vary over very wide ranges, the effect of this factor on the overall catalytic reactor performance is generally attributable to a single activated step.

The quantitative estimation of these factors presents a formidable mathematical problem complicated by the effect of changing gas composition through the reactor, temperature effects resulting from the release or absorption of heat by the reactions occurring on the surface of the catalyst particles, and a lack of fundamental rate data. The frequent occurrence of catalyst poisoning is a further complicating factor. Discussions of design theory and design equations which have been developed are presented in specialized texts such as Hougen and Watson (42), Smith (56), Levenspiel (57), Hill (58), and Rase (59).

Although a great deal of study has gone into the development of quantitative expressions for relating the various phenomena occurring during catalytic reactions, techniques which are applicable to everyday design problems involve numerous simplifying assumptions. Commercial reactors are, in fact, generally designed on the basis of pilot- or full-scale plant test data obtained with the identical catalyst and reactants. A convenient method of relating the design capacity of reactors with test data is in terms of the space-velocity. This is usually defined as the volume of gas measured at standard conditions passed through the bed per unit time per unit volume of catalyst. Typical units of space velocity are standard cubic feet per hour per cubic foot of catalyst reactor volume (reciprocal hours). Care must be exercised in using published space velocity data as some investigators report it in terms of actual rather than standard cubic feet per hour.

For cases where mass-transfer effects are negligible, operation of two reactors at the same space-velocity should result in the same conversion (assuming equal pressure and temperature). If mass-transfer effects are important, however, the gas velocity, particle size, and other physical factors must be taken into consideration. As in adsorption, mass transfer is generally improved by operating a given catalyst volume with a maximum L/D (length/diameter) ratio which, of course, involves the use of a high gas velocity. As in adsorber design, the gas velocity is limited by considerations of bed movement, attrition, and pressure drop. Evaluation of these factors for catalyst beds is similar to that for fixed-bed adsorbers, and a more detailed discussion with design data (for typical adsorbent particles) is presented in Chapter 12.

References

1. Anon. 1952. *Petrol. Refiner* 31(Nov.):105.
2. Bowles, V.O. 1954. *Chem. Eng.* 61(May):174.
3. Nutter, I.E. 1954. *Chem. Eng.* 61(May):176.
4. Thrift, G.C. 1954. *Chem. Eng.* 61(May):177.

5. Pfeiffenberger, C.A. 1953. *Chem. Eng.* 60(Apr.):242.

6. Brief, R.S., and Oiestand, A. 1964. "Impingement baffle plate scrubber for flue gas." *J. Air Pollution Assoc.* 14(Sept.):372.

7. Kopita, R., and Gleason, F.G. 1968. "Wet scrubbing of boiler flue gas." *Chem. Eng. Progr.* 64(Jan.):74.

8. Pollock, W.A.; Tomany, J.P.; and Frieling, G. 1966. *Removal of Sulfur Dioxide and Fly Ash from Coal Burning Power Plant Flue Gases.* ASME Paper 66-WA/CD-4.

9. Douglas, H.R.; Snider, I.W.A.; and Tomlinson, G.H. 1963. *An Evaluation of Mass and Heat Transfer in the Turbulent Contact Absorber.* Paper presented at 50th National Meeting of AIChE.

10. Perry, R.H., and Chilton, C.H. (ed.). 1973. *Chemical Engineers' Handbook,* 5th ed. New York: McGraw-Hill Book Company, Inc.

11. Sherwood, T.K., and Pigford, R.L. 1952. *Absorption and Extraction,* 2nd ed. New York: McGraw-Hill Book Company, Inc.

12. Treybal, R.E. 1968. *Mass-Transfer Operations,* 2nd ed. New York: McGraw-Hill Book Company, Inc.

13. Whitman, W.G. 1923. *Chem. & Met. Eng.* 29:147.

14. Walker, W.H.; Lewis, W.K.; McAdams, W.H.; and Gilliland, E.R. 1937. *Principles of Chemical Engineering,* 3rd ed. New York: McGraw-Hill Book Company, Inc.

15. Chilton, T.H., and Colburn, A.P. 1935. *Ind. Eng. Chem.* 27:255.

16. Colburn, A.P. 1939. *Trans. Am. Inst. Chem. Engrs.* 35:211.

17. Colburn, A.P. 1941. *Ind. Eng. Chem.* 33:459.

18. White, G.E. 1940. *Trans. Inst. Chem. Engrs.* 36:359.

19. Sherwood, T.K.; Pigford, R.L.; and Wilke, L.G. 1975. *Mass Transfer.* New York: McGraw-Hill Book Company, Inc.

20. Vivian, J.E., and King, C.J. 1963. *Modern Chemical Engineering,* vol. 1. Edited by A. Acrivos. New York: Reinhold Publishing Corporation.

21. Eckert, J.S.; Foote, E.H.; Rollison, L.R.; and Walter, L.F. 1967. *Ind. Eng. Chem.* 59:41.

22. Murphree, E.V. 1935. *Ind. Eng. Chem.* 17:747.

23. O'Connell, H.E. 1946. *Trans. Am. Inst. Chem. Engrs.* 42:741.

24. American Institute of Chemical Engineers. 1958. *Bubble Tray Design Manual, Prediction of Fractionation Efficiency.* New York: AIChE.

25. Hatta, S. 1929. *Technol. Repts. Tôhoku Univ.* 8:1.

26. Hatta, S. 1932. *Technol. Repts. Tôhoku Univ.* 10:119.

27. Van Krevelen, D.W., and Hoftijzer, P.J. 1948. *Chem. Eng. Progr.* 44:529.

28. Perry, R.H., and Pigford, R.L. 1953. *Ind. Eng. Chem.* 45:1247.

29. Brian, P.L.T.; Hurley, J.F.; and Hasseltine, E.H. 1961. *AIChE J.* 7:226.

30. Gilliland, E.R.; Baddour, R.F.; and Brian, P.L.T. 1958. *AIChe J.* 4:223.

31. Brian, P.L.T. 1964. *AIChE J.* 10:5.

32. Danckwertz, P.V., and Gillham, A.J. 1966. *Trans. Inst. Chem. Engrs. (London)* 44:T42.

33. Sherwood, T.K.; Shipley, G.H.; and Holloway, F.A.L. 1938. *Ind. Eng. Chem.* 30:765.

34. Elgin, J.C., and Weiss, F.B. 1939. *Ind. Eng. Chem.* 31:435.

35. Lobo, W.E.; Friend, L.; Hashmall, F.; and Zenz, F. 1945. *Trans. Am. Inst. Chem. Engrs.* 41:693.

36. Eckert, J.S. 1970. *Chem. Eng. Progr.* 66(3):39.

37. Eckert, J.S. 1970. *Oil Gas J.* 60(Aug. 24):58.

38. Fair, J.R. 1963. Chapter 15 in B.D. Smith, *Design of Equilibrium Stage Processes.* New York: McGraw-Hill Book Company, Inc.

39. Zenz, F.A. 1972. *Chem. Eng.* (Nov. 13):120-138.

40. Souders, M., and Brown, G.G. 1934. *Ind. Eng. Chem.* 26:98.

41. Hougen, O.A., and Marshall, W.R. 1947. *Chem. Eng. Progr.* 43:197.

42. Hougen, O.A., and Watson, K.M. 1947. *Chemical Process Principles,* vol. 3. New York: John Wiley & Sons, Inc.

43. Hougen, O.A., and Watson, K.M. 1946. *Chemical Process Principles Charts.* New York: John Wiley & Sons, Inc.

44. Marshall, W.R., and Pigford, R.L. 1947. *The Application of Differential Equations in Chemical Engineering Problems.* Newark: University of Delaware Press.

45. Vermuelen, T. 1958. *Advances in Chemical Engineering,* vol. 2. Edited by T.B. Drew and J.W. Hoopes, Jr. New York: Academic Press, pp. 148-205.

46. Acrivos, A. 1956. *Ind. Eng. Chem.* 48:703.

47. Leavitt, F.W. 1962. *Chem. Eng. Progr.* 58:54.

48. Carter, J.W. 1966. *Trans. Am. Inst. Chem. Engrs.* 44:T253-T259.

49. Meyer, D.A., and Weber, T.W. 1967. *AIChE J.* 13:457.

50. Chi, C.W., and Wasan, D.T. 1970. *AIChE J.* 16:23.

51. Lee, H., and Cummings, W.P. 1967. *Chem. Eng. Progr. Symposium Series* 63(74):42.

52. Lee, H., and Chi, C.W. 1971. *CO_2 Removal from Air by Na-X Molecular Sieve under Nonisothermal Conditions.* Paper presented at AIChE 68th National Meeting, Houston, Texas (Feb. 28-Mar. 4).

53. Gamson, B.W.; Thodos, G.; and Hougen, O.A. 1943. *Trans. Am. Inst. Chem. Engrs.* 39:1.

54. Wheeler, A. 1955. *Catalysis,* vol. 2. Edited by P.H. Emmett. New York: Reinhold Publishing Corporation, pp. 105-165.

55. Hougen, O.A., and Watson, K.M. 1947. *Chemical Process Principles,* vol. 3. New York: John Wiley & Sons, Inc., p. 906.

56. Smith, J.M. 1970. *Chemical Engineering Kinetics,* 2nd ed. New York: McGraw-Hill Book Company, Inc.

57. Levenspiel, O. 1972. *Chemical Reactor Engineering,* 2nd ed. New York: John Wiley & Sons.

58. Hill, C.G., Jr. 1977. *An Introduction to Chemical Engineering Kinetics and Reactor Design.* New York: John Wiley & Sons.

59. Rase, H.F. 1977. *Chemical Reactor Design for Process Plants.* Volume One, *Principles and Techniques;* Volume Two, *Case Studies and Design Data.* New York: John Wiley & Sons.

60. Maddox, R.N. 1974. *Gas and Liquor Sweetening,* 2nd ed. Norman, Oklahoma: John M. Campbell (Campbell Petroleum Series).

61. Sorenson, P.; Takvoryan, N.E.; and Jaworowski, R.J. 1976. *Proc. Symp. on Flue Gas Desulfurization,* New Orleans, March 1976, Vol. 1. EPA-600/2-76-136a (May 1976):999.

62. Frank, O. 1977. *Chem. Eng.* 84,6(March 14):111-128.

63. Blecker, H.G., and Nichols, T.M. 1973. "Capital and Operating Costs of Pollution Control Equipment Modules," *Data Manual,* Vol. 2. EPA-R5-73-023b (July), PB-224536.

64. Chase, J.D. 1967. *Chem. Eng.* Part 1, (July 31):105-106. Part 2, (Aug. 28):139-146.

65. Bolles, W.L. 1976. *Chem. Eng. Progr.* 72,9(Sept.):43-49.

66. Smith, V.C., and Delnicki, W.V. 1975. *Chem. Eng. Progr.* 71,8(Aug.):68-73.

67. Collins, H.W., Jr., and Chao, K.C. 1973. "Gas Pur:fication by Adsorption," *AIChE Symposium Series,* No. 134, Vol. 69:9-17.

68. Michaels, A.S. 1952. *Ind. Eng. Chem.* 44,8(Aug.):1922-30.

69. Lukchis, G.M. 1973. *Chem. Eng.* 80,13(June 11):111-116.

70. Onda, K., Takeuchi, H., and Okumoto, Y., 1968. *Journal ChE, Japan,* Vol 1, No. 1, p 56.

71. Bolles, W. L. and Fair, J. R., 1982. *Chem. Eng,* Vol 89, July 12, p 109.

72. Bravo, J. L. and Fair, J. R., 1982. *Ind. Eng. Chem., Process Des. Dev.,* Vol 21, No. 1, p 162.

73. Zuiderweg, F. J., *Chemical Engineering Science,* 1982. Review Article No. 9, Vol 37, No. 10, p 1441.

74. Yanagi, T. and Sakata, M., 1981. 90th National AIChE Meeting, Symp 44, Houston, Texas, April.

75. Decoursey, W. J., 1974. *Chemical Engineering Science,* Vol 29, p 1867.

76. Matheron, E. R. and Sandall, O. C., 1978. *Am. Inst. Chem. Engrs. Journal,* Vol 24, No. 3, p 552.

77. Olander, D. R., 1960. *Am. Inst. Chem. Engrs. Journal,* Vol 6, No. 2, p 233.

78. Joshi, S. V., Astarita, G., and Savage, D. W., 1981. *Transport with Chemical Reactions,* AIChE Symposium Series No. 202, Vol 77, p 63.

79. Ouwerkerk, C., *Hydrocarbon Processing,* 1978. Vol 57, No. 4, p 89.

80. Astarita, G. and Savage, D. W., 1980. *Chemical Engineering Science,* Vol 35, p 649.

81. Savage, D. W., Astarita, G., and Joshi, S., 1980. *Chemical Engineering Science,* Vol 35, p 1513.

82. Weiland, R. H., Rawal, M., and Rice, R. G., 1982. *Am. Inst. Chem. Engrs. Journal,* Vol 28, No. 6, p 963.

83. Astarita, G., 1967. *Mass Transfer with Chemical Reactions,* Elsevier, Amsterdam.

84. Danckwerts, P. V., 1970. *Gas Liquid Reactions,* McGraw-Hill Book Company, New York.

85. Astarita, G., Savage, D. W., and Bisio, A., 1983. *Gas Treating with Chemical Solvents,* John Wiley & Sons, New York.

86. Ufrecht, R. H., Sommerfield, J. T., and Lewis, H. C., 1980. *Journal of the Air Pollution Control Association,* Vol 30, No. 12, p 1348.

2

Alkanolamines for Hydrogen Sulfide and Carbon Dioxide Removal

Credit for the development of alkanolamines as absorbents for acidic gases goes to R.R. Bottoms, (1) who in 1930 was granted a patent covering this application. Triethanolamine (TEA), which was the first to become commercially available, was used in the early gas-treating plants. As other members of the alkanolamine family were introduced into the market, they were also evaluated as possible acid-gas absorbents. As a result, sufficient data are now available on several of the alkanolamines to enable design engineers to choose the most suitable compound for each particular requirement.

The two amines which have proved to be of principal commercial interest for gas purification are monoethanolamine (MEA) and diethanolamine (DEA). Triethanolamine has been displaced largely because of its low capacity (resulting from higher equivalent weight), its low reactivity (as a tertiary amine), and its relatively poor stability. Diisopropanolamine (DIPA) (2, 3) is being used to some extent in the Adip process and in the Sulfinol process (see Chapter 14) as well as in the SCOT process for Claus plant tail gas purification (see Chapter 13). Although methyldiethanolamine (MDEA) was described by Kohl et al. (4, 5, 6) as a selective absorbent for H_2S in the presence of CO_2 as early as 1950, its use in industrial processes has only become important in recent years. A somewhat different type of alkanolamine, β,β'-hydroxy-aminoethyl ether, (2-(2-aminoethoxy) ethanol) commercially known as Diglycolamine* (DGA) was first proposed by Blohm and Riesenfeld (7). This compound couples the stability and reactivity of monoethanolamine with the low vapor pressure and hygroscopicity of diethylene glycol and, therefore, can be used in more concentrated solutions than monethanolamine.

*Registered trade name

Figure 2-1. High-pressure natural-gas treating plant using diethanolamine solution (S.N.P.A.-DEA process). *The Ralph M. Parsons Company*

Several proprietary formulations of ethanolamine solutions containing, besides the amine, corrosion inhibitors, foam depressants, and activators are being offered under various trade names such as UCARSOL, Amine Guard (Union Carbide Corporation) (83, 84, 85); GAS/SPEC IT-1 Solvents (Dow Chemical Company) (86); and Activated MDEA (BASF Aktiengesellschaft) (87, 88, 89). EXXON Research and Engineering Company offers an improved process using MEA which is reportedly capable of effecting savings in capital and operating costs, on the order of 30 percent and 60 percent respectively (93).

A different class of acid gas absorbents, the sterically hindered amines have recently been disclosed, also by EXXON Research and Engineering Company (90, 91, 93).

Typical ethanolamine gas-treating plants are shown in Figures 2-1, 2-2, and 2-3. Figure 2-1 is a photograph of a unit treating natural gas at high pressure to pipeline specifications using an aqueous diethanolamine solution (S.N.P.A.-DEA process). Figure 2-2 shows an installation for purifying and simultaneously dehydrating natural gas for pipeline transmission (glycol-amine pro-

Figure 2-2. Glycol-amine plant for purifying and dehydrating high-pressure natural gas. Stripping column and associated reboiler and heat exchangers (beneath reboiler) are located in the foreground with the absorber and purified-gas scrubber slightly behind and to the right of the stripping column. *Fluor Corporation*

cess). Figure 2-3 depicts a natural gas-treating plant using Diglycolamine as the solvent.

Basic Chemistry of the Process

Structural formulas for the alkanolamines mentioned above are presented in Figure 2-4.

Each has at least one hydroxyl group and one amino group. In general, it can be considered that the hydroxyl group serves to reduce the vapor pressure and increase the water solubility, while the amino group provides the necessary alkalinity in water solutions to cause the absorption of acidic gases.

The principal reactions occurring when solutions of a primary amine, such as monoethanolamine, are used to absorb CO_2 and H_2S may be represented as follows:

Figure 2-3. High-pressure gas-treating plant using Diglycolamine solution (Fluor Econamine process). *The Fluor Corporation*

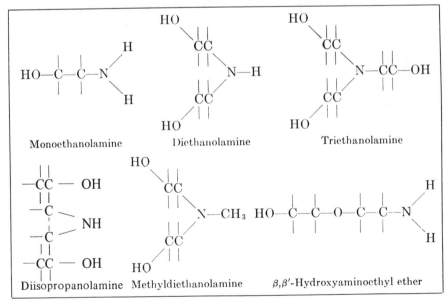

Figure 2-4. Structural formulas for different alkanolamines.

$$2RNH_2 + H_2S \rightleftharpoons (RNH_3)_2S \tag{2-1}$$

$$(RNH_3)_2S + H_2S \rightleftharpoons 2RNH_3HS \tag{2-2}$$

$$2RNH_2 + CO_2 + H_2O \rightleftharpoons (RNH_3)_2CO_3 \tag{2-3}$$

$$(RNH_3)_2CO_3 + CO_2 + H_2O \rightleftharpoons 2RNH_3HCO_3 \tag{2-4}$$

$$2RNH_2 + CO_2 \rightleftharpoons RNHCOONH_3R \tag{2-5}$$

Although these products are definite chemical compounds (some have been isolated and crystallized), they have appreciable vapor pressures under normal conditions so that the composition of the equilibrium solution varies with the partial pressure of the acidic gases over it. As the vapor pressures of these compounds increase rapidly with temperature it is possible to strip the absorbed gases from the solution by the application of heat.

As can be seen from equations 2-1 and 2-2, H_2S reacts directly and rapidly with MEA (and other primary amines) forming the amine sulfide and hydrosulfide. The same reactions occur between H_2S and secondary and tertiary amines, e.g., DEA, DIPA, and MDEA. Carbon dioxide undergoes the reactions shown in equations 2-3, 2-4, and 2-5. The reaction shown in Equation 2-5, resulting in the formation of the amine salt of a substituted carbamic acid, is also direct and relatively rapid but can only take place between CO_2 and a primary or secondary amine. The reactions symbolized by Equations 2-3 and 2-4 are slow because, before reacting with the amine, CO_2 must first react with water to form carbonic acid (a slow reaction) which subsequently reacts with the amine.

If the reaction according to equation 2-5 is predominant, as it is with primary amines, the capacity of the solution for CO_2 is limited to approximately 0.5 mole of CO_2 per mole of amine, even at relatively high partial pressures of CO_2 in the gas to be treated. The reason for this limitation is the high stability of the carbamate and its low rate of hydrolysis to bicarbonate. With tertiary amines, which are unable to form carbamates, a ratio of one mole of CO_2 per mole of amine can theoretically be achieved. However, as just pointed out, the CO_2 reactions involved are very slow. In a recently described process this problem was overcome (for MDEA) by the addition of a catalyst which increases the rate of hydration of dissolved CO_2 (see following section).

The effectiveness of any amine for absorption of both acid gases is due primarily to its alkalinity. The magnitude of this factor is illustrated in Figure 2-5 which shows pH values on titration curves for approximately 2 N solutions of several amines when they are neutralized with CO_2. The curves were obtained by bubbling pure CO_2 through the various solutions and periodically determining the concentration of the solution and pH. The curve for an equivalent KOH solution is included for comparison. The relatively smooth curves for the amines, as compared to the sharp breaks in the KOH curve, may be interpreted as an indication of the presence of non-ionized species during neutralization of the former compounds.

The curves for the tertiary amines MDEA and TEA are seen to cross the DEA and MEA curves at a mole ratio near 0.5 indicating that the tertiary amines, while initially less alkaline, may be expected to attain higher ultimate CO_2/amine ratios. Figure 2-6 shows a comparison of pH values versus temperature curves of 20 percent solutions of monoethanolamine and diethanolamine (8).

In view of the difference in the rates of reaction of H_2S and CO_2 with tertiary amines, partially selective H_2S absorption would be expected with these compounds. The kinetics of H_2S and CO_2 absorption into aqueous solutions of MDEA has been studied by a number of investigators (94, 101, 105). Savage, Funk, and Astarita (94) found that although the rate of H_2S absorption could be thermodynamically predicted, the rate of CO_2 absorption, measured experimentally, appreciably exceeded that predicted on the basis of thermodynamic considerations, and they concluded that MDEA apparently acts as a base catalyst for hydration of CO_2.

The chemistry of acid gas reactions with sterically hindered amines is discussed in some detail by Sartori and Savage (92) and by Weinberg et al (93). A

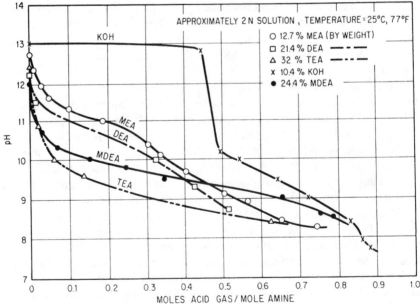

Figure 2-5. Titration curves showing pH during neutralization of ethanolamine and KOH solutions with CO_2.

sterically hindered amine is defined structurally as a primary amine in which the amino group is attached to a tertiary carbon atom or a secondary amine in which the amino group is attached to a secondary or tertiary carbon atom. Typical sterically hindered amines are shown in Figure 2-7 (92).

The key to the concept of CO_2 absorption by such amines is that, by control of the molecular structure, amines can be synthesized which form either a stable carbamate ion, an unstable carbamate ion, or no carbamate ion. For example, by an appropriate molecular configuration an unstable carbamate would be formed with CO_2 which is readily hydrolyzable, resulting in the formation of bicarbonate as the end product. This would result in a theoretical ratio of one mole of CO_2 per mole of amine. For selective H_2S absorption, a molecular structure would be selected which suppresses carbamate formation and, consequently, the rate of CO_2 absorption, without affecting the rate of H_2S absorption. It is claimed that better selectivity can be obtained with sterically hindered amines than with the presently used tertiary or secondary alkanolamines (93).

The lower stability of the carbamate formed between CO_2 and DEA, as compared with that formed with MEA, may explain the high solution loadings observed in plants employing DEA as for example in the SNPA process.

Selection of Process Solutions

The choice of the process solution is determined by the pressure and temperature conditions at which the gas to be treated is available, its composition with respect to major and minor constituents, and the purity requirements of the treated gas. In addition, consideration must, of course, be given to whether simultaneous H_2S and CO_2 removal or selective H_2S absorption is desired. Although no ideal solution is available to give optimum operating conditions for each case, sufficient data and operating experience with several alkanolamines are on hand to permit a judicious selection of the treating solution for a wide range of conditions. A comparison of alkanolamines used for gas purification, based on selected physical properties and approximate cost, is shown in Table 2-1.

Aqueous monoethanolamine solutions which were used almost exclusively for many years for the removal of H_2S and CO_2 from natural and certain synthesis gases are rapidly being replaced by other more efficient systems, particularly for the treatment of high-pressure natural gases. However, monoethanolamine is still the preferred solvent for gas streams containing relatively low concentrations of H_2S and CO_2 and essentially no minor contaminants such as COS and CS_2. This is especially true when the gas is to be treated at low pressures, and maximum removal of H_2S and CO_2 is required. The low molecular weight of monoethanolamine, resulting in high solution capacity at moderate concentrations (on a weight basis), its high alkalinity,

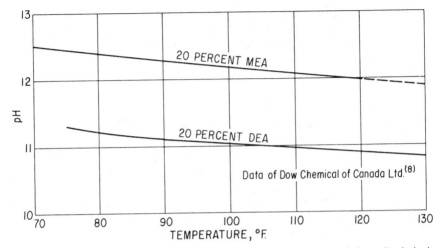

Figure 2-6. pH values of aqueous mono- and diethanolamine solutions (technical grade).

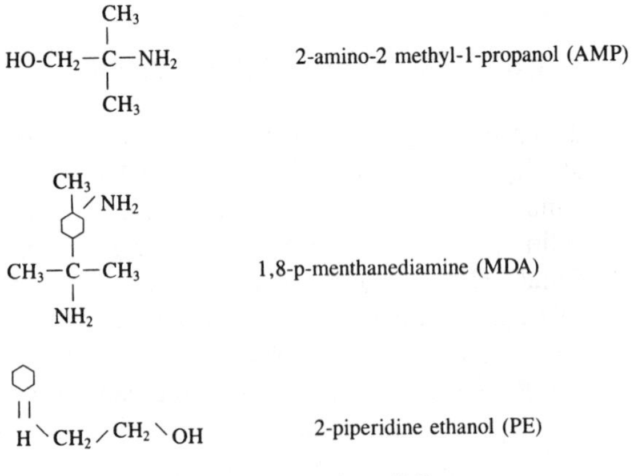

$$HO-CH_2-\underset{\underset{CH_3}{|}}{\overset{\overset{CH_3}{|}}{C}}-NH_2 \qquad \text{2-amino-2 methyl-1-propanol (AMP)}$$

1,8-p-menthanediamine (MDA)

2-piperidine ethanol (PE)

Figure 2-7. Examples of sterically hindered amines (92).

and the relative ease with which it can be reclaimed from contaminated solutions (see Chapter 3) are advantages which in many cases more than counterbalance inherent disadvantages. Among the latter, the most serious one is the formation of irreversible reaction products with COS and CS_2, resulting in excessive chemical losses if the gas contains significant amounts of these compounds. Furthermore, monoethanolamine solutions are appreciably more corrosive than solutions of most other amines, particularly if the amine concentrations exceed 20 percent and the solutions are heavily loaded with acid gas. This feature limits the capacity of monoethanolamine

Table 2-1. Physical Properties of Alkanolamines

Property	MEA*	DEA*	TEA*	MDEA*	DIPA*	DGA**
Mol. weight	61.09	105.14	149.19	119.17	133.19	105.14
Specific gravity, 20/20°C	1.0179	1.0919 30/20°C	1.1258	1.0418	0.9890 45/20°	1.0550
Boilng point, °C						
760 mmHg	171	decomp.	360	247.2	248.7	221
50 mmHg	100	187	244	164	167	—
10 mmHg	69	150	208	128	133	—
Vapor pressure, mmHg						
at 20°C	0.36	0.01	0.01	0.01	0.01	0.01
Freezing point, °C	10.5	28.0	21.2	−21.0	42	−9.5
Solubility in water,						
% by weight at 20°C	Complete	96.4	Complete	Complete	87	Complete
Absolute viscosity,						
cps at 20°C	24.1	380(30°C)	1,013	101	198(45°C)	26(24°C)
Heat of vaporization,						
Btu/lb at 1 atm	355	288(23 mm) (168.5°C)	230	223	184.5	219.1
Approximate cost, $/lb	0.47	0.48	0.49	0.97	0.44	0.68

*Data of Union Carbide Chemicals Company
**Data of Texaco Chemical Company, Inc.

solutions in cases where high partial pressures of the acid gases would permit substantially higher loadings. However, several systems, using effective corrosion inhibitors, reportedly overcome these limitations (see Chapter 3). Such systems include Dow Chemical Company's GAS/SPEC IT-1 technology which is suitable for CO_2 removal in ammonia and hydrogen plants, as well as from sweet natural gas streams (86) and Union Carbide Corporation's Amine Guard Systems (83, 84, 85). It is claimed that use of these systems permits substantially higher amine concentrations, on the order of 30 percent MEA and 50 percent DEA, without excessive corrosion, with attendant reduction in circulation rate and heat consumption for solution stripping. Finally, the relatively high vapor pressure of monoethanolamine causes significant vaporization losses, particularly in low-pressure operations. This difficulty can, however, be overcome by a simple water wash treatment of the purified gas.

Mixtures of monoethanolamine with di- or triethylene glycol, as first described by Hutchinson (9) have been used extensively for simultaneous acid-gas removal and dehydration of natural gases. This process, commonly known as the glycol-amine process, has as its principal advantages the features of simultaneous purification and dehydration and somewhat lower steam consumption when compared to aqueous systems. Furthermore, glycol-amine solutions can be stripped almost completely of H_2S and CO_2, resulting in the capability of producing extremely high purity treated gas. However, the glycol-amine process has a number of drawbacks which have seriously limited its usefulness. Probably the most important of these is the fact that, in order to be effective as a dehydrating agent, the water content of the solution has to be kept at or below 5 percent, requiring relatively high reboiler temperatures. At these temperatures rather severe corrosion occurs in the amine to amine heat exchangers, the stripping column, and, under certain operating conditions, the reboiler. The only practical solution to the corrosion problem is the utilization of corrosion-resistant ferrous alloys or non-ferrous metals (see Chapter 3). Another undesirable feature of the glycol-amine process is a high vaporization loss, especially of the amine. This loss can be materially reduced by a glycol wash of the purified gas, but such an operation is not quite as simple as the water wash used in aqueous systems. Furthermore, because of the very low vapor pressure of the glycol, a contaminated glycol-amine solution cannot be reclaimed by simple distillation as is possible with the aqueous system. Finally, hydrocarbons, especially aromatics, are substantially more soluble in glycol-amine than in aqueous amine solutions. This feature is of major importance if the acid gas is to be further processed in a Claus type sulfur plant, as the presence of high molecular weight hydrocarbons usually leads to rapid catalyst deactivation and production of discolored sulfur.

Aqueous solutions of diethanolamine have been used for many years for the treatment of refinery gases which normally contain appreciable amounts of COS and CS_2, besides H_2S and CO_2. As discussed in Chapter 3, secondary amines are much less reactive with COS and CS_2 than primary amines, and the reaction products are not particularly corrosive. Consequently, diethanolamine and other secondary amines are the better choice for treating gas streams containing COS and CS_2. The low vapor pressure of diethanolamine makes it suitable for low-pressure operations as vaporization losses are quite negligible. One disadvantage of diethanolamine solutions is the fact the reclaiming of contaminated solutions may require vacuum distillation. Application of diethanolamine solutions to the treatment of natural gas was first disclosed by Bertheir (10) and later described in more detail by Wendt and Dailey (11), Bailleul (12), Dailey (13), and in Canadian Patent No. 651,379. This process, which is commonly known as the S.N.P.A.-DEA process, was developed by Societe Nationale des Petroles d'Aquitaine (S.N.P.A.)* of France in the gas field at Lacq in southern France. S.N.P.A. recognized that relatively concentrated aqueous diethanolamine solutions (25 to 30 percent by weight) can absorb acid gases up to the stoichiometric molar ratio typically 1.0 to 1.3 mole of DEA per mole of acid gas, provided the partial pressure of the acid gases in the feed gas to the plant is sufficiently high. If the regenerated solution is well enough stripped when returned to the absorber and the operating pressure is high, a purified gas satisfying pipeline specifications can be produced. The presence of impurities such as COS and CS_2 is not injurious to the solution. Under normal operating conditions decomposition products which may be formed are removed quite easily by filtration through activated carbon. In general, diethanolamine solutions are less corrosive than monoethanolamine solutions because the acid gases are stripped easier and less vigorous reboiling is required. In addition, the decomposition products from side reactions are essentially noncorrosive (see Chapter 3).

As a result of S.N.P.A.'s experience in Lacq, the S.N.P.A.-DEA process has been widely accepted and is at present, the preferred choice for the treatment of high-pressure natural gases with high concentrations of acidic components, especially if COS and CS_2 are also present in appreciable amounts. Beddome (14) reports that in 1969 the S.N.P.A.-DEA process predominated for the recovery of sulfur from natural gas in Alberta, Canada. Comparative operating data for mono- and diethanolamine systems as reported by Beddome (14) for typical Canadian gas-treating plants are shown in Table 2-2. Although not stated in the article, it is assumed that all plants operated at a pressure of about 1,000 psig, which is typical for Canadian operation.

The use of aqueous solutions of Diglycolamine was commercialized jointly by the Fluor Corporation, the El Paso Natural Gas Company, and the Jefferson Chemical Company Inc. (15, 16). The process employing this solvent has

---

*Now Societe Nationale Elf Aquitaine (Production) (SNEAP)

Table 2-2. Comparative Operating Data for MEA and DEA Systems (14)

Gas Plant	A	B	B	C	D
Feed gas composition					
Mole % H_2S	2.1	7.1	7.1	2.4	16.5
Mole % CO_2	0.7	5.9	5.9	4.9	8.0
Solvent					
(% active reagent in	18%	15%	24% SNPA-	22.5%	27.5% SNPA-
water solution)	MEA	MEA	DEA	DEA	DEA
Solvent circulation					
Moles amine per mole acid gas	1.8	2.5	1.3	1.5	1.0
Gallons solvent per mole					
acid gas	74	123	68	84	44
Reboiler steam					
lbs steam/gal solvent	1.0	1.2	1.5	1.2	1.0
lbs steam/mole acid gas	74	148	72	101	44

Table 2-3. Comparison of Typical Operating Data of MEA-DEG and DGA Systems (15)

	MEA-DEG	DGA
Gas volume, MMSCFD	121.2	121.3
Solution rate, gpm	714	556
Reboiler steam, lb/hr	50,700	40,100
Solution loading		
SCF acid gas/gal	4.0	5.5
H_2S in treated gas		
grain/100 SCf	0.25	0.25
CO_2 in treated gas, Mol %	0.01	0.01

been named the Fluor Econamine process. The solvent is in many respects similar to monoethanolamine, except that its low vapor pressure permits its use in relatively high concentrations, typically 40 to 60 percent, resulting in appreciably lower circulation rates and steam consumption when compared to typical monethanolamine solutions. A comparison of operating data, for glycol-monoethanolamine and Diglycolamine solutions in a commercial installation treating natural gas containing 2 to 5 percent total acid gas at a pressure of 850 psig was given by Holder (15) and is shown in Table 2-3. Additional information on commercial applications of the Fluor Econamine process has been presented by Dingman (95), Mason and Griffith (96), Huval and van de Venne (97), Bucklin (98) and Weber and McClure (99). Comparison of the process with systems using MEA solutions indicates some capital and operating cost savings, as well as improved operation at relatively low pressures (97). An additional advantage is partial removal of mercaptans and COS by the DGA solution. Furthermore, a substantial portion of DGA can be reclaimed from degradation products, resulting from reactions of DGA with CO_2 and COS, by steam distillation (see Chapter 3). At present, Diglycolamine solutions are being used for the treatment of more than three billion cubic feet of natural gas per day.

Diisopropanolamine (DIPA) is being used in the Adip process and the Sulfinol process, both licensed by the Shell International Petroleum Company. In the Sulfinol process diisopropanolamine is used in conjunction with a physical organic solvent; a more detailed discussion of this process is given in Chapter 14. The Adip process which employs relatively concentrated aqueous solutions of diisopropanolamine, about 30 to 40 percent, has been described by Bally (2) and by Klein (3). It has been widely accepted, primarily in Europe, for the treatment of refinery gases and liquids which, besides H_2S and CO_2, also contain COS. It is claimed that substantial amounts of COS are removed without detrimental effects to the solution. Furthermore, diisopropanolamine solutions are reported to have low regeneration steam requirements and to be noncorrosive (3). A more recent development by Shell International Petroleum Company is the application of the ADIP Process to the selective absorption of H_2S from refinery gas streams (100) and, as part of the SCOT process, to selective absorption of H_2S from Claus plant tail gas (see Chapter 13). A theoretical study of the absorption kinetics involved in the selective absorption of H_2S in DIPA has been presented by Ouwerkerk (101). Equations for mass transfer with chemical reaction are utilized in the study to develop a computer program which takes into account the competition between H_2S and CO_2 when absorbed simultaneously.

Selective absorption of hydrogen sulfide in the presence of carbon dioxide, especially in cases where the ratio of carbon dioxide to hydrogen sulfide is very high, has recently become the subject of considerable interest, particularly in the purification of non-hydrocarbon gases such as coke-oven gas, the products from coal gasification processes and Claus plant tail gas. The early work of Kohl et al. (4, 5, 6) has shown that tertiary amines, especially methyldiethanolamine, can absorb hydrogen sulfide reasonably selectively under proper operating conditions involving short contact times. A study by Vidaurri and Kahre (71) in which selective absorption with several ethanolamines was investigated in a pilot and commercial plant demonstrated that purified gas containing as little as 5 parts per million of hydrogen sulfide could be obtained with absorption of only about 30 percent of the carbon dioxide contained in the feed gas. The most selective solvent was methyldiethanolamine although other amines also showed some selectivity.

Additional information on selective H_2S absorption with MDEA or MDEA based solutions is presented by Pearce (102), Crow and Baumann (103), Goar (104), Blanc and Elgue (105), Sigmund, Buttwell and Wussler (108) and Dibble (82). A proprietary selective MDEA solvent was disclosed by Gas/Spec Division of Dow Chemical USA under the trade name of ST-1 Technology (106). The papers by Sigmund et al and Dibble give a description of Union Carbide Corporation's proprietary HS Process using MDEA based solutions under the trade name of UCARSOL Solvents. These solvents are claimed to be more selective than conventional MDEA and DIPA solutions and, consequently, more economical with respect to energy consumption. A comparison of UCARSOL with DIPA for recovering H_2S from Claus plant tail gas (after hydrogenation) is shown in Table 2-4.

The data from these studies indicate that, with proper design, selective solvents can yield H_2S concentrations as low as 4 ppmv in the treated gas while permitting a major fraction of the CO_2 to pass through unabsorbed.

Because of its low vapor pressure MDEA can be used in concentrations up to 60 percent in aqueous solutions without appreciable evaporation losses. Furthermore, MDEA is highly resistant to thermal and chemical degradation, is essentially noncorrosive (see Chapter 3), has low specific heat and heats of reaction with H_2S and CO_2 and, finally, is only sparingly miscible with hydrocarbons.

Use of MDEA as a nonselective solvent for removing acid gases, particularly CO_2, from synthesis and natural gases has been disclosed by BASF Aktiengesellschaft and described by Meissner and Wagner (87, 88). This process, which is licensed in the USA by The Ralph M. Parsons Company, employs a 2.5 to 4.5 molar MDEA solution containing 0.1 to 0.4 mole of monomethyl-monoethanolamine or up to 0.8 mole of piperazine as absorption activators (89). The activators apparently increase the rate of hydration of CO_2 in a manner analogous to activators used in hot potassium carbonate solutions (see Chapter 5) and thus increases the rate of absorption.

The process can be operated with one or two absorption stages, depending on the required gas purity. In the single absorption stage version, which is suitable for bulk CO_2 removal from high pressure gases, the rich MDEA solution is regenerated by simple flashing at reduced pressure. In the two-stage version, when essentially complete CO_2 removal is required, a small stream of steam stripped MDEA solution is used in the second stage.

The comparative capacities of MDEA and MEA for CO_2 recovery in an absorption/flash process is illustrated by Figure 2-8 (88). If it is assumed that equilibrium is attained in both the absorption and stripping steps and that isothermal conditions are maintained, the maximum net capacity is simply the difference between equilibrium concentrations at the absorption and stripping

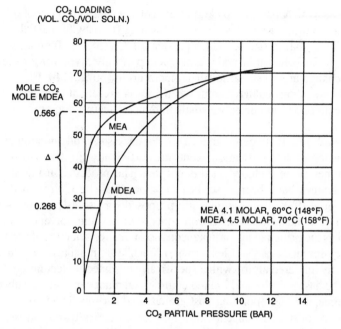

Figure 2-8. CO_2 Solution Isotherms in MEA and MDEA at 0°C, (87). Reprinted with permission from *Oil & Gas Journal*, February 7, 1983. Copyright Penn Well Publishing Company.

Table 2-4. UCARSOL HS Solvent 101 VS. Dipa in Claus Tail Gas Cleanup Unit (82)

	27% Dipa	50% Ucarsol HS
Circulation Rate, gpm	76	28
CO_2 Slippage, %	84	95
H_2S Content in Recycle Stream to Claus Unit, %	35	66
Reboiler Steam Consumption MLbs/Hr	4,469	2,284
Steam Cost @ $5.50/MLbs $/Year	212,400	108,500
Savings with UCARSOL HS $/Year		103,900

partial pressures. A net CO_2 pickup of 30 vol/vol (0.297 mole/mole) is indicated for a 4.5 molar MDEA solution by flashing from a CO_2 partial pressure of 5 bar (72.5 psia) to one bar (14.5 psia) at 70°C (158°F). By comparison a 4.1 molar MEA solution provides a net pickup of only 5 vol/vol for the same pressure change at somewhat lower temperature of 60°C (140°F).

The process is particularly useful when CO_2 is present at high partial pressures, as either no steam or only a small amount of steam is required for regeneration.

Although sterically hindered amines are not necessarily alkanolamines, their characteristics as gas purification agents are sufficiently similar to those of the alkanolamines to be included in this chapter. The processes using sterically hindered amines have been described in some detail by Goldstein (91) and Weinberg et al (93). These amines are proposed for use as promoters in hot potassium carbonate systems; as components of organic solvent/amine systems; and as the principal agent in aqueous solutions for selective absorption of H_2S in the presence of CO_2 (Flexsorb SE). Each system makes use of a different sterically hindered amine with a specifically designed molecular configuration. On the basis of pilot and limited commercial plant experience, substantial savings in capital and operating cost are claimed for this technology.

The choice of amine concentration may be quite arbitrary and is usually made on the basis of operating experience. Typical concentrations of monoethanolamine in aqueous or glycol solutions range from 12 percent to a maximum of 25 percent. On the basis of operating experience in five plants, Feagan et al. (17) recommend the use of a design concentration of 15 percent monoethanolamine in water. The same solution strength is recommended by Connors (18). However, it should be noted that higher amine concentrations may be used when corrosion inhibitors are added to the solution.

Diethanolamine solutions used for treatment of refinery gases range in concentration from 10 to 25 percent, while concentrations of 25 to 30 percent are commonly used for natural gas purification. Diglycolamine solutions typically contain 40 to 60 percent amine in water and MDEA solution concentrations may range from 35 to 55 percent.

It should be noted that increasing the amine concentration will generally reduce the required solution circulation rate and therefore the plant cost. However, the effect is not as great as might be expected, the principal reason being that the acid-gas vapor pressure is higher over more concentrated solutions at equivalent acid-gas/amine mole ratios (see Figure 2-16). In addition, when an attempt is made to absorb the same quantity of acid gas in a smaller volume of solution, the heat of reaction results in a greater increase in temperature and a consequently increased acid-gas vapor pressure over the solution.

Data on the economics of CO_2 removal from very concentrated gas streams by monoethanolamine and diethanolamine solutions are presented in Chapter 5 in connection with an evaluation of the hot potassium carbonate process for this service. The evaluation indicates that the conventional amine solutions are not competitive in this case although they are applicable as a final stage purification after bulk CO_2 removal by the hot potassium carbonate or water-wash systems.

Flow Systems

Basic Flow Scheme. The basic flow for all alkanolamine acid-gas absorption-process systems is shown in Figure 2-9. Gas to be purified is passed upward through the absorber, countercurrent to a stream of the solution. The rich solution from the bottom of the absorber is heated by heat exchange

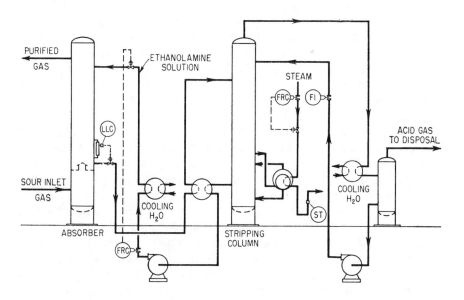

Figure 2-9. Basic flow scheme for alkanolamine acid-gas removal processes.

with lean solution from the bottom of the stripping column and then fed to the stripping column at some point near the top. In units treating sour hydrocarbon gases at high pressure, it is customary to flash the rich solution in a drum kept at an intermediate pressure to remove dissolved and entrained hydrocarbons before stripping (see Chapter 3). The lean solution, after partial cooling in the heat exchanger, is further cooled by exchange with water or air and fed into the top of the absorber to complete the cycle. Acid gas which is removed from the solution in the stripping column is cooled to condense a major portion of the water vapor. This condensate (or other pure water source) is continually fed back to the system to prevent the amine solution from becoming progressively more concentrated. Generally, all this water, or a portion of it, is fed back at the top of the stripping column at a point above the rich-solution feed and serves to force back amine vapors carried by the acid-gas stream.

Water Wash for Amine Recovery

The simplest modification of the flow system shown in Figure 2-9 is the inclusion of a water wash at the top of the absorber to reduce losses of amine with the purified gas. If acid gas condensate is used for this purpose, no draw-off tray is required because it is necessary to readmit this water to the system at some point. It should be noted that the condensate is saturated with acid gas under the conditions of the condenser and that this acid gas will be reintroduced into the gas stream if the water is used "as is" for washing. If the gas volume is very large, compared to the amount of wash water, this may be of no consequence. However, if calculations indicate that the quantity of acid gas so introduced is excessive, a water stripper can be included in the process. In its simplest form, this unit consists of a small, packed column; the water is admitted at the top and withdrawn at the bottom, with process steam fed into the bottom to provide heat and stripping vapor. In some cases, insufficient condensate is available from the acid-gas stream to provide water reflux at the top of both the stripper and the absorber. In this event, water from some other source may be used to wash the purified gas and a draw-off tray provided to prevent buildup of water in the system.

A water wash is primarily used in monoethanolamine systems, especially at low operating pressures, as the relatively high vapor pressure of monoethanolamine may cause appreciable vaporization losses. The other amines have sufficiently low vapor pressures to make water washing unnecessary, except in rare cases when the purified gas is used in catalytic processes and the catalyst is sensitive even to traces of amine vapors.

The number of trays used for water wash varies from two to five in commercial installations. Preliminary pilot-plant experience (19) has indicated that an efficiency of 40 or 50 percent can be expected per tray under typical absorber operating conditions. From this, it would appear that four trays would be ample to remove over 80 percent of the vaporized amine from the purified gas and, incidentally, a major portion of the amine carried as entrained droplets in the gas stream.

It is probable that an even greater tray efficiency is obtained in the water-wash section of the stripping column. However, because of the higher temperature involved, the amine content of the vapors entering this section may be quite high. Four to six trays are commonly used for this service.

Glycol Wash

Obviously, a water wash cannot be used with the glycol-amine system to recover amine vapors from the purified gas stream if dehydration is desired. Because of this, several novel schemes have been devised to wash the gas with glycol (20-22). A flow diagram of one of these is shown in Figure 2-10. In this arrangement, a small stream of solution which is relatively low in amine content is used to wash the gas stream after it leaves the primary absorber. This solution is stripped in a separate small reboiler, the vapors from which may be vented directly into the main stripping column as shown. In one modification, the primary contacting is done at a low pressure to reduce the H_2S and

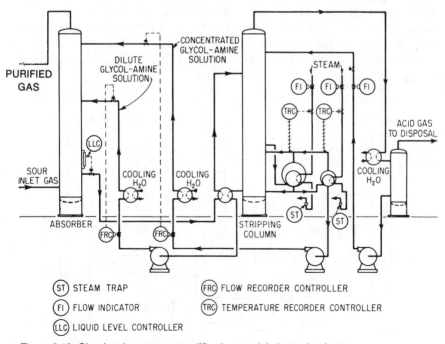

Figure 2-10. Glycol-amine process modification to minimize amine losses.

CO_2 contents of the gas flowing to the compressors, and the final wash with a solution which is low in both amine and water serves to remove the last traces of H_2S and CO_2 and provide maximum dehydration with negligible loss of amine by vaporization. In a commercial operation using an arrangement of this type, the primary solution contains 30 percent monoethanolamine and 13 percent water with the balance glycol, while the wash solution contains about 10 percent monoethanolamine and 8 percent water.

Split-Stream Cycle

A flow modification which has been proposed for aqueous amine solutions to reduce the steam requirement is shown in Figure 2-11. The split-stream scheme, in which only a portion of the solution is stripped to a low acid-gas concentration, has been applied to several gas-purification processes, including the Shell tripotassium phosphate process and the U.S. Bureau of Mines hot potassium carbonate process (see Chapter 5), and was first disclosed by Shoeld in 1934 (23). The rich solution from the bottom of the absorber is split into two streams, one being fed into the top of the stripping column and one at the midpoint. The top stream flows downward countercurrent to the stream of vapors rising from the reboiler and is withdrawn at a point which is above the inlet of the second portion of the rich solution. The liquid withdrawn from the upper portion of the stripping column is not completely stripped and is recycled back to the absorber to absorb the bulk of the acid gases in the lower portion of the absorber column. The portion of solution which is introduced near the midpoint of the stripping column flows through

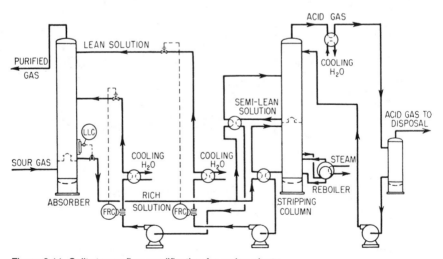

Figure 2-11. Split-stream flow modification for amine plants.

the reboiler and is very thoroughly stripped of absorbed acid gases. This solution is returned to the top of the absorber where it serves to reduce the acid-gas content of the product gas to the desired low level. In this system, the quantity of vapor rising through the stripping column is somewhat less than that in a conventional stream plant. However, the ratio of liquid to vapor is lower in both sections because neither carries the total liquid stream.

The obvious drawback of this process modification is the fact that it increases appreciably the initial cost of the treating plant. The stripping column is taller and somewhat more complex, and the two streams require separate piping systems with two sets of pumps, heat exchangers, and coolers. Commerical units utilizing a system of this type have been described by Bellah et al. (24) and by Estep, McBride, and West (25).

A simplified form of the split-stream cycle consists of dividing the lean solution before introduction into the absorber into two unequal streams. The larger stream is fed to the middle of the absorber, while the smaller stream is introduced at the top of the column. In cases where gases of high acid-gas concentration are treated, this scheme may be more economical than the basic flow scheme, as the diameter of the top section of the absorber may be appreciably smaller than that of the bottom section. Furthermore, the lean-solution stream fed to the middle of the absorber may not have to be cooled to as low a temperature as the stream flowing to the top of the column, resulting in reduction of heat exchange surface.

Heat Exchanger-Absorber

To permit more concentrated solutions to be used to their full capacity, a process scheme has been proposed in which initial absorption takes place in a heat exchanger as shown in Figure 2-12 (26). The sour-gas stream is passed concurrently with partially rich solution through a heat exchanger in which cooling water is circulated to remove the heat of reaction. The mixture is separated, and the resulting gas is passed countercurrent to fresh solution in an absorber to provide final purification of the gas. Solution from the bottom of the absorber contains some acid gas; this is fed to the heat exchanger with the sour-gas stream. The process has found only limited application to date for reabsorbing acid-gas streams from conventional-plant stripping columns in cases where it is necessary to transport the acid gas absorbed in the solution stream and a minimum volume of solution is required.

Design of Absorption Column

Tray Versus Packed Columns

While absorbers in most of the older plants contain either bubble-cap trays or raschig ring packing, alternate tray designs (e.g., sieve trays and valve type

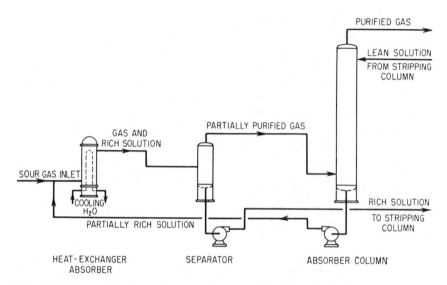

Figure 2-12. Flow system employing heat exchanger for initial gas contact.

trays), and different packing shapes (saddles and Pall rings) are being used extensively in modern plants. The choice between trays and packed columns may be somewhat arbitrary because generally either can be designed to do a satisfactory job and the overall economics are seldom decisively in favor of one or the other. There have been several instances reported where trays have been replaced with packing to alleviate difficult foaming problems (27). On the other hand, the authors know of several more recent cases of foaming in bubble-cap columns and other tray type columns which have been quite satisfactorily relieved by the addition of anti-foaming agents.

Packed columns are normally used when a very high degree of CO_2 removal is required, and the low tray efficiencies obtained with plate columns may result in objectionably tall columns. Although it has been generally assumed that packed columns require a larger diameter than plate columns for the same gas and liquid volumes processed, use of the recently introduced high efficiency packings, such as Pall rings and Hy-pack, has in many instances, resulted in columns of equal or even smaller diameter than equivalent plate columns.

Estimation of Solution Rate

Before detailed design of the absorber can be undertaken, it is of course necessary to know the flow rates and physical properties of the gas and solution streams. Data on the physical properties of several typical ethanolamine solutions are presented in Figures 2-11 through 2-39.

(text continued on page 61)

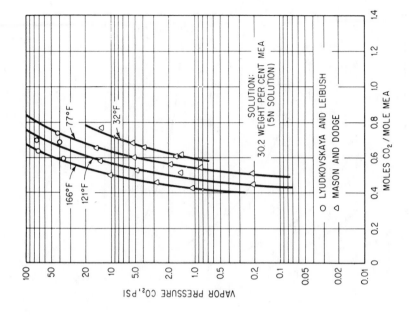

Figure 2-14. Vapor pressure of CO₂ versus CO₂ concentration in 5 N monoethanolamine solution.

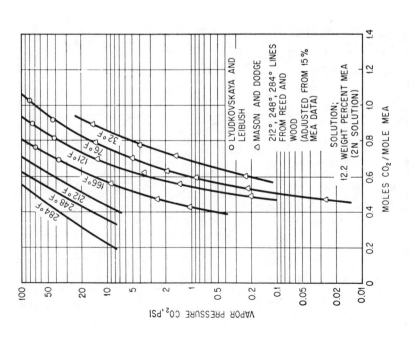

Figure 2-13. Vapor pressure of CO₂ versus CO₂ concentration in 2 N monoethanolamine solution.

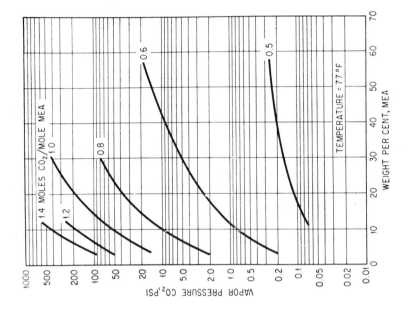

Figure 2-16. Effect of amine concentration on CO_2 vapor pressure.

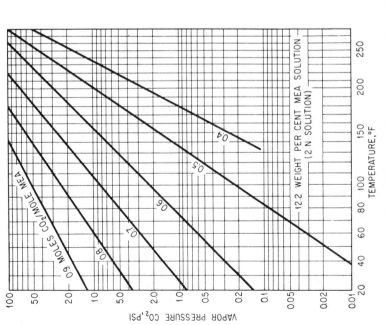

Figure 2-15. Effect of temperature on CO_2 vapor pressure for various CO_2 concentrations in 2 N monoethanolamine solution.

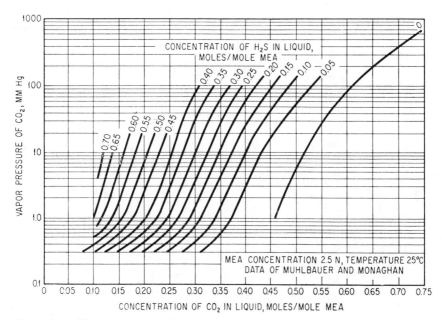

Figure 2-17. Effect of dissolved hydrogen sulfide on vapor pressure of CO_2 over 2.5 N monoethanolamine solution at 25°C. *Data of Muhlbauer and Monaghan* (30)

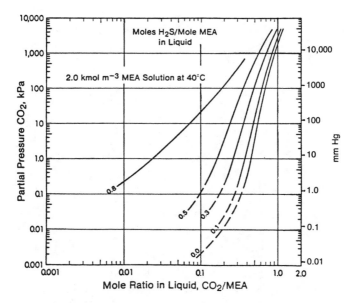

Figure 2-18. Effect of H_2S on the Solubility of CO_2 in 2.5 kmol/m³ MEA Solution at 40°C (110).

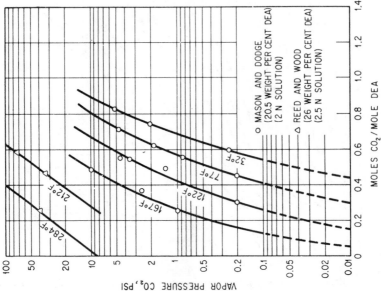

Figure 2-20. Vapor pressure of CO_2 versus CO_2 concentration in 2 and 2.5 N diethanolamine solutions.

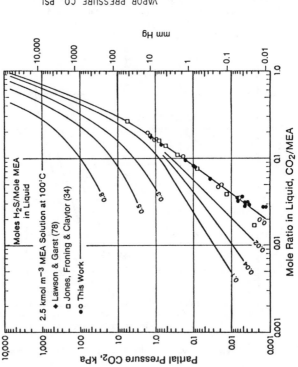

Figure 2-19. Effect of H_2S on the Solubility of CO_2 in 2.5 kmol/m³ MEA Solution at 100°C (110).

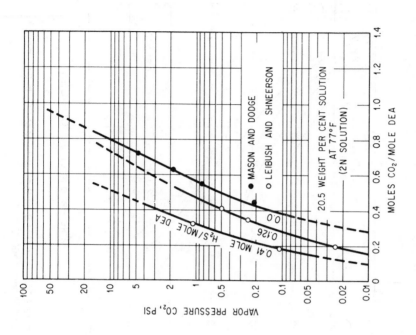

Figure 2-22. Effect of dissolved H_2S on CO_2 vapor pressure over 2 N diethanolamine solution containing both CO_2 and H_2S.

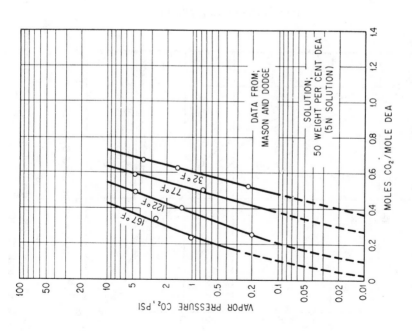

Figure 2-21. Vapor pressure of CO_2 versus CO_2 concentration in 5 N diethanolamine solution.

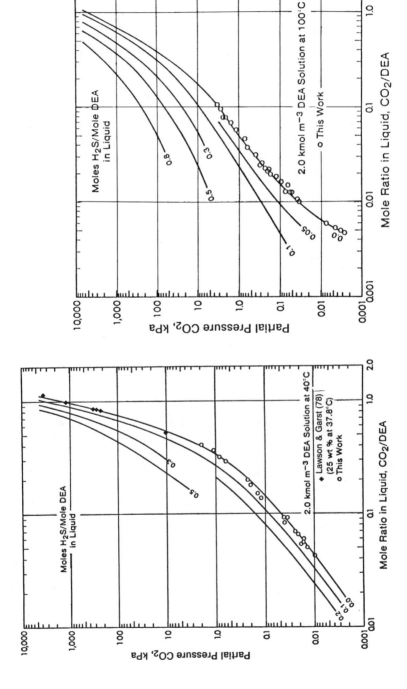

Figure 2-24. Effect of H_2S on the Solubility of CO_2 in 2.0 kmol/m^3 DEA Solution at 100°C (110).

Figure 2-23. Effect of H_2S on the Solubility of CO_2 in 2.0 kmol/m^3 DEA Solution at 40°C (110).

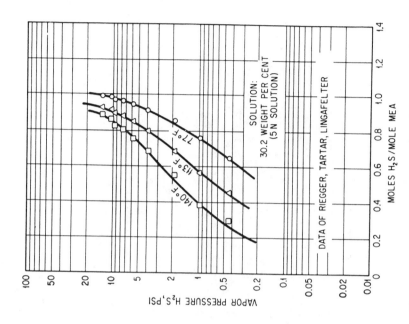

Figure 2-26. Vapor pressure of H_2S versus H_2S concentration in 5 N monoethanolamine solution.

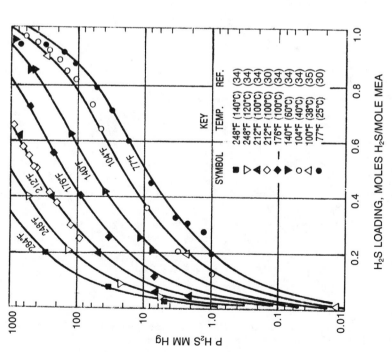

Figure 2-25. Comparison of Calculated Curves for H_2S Partial Pressure over 15.3 Wt. % MEA with Experimental Points (77). (Courtesy *Hydrocarbon Processing*, Feb. 1979.)

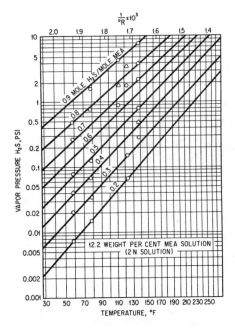

Figure 2-27. Effect of temperature on vapor pressure of H_2S for various H_2S concentrations in 2 N monoethanolamine solution.

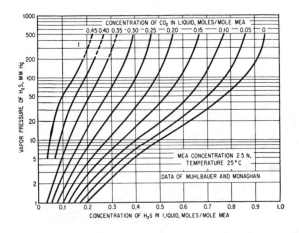

Figure 2-28. Effect of dissolved carbon dioxide on vapor pressure of H_2S over 2.5 N monoethanolamine solution at 25°C. *Data of Muhlbauer and Monaghan* (30)

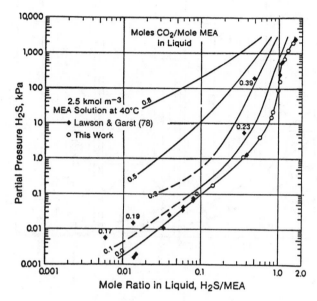

Figure 2-29. Effect of CO_2 on the Solubility of H_2S in 2.5 kmol/m³ MEA Solution at 40°C (110).

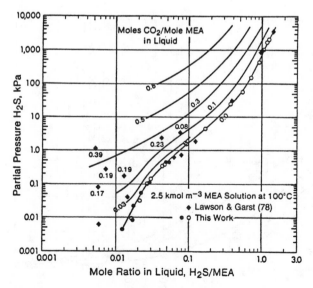

Figure 2-30. Effect of CO_2 on the Solubility of H_2S in 2.5 kmol/m³ DEA Solution at 100°C (110).

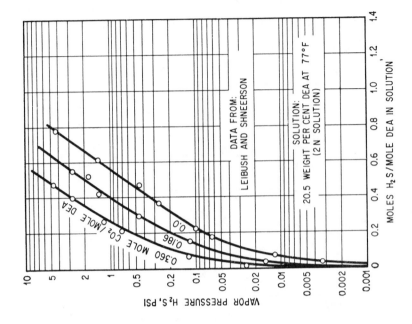

Figure 2-32. Effect of CO_2 on vapor pressure of H_2S over 2 N diethanolamine solution containing both CO_2 and H_2S.

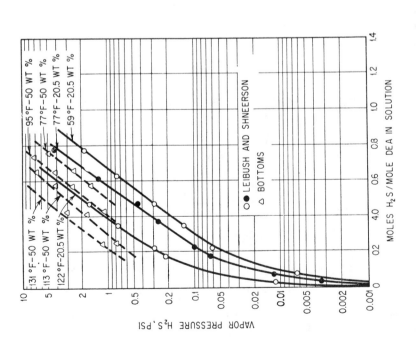

Figure 2-31. Vapor pressure of H_2S versus H_2S concentration in 2 and 5 N diethanolamine solutions.

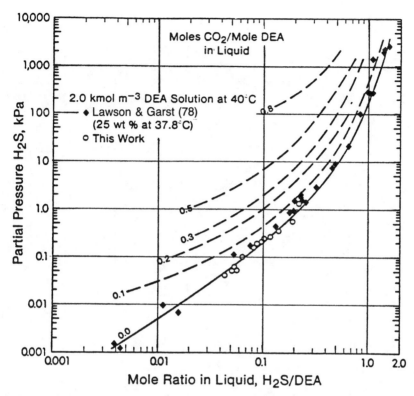

Figure 2-33. Effect of CO_2 on the Solubility of H_2S in 2.0 kmol/m³ DEA Solution at 40°C (110).

Vapor pressure data for CO_2 and H_2S in solutions of monoethanolamine (MEA), diethanolamine (DEA), diisopropanolamine (DIPA), methyldiethanolamine (MDEA), and diglycolamine (DGA) are given in Figures 2-13 through 2-43. The curves cover the conditions and components of greatest commercial interest, but represent only a small fraction of the equilibrium data available in the literature. A considerable number of studies have been conducted during the past ten years corroborating and extending previous studies and covering amines of more recent interest. Especially noteworthy is the work conducted at the University of Alberta at Edmonton, Canada by Lee, Otto, Mather, Isaacs, Jou, Lal and Martin, as well as the studies of Lawson and Garst of Sunoco Production Co. (72, 73, 74, 75, 76, 78, 79, 109, 110, 111, 122). Data were also presented by Jones, Froning and Claytor (34) and by Atwood et al who proposed a calculation procedure for hydrogen sulfide equilibria (35). These investigations have greatly extended the range of equilibrium data with regard to both physical conditions and types of amines covered. A large amount of experimental vapor-liquid equilibrium data on the sys-

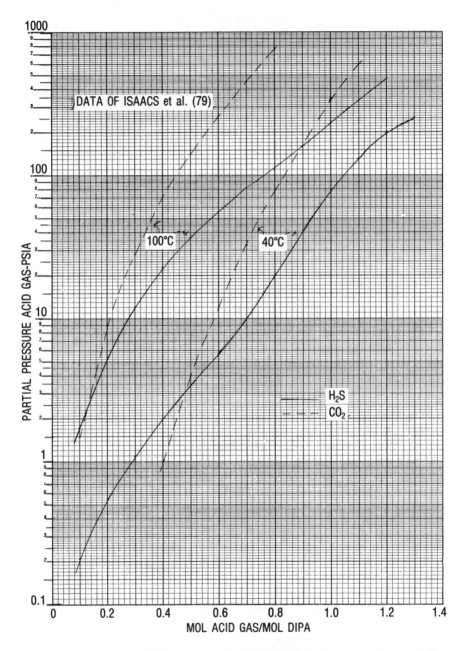

Figure 2-34. Equilibrium Solubility of H_2S and CO_2 in 2.5 M diisopropanolamine solution (79).

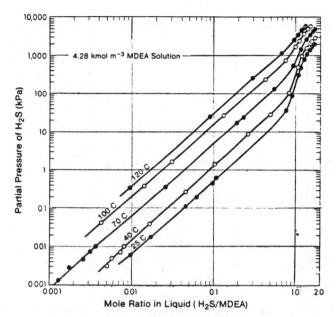

Figure 2-35. Effect of Temperature on the Solubility of H_2S in 4.28 kmol/m³ MDEA Solution
(111). Reprinted with permission from *Industrial and Engineering Chemistry, Process Design
and Development*, Vol 21, No. 4. © 1982, American Chemical Society.

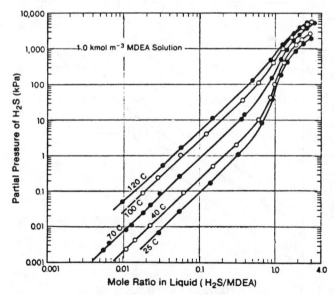

Figure 2-36. Effect of Temperature on the Solubility of H_2S in 1.0 kmol/m³ MDEA Solution
(111). Reprinted with permission from *Industrial and Engineering Chemistry, Process Design
and Development*, Vol 21, No. 4. © 1982, American Chemical Society.

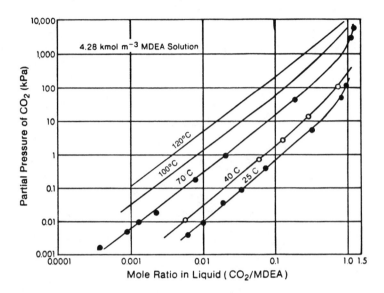

Figure 2-37. Effect of Temperature on the Solubility of CO_2 in 4.28 kmol/m³ MDEA Solution (111). Reprinted with permission from *Industrial and Engineering Chemistry, Process Design and Development,* Vol 21, No. 4. Copyright 1982, American Chemical Society.

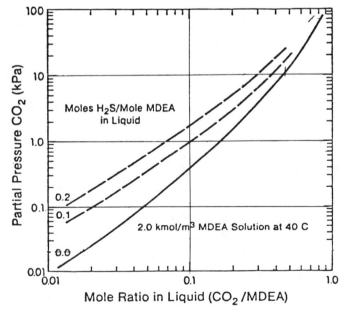

Figure 2-38. Effect of H_2S on the Solubility of CO_2 in a 2.0 kmol/m³ MDEA Solution at 40°C (122).

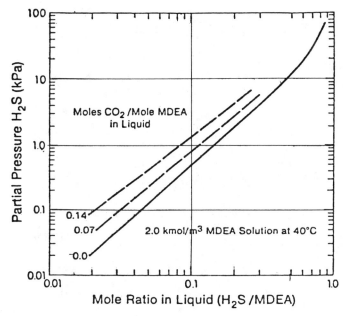

Figure 2-39. Effect of CO_2 on the Solubility of H_2S in a 2.0 kmol/m³ MDEA Solution at 40°C (122).

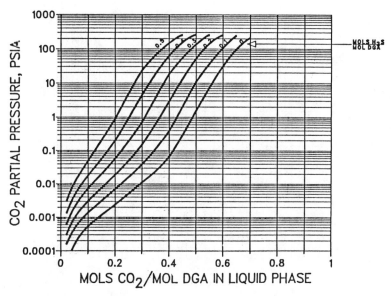

Figure 2-40. CO_2 Partial Pressure Curves for the Diglycolamine Agent - H_2S - CO_2 - H_2O System at 100°F, 65/35 Amine-Water Weight Ratio (112).

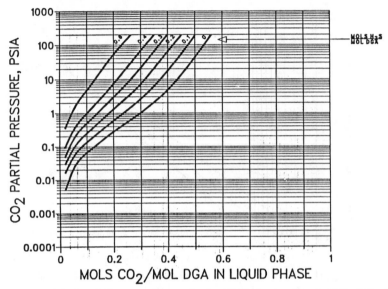

Figure 2-41. CO_2 Partial Pressure Curves for the Diglycolamine Agent - H_2S - CO_2 - H_2O System at 180°F, 65/35 Amine-Water Weight Ratio (112).

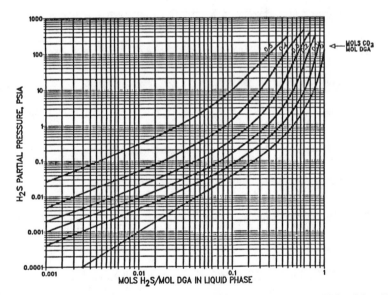

Figure 2-42. H_2S Partial Pressure Curves for the Diglycolamine Agent - H_2S - CO_2 - H_2O System at 100°F. 65/35 Amine-Water Weight Ratio (112).

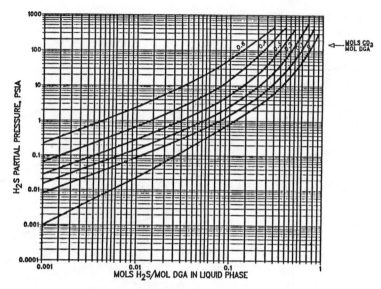

Figure 2-43. H₂S Partial Pressure Curves for the Diglycolamine Agent - H_2S - CO_2 - H_2O System at 180°F, 65/35 Amine-Water Weight Ratio (112).

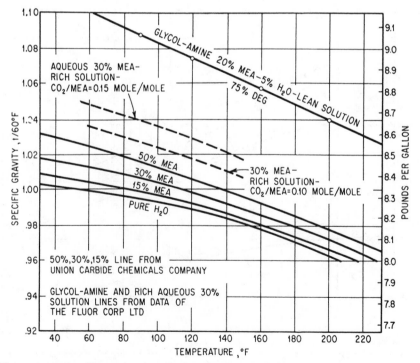

Figure 2-44. Specific gravity of monoethanolamine solutions.

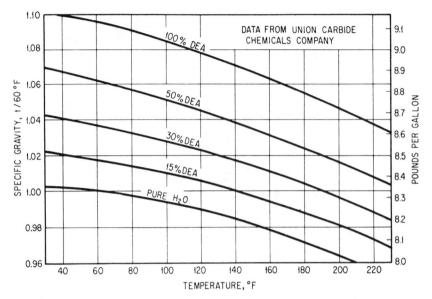

Figure 2-45. Specific gravity of diethanolamine solutions.

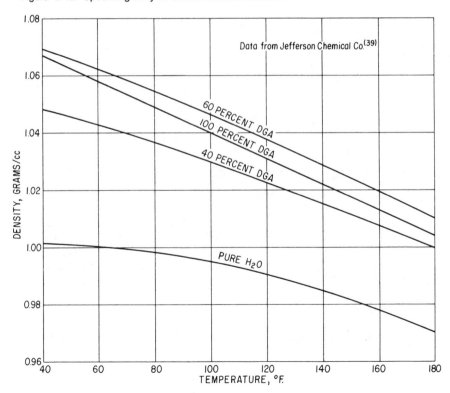

Figure 2-46. Density of Diglycolamine solutions.

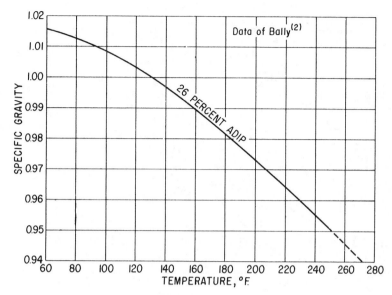

Figure 2-47. Specific gravity of Adip solution.

tem $DGA-H_2S-CO_2$, with a correlation of the data for extrapolation over a wide range of conditions was presented by Dingman et al of Texaco Chemical Co. (112). A limited amount of vapor-liquid equilibrium data for sterically hindered amines was reported by Sartori and Savage of EXXON Research and Engineering Company (92).

Physical data on alkanolamines beyond those shown (Figures 2-44 to 2-58) are available from the references cited above, chemical manufacturers, and from process licensors. The volume of such data is so large, that it would be impractical to incorporate them in this text.

Kent and Eisenberg (77) presented a method for correlating published vapor-liquid equilibrium data for the systems H_2S-CO_2-MEA and H_2S-CO_2-DEA, resulting in a model which can be used to extrapolate outside the range of existing data. The work is an expansion and modification of the studies of Dankwerts and McNeil (113). The curves calculated by the proposed method are in excellent agreement with published experimental data (Figure 2-25). Equilibrium constants given by Kent and Eisenberg (77) were used by Chen and Ng (114) for the construction of a series of monographs predicting equilibrium behavior of simultaneous H_2S and CO_2 absorption in MEA and DEA solutions. Another study extending the work of Kent and Eisenberg was presented by Moshfeghian, Bell, and Maddox (115).

As a result of better understanding of the reaction mechanism and the availability of more complete vapor-liquid equilibrium data, several computer models for plant design have been developed. Examples are described by Vaz

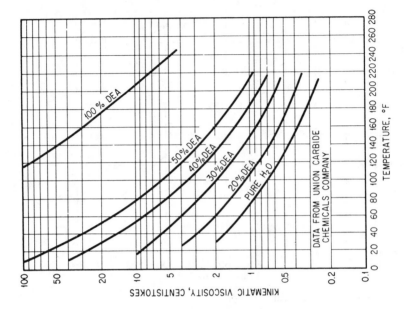

Figure 2-49. Viscosity of diethanolamine solutions.

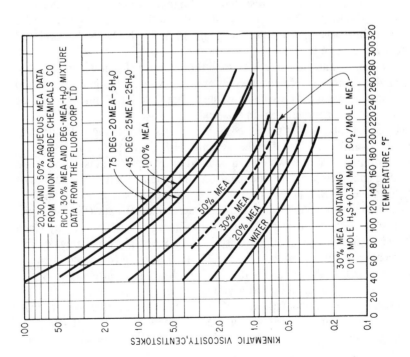

Figure 2-48. Viscosity of monoethanolamine solutions.

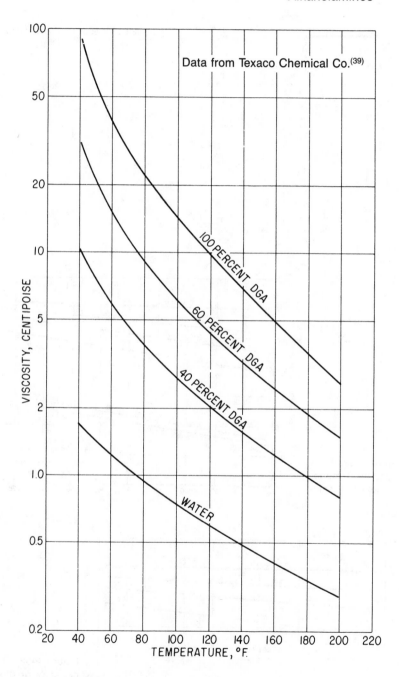

Figure 2-50. Viscosity of Diglycolamine solutions.

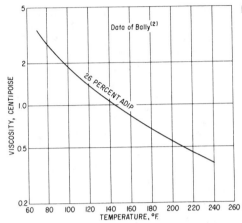

Figure 2-51. Viscosity of Adip solution.

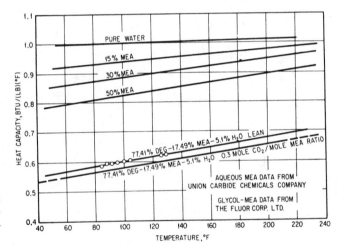

Figure 2-52. Heat capacity of mono-ethanolamine solutions.

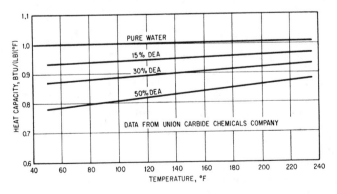

Figure 2-53. Heat capacity of diethanolamine solutions.

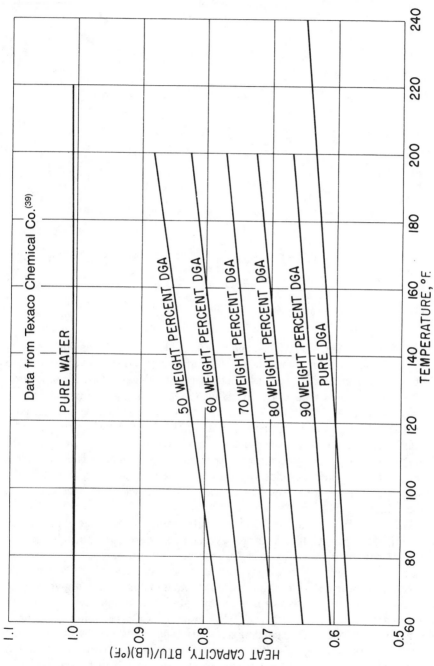

Figure 2-54. Heat capacity of Diglycolamine solutions.

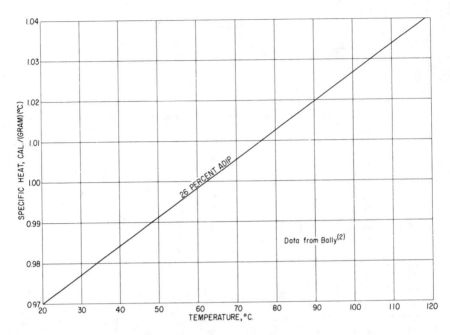

Figure 2-55. Specific heat of Adip solution.

and Maddox (116), Vaz, Maius, and Maddox (117) of Oklahoma State University and Tomcej, Otto, and Nolte (119). Computer programs are also available from commercial firms, such as ChemShare, Simulation Science, and others.

As a first approximation for calculating the required liquid-flow rate in non-selective systems (MEA, DEA, DGA) it can be assumed that the acid-gas content of the rich solution will approach to within about 75 to 80 percent of the equilibrium value with respect to the feed gas at the temperature at the absorber bottom. In many commercial operations, the approach is not even this close, and an appreciably closer approach is quite difficult to attain because the rate of absorption decreases rapidly with increased acid-gas concentration in the solution.

Precise calculation of the liquid-flow rate requires knowledge of the degree of stripping obtained in the stripping column. However, an approximate rate can be estimated on the basis of the absorber alone by assuming a lean solution composition and making a correction, if necessary, after the stripping column is designed. For aqueous monoethanolamine and diglycolamine solutions used for CO_2 absorption, plant data indicate that the lean solution will contain between 0.05 and 0.2 mole of CO_2 per mole of mono-ethanolamine or DGA, depending upon stripping conditions. A value of 0.15 may be considered typical if a low-pressure stripping column is used. For material-balance purposes, it is safe to assume that all of the H_2S will be removed from the solution in the stripping column. Monoethanolamine-glycol mixtures of low water-content and solutions of the weaker amines (diethanolamine, triethanolamine, and methyldiethanolamine) can be assumed to be stripped essentially free of CO_2 and H_2S. Note, however, that in all amine systems, the small amount of H_2S left in the lean solution must be considered with respect to its effect on the purity of the product gas.

Estimation of the liquid flow rate in systems for selective H_2S absorption is considerably more complex than in nonselective systems. While in the latter the amount of liquid circulation required to absorb the acid gases is determined by the vapor-liquid relations at the bottom of the absorber and the degree of stripping of the lean solution, selective absorption relies on the difference in the rate of reaction of the selective amine, e.g., MDEA, with H_2S and CO_2. A number of factors, such as the ratio of H_2S to CO_2 in the feed gas, absorption pressure and temperature, mechanical design of contact equipment, height of packing (or number of trays), liquid/gas ratio and solution physical properties have to be taken into account. The rate of H_2S absorption in MDEA solutions has been found to decrease appreciably faster with solution loading than the rate of CO_2 absorption (108).

Experimental solution loadings reported in the literature vary over a wide range, from less than 0.1 mole of acid gas per mole of MDEA to more than 0.9 mole of acid gas per mole of MDEA. The same is true of reported selectivity which ranges from less than 40 percent to more than 80 percent CO_2 slippage. Table 2-5 shows a selection of published data on pilot and commercial plant operations with MDEA solutions.

The key to effective high selectivity is gas-liquid contact time, or the number of contact stages. This effect is shown in the work of Miller and Kohl (6), Viddauri and Ferguson (120), Sigmund et al (108), Viddauri and Kahre (71), and Pierce (102). Sigmund et al propose a flow scheme, shown in Figure 2-59, with a multiple stage absorber with fresh solution feed going to each stage. The absorber contains specially designed proprietary trays.

For a theoretical treatment of selective absorption the reader is referred to the articles by Savage, Funk and Astarita (94), Blanc and Elgue (105) and Ouwerkerk (101).

Column Temperatures

To estimate the temperature of the solution leaving the absorber, it is necessary to know the heat of reaction and heat capacity of the solution. Approximate data for heats of reaction are given in Table 2-6, and data on heat capacity are presented in Figures 2-52 to 2-55 (pages 72–74). A method of determining heats of reaction from partial pressure data has been proposed by Crynes and Maddox (36).

When gas streams containing relatively large proportions of acid gases (over 5 percent) are purified, the quantity of solution required is normally so

Table 2-5. Selective Absorption of H_2S and CO_2 in MDEA Solutions

	Absorber			Solvent	Feed Gas		Product Gas		MDEA Loading	
Number of Trays	Press. Psig	Temp. °F		MDEA Wt-%	H_2S %	CO_2 %	H_2S ppmv	CO_2 % Passed	Mole Acid Gas/ Mole MDEA	Reference
16[1]	50	85		20	8.8	2.1	6	72	0.542	117
8[1]	250	85		15	2.0	24.0	6	68	0.651	117
20[2]	78	< 110		14.8	8.5	1.4	25	70	0.489	117
8[3]	150			16-19	5.3	1.2	5,824	75	0.43	6
20[3]	150			16-19	5.1	1.2	80	32	0.57	6
Packing[4]	585	~ 100		35	0.7	8.0	4	41	0.508	5
½ Packing[4]	585	~ 100		35	0.74	9.5	46	59	0.555	5

(1) Pilot Plant, 2 inch I.D. Absorber
(2) Refinery Absorber, 6 ft. I.D., 20 Type A Koch Flexitrays.
(3) Refinery Absorber, Bubblecap Trays.
(4) Pilot Plant Absorber, 4-inch I.D., 11 ft. 3/8 inch Porcelain Raschig Rings.

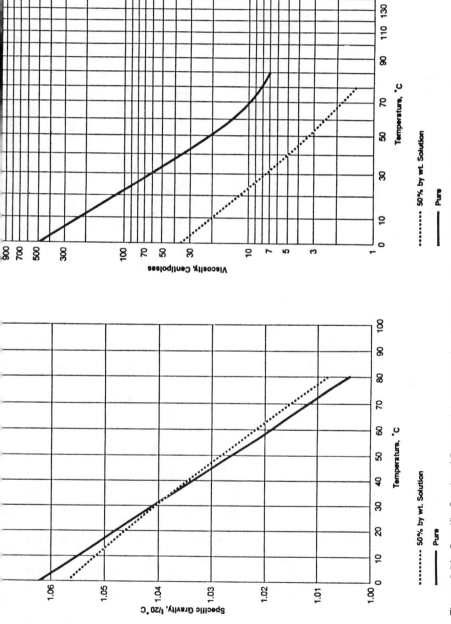

Figure 2-57. Viscosity of Pure and Aqueous UCARSOL™ HS Solvent 101 (82).

Figure 2-56. Specific Gravity of Pure and Aqueous UCARSOL™ HS Solvent 101 (82).

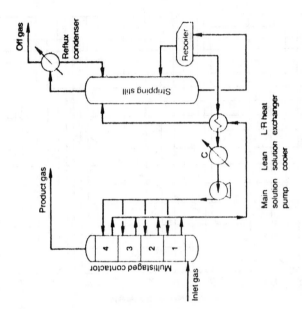

Figure 2-59. Typical HS Process Flow Diagram (108). Courtesy of *Hydrocarbon Processing*, May 1981.

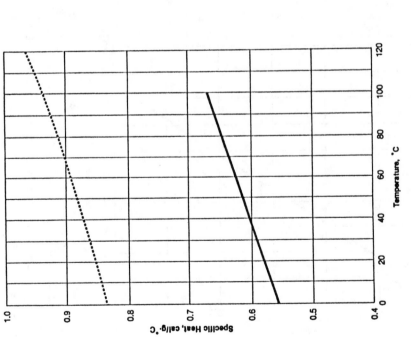

Figure 2-58. Specific Heat of Pure and Aqueous UCARSOL™ Solvent 101 (82).

Table 2-6. Heats of Reaction for Absorption of H₂S and CO₂ in Alkanolamines Solutions

Acid Gas	Amine	Heat of Reaction btu/lb gas
H₂S (37)	MEA	820
H₂S (38)	DEA	511
H₂S (38)	TEA	400
H₂S (3)	DIPA	475*
H₂S (39)	DGA	674
CO₂	MEA	825**
CO₂ (38)	DEA	653
CO₂ (38)	DEA	653
CO₂ (82)	TEA	425
CO₂ (39)	DGA	850
CO₂ (82)	PIPA	720
CO₂ (82)	MDEA	475

*At 40°C and 0.4 mole of H₂S per mole of amine
**Calculated for 0.4 mole of CO₂ per mole of MEA from Figure 2-12

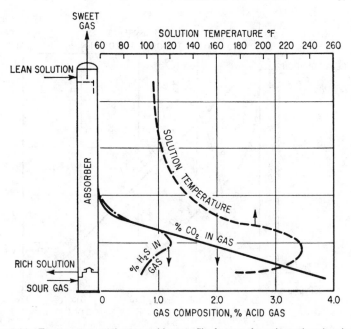

Figure 2-60. Temperature and composition profile for treating plant absorber handling a gas stream containing a high concentration of acid gas. *Data from Plant A of Table 2-10.*

large that the purified gas is cooled by the lean solution at the top of the column to within a few degrees of the temperature of the lean solution. In such cases, essentially all of the heat of reaction is taken up by the solution, which leaves the column at an elevated temperature. A typical temperature profile for an absorber of this type is shown in Figure 2-60. The temperature "bulge" is a result of the cool inlet-gas absorbing heat from the rich solution at the bottom of the column and later losing this heat to the cooler solution near the upper part of the column. The effect is similar to that of preheating air and fuel for a burner with the combustion products so that an internal temperature may be attained which considerably exceeds that calculated on the basis of heat of reaction and heat capacities.

When more dilute gases are purified, the quantity of gas may be so large, relative to the mass of solution, that the gas leaving the contact zone will carry more of the heat which is generated than will the solution. In the extreme case, illustrated in Figure 2-61, the solution is cooled to approximately the temperature of the incoming gas before it leaves the column, and essentially all of the heat of reaction is taken out of the column by the gas stream, which leaves at an appreciably higher temperature than that of the incoming lean solution.

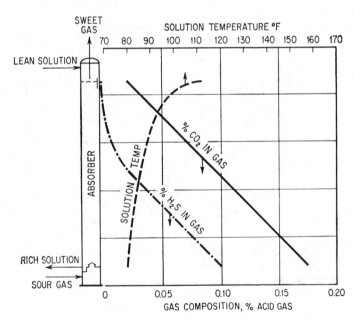

Figure 2-61. Temperature and composition profile of gas-treating plant absorber handling a gas stream containing a low concentration of acid gas. *Data from Plant E of Table 2-8.*

Absorber Diameter

After establishing the approximate liquid- and gas-flow rates and physical properties of the two streams, the required column diameter can be calculated by conventional techniques. For packed towers, correlations of the type proposed by Sherwood et al. (40), Elgin and Weiss (41), Lobo et al. (42), and Zenz and Eckert (43) are satisfactory (see Chapter 1). Pressure-drop data for specific packing shapes are available in the trade literature supplied by packing manufacturers.

The use of flooding correlations of this type for designing commercial columns has been described by Reed and Wood (27). In the operation of relatively small pilot-scale absorption equipment, the authors have found that flooding frequently occurs well below the values predicted by such correlations, presumably due to the deposition of iron sulfide on the packing and possibly some tendency for foaming to occur. Because of these factors, it is recommended that a conservative safety factor be utilized in establishing the diameter of packed columns. Design rates not over 60 percent of those of flooding are recommended. Reed and Wood (27) also discuss the design of bubble-cap columns for amine plant absorbers and conclude that the Souders and Brown (44) correlation based upon entrainment can be satisfactorily used. The detailed design of bubble-cap trays is discussed by most texts on chemical engineering and need not be reviewed in this work. The reader is referred to Brown and associates (45) for a particularly complete discussion of this problem.

Design data for sieve trays are available from Fractionation Research Inc. to members of that organization. Manufacturers of proprietary trays, such as valve type trays, usually reserve the right to set the column diameter for guarantee purposes. However, most of these manufacturers supply shortcut methods to enable the designer to arrive at preliminary column sizes. The possibility of foaming and iron sulfide deposition on the trays in amine plant absorbers should not be overlooked, however. The use of a conservative gas velocity is therefore recommended.

Absorber Height

The problem of estimating the column-height requirement is considerably more difficult than determining that of the diameter as the mathematical formulation of absorption with chemical reaction is quite complex. It is therefore common practice to correlate absorption coefficients and tray-efficiency data on an empirical basis for application to practical problems in plant design. Studies of the rates of absorption of CO_2 and H_2S in various ethanolamines in packed columns have been reported by Cryder and Maloney (46), Gregory and Scharmann (47), Wainwright et al. (48), Benson et al. (49, 50), Teller and

Ford (51), and Shneerson and Leibush (52, 53). An attempt at correlating the data of Cryder and Maloney for diethanolamine on a theoretical basis has been made by van Krevelen and Hoftijzer (54). Packed-column performance data for the absorption of H_2S in methyldiethanolamine solutions in the presence of CO_2 have been presented by Frazier and Kohl (4) and Kohl (5). Some data on absorption coefficients for CO_2 in aqueous mono- and diethanolamine solutions with Intalox saddles and Pall rings have been reported by Eckart, Foote, Rollison, and Walter (55). In addition to these studies, numerous descriptions of commercial operations have appeared in the literature.

A comprehensive theoretical treatment of the absorption of CO_2, H_2S, and carbonyl sulfide (COS) in solutions of alkalis and amines has been presented by Danckwerts and Sharma (56). Much data on the subject, available up to 1966 are reviewed and design procedures based on fundamental concepts are proposed. Although the design methods are considered to be sound in principle, Danckwerts and Sharma recognize that additional fundamental information will have to be obtained before rigorous design procedures can be developed. A methodology for absorber design, based on the use of chemical reaction rate data has been described by Ouwerkerk for the case of H_2S and CO_2 absorption in DIPA. (101)

Absorption of CO_2 in Packed Columns by Monoethanolamine Solutions

The published data clearly indicate that the liquid film is the controlling factor in the absorption of CO_2 by any of the commonly used amines. Correlation is generally made on the basis of K_Ga values, however, rather than those for K_La because the former can be more readily calculated from experimental data and are more directly applicable to the design of commercial columns where the efficiency of CO_2 removal is usually of primary interest.

In general, it has been found that K_Ga increases with increased liquid loading and decreases with increased concentration of CO_2 in the solution. Increasing temperature or amine concentration causes K_Ga to rise to a maximum and then decrease. Several investigators have reported K_Ga to decrease with increased partial pressure of CO_2 over the solution. This phenomenon is explainable on the basis of theoretical equations developed to express the process of absorption followed by chemical reaction.

The most complete data for the absorption of CO_2 in aqueous monoethanolamine solutions are those of Shneerson and Leibush (52), although the utility of these data for design is decreased by the fact that they were obtained in a small laboratory column only 1 in. in diameter and packed with 5-to 6-mm glass rings. A correlation of these data has been made to include the observed effects of temperature, partial pressure, viscosity, CO_2

content of solution, and amine strength, and the following equation was proposed (57):

$$K_Ga = 0.56/\mu^{0.68}[1 + 5.7(0.5 - C)Me^{0.0067T-3.4p}] \qquad (2\text{-}6)$$

where K_Ga = overall gas-film coefficient, lb moles/(hr)(cu ft)(atm)

μ = viscosity, centipoises

C = concentration of CO_2 in the solution, moles/mole monoethanolamine

M = amine concentration of solution (molarity, g moles/liter)

T = temperature, °F

p = partial pressure, atm

The equation was developed on the basis of runs made at a liquid-flow rate of 695 lb/(hr)(sq ft) and gas velocities varying from 0.09 to 0.55 ft/sec. Gas velocity has not been found to have an appreciable effect; however, liquid-flow rate is quite important. If all of the resistance is in the liquid phase, it would be expected that K_Ga could be extrapolated to other liquid-flow rates by assuming it to vary approximately as $L^{2/3}$, although there is some disagreement as to the proper exponent. Extrapolation of the data of Shneerson and Leibush to other sizes of packing is even more doubtful, however, as the effect of this variable on the absorption coefficient has not been clearly defined. Koch et al. (58) studied the absorption of CO_2 in water and concluded that for packings from ⅜ to 1¼ in. in diameter, the size had no effect on K_La (which they found to vary as $L^{0.96}$). Sherwood and Holloway (59) on the other hand, found that packing size affected K_La and, in fact, that the liquid-flow rate exponent was also affected by the packing used. Their data indicate, for example, that at a liquid-flow rate comparable to that used by Shneerson and Leibush [695 lb/(hr)(sq ft)], K_La (for O_2 desorption) with ½-in. rings was almost twice that with 2-in. rings.

Comparison of the small amount of available data on monoethanolamine absorption in large columns with the Shneerson and Leibush values seems to indicate that the very small packings which they used resulted in values for K_Ga two to ten times greater than could be expected with columns using ⅜-in. rings or larger.

To take into account the effect of liquid-flow rate and packing size, as well as the variables defined above, the following general equation* is suggested for the absorption of CO_2 by aqueous monoethanolamine solutions in packed columns:

$$K_Ga = F(L/\mu)^{2/3}[1 + 5.7(C_e - C)Me^{0.0067T-3.4p}] \qquad (2\text{-}7)$$

---

*The equation is believed to be valid only for CO_2 partial pressures below ½ atm, temperatures below 125°F, and CO_2/amine ratios below 0.5 moles/mole.

where L = liquid-flow rate, lb/(hr)(sq ft)
      C_e = equilibrium concentration of CO_2 in solution, moles/mole monoethanolamine
      F = factor to correct for size and type of packing which has approximately the values given in Table 2-7.

As K_Ga is very sensitive to both CO_2 partial pressure and degree of solution saturation, it is found to vary appreciably from the top to the bottom of any absorber. In attempting to use an average K_Ga for an entire column, it has been found that a logarithmic mean of these two variables [p and ($C_e - C$)] is most satisfactory.

Application of Equation 2-7 to data for several packings is shown in Figure 2-62. Values of F can be estimated from such a chart by noting the intercept of the various packing lines with a common ordinate (e.g., 1,000). Only a limited number of the runs published by Gregory and Scharmann are used in Figure 2-62 because many showed zero CO_2 concentration in the product gas. This value obviously cannot be used in connection with a logarithmic-mean driving force. Since material balances for these plant tests were poor, average values for observed values of K_Ga, based on gas and liquid data, are used.

The performances of several commercial packing designs for CO_2 absorption by aqueous monoethanolamine solutions at atmospheric pressure have been compared by Teller and Ford (51), and data from all of the runs by these investigators are included in Figure 2-62. These authors studied nominal 1-in. ceramic berl saddles, steel raschig rings, and a polyethylene packing shape referred to as "Tellerettes." The results of the investigation indicated that the Tellerettes have an absorption efficiency from 23 to 72 percent higher than that of the more conventional packings and a pressure drop from 46 to 78 percent that of the raschig rings.

Table 2-7. Values of F for Different Packings

Packing	F	Basis for calculation of F
5- to 6-mm glass rings	7.1×10^{-3}	Shneerson and Leibush (52) data, 1-in. column, atmospheric pressure
⅜-in. ceramic rings	3.0×10^{-3}	Unpublished data for 4-in. column, atmospheric pressure (19)
¾- by 2-in. polyethylene Tellerettes	3.0×10^{-3}	Teller and Ford (51) data, 8-in. column, atmospheric pressure
1-in. steel rings		
1-in. ceramic saddles	2.1×10^{-3}	
1½- and 2-in. ceramic rings	$0.4\text{-}0.6 \times 10^{-3}$	Gregory and Scharmann (47) and unpublished data for two commercial plants (19), pressures 30 to 300 psig

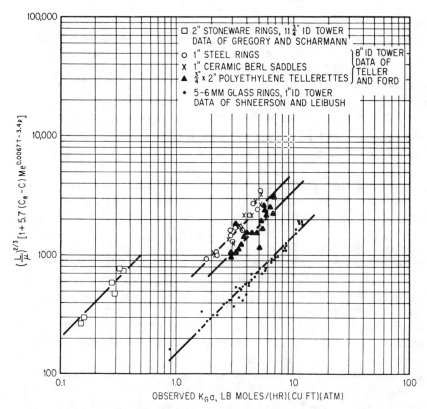

Figure 2-62. Correlation of K_Ga for CO_2 absorption by aqueous monoethanolamine solutions in packed towers employing various packings. *Based on data of Gregory and Scharman (47) Teller and Ford (51) and Shneerson and Leibush (52)*

Additional data on CO_2 absorption in aqueous monoethanolamine solutions are presented in two papers describing work by the U.S. Bureau of Mines (49,50). However, no attempt has been made to correlate these data by Equation 2-7 as CO_2 partial pressure and CO_2 saturation values are generally higher than the limits set above. The Bureau of Mines data were obtained to compare the steam economy of the hot potassium carbonate process (see Chapter 5) with that of monoethanolamine absorption. The columns were therefore operated with the solution and gas more nearly at equilibrium than they are in most commercial operations. This resulted in relatively good steam economy but generally poor efficiency of CO_2 removal. In the work done by the U.S. Bureau of Mines, K_Ga values for CO_2 absorption in 15 percent monoethanolamine solutions varied from 0.061 to 0.11 lb mole/(hr)(cu ft)(atm) with operation at 300 psig, 20 percent CO_2 feed gas, 0.4 to 2.2 percent CO_2 product, and liquid-flow rates from 1,800 to 5,000 lb/(hr)(sq ft) (49). The column for these studies was a 6-in. pipe packed to a height of 20.8 ft with 1-in. porcelain raschig rings. Values of K_Ga are also presented for a 30 percent monoethanolamine solution; however, these are even more variable, ranging from 0.05 to 1.6 lb moles/(hr)(cu ft) (atm), depending upon liquid-flow rate, degree of regeneration, and other factors (49, 50).

Absorption of CO_2 in Packed Columns by Diethanolamine and Triethanolamine Solutions

Diethanolamine and triethanolamine solutions are appreciably less efficient for the absorption of CO_2 than monoethanolamine solutions, although, under some circumstances, they are used for this purpose. The data of Shneerson and Leibush (52) are of value in comparing the absorption coefficients of the three amines because they conducted studies with each amine in the same apparatus. They concluded that under the same conditions, K_Ga values for monoethanolamine solutions are two to two and one-half times greater than they are for diethanolamine and twenty to thirty times greater than for triethanolamine. Data on the absorption of CO_2 by diethanolamine solutions in an apparatus more nearly appraoching commercial size have been presented by Cryder and Maloney (46), who used an 8-in. diameter column packed with ¾-in. raschig rings. They found K_Ga to decrease with increased saturation of the solution, increase with increased liquid-flow rate, and decrease with increased CO_2 partial pressure. The K_Ga values which they observed for 1 N and 2 N diethanolamine solutions are plotted in Figure 2-63. The coefficients are plotted against $(L^{2/3})$ $(0.5-C)$ to indicate the combined effect of these factors. To minimize scatter resulting from wide variations in other variables, runs at temperatures above 40°C, feed-gas CO_2 contents above 20 percent, and liquid-flow rates above 2,100 lb/(hr)(sq ft) are not included. Runs were also made with 3 N and 4 N diethanolamine solutions. At these higher normalities, the coefficients were found to decrease, presumably due to the increased viscosity. Curves for the 3 N or 4 N solutions would fall in an intermediate position between those for 1 N and 2 N solutions shown in Figure 2-63.

Absorption of CO_2 in Plate Columns

The problem of estimating plate efficiency when CO_2 is absorbed by aqueous monoethanolamine solutions has been analyzed by Kohl (57). The study is based on the use of an equation similar to Equation 2-6 to correlate the effects of viscosity, amine strength, CO_2 concentration, temperature, and partial pressure of CO_2 on the absorption coefficient and on the use of equation 2-8 to relate the absorption coefficient to plate efficiency.

$$E_{MV} = 1 - e^{-K_G(A/V)RT_A} \qquad (2\text{-}8)$$

where E_{MV} = Murphree vapor efficiency for single plate
   A = interfacial contact area, sq ft/sq ft of tray
   = ah, where a = interfacial area, sq ft/cu ft of contact volume
      h = height of contact zone
   V = actual gas volume, cu ft/(hr)(sq ft of tray)
   R = gas constant
   T_A = absolute temperature, °R

For the specific column tested at design gas rate, $K_G((A/V)$ was found to be represented by the following equation:

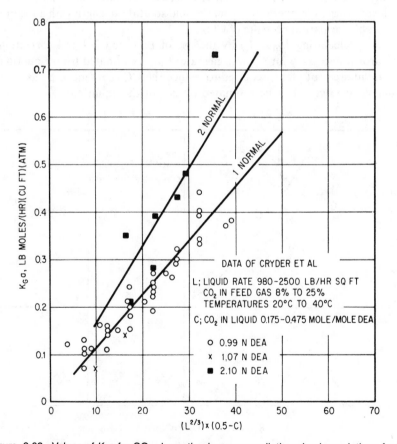

Figure 2-63. Values of $K_G a$ for CO_2 absorption in aqueous diethanolamine solutions for 8-in. column packed with 3/4-in. raschig rings.

$$K_G(A/V) = 1.2 \times 10^{-4}/\mu^{0.68}[1 + 1.2(0.5 - C)Me^{0.0067T-3.4p}] \qquad (2\text{-}9)$$

The equation was developed from data on a commercial column operating at atmospheric pressure with a gas rate which resulted in an approximate superficial velocity of 35 ft/sec through the bubble-cap slots. It was observed that, at reduced velocity, an increased plate efficiency was obtained presumably due to an increase in (A/V). If the factor to correct for (A/V) variation is taken to be unity at 35 ft/sec, the value which it would have at lower gas velocities can be estimated from Figure 2-64.

As in Equation 2-7, the factor $(0.5-C)$ should probably be replaced by (C_e-C) if the equilibrium concentration of CO_2 in the amine solution is appreciably above 0.5 moles CO_2 per mole of monoethanolamine. The equation can be applied to single plates or to groups of plates where concentration changes are not extreme. In the latter case, average values should be used for such variables as temperature, viscosity and solution strength with a log mean preferable for partial pressure of CO_2.

After calculating $K_G(A/V)$ by the use of Equation 2-9 and correcting it upwards if necessary for a lower gas rate, E_{MV} is estimated from Equation 2-8. The number of plates required to reduce the CO_2 content of a gas stream from y_1 to y_2 can then be estimated by use of the equation.

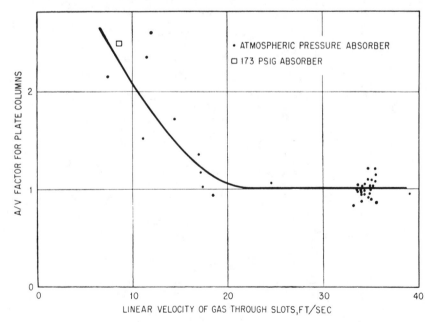

Figure 2-64. Correction factor to adjust $K_G(A/V)$ for lower gas rates. Value of $K_G(A/V)$ is to be calculated by Equation 2-9 and multiplied by indicated factor.

$$n = - \frac{\ln (y_1/y_2)}{\ln (1 - E_{MV})}$$ (2-10)

This assumes constant plate efficiency and zero CO_2 vapor pressure over the solution. If the plate efficiency varies appreciably over the column or plate section under consideration, a conventional graphical method may be used as shown in Figure 2-65.

No correlation is available for estimating the plate efficiency of columns operating with diethanolamine solutions. However, if it is assumed that a similar relationship holds, and one accepts the findings of Shneerson and Leibush (52) that $K_G a$ for absorption of CO_2 in diethanolamine solutions is approximately one-half that for monethanolamine, it would be expected that approximate plate efficiencies from 5 to 15 percent would be obtained.

Absorption of H₂S and H₂S—CO₂ Mixtures

Less work has been done on the rate of absorption of H_2S by amine solutions than on the rate of CO_2 absorption. However, the lack of absorption-rate data is normally of little concern in the design of H_2S absorbers because of the relatively high efficiencies obtained. In fact, it is not uncommon for purified-gas streams leaving the top of amine absorbers to be practically in equilibrium with the inlet solution, with regard to H_2S content. This has the effect of making the stripper performance more important than the absorber height if very pure gas is required. The effect is graphically shown in Figure 2-66, which presents the data of Wainwright et al. (48) for the absorption of H_2S by triethanolamine from a gas containing a high percentage of CO_2.

Leibush and Shneerson (53) conducted a study of H_2S absorption in aqueous monoethanolamine and diethanolamine solutions using the apparatus described in connection with their CO_2 absorption studies (52). They found, in general, that the coefficients for H_2S absorption were three to five times those for CO_2 absorption under equivalent conditions. They found H_2S absorption to resemble that of CO_2 in that raising the degree of conversion of ethanolamine, increasing the acid-gas content of the gas, or decreasing the liquid-flow rate resulted in a decrease of the absorption coefficient. One point of difference, however, was in the effect of temperature. Raising the temperature was found to decrease the absorption coefficient even in the low-temperature range. Because of the higher absorption coefficient, H_2S was found to be absorbed somewhat selectively by either of the amines studied. With a mixed gas containing two and one-half to twenty times as much CO_2 as H_2S they found that H_2S has an absorption coefficient six to ten times greater than that of CO_2. Typical values which they determined for H_2S absorption alone are presented in Table 2-8. All runs are at a temperature of about 77°F and a liquid-flow rate of 400 lb/(hr)(sq ft).

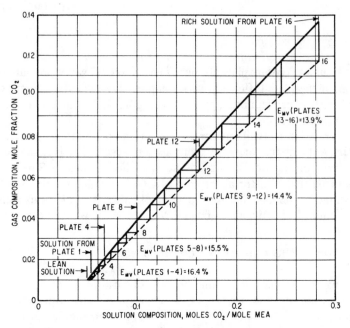

Figure 2-65. Graphical analysis of plate-efficiency data for atmospheric-pressure CO_2 absorption with 14.5 percent aqueous monoethanolamine in atmospheric pressure bubble-cap column. *Data of Kohl* (57)

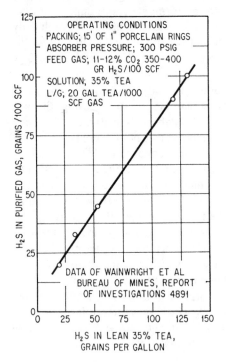

Figure 2-66. Effect of H_2S concentration in lean solution on gas purity obtainable with aqueous triethanolamine solution in packed absorber.

Purity requirements for high pressure natural gas, to make it suitable for transmission and utility use, are very high. A maximum H_2S content of 0.25 grain/100 scf (standard cubic feet) is frequently specified. Because of difficulties in analyzing for H_2S in concentrations appreciably below this value, as well as the questionable accuracy of analyses of lean solutions and equilibrium data at this level, H_2S absorption coefficients (or tray efficiencies) cannot be satisfactorily calculated from data on commercial columns, in which terminal conditions only are normally presented.

The available commercial data are of value, however, in indicating the degree of purity obtainable.

In general, it can be stated that H_2S concentrations of 0.25 grain/100 scf or less in the treated gas are obtainable with properly stripped monoethanolamine solutions at operating pressures above 200 psig, provided, of course, that sufficient packing height or number of trays are available for gas-liquid contact. The same gas purities can be attained with diethanolamine solutions at somewhat higher operating pressures, on the order of 500 psig or above.

Design of Stripping Column

As in the case of absorbers, stripping columns may be designed to use trays or packing, and, again, column diameters are established on the basis of conventional design equations. Principal design problems are the calculation of reboiler heat load and establishment of the column height which is required.

Estimation of Heat Load

In conventional stripping operations, heat is supplied to the column by steam or by a heat medium within tubes in the reboiler. Sufficient heat must

Table 2-8. Absorption Coefficients of H_2S Absorption in Monoethanolamine and Diethanolamine Solutions

Absorbent	Per cent H₂S in gas		H₂S in solution, moles/mole amine		K_Ga, lb moles/(hr) (cu ft)(atm)
	Feed	Exit	Feed	Exit	
2.5 N MEA	1.08	0.11	0.007	0.107	39.9
2.5 N MEA	1.08	0.40	0.207	0.275	25.6
2.5 N MEA	8.80	3.86	0.007	0.378	11.2
2 N DEA	2.62	0.39	0.007	0.101	12.8
2 N DEA	2.56	1.64	0.266	0.315	5.7

Source: Data of Shneerson and Leibush (52) for 1-in.-diameter laboratory column packed with 5- to 6-mm glass rings.

be supplied (a) to provide sensible heat to raise the temperature of the feed solution to that of the lean solution leaving the reboiler, (b) to provide heat of reaction to dissociate the amine acid-gas compounds, and (c) to evaporate water which leaves the stripping section of the column as vapor with the acid-gas stream.

Heat-capacity data for typical alkanolamine solutions are presented in Figures 2-52 to 2-55. Data are included for one rich MEA solution to illustrate the effect of acid gas on the specific heat. It will be noted that the glycol-amine solutions has an appreciably lower specific heat than the aqueous solutions because of the effect of the glycol.

Approximate heats of reaction for several amines are presented in Table 2-6. The heat of reaction for CO_2 and monoethanolamine is approximately constant at the given value when the CO_2 concentration in the solution is increased, up to a concentration of about 0.5 moles CO_2 per mole monoethanol-amine (representing the carbonate or carbamate compounds). At higher concentrations of CO_2, the heat of reaction falls off rapidly. Heat of reaction for H_2S and monoethanolamine solutions also varies somewhat with H_2S concentration in the solution, although no sharp break occurs. A similar effect of acid-gas concentration can be assumed for other alkanolamines. In general, however, for the CO_2 and H_2S concentrations encountered in most engineering-design calculations, the value given in Table 2-6 are of sufficient accuracy.

The quantity of stripping vapor required depends upon the solution purity needed to produce the required product gas, the stripping-column height, and the nature of the solution. The stripping vapor which passes out of the column with the acid gas is normally condensed and returned to the column as reflux, resulting in a heat load additional to the sensible and latent heat requirements of the operation. The ratio, moles water in the acid gas from the stripping column to moles acid gas stripped, is commonly referred to as the "reflux ratio" and is used in design as a convenient measure of the quantity of stripping vapor provided.

Typical reflux ratios in commercial columns range from 3:1 to less than 1:1, depending on the gas purity required, the gas composition (i.e., the ratio of $H_2S:CO_2$), the amine used, and the height of the stripping column. In general, aqueous monoethanolamine solutions require the highest reflux ratios while the other amines, especially diethanolamine, are stripped to a satisfactory degree at substantially lower reflux ratios. Experience with DGA solutions indicates that reflux ratios of 1.5 mole of water per mole of acid gas are usually satisfactory for adequate stripping.

Methyldiethanolamine being less basic than MEA, DEA and DGA is stripped more easily to a satisfactory level than these primary and secondary amines. Reflux ratios ranging from 0.3 to 1.0 mole water per mole of acid gas have been reported (95). Figure 2-67 shows the effect of reflux ratio on the H_2S content of the purified gas in an MDEA system (120).

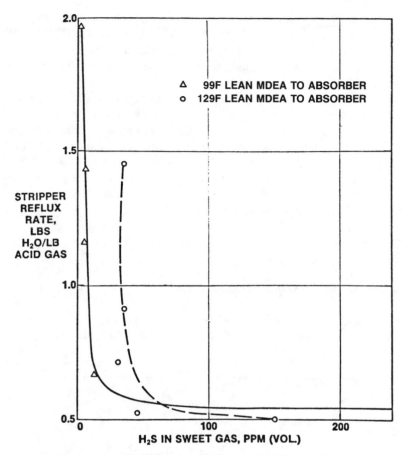

Figure 2-67. Effect of Stripper Reflux on H₂S in Sweet Gas (120).

In commercial design it is customary to express the heat requirements for solution stripping in terms of pounds of steam per gallon of circulated solution. Depending again on the required gas purity, the ratio of H_2S to CO_2 in the gas, the amine used, the height of the stripping column, and, in addition, the temperature at the bottom of the column, this value may range from less than 1 lb/gal to 1.5 or even 2 lb/gal. The critical parameter for high-pressure natural gas purification is the residual H_2S in the lean solution, as it determines the equilibrium conditions at the top of the absorber and thus the H_2S content of the treated gas. Buskel (60) reports stripping steam requirements, as related to residual H_2S in the solution in a large commercial unit. The feed gas contained 11.0 percent H_2S and 6.0 percent CO_2 and was treated at a pressure of 1,000 psig with a 20 percent aqueous monoethanolamine solution.

The stripping column contained 19 stripping trays. The data are shown in Figure 2-68. Data presented by Estep et al. (25) are included in Figure 2-68. These data were obtained in a commercial installation using a 14.4 percent aqueous monoethanolamine solution for treating natural gas containing 19.90 percent H_2S and 1.84 percent CO_2.

A correlation of residual H_2S versus the ratio of $H_2S:CO_2$ in the gas treated for different steam rates has been presented by Fitzgerald and Richardson (61) and is shown in Figure 2-69. These data are based on a study of thirteen Canadian plants using aqueous monoethanolamine solutions ranging in concentration from about 11 to 22 percent, for the treatment of gases of widely varying composition. It is interesting to note that the amount of steam required to obtain a given H_2S residual decreases with decreasing ratio of $H_2S:CO_2$, indicating that CO_2 acts as stripping vapor for H_2S.

The available operating data indicate that in most cases a steam rate of about 1 lb/gal is sufficient to obtain a satisfactory treated gas purity when mono- or diethanolamine solutions are used.

Stripping-Column Height Requirements

The number of trays (or height of packing) required for stripping columns is generally established on the basis of experience, rather than on rigorous

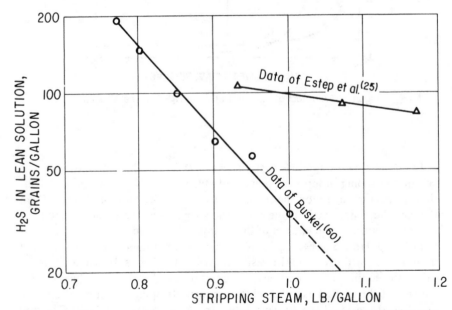

Figure 2-68. Effect of stripping steam rate on H_2S residual in lean monoethanolamine solutions.

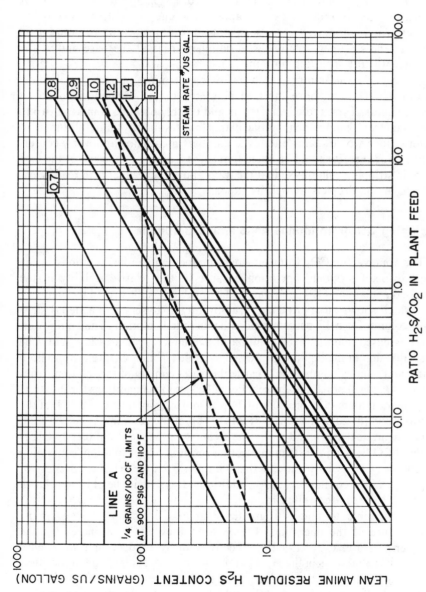

Figure 2-69. Effect of ratio H_2S:CO_2 on residual H_2S in MEA solutions. Data of Fitzgerald, et al. (61)

column calculations. Typical stripping columns used with conventional aqueous monoethanolamine and diethanolamine plants or glycol-amine units contain 12 to 20 trays below the feed point for stripping and two to six trays above the feed to prevent loss of vaporized amine. Split-stream plants obviously require more trays. One such unit, for example, employs a total of 33 trays in the stripping column, with chimney trays at the base and at the seventeenth tray from the bottom (24).

The less volatile amines, such as diethanolamine and triethanolamine, require fewer trays above the feed point to achieve adequate recovery of amine vapors. Typical diethanolamine and triethanolamine plants use two to four such trays; monoethanolamine plants four to six trays. Equilibrium conditions alone would indicate that the numbers mentioned are overly conservative. However, it should be noted that some foaming may occur at the feed point in stripping columns so that the trays above the feed serve to remove entrained amine droplets as well as vapor.

Stripping CO_2 from Monoethanolamine Solutions

The removal of CO_2 from aqueous monoethanolamine solutions is the most difficult stripping operation because of the relative stability of the monoethanolamine-CO_2 compounds. The problem is illustrated by Figure 2-70, which shows an approximate tray-by-tray analysis of an aqueous stripping column for a 17 percent monoethanolamine solution. The equilibrium curve is based upon an extrapolation of available vapor-pressure data. The concentration in the liquid is expressed as mole fraction CO_2 relative to both water and monoethanolamine because the water content of the solution varies between the feed point and the reboiler. The assumed conditions for the stripping operation are (a) a pressure of 24 psia and a temperature of 240°F at the reboiler and (b) 20 psia and 208 °F at the top of the column. As can be seen, eight theoretical trays and sufficient steam to produce 2.1 moles H_2O per mole of CO_2 leaving the stripping section result in a lean solution, containing 0.14 moles CO_2 per mole monoethanolamine, from the reboiler. Because of the shape of the equilibrium curve, additional trays would be of little value in reducing the required reflux ratio.

It is possible to obtain a greater degree of stripping by raising the temperature of the operation. This can be done either by increasing the stripper pressure, as described by Reed (62), or by a reduction in water content of the solution as in the glycol-amine process. A comparison of the effect of raising the temperature by increasing the pressure with the effect of glycol addition is shown in Figure 2-71, which is based upon data presented by Chapin (63). As these curves were derived from plant data in which reflux ratios, number of stripping trays, and other variables were not necessarily constant, they can be taken only as an indication of expected performance.

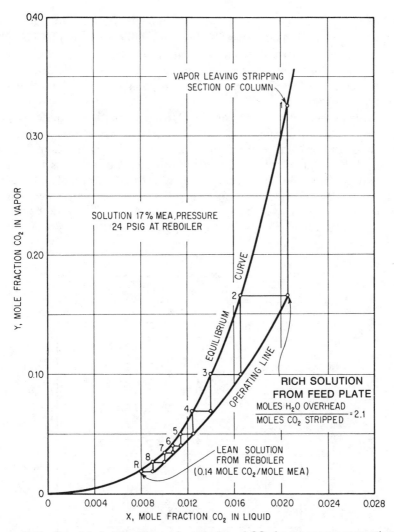

Figure 2-70. Calculated plate diagram for stripping of CO_2 from aqueous monoethanolamine solutions.

Data on the stripping effectiveness of an 8-in. ID stripping column packed with 25 ft of 1/2-in. raschig rings have been presented by Benson et al.(49). The column was operated at 10 psig with 15 and 30 percent monoethanolamine solutions (as well as potassium carbonate solutions; see Chapter 5). A maximum regeneration efficiency of about 5.2 scf of CO_2 absorbed per pound of steam was attained when using the 30 percent monoethanolamine solution and operating the plant in such a manner that the solution carried about 5.5 scf/gal of CO_2 (net). Compositions of lean solution varied from 0.1 to 0.5 moles CO_2 per mole monoethanolamine during the series of tests. Best overall steam economy was obtained when the solutions were stripped to values intermediate between these values. No attempt was made to correlate the effectiveness of the stripping operation independent of absorber performance.

A study of the rate of desorption of CO_2 from monoethanolamine solutions has been reported by Ellis et al. (64). The data obtained show that the approach to equilibrium is relatively slow and that, even with substantial driving forces, the time required for going half way toward equilibrium is of the order of minutes. The results of the study also indicate that the rate of desorption is not materially affected by the rate of stripping steam flow and by temperature variation over the range of 244° to 282°F. This work would indicate the usefulness of high weirs on stripping column trays.

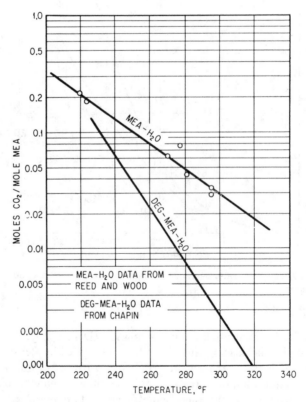

Figure 2-71. Effect of temperature on degree of CO_2 stripping obtainable for monoethanolamine solutions. Temperature increased for aqueous amine solution by increasing pressure; for glycolamine by increasing glycol content.

A study on stripping of CO_2 from MEA solutions in a packed column, involving experimental data and construction of a mathematical model, was presented by Weiland, Rawal, and Rice (121). The authors concluded that the measured mass transfer coefficients were predictable within ± 25 percent by use of the model.

Commercial Data

Design and performance data of several commercial units, using mono- and diethanolamine, Diglycolamine, and Adip solutions are available in the literature. A complete description of the performance of two installations using aqueous monoethanolamine (one with a split-flow cycle) has been presented by Estep, McBride, and West (25). Obviously, only a limited amount of such data can be included in this text, and the reader is referred to the original publications for complete information. Typical design and operating data for a commercial installation using an aqueous monoethanolamine solution are presented in Table 2-9.

More complete analytical data are available on plants operating with glycolamine solutions containing from 10 to 30 percent monoethanolamine, 5 to 10 percent water, and the balance diethylene glycol. Operating data of absorbers from seven such plants are presented in Table 2-10 (66).

Table 2-9. Design and Operating Data for Large Commercial Natural-Gas Treating Plant (65)

Absorber design:
    5 units, each 7-ft ID by 68 ft high
    23 trays, with solution to 4th tray
    water wash to top tray
    pressure, 200 psig
Feed gas:
    50 MMSCF per absorber
    H_2S content, 160 to 180 grains/100 SCF
    CO_2 content, 0.3 to 0.4 per cent (by volume)
Outlet gas:
    H_2S content, 0.02 to 0.3 grain/100 SCF
Solution:
    17 per cent monoethanolamine, 83 per cent water
    rate of flow, 2 to 3 gal/MSCF of gas
Stripping columns:
    2 units, each 7-ft ID
    20 trays
    pressure, 12 psig
    still top temperature, 240°F
    feed, 200°F; bottom, 250°F
    steam consumption, 3.2 lb/MSCF gas (maximum)

Table 2-10. Operating Data for Commercial Glycol-Amine Plant Absorbers

Plant	Run	Gas flow rate MMSCF day	Absorber pressure, psia	Solution flow rate, in gpm	Inlet gas composition, %		Outlet-gas composition	
					H_2S	CO_2	H_2S grain/100 SCF	CO_2, %
A	1	50.9	542	450	0.74	3.9	0.06	0.00
A	2	63.9	537	465	0.72	4.0	0.03	0.00
A	3	63.5	533	445	0.71	4.1	0.05	0.00
A	4	63.5	536	441	0.76	3.8	0.04	0.00
A	5	63.5	533	450	0.84	3.9	0.05	0.00
A	6	63.5	533	451	0.76	3.8	0.06	0.00
A	7	64.0	533	451	0.76	3.8	0.06	0.00
B	8	86.5	533	347	0.34	1.66	0.02	0.00
B	9	86.0	535	341	0.35	1.45	0.04	0.00
B	10	87.8	537	354	0.34	1.68	<0.25	
C	11	68.5	523	123	0.12	0.42	0.06	0.00
D	12	41.6	518	169	0.12	0.43	0.04	0.00
E	13	50.7	538	32.2	0.10	0.18	0.09	0.01
F	14	44.9	596	191	0.10	0.46	0.06	0.00
F	15	45.1	603	61.8	0.09	0.41	<0.25	
F	16	44.2	596	191	0.11	0.38	<0.25	
G	17	34.5	558	146	0.86	0.90	0.20	0.00

Table 2-11. Performance Data for Typical Aqueous Diethanolamine Plant Absorbers used to Remove H_2S from Refinery-Gas Streams.

Absorber	Pressure, psig	Temperature, °F	H_2S grains/100 SCF	
			In	Out
16 trays (27)	101	66	3,196	15
26 ft of 3-in. rings (27)	225	140	1,490	26
2-in. rings (27)	250	140-155	260	6
51 ft packing (67)	150	95-100	1,500	5
30 ft of ¾-in. rings (68)	175	125-130	2,500	15

In all cases, the solution fed to the absorber had been stripped to below 20 grains H_2S/gal.

Diethanolamine is frequently used for purifying refinery-gas streams because of its resistance to COS, which reacts nonregenerably with monoethanolamine. In refinery operations, the gas is often used for fuel within the refinery so that a high degree of purity is not required. Typical performance data for aqueous diethanolamine plant absorbers are presented in Table 2-11.

It will be noted that gas purities from 5 to 26 grains/100 scf are reported. Other refinery diethanolamine plants are known to produce gas containing up to 50 grains H_2S/100 scf and are considered to be performing quite satisfactorily.

Operating data for an aqueous diethanolamine plant in high-pressure natural gas service have been presented by Berthier (10) and are presented in Table 2-12. These data were obtained in the early phases of development of the S.N.P.A.-DEA process at Lacq and are not quite representative of the process, as substantially lower amine circulation rates are currently being used (see Table 2-2).

Where large quantities of CO_2 are absorbed together with H_2S, it has been found that lower H_2S levels are obtainable in the product gas. This is a result of the stripping effect of CO_2 on H_2S in the stripping column, which produces

Table 2-12. Operating Data for Aqueous Diethanolamine Plant in High-Pressure Natural-Gas Service (10)	
Gas feed, MMSCFD	35.5
Feed-gas analysis:	
H_2S, percent	15.0
CO_2, precent	10.0
COS, ppm	300
CS_2, ppm	600
Outlet gas analysis:	
H_2S, gr/100 SCF	0.28
CO_2, gr/100 SCF	1.6
COS	0
DEA—20%, gpm	1,540
Steam, lb/hr	92,000
Absorber:	
Number of trays	30
Pressure, psig	1,000
Stripper:	
Number of trays	20
Pressure, psig	25
Reboiler Temp. °F	272

a lean solution containing very little H_2S. Data on an absorber utilizing an aqueous diethanolamine solution to absorb both CO_2 and H_2S have been presented by Willmott et al. (69). The absorber used in this study consisted of a 30-in. diameter tower packed with 35 ft of 1¼-in. stoneware raschig rings. Pertinent operating data for two runs are presented in Table 2-13.

Typical design and performance data for plants employing the Fluor Econamine process and the Adip process are shown in Tables 2-14 and 2-15.

Removal of Organic Sulfur Compounds

Most natural and refinery gases contain, besides H_2S and CO_2, relatively small quantities of mercaptans, carbonyl sulfide (COS), and carbon disulfide (CS_2). These compounds, if present in sufficiently large concentrations, must be removed from the gas stream if stringent specifications for total sulfur content of the treated gas are to be satisfied. The problem associated with the presence of organic sulfur compounds is two-fold: (a) the degree of removal obtainable and (b) the effect on the treating solution used. The latter is discussed in some detail in Chapter 3. Mercaptans are not sufficiently acid to form heat regenerable compounds, analogous to those formed by H_2S and CO_2 with ethanolamines. They are, therefore, only removed to the extent of

Table 2-13. Performance Data for Aqueous Diethanolamine Plant Absorber Used to Remove CO_2 and H_2S from Synthesis Gas (69)

Absorber variables	Run	
	A	B
Gas feed, SCF/hr	87,000	71,900
Absorber pressure, psig	350	340
Feed-gas analysis:		
CO_2, per cent	15.0	19.4
H_2S, grains/100 SCF	130	75
Outlet-gas analysis:		
CO_2, per cent	2.5	4.2
H_2S, grains/100 SCF	12.0	2.1
Solution rate, gpm	36	41
Diethanolamine, per cent	35	41
Lean-solution analysis:		
H_2S, grains/gal	16.5	38.0
CO_2, cu ft/gal	1.3	0.5
Rich-solution analysis:		
H_2S, grains/gal	55	55
CO_2, cu ft/gal	6.2	5.2

Table 2-14. Design Data for Fluor Econamine Process (15)
(Diglycolamine Solution)

Gas feed, MMSCFD	100
H_2S, percent	5
CO_2, percent	5
Outlet gas analysis:	
H_2S, gr/100 SCF	0.25
CO_2, percent	0.01
Solution circulation	
(60 wt % DGA), gpm	1,405
Stripping steam, lb/Hr	127,000
Absorber:	
No. of trays	20
Temperature in, °F	110
Temperature out, °F	180
Pressure, psig	900
Stripper:	
No. of trays:	
Stripping	18
Reflux	4
Pressure, psig	8
Temperature Top, °F	220
Reboiler temperature, °F	250

Table 2-15. Operating Data of ADIP Plants (3)

Plant	1	2	3
Gas feed, cu ft/hr	700,000	85,000	1,200,000
H_2S, percent	0.5	10.4	15.6
CO_2, percent	5.5	2.5	—
COS, ppm	200	—	—
Absorber pressure, psig	350	280	59
Absorber temp., °F	104	95	104
No. of trays in absorber	25	20	15
Outlet gas:			
H_2S, ppm	2	10	100
COS, ppm	100	—	—
Steam, lb/lb acid gas			
Removed	1.3	1.8	2.3
Power, KWH per ton of acid			
Gas removed	14	15	6

Plant 1: Synthesis gas from oil gasification unit
Plant 2: Gases from catalytic cracking unit
Plant 3: Off gases from gas oil hydrodesulfurizer

their physical solubility in the solution, which is quite small. If the gas to be treated contains appreciable amounts of mercaptans, an additional processing step, such as adsorption, caustic wash, or condensation by refrigeration (usually together with liquid hydrocarbon recovery), is required for complete removal. Since mercaptans are chemically inert toward ethanolamines, they have no deleterious effect on the solution.

Carbonyl sulfide and carbon disulfide undergo chemical transformation when contacted with primary and secondary ethanolamines under typical gas-purification operating conditions. The reactions involved are hydrolysis to H_2S and CO_2 and formation of well-defined compounds which are not regenerable by application of heat (see Chapter 3). The studies of Pearce, Arnold, and Hall (70) indicate that both types of reactions do occur in plants using mono- and diethanolamine solutions. However, while monoethanolamine is degraded fairly rapidly, diethanolamine is largely unaffected. Consequently, the use of diethanolamine solutions is indicated if the gas to be treated contains appreciable quantities of COS and CS_2. Both mono- and diethanolamine solutions are capable of fairly complete removal of COS and CS_2, so that in most cases no additional treatment of the gas is required. According to Klein (3), the solution used in the Adip process (diisopropanolamine) is suitable for partial COS removal from gas streams.

The Sulfiban Process

This process which was developed jointly by Bethlehem Steel Corporation and BS&B Process Systems Inc. (now Applied Technology Corporation), a subsidiary of International Systems and Control Corporation, is an adaptation of the ethanolamine process to the purification of coke-oven gases. Although early studies indicated that ethanolamines are not suitable for this service because of the presence of impurities such as carbonyl sulfide, carbon disulfide, and hydrogen cyanide in coke-oven gas which would react irreversibly with the amine and thus lead to excessive chemical consumption, pilot plant tests and commercial operations conducted by the co-developers of the process have shown that the anticipated deleterious effects of the impurities are less than expected.

The process has been described by Massey and Dunlap (80), and by Williams and Homberg who report operating experience in a commercial unit installed by the Coke Oven Division of the Bethlehem Plant of Bethlehem Steel Corporation (81).

The flow scheme used in the Sulfiban process is identical with that of a typical amine gas treating plant using monoethanolamine as the active agent. It contains a solution reclaimer, processing a side stream of about two percent of the circulating solution to remove undesirable products accumulating in the amine solution. The operating conditions are such that sulfur com-

Table 2-16. Coke-Oven Gas Desulfurization with Sulfiban Process (81)			
Inlet Gas, MM scf/day	60	86	63
H_2S, grains/100 scf	325-375	325-375	280-360
CO_2, Mol %	—	1.8	2.5
MEA (13-18%), gpm	275	310	440
Steam, M lbs/hr.	37.8	44.3	39.6
Treated Gas:			
H_2S, grains/100 scf	14-19	19-20	14-20
Acid Gas:			
H_2S, Mol%	42.6	36.9	34.7
CO_2, Mol %	52.8	58.2	62.7
HCN, Mol %	3.9	2.8	1.6

pounds are sufficiently removed to make the purified gas acceptable under Pennsylvania state regulations. Only a relatively small portion of the CO_2 contained in the gas is absorbed, thus making it possible to process the acid gas stream in a Claus unit. Essentially all of the hydrogen cyanide is absorbed, and a large portion of it is expelled in the regenerator, together with the acid gases. Before entering the Claus unit, the acid gases are treated catalytically to destroy hydrogen cyanide.

Typical performance data reported by Williams and Homberg are shown in Table 2-16.

The most serious operating problem reported was corrosion in the regenerator which was originally constructed partly of carbon steel and partly of 316L stainless steel. The carbon steel section of the vessel corroded badly and the entire column had to be constructed of 316L stainless steel. No serious corrosion problems were reported in other sections of the plant which, apparently are all constructed of carbon steel. However, continuous addition of corrosion inhibitor is practiced. Although there is no mention of consumption of monoethanolamine as a serious problem, the reported losses of two gallons of amine per million cubic feet of treated gas are rather high. Since the amine is lost as a result of chemical reactions with impurities, especially hydrogen cyanide, disposal of the waste liquid requires additional processing to avoid environmental pollution.

References

1. Bottoms, R.R. 1930. U.S. Patent 1,783,901; Re. 1933. 18,958.
2. Bally, A.P. 1961. *Erdöl und Kohle* 14:921-923.
3. Klein, J.P. 1970. *Oil and Gas Int.* 10(Sept.):109-112.
4. Frazier, H.D., and Kohl, A.L. 1950. *Ind. Eng. Chem.* 42(Nov.):2282-2292.

5. Kohl, A.L. 1951. *Petrol. Processing* 6(Jan.):26-31.
6. Miller, F.E., and Kohl, A.L. 1953. *Oil Gas J.* 51(Apr. 27):175-183.
7. Blohm, C.L., and Riesenfeld, F.C. 1955. U.S. Patent 2,712,978.
8. Dow Chemical of Canada, Limited. 1962. *Gas Conditioning Fact Book.*
9. Hutchinson, A.J.L. 1939. U.S. Patent 2,177,068.
10. Berthier, P. 1959. *Science et Technique* 81(Jan.):49-55.
11. Wendt, C.J., and Dailey, L.W. 1967. *Hydro. Process.* 46(Oct.):155-157.
12. Bailleul, M. 1969. *Gas Processing Canada* 61(3):34-38.
13. Dailey, L.D. 1970. *Oil Gas J.* 68(May 4):136-141.
14. Beddome, J.M. 1969. *Current Natural Gas Sweetening Practice.* Paper presented at 19th Canadian Chemical Engineering Conference (Oct.).
15. Holder, H.L. 1966. *Oil Gas J.* 64(May 2):83-86.
16. Dingman, J.C., and Moore, T.F. 1968. *Hydro. Process.* 47(July):138-140.
17. Feagan, R.A.; Lawler, H.L.; and Rhames, M.H. 1954. *Petrol. Refiner* 33(June):167.
18. Connors, J.S. 1958. *Oil Gas J.* 56(Mar. 3):100-120.
19. The Fluor Corporation, unpublished data.
20. McCartney, E.R. 1948. U.S. Patent 2,435,089.
21. Chapin, W.F. 1950. U.S. Patent 2,518,752.
22. McCartney, E.R. 1951. U.S. Patent 2,547,278.
23. Shoeld, M. 1934. U.S. Patent 1,971,798.
24. Bellah, J.S.; Mertz, R.V.; and Kilmer, J.W. 1949. *Petrol. Refiner* 28(June):154.
25. Estep, J.W.; McBride, J.T.; and West, J.R. 1962. *Advances in Petroleum Chemistry and Refining,* vol. 6. New York: Interscience Publishers, pp. 315-466.
26. Kohl,A.L., and Bechtold, I.C. 1952. U.S. Patent 2,607,657.
27. Reed, R.M., and Wood, W.R. 1941. *Trans. Am. Inst. Chemical Engrs.* 37(June 25):363-383.
28. Lyudkovskaya, M.A., and Leibush, A.G. 1949. *J. Appl. Chem. (U.S.S.R.)* 22(6):558-567.
29. Mason, J.W., and Dodge, D.F. 1936. *Trans. Am. Inst. Chemical Engrs.* 32(1):27-48.
30. Muhlbauer, H.G., and Monaghan, P.R. 1957. *Oil Gas J.* 55(Apr. 29):139-145.
31. Leibush, A.G., and Shneerson, A.L. 1950. *J. Appl. Chem. (U.S.S.R.).* 23(2):145-152.
32. Riegger, E.; Tartar, H.V.; and Lingafelter, E.C.1944. *J. Am. Chem. Soc.* 66(12):2024-2027.
33. Bottoms, R.R. 1931. *Ind. Eng. Chem.* 25(May):501-504.
34. Jones, J.H.; Froning, H.R.; and Claytor, E.E. 1959. *J. Chem. and Eng. Data* 4(Jan.):85-92.
35. Atwood, K.; Arnold, M.R.; and Kindrich, R.C. 1957. *Ind. Eng. Chem.* 49(Sept.):1439.
36. Crynes, B.L., and Maddox, R.N. 1969. *Oil Gas J.* 67(Dec.):65-67.
37. Union Carbide Chemicals Company. 1957. *Gas Treating Chemicals,* vol. 1.
38. Bottoms, R.R. 1931. *Am. Gas Assoc., Proc. (Technical Section)* 13:1071-1082.
39. Jefferson Chemical Company, Inc. 1969. *Gas Treating Data Book.*
40. Sherwood, T.K.; Shipley, G.H.; and Holloway, F.A.L. 1938. *Ind. Eng. Chem.* 30(July):765-769.
41. Elgin, J.C., and Weiss, F.B. 1939. *Ind. Eng. Chem.* 31(Apr.):435-445.
42. Lobo, W.E.; Friend, L.; and Skaperdas, G.T. 1944. *Ind. Eng. Chem.* 34(July):821-823.
43. Zenz, F.A., and Eckert, R.A. 1961. *Petrol. Refiner* 40(Feb.):130-132.

44. Souders, M., Jr., and Brown, G.G. 1934. *Ind. Eng. Chem.* 26(Jan.):98-103.
45. Brown, G.G., and Associates. 1950. *Unit Operations.* New York: John Wiley & Sons, Inc., p. 346.
46. Cryder, D.S., and Maloney, J.O. 1941. *Trans. Am. Inst. Chem. Engrs.* 37(Oct.):827-852.
47. Gregory, L.B., and Scharmann, W.G. 1937. *Ind. Eng. Chem.* 29(May):514-519.
48. Wainwright, H.W.; Egelson, G.C.; Brock, C.M.; Fisher, J.; and Sands, A.E. 1952. *U.S. Bur. Mines., Rept. Invest., No. 4891* (Oct.).
49. Benson, H.E.; Field, J.H.; and Haynes, W.P. 1956. *Chem. Eng. Progr.* 52:33.
50. Benson, H.E.; Field, J.H.; and Jimeson, R.M. 1954. *Chem. Eng. Progr.* 50(July):356-364.
51. Teller, A.J., and Ford, H.E. 1958. *Ind. Eng. Chem.* 50(Aug.):1201.
52. Shneerson, A.L., and Leibush, A.G. 1946. *J. Appl. Chem. (U.S.S.R.)* 19(9):869-880.
53. Leibush, A.G., and Shneerson, A.L. 1950. *J. Appl. Chem. (U.S.S.R.)* 23:1253-1263.
54. Van Krevelen, D.W., and Hoftijzer, P.J. 1948. *Chem. Eng. Progr.* 44(7):529-536.
55. Eckart, J.S.; Foote, E.H.; Rollison, L.R.; and Walter, L.F.; 1967. *Ind. Eng. Chem.* 59(Feb.):41-47.
56. Danckwertz, P.V., and Sharma, M.M. 1966. *Chem. Engr.* (Oct.):CE244-280.
57. Kohl, A.L. 1956. *AIChE J.* 2(June):264.
58. Koch, H.A., Jr.; Stutzman, L.F.; Blum, H.A.; and Hutchings, L.E. 1949. *Chem. Eng. Progr.* 45(11):677.
59. Sherwood, T.K. and Holloway, F.A.L. 1940. *Trans. Am. Inst. Chem. Engrs.* 36(Dec. 25):39.
60. Buskel, C. 1959. *Oil Gas J.* 57(Nov. 30):67-70.
61. Fitzgerald, K.J., and Richardson, J.A. 1966. *Hydro. Process.* 45(July):125-129.
62. Reed, R.M. 1946. U.S. Patent 2,399,142.
63. Chapin, W.F. 1947. *Petrol. Refiner* 26(6):109-112.
64. Ellis, G.C.; Leachman, G.S.; Formaini, R.E.; Hazelton, R.F.; and Smith, W.C. 1963. Paper presented at the 13th Annual Gas Conditioning Conference, University of Oklahoma (Apr.).
65. Carney, B.R. 1947. *Oil Gas J.* 46(Aug. 30):56-63.
66. Kohl, A.L., and Blohm, C.L. 1950. *Petrol. Engr.* 22(June):C-37.
67. Love, F.H. 1941. *Petrol. Engr.* 13(Nov.):31-32.
68. Anon. 1946. *Petrol. Refiner* 25(10):121.
69. Willmott, L.F.; Batchelder, H.R.; Wenzell, L.P. Jr.; and Hirst, L.L. 1956. *U.S. Bur. Mines, Rept. Invest., No. 5196* (Feb.).
70. Pearce, R.L.; Arnold, J.L.; and Hall, C.K. 1961. *Hydro. Process.* 40(Aug.):121-129.
71. Vidaurri, F.C., and Kahre, L.C. 1977. *Hydro. Process.* 56(Nov.):333-337.
72. Lee, J.I.; Otto, F.D.; and Mather, A.E. 1972. *J. of Chem. and Eng. Data* 52(Dec.):803-805.
73. Lee, J.I.; Otto, F.D.; and Mather, A.E. 1973. *J. of Chem. and Eng. Data* 17(41):465-468.
74. Lee, J.I.; Otto, F.D.; and Mather, A.E. 1973. *J. of Chem. and Eng. Data* 18(1):71-73.
75. Lee, J.I.; Otto, F.D.; and Mather, A.E. 1974. *Canadian J. of Chem. Eng.* 52(Feb.):125-127.

76. Lee, J.I.; Otto, F.D.; and Mather, A.E. 1976. *J. of Chem. and Eng. Data* 21(2):207-208.

77. Kent, R.L., and Eisenberg, B. 1976. *Hydro. Process.* 55(Feb.):87-90.

78. Lawson, J.D., and Garst, A.W. 1976. *J. of Chem. and Eng. Data* 21(1):20-30.

79. Isaacs, E.E.; Otto, F.D.; and Mather, A.E. 1977. *J. of Chem. and Eng. Data* 22(1):71-73.

80. Massey, M.J., and Dunlap, R.W. 1975. *Journal of the Air Pollution Control Association* 25(10):1019-1027.

81. Williams, J.A., and Homberg, O.A. 1977. McMaster Symposium, "Treatment of Coke Oven Gas," (May 1977), McMaster University, Hamilton, Ontario, Canada.

82. Dibble, J.H. 1983 "UCARSOL Solvents for Acid Gas Removal," paper presented at Petroenergy '83, Houston, Texas, September 14.

83. Butwell, K.F., Hawkes, E.N., and Mago, B.F. 1973. *Chem. Eng. Progress.* February. p. 57.

84. Butwell, K.F., Kubek, D.J., and Sigmund, P.W. 1979. *Chem. Eng. Progress.* Feb. p. 75.

85. Kubek, D.J. and Butwell, K.F. 1979. "Amine Guard Systems in Hydrogen Production," paper presented at AIChE National Meeting, April 1-5.

86. Dow Chemical U.S.A. Gas/Spec Technology Group, Technical Bulletin IT-1 Technology.

87. Meissner, R. E. and Wagner, U. 1983. *Oil and Gas Journal.* Feb. 7, p. 55.55.

88. Meissner, R. E. 1983. "A Low Energy Process for Purifying Natural Gas," paper presented at 1983 Gas Conditioning Conference, University of Oklahoma.

89. U.S. Patent 3,622,267; Canadian Patent 1,090,098.

90. Anon. 1981. *Chemical & Engineering News,* Sept. 7, p. 58.

91. Goldstein, A.M. 1983. "Commercialization of a New Gas Treating Agent," paper presented at Petroenergy '83 Conference, Houston, Texas, Sept 14.

92. Sartori, G., and Savage, D. W. 1983. I.& E.C. Fundamentals, 22, 239.

93. Weinberg, H.N., Eisenberg, B., Heinzelmann, F.J., and Savage, D.W. 1983. "New Gas Treating Alternatives for Saving Energy in Refining and Natural Gas Processing," paper presented at 11th World Petroleum Congress, London, England, August 31.

94. Savage, D.W., Funk, E.W., and Astarita, G. 1981. "Selective Absorption of H_2S and CO_2 into Aqueous Solutions of Methyldiethanolamine," paper presented at AIChE Meeting, Houston, Texas, April 5-9.

95. Dingman, J. C. 1977, "Gas Sweetening with Diglycolamine Agent," paper presented at Third Iranian Congress of Chemical Engineering, Shiraz, Iran, November 6-11.

96. Mason, J. R., and Griffith T. E. 1969. *Oil and Gas Journal.* June 5. p. 67.

97. Huval, M., and Van de Venne, H. 1981. *Oil and Gas Journal.* August 17, p. 91.

98. Bucklin, R. W. 1982. *Oil and Gas Journal.* November 8. p. 204.

99. Weber, S., and McClure, G. 1981. *Oil and Gas Journal.* 79 (23),:160-163.

100. Abe, T., and Peterzan, P. 1980. *Oil and Gas Journal.* March 31. p. 139.

101. Ouwerkerk, C. 1978. *Hydrocarbon Processing.* 57 (4) April. p. 89.

102. Pearce, R. L. 1978. "Hydrogen Sulfide Removal with Methyldiethanolamine," paper presented at 57th Annual GPA Convention, New Orleans, Louisiana, March 20-22.

103. Crow, J. H., and Baumann, J.C. 1974. *Hydrocarbon Processing.* J3(10) October p. 131.

104. Goar, B. G. 1980. "Selective Gas Treating Produces Better Claus Feed," paper presented at 1980 Gas Conditioning Conference, University of Oklahoma.

105. Blanc, C., and Elgue, J. 1981. *Hydrocarbon Processing.* 60(8) August p. 111.

106. Dow Chemical USA. Gas/Spec ST-1 Technology, Technical Bulletin.

107. Dow Chemical USA. Gas/Spec FT-1 Technology, Technical Bulletin.

108. Sigmund, P. W., Butwell, K. F., and Wussler, A. J. 1981. *Hydrocarbon Processing*, 60(5) May, p. 118.

109. Martin, J.L., Otto, F.D., and Mather, A.E. 1978. Journal of Chemical and Engineering Data 23:2, 163.

110. Lal, D., Isaacs, E.E., Mather, A.E., and Otto, F.D. 1980. "Equilibrium Solubility of Acid Gases in Diethanolamine and Monoethanolamine Solutions at Low Partial Pressures," paper presented at 1980 Gas Conditioning Conference, University of Oklahoma.

111. Jou, F., Mather, A.E., and Otto, F.D. 1982. Ind. Eng. Chem. Process Des. Dev. 21, 539-544.

112. Dingman, J.C., Jackson, J.L., Moore, T.F., and Branson, J.A. 1983. "Equilibrium Data for the H_2S-CO_2-Diglycolamine-Water System," paper presented at 62nd Annual Gas Processors Assoc. Convention, San Francisco, March 14–16.

113. Dankwerts, P.V., and McNeil, K.M. 1967. Trans. Institute Chemical Engineering 45, T32-T38.

114. Chen, C., and Ng, A. 1980. *Hydrocarbon Processing*. 59(4) April. p. 122.

115. Moshfeghian, M., Bell, J.K., and Maddox, R.N. 1977. "Reaction Equilibria for Acid Gas Systems," paper presented at 1977 Gas Conditioning Conference, University of Oklahoma.

116. Vaz, R.N., and Maddox, R.N. 1981. "Ethanolamine Sweetening Process Calculations Using Reaction Equilibrium Model," paper presented at AIChE Spring Meeting, Houston, Texas, April 5–9.

117. Vaz, R.N., Mains, G.J., and Maddox, R.N. 1981. *Hydrocarbon Processing*. 81(4) April. p. 139.

118. Majeed, A., Diab, S., Mains, G.J., and Maddox, R.N. 1982. "Computer Calculations of Ethanolamine Sweetening," paper presented at 1982 Gas Conditioning Conference, University of Oklahoma.

119. Tomcej, R.A., Otto, F.D., and Nolte, F.W. 1983. "Computer Simulation of Amine Treating Units," paper presented at 1983 Gas Conditioning Conference, University of Oklahoma.

120. Viddauri, F.C., and Ferguson, R.G. 1977. "MDEA Used in Ethane Purification," paper presented at 1977 Gas Conditioning Conference, University of Oklahoma.

121. Weiland, R.H., Rawal, M., and Rice, R.G. 1982 AIChE Journal November 28:6, 963-973.

122. Jou, F., Lal, D., Mather, A.E., and Otto, F.D. 1981. "Solubility of Acid Gases in MDEA Solutions," paper presented at CGPA Meeting, November 12, 1981, Calgary, Alberta.

3

Mechanical Design and Operation of Alkanolamine Plants

One of the reasons the alkanolamine processes have displaced to a large extent such operations as the iron oxide and sodium carbonate processes for natural-gas purification is their comparative freedom from operating difficulties. Nevertheless, several factors can result in undue expense and cause difficulty in the operation of alkanolamine units. Chief among these, from an expense standpoint, are corrosion and amine loss. Operating difficulties which occasionally limit the capacity of a plant for gas purification include foaming and the plugging of equipment.

Corrosion

The most serious operating problem encountered with alkanolamine gas-purification plants is corrosion, and, as would be expected, this problem has been given the widest attention. Several theories have been advanced to explain the corrosion mechanism, patents describing measures to eliminate or alleviate corrosion have been issued, and a considerable number of technical articles have been published on the subject (1-16). In addition, numerous case histories have been presented in the literature (17-25). Based on this appreciable amount of information and experience, the corrosion phenomena observed in a large number of plants operating under a wide variety of conditions can be reasonably well explained, and certain guidelines can be established to minimize corrosion.

Corrosion Mechanisms

It is quite evident that no single mechanism can be blamed for all corrosion in alkanolamine systems. The extent and type of corrosion has been observed to depend on such factors as the amine used, the presence of contaminants in the solution, the solution loading with acid gas, the temperatures and pressures prevailing in various parts of the plant, the velocity with which the solution flows, and others. However, it appears that the principal corroding agents are the acid gases themselves. This premise is borne out by the generally observed increase in corrosion with increased acid-gas concentration in the solution, especially in plants using monoethanolamine (MEA).

It is known that free or "aggressive" CO_2 causes severe corrosion, particularly at elevated temperatures and in the presence of water. It is believed that the mechanism involved in such systems consists of the reaction of metallic iron with carbonic acid which results in the formation of soluble iron bicarbonate (26). Further heating of the solution may cause the release of CO_2 and precipitation of the iron as the relatively insoluble carbonate. The iron may also be removed from solution by hydrolysis to basic carbonates or hydrated oxides, further oxidation to the less soluble ferric compounds, or by precipitation as sulfide by H_2S. The darkening of the solution which frequently occurs in absorbers purifying gas containing both CO_2 and H_2S may be attributed to the latter reaction. Upon resaturation with CO_2 and subsequent heating, the solution is capable of dissolving more iron and repeating the corrosion cycle. This mechanism can cause rather rapid corrosion of carbon steel, particularly at locations of high temperature and high acid-gas concentration.

Hydrogen sulfide attacks steel as an acid, with the subsequent formation of insoluble ferrous sulfide. This compound forms a coating on the metal surface which does not adhere tightly and therefore affords little protection from further corrosion. There is no satisfactory correlation available for CO_2-H_2S mixtures which relates the corrosive attack to be expected with any given ratio of H_2S to CO_2. However, certain generalized observations have been made. It appears that in plants handling predominantly CO_2 very small quantities of H_2S may actually reduce corrosion. On the other hand, the severe corrosive attack which has been noted at other ratios indicates that under some conditions a synergism may exist, i.e., each of the acid gases increases the corrosive attack of the other.

This effect has been observed experimentally by Lang and Mason (8) for monel in contact with a 20 percent aminoethylethanolamine solution saturated with acid gases at 240°F. The corrosion rate for this metal was found to be negligible with pure CO_2 and pure H_2S but peaked up to a maximum of 0.137 in./year with a 50:50 H_2S-CO_2 mixture. However, under the same conditions, most of the other metals tested, including steel showed less corrosion from all of the H_2S-CO_2 mixtures than from pure CO_2.

Froning and Jones (7) conducted a laboratory study of the corrosion of mild steel in boiling aqueous monoethanolamine solutions and concluded that corrosion by carbon dioxide is most severe at concentrations of CO_2 (in the saturating gas) of 20 to 30 percent. Under these conditions corrosion rates up to 0.030 in./year were measured. They found corrosion due to H_2S to decrease with increased H_2S concentration from a maximum of 0.006 in. /year at a concentration of 0.01 to 0.5 percent H_2S in the saturating gas (about 200 ppm H_2S in the liquid) to only 0.001 in./year with saturating gas atmospheres containing more than 2 percent of H_2S and CO_2. Mixtures of H_2S and CO_2 were found to behave like H_2S alone, corrosion being lower than for the corresponding CO_2 concentration and decreasing with increased H_2S concentration.

Dingman (56) reports that in systems employing diglycolamine (DGA) solutions high CO_2 to H_2S ratios in the feed gas represent much more corrosive conditions than low CO_2 to H_2S ratios, requiring alloy steel in several locations of the plant.

The generally encountered corrosion increase with increased amine concentration may also be attributed to the effect of acid gases, as plants with high amine concentrations are ordinarily operated to absorb more acid gas per unit volume of solution. An additional factor which may be of some importance at very high amine concentrations is the possible formation of soluble ethanolamine-iron compounds. Isolation and identification of compounds of this type have been reported by Dixon and Williams (27).

Next to the acid gases, in the order of importance as corroding agents, are degradation products of the solvents, resulting from irreversible reactions between the solvents and constituents of the feed gas. Although the early theory that glycine or glycolic acid form as a result of monoethanolamine decomposition (1,2) has not been substantiated by later investigators, a variety of compounds, notably polyamines, have been isolated and shown to have a marked effect on corrosion. The chemistry of the formation of such compounds has been studied by Polderman and Steele (28), Polderman, Dillon and Steele (29), and Hakka et al. (16). A comprehensive study of the effect of diethanolamine (DEA) and methyldiethanolamine (MDEA) degradation products on corrosion in treating plants has been presented by Blanc, Grall, and Demarais (58). A large number of basic and acid compounds were identified in solutions used in laboratory experiments and in samples of solutions from operating plants. The authors conclude that DEA and MDEA degradation products have little or no effect on corrosion of carbon steel equipment. Contact of the solution with oxygen at some point in the process cycle has also been blamed for certain cases of unusually severe corrosion. Possible mechanisms for amine degradation and oxidation are discussed in more detail later in this chapter. Comeaux (15) presents data showing the effect of polyamines on corrosion in ethanolamine systems. It is proposed that such compounds—for example hydroxyethylethylenediamine, which is a reaction product of monoethanolamine and CO_2—act as chelating agents for iron in the hot sections of the process. The iron chelates become unstable when cooled, and iron reacts with hydrogen sulfide in the absorber, thus regenerating the chelating agents for further corrosive attack. This mechanism would to some extent explain the difference in corrosive behavior of systems using different amines. For example, the principal degradation product of diethanolamine resulting from reaction with CO_2 is a heterocyclic compound (see later in this chapter) which is a much poorer chelating agent than the diamine formed from monoethanolamine. Experience has shown that plants using diethanolamine solutions have appreciably fewer corrosion problems than those using monoethanolamine. The reported relatively low rate of corrosion in plants using diisopropanolamine (DIPA) (Adip and Sulfinol) may be at least partially due to a similar effect. The limited amount of information available on systems employing DGA and sterically hindered amines indicates that amine degradation products play only a minor part in the corrosion mechanism (56, 57, 59, 60).

Rapid corrosion can occur as the result of erosion due to suspended solids, i.e., iron sulfide in the solution and high velocity flow through heat exchanger tubes and process lines. Under such conditions a rate limiting film of iron sulfide is prevented from forming, and iron is continuously removed by reaction with H_2S.

Isolated cases of stress-corrosion cracking, typical for alkaline systems, have been observed, especially in contactors and stripping columns. The alkanolamines themselves are not corrosive to carbon steel even at relatively high temperatures; in fact, they act somewhat as corrosion inhibitors in the presence of CO_2. A comparison of the rates of corrosion of carbon steel in water, aqueous monoethanolamine solutions, and glycol-monoethanolamine solutions in the presence of CO_2 shows the definite corrosion-depressing effect of the monoethanolamine (Figure 3-1).

In the light of the foregoing explanation of the corrosion mechanism, the location and severity of the attack in gas-purification units using alkanolamine solutions can be fairly well predicted. In systems operating with uncontaminated solutions, free of solids, corrosion will be most prevalent in places where the highest concentration of acid gases is accompanied by the highest temperature. It therefore becomes apparent that in aqueous amine systems, especially monoethanolamine systems, the area most susceptible to corrosive attack is the reboiler. In systems containing di-, tri-, methyldiethanolamine or diisopropanolamine which are much more easily stripped of acid gas, the amine-to-amine heat exchanger will be most severely attacked. In systems containing glycol and an amine, in which essentially total acid-gas removal is achieved in the stripping column, the highest concentration of acid gas at the highest temperature is the hottest pass of the heat exchanger on the rich-solution side.

Corrosion is more widely distributed throughout the system in plants using aqueous monoethanolamine solutions than in those employing either mix-

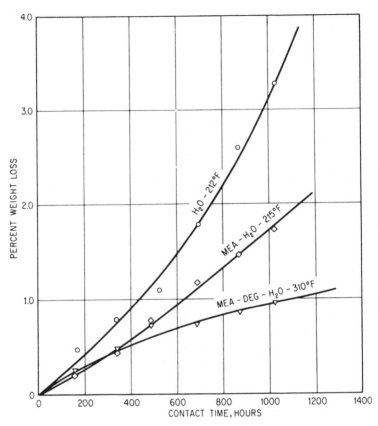

Figure 3-1. Corrosion rate of carbon steel in water, monoethanolamine-water, and monoethanolamine-diethylene glycol-water solutions at their boiling points in the presence of carbon dioxide.

tures of glycol and monoethanolamine or other amines such as DEA, MDEA and DIPA. This is due to the incomplete stripping obtained in aqueous monoethanolamine systems where the regenerated solution still contains appreciable quantities of acid gas, especially CO_2. In other systems the regenerated solution is essentially free of acid gases, and corrosion is generally confined to the portion of the plant where rich solutions are in contact with hot metallic surfaces.

Comparative corrosion rates of carbon steel in an aqueous methyldiethanolamine solution and a monoethanolamine—diethylene glycol—water solution are shown in Figure 3-2. The data presented in Figure 3-2 were obtained in a continuous pilot plant in which the metal specimens were exposed to amine solutions containing three different concentrations of acid gas at three different temperatures. The specimens placed in the monoethanolamine—

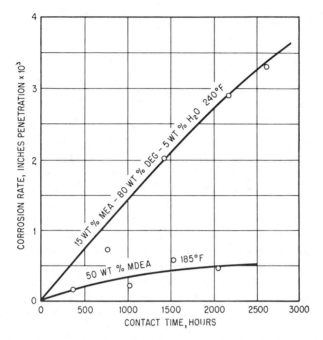

Figure 3-2. Corrosion rate of carbon steel in monoethanolamine-diethylene glycol-water and methyldiethanolamine-water solutions containing acid gas.

diethylene glycol—water solution corroded both at 240° and 320°F (see Figure 3-5) while the samples in the methyldiethanolamine solution showed corrosion only at 185°F. This is understandable if it is considered that methyldiethanolamine solutions do not contain appreciable amounts of acid gas above 185°F while monoethanolamine solutions retain acid gas at considerably higher temperatures.

Treating-Plant Corrosion Experience

Contactors are normally free of corrosion, although stress-corrosion cracking has been reported in several plants using aqueous alkanolamine solutions. Relieving of stress on vessels has been found to prevent this type of attack (30).

Considerable corrosion of carbon steel heat-exchanger tubes has been encountered in systems using alkanolamine solutions. In aqueous monoethanolamine systems the attack frequently occurs in both the rich- and lean-solution sides and is normally most severe in the hottest exchanger pass. In plants using mixtures of monoethanolamine and glycol, the most severe corrosion is generally found on the rich-solution side of the hottest exchanger pass. The same experience has been reported for aqueous diethanolamine systems. Normally, only insignificant corrosion occurs on the carbon-steel heat-exchanger shells.

Carbon-steel stripping-still columns in both glycol-amine and aqueous amine systems are subject to various degrees of corrosion depending on operating conditions. High regeneration temperatures, especially in aqueous systems, are particularly conducive to corrosion. In general, still-column corrosion is more severe in aqueous than in glycol-amine systems. Again, the corrosion observed in plants using diethanolamine appears to be much milder than in those using monoethanolamine as the active agent. Corrosion occurs in both the liquid and vapor space and is most pronounced below the solution-feed point. Stress-corrosion cracking of vessels which are not stress-relieved is more frequent than in contactors.

Corrosion occurs in various degrees on carbon-steel reboiler tubes in both aqueous and glycol-amine systems. Instances of very severe corrosion of reboiler tubes in aqueous monoethanolamine plants have been found to be due to the use of too high a still pressure (and thus temperature) or the use of high-temperature steam or heat medium in the reboiler tubes. In glycol-amine systems it appears that the beneficial effect of low acid-gas level more than compensates for the effect of the higher temperatures employed in the reboiler.

Solution coolers and the overhead condenser tubes of regenerators are not particularly susceptible to corrosive attack from the solution or acid-gas sides, respectively. However, solution-cooler corrosion from the solution side, especially in cases where the solution is incompletely stripped, can occur. In general, corrosion from the water side is much more critical. Plants using aqueous solutions of diisopropanolamine (Adip process) are reported to have essentially no corrosion problems, even if they are entirely constructed from carbon steel (31).

Methods of Minimizing Corrosive Attack

Corrosion can be reduced or eliminated by various means, including certain practices in the design and operation of purification plants, use of more expensive corrosion-resistant materials, continuous or periodic removal of corrosion-promoting agents and suspended solids from the solution, and addition of corrosion inhibitors. A combination of several of these measures usually leads to the most satisfactory and economical solution.

The following operating practices are recommended to minimize corrosive attack:

1. The temperature of the solution in the reboiler and the temperature of the steam should be kept as low as possible.

2. Use of high-temperature heat-carrying media, such as oil, in reboilers should be avoided to maintain the lowest possible metal skin temperature.

3. Solution regeneration under pressure with its accompanying high temperatures results in severe corrosion of reboiler tubes; it is, therefore, good practice to maintain the lowest possible pressure on stripping columns and reboilers.

4. It is advisable to exclude oxygen from the system by maintaining an inert gas blanket over all portions of the solution which could be exposed to the atmosphere and by ensuring positive pressure on the suction side of all pumps.

5. Continuous removal of suspended solids (by filtration) and of amine degradation products (by distillation of a side stream or by filtration through activated carbon) generally helps to reduce corrosion.

6. In some cases addition of caustic to the circulating amine solution

7. Corrosion inhibitors may be used including filming high molecular weight amines and salts of heavy metals such as vanadates. The success of corrosion inhibitors depends on many factors in the operation of the plant. It may, therefore, be necessary to experiment with several inhibitors until the proper compound is found. Several proprietary inhibitors and solution formulations are commercially available.

It should also be mentioned that proper practices in cleaning of plant equipment can influence corrosion quite critically. The use of acid, even if inhibited, for the removal of iron sulfide and other sludge deposits may lead to increased corrosion by exposing bare metal surfaces to the attack. In most cases, plants can be cleaned quite successfully by employing a good detergent. Sometimes it is necessary to boil the cleaning solution for several hours, especially in major vessels such as contactors and stripping stills. In plants where very severe fouling occurs, it may even be necessary to remove the deposits mechanically.

Plant design features to minimize corrosion are usually accompanied by increased operating and investment costs and are therefore limited by these considerations. For example, it has been proposed to design plants with large amine circulation rates to maintain low acid-gas concentration in the rich solution flowing from the absorber (9,14,22). However, recent experience has shown that substantially higher solution loadings, in terms of moles of acid gas per mole of amine, can be tolerated without leading to excessive corrosion in properly designed and operated plants. This is particularly evident in plants using the S.N.P.A.-DEA process where diethanolamine solutions are saturated with acid gas to almost stoichiometric proportion.

In spite of the limitations imposed on the process design, certain features appear to be quite practical. These include (a) maintaining low velocities in heat exchangers (not exceeding 3 ft/sec); (b) placing the rich solution in the heat-exchanger tubes rather than in the shell; and (c) maintaining pressure on heat exchangers to reduce corrosive attack by flashing of acidic gases. Furthermore, the use of low-pressure steam (40 to 75 psig) and low reboiler temperatures (below 240°F in aqueous and 300°F in glycol-amine systems) is also indicated to minimize corrosion. There are several features which can be incorporated in the design of reboilers to alleviate corrosion. To maintain the minimum steam temperature at all times, the steam-rate control valve should be located in the steam line ahead of the reboiler instead of in the condensate line from the reboiler. Vibration of the tubes can be reduced by using a square-pitch design permitting evolved gas to escape freely. This measure will also reduce accumulation of sediments. Finally, the level in the reboiler should be maintained high enough so that all of the tubes are always covered with liquid.

As stated earlier, several corrosion inhibitors have proved successful in alleviating or eliminating corrosion of carbon steel in amine units. The composition of these materials is usually proprietary and they are commercially available under trade names, such as Petromeen 52 (Betz Laboratories, Inc.), Drewgard 100 (Drew Chemical Corporation) and others. Union Carbide Corporation and Dow Chemical Company are offering process systems which basically are built around corrosion control but are claimed to incorporate additional benefits such as reduced energy requirements and low chemical consumption.

Union Carbide's UCAR Amine Guard system has been described by Kelly (51) as applied to MEA units used for CO_2 removal in ammonia synthesis plants. The author cites specific cases where appreciable energy savings—on the order of 60,000 pounds per hour of steam in a 1,000 ton per day ammonia plant—could be achieved by use of the UCAR Amine Guard system. The energy savings are due to the fact that amine solutions of relatively high concentrations, up to 30 percent in the case of monoethanolamine, and consequently low circulation rates can be used without incurring excessive corrosion and solution degradation.

The Amine Guard system, as used in ammonia and hydrogen manufacture, has been further described by Butwell, Hawkes, and Mago (61), and Butwell and Kubek (62, 63). An improved system, named Amine Guard III, has been described by Butwell, Kubek and Sigmund (64). In this system the lean/rich heat exchanger is eliminated and substantial operating advantages are claimed over conventional MEA systems for CO_2 removal (65). One of the principal advantages claimed for the Amine Guard system is that it permits use of high MEA concentrations, about 30 to 40 percent, without corrosion of carbon steel and, consequently, appreciable savings in energy consumption. In 1979, over 100 Amine Guard systems were reportedly in use in ammonia, hydrogen, methanol, synthesis gas and natural gas plants.

The recently disclosed UCARSOL solvents (66) are reported to be noncorrosive special formulations for the removal of both H_2S and CO_2. UCARSOL HS Solvent 101 is based on MDEA and proposed for selective H_2S removal in the presence of CO_2.

A group of proprietary systems is offered by Gas/Spec Technology Group of Dow Chemical U.S.A. under the trade names Gas/Spec IT-1 Technology, Gas/Spec ST-1 Technology and Gas/Spec FT-1 Technology. IT-1 is characterized by the feature that it permits use of highly concentrated ethanolamine solutions—30 percent MEA and 50 percent DEA—with corresponding high carrying capacities for CO_2; ST-1 Technology is a low corrosion, selective H_2S removal process; and FT-1 Technology is used for very low cost CO_2 removal from flue gases (67). In addition, a corrosion inhibitor, IT-67L, which can be used in amine systems for CO_2 removal, is available.

Substantial savings in capital and operating costs as compared to conventional systems are claimed for all Gas/Spec Technologies. For example, FT-1 Technology is reported to save about 15 percent and 25 percent respectively per ton of CO_2 produced as compared with 30 percent DEA and 60 percent DGA (67). A commercial application of this process has been described by Wiggins and Bixler (68).

Corrosion inhibitors of the film-forming amine type have been proposed for systems using MEA and DEA solutions. Liebermann (69) reports that addition of such inhibitors reduced reboiler corrosion from 50 to 3 mils per year, in a refinery gas purification unit employing a 25 percent MEA solution. The values given are based on results obtained with probes inserted in the reboiler outlet line.

Another approach to the alleviation of corrosion is the use of more expensive corrosion-resistant alloys in place of carbon steel. Generally, these alloys are used only in places especially exposed to corrosive attack, such as the heat exchangers, reboiler, certain portions of the stripping column, and certain portions of the piping.

A comparison of several metals and alloys for various treating-plant environments is presented in Table 3-1; this is based on the data of Lang and Mason (8) for coupons supported in plant-solution lines by spool holders. Although tests of this type are of some value for comparison purposes, they do not always provide meaningful data for estimating plant-equipment corrosion. This is particularly true for equipment used to transfer heat where the temperature of the metal wall may be appreciably higher than that of the solution (and coupon). In test E of Table 3-1, for example, the spool was exposed in the exit head of the reboiler, in contact with flowing liquid at 230° to 240°F. The results indicated that all materials except aluminum would be satisfactory. In actual experience in this plant, mild-steel reboiler-tubes had a

Table 3-1. Plant-Corrosion Test Results

	Test A	Test B	Test C	Test D	Test E
Test variables					
Amine	MEA	DEA	MEA	MEA	MEA
Concentration, per cent	90–95	11–15	15–20	15	17
CO_2	0	0	present	present	2% by wt
H_2S	0	10–50 grains/gal	0	present	0
Temperature, °F	338–374	225–231	180–220	230	230–240
Duration of test, days	36	483	100	270	293
Service	Reclaimer	Stripping column	Refinery-gas treater	Natural-gas treater	Synthesis-gas treater reboiler
			Indicated corrosion rate, in./year		
Metal					
Monel	0.0003	0.0021	0.0016	0.0013	0.001
Nickel	0.0004	0.0049	0.0021	0.0033	0.0003
Inconel	0.0002	<0.0001	....	<0.0001	<0.0001
T-302 and 304 SS	0.0005	<0.0001†	<0.0001	<0.0001	<0.0001
T-416 SS	0.0004				
T-410 SS	0.008	0.0001†	....	....	0.0008
T-502 SS	0.004	*			
Aluminum 2S and 3S	*	....	....	....	*
Mild steel	0.003	*	0.0014	0.0054	0.0002
70-30 Cu-Ni	....	*	0.0019	0.010	
Cast Iron	....	0.017	0.0082	0.0021	
T-316 SS	....	....	....	<0.0001	<0.0001

Source: Data of Lang and Mason (8)
*Specimens originally 0.031 in. thick destroyed during test period.
†Pitted.

life of 6 weeks to 6 months, and Type 304 stainless steel had a life varying from 6 months to 1 year. Monel was found to be quite resistant in this environment and had not failed after 2 years' service.

A study of corrosion of various ferrous and nonferrous alloys under actual operating conditions in a monoethanolamine system used for carbon dioxide removal from crude hydrogen produced by steam reforming of natural gas has been reported by Montrone and Long (32). The plant is of conventional design for such applications, using the hot shift converter effluent gases as the heat supply in the reboiler. A monoethanolamine reclaimer is included in the system.

Corrosion spools were placed in 11 locations of the plant, and, in addition, single exchanger tubes of six different alloys were inserted in the reboiler. The materials tested included, besides carbon steel, five types of stainless steel, two Incoloys, Monel, three copper-nickel alloys, and two aluminum alloys. The test was conducted over a period of 5,000 hours. The results of this study are shown in Table 3-2.

The authors concluded from the results of this study that all vessels and exchanger shells in the treating plant can be constructed of carbon steel. However, stainless steel, Type 304, is recommended for the hottest amine exchanger pass, the reboiler, the amine cooler, the amine reclaimer, and certain sections of the piping.

In aqueous amine systems using monoethanolamine, diethanolamine, triethanolamine, and methyldiethanolamine, stainless steels of the 304 and 316 types show appreciable resistance to corrosive attack if either CO_2 alone or both CO_2 and H_2S are present. Monel is generally less resistant than types 304 and 316 stainless steel. If the lean solution contains appreciable quantities of CO_2 or H_2S, aluminum 1100, 3003 and 6061 can be used successfully in heat exchangers.

It appears that the residual CO_2 and H_2S protect aluminum against corrosive attack by the alkaline amines (see Figure 3-3). In glycol-amine systems, the use of aluminum-alloy heat-exchanger tubes is entirely satisfactory. Aluminum is not only protected against attack by the amine by the presence of residual CO_2 and H_2S, but also by the presence of glycol in the solution. A minimum of 40 percent glycol is necessary to obtain satisfactory protection against the alkaline attack exerted by lean, aqueous, monoethanolamine solutions. This effect is shown in Figure 3-4. In addition, steel alloys containing a minimum of 5 percent chromium and ½ percent molybdenum are quite resistant to corrosive attack (see Figure 3-5). However, where aluminum can be safely used, it generally has an economic advantage over other corrosion-resistance alloys. One problem which is sometimes encountered with aluminum, and to a minor extent with carbon-steel heat-exchanger tubes, is baffle cutting caused by vibration of the tubes in the baffle holes. This difficulty can be reduced somewhat by removal of the hydrocarbons and acid gas evolved ahead of the heat exchangers and between the various heat-exchanger passes. Another protective measure against baffle cutting on aluminum is the use of anodized tubes.

Several materials are suitable for protecting stripping-still columns in aqueous monoethanolamine systems. Lining the vessel with stainless steel and using stainless steel trays and bubble caps, primarily of the 304 and 316 types, have been quite satisfactory in a number of plants. Coating the stripping-column wall with concrete and using ceramic packing have also

Table 3-2. Test Spool Corrosion Rate Date for Aqueous MEA Plant

Spool No.	Location	Process Stream	Temp. %	Carbon Steel	Stainless Steels					Incoloys		Monel 400	Copper Nickels			Aluminum	
					405	410	304	304†	316	800	600		70-30	60-40	65-30-5 Iron	6063	5052
1.	MEA reclaimer	MEA vapor**	270	0.55	0.01	0.07	0.36	0.81	0.91	0.53	1.02	0.71	0.74	0.55	0.55	0.45	0.66
2.	Hot cond. drum	H_2CO_3, H_2O vapor	310	Lost	0.56	4.5	0.02	0.02	0.01	0.02	0.02	0.99	0.12	0.54	0.28	0.05	0
3.	Hot cond. drum	H_2CO_3, H_2O liquid	260	Lost	0.48	4.3	0.02	0.02	0	0.02	0.02	0.22	0.10	0.12	0.11	0	0
4.	Cold cond. drum	H_2CO_3, H_2O liquid	110	6.5	0	0	0	0	0.02	0.17	0.05	0.13	0.11	0.16	0.10	3.1	2.9
5.	Regen. Tray 24	MEA solution	207	5.5	0.4	5.4	1.6	5.5	0.7	6.2	3.0	2.3	5.5	3.2	4.2	1.8	2.4
6.	Regen. Tray 21	MEA solution	207	0.9	0	1.9	3.3	4.6	0.6	9.0	3.8	2.2	2.3	2.3	1.9	2.0	2.0
7.	Lean/rich exchanger	Rich MEA solution	200	3.0	0	0	0.02	0.04	0.02	0.02	0.03	0.04	0.20	0.11	0.12	0.74	1.0
8.	Reboiler	Lean MEA vapor	240	0.05	0	0.02	0.07	0.11	0.16	0.22	0.41	0.57	0.64	0.58	0.52	0.49	0.68

Indicated Corrosion Rate, Mils/Year

Source: Data of Montrone and Long.
*Eighteen percent MEA solution containing about 5.2 scf CO_2/gal (rich) and 1.4 scf CO_2/gal (lean).
†Sensitized Type 304.
‡ Indicates pitting attack.
**Entrained Na_2CO_3 also present in reclaimer vapor.

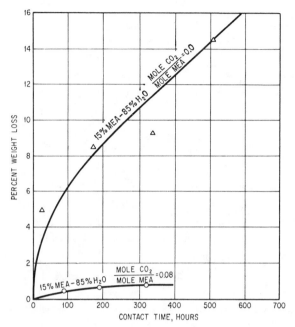

Figure 3-3. Corrosion rate of aluminum 1100 in aqueous monoethanolamine solutions with and without carbon dioxide at 77°F.

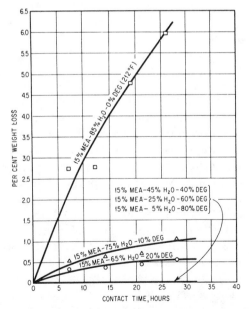

Figure 3-4. Corrosion rate of aluminum 1100 in aqueous monoethanolamine and monoethanolamine-diethylene glycol-water solutions containing no acid gas at 200°F.

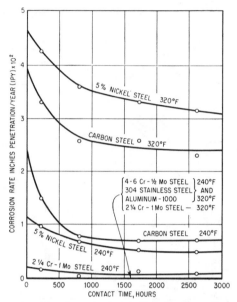

Figure 3-5. Corrosion rates of carbon steel and alloys in monoethanolamine-diethylene glycol-water solutions containing carbon dioxide and hydrogen sulfide.

been effective measures where severe corrosion was encountered. In glycol-monoethanolamine systems where stripping-column corrosion has been encountered, it has been found sufficient to protect the zone near the feed point with an aluminum liner and to use aluminum trays and bubble caps in this section. The zone requiring such protection extends approximately six trays downward and two trays upward from the feed point.

Stainless steel tubes (types 304 and 316) are quite resistant to corrosion by aqueous amine solutions when used in reboilers, while copper-bearing alloys such as cupronickel and Monel do not perform consistently better than carbon steel. Aluminum alloys 1100, 3003, and 6061 appear to be resistant to corrosion by glycol-monoethanolamine solutions in reboiler service. However, the problem of baffle cutting has to be resolved in order to design a satisfactorily operating reboiler containing aluminum tubes.

To eliminate or reduce corrosion of steel by water in the overhead condensers and solution coolers of stripping stills, bimetallic tubes containing admiralty metal on the outside and steel on the inside can be used successfully. In glycol-monoethanolamine systems, admiralty-metal tubes can be used for solution coolers since the lean solution does not attack admiralty metal to an appreciable extent.

Summarizing, it can be said that gas-purification plants employing alkanolamine solutions can be constructed for the most part of carbon steel. A variety of corrosion inhibitors and specially formulated solutions have proven to be quite successful and economical in alleviating or practically eliminating corrosion. However, the use of corrosion-resistant alloys such as the stainless steels, low-chromium steels, and aluminum alloys in areas especially susceptible to corrosion may prove to be economical in many cases. This, in combination with proper process-design and operational practices, results in satisfactory control of corrosion problems.

Foaming

Foaming of alkanolamine solutions is most frequently encountered in the contactor but may also occur in the stripping column. Foaming is usually caused by contamination of the solution by condensed light hydrocarbons, finely divided suspended solids (e.g., iron sulfide), amine degradation products, or surface active agents carried by the feed gas stream. Foam problems and some remedies are discussed in detail by Smith (70) and analytical methods are described which can be used to determine whether treating solutions are susceptible to foaming.

Condensation of light-hydrocarbon constituents of the feed gas can be avoided by keeping the temperature of the lean-amine solution about 10° to 15°F above that of the feed gas, thus ensuring that no phase change occurs. Suspended solids can be removed from the treating solution by continuous filtration of a side stream. Although several types of filters have been used successfully, the best results have generally been obtained with precoat filters. Except for cases of very severe fouling of the solution, continuous filtration of 5 percent of the circulating solution is adequate. In aqueous monoethanolamine plants, suspended solids may be effectively removed by the solution reclaimer (see later in this chapter).

When foaming is caused by dissolved or emulsified high molecular weight organic compounds, these agents may be eliminated by passage of the solution through a bed of activated carbon as described by Fife (33). Continuous treatment of 5 to 10 percent of the circulating solution by passage through a carbon bed is usually sufficient to avoid foaming. The activated carbon filters contain beds of a depth of 8 to 12 ft. and are sized for flow rates ranging from 2 to 4 gallons per minute per square foot of cross-sectional area (34). In most cases the carbon beds are not regenerated. Use of filter aid and activated carbon in a precoat filter of the pressure leaf type has been reported by Dailey (21). Such a system removes suspended solids and undesirable dissolved compounds in one operation. A review of experience with different kinds of filters in DEA plants located in Canada and the United States has been presented by Scheirman (54).

Several types of activated carbon filters for the purification of amine solutions have been described by Perry (71). The author concludes that "Deep Bed" filters (72) with a minimum bed depth of 5 feet, containing relatively large carbon particles (optimum 4 × 10 mesh) are capable of removing all contaminants, both solid and liquid, from amine and glycol systems and that no further purification of the circulating solution is required. The filters, when properly designed and operated, are reported to have a long service life and can be regenerated with low pressure steam. Operating results from two plants using about 28 weight percent DEA solutions indicate excellent solution quality with no foaming tendency, less than 1 ppm of iron and about 0.2 weight percent DEA as heat stable salts.

Another activated carbon filtration system in a plant using a DEA solution was reported by Keaton and Bourke (73). The carbon filter, which is used on a 10 percent slipstream of the lean solution at the pump discharge is a 10 feet diameter by 11 feet carbon steel vessel with a 316 stainless steel liquid collection system containing 20,000 pounds of Calgon SGL 8 × 30 granular activated carbon. Surface loading is 0.6 gpm per square foot and carbon contact is 120 minutes.

Foaming can in many cases be controlled by the addition of foam inhibitors or by the removal of foam-generating substances such as fine precipitates. The most widely used foam inhibitors are either silicone compounds or high-boiling alcohols such as oleyl alcohol or octylphenoxyethanol. The silicones are commercially available either as water emulsions or in their pure form. In general, the pure silicones are preferable. Since foam inhibitors are continuously lost from the system either mechanically or by steam distillation in the stripping still, it is necessary to add them continuously. The concentrations desirable for satisfactory foam inhibition are of the order of 10 to 15 ppm. Wainwright et al. (35) reported experiments with several commercial foam-depressing compounds in concentrations ranging up to 0.6 percent by volume. Alkaterge-C and 2-ethylhexanol were found most effective in concentrations of 0.4 percent by volume with the di- and triethanolamine solutions used in this work. However, 2-ethylhexanol had to be eliminated because of its high volatility. Ucon HB 5100 and Hercules defoamer 4, in small concentrations, increased the foaming tendencies of the solutions. A theoretical approach to the selection of foam depressants in gas processing systems using cohesive energy density has been presented by Meusburger and Segebrecht (74). It is of interest that both diethylene glycol and triethylene glycol are very effective foam depressants, and therefore foaming does not occur in plants using glycol-amine solutions.

Chemical Losses

The loss of solvent can be a serious operating difficulty in alkanolamine gas-purification plants. Losses can be incurred by entrainment of the solution in the gas stream, vaporization, or chemical degradation of the amine. Loss of solvent by entrainment or vaporization is undesirable, not only because of the cost of chemicals, but also because of the contamination of pipelines by liquids deposited on their walls. In addition, when alkanolamine solutions are used to purify gas to be used in catalytic processes, entrainment or vaporization of solvents may result in serious poisoning of the catalyst.

Entrainment

Entrainment losses are caused either by inefficient mist extraction or by foaming and subsequent carry-over of the solution. This problem can be eliminated by use of efficient mist-eliminating equipment and the application of foam inhibitors.

Vapor Losses

Although the vapor pressures of alkanolamine are relatively low, vaporization losses are significant because of the tremendous volumes of gas passed through the solution. The magnitude of losses from aqueous monoethanolamine and diethanolamine solutions can be estimated from Figure 3-6, which shows the amine vapor pressure over a number of typical solution concentrations. Vaporization losses from systems using MDEA or DGA solutions are comparable to those from DEA solutions, as the vapor pressures of these compounds are quite similar to that of DEA. Chemical losses due to vaporization can be combated by various methods. Probably the most simple procedure consists of washing the purified gas in a short section of packed or tray column with water or glycol (as described in Chapter 2). Vaporized amine can also be recovered by adsorption on bauxite or similar solids and subsequent regeneration of the saturated adsorbent by heating and steam addition (36). This process is very effective, and gas containing extremely low amounts of solvent vapor can be obtained by this method. In addition, complete recovery of the adsorbed amine can be achieved. Many adsorbents have high capacity of a very satisfactory life; thus this method is economically sound. The design of such plants is identical to solid-bed dehydration installations. When the inlet gas is saturated with water and dehydration is desirable, the capacity of the solid for water will determine the size of the adsorber because no amine will pass through the bed at the point of its saturation with water. However, in cases where partially dehydrated gas, such as is pro-

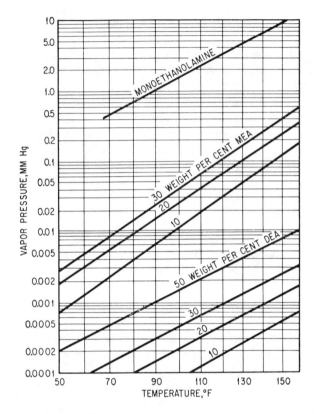

Figure 3-6. Vapor pressure of monoethanolamine and diethanolamine over aqueous solutions. *Chemetron Corporation and Texaco Chemical Company.*

duced in glycol-amine plants, is fed to the solid bed and further dehydration is not required, the equilibrium water-saturation point of the bed can be far surpassed before amine will "break through" the bed. The results obtained in a pilot-plant study (37) of such a system are shown in Table 3-3.

Amine vapors can be precipitated from the purified gas stream by intimate contact with small amounts of CO_2. Ethanolamines, especially monoethanolamine, react in the vapor phase with CO_2 to form a salt of the carbamic acid corresponding to the ethanolamine used. This compound condenses because of its extremely low vapor pressure and can be recovered in efficient mist-eliminating equipment (38,39).

Vaporization losses can also be minimized by the use of amines and glycols of extremely low vapor pressure such as diethanolamine, triethanolamine, methyldiethanolamine, and diglycolamine (40). However, these amines are somewhat less efficient absorbers of acid gases, and the purified gas may

Table 3-3. Pilot-plant Operation—Adsorption of Monoethanolamine Vapor from Natural Gas on Driocel (Bauxite)

Plant variables	Run A	Run B	Run C
Physical properties of bed:			
Bed depth, in	30	30	30
Bed diameter, in	1.875	1.875	1.875
Weight of Driocel in bed, lb	2.55	2.55	2.55
Superficial gas velocity, cu ft/(sq ft)(min)	36	33	33
Average operating conditions:			
Temperature of gas from contactor, °F	125	95	95
Temperature of gas to bed, °F	125	95	95
Temperature of solution to contactor, °F	133	98	98
Gas flow rate, SCF/hr	1,390	1,340	1,340
Solution flow rate, gpm	0.74	0.64	0.65
Contactor pressure, psia	555	555	555
Bed pressure, psia	555	555	555
Total operating time, hr	5	15	14
Analyses:			
Solution:			
Monoethanolamine, % by weight	30.6	31.0	31.0
H_2O, % by weight	9.2	5.6	6.2
Diethylene glycol, % by weight	60.2	63.4	62.8
Inlet gas:			
H_2O, lb/MMSCF	58.5	13.2	13.9
Monoethanolamine, lb/MMSCF	8.3	2.4	2.6
Outlet gas:			
H_2O break (after SCF)	1,740	6,700	6,820
Monoethanolamine break (after SCF)	3,590	13,400	13,200
H_2O in bed at H_2O break, % by weight	3.9	3.5	3.7
Monoethanolamine in bed at monoethanolamine break, % by weight	1.17	1.29	1.35

The Fluor Corporation

contain larger amounts of acid gases than gas treated with monoethanolamine.

Solution Degradation

Probably the most serious solution loss is caused by chemical degradation of the amine or glycol. In general, the amines and glycols are thermally stable at the temperatures normally used in the regeneration step.

The alkanolamines are somewhat subject to oxidative degradation. Several reaction mechanisms are possible. For example, the oxygen may react with H_2S removed from the gas to form free sulfur which, upon heating, reacts with the amine and forms dithiocarbamates, thioureas, and further decomposition products which cannot be reconverted to the free amine by application of heat. In addition, thiosulfuric acid may be formed; this combines with the amine to form a stable salt which also cannot be recovered for gas absorption by heating. Hofmeyer, Scholten, and Lloyd (41) have shown that monoethanolamine is subject to oxidative deamination which results in the formation of formic acid, ammonia, substituted amides, and high-molecular-weight polymers. Monoethanolamine appears to be considerably more vulnerable to oxidative degradation than the other ethanolamines. Glycols are also subject to oxidative degradation resulting in the formation of acids (42). Formic acid has been isolated from deteriorated diethylene glycol solutions. The effect of the pH of the solution on the oxidative degradation of glycols is fairly sharp. The rate of oxidation is lowered markedly if a pH of 6.0 or higher is maintained (43). The obvious remedy for oxidative degradation of the solution is exclusion of oxygen from the system, although the use of oxidation inhibitors may also be beneficial.

Oxidative degradation of DEA and MDEA was investigated by Blanc, Grall and Demarais (58) in their comprehensive work on degradation of DEA and MDEA solutions. The authors working with synthetic solutions prepared in the laboratory and solutions taken from operating plants were able to identify acetic, propionic, formic and oxalic acids, as well as thiosulfuric acid, in solutions which had been exposed to air. Concentrations of amine salts of these acids found in solution samples from nine plants ranged from 0.35 to 3.0 weight percent in DEA units and 0.24 to 0.53 weight percent units using MDEA solutions.

Gases subject to treatment by alkanolamine solutions may contain a great variety of impurities which react irreversibly with the amines. These compounds include carboxylic acids such as formic, acetic, and butyric acids; sulfur compounds such as carboxyl sulfide and carbon disulfide; hydrochloric acid; and hydrogen cyanide. The amines form simple thermally stable salts with all these compounds except hydrogen cyanide. The reaction of alkanolamines with hydrogen cyanide appears to result in salts of complex inorganic acids. One such compound, the diethanolamine salt of hydroferrocyanic acid, has been isolated and identified (28).

The stable acid salts of the amines can be dissociated by the addition of strong bases, such as sodium carbonate and sodium hydroxide, and the amine is recovered by subsequent distillation.

The most common loss of solvent is caused by the irreversible reactions of mono- and diethanolamine with the CO_2 present in the gas, as described by Polderman et al. (28, 29), Hakka et al. (16), Kennard and Meisen (75, 76), and Blanc et al. (58). This reaction is relatively slow but does occur at a significant rate under the conditions prevailing in the regeneration section of a purification plant. Monoethanolamine carbonate (or monoethanolamine carbamate) is first converted to oxazolidone-2 which reacts with one molecule of monoethanolamine, yielding 1-(2-hydroxyethyl)-imidazolidone-2. The substituted imidazolidone hydrolyzes to N-(2-hydroxyethyl)-ethylenediamine and CO_2. The hydrolysis of the substituted imidazolidone to the diamine restores part of the lost alkalinity and acid-gas absorption capacity of the solution.

$$HOCH_2CH_2NH_2 + CO_2 \rightarrow \begin{array}{c} CH_2 \text{---} CH_2 \\ | \qquad | \\ O \qquad NH \\ \diagdown \diagup \\ C \\ \| \\ O \end{array} + H_2O \qquad (3\text{-}1)$$

Monoethanolamine   Oxazolidone-2

$$\begin{array}{c} CH_2 \text{---} CH_2 \\ | \qquad | \\ O \qquad NH \\ \diagdown \diagup \\ C \\ \| \\ O \end{array} + HOCH_2CH_2NH_2 \rightarrow HOCH_2CH_2N \begin{array}{c} CH_2 \text{---} CH_2 \\ | \qquad | \\ \qquad NH \\ \diagdown \diagup \\ C \\ \| \\ O \end{array} + H_2O \qquad (3\text{-}2)$$

Oxazolidone-2   1-(2-Hydroxyethyl) imidazolidone-2

$$HOCH_2CH_2N \begin{array}{c} CH_2 \text{---} CH_2 \\ | \qquad | \\ \qquad NH \\ \diagdown \diagup \\ C \\ \| \\ O \end{array} + H_2O \rightarrow HOCH_2CH_2NHCH_2CH_2NH_2 + CO_2 \qquad (3\text{-}3)$$

1-(2-Hydroxyethyl)-imidazolidone-2   N-(2-Hydroxyethyl)-ethylenediamine

However, because the ethylenediamine derivative is a stronger base than monoethanolamine, it is more difficult to regenerate its sulfide and carbonate salts, and a significant portion of the diamine remains unregenerated. This may contribute somewhat to the corrosion of plant equipment. Since the hydrolysis of 1-(2-hydroxyethyl)imidazolidone-2 to N-(2-hydroxyethly)ethylenediamine is an equilibrium reaction, it is possible to prevent the presence of corrosion-promoting diamine in the solution by removal of the imidazolidone. It appears that high partial pressures of CO_2 favor

Reactions 3-1 and 3-2, which would explain the high rates of corrosion in plants using high acid-gas loadings of monoethanolamine solutions. Diethanolamine carbonate or carbamate is probably converted to 3-(2-hydroxyethyl)oxazolidone-2 which, at elevated temperatures and in the presence of base, loses two molecules of CO_2 to form N, N'-di(2-hydroxyethyl)piperazine.

$$(HOCH_2CH_2)_2NH + CO_2 \rightarrow \quad \underset{\substack{\text{HOCH}_2\text{CH}_2\text{N} \diagdown \quad \diagup \text{O} \\ \text{C} \\ \parallel \\ \text{O}}}{\overset{\text{CH}_2\text{---CH}_2}{|\qquad|}}$$

Diethanolamine   3-(2-Hydroxyethyl) oxazolidone-2(HEOZD) (3-4)

$$2HOCH_2CH_2N \quad \overset{\text{CH}_2\text{---CH}_2}{\underset{\substack{\diagdown \quad \diagup \\ \text{C} \\ \parallel \\ \text{O}}}{|\qquad|}} O \quad \rightarrow$$

3-(2-Hydroxyethyl) oxazolidone-2

$$HOCH_2CH_2N \overset{\text{CH}_2\text{---CH}_2}{\underset{\text{CH}_2\text{---CH}_2}{\diagup \diagdown \diagup \diagdown}} N\ CH_2CH_2OH + 2CO_2$$

N,N'-Di(2-hydroxyethyl)piperazine (HEP)   (3-5)

N, N'-Di(2-hydroxyethyl)piperazine, which has been identified in commercial diethanolamine solutions, is basic and capable of absorbing both H_2S and CO_2; this results in only a partial loss of acid-gas removing capacity. In addition to this compound, other nitrogen-containing degradation products have been found in diethanolamine solutions.These compounds are quite complex and probably of a linear carbamide structure. One such compound has been isolated by Hakka et al. (16) and identified as N, N, N'-tris (2-hydroxyethyl) ethylene diamine.

The DEA degradation products shown in Equations 3-4 and 3-5 and N,N,N'-tris (2-hydroxyethyl) ethylenediamine (THEED) have also been identified by Kennard and Meisen (75,76) and by Blanc et al. (58). However, a number of additional compounds were found by these investigators and rather complex reaction mechanisms for their formation are proposed. Kennard and Meisen (75) suggest that the most important measures to control DEA degradation are to keep reboiler and heat exchanger temperatures at the lowest practical levels and to operate with reasonably low DEA concentrations. It is interesting to note that they found that the presence of degradation products retards further degradation of DEA. Furthermore, they found in laboratory experiments that filtration through activated charcoal did not remove major degradation products. This observation is contrary to the experience of other investigators, notably that reported by Perry (71).

Regarding MDEA degradation, Blanc et al. state that acids have been found in plant solutions but the only basic compound mentioned as theoretically possible is Bis (hydroxyethylaminoethyl) ether (BHEAE).

Degradation products of diglycolamine (DGA) are the result of reactions with CO_2, COS or CS_2. The principal compounds are N,N' bis (hydroxyethoxy) urea (BHEEU) and N,N' bis (hydroxyethoxy) thiourea. These reactions are reportedly reversible at temperatures used in a reclaimer, so that a major portion of the degraded diglycolamine is recoverable (56).

Polderman et al (28,29) report that under certain conditions, the rate of conversion of diethanolamine is significantly faster than that of monoethanolamine. For example, when an aqueous solution containing 20 percent by weight of either amine saturated with CO_2 was heated for 8 hours at a pressure of 250 psi and 258°F, 22 percent of the diethanolamine was converted while practically no conversion of monoethanolamine was observed. In actual plant operation, lower rates of diethanolamine conversion have been observed as compared with monoethanolamine. This seeming inconsistency with the reported laboratory results may be explained by the fact that the vapor-liquid equilibrium relationships of the systems monoethanolamine-CO_2 and diethanolamine-CO_2 are such that in the reboiler (the point of highest temperature) of a plant using diethanolamine the solution is essentially free of CO_2. On the other hand, appreciable amounts of CO_2 are left in monoethanolamine solutions at the same point of the plant. The presence of degradation products does not, in general, impair the absorption characteristics of the free amine contained in the solution. However, accumulation of large amounts of degradation products results in increased viscosity of treating soutions and consequent decrease of absorption efficiency.

Sulfur-containing impurities such as mercaptans, disulfides, and thiophenes do not react with amines and therefore do not contribute to solution loss. Carbonyl sulfide and carbon disulfide do react with primary and secondary amines and therefore cause solvent losses.

The reaction of carbonyl sulfide with alkanolamines, as applicable to gas-purification operations, has been studied by Pearce, Arnold and Hall (44), Berlie, Estep and Ronicker (45), and Orbach and Selleck (46). Pearce et al. found that the reactions between carbonyl sulfide and monoethanolamine are essentially analogous to those of CO_2 with monoethanolamine as shown in Equations 3-1 and 3-2, except that they take place readily in ambient temperatures. In addition to oxazolidone and imidazolidone, the presence of diethanol urea was reported. Orbach and Selleck conducted experiments in a continuous bench scale pilot plant simulating an absorption-regeneration cycle. When contacting essentially pure carbonyl sulfide with a 20 weight percent monoethanolamine solution, they found rapid disappearance of alkalinity. A compound identified as 2-oxazolidone was isolated from the solution. The same experiment conducted with a 35 weight percent diethanolamine solution indicated no loss of alkalinity with time, and no degradation product could be found in the solution. The results from this study are shown graphically in Figure 3-7. These data are in agreement with the finding of Pearce et al. who also report essentially no degradation of diethanolamine by carbonyl sulfide, both in laboratory experiments and in field tests.

In commercial plants all of the carbonyl sulfide present in the feed gas does not react with monoethanolamine. Pearce et al. (44) and Berlie et al. (45) report that the major portion of the COS undergoes hydrolysis, forming H_2S and CO_2, and only about 15 to 20 percent of the COS reacts irreversibly with the monoethanolamine. The addition of strong alkalis, such as sodium carbonate or sodium hydroxide, to monoethanolamine solutions reduces losses due to reaction with COS substantially, probably by increasing the rate of COS hydrolysis (44). Although monoethanolamine losses can be reduced by this method, use of diethanolamine is preferred if the gas to be treated contains appreciable amounts of COS.

Tertiary amines such as triethanolamine and methyldiethanolamine do not react with carbonyl sulfide.

Carbon disulfide reacts with primary and secondary amines, first forming substituted dithiocarbamates and, subsequently, thiocarbamides.

$$2(HOCH_2CH_2)NH + CS_2 \rightarrow (HOCH_2CH_2)_2N\overset{\displaystyle S}{\overset{\|}{C}}SH_2N(CH_2CH_2OH)_2$$

Diethanolamine   Diethanolammonium N,N-di-
                            (2-hydroxyethyl)dithiocarbamate

$$\text{(3-6)}$$

$$(HOCH_2CH_2)_2N\overset{\displaystyle S}{\overset{\|}{C}}SH_2N(CH_2CH_2OH)_2$$

Diethanolammonium N,N-di-
(2-hydroxyethyl)dithiocarbamate

$$\rightarrow (HOCH_2CH_2)_2N\overset{\displaystyle S}{\overset{\|}{C}}N(CH_2CH_2OH)_2 + H_2S$$

N,N,N′,N′-Tetra(2-hydroxyethyl)-
thiocarbamide   (3-7)

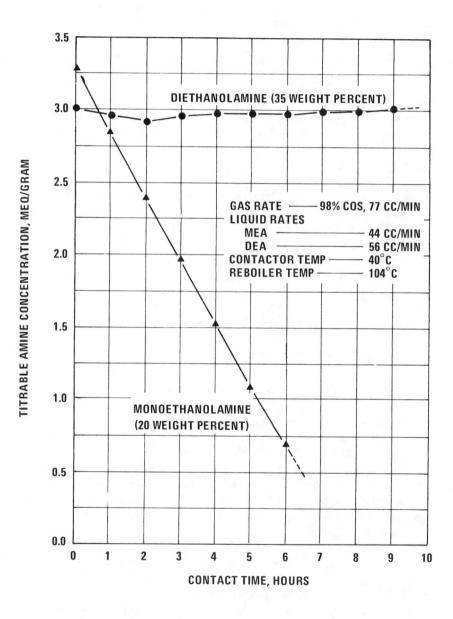

Figure 3-7. Effect of COS on mono- and diethanolamine. Data of Orbach and Selleck (46).
Fluor Corporation

A laboratory study of the reactions of carbon disulfide with monoethanolamine, diethanolamine, and diisopropanolamine has been reported by Osenton and Knight (47). The results of this study indicate that all three amines form dithiocarbamic acid salts quite rapidly. While the salts formed from diethanolamine and diisopropanolamine appear to be thermally stable, the monoethanolamine salt is less stable and forms one mole of oxazolidone 2-thione for each mole of monoethanolamine.

The field tests reported by Pearce et al. (43) showed that, under the operating conditions investigated, more than half of the carbon disulfide could be regenerated in the solution stripper, with the balance being hydrolyzed to H_2S and CO_2. This would indicate that the rate of reaction of monoethanolamine with carbon disulfide is slow and that amine losses resulting from this reaction are not serious. Again, tertiary amines do not react with carbon disulfide.

Amine deactivation by relatively strong acids present in the gas can be prevented by the use of a water or caustic wash prior to feeding it into the amine absorber. Since the acids most commonly occurring in gas streams are quite soluble in water, their removal can be carried out in simple equipment. The water wash will also remove ammonia which is present in certain gas streams. Ammonia itself is not harmful to alkanolamines, and therefore it is not necessary to take special precautions to remove this constituent from the gas before it enters the contactor. When ammonia is present, most of it is dissolved in the alkanolamine and expelled in the regenerating step.

Purification of Degraded Solutions

Precipitates and heavy sludge are removed from solutions by settling or by filtration. Settling is quite effective in aqueous solutions provided sufficient time is allowed for the precipitates and sludges to separate. Filtration involves a great variety of problems, primarily because iron sulfide, formed as a corrosion product, is very difficult to handle in conventional inexpensive filtration equipment. The most effective filter for this application appears to be the leaf-type precoat filter, although simpler equipment such as hay filters and cartridge-type filters are also used. As stated earlier in this chapter, high-molecular-weight degradation products of the amine can be eliminated from the solution by filtration through activated carbon.

High-boiling deterioration products of the amines, as well as sludge, can also be removed from the solutions by distillation of a small side stream, usually ½ to 2 percent of the main flow. This stream is withdrawn from the solution leaving the reboiler and fed to a small steam-heated or direct-fired kettle which operates either at atmospheric or reduced pressure. Since there is little acid gas present in the solution, the kettle is usually constructed of carbon steel although stainless-steel tubes are commonly used.

In plants using aqueous monoethanolamine solutions, purification is effected by semicontinuous steam distillation   as shown in Figure 3-8. Sodium carbonate or hydroxide is added, if necessary, to liberate the amine from the acid salts and minimize corrosion. This may be added directly to the kettle before charging. After the inital charging of the kettle with solution, fresh solution is fed at a slow rate, and then, by gradual increase of the steam pressure, it is brought to such a concentration that the equilibrium vapor has the same amine content as the feed to the kettle. As shown in Figure 3-9, for example, if the solution fed to the purification kettle contains 20 percent (by weight) MEA in water, the solution in the kettle has to be concentrated to an MEA content of 76 percent (by weight) in order to obtain an MEA concentration of 20 percent (by weight) in the vapor. As the distillation continues, the temperature in the kettle increases because of the accumulation of solids and high-boiling constituents. The addition of feed is stopped when the temperature reaches about 280°F. However, distillation is continued for a short time with the addition of water as liquid feed or as steam to remove as much of the residual MEA as economically possible. The kettle is then cleaned and recharged, and the cycle is repeated. This simple system can be used with aqueous monoethanolamine solutions because the vapor pressure of monoethanolamine is sufficiently high to permit distillation at temperatures at which the amine does not decompose thermally.

Reclaiming of diglycolamine from degraded solutions, as described by Dingman (56), is quite similar to reclaiming of MEA by semicontinuous steam distillation. A diagram of a reclaimer is shown in Figure 3-10. The reclaimer should be sized for a side stream of about 1 to 2 percent of the circulating solution and distillation is conducted at a kettle temperature of 360 to 380°F maximum. When steam is available, steam is usually sparged in below the tube bundle. As can be seen in Figure 3-11, at a temperature of 360°F and a pressure of 20 psia, the vapor from the reclaimer contains about 50 weight percent diglycolamine. This is returned directly to the stripping column.

Different techniques have to be employed with solutions containing higher-boiling amines such as diethanolamine and triethanolamine. One such method consists of withdrawing a small side stream of the ethanolamine solution, adding sodium or potassium hydroxide, and then removing most of the water by distillation. After this, the solution, containing 5 to 10 percent of water, is mixed with an alcohol—preferably isopropyl alcohol—which causes precipitation of the sodium or potassium salts of volatile organic and inorganic acids. After filtration of these salts, the alcohol is removed from the mixture by distillation, and the amine is returned to the main solution stream (48).

Another method for purifying contaminated diethanolamine solutions, covered by U.S. Patents 2,892,775 and 2,914,469, has been disclosed in the

138   Gas Purification

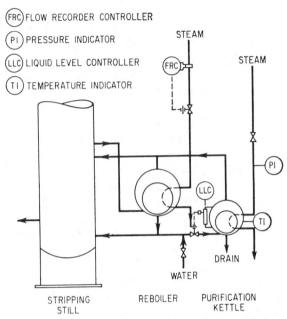

Figure 3-8. Purification system for aqueous monoethanolamine solutions at atmospheric pressure.

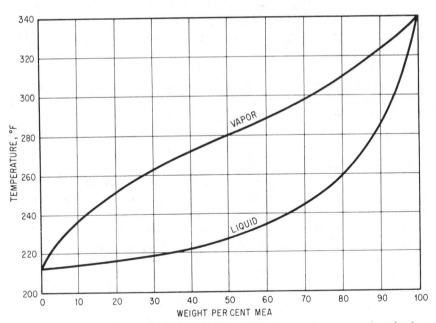

Figure 3-9. Vapor/liquid equilibrium composition of aqueous monoethanolamine solutions at boiling points. *Jefferson Chemical Company, Inc.*

literature (49). This system has been developed for reclaiming solutions treating refinery gases containing small amounts of acids. In order to maintain the activity of the diethanolamine solutions in such cases, continuous addition of sodium hydroxide is necessary to dissociate the amine-acid salts. This results in a buildup of sodium salts, requiring periodic disposal of the solution. By adding potassium salts to the contaminated solution, the melting point of the salt mixture is lowered sufficiently to permit distillation of diethanolamine at essentially atmospheric pressure without thermal decomposition.

A schematic flow diagram of the unit is shown in Figure 3-12. A solution of 48 percent potassium hydroxide is added to the contaminated diethanolamine before it enters the top of a packed tower where it is contacted with superheated steam rising from the reboiler. Essentially all of the amine and water is vaporized, and the product is withdrawn at a point below the packed section into a condenser which is held at a slightly lower pressure than that prevailing in the reboiler. A steam jet is used to maintain the required pressure. The salts drop into the reboiler and melt at 420° to 440°F, and any residual diethanolamine is flashed off. The temperature of the diethanolamine water vapor withdrawn is 300° to 375°F. The temperatures in the column and reboiler are sufficient to vaporize practically all the amine without decomposition. Heat is supplied by circulating oil at 600°F through the reboiler tubes. Recovery of 90 percent of reusable diethanolamine is claimed. As pointed out earlier in this chapter, experience with diethanolamine solutions in operating plants has demonstrated that degradation products can satisfactorily be removed by filtration through activated carbon.

A method for purifying DEA solutions contaminated with sodium chloride has been described by Morgan and Klare (77). A side stream of the circulating solution flows downward through a bed of Rohm and Haas Amberlite IRA-910, a strong base anion exchange resin, and the chloride ion is replaced with hydroxide ions. The hydroxide ions then react with CO_2 and sodium bicarbonate precipitates and is collected in the amine filter.

Solutions containing both glycol and amine cannot be purified by application of either of these methods. These solutions are purified by subjecting them to distillation under reduced pressure (approximately 10 in. Hg) and temperatures ranging from 300° to 375°F. The equipment used for this distillation, which consists of a small, steam-heated or direct-fired kettle, is shown in Figure 3-13. Proper heat densities to avoid excessive skin temperatures have to be used in direct-fired kettles. The feed to the kettle is separated from the main solution stream at the reboiler outlet; overhead vapors are condensed, collected in an accumulator, and returned to the plant system. The entire purification unit can be constructed of carbon steel.

By proper operation of side-stream purification units, it is possible to maintain a constant concentration of active amine in the treating solution and

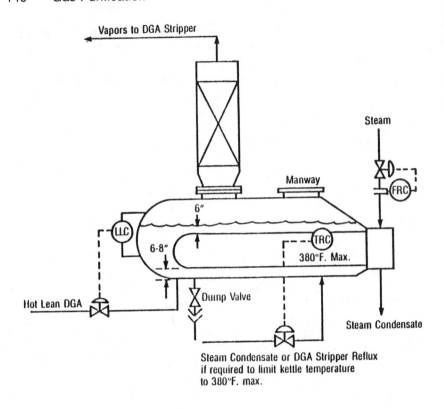

Figure 3-10. Diglycolamine Reclaimer (56).

to prevent undue accumulation of undersirable materials. In commercial plant operations, the concentration of amine-decomposition products in the solution should not be allowed to exceed 10 percent of the concentration of active amine.

Purification of the solution by distillation is also an effective method of removing precipitates such as iron sulfide and can therefore obviate the need for cumbersome and expensive filtration equipment. The effectiveness of the purification is determined by comparison of solution analyses for primary amino nitrogen, total alkalinity, and total nitrogen.

Nonacidic-gas Entrainment In Solution

In certain operations, especially when acid-gas removal is carried out at high pressure, appreciable amounts of nonacidic gases are carried by the solu-

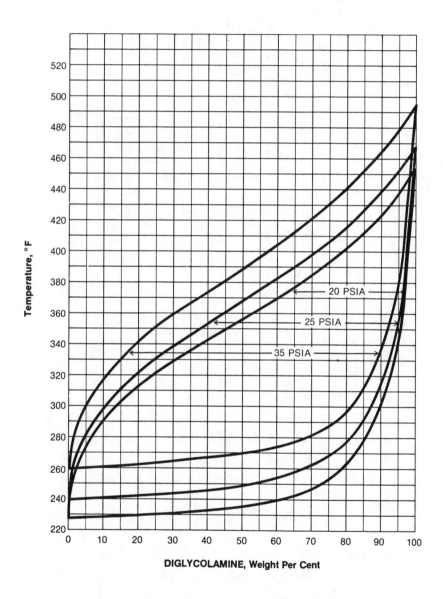

Figure 3-11. Vapor - Liquid Equilibrium for Aqueous Diglycolamine Solutions at Various Pressures (57).

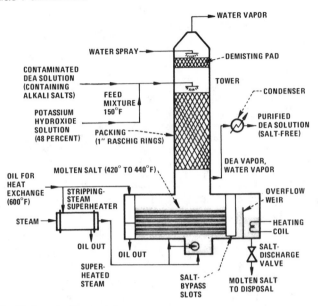

Figure 3-12. Diethanolamine reclaimer.

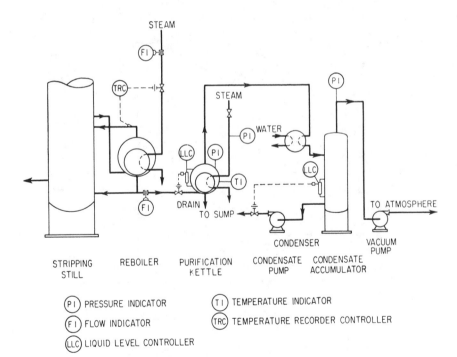

Figure 3-13. Monoethanolamine-diethylene glycol-water solution purification system at reduced pressure.

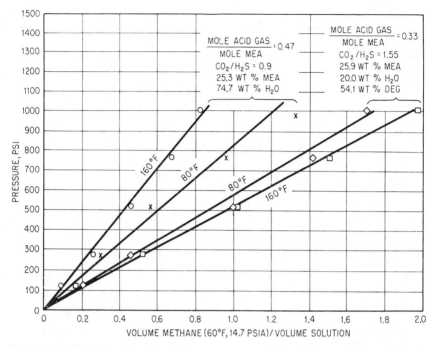

Figure 3-14. Solubility of methane in aqueous monoethanolamine and monoethanol-amine-diethylene glycol-water solutions containing acid gas. *Fluor Corporation*

tion from the contactor to the regeneration section of the plant. This is particularly undesirable if the acid gases are intended to be used further, as for instance, for the production of dry ice or elemental sulfur. The nonacidic gas may be carried both in solution and as entrained bubbles (or drops of liquid hydrocarbon). Mechanical carry-over of nonacidic gases can be minimized by proper design of the bottom section of the contactor. Both splashing and the free fall of liquids through the vapor space (which result in froth in the bottom of the contactor) should be avoided by the installation of a properly designed downcomer, and the outlet line should be designed to prevent the formation of a vortex. Even when the contactor bottom is properly designed, it is in most cases impossible to eliminate totally mechanical carry-over. In addition, the solubility of most nonacidic gases in ethanolamine solutions becomes appreciable at high pressures. Provisions, therefore, must be made to separate these gases from the solution after it leaves the contactor and before it enters the regenerating section. Depending on the operating conditions of the plant and the purity requirements of the acid gas, one or several disengaging drums can be installed at several locations between the solution outlet at the contactor and the solution inlet at the stripping still. To provide

Table 3-4. Solubility of Methane and Ethane In Aqueous MEA and DEA Solutions (55)

Amine Type	Wt. % Amine	Temperature °F	Vapor Pressure—psi Methane	Vapor Pressure—psi Ethane	Solubility Vol./Vol.
MEA	15	100	954	—	1.55
MEA	15	150	498	—	0.73
MEA	15	150	993	—	1.37
MEA	15	100	—	491	0.92
MEA	15	100	—	868	1.19
MEA	15	150	—	501	0.74
MEA	15	150	—	955	1.18
DEA	25	100	510	—	0.76
DEA	25	100	968	—	1.42
DEA	25	150	511	—	0.68
DEA	25	150	982	—	1.28
DEA	25	200	516	—	0.61
DEA	25	200	920	—	1.19
DEA	25	100	—	480	0.91
DEA	25	100	—	868	1.19
DEA	25	150	—	498	0.74
DEA	25	150	—	972	1.13

Gas volume at 60°F, 14.7 psi.

a maximum of vapor-disengaging area, horizontal vessels are frequently used.

The gases evolved in the disengaging drums contain acid gas in varying concentrations. This acid gas can be recovered by contacting the flashed gases with a small stream of lean amine solution in a small column usually installed at the top of the disengaging vessel (see Chapter 2).

The solubility of methane in rich aqueous and glycol-monoethanolamine solutions is presented in Figure 3-12 for two temperatures and varying pressures (50). This chart can be used in estimating the amount of methane (or dry natural gas) which will be dissolved in a treating-plant solution and the quantity which will remain after an equilibrium flash separation. Data on the solubility of methane and ethane in aqueous monoethanol and diethanolamine solutions either free of acid gas or partially saturated with H_2S or CO_2 were reported by Lawson and Garst (55). Selected values from this study are shown in Tables 3-4 and 3-5. It should be noted that hydrocarbon gases are appreciably more soluble in amine solutions containing no dissolved acid gases than in solutions containing either one or both of the acid gases.

Table 3-5. Solubility of Methane in 25 Wt.% Aqueous DEA Containing H_2S or CO_2 (55)				
Temperature °F	Vapor Pressure psi	Mol/Mol DEA H_2S	Mol/Mol DEA CO_2	Solubility Vol./Vol.
100	973	0.20	—	1.27
100	862	1.10	—	1.07
150	456	0.19	—	0.47
150	495	0.25	—	0.54
150	951	0.20	—	1.03
100	500	—	0.27	0.65
100	956	—	0.27	1.16
150	489	—	0.25	0.55
150	970	—	0.25	0.97

Gas volume at 60°F, 14.7 psi.

References

1. Bottoms, R.R. 1936. U.S. Patent 2,065,112.
2. Bottoms, R.R. 1936. U.S. Patent 2,031,632.
3. Reed, R.M. 1945. U.S. Patent 2,377,966.
4. Riesenfeld, F.C., and Blohm, C.L. 1950. *Petrol. Refiner* 29(Apr.):141-150.
5. Riesenfeld, F.C., and Blohm, C.L. 1951. *Petrol. Refiner* 30(Feb.):97-106.
6. Riesenfeld, F.C., and Blohm, C.L. 1951. *Petrol. Refiner* 30(Oct.):107-115.
7. Froning, H.R., and Jones, J.H. 1958. *Ind. Eng. Chem.* 50(Dec.):1737.
8. Lang, F.S., and Mason, J.F. Jr. 1958. *Corrosion* 14(Feb.):65-68.
9. Connors, J.S. 1958. *Oil Gas J.* 56(Mar. 3):100-110.
10. Graff, R.A. 1959. *Refining Eng.* 31(3):C12-14.
11. Leachman, G.S., and Ellis, J.C. 1961. *Oil Gas J.* 59(July 24):84-88.
12. Mottley, J.R., and Fincher, D.R. *Materials Protection* 2(8):26-30.
13. Ballard, D. 1966. *Hydro. Process.* 45(Apr.):137-144.
14. Dingman, J.C.; Allen, D.L.; and Moore, T.F. 1966. *Hydro. Process.* 45(Sept.):285-290.
15. Comeaux, R.V. 1962. *Hydro. Process.* 41(May):141-143.
16. Hakka, L.E.; Singh, K.P.; Bata, G.L.; Testart, A.C.; and Andrejehyshyn, W.M. 1968. *Some Aspects of Diethanolamine Degradation in Gas Sweetening.* Paper presented at Canadian Natural Gas Processing Association meeting (May 9).
17. Hall, G.D. 1959. *Can. Oil Gas Ind.* (May):84-89.
18. Hall, G.D., and Polderman, L.D. 1960. *Chem. Eng. Progr.* 56(Oct.):52-58.
19. Petrunic, A. 1968. *Wildcat Hills Complex MEA-gas Treating Sulphur Production System.* Paper No. 6803 presented at 19th Annual Technical Meeting of the Petroleum Society of CIM, Calgary (May 7-10).
20. Bailleul, M. 1969. *Gas Process. (Can.)* 61(3):34-38.
21. Dailey, L.W. 1970. *Oil Gas J.* 68(May 4):136-141.
22. Moore, K.L. 1960. *Corrosion* 16(Oct.):503t-506t.
23. Estep, J.W.; McBride, J.T.; and West, R.J. 1962. *Advances in Petroleum Chemistry and Refining,* vol. 6. New York: Interscience Publishers, pp. 315-466.

24. Dow Chemical of Canada, Ltd. 1962. *Gas Conditioning Fact Book.*
25. Wendt, C.J., and Dailey, L.W. 1967. *Hydro. Process.* 46(Oct.):155-157.
26. Evans, U.R. 1948. *Metallic Corrosion, Passivity and Protection.* New York: Longmans, Green and Co., Inc., pp. 205-287.
27. Dixon, B.E., and Williams, R.A. 1950. *J. Soc. Chem. Ind. (London)* 69(Mar.):69-71.
28. Polderman, L.D., and Steele, A.B. 1956. *Oil Gas J.* 54(July 30):206-214.
29. Polderman, L. D.; Dillon, C.P.; and Steele, A.B. 1955. *Oil Gas J.* 53(May 16):180-183.
30. Garwood, G.L. 1953. *Oil Gas J.* 51(July 27)334-340.
31. Klein, J.P. 1970. *Oil Gas Int.* 10(Sept.):109-112.
32. Montrone, E.D., and Long, W.P. 1971. *Chem. Eng.* 78(Jan.):94-99.
33. Fife, H.R. 1932. U.S. Patent 1,944,122.
34. Jefferson Chemical Company, Inc. 1969. *Gas Treating Data Book.*
35. Wainwright, H.W.; Egelson, G.C.; Brock, C.M.; Fisher, J.; and Sands, A.E. 1952. *U.S. Bur. Mines, Rept. Invest., No. 4891* (Oct.).
36. Nevens, T.D. 1954. *Petrol. Engr.* 26(June):D-33.
37. The Fluor Corporation. Unpublished data.
38. Frazier, H.D. 1952. U.S. Patent 2,608,461.
39. Frazier, H.D., and Riesenfeld, F.C. 1952. U.S. Patent 2,608,462.
40. Blohm, C.L., and Riesenfeld, F.C. U.S. Patent 2,712,978.
41. Hofmeyer, B.G.; Scholten, H.G.; and Lloyd, W.G. 1956. *Contamination and Corrosion in Monoethanolamine Gas Treating Solutions.* Paper presented at National Meeting of American Chemical Society, Dallas, Texas (April 8 to 13).
42. Lloyd, W.G., and Taylor, F.C. 1954. *Ind. Eng. Chem.* 46(Nov.):2407-2415.
43. Lloyd, W.G. (Dow Chemical Company). Private communication.
44. Pearce, R.L.; Arnold, J.L.; and Hall, C.K. 1961. *Hydro. Process.* 40(Aug.):121-129.
45. Berlie, E.M.; Estep, J.W.; and Ronicker, F.J. 1965. *Chem. Eng. Progr.* 61(Apr.):82-85.
46. Orbach, H.K., and Selleck, F.T. "The Effect of Carbonyl Sulfide on Ethanolamine Solutions." Unpublished paper by permission of The Fluor Corporation.
47. Osenton, J.B., and Knight, A.R. 1970. *Reaction of Carbon Disulphide with Alkanolamines used in the Sweetening of Natural Gas.* Paper presented at the Canadian Gas Processing Association, Fourth Quarterly Meeting, Calgary (Nov. 20).
48. Paulsen, H.C.; Holtzelaw, J.B.; and McNamara, T.P. 1955. U.S. Patent 2,-701,750.
49. Anon. 1963. *Chem. Engr.* 70(Mar. 4):40-41.
50. The Fluor Corporation. Unpublished data.
51. Kelly, W. 1977. *Hydro. Process.* 56(July):118-119.
52. Dow Chemical Company, Private Communication 1978.
53. Martin, C.W.; Krawczyk, L.S.; Fowler, A.E.; and Clouse, R.C. 1978. "A Versatile New Inhibited Amine Process," Paper presented at Gas Conditioning Conference, Norman, Oklahoma (March 7, 1978).
54. Scheirman, W.L. 1973. *Hydro. Process.* 52(Aug.):95-96.
55. Lawson, J.D., and Garst, A.W. 1976. *J. of Chem. and Eng. Data* 21(1):30-32.
56. Dingman, J.C. 1977. "Gas Sweetening with Diglycolamine Agent," paper presented at Third Iranian Congress of Chemical Engineering, Shiraz, Iran, Nov. 6–10.
57. Texaco Chemical Company. "Diglycolamine Agent."

58. Blanc, C., Grall, M., and Demarais, G. 1982. "The Part Played by Degradation Products in the Corrosion of Gas Sweetening Plants Using DEA and MDEA," paper presented at 32nd Gas Conditioning Conference, University of Oklahoma.

59. Goldstein, A.M. 1983. "Commercialization of a New Gas Treating Agent," paper presented at Petroenergy '83 Conference, Houston, Texas, Sept. 14.

60. Weinberg, H.N., Eisenberg, B., Heinzelmann, F.J. and Savage, D.W. 1983. "New Gas Treating Alternatives for Saving Energy in Refining and Natural Gas Production," paper presented at the 11th World Petroleum Congress, London, England, Aug. 31.

61. Butwell, K.F., Hawkes, E.N., and Mago, B.F. 1973. *Chemical Eng. Progress,* 69(2):57, Feb.

62. Butwell, K.F., and Kubek, D.J. 1977. *Hydrocarbon Processing,* 56(10), Oct. p 173.

63. Kubek, D.J., and Butwell, K.F. 1979. "Amine Guard Systems in Hydrogen Production," paper presented at 86th AIChE National Meeting, Houston, Texas, April 1–5.

64. Butwell, K.F., Kubek, D.J., and Sigmund, P.W. 1979. *Chemical Eng. Progress,* February. p 75.

65. Butwell, K.F. 1978. U.S. Patent 4,079,117.

66. Dibble, H.J. 1983. "UCARSOL Solvents for Acid Gas Removal," paper presented at Petroenergy '83, Houston, Texas, Sept. 14.

67. Dow Chemical U.S.A. 1983. "Gas/Spec, A New Kind of Gas Processing Intelligence from Dow," Form 112-1815-83.

68. Wiggins, W.R., and Bixler, R.L. 1983. Energy Progress 3:3, Sept. p 132.

69. Liebermann, N.P. 1980. *Oil and Gas Journal,* May 12, p 113.

70. Smith, R.F. 1978. Travis Chemicals, Calgary, Alberta.

71. Perry, C.R. 1980. "Activated Carbon Filtration of Amine and Glycol Solutions," paper presented at 1980 Gas Conditioning Conference, University of Oklahoma.

72. Perry, C.R. 1971. U.S. Patent 3,568,405.

73. Keaton, M.M., and Bourke, M.J. 1983. *Hydrocarbon Processing,* 62(8) Aug. p 71.

74. Meusburger, K.E., and Segebrecht, E.W. 1980. "Foam Depressants for Gas Processing Systems," paper presented at 1980 Gas Conditioning Conference, University of Oklahoma.

75. Kennard, M.L., and Meisen, A. 1980. *Hydrocarbon Processing,* 59(4) April p 103.

76. Meisen, A., and Kennard, M.L. 1982. *Hydrocarbon Processing,* 61(10) October p 105.

77. Morgan, C., and Klare, T. 1977. "Chloride Removal from DEA by Ion Exchange," paper presented at 1977 Gas Conditioning Conference, University of Oklahoma.

4

Hydrogen Sulfide and Carbon Dioxide Removal with Ammonia

The alkanolamine processes, discussed in the two preceding chapters, are ideally suited for the treatment of natural, refinery, and synthesis gases which contain hydrogen sulfide and carbon dioxide as the only impurities to be removed. For treating gas streams containing carbonyl sulfide, carbon disulfide, hydrogen cyanide, organic acids, nitrogen bases, and other impurities, the alkanolamines are of limited application because (a) they react nonregenerable with certain impurities, and (b) they are not readily recoverable from fouled solutions. The purification of coal gas, which contains such impurities and which in most countries outside the United States is still an important industrial and domestic fuel, requires processes which do not have the above limitation. The presence of ammonia in coal gases naturally led to consideration of its utilization for removal of acid gases, ideally in such a manner that the maximum amount of ammonia and the acid gases could be recovered. Typical concentrations of impurities, other than hydrocarbons, present in coal gases are shown in Table 4-1.

The principal objectives of coal-gas purification processes are the removal of hydrogen sulfide and of nitrogen compounds, primarily ammonia. These compounds are undesirable because of their corrosive nature and are present in such high concentrations that their removal from the gas is an absolute necessity. In addition, both H_2S and NH_3 are relatively valuable chemicals, and their recovery and conversion to elemental sulfur and ammonium sulfate, nitrate, or phosphate is in many cases of considerable economic significance. This was particularly true before the advent of synthetic ammonia, when coal gas constituted the largest source of fixed nitrogen.

As can be seen from the analysis in Table 4-1, carbon dioxide is always a major impurity in coal gas. Although it is not generally necessary to remove

Table 4-1. Nonhydrocarbon Impurities in Coal Gases

Type of Impurity	Typical Concentration, Percent by Volume
Hydrogen sulfide	0.3–3.0
Carbon disulfide	0.016
Carbonyl sulfide	0.009
Thiophene	0.010
Mercaptans	0.003
Ammonia	1.1
Hydrogen cyanide	0.10–0.25
Pyridine bases	0.004
Nitric oxide	0.0001
Carbon dioxide	1.5–2.0

CO_2 from gases used as fuels, partial removal is sometimes desirable to improve the heating value, and complete CO_2 elimination is required for gases undergoing processing at very low temperatures—as for example in coke-oven gas purification to provide hydrogen for ammonia synthesis.

The technology of coal-gas purification has its beginnings in the early nineteenth century, and the literature covering the subject is voluminous. Complete coverage of the many coal- and synthesis-gas purification processes which have been proposed is not possible in a text of this scope. However, the most important techniques presently used on a commercial scale are discussed in this and some of the following chapters.

Although the idea of using aqueous ammonia solutions for the removal of acidic constituents from gases is quite old, commerical application of such processes, especially for selective removal of hydrogen sulfide from coal gases and complete removal of carbon dioxide from synthesis gases, has only recently been accomplished. Availability of corrosion-resistant materials and better understanding of certain physical phenomena have led to considerable improvement of the economics of the process and, consequently, to the acceptance of these methods by industries concerned with gas-purification problems.

The recent strong emphasis on environmental protection, especially in heavily industrialized areas, gave new impetus to purification of coke-oven gases which in many instances have been burned without treatment as an industrial fuel. Several ammonia based processes, developed primarily in Japan and Germany, are now either in use or projected for early operation. Reviews of the state of the art of coke-oven gas purification have been presented by Massey and Dunlap (31) and at a symposium on treatment of coke-oven gas held at McMaster University, Ontario, Canada, in May, 1977 (32).

Basic Chemistry

The reactions occuring in the system comprised of ammonia, hydrogen sulfide, carbon dioxide, and water can be represented by the following equations:

$$NH_3 + H_2S = NH_4HS \tag{4-1}$$

$$2NH_3 + H_2S = (NH_4)_2S \tag{4-2}$$

$$2NH_3 + CO_2 = NH_2COONH_4 \tag{4-3}$$

$$NH_3 + CO_2 + H_2O = NH_4HCO_3 \tag{4-4}$$

$$2NH_3 + CO_2 + H_2O = (NH_4)_2CO_3 \tag{4-5}$$

$$NH_2COONH_4 + H_2O = (NH_4)_2CO_3 \tag{4-6}$$

$$(NH_4)_2CO_3 + H_2S = NH_4HCO_3 + NH_4HS \tag{4-7}$$

$$(NH_4)_2S + H_2CO_3 = NH_4HCO_3 + NH_4HS \tag{4-8}$$

$$NH_4HS + H_2CO_3 = NH_4HCO_3 + H_2S \tag{4-9}$$

Van Krevelen, Hoftijzer, and Huntjens (1) have shown that under equilibrium conditions the ionic species NH_4^+, HCO_3^-, NH_2COO^-, and $CO_3^=$, as well as undissociated NH_3, are present in aqueous solution in measurable quantities. The same authors state that under most conditions H_2S is present in the form of SH^- ions. When the pH of the solution is below 12 the sulfide ion concentration is negligible, and even at pH 12 this ionic species amounts to only 0.1 percent of the fixed hydrogen sulfide.

The following equations present heats of reaction (based on published heats of formation) for the four reactions most predominant in processes using ammonia for the removal of hydrogen sulfide and carbon dioxide from gases:

$$2NH_3(g) + H_2S(g) \rightarrow (NH_4)_2S \ (aq.) \tag{4-10}$$
$$-\Delta H = 39,600 \ Btu/lb \ mole \ (77°F)$$

$$NH_3(g) + H_2S(g) \rightarrow (NH_4)HS \; (aq.) \tag{4-11}$$
$$-\Delta H = 34,800 \; \text{Btu/lb mole (77°F)}$$

$$2NH_3(g) + CO_2(g) + H_2O(l) \rightarrow (NH_4)_2CO_3 \; (aq.) \tag{4-12}$$
$$-\Delta H = 73,200 \; \text{Btu/lb mole (77°F)}$$

$$2NH_3(g) + CO_2(g) \rightarrow NH_4CO_2NH_2 \; (aq.) \tag{4-13}$$
$$-\Delta H = 61,100 \; \text{Btu/lb mole (77°F)}$$

Vapor Pressures

The vapor pressures of H_2S and NH_3 over solutions containing the two compounds in various concentrations and mole ratios were determined by Lohrmann and Stoller (2), Pexton and Badger (3), Badger and Silver (4), Badger (5), Badger and Wilson (6), Dryden (7), and van Krevelen, Hoftijzer, and Huntjens (1). Terres, Attig, and Tscherter (8) determined the vapor pressures of H_2S, NH_3, and water over aqueous ammonium sulfide and ammonium hydrosulfide solutions, containing up to 9 percent total ammonia, over the temperature range of 20° to 60°C. These data are shown in Tables 4-2 and 4-3.

Van Krevelen et al. (1) developed a method for calculating the vapor pressures of H_2S and NH_3 over aqueous solutions on the basis of the chemical equilibria in the solutions. The numerical values of the equilibrium coefficients were derived from experimental data.

Since hydrogen sulfide is present as HS^-, the equilibrium to be considered is

$$NH_3 + H_2S \rightleftharpoons NH_4^+ + HS^- \tag{4-14}$$

For a total ammonia content of A moles/liter, a total H_2S content of S moles/liter and an equivalent anion concentration of other ammonium salts present of Z equivalents/liter, the following equations can be written:

$$(NH_3) = A - S - Z \tag{4-15}$$

$$(HS^-) = S \tag{4-16}$$

$$(NH_4^+) = S + Z \tag{4-17}$$

Table 4-2. Vapor Pressures of H_2S, NH_3 and H_2O over Aqueous $(NH_4)_2S$ Solutions

Temperature, °C	NH_3, % by wt.	H_2S, % by wt.	Vapor Pressure, mm Hg		
			H_2S	NH_3	H_2O
20	1.38	1.40	53.3	0	16.3
20	2.68	2.68	65.6	0	16.2
20	6.77	6.81	130.1	0	15.8
20	8.05	8.03	154.8	5.6	15.1
20	9.41	9.40	190.0	13.0	14.3
40	2.31	2.33	92.0	8.2	49.1
40	5.88	5.86	183.4	22.2	44.7
40	7.21	7.26	220.6	26.9	48.4
40	9.34	9.36	293.2	30.4	47.9
60	3.22	3.25	134.1	25.5	125.1
60	5.54	5.56	205.0	48.5	127.3
60	6.80	6.81	250.9	76.1	113.5
60	9.00	9.05	365.0	145.0	105.0

Source: Data of Terres, Attig, and Tscherter. (8)

Table 4-3. Vapor Pressures of H_2S, NH_3, and H_2O over Aqueous NH_4HS Solutions

Temperature, °C	NH_3, % by wt.	H_2S, % by wt.	Vapor Pressure, mm Hg		
			H_2S	NH_3	H_2O
20	0.80	1.58	115.7	0	13.9
20	2.14	4.23	163.0	0	15.2
20	2.97	5.93	197.6	0	16.1

Source: Data of Terres, Attig, and Tscherter. (8)

Table 4-4. Henry Coefficient for Ammonia in Pure Water

Temperature, °C	H_0, kg mole/ cu m/mm Hg
20	0.099
40	0.0395
60	0.017
80	0.0079
90	0.0058

An equilibrium coefficient K can be expressed as follows:

$$K = \frac{(NH_4^+)(HS^-)}{(NH_3)P_{H_2S}} = \frac{(S + Z)S}{(A - S - Z)P_{H_2S}} \qquad (4\text{-}18)$$

from which the vapor pressure of H_2S is

$$P_{H_2S} = \frac{(S + Z)S}{(A - S - Z)K} \qquad \text{mm Hg} \qquad (4\text{-}19)$$

and from Henry's law,

$$P_{NH_3} = \frac{A - S - Z}{H_{NH_3}} \qquad \text{mm Hg} \qquad (4\text{-}20)$$

The authors (1) propose the following equation for estimating Henry coefficients for solutions of the type commonly encountered in H_2S absorption:

$$- \log \frac{H_{NH_3}}{H_0} - 0.025(NH_3) \qquad (4\text{-}21)$$

where H_0 = Henry coefficient for ammonia in pure water (see Table 4-4)

(NH_3) = concentration of free ammonia

The equilibrium coefficient K is not a true constant and varies with the concentration of dissolved salts in the solution. The authors found that this variation can be expressed as a function of the ionic strength I

$$I = \tfrac{1}{2}\Sigma C_i Z_i \qquad (4\text{-}22)$$

Where C_i = concentration of a given ion
Z_i = corresponding valency,
by the equation

$$\log K = a + 0.089I \qquad (4\text{-}23)$$

where a has the following values:

$t,°C$	a
20	−1.10
40	−1.70
60	−2.19

In aqueous solutions containing only ammonia and H_2S, I equals S, the total H_2S concentration, and Equation 4-23 becomes

$$\log K = a + 0.089S \qquad (4\text{-}24)$$

Vapor pressures of H_2S and NH_3 for 0.5, 1.0, 2.0 N aqueous ammonia solutions at 20°C (68°F) as a function of the mole ratio H_2S: NH_3, calculated by this method are shown in Figure 4-1.

The calculated values agree quite well with most experimental data presented in the literature. However, the data of Terres, Attig, and Tscherter (8) for ammonium sulfide and ammonium hydrosulfide solutions are in considerable diagreement. The reason for this discrepancy may be the dynamic method employed by Terres et al. and the analytical difficulties reported in their paper. Van Krevelen and coworkers used both a dynamic and static method which yielded practically identical results.

Vapor pressures of ammonia and carbon dioxide over aqueous solutions containing these two gases are presented by Pexton and Badger (3), Badger and Wilson (6), Terres and Weiser (9), Terres, Attig, and Tscherter (8), and van Krevelen, Hoftijzer, and Huntjens (1). The latter authors propose a method, similar to the one described for hydrogen sulfide-containing ammonia solutions, for calculating vapor pressures of ammonia and carbon dioxide. Figures 4-2 to 4-4 show calculated vapor pressures of ammonia and carbon dioxide for 0.125, 0.5, 1.0, and 2.0 N ammonia solutions as a function of the mole ratio CO_2:NH_3 at various temperatures. These values agree very well with the experimental data obtained by van Krevelen et al. (1) and those of Pexton and Badger (3), Badger and Wilson (6), and Terres and Weiser (9). However, there is considerable disagreement between the calculated data and the data presented by Terres, Attig, and Tscherter for ammonium bicarbonate solutions. The calculated values for the vapor pressures of CO_2 at mole ratios CO_2:NH_3 approaching 1.0 are very high, and, judging from the shape of the curves, it appears that the calculation method may not be reliable for solutions in which essentially all the ammonia and carbon dioxide are present as ammonium bicarbonate. On the other hand, the values given by Terres et al. for ammonium bicarbonate solutions seem to be lower than could be expected from extrapolation of other experimental data reported in the literature.

A method for the approximate calculation of the vapor pressures of NH_3, H_2S, and CO_2 in solutions containing all three of these compounds is presented by Badger and Silver (4). These authors postulate that, in solutions containing A moles of ammonia, C moles of carbon dioxide, and S moles of hydrogen sulfide, the vapor pressures of ammonia and carbon dioxide are essentially the same as those for solutions containing $(A - S)$ moles of am-

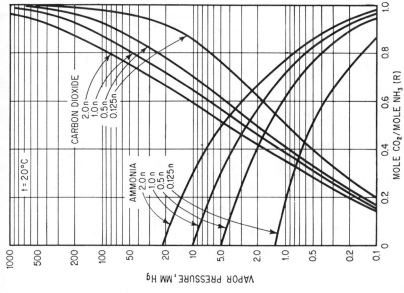

Figure 4-2. Equilibrium vapor pressures of carbon dioxide and ammonia over 0.125, 0.5, 1.0, and 2.0 N aqueous ammonia solutions at 20°C (68°F). Data of van Krevelen, Hoftijzer, and

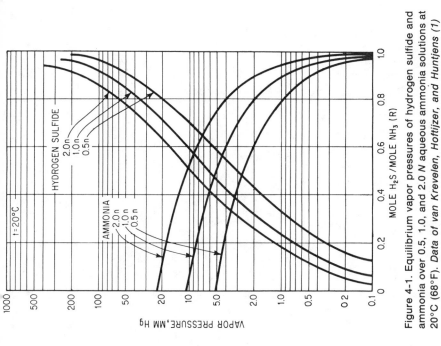

Figure 4-1. Equilibrium vapor pressures of hydrogen sulfide and ammonia over 0.5, 1.0, and 2.0 N aqueous ammonia solutions at 20°C (68°F). Data of van Krevelen, Hoftijzer, and Huntjens (1)

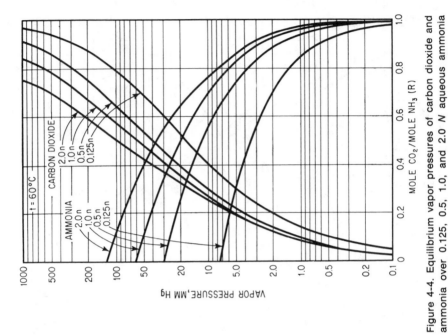

Figure 4-4. Equilibrium vapor pressures of carbon dioxide and ammonia over 0.125, 0.5, 1.0, and 2.0 N aqueous ammonia solutions at 60°C (140°F). *Data of van Krevelen, Hoftijzer, and*

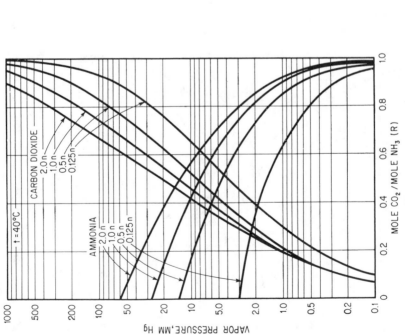

Figure 4-3. Equilibrium vapor pressures of carbon dioxide and ammonia over 0.125, 0.5, 1.0, and 2.0 N aqueous ammonia solutions at 40°C (104°F). *Data of van Krevelen, Hoftijzer, and*

monia and C moles of CO_2. The vapor pressure of hydrogen sulfide in such systems can be estimated in the same manner because

$$P_{H_2S} = \frac{(C + S)S}{P_{NH_3}H_{NH_3}} \tag{4-25}$$

A comparison of values obtained by this method with the experimental data of Badger and Silver (4) is shown in Table 4-5. Van Krevelen, Hoftijzer, and Huntjens (1) propose a considerably more involved method for systems containing H_2S, CO_2, and ammonia. However, the simple procedure of Badger and Silver (4) gives practically the same results for relatively dilute ammonia solutions $(0-2.0\ N)$.

Selectivity for Hydrogen Sulfide

Observations of Bähr (10) showed that selective removal of H_2S from CO_2 by means of aqueous ammonia solutions can be achieved if short contact

Table 4-5. Comparison of Experimental and Calculated Vapor Pressures of CO_2 and H_2S over NH_3 Solutions at 20°C

Concentration, moles/liter			P_{NH_3}, mm Hg		P_{CO_2}, mm Hg		P_{H_2S}, mm Hg	
NH_3	CO_2	H_2S	Exp.	Calc.	Exp.	Calc.	Exp.	Calc.
1.189	0.41	0.189	4.11	3.8	1.45	1.8	3.21	3.6
1.194	0.49	0.194	2.92	2.8	3.50	3.7	5.12	5.7
1.390	0.495	0.390	2.42	2.7	3.70	4.4	12.56	14.3
1.045	0.50	0.045	2.72	2.7	3.65	3.6		
1.380	0.64	0.380	1.54	1.44	13.10	15.0	26.97	29.3
1.097	0.66	0.097	1.35	1.34	12.15	14.9	5.29	6.3
1.192	0.67	0.192	1.31	1.26	13.10	17.7	11.14	14.8
1.196	0.69	0.196	1.30	1.16	19.00	20.6	15.25	16.7
1.192	0.70	0.192	0.89	1.11	20.45	21.9	15.95	17.2
1.193	0.745	0.193	0.90	0.87	29.2	32.2	27.38	22.8
1.092	0.77	0.092	0.71	0.76	35.15	38.8	12.20	11.7
1.088	0.79	0.088	....	....	42.45	44.7		
1.090	0.80	0.090	....	....	45.1	48.9		
1.088	0.815	0.088	0.55	0.59	.....	....	13.06	14.8
1.090	0.82	0.090	0.57	0.56	.....	....	16.91	16.2

Source: Data of Badger and Silver (4).

times—on the order of 5 sec (based on gas)—between gas and liquid are used. Lohrmann and Stoller (2) proposed that the difference in the rate of absorption of the two acid gases is due to different diffusion velocities in solutions of ammonium sulfide, ammonium carbamate, and ammonium carbonate. However, a more probable explanation can be based on the fact that hydrogen sulfide ionizes immediately to HS^- and H^+, which reacts rapidly with the hydroxyl ions in the solution, while carbon dioxide must first react with water, forming carbonic acid, before it can react ionically with ammonia. The rate of the hydration reaction is quite low and presumably is the controlling step in the overall reaction.

The selectivity is influenced by the method of contact between gas and liquid. For example, Eymann (11) and Bähr (12) measured the rate of solution of hydrogen sulfide and carbon dioxide in dilute aqueous ammonia solutions (0.5 to 2.0 percent) with quiescent surfaces and found that at a gas pressure of 1 atm and at room temperature, H_2S dissolves about twice as fast as CO_2. Eymann also found experimentally that, if the gases are absorbed by falling drops of liquid under the same conditions of pressure and temperature, the hydrogen sulfide is dissolved 85 times faster than the carbon dioxide. Experiments conducted by the same author with typical coke-oven gas containing about 0.5 percent H_2S and 2.0 percent CO_2 showed that H_2S was dissolved approximately 17 times faster than CO_2 when the gas was contacted in a spray tower with an excess of dilute aqueous ammonia at 21°C (70°F).

The absorption of ammonia into water is quite rapid and governed almost entirely by the gas-film resistance; in fact, this is the classical system for chemical-engineering studies of gas-film resistance. The rate of absorption of hydrogen sulfide into aqueous ammonia solutions is also rapid although it is dependent upon the ammonia concentration. In the presence of an adequate concentration of ammonia at the interface, it is probable that the rate of this absorption is also governed by the gas-film resistance. On the other hand, absorption of carbon dioxide into water or weak alkaline solutions is considered typical of liquid-film controlled systems, not because its gas-film resistance is any lower than that of H_2S and ammonia, but because its liquid-film resistance is very much greater. The net result is that when gases containing H_2S, ammonia, and CO_2 are contacted with water, the ammonia and H_2S are absorbed much more rapidly than the CO_2 and this difference can be accentuated by operating under conditions which reduce the gas-film resistance or increase the liquid-film resistance.

As mentioned previously, the absorption of CO_2 into water and dilute alkaline solutions is hindered by a slow chemical reaction which is required to convert the dissolved carbon dioxide molecules into the more reactive ionic species. In effect, the CO_2 molecules are absorbed by physical solubility (which is not high), and more cannot be absorbed until some molecules are removed by the hydration reaction. The efficiency of CO_2 absorption can, therefore, be improved by turbulence in the liquid film (this aids diffusion of unreacted molecules into the body of the liquid) and by an extended hold-time of liquid in the absorption zone (this provides for the continuous reaction of CO_2 molecules which do enter the liquid phase). These conditions can be met, for example, by a tall, packed column operating at a relatively high liquid-flow rate or by a liquid-filled column through which bubbles of gas are made to pass.

The hydrogen sulfide and ammonia-absorption rates can be increased by inducing turbulence in the gas phase at the interface, a condition which requires a high relative velocity between gas and liquid. This can be achieved by the use of high-pressure spray nozzles which produce a much higher relative velocity than can be realized with gravity flow devices. If maximum selectivity for H_2S and ammonia is desired, the use of spray columns in combination with relatively short contact-time is indicated.

An additional factor which bears on the selectivity of the process is the fact that, once in solution, CO_2 is an appreciably stronger acid than H_2S and, under equilibrium conditions, the process would actually be expected to be selective for CO_2. If the ammonia is added with the wash water, this is indeed possible providing that a sufficiently tall column is used. When the ammonia enters with the feed gas, however, it is absorbed near the bottom of the column and, since very little absorption of CO_2 can occur in pure water, additional column height will have little effect on the selectivity or on the total amount of acid gases absorbed.

Selective Hydrogen Sulfide Removal

For many years the iron oxide dry-box process was the most commonly used method for the removal of H_2S from gases. This process, which is, discussed in Chapter 8, is still employed to a very large extent in Europe. However, liquid processes utilizing the ammonia contained in coal gas for H_2S removal were proposed as early as the end of the nineteenth century. The first process of this type was developed by Claus (13, 14) in 1883 and used for the treatment of coke-oven gas. The process consisted of washing the gas with sufficient aqueous ammonia to absorb essentially all of the H_2S and CO_2. The acid gases were subsequently expelled from the solution by heating, and the regenerated solution was recirculated to the absorber. Maximum removal of the CO_2 required circulation of large volumes of liquid and consumption of large amounts of steam for solution regeneration; these factors made the process economically unattractive. Several subsequent attempts to develop similar processes were equally unsuccessful, mainly because of the same economic shortcomings.

Feld and Burkheiser, the two pioneers of German gas-purification technology, developed complicated processes for the simultaneous absorption of H_2S and ammonia and subsequent conversion to ammonium sulfate and elemental sulfur. These processes, which involve oxidation of the H_2S are discussed in Chapter 9. Innumerable modifications, especially of the Feld processes, were proposed and, in some cases, commercialized. An excellent review of these processes is presented by Hill (15). In spite of the tremendous amount of work done in this direction, no satisfactory process along the lines envisioned by Feld was ever developed. Terres (16) examined the problem of H_2S and ammonia recovery from coal gases and concluded that simultaneous removal and subsequent conversion of the two compounds to ammonium sulfate constitute the least desirable route.

New interest in the process proposed by Claus was aroused after it was discovered that under certain operating conditions aqueous ammonia solutions absorb H_2S selectively from gas streams which also contain CO_2. A fairly large number of processes based on this principle have been proposed, and some have been applied on an experimental or industrial scale. These developments originated primarily in Germany. An indication of the rapid increase in the popularity of these processes is the fact that between 1949 and 1954 the daily volume of gas treated with ammonia solutions in Germany increased form 85 to 260 million cu ft (17). A review of the status of this technology at the beginning of 1957 is presented by Rühl (18).

Process Description

Selective H_2S removal processes can be operated without recycle, with partial recycle, and with total recycle of the wash solution, and a clearly distinct classification on the basis of the flow scheme is not possible. The first two types of processes are usually integrated with the ammonia-removal system of a coke-oven or gas plant. Most of the ammonia contained in the gas is absorbed simultaneously with the H_2S and thus serves as the active agent in the solution.

The conventional indirect process for the recovery of ammonium sulfate (see Chapter 10) has the disadvantage that, because of the conditions prevailing in conventional scrubbers, a large portion of the CO_2 and only a relatively small portion (15 to 20 percent) of the H_2S is absorbed with the ammonia, and the bulk of the H_2S has to be removed subsequently in dry-box purifiers. Installation of a selective H_2S absorber, which permits high relative velocity between gas and liquid, ahead of the ammonia scrubbers (or replacement of one scrubber with a selective absorber), results in more complete H_2S removal and better utilization of the available ammonia to combine with H_2S rather than with CO_2. Furthermore, the ammonia available from the raw gas can be supplemented by partial recycle of ammonia solution, which has been freed of acid gases in a separate stripping column, or by addition of gaseous ammonia to the feed gas. By proper operation of such a process, most of the

H₂S can be eliminated from the gas in the selective absorber. Removal of H₂S, CO₂, and HCN from the ammonia solution in a stripping column located ahead of the ammonia-distillation column permits completely separate processing of the ammonia and the acid gases. This serves to prevent some difficulties in the operation of the saturator or, if strong ammoniacal liquor is produced, results in much higher purity of the crude liquor. Finally, selective absorption of H₂S yields an acid-gas stream of high H₂S concentration which is desirable in the further processing to sulfur or sulfuric acid. Most of these advantages apply also to the semidirect ammonia-removal process (see Chapter 10).

Processes with total solution recycle do not require the presence of ammonia in the gas for absorption of H₂S and can therefore be operated independently of the ammonia-removal plant. The principal advantage of such processes is their high flexibility, which is due to the separate processing of acid gases and ammonia.

Selective H₂S-removal processes do not result in complete elimination of H₂S and, if the treated gas is intended for use as a domestic fuel or in synthesis processes, a final purification step is required. The degree of H₂S removal depends on several operating variables, but it appears that elimination of about 90 percent of the H₂S is the maximum which can be attained economically. Substantial amounts of hydrogen cyanide are also removed in the selective absorber.

Processes without Recycle and with Partial Recycle

The basic flow scheme of the selective H₂S-removal process without liquid recycle is quite simple. The cooled gas is contacted countercurrently in an absorber with water or a mixture of water and cooled gas condensate. The absorber effluent is sent to the ammonia plant for further processing. The principal difference in the various processes presently used lies in the design of the absorber, a matter discussed in some detail later in this chapter.

The ammonia present in most coal gases is insufficient to remove more than about 30 to 50 percent of the hydrogen sulfide and, as pointed out above, recycling of regenerated acid-gas free ammonia solution is necessary if more complete H₂S elimination is desired. Figure 4-5 shows a schematic flow diagram of a typical selective H₂S-removal process without recycle (solid lines) and with partial liquid recycle (dotted lines) combined with an ammonia-removal plant using the indirect process for ammonium sulfate recovery. The cooled gas enters the bottom of the H₂S absorber where it is contacted countercurrently with the absorption liquid, which may be a mixture of fresh water, gas condensate, and recirculated, regenerated aqueous ammonia. From the H₂S absorber the gas flows to the ammonia scrubber where residual ammonia is removed with water, and from there to the final

H2S purifier (usually iron oxide dry boxes), and the benzene-recovery column. In the once-through process, the liquid effluent from the H2S absorber is combined with the effluent from the ammonia scrubber and fed to an ammonia-distillation column. The ammonia contained in the overhead vapors from this column is converted to ammonium sulfate by reaction with sulfuric acid in a conventional saturator. The acid gases are processed to elemental sulfur or sulfuric acid. Of course, it is also possible to condense the overhead vapors from the distillation column to strong annoniacal liquor. If partial liquid recycle is used, the rich ammonia solution which leaves the H2S absorber is fed to the middle of the acid-gas stripper, after heat exchange with a portion of the stripped solution. In order to control the temperature at the top of the stripper and to avoid losses of ammonia in the overhead vapors, a small stream of the cold, rich solution is bypassed around the heat exchanger and fed into the top of the stripping column. The acid gases are expelled from the solution by indirect heating with steam, and the regenerated solution is divided into two portions. The recycle stream is subjected to heat exchange with the incoming solution, cooled, and returned to the H2S absorber. The remaining portion of the acid-gas free solution is combined with the effluent liquid from the ammonia scrubber and fed into the ammonia-distillation

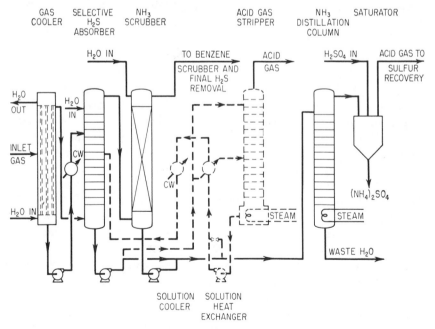

Figure 4-5. Typical flow diagram of selective hydrogen sulfide-removal process without solution recycle (solid lines) and with partial solution recycle (dotted lines); indirect ammonium sulfate-recovery process.

column. The overhead vapors from this column are absorbed in acid in a conventional saturator. Several other flow schemes, depending on the desired end products of the process (strong ammoniacal liquor, ammonium sulfate, ammonium carbonate, ammonium sulfide, ammonium hydroxide, ammonia, hydrogen sulfide), are proposed by Bähr (12). Variations of this process flow-scheme are also discussed by Bayerlein (19) and Rühl (18).

A typical selective hydrogen sulfide-removal system with partial solution recycle, combined with a semidirect ammonium sulfate-recovery plant, is shown in Figure 4-6. The cooled gas is contacted countercurrently in the H_2S absorber with ammonia solution and, after leaving the absorber, flows to the saturator for conversion of the ammonia to ammonium sulfate. The rich ammonia solution leaving the bottom of the H_2S absorber flows to the acid-gas stripper in which the H_2S and CO_2 are expelled by heating. A portion of the acid-gas free solution is recycled to the H_2S absorber while the remainder is stripped of ammonia in a distillation column. The overhead vapors from this column, which consist essentially of ammonia, are recycled to the H_2S absorber. The amount of ammonia available for H_2S removal is controlled by the addition of fresh water to the H_2S absorber. Different flow arrangements of this process are also described by Bähr (12).

The ammonia and the acid gases removed from the gas can be processed to yield a variety of products by addition of a Claus-type sulfur plant, a sulfuric

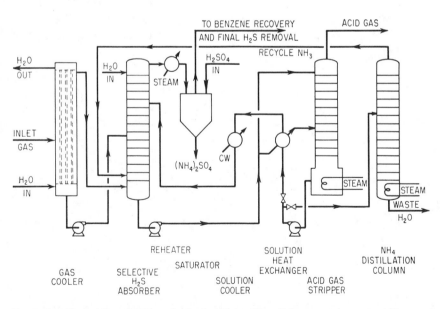

Figure 4-6. Typical flow diagram of selective hydrogen sulfide-removal process with partial solution recycle; semidirect ammonium sulfate-recovery process.

acid plant, or a sulfite-sulfate process (12). An interesting method for the chemical separation of the acid gases and ammonia, the so-called "Chemo-Trenn" process, is described by Bähr (20). In this process, which operates in conjunction with a once-through selective H_2S absorber, the overhead vapors from the ammonia-distillation column which contain ammonia, H_2S, CO_2, and HCN are first contacted in a spray column with a weak acidic solution at 40°C (104°F). The ammonia is absorbed quantitatively by the liquid, and the ammonia-free acid gases are processed further as desired. The ammonia is removed from the solution by heating to 130°C (266°F) in a second tower provided with a reboiler, and the cooled solution is recycled to the absorber. Figure 4-7 shows a schematic flow diagram of the process with the absorption and stripping section located in one vessel. Phenol, xylenols, and amino acids are suitable for use as the acidic absorbent. Operating data presented by Bähr (12) indicate that 99.5 percent separation can be obtained.

A similar process for the separation of H_2S and ammonia has recently been introduced by Howe-Baker Engineers, Inc. (21, 22). This process, which has been named the Ammonex process, uses an undisclosed solvent which is highly selective for ammonia in the presence of H_2S, CO_2, HCN, and light hydrocarbons.

The process flow scheme is quite simple; the feed gas, which may be the overhead stream from a refluxed sour water stripper commonly used in petroleum refineries, is contacted countercurrently with the solvent in an absorption tower, and essentially all of the ammonia is removed. The effluent gas which contains only a very small amount of ammonia can be treated without difficulty in a conventional Claus-type unit for the production of elemental sulfur.

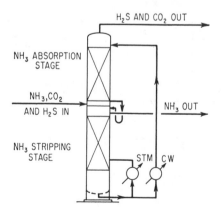

Figure 4-7. Schematic flow diagram of Chemo-Trenn process for separation of ammonia and carbon dioxide.

The rich solvent is regenerated by flashing and, after heat exchange and final cooling, is returned to the absorption column. The regenerator overhead stream is pure ammonia which can be recovered as anhydrous liquid ammonia.

The process is not suited for overhead streams from nonrefluxed sour water strippers or other gas streams containing large amounts of water vapor. The excessive water content of such gas streams leads to operating problems, increased utilities consumption, and increased plant investment.

The problem of separating H₂S and ammonia is of particular significance in the processing of aqueous condensates containing the two compounds combined as ammonium sulfide. Typical of such streams are aqueous condensates from catalytic crackers, hydrotreaters and hydrocrackers in petroleum refineries and condensates from coal, oil shale, or tar-sand hydrotreaters in synthetic fuel plants. In view of air and water pollution problems, these streams have to be freed of H₂S and ammonia prior to disposal. The conventional method is steam stripping of the condensate in a "sour water stripper," followed by further processing of the overhead vapors in a Claus-type sulfur plant. However, in certain cases this leads to serious problems in the sulfur plant, and at least partial separation of H₂S and ammonia is required. This separation can be achieved either by a process such as the Chemo-Trenn or the Ammonex process or by separate stripping of H₂S and ammonia in a system of the type described by Annessen and Gould (23). In this system which has been developed by Chevron Research Company and which is known as the Waste Water Treatment (WWT) process, streams of essentially pure H₂S and ammonia are produced.

A schematic flow diagram of this process is shown in Figure 4-8. The sour water containing typically about 3 to 5.5 weight percent ammonia and 4 to 5.5 weight percent H₂S flows first to a degassing unit to remove dissolved inert gases and from there to a surge tank. In the surge tank liquid hydrocarbons are skimmed from the water and sent to recovery facilities. The water then flows to a column operating typically at a pressure of 100 psig, where H₂S is expelled overhead while the water containing ammonia flows from the bottom of this column to a second column operating at 50 psig, for ammonia stripping. The overhead of this column is crude ammonia of about 98 percent purity which has to be further purified in a scrubbing system before being compressed and condensed as liquid ammonia. The water effluent from the bottom of the ammonia stripping tower is sufficiently pure to cause no problems in its disposal.

Processing of sour aqueous condensates from petroleum refining operations has been the subject of many studies. A particularly good treatment of the subject has been presented by Beychok (24). Design methods for sour water strippers, based on the vapor-liquid equilibrium data of Van Krevelen, Hoftijzer and Huntjens (1) are given for a variety of conditions.

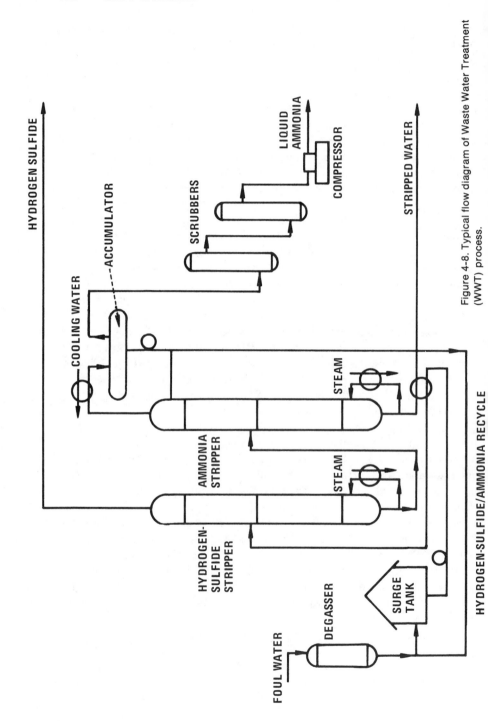

Figure 4-8. Typical flow diagram of Waste Water Treatment (WWT) process.

Processes with Total Recycle

As mentioned earlier, processes using total solution recycle are entirely independent of the ammonia-recovery plant. The ammonia in the wash solution is maintained at such a concentration that the vapor pressure of ammonia over the solution is essentially the same as the partial pressure of ammonia in the gas, and, therefore, practically no ammonia is removed from the gas in the H_2S absorber. This mode of operation permits completely separate H_2S- and ammonia-removal systems. One such process is the Collin (25, 26, 27) process which is used in several commercial installations in England and on the European continent.

A schematic flow diagram of the Collin process is shown in Figure 4-9. The feed gas is contacted countercurrently with the wash solution in a spray tower containing six stages. The tower is constructed in such a manner that the solution passes from stage to stage by flowing over a weir. Solution is withdrawn from the bottom of each stage and pumped to the top of the same stage where it is atomized by small spray nozzles. The final rich ammonia solution flows from the bottom of the spray tower to a surge tank, from which it is pumped to a heat exchanger where it is preheated by exchange with the regenerated solution. The preheated solution enters the stripper at a point located near the middle of the upper section of the column. The upper section

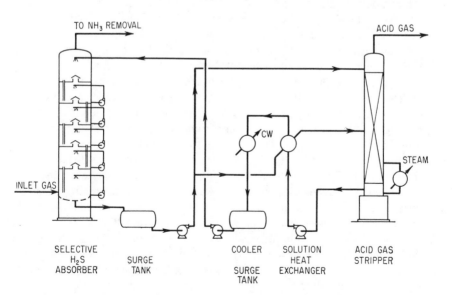

Figure 4-9. Schematic flow diagram of Collin process for selective absorption of hydrogen sulfide with total liquid recycle.

of the stripper contains bubble-cap trays; ceramic packing is used in the lower section of the column. The solution is regenerated by indirect steam-heating in a reboiler and is pumped back to the absorber after passage through the heat exchanger, a cooler, and a surge tank. The temperature at the top of the stripping column as well as the ammonia losses from the vessel are controlled by admitting a small stream of cold, rich solution into the top of the stripper. The acid-gas stream leaving the top of the stripping column contains H_2S, CO_2, HCN, and a trace of ammonia. The ammonia lost from the system is replaced by absorption of ammonia from the inlet gas.

Two other processes falling into the category of selective hydrogen sulfide removal with totally recycled ammonia solutions have recently been developed and commercialized. The first process, referred to as the DIAMOX process, was developed jointly by Mitsubishi Chemical Industries (MCI) and Mitsubishi Kakoki Kaisha (MKK) in Japan; the second process, referred to as the Carl Still Process, was developed by Firma Carl Still in Germany. Both processes almost completely remove hydrogen sulfide from coke-oven gas (COG); however, neither process will remove organic sulfur compounds such as carbonyl sulfide or carbon disulfide.

The DIAMOX process has been described by H. Hiraoka, E. Tanaka, and F. Sudo (33) who report operating experience in commercial units installed at MCI plants in Japan.

In the DIAMOX process shown in Figure 4-10, the COG, after initial cooling and detarring in the primary coolers, is further cooled by direct contact with a circulating cold stream of gas liquor. As the COG is cooled, water and naphthalene are condensed. Sufficient naphthalene is removed so that it will not condense from the COG downstream in the hydrogen sulfide absorber where it may solidify and plug the absorption liquor circuit.

The cooled COG next enters the hydrogen sulfide absorber where the hydrogen sulfide is selectively absorbed from the COG by countercurrent contacting with the aqueous ammoniacal liquor. The rich liquor containing hydrogen sulfide, carbon dioxide, and ammonia enters the acid gas stripper after recovering heat from the stripper bottoms solution. In the acid gas stripper, the rich liquor is heated to expel the acid gases; however, the ammonia is retained in the liquor due to the temperature and reflux at the top of the stripper. The regenerated solution is cooled and recycled to the hydrogen sulfide absorber. The stripped acid gas is sent to a typical Claus unit where the hydrogen sulfide is recovered as 99.98 percent pure sulfur.

Impurities such as thiocyanates and formates accumulate in the circulating liquor and will reduce the desulfurization efficiency causing corrosion if they are not continuously purged from the system. Consequently, a part of the liquor must be continuously purged. In the existing plants, this purge which is similar in composition to the gas liquor produced during the destructive distillation of the coal is either used for coke quenching or treated directly in a biological oxidation unit.

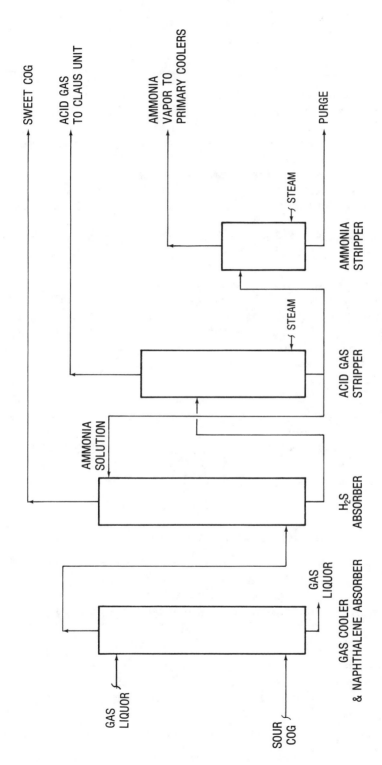

Figure 4-10. Flow diagram of DIAMOX process which selectively removes hydrogen sulfide from coke-oven gas.

More than 98 percent of the hydrogen sulfide can be removed from the COG which will contain less than 8 grains of hydrogen sulfide per 100 SCF. Although there is about seven times as much carbon dioxide as hydrogen sulfide in the COG, the DIAMOX process selectively removes the hydrogen sulfide. Operating data from the MCI Kurosaki plant in Table 4-6 show the amount of ammonia, hydrogen sulfide, and carbon dioxide present in the rich and lean liquors as a function of absorption liquor temperature and percent hydrogen sulfide removed from the COG. The data show that the lower the operating temperature, the greater the percent of desulfurization. Equal concentrations of hydrogen sulfide and carbon dioxide exist in the rich and lean solutions indicating that hydrogen sulfide is selectively removed over carbon dioxide. Only about 30 percent of the hydrogen cyanide present in the COG is absorbed during desulfurization and hydrolyzed into ammonia and other by-products.

The acid gas produced contains more than 45 volume percent hydrogen sulfide with the balance being carbon dioxide and a small amount of other impurities such as 0.1 volume percent hydrogen cyanide, .01 volume percent ammonia, and 1.1 volume percent aromatics. Consequently, the acid gas is a suitable feed for a typical Claus unit. Operating problems such as plugging and corrosion associated with larger concentrations of these impurities are

Table 4-6. DIAMOX Operating Data from MCI Kurosaki Plant (33)

Absorption Liquor Temp., °F	% Removal H_2S	Solution Analysis, Grains/U.S. Gal.					
		NH_3		H_2S		CO_2	
		Rich	Lean	Rich	Lean	Rich	Lean
90	96.2	473	473	118	12.3	117	2.3
98	91.1	436	417	108	14.6	123	2.3
104	86.0	418	386	102	16.4	106	5.2

Table 4-7. Utilities for DIAMOX Process for Two COG Desulfurization Levels (Basis: 500 grains H_2S/100 SCF at inlet, 90 MMSCFD capacity) (33)

Utility $\quad$ H_2S at Absorber Outlet	8 Grains/100 SCF	30 Grains/100 SCF
L.P. Steam, Lb/Hr	34,180	20,950
Cooling Water, GPM	2,910	1,590
Boiler Water, GPM	7	33
Electric Power, kW	1,000	570

avoided. As an alternative, the acid gas is suitable feed for a sulfuric acid plant.

Table 4-7 shows the utility requirements for the DIAMOX process for two levels of COG desulfurization. The DIAMOX process does not require any chemicals since the absorption liquor is generated from the ammonia present in the COG.

The basic flow scheme of the Carl Still process is similar to that of the DIAMOX process. The Carl Still process and operation of a plant at ARMCO Steel Corporation, Middletown, Ohio, are described by Snook (34) who reports that the concentration of ammonia in the Carl Still circulating liquor is about 2 weight percent. In the DIAMOX process, the ammonia content is about 0.7 weight percent. As a result of the higher ammonia concentration, the Carl Still process is not as selective for hydrogen sulfide as the DIAMOX process. A substantial quantity of hydrogen cyanide is present in the acid gas. In addition, ammonia is stripped from the rich liquor when it is regenerated. At the ARMCO plant, ammonia is removed from the acid gas with the USS Phosam process, and the remaining acid gases are fed to a sulfuric acid plant. Snook reports that the Carl Still process can reduce the hydrogen sulfide content of the COG from 350 grains per 100 SCF to 50 grains per 100 SCF.

Design and Operation

Absorption

The degree of selectivity for H_2S obtainable in any one of the selective absorption processes described is largely determined by the method of gas-liquid contact. The basic requirements for maximum selectivity are high relative velocities (contact times on the order of 5 sec, based on the gas) and intimate contact between gas and liquid. Unfortunately, insufficient basic data are available to permit a rigorous analysis of absorber design. However, operating data with different types of columns are available in the literature and may be used as guides for the design of selective H_2S absorbers. Bähr (12) and Bayerlein (19) report experimental and operating data with a variety of absorption columns. Some of these data are the results of experiments conducted by these authors while others are based on experience gained in industrial installations.

Operating results obtained with hurdle-packed, bubble-cap, and Kittel-tray columns in the treatment of typical coal gas are shown in Table 4-8.

These data, which are based on experiments conducted by Bähr (12), indicate clearly the superior performance of the shorter bubble-cap and Kittel-tray columns as compared to the packed column. Both the degree of H_2S removal and the selectivity for H_2S are markedly higher in the bubble-cap

Table 4-8. Selective H₂S Removal with Different Absorption Columns

Column Characteristics	Type of Column		
	Hurdle Packed	Bubble-Cap	Kittel-Tray
Diameter, ft	10.5	10.5	6.6
Height, ft	100	19.5	18.0
Pressure, psig	1.5	1.5	1.5
Gas-liquid contact time, sec (based on gas)	47.5	6.0	3.5
Gas velocity, ft/sec (based on empty column)	2.1	2.1	5.3
Gas rate, MSCF/hr	637	637	637
Gas composition:			
NH₃, mole %	0.775	0.775	0.775
H₂S, mole %	0.605	0.605	0.605
CO₂, mole %	2.73	2.73	2.73
Moles NH₃/mole H₂S	1.28	1.28	1.28
Water rate, gpm	53.0	46.5	48.5
Liquid composition (bottom of absorber):			
NH₃, g/liter	8.8	9.8	9.4
H₂S, g/liter	2.2	6.7	7.8
CO₂, g/liter	10.6	3.4	1.6
H₂S absorbed, % of inlet	16.0	42.5	52.0
CO₂ absorbed, % of inlet	13.1	3.7	1.8
Inlet gas, moles H₂S/mole CO₂	0.24	0.24	0.24
Outlet liquid, moles H₂S/mole CO₂	0.27	2.50	6.25
Outlet liquid, moles acid gas/mole NH₃	0.59	0.48	0.48

and Kittel-tray columns than in the column packed with hurdles. The data are typical for a process without liquid recycle where only the ammonia contained in the gas is available for H₂S removal. The low mole-ratio of NH₃ to H₂S in the gas accounts for the incomplete H₂S removal. A minimum ratio of two moles of available ammonia (either in the gas or in recycled solution) per mole of H₂S is required to obtain about 70 to 80 percent H₂S removal, even under the most favorable contact conditions. Somewhat better removal—up to 90 percent—is possible with a large excess of available ammonia: four moles of NH₃ per mole of H₂S or more.

Although the number of trays in the bubble-cap and Kittel-tray columns used in these experiments is not given, other data presented by Bähr indicate that eight bubble-cap trays and 15 Kittel trays are sufficient for maximum H₂S removal. The selectivity does not change appreciably with the number of trays, probably because the bulk of the ammonia is removed in the lower portion of the column and additional trays at the top are not supplied with an active solution for absorption. The completeness of H₂S removal in the bubble-cap column is relatively unaffected by changes in gas velocity ranging from 0.85 to 2 ft/sec and is primarily determined by the ammonia content of the gas.

Bähr (12) also reports operating data obtained with a Kittel-tray absorber, 76 in. in diameter and 20 ft high, containing 16 double trays, which was used for treating 760,000 cu ft/hr of coal gas with fresh water and without solution recycle. The inlet gas contained 0.71 mole percent NH_3 and 0.63 mole percent H_2S, 54 percent of which was removed. The mole ratio $H_2S:CO_2$ in the saturated solution was about 10; this indicates extremely high selectivity if an inlet CO_2 concentration of 2 to 3 mole percent is assumed. Essentially complete ammonia removal was obtained in all the above experiments. A photograph of a Kittel tray is shown in Figure 4-11.

Bayerlein (19) describes the operation of a spray tower, containing six stages, each approximately 10 ft high, used for treating 390,000 cu ft/hour of coal gas which contained about 0.53 mole percent NH_3 and 0.53 mole percent H_2S. Fresh water was used as the wash liquid in a once-through process. The data obtained are presented in Table 4-9, and the enrichment of the water with NH_3, H_2S, CO_2, and HCN after successive stages is shown graphically in Figure 4-12.

Figure 4-11. Kittel tray used for selective removal of hydrogen sulfide. *John Brown, Ltd. and Fractionating Towers, Inc.*

Table 4-9. Operating Data of Six-Stage Spray Tower

Tower Characteristics	Stages	
	1 to 3	4 to 6
Column diameter, in	63.0	93.0
Height of stage, ft	10.0	10.0
Gas velocity, ft/sec	5.0	2.0
Gas rate, MSCF/hr	390	390
Inlet gas composition, mole %:		
NH_3	0.53	
H_2S	0.53	
HCN	0.08	
Outlet gas composition, mole %:		
NH_3	. . . . . .	0.003
H_2S	. . . . . .	0.32
HCN	. . . . . .	0.04
Solution composition, from first stage, moles H_2S/mole CO_2.	6.3	

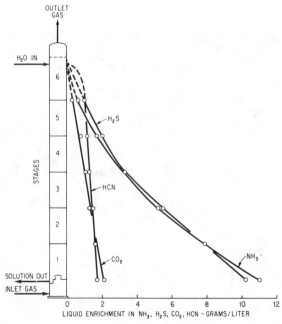

Figure 4-12. Change of liquid composition during absorption of ammonia, hydrogen sulfide, carbon dioxide, and hydrogen cyanide with water in a six-stage spray tower. *Data of Bayerlein (19)*

As can be seen, good selectivity was obtained; this is evidenced by the mole ratio, $H_2S:CO_2$, of 6.3 in the rich solution. It is stated that this ratio does not change at operating temperatures ranging from 65° to 77°F but that it decreases markedly at temperatures above 80°F. This could be expected on the basis of the known improvement of CO_2 absorption in aqueous solutions with moderate temperature-increase and the more pronounced increase in the equilibrium vapor pressure of H_2S, as compared to that of CO_2, with increasing temperature.

Operating data of a plant with total-solution recycle (Collin process) using two three-stage spray towers in series as the absorbers are shown in Table 4-10 (26).

The concentration of hydrogen cyanide in the inlet gas appears to have a marked effect on the capacity of the ammonia solution for absorbing H_2S. The HCN removed in the selective absorber (about 50 percent) is not completely stripped in the regenerator and accumulates in the circulating solution in the form of cyanide, thiocyanate, and iron-cyanide complexes until an equilibrium condition is reached. No particular operating difficulties are reported with gases containing about 40 grains/100 scf of HCN or less (normal for most coal gases). Bayerlein (19) states that solutions containing 2g/liter of HCN, 0.8 g/liter of HSCN, and 8 g/liter of iron-cyanide complex-

Table 4-10. Typical Operating Data of Collin Process

Gas rate, MSCF/hr	1,125
Liquid rate, gpm	172
Gas temperature, °F	70
Liquid temperature, °F	76
Inlet gas, grains/100 SCF:	
NH_3	250
H_2S	555
Outlet gas, grains/100 SCF:	
H_2S	125
Acid gas composition, %:	
H_2S	74–77
CO_2	12–14
HCN	6–9
NH_3	0.4–0.5
H_2S removal, %	78
Inlet liquid, g/liter:	
NH_3	11.7
H_2S	1.0
Outlet liquid, g/liter:	
H_2S	7.2

Source: Data of Williams (26).

es may be obtained when typical coal gases are treated, and they may be used without loss of treating efficiency. However, higher HCN concentrations in the inlet gas result in a much higher contaminant content of the solution and in considerably lower solution capacity for H_2S. Operating data (19) of a plant using the Collin process to treat coke-oven gas containing 147 grains/100 scf of HCN show that after several weeks of operation the solution capacity decreased from an original value of 10 g/liter of H_2S to 3 g/liter and the HCN content increased from about 2 to 7 g/liter. The thiocyanate concentration rose rather slowly to a maximum to 2 g/liter while the iron cyanide content rose very rapidly to 12 g/liter. The difficulty could be remedied by continuously discarding a small portion of the solution and adding an equivalent volume of steam condensate. The formation of iron-cyanide complexes was minimized by installing aluminum equipment in the high-temperature zones in place of the cast iron and carbon steel originally used.

Regeneration

Regeneration of the rich ammonia solution is carried out in conventional manner. Bubble-cap tray columns or columns containing bubble-cap trays in the upper section and ceramic packing in the bottom part are used. The heat required for expelling the acid gases is supplied either by direct or indirect steam or a combination of both. Steam coils installed in the bottom of the stripping column or thermosiphon reboilers, as in the case of the Collin process (27), can be used to provide indirect steam heating.

The rich solution is admitted at a point near the middle of the column and stripped in the lower portion of the vessel by contact with the hot vapors rising from the bottom. The upper part of the regenerator serves as a rectification section for the ammonia. If the proper temperature is maintained at the top of the column (by proper regulation of the cold-solution reflux rate), the acid-gas stream contains practically no ammonia. The regenerated solution leaving the bottom of the stripping column contains essentially all the ammonia present in the stripper feed and 0.5 to 1.0 g/liter of hydrogen sulfide.

Sufficient heat must be added in the regeneration column (a) to raise the temperature of the components of the two solution-feed streams to the temperature at which they leave the column (in the product solution or acid-gas stream); (b) to provide heat of reaction to dissociate the chemical compounds present in the solution; (c) to evaporate the water vapor which leaves the column with the acid-gas stream; and (d) to compensate for heat loss from the vessel.

Steam requirements for solution regeneration in commercial plants, as reported in the literature, vary from 10 to 20 lb per pound of hydrogen sulfide, depending on the ammonia content of the solution. Williams (26) reports a steam consumption of about 11 lb/lb H_2S in the total-recycle process (Collin process) for the operating conditions given in Table 4-10. Bähr (12) presents steam-requirement values of 20.2, 14.8, and 9.8, lb/lb H_2S for solutions containing 1.0, 1.36, and 2.15 percent ammonia respectively. Although not specifically so stated, it is assumed that all three solutions were saturated with H_2S in the same molar proportions.

Since a major portion of the heat required in the process is used to raise the temperature of the rich solution during regeneration, significant heat economy can be effected by operating with ammonia solutions of relatively high concentration and, therefore, with correspondingly low solution-circulation rates. Hoffmann (28) reports the use of concentrated solutions (7 to 20 percent ammonia) for the purification of coke-oven gas containing about 650 grains $H_2S/100$ scf. Solution regeneration is carried out at a pressure of about 150 psig and regenerator-bottom temperatures ranging from 295° to 310°F. Considerable steam savings are claimed over conventional processes using 1 to 2 percent ammonia solutions.

A serious disadvantage in the use of concentrated ammonia solutions is their extreme corrosiveness which necessitates the use of expensive alloys as materials of construction.

Materials of Construction

Corrosion problems in selective H_2S-removal plants are entirely confined to the hot portions of the systems and largely determined by the H_2S and HCN contents of the ammonia solution. Carbon steel appears to be quite satisfactory for the construction of absorbers and solution coolers. Regenerators and solution heat exchangers are constructed from cast iron in installations designed for treating gas of relatively low HCN content. A service life of several years can be expected with rich solutions containing 5 to 6 g/liter of H_2S and 0.5 g/liter of HCN. At higher H_2S and HCN concentrations in the ammonia solution, aluminum (99.5 percent) and ceramic materials have been found quite satisfactory.

Carbon Dioxide Removal

Aqueous ammonia solutions are used to some extent for the removal of carbon dioxide from synthesis gases. The best-known application of this process is probably the purification of the hydrogen used in ammonia synthesis. Several installations of this type are operating in Europe, and one plant treating coke-oven gas has been built in the United States. The process is probably most economical for the treatment of partially desulfurized coke-oven gases of relatively low carbon dioxide content, although it is also useful for the purification of synthesis gases containing about 30 percent carbon dioxide. Mullowney (29) presents a comparison of the economics of CO_2 removal, from a gas stream containing 34 percent CO_2 by seven different combinations of hot potassium carbonate, ethanolamine, water-wash, and ammonia processes. The analysis shows that a combination involving bulk CO_2 removal with hot potassium carbonate (from 34 to 2 percent CO_2), followed by treatment with aqueous ammonia (reduction of CO_2 from 2 percent to 150 ppm), and final CO_2 removal (to 10 to 25 ppm) by scrubbing with caustic, results in much more favorable economics than a process in which the CO_2 is reduced from 34 percent to 150 ppm by treatment with aqueous ammonia followed by caustic scrubbing for final purification. A summary of the cost comparison between these two schemes is shown in Table 4-11.

The main advantage of the process is the low cost of the treating agent and the fact that the efficiency of the process is essentially unaffected by the presence of carbonyl sulfide, carbon disulfide, and relatively small amounts of H_2S and HCN in the gas. The principal disadvantage is the somewhat corrosive nature of the carbonated solution (especially if the gas contains appreciable quantities of HCN) which requires special materials of construction in the solution-regeneration section. Another disadvantage is the somewhat more complex flow scheme as compared to either the ethanolamine or the hot potassium carbonate processes.

Table 4-11. Cost Comparison of Two Cases* of CO₂ Removal with Ammonia Solutions

Cost Variables	Case†	
	I	II
Plant investment, $	619,000	793,000
Direct operating costs, $/year	286,000	536,000
Indirect operating costs, $/year	77,000	99,000
Cost of profit (before income tax) @ 25% of plant cost, $/year	155,000	198,000
Total treating cost, $/year	518,000	833,000
Treating cost, $/MSCF	0.12	0.19

*Case I: hot potassium carbonate-ammonia—caustic; Case II: ammonia-caustic.
†Constant data for each case: gas-flow rate, 12 MMSCF/day; pressure, 350 psig; gas temperature, 100°F; CO_2 content, 34.3 mole percent.

Process Description

A schematic flow diagram of the process as used for CO_2 removal from coke-oven gas (30) is shown in Figure 4-13. Compressed coke-oven gas, from which most of the hydrogen sulfide, benzene, and high-molecular-weight unsaturated hydrocarbons have been removed, enters the bottom of the carbon dioxide absorber where it is contacted countercurrently with a 2 to 5 percent aqueous ammonia solution. The treated gas contains about 150 ppm CO_2, practically no H_2S, and a small amount of NH_3 which is recovered in the water-wash column. The CO_2 content of the gas is subsequently reduced to a final value of 10 to 25 ppm by treatment with sodium hydroxide solution.

The carbonated ammonia solution withdrawn at the bottom of the CO_2 absorber flows to a heat exchanger, where it is heated by exchange with the regenerated solution, and from there to the regenerator (acid gas stripper). Carbon dioxide is expelled from the solution by heating with vapors from the ammonia stripper. The regenerated ammonia solution leaving the bottom of the regenerator is split into two unequal streams. The larger stream flows through the heat exchanger to a surge tank, from which it is pumped to the top of the absorber after cooling to about 100°F in the solution cooler. The smaller stream flows to the ammonia stripper where the ammonia is removed from the liquid by contact with direct steam. The mixture of steam and am-

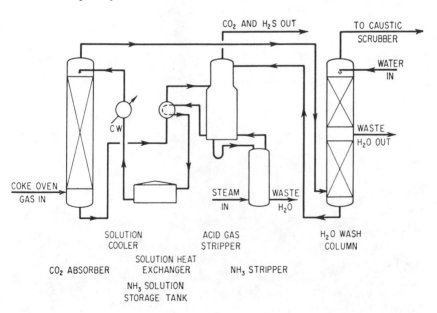

Figure 4-13. Schematic flow diagram of carbon dioxide removal with aqueous ammonia.

monia passes from the top of the ammonia stripper to the regenerator, where the ammonia is absorbed and the steam is used for removing CO_2 from the carbonated solution. The ammonia-free water, which contains inorganic salts such as thiocyanates and iron-cyanide complexes, is discarded, and excessive build-up of undesirable salts in the circulating solution is thus avoided.

The water rejected from the system in the ammonia stripper is replaced by adding an equivalent volume of cold reflux to the top of the regenerator. The effluent from the bottom of the water-wash column, which contains essentially all the ammonia removed from the treated gas, is used as reflux. Since in most instances a larger quantity of water is used in the water-wash column than is required for regenerator reflux, the excess wash water is withdrawn at the middle of the wash column and discarded. This stream contains only a trace of ammonia, and the losses incurred are negligible.

The absorber, water-wash column, and solution cooler are constructed from carbon steel while cast iron is normally used for the solution heat exchanger, regenerator, and ammonia stripper.

Corrosion of plant sections operating at elevated temperatures is aggravated by high H_2S and HCN contents of the gas to be treated. In cases

Figure 4-14. Plant for removal of carbon dioxide from coke-oven gas using aqueous ammonia solution. *Ketona Chemical Company*

where such gases are processed, use of special alloys or ceramic linings may be required.

Figure 4-14 shows a photograph of a commercial installation for treating coke-oven gas prior to its use in ammonia synthesis.

References

1. Van Krevelen, D.W.; Hoftijzer, P.J.; and Huntjens, F.Y. 1949. *Rec. Trav. Chim.* 68:191.
2. Lohrmann, H., and Stoller, P. 1942. *Arch. bergbauliche Forsch.* 3:43.
3. Pexton, S., and Badger, E.H.M. 1938. *J. Soc. Chem. Ind. (London)* 57:106.
4. Badger, E.H.M., and Silver, L. *J. Soc. Chem. Ind. (London)* 57:110.
5. Badger, E.H.M. 1938. *J. Soc. Chem. Ind. (London)* 57:112.
6. Badger, E.H.M., and Wilson, D.S. 1947. *J. Soc. Chem. Ind. (London)* 66:84.
7. Dryden, J.G.C. 1947. *J. Soc. Chem. Ind. (London)* 66:59.
8. Terres, E.; Attig, W.; and Tscherter, F. 1957. *Gas-u. Wasserfach* 98(21):512.
9. Terres, E., and Weiser, J. 1920. *J. Gasbelaucht* 63:705.
10. Bähr, H. 1940. British Patent 520,327.
11. Eymann, C. 1948. *Gas-u. Wasserfach* 89(1):10.
12. Bähr, J. 1955. *Brennstoff-Chem.* 36(9/10):129.
13. Claus, C.F. 1883. German Patent 23,763.
14. Claus, C.F. 1883. German Patent 39,277.
15. Hill, W.H. 1945. *Chemistry of Coal Utilization.* Edited by H.H. Lowry. New York: John Wiley & Sons, Inc., p. 1060.
16. Terres, E. 1953. *Gas-u. Wasserfach* 94:260-265; 311-315.
17. Reinhardt, K. 1956. *Energietechnik* 6(10):454.
18. Rühl, G. 1957. *Brennstoff-Chem.* 38(1/2):27.
19. Bayerlein, K. 1954. *Gas, Wasser, Wärme* 8(2):25.
20. Bähr, H. 1943. German Patent 741,222. 1942. German Patent 728,102.
21. Howe-Baker Engineers, Inc. 1971. Private communication.
22. Goar, G. 1971. *Today's Gas Treating Processes.* Paper presented at Gas Conditioning Conference, University of Oklahoma (Apr.).
23. Annessen, R.J., and Gould, G.D. 1971. *Chem. Eng.* 78(Mar.22):67.
24. Beychok, M.R. 1967. *Aqueous Wastes from Petroleum and Petrochemical Plants.* New York: John Wiley & Sons.
25. Collin, F.J. 1943. German Patent 743,088.
26. Williams, T.H. 1954. *Coke and Gas* 16:61.
27. Doerges, A. 1956. *Gas-u. Wasserfach* 97(21):893.
28. Hoffmann, G. 1958. *Gas-u. Wasserfach* 99(21):509.
29. Mullowney, J.F. 1957. *Petrol. Refiner* 36(Dec.):149.
30. Anon. 1956. *Chem. Eng.* 63(June):400.
31. Massey, J.Y., and Dunlap, R.W. 1975. *Journal of the Air Pollution Control Association* 25(1):1019-1027.
32. Symposium on Treatment of Coke-Oven Gas, May, 1977. McMaster University Press, McMaster University, Hamilton, Ontario, Canada.
33. Hiraoka, H.; Tanaka, E.; and Sudo, H. 1977. "DIAMOX Process for the Removal of H_2S in Coke Oven Gas." ibid.
34. Snook, R.D. 1977. "The Carl Still Coke Oven Gas Desulfurization Plant at the ARMCO Steel Corp. No. 3 Coke Plant." ibid.

5

Alkaline Salt Solutions for Hydrogen Sulfide and Carbon Dioxide Absorption

A prime requirement for absorptive solutions to be used in regenerative CO_2 and H_2S removal processes is that any compounds formed by reactions between the acid gas and the solution must be readily dissociated. This precludes the use of strong alkalies; however, the salts of these compounds with weak acids offer many possibilities, and a number of processes have been developed which are based on such salts. Typically the processes employ an aqueous solution of a salt containing sodium or potassium as the cation with an anion so selected that the resulting solution is buffered at a pH of about 9 to 11. Such a solution, being alkaline in nature, will absorb H_2S and CO_2 (and other acid gases), and, because of the buffering action of the weak acid present in the original solution, the pH will not change rapidly as the acid gases are absorbed. Salts which have been proposed for processes of this type include sodium and potassium carbonate, phosphate, borate, arsenite, and phenolate, as well as salts of weak organic acids. Major industrial processes based on such solutions are described in subsequent sections of this chapter. The processes discussed are divided into two groups, i.e., those in which absorption of acid gases takes place at ambient temperature, typically 70 to 100°F, and those where absorption is carried out at elevated temperatures, approximating the temperature at which the solution is regenerated. The latter types of processes are primarily suitable for the treatment of gases under pressure.

Sodium and potassium carbonate solutions have been used extensively for the absorption of CO_2 and H_2S from gas streams because of their low cost and ready availability. Probably for similar reasons, a very considerable amount of laboratory data on absorption has been obtained with the system CO_2—sodium carbonate. In fact, the extent of theoretical research on this system is probably out of proportion to its industrial importance. Less basic design information is available on the absorption of H₂S in sodium and potassium carbonate solutions than there is on CO_2 absorption, although H₂S removal by such solutions is the basis for several commercial processes. A comprehensive theoretical study of absorption of hydrogen sulfide and carbon dioxide in alkaline solutions has been presented by Dankwerts and Sharma (1), and experimental data, obtained in laboratory scale equipment, on the selective absorption of hydrogen sulfide in carbonate solutions have been reported by Garner, Long, and Pennell (2).

Carbon dioxide is a slightly stronger acid in solution than hydrogen sulfide. Its ionization constant for the first-step ionization,

$$H_2CO_3 = H^+ + HCO_3^-$$

is approximately 4×10^{-7} at 25°C as compared to about 1×10^{-7} for the corresponding hydrogen sulfide ionization. Nevertheless, H₂S is absorbed more rapidly than CO_2 by aqueous alkali-salt solutions, and partial selectivity can be obtained when both acid gases are present. The data of Garner, Long, and Pennell (2) indicate that selectivity is favored by short gas-liquid contact times and low temperatures.

It is generally believed that the low rate of absorption of CO_2 is due to a slow chemical reaction occurring in solution between dissolved molecular CO_2 and hydroxyl ions. If the proposed mechanism is correct and the rate of absorption of carbon-dioxide is governed by a chemical reaction rather than by diffusion, it can be expected that the reaction rate and therefore the absorption coefficient can be increased by the addition of catalysts to the carbonate solution. Some work with additives to sodium and potassium carbonate solutions has, in fact, indicated such an effect. Riou and Cartier (3,4) investigated numerous additives including glycerol, dextrose, sucrose, ethylene glycol, levulose, methyl alcohol, ethyl alcohol, formaldehyde, and lactose and found that many of these produced a marked increase in the rate of absorption of CO_2 by sodium carbonate solutions. The addition of about one percent sucrose, for example, more than doubled the rate of absorption. At present, several processes using carbonate solutions containing catalysts are in commercial operation. Such processes are discussed later in this chapter.

Absorption at Ambient Temperature

CO₂ Absorption in Alkali-Carbonate Solutions

Sodium and potassium carbonate solutions have been used to a considerable extent for the absorption of CO_2 from flue gases for the manufacture of dry ice. The operation, which has been adequately described by Quinn and Jones (5) and by Reich (6), will not be discussed in detail as it cannot properly be classified as a gas-purification process; the primary reason for removing CO_2 is to provide a raw material for subsequent processing. In the conventional CO_2-recovery process, the alkali carbonate is partially converted to bicarbonate in the absorber and back to the carbonate again in the regenerator, which is heated by steam. Absorption is very slow because of the low alkalinity of the solution, and two packed absorbers in series are commonly used. Major drawbacks of the process are low efficiency of CO_2 recovery and the high steam requirement for regeneration. Because of these drawbacks, most of the more modern CO_2 plants employ aqueous monoethanolamine to remove CO_2 from flue gas (see Chapters 2 and 3).

Sodium carbonate (with free hydroxyl due to excess caustic) is also frequently used to remove the last traces of carbon dioxide from hydrogen (or other gases) from which carbon dioxide has been removed by the use of a water-wash or other relatively inefficient process. In this operation, the sodium carbonate in the solution is formed by the reaction of CO_2 with free hydroxyl, and the alkalinity is maintained at a high level by the periodic addition of fresh caustic. No attempt is made to regenerate the solution, which is discarded or used elsewhere when its alkalinity is reduced to the point where it is no longer effective for CO_2 removal. This type of process is also used to remove traces of carbon dioxide from air which is used as feed to low-temperature air-separation plants, or in other processes requiring high purity. Some increase in the efficiency of caustic utilization can be realized by using two or more stages of absorption.

A flow sheet of a typical unit for removing CO_2 in trace quantities is shown in Figure 5-1. This type of unit has been found to be capable of reducing the CO_2 content of air from about 300 ppm to as low as 5 ppm (7).

Design and Operating Data

The vapor pressure of carbon dioxide over sodium carbonate bicarbonate solutions can be related by the following equation presented by Harte, Baker, and Purcell (8).

$$P_{CO_2} = \frac{137 f^2 N^{1.29}}{S(1 - f)(365 - t)} \tag{5-1}$$

where N = sodium normality

f = fraction of total base present as bicarbonate

t = temperature, °F

P_{CO_2} = equilibrium partial pressure of CO_2, mm Hg

S = solubility of CO_2 in water under a pressure of 1 atm, moles CO_2/liter

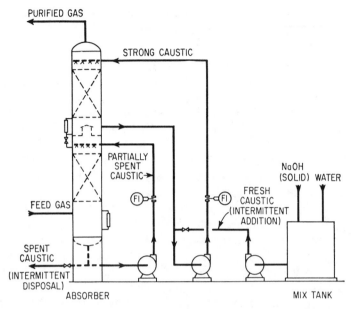

Figure 5-1. Flow diagram of caustic-wash process to remove traces of CO_2 from air or other gases.

Table 5-1. Solubility of CO_2 in Water with Changing Temperature

Temperature	S
15°C (59°F)	0.0455
25°C (77°F)	0.0336
45°C (113°F)	0.0215
75°C (167°F)	0.0120

Values for S are available in standard texts. Seidell (9) gives the values listed in Table 5-1.

When free hydroxyl is present, the vapor pressure of CO_2 over the solution is so low that, in most cases, it can be neglected.

As mentioned above, numerous theoretical studies have been made of the absorption of CO_2 in sodium carbonate, bicarbonate, and hydroxide. An excellent review of this work is presented by Sherwood and Pigford (10). For the carbonate bicarbonate region, the data of Furnas and Bellinger (11) are of particular value in that they provide an equation relating the overall gas-film coefficient with packing characteristics and liquid rate, as follows.

$$K_G a = C'L^{1-n}a_d n \qquad (5-2)$$

where $K_G a$ = overall gas-film coefficient, lb moles/(hr)(cu
     ft)(atm)
     L = liquid rate; lb/(hr)(sq ft)
     a_d = surface area of packing (dry), sq ft/cu ft
C' and n = constants

At a temperature of 77°F, a sodium carbonate molality of 0.5, and with 20 percent of the sodium in the form of the bicarbonate, the constants for Equation 5-2 are as in Table 5-2.

At temperatures above 77°F, the absorption coefficient goes up rapidly. An increase to 130°F, for example, almost doubles the coefficient.

In the carbonate-hydroxide region, the rate of absorption of CO_2 is found to be much higher than in the carbonate-bicarbonate region. It is theorized that this is because the molecular carbon dioxide can react directly with hydroxyl ions to form the carbonate ion and that this reaction proceeds rapidly in the presence of free hydroxyl ions. Useful design data on the absorption of carbon dioxide by sodium hydroxide solutions have been presented by Tepe and Dodge (12). The data were obtained in an absorber of 6-in. diameter packed with 36 in. of 0.5-in. carbon raschig rings. These observers found that $K_G a$ varies with sodium-ion normality and percentage conversion to carbonate as shown in Figure 5-2. They propose that $K_G a$ values taken from the chart be adjusted to other liquid rates and temperatures by the use of the relation:

$$K_G a \propto L^{0.28} T^6 \tag{5-3}$$

where T is the absolute temperature and L is the liquid rate. They found $K_G a$ to be essentially independent of gas rate, a condition which would normally indicate that the liquid film was controlling absorption. However, the exponent relating the effect of liquid rate is not as high as would be expected in a simple liquid-film-controlled absorption. As shown in the figure, the absorption coefficient increases with increased sodium hydroxide concentrations up to about 2 N and then decreases. The decrease is presumably due to the higher viscosity of more concentrated solutions—a phenomenon also observed for alkanolamine solutions.

The absorption of carbon dioxide in potassium hydroxide and carbonate solutions generally parallels that in solutions of the corresponding sodium compounds. However, Spector and Dodge (13) found that slightly larger coefficients are obtained for KOH than for NaOH. The absorption of carbon dioxide in hot potassium carbonate solutions is discussed in a separate section.

Table 5-2. Constants for Equation 5-2.

Packing	C'	n	a_d, sq ft/cu ft
Raschig rings, ⅜-in.	1.53×10^{-4}	0.56	148
Raschig rings, 1-in.	0.81×10^{-4}	0.36	58
Berl saddles, 1-in.	1.00×10^{-4}	0.42	79

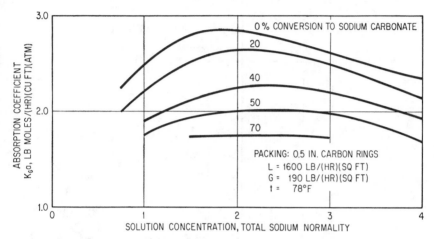

Figure 5-2. Absorption of CO_2 in sodium hydroxide-sodium carbonate solutions. *Data of Tepe and Dodge* (12)

Seaboard Process

The Seaboard process is based upon the absorption of hydrogen sulfide by a dilute sodium carbonate solution and regeneration by air. Although in many instances, this process has been supplanted by newer developments, it is of interest as being the first regenerative liquid process for H_2S removal applied commerically on a large scale. The process was introduced by Koppers Company, Inc., in 1920, and has been thoroughly described by Sperr (14). Principal advantages of the process are extreme simplicity and economy. Its chief drawbacks are (a) the occurrance of side reactions caused by the introduction of oxygen in the air-regeneration step and (b) the disposal of the foul air containing H_2S. The process is capable of H_2S-removal efficiencies of 85 to 95 percent in single-stage plants.

Process Description

A flow sheet of a typical Seaboard-process installation is presented in Figure 5-3. The circulating solution normally contains 3.0 to 3.5 percent

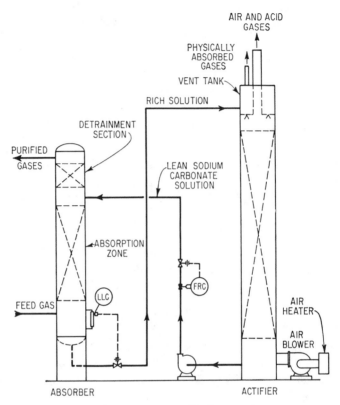

Figure 5-3. Flow diagram of Seaboard gas-purification process.

sodium carbonate. This is used to wash the gas in a countercurrent absorber column and is regenerated in a separate actifier column by a countercurrent flow of low-pressure air. The principal chemical reaction in the process is as follows:

$$Na_2CO_3 + H_2S \rightleftharpoons NaHCO_3 + NaHS \qquad (5\text{-}4)$$

According to Gollmar (15), the solution flow rate is between 60 and 150 gal/1,000 cu ft of gas, depending upon the H_2S and CO_2 concentrations in the ·gas. For typical coke-oven gas containing 300 to 500 grains of H_2S/100 scf and 1.5 to 2.0 percent CO_2 the solution flow rate will be in the range of 60 to 80 gal/1,000 scf, a carrying capacity of around 50 grains/gal. For higher H_2S concentrations and higher than normal CO_2, particularly the latter, solution flow rates up to 150 gal/1,000 scf may be required.

The air rate required for solution regeneration is usually from 1.5 to 3.0 times that of the gas, depending upon the H_2S removal efficiency desired. The air from the actifier, therefore, contains the H_2S removed at a concentration on the order of one-half that in the feed-gas stream. Since such concentrations of H_2S are extremely objectionable, it is necessary either to vent the air through a tall stack or, preferably, to use it as combustion air for some other plant operation. Disposal by combustion offers the advantages that the sulfur is vented to the atmosphere as the less objectionable SO_2, and better disposal is obtained because of the high temperature and velocity of the combustion-stack gases.

Plant Operation

As shown by the flow diagram, the process is quite simple. In some cases, plants have been built with but one tall tower, half of this being used for absorption and the other half for stripping. Principal auxiliary equipment items are the air fan, solution pump, and solution and air heaters. Since air is required in large quantities for regeneration, it is important that the actifier column be designed for very low pressure drop to minimize horsepower requirements.

The use of air as stripping vapor has the disadvantage of promoting oxidation in the system. Approximately 5 percent of the hydrogen sulfide which is absorbed is oxidized to thiosulfate. As sodium thiosulfate is not regenerable by simple stripping, buildup of this salt results in a loss of solution activity, and it becomes necessary to replace a portion of the solution periodically.

Hydrogen cyanide is also absorbed by the sodium carbonate solution and, to a considerable extent, is oxidized by the oxygen in the air used for stripping according to the following reaction:

$$2NaHS + 2HCN + O_2 \rightarrow 2NaSCN + 2H_2O \qquad (5\text{-}5)$$

Although it has been reported that the spent Seaboard process solutions have some value because of their NaSCN content, in most cases their disposal is a serious problem from the standpoint of water pollution.

Vacuum Carbonate Process

The use of vacuum distillation for regeneration of the alkali-carbonate solutions used in H_2S absorption is a more recent development of the Koppers Company, Inc. (16, 17). The process was an outgrowth of the early Seaboard process (which used air reactivation) and offered the advantage over this process of recovering the H_2S in a concentrated, usable form. The use of vacuum was found to reduce the steam requirement to about one-sixth that which would be required for steam-stripping at atmospheric pressure (18). The first installation of the vacuum carbonate process was built in Germany in 1938 and used a potassium carbonate solution. Plants constructed in the United States have in general utilized sodium carbonate solutions. The process has been primarily applied to coke oven gas streams, which generally contain 300 to 500 grains $H_2S/100$ scf. This type of gas contains HCN and other impurities which may cause difficulties with other H_2S-absorption systems. A photograph of a vacuum-carbonate-process installation in a large coke-oven plant is shown in Figure 5-4.

In recent years the vacuum carbonate process has been replaced in many coke-oven gas treating installations by processes based on oxidation of hydrogen sulfide to elemental sulfur in the liquid phase (see Chapter 9).

Process Description

A simplified flow diagram of the process is shown in Figure 5-5. The impure gas is contacted with dilute sodium carbonate in a packed countercurrent absorber, and the rich solution is passed to the top of an actifier column, where it is regenerated by vacuum distillation. The regenerated solution is then pumped from the bottom of the actifier through a solution cooler and back to the absorber. Gases from the top of the actifier, which consist of H_2S, HCN, CO_2, and water vapor, pass through condensers and a vacuum-pump system. In the process as illustrated in Figure 5-5, the heat required for activation is provided by low-pressure steam in a reboiler at the base of the actifier.

In a modification of the process (19), a major portion of the heat is supplied to the solution in the actifier by heat exchange with a source of low-level waste heat. Such heat is available in most coke-oven plants in the form of "flushing liquor," which is circulated through the hot gas-collecting mains and reaches a temperature of 75 to 80°C (167 to 176°F) (20). Solution is pumped from the bottom of the actifier through the heat exchangers, where it is heated by the flushing liquor, then back to the lower portion of the actifier, where stripping vapors are flashed off because of the relatively low boiling temperature in the vacuum actifier.

Basic Data

Principal reactions occurring in the vacuum carbonate process are given below.

$$Na_2CO_3 + CO_2 + H_2O \rightleftharpoons 2NaHCO_3 \tag{5-6}$$

$$Na_2CO_3 + H_2S \rightleftharpoons NaHS + NaHCO_3 \tag{5-7}$$

$$Na_2CO_3 + HCN \rightleftharpoons NaCN + NaHCO_3 \tag{5-8}$$

Figure 5-4. Photograph of vacuum carbonate (hot-activation) gas-purification plant. *Koppers Company, Inc.*

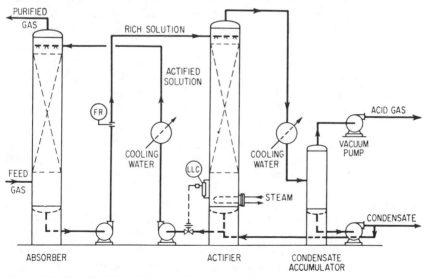

Figure 5-5. Simplified flow diagram of vacuum carbonate process.

Side reactions, which may occur if air contacts the solution, include Equations (5-9) and (5-10):

$$2NaHS + HCN + O_2 \rightarrow 2NaSCN + 2H_2O \qquad (5\text{-}9)$$

$$2NaHS + 2O_2 \rightarrow Na_2S_2O_3 + H_2O \qquad (5\text{-}10)$$

The absorption-desorption reaction (Equation 5-6) is not completely reversed in the stripping operation; thus, a considerable proportion of the soda ash remains as sodium bicarbonate and very little net CO_2 removal is realized. An analysis of the actifed solution from an operating plant, Table 5-3, has been given by Kurtz (21).

The high values for NaSCN and $Na_2S_2O_3$ given in Table 5-3 are not necessarily typical, as these will depend upon the extent to which oxygen is admitted to the system as well as on the age of the solution. In the plant from which the solution shown in Table 5-3 was taken, a specific gravity of 1.125 at 60°F has been adopted as the maximum allowable in the actified solution. The solution is discarded when dissolved solids increase to the point where this specific gravity is attained—usually after 6 to 8 months' operation.

The active ingredient of the solution can be considered to be the Na_2CO_3, which provides alkalinity for the absorption of acid gases. The reactions between acid gas and Na_2CO_3 are not complete, however, and can only approach the equilibrium condition corresponding to the gas and liquid composition. When H_2S is absorbed by sodium carbonate solution, NaHS and $NaHCO_3$ are formed by the reaction given in Equation 5-7. The vapor in equilibrium with such a solution thus contains both H_2S and CO_2 in accordance with the decomposition pressure of these two compounds. Carbon dioxide is the stronger acid and is held more firmly. However, its rate of absorption is much lower. As previously stated, the low rate of absorption of carbon dioxide is believed to be due to a slow chemical reaction which must occur before combination of CO_2 and Na_2CO_3 can take place. The rate-controlling reaction is most probably that of the dissolved molecular CO_2 with OH^- ions to form HCO_3^- ions. Hydrogen sulfide does not require hydration to form an acid and can, therefore, react rapidly with alkaline solutions.

Litvinenko (22) studied the absorption of H_2S in solutions of sodium and potassium carbonate on a laboratory scale and found that at low partial pressures of H_2S (1 percent H_2S in nitrogen at 1 atm total pressure) the gas film presented the major resistance. Solutions containing 5 percent Na_2CO_3 and 15 percent K_2CO_3 were studied. Results with both solutions were similar, although the potassium carbonate produced slightly higher absorption coefficients. K_Ga was found to vary as $G^{0.8}$ (in a 13-mm ID column packed with 5-mm glass rings) and as $L^{0.34}$ (in a 9.5-mm ID wetted-wall column). These results indicate that the rate of reaction of H_2S in the solution is sufficiently

Table 5-3. Solution Analysis

Constituent	Concentration g/liter
Na_2CO_3	20.25
$NaHCO_3$	25.38
NaHS	0.67
$Na_2S_2O_3$	9.48
NaSCN	66.62
Total solids	163.86 g/liter
Specific gravity (60°F)	1.103
Boiling point	
(30.38 in. Hg)	103.0°C (217.4°F)

rapid to make the resistance in the liquid film a minor portion of the total resistance. Similar conclusions were reached by Garner, Long, and Pennell (2).

When both H_2S and CO_2 are present in the gas stream contacting a sodium carbonate solution, a higher percentage of the H_2S will be absorbed (in a tower of reasonable height) because of the difference in rates of reaction. Unfortunately, however, the CO_2 which is absorbed will not be completely stripped and the bicarbonate concentration of the solution will gradually build up until a steady-state condition is attained. At this point, the quantity of CO_2 absorbed (which decreases with increased $NaHCO_3$ concentration) will equal the quantity stripped in the desorption step. The net effect of increased CO_2 concentration in the gas (with the same H_2S concentration) is thus to increase the $NaHCO_3$ content of the circulating solution, thereby decreasing its absorptive capacity and resulting in an increased solution-circulation requirement.

Plant Operation

Absorption. Packed absorbers are generally employed. In one commercial unit (plant A of Table 5-4) the packing is in the form of hurdles made up of numerous small wood slats (West Virginia spruce) (23). All of the plants which have been described are used for treating coke-oven gas, which contains 250 to 500 grains H_2S/100 cu ft and 1½ to 3 percent CO_2. This type of gas is usually treated at a relatively low pressure (below 25 psig).

According to Smith (24), up to 93 percent of the HCN and 5 to 7 percent of the CO_2 are also removed with the H_2S. When the process is operated to attain very high H_2S-removal efficiency, more CO_2 is also removed. Since the absolute quantity of H_2S absorbed does not increase materially, the resulting acid-gas stream then contains a higher percentage of CO_2.

Table 5-4. Vacuum Carbonate Process Absorber Data		
Absorber	Plant A (23)	Plant B (21)
Gas rate, SCF/hr.....................	2,300,000	$\begin{cases} 1,000,000, \text{avg.} \\ 1,300,000, \text{max.} \end{cases}$
Gas composition:		
H₂S, grains/100 SCF...............	300	406 avg.
H₂S, per cent......................	0.48	0.64
HCN, per cent.....................	0.13	
Solution rate, gpm....................	533	375–400
Absorber pressure, psig................	2	15–20
Purification efficiency, % removed:		
H₂S.............................	93	85 avg.
HCN............................	85–90	

Operating data on the absorption step in two commercial installations are presented in Table 5-4.

Desorption. The solution is stripped in a column called the "actifier" which is operated at a pressure of about 2.0 to 2.5 psia. Steam is used as stripping vapor, and this is generated by boiling the solution in the reboiler at the base of the actifier. Because of the reduced pressure, boiling occurs at a temperature of about 140°F (60°C). Both wood hurdle packed and bubble-plate columns have been employed for actifying. In plant B of Table 5-4, which was built in 1950, the actifier is equipped with 15 stainless-steel bubble-cap trays.

Since the reboiler heat can be supplied at a very low level, it is frequently possible to utilize waste heat from other plant streams. As mentioned above, one such source of heat is the flushing liquor which is circulated in the hot-gas mains of coke-oven plants.

The quantity of heat required for regeneration depends to a considerable extent upon the degree of H₂S removal desired. In plant A of Table 5-4, flushing liquor is used as the source of heat. In plant B, it is reported (21) that 13,000 to 15,000 lb/hr of steam are used for reboiling at the base of each actifier. Since 1 to 1.3 MMscf/hr of gas are treated per unit, the steam requirement for this plant is indicated to be about 12.0 lb/Mscf of gas purified.

Acid gas leaving the actifier of plant B contains 5 to 9 percent HCN. This is removed by a water wash prior to combustion of the H₂S for use in sulfuric acid manufacture. Acid-gas analyses for two other plants are presented in Table 5-5.

**Table 5-5. Analysis of Acid Gas from Plants Using
Vacuum Carbonate Process**

Gas	Gas content, %	
	Plant A(23)	Plant C (24)
H_2S	55	70
HCN	15	10
CO_2	25	15
N_2, etc.	5	5

At plant A, the HCN which is removed in the vacuum carbonate process is recovered as a by-product. In most other plants, this substance is either removed and burned (to permit the H_2S to be utilized in conventional acid plants), or it is destroyed in a specially designed sulfuric acid plant which utilizes the H_2S.

Operating Problems. Minor constituents in coke-oven gas, particularly naphthalene, can cause difficulty in the operation of vacuum carbonate process plants. Smith (24) recommends that the process be limited in application to gases containing not more than 2 grains napthalene/100 scf; 1.5 grains ammonia (including pyridine)/100 scf; and 1 grain tar fog/100 scf.

Coke-oven gas normally meets the above specifications after it has passed through benzene washers. If the vacuum carbonate process is located ahead of the benzene washers, which is sometimes done to improve HCN yield, special precautions are necessary to minimize difficulties caused by naphthalene.

Naphthalene is only very slightly soluble in the sodium carbonate solution; however, a small amount is absorbed by the solution when it contacts the coke-oven gas. The napthalene which is absorbed is stripped from the solution in the actifier and carried by the acid-gas stream to the condensers. The concentration of naphthalene in the acid-gas stream will thus be a function of the initial concentration in the coke-oven gas, and, if this is too high, condensation can occur when the acid gas is cooled. Since naphthalene condenses as a solid, this can cause plugging of condensers and other equipment.

Ammonia dissolves readily in the Na_2CO_3 solution and is also stripped in the actifier. Cooling the actifier effluent gas causes the condensation of water, which readily takes up ammonia from the acid-gas stream. The ammonia dissolved in the condensate increases the solubility of H_2S and, as this is returned to the system, the overall H_2S-removal efficiency of the process is reduced. Pyridine behaves in a similar manner; however, the quantity of this compound is usually too small to have an appreciable effect on the efficiency of the process.

Corrosion is not considered to be a major difficulty with vacuum carbonate plants, probably because of the relatively low temperatures employed. Mild steel is the primary material of construction, although, as mentioned above, wooden absorber packings and stainless-steel actifier trays are used. Stainless steel is also used for minor components of the vacuum and other pumps, thermometer wells, and instrument diaphragms (24).

One problem common to all processes using liquids for the purification of coke oven gases or other coal derived gases is the handling of contaminated waste streams within the framework of existing environmental pollution control regulations. The principal offender in coke-oven gas is hydrogen cyanide which is readily soluble in most solvents and quite reactive with chemicals used as the active treating agents. The reaction products such as cyanides, thiocyanates, and iron-sulfur-cyanide compounds have to be eliminated before liquid waste streams can be disposed of in sewage systems. A considerable effort has been made to develop economical methods for acceptable waste disposal and some are in industrial use. A good review of such methods including those used in vacuum carbonate units has been presented by Massey and Dunlap (77).

Operating Costs. Daily utility and chemical requirements and operating costs for a typical vacuum carbonate process plant, treating 55 MMscf/day of coke-oven gas, have been made available by Koppers Company, Inc. (25) and are presented in Table 5-6. In order to point out the steam savings possible by the use of waste heat in the actifier reboiler, data are presented for two plant designs; one using external steam to provide reboiler heat and the other using the heat-exchanger system. The operating-cost savings resulting from the use of the heat-exchanger systems are appreciable. However, it should be realized that the plant investment for this design would be somewhat higher.

Tripotassium Phosphate Process

The use of a tripotassium phosphate solution for H_2S removal was introduced by the Shell Development Company (26). The process has been displaced to a considerable extent by alkanolamine processes for refinery and natural-gas purification; however, it offers some advantages for special applications. The principal advantages are nonvolatility of the active component in the solution, insolubility in hydrocarbon liquids and nonreactivity with COS and other trace impurities. These advantages make the tripotassium phosphate process suitable for high-temperature application, the treatment of liquid hydrocarbons, and the purification of refinery or synthesis gases. In common with the other alkali-salt processes, the tripotassium phosphate process offers some selectivity for H_2S in the presence of CO_2. This selectivity gives the phosphate process an economic advantage over

Table 5-6. Operating Costs for Vacuum Carbonate Process (25)

Operating variable	Quantity considered
Coal carbonized, tons/day	5,000
Gas produced, SCF/day	55,000,000
Average H_2S concentration, grains/100 cu ft:	
Inlet	500
Outlet	50
Average H_2S removal, %	90
Average HCN concentration, grains/100 cu ft:	
Inlet	40
Outlet	6
Average HCN removal, %	85
HCN recovered, lb/day	nil
H_2S removed, lb/day	35,400
Equivalent available sulfur, tons/day	16.3

Part B. Comparative Requirements for Plant Designs I and II

Daily utility and chemical requirements	Plant design I, with Koppers heat-exchange system		Plant design II, without Koppers heat-exchange system	
Cooling water, gal/day:	Max (80°F)	Avg (65°F)	Max (80°F)	Avg (65°F)
Vapor condensers	1,700,000	1,055,000	1,700,000	1,055,000
Vapor coolers	470,000	187,000	470,000	187,000
Solution coolers	1,440,000	907,000	1,440,000	907,000
Vacuum-pump jackets	86,500	48,000	86,500	48,000
Cyanogen scrubber	55,200	55,200	55,200	55,200
Fresh solution	2,400	2,400	2,400	2,400
Total for sulfur-recovery plant	3,754,100	2,254,600	3,754,100	2,254,600
Less water saved at primary coolers	1,119,000	981,000	nil	nil
Net additional water required	2,635,100	1,273,600	3,754,100	2,254,600
Electric power, kwhr/day:				
Flash-solution pumps	2,017		nil	
Vacuum pumps	4,200		nil	
Actified-solution pumps	1,167		1,167	
Foul-solution pumps	413		413	
Cyanogen-scrubber pump	34		34	
Lighting	48		48	
Total	7,879		1,662	

Table 5-6 continued

Table 5-6 continued

Steam, lb/day:		
Live, for vacuum pumps.............	nil	175,000
Low-pressure steam (15 psig)		
Actifier reboilers..................	nil	696,000
Miscellaneous...................	9,600	9,600
Total low-pressure steam required.	9,600	705,600
Less steam available from		
vacuum-pump exhaust.......	nil	158,300
Net low-pressure steam required		
for make-up.................	9,600	547,300
Sodium carbonate, lb/day.............	400	400

Part C. Comparative Daily Operating Costs for Plant Designs I and II

Daily cost items		Total daily cost per item	
Requirement	Unit price	Plant design I, with Koppers heat-exchange system	Plant design II, without Koppers heat-exchange system
Cooling water.............	$0.015/1,000 gal	$ 19.10	$ 33.80
Electric power.............	0.01/kwhr	78.80	16.60
Steam:			
Live....................	0.70/1,000 lb	....	122.50
Low-pressure...........	0.50/1,000 lb	6.70	273.70
Sodium carbonate..........	0.02/lb	8.00	8.00
Operating labor............	1.75/man-hr	84.00	84.00
Laboratory supervision...................		6.00	6.00
Maintenance..........................		75.00	75.00
Total daily operating cost................		$277.60	$619.60

monoethanolamine or diethanolamine systems when it is desired to remove H_2S with a minimum extraction of CO_2 from gas streams containing both at CO_2/H_2S ratios exceeding about 4:1.

Process Description

The absorption of hydrogen sulfide by tripotassium phosphate may be represented by Equation 5-11.

$$K_3PO_4 + H_2S \leftrightarrows K_2HPO_4 + KHS \qquad (5\text{-}11)$$

The basic flow-cycle used is similar to that of alkanolamine processes and other heat-regenerative systems. Where a high degree of purification of a gas is required, a double-stream system such as that shown in Figure 5-6 is suggested.

If no CO_2 is present, a 40 to 50 percent (by weight) solution of tripotassium phosphate is used; however, if CO_2 is present, a 35 percent solution is generally used to avoid precipitation of bicarbonate. In the split-stream cycle, a portion of the solution is passed through a second stripping zone and aqueous condensate is added to this portion after final stripping. This more completely regenerated (and more dilute) stream is fed into the top of the absorber to provide final cleanup of the gas stream. Higher purity is attainable with the dilute solution because, for a given H_2S/K_3PO_4 ratio, the vapor pressure of H_2S is lower over solutions containing less tripotassium phosphate. Figure 5-7 shows a photograph of a plant which removes H_2S from refinery-gas streams by the tripotassium phosphate process.

Design and Operating Data

The equilibrium vapor pressure of hydrogen sulfide over a 50 percent tripotassium phosphate solution at several temperatures is shown in Figure 5-8, which is based on the data of Rosebaugh (27). One isotherm for a 20 percent solution is also included to point out the effect of dilution on the H_2S vapor pressure.

Detailed data on a pilot-plant study of the applicability of tripotassium phosphate solutions to the selective absorption of H_2S in the presence of CO_2 have been presented by Wainwright et al. (28). Results of this study with regard to the effect of CO_2, and H_2S on solution circulation rate requirements are shown in Figure 5-9. As would be expected, an increase in either the CO_2 or H_2S content of the feed gas increases the quantity of 35 percent potassium phosphate solution required to make a product gas containing 25 grains $H_2S/100$ scf. The data were obtained with a column made of 10-in. standard iron pipe packed with 15 ft of 1-in. porcelain raschig rings and operated at 300 psig pressure. All runs were made with relatively high CO_2/H_2S ratios in the feed gas. The effect of varying this ratio on the quantity of H_2S absorbed per gallon of solution is shown in Figure 5-10.

Wainwright, et al., found that their initial solution, which contained 43.7 percent tripotassium phosphate, could not be used on gases of high CO_2 content because of the formation of potassium bicarbonate, which precipitated and plugged the absorber packing. The solution was therefore diluted to 32 to 35 percent and no further difficulty was encountered as a result of precipitation.

Reactivation was accomplished in a column also made from 10-in. steel pipe and packed with 15 ft of 1-in. porcelain raschig rings. The reboiler

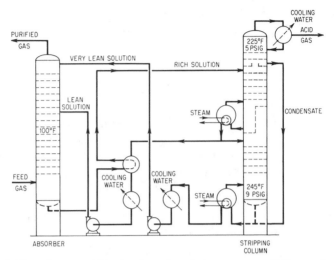

Figure 5-6. Flow diagram of tripotassium phosphate process, split-flow cycle.

Figure 5-7. Refinery-gas purification plant employing Shell tripotassium phosphate process. Three columns on right are used to remove H_2S from various high- and low-pressure gas streams originating from thermal cracking operations. Stripper column is located in left center background; overhead condenser and associated structure are visible. *Shell Oil Company*

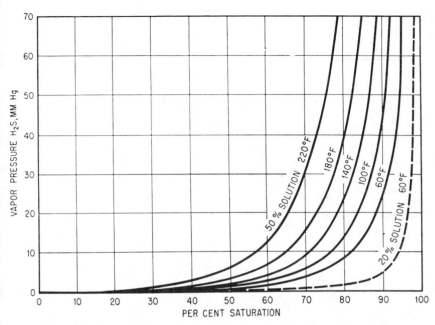

Figure 5-8. Equilibrium vapor pressure of hydrogen sulfide over 50 and 20 per cent tripotassium phosphate solutions. *Data of Rosebaugh* (27)

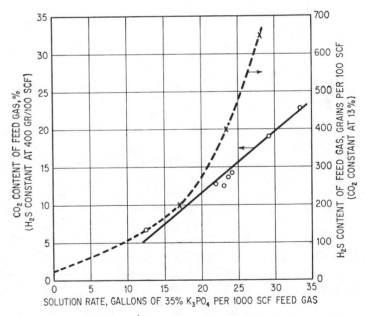

Figure 5-9. Effect of CO_2 and H_2S contents of feed gas on tripotassium phosphate—solution rate required to produce exit gas containing 25 grains H_2S/100 SCF. Absorber of 10-in. ID packed with 15 ft of 1-in. porcelain raschig rings; solution regenerated with steam, approximately 1 lb/gal. *Data of Wainwright et al* (28)

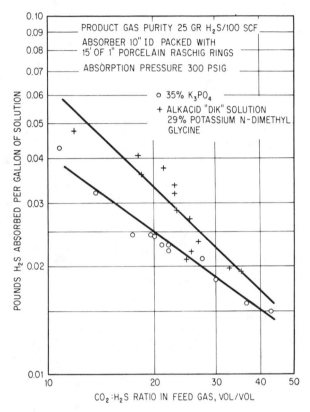

Figure 5-10. Effect of CO2/H2S ratio on capacity of K3PO4 and Alkacid "dik" solutions. *Data of Wainwright et al (28)*

pressure was generally maintained at about 3 psig. Steam consumption averaged about 1 lb/gal of circulating solution. Because of relatively high heat losses from the pilot-scale equipment, it was estimated that the quantity of steam actually used for reactivation was somewhat lower, about 0.9 lb/gal of solution. Increasing the steam flow rate by 20 percent was found to have only a slight effect on the degree of stripping. It has been reported (29) that commercial plants operating under about the same conditions have substantially lower steam consumption than that observed by Wainwright et al (28). This higher stripping efficiency is believed due to (a) more efficient stripping columns, (b) more nearly optimum solution circulation rate, and (c) less heat loss from large-scale equipment.

A commercial plant which removes H_2S and CO_2 from a stream of light hydrocarbons in the liquid state has been described by Zahnstecher (30). Although this cannot be considered to be gas purification, the regeneration step is identical whether the absorber treats liquid or gaseous hydrocarbons and the data presented are of interest in this regard. The stripping column consists of a unit 2 ft 6 in. in diameter with 10 bubble cap trays installed directly on top of a steam-heated reboiler. With a solution-circulation rate of 25 gpm, approximately 1.5 lb of steam are required per gallon of solution circulated. The regenerator operates at a pressure of 2.5 psig and a temperature of about 226°F. Under these regenerator conditions, the solution concentration is reduced from 0.187 mole H_2S and 0.640 mole CO_2 to 0.090 moles H_2S and 0.493 mole CO_2 per mole of K_3PO_4. In this plant, the feed to the absorption unit (liquid phase) contains 0.50 mole percent CO_2 and about 0.4 mole percent H_2S.

Sodium Phenolate Process

The sodium phenolate process is of more interest historically than for practical application as it has been largely supplanted by other processes. Because of this, no detailed operating or design data are presented. The process, which was developed by Koppers Company, Inc,. employs a fairly concentrated solution of sodium phenolate in a conventional heat-regenerative flow cycle. The solution contains approximately three moles of NaOH (120 g) and two moles phenol (188 g) per liter (31). Its capacity for H_2S is quite high, up to about 35 scf/cu ft for an extremely sour gas (32). The basic difficulty of the process is the low efficiency of H_2S removal (on the order of 90 percent) and the high steam consumption. The steam consumption can be cut almost in half by the use of a two-stage stripping system; however, this complicates the plant design considerably. Phenolate plants have also reportedly encountered corrosion difficulties.

Alkacid Process

The Alkacid (Alkazid) process, which was developed by I.G. Farbenindustrie (33) in Germany, could be classified as three separate processes in that three different absorption solutions are used. However, all of the process variations use a solution of the salt of a strong inorganic base and weak organic nonvolatile acid and all use a conventional heat-regenerative cycle such as employed by the phenolate, tripotassium phosphate, and ethanolamine processes. The solutions are designated as Alkacid solution "M," Alkacid solution "dik," and Alkacid solution "S"; and each has a specific field of application. The "M" solution contains sodium alanine and is used for absorbing either H_2S or CO_2 when present alone, or for absorbing both gases simultaneously. The "dik" solution contains the potassium salt of diethyl- or dimethylglycine and is used for the selective removal of hydrogen sulfide from gases containing carbon dioxide and also from gases containing small quantities of carbon disulfide or hydrogen cyanide. Solution "S," which is reported to be a solution of sodium phenolate (34), was developed for gases containing an appreciable amount of other impurities such as HCN, ammonia, carbon disulfide, mercaptans, dust, and tar. However, the version of the process using solution "S" has not been commercialized.

Design and Operating Data

The Alkacid process was described in considerable detail by Hans Bähr in 1938 (35); at that time he reported that a number of Alkacid plants for the removal of hydrogen sulfide and carbon dioxide were operating in Germany and that several more plants were under construction. More recently industrial applications of the Alkacid process have been described by Pasternak (36) and by Leuhddemann, Noddes, and Schwarz (37). The latter authors report that in 1959 more than 50 Alkacid plants were in operation, in Europe, the Middle East, and Japan. The process is suitable for treating synthesis gases, water gas, natural gas, and hydrocarbon liquids at atmospheric as well as at elevated pressures (36,37,38). No commercial installations are known to have been operated in the United States, although the process has been studied on a pilot-plant scale (28). The "dik" solution is of particular interest as the more selective solvent for hydrogen sulfide in the presence of carbon dioxide. A comparison of the selectivity of the two solvents, based on the data of Leuhddemann et al. (37), is shown in Figure 5-11. These data were obtained by shaking the "dik" and "M" solutions at ambient temperatures with

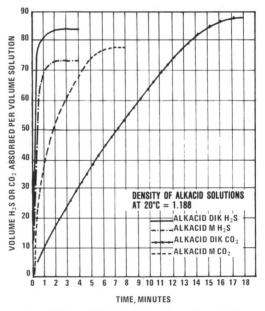

Figure 5-11. Absorption of H_2S and CO_2 by Alkacid solutions (37).

pure hydrogen sulfide and carbon dioxide and measuring the volume of gas dissolved per volume of solution every minute, until saturation was reached. Vapor pressures of hydrogen sulfide and carbon dioxide over Alkacid "dik" and "M" solutions reported by the same authors (37) are shown in Figures 5-12 to 5-15.

These pilot-plant studies, conducted by the U.S. Bureau of Mines, were made to evaluate the Alkacid "dik" solution as a selective absorbent for hydrogen sulfide in the presence of carbon dioxide, and the results were compared with those of other absorbents tested in the same apparatus. The solution capacity of the Alkacid solution tested has been compared with that of the 35 percent tripotassium phosphate solution in Figure 5-10 which demonstrates the effect of CO_2/H_2S ratio on the absorptive capacity for both solutions. It will be noted that the Alkacid solution takes up appreciably more H_2S particularly at low CO_2/H_2S ratios. Although its capacity is higher, the Alkacid solution was found to be somewhat less selective than tripotassium phosphate.

An indication of the capacity of the Alkacid solution for absorbing H_2S is shown in Figure 5-16, which presents the vapor pressure of H_2S over Alkacid solutions containing two concentrations of CO_2. This chart is based on U.S. Bureau of Mines data (28) and represents results obtained with a solution of the following composition and characteristics:

Total nitrogen, %	2.60
Sulfated ash, %	19.26
NH_3, nitrogen %	0.01
Solids, %	29.2
Specific gravity	1.138

Another pilot-plant investigation comparing the selectivity of Alkacid "dik" with that of diethanolamine (DEA) and methyldiethanolamine (MDEA) has been presented by Pasternak (39). The feed gas to the pilot plant was water gas of the following composition:

CO_2—27.5 volume percent
H_2S—0.7 volume percent
CO—31.4 volume percent
CH_4—1.1 volume percent
N_2—1.0 volume percent

During the experiments the hydrogen sulfide content was varied from 0.7 to 2.0 percent. All solvents used consisted of aqueous solutions containing the active chemicals in a concentration of 40 percent. The results from this study

(text continued on page 210)

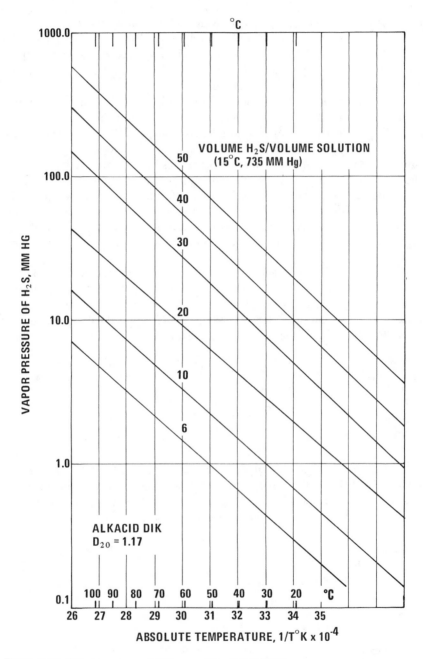

Figure 5-12. Vapor pressure of H₂S over Alkacid "dik" (37).

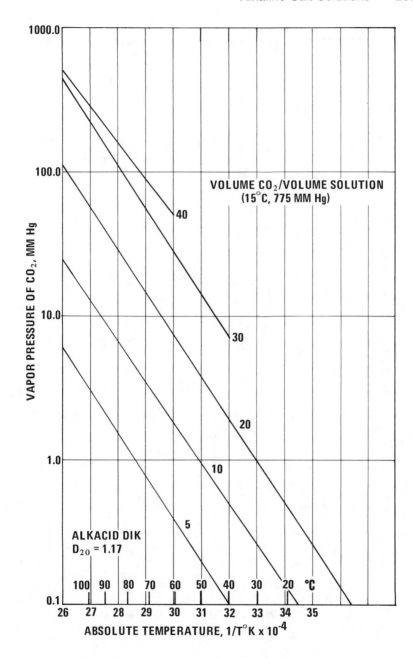

Figure 5-13. Vapor pressure of CO₂ over Alkacid "dik" (37).

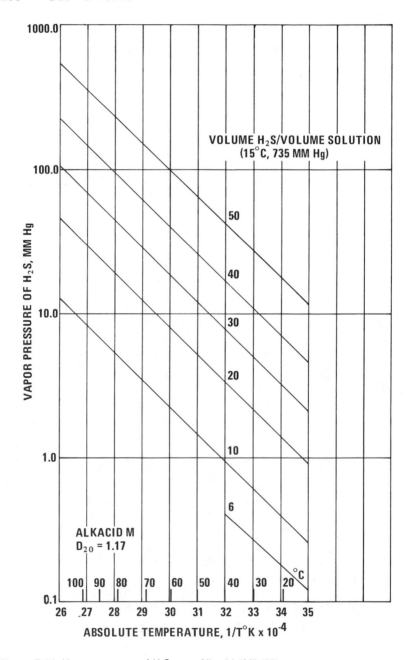

Figure 5-14. Vapor pressure of H₂S over Alkacid "M" (37).

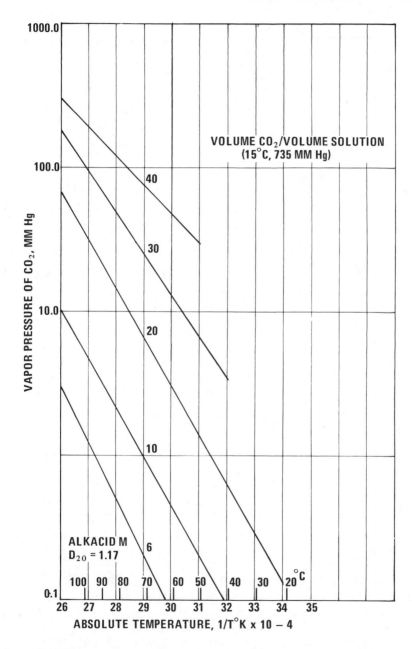

Figure 5-15. Vapor pressure of CO_2 over Alkacid "M" (37).

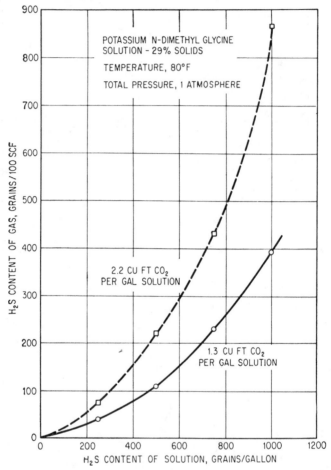

Figure 5-16. Vapor pressure of H_2S over potassium N-dimethylglycine solutions containing CO_2. *Data Wainwright et al (28)*

show that the "dik" solution is the most selective, followed by MDEA and DEA. Also, the "dik" solution has the highest capacity for hydrogen sulfide for comparable partial pressures of hydrogen sulfide and carbon dioxide in the feed gas. It should be noted that no attempt was made in this study to attain complete hydrogen sulfide removal, the maximum being on the order of 70 percent. Another interesting result was that the steam requirement for solution regeneration was appreciably lower for the "dik" solution than for the two ethanolamines.

The "dik" solution is reported to be quite stable although some aging and loss of selectivity have been observed, especially when the gas to be treated contained HCN and oxygen (39).

Operating results from an Alkacid plant using "M" solution for the removal of hydrogen sulfide and carbon dioxide from natural gas are shown in Table 5-7 (37).

It is claimed that the Alkacid solutions are relatively noncorrosive, and the equipment requires no special materials of construction except in a few parts which are particularly subjected to wear (35); however, patents have been obtained on methods of avoiding corrosion (40). It is reportedly German practice to use aluminum and special alloys for the hot-solution pumps and lines, the reactivator, and the reboiler (41).

Absorption at Elevated Temperature

Hot Potassium Carbonate (Benfield) Process

This process was developed by the U.S. Bureau of Mines, at Bruceton, Pennsylvania, as part of a program on the synthesis of liquid fuel from coal. Research on CO_2 removal was conducted with the objective of reducing the cost of synthesis-gas purification by designing a process which would take maximum advantage of the synthesis-gas conditions; i.e., high CO_2 partial pressure and high temperature. A flow sheet of the basic process is shown in Figure 5-17 and a photograph of a commercial unit is shown in Figure 5-18. The process has been described in considerable detail by publications of Benson and coworkers (42, 43, 44).

In recent years the hot potassium carbonate process has been developed further by Benson and Field who conducted much of the original work at the U.S. Bureau of Mines, and many improvements have been made. Among these, development of activating additives to the potassium carbonate solution, resulting in substantial lowering of capital and operating costs and higher treated gas purity, is probably the most important (44). Furthermore, major improvements have been made in energy economy (80, 81) and the process has been demonstrated to be suitable for partially selective removal

Table 5-7. Removal of H_2S and CO_2 with Alkacid "M"	
Feed gas volume, SCF per hour	650,000
Pressure, psia	85
Temperature, °F	78
H_2S, vol. percent	0.5
CO_2, vol. percent	3.0
Treated Gas:	
H_2S, grains/100 SCF	4.4
CO_2, vol. percent	0.04
Steam, lb/lb CO_2	4

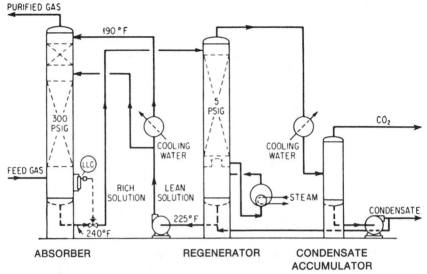

Figure 5-17. Flow diagram of the hot potassium carbonate process for absorption of CO_2 (split-stream absorption).

Figure 5-18. Photograph of hot potassium carbonate gas-purification plant employed for removal of CO_2 from natural gas. *Petrocon Engineering Company*

of hydrogen sulfide in the presence of carbon dioxide (82). The process is practiced in several hundred plants located all over the world for the removal of carbon dioxide and hydrogen sulfide from ammonia synthesis gas, crude hydrogen, natural gas, town gas, and others (44). The process is licensed under the name of "Benfield Process" by the Benfield Corporation of Pittsburgh, Pennsylvania.

The applicability of potassium carbonate to CO_2 removal has been known for many years. A German patent, granted as early as 1904, described a process for absorbing CO_2 in a hot solution of potassium carbonate and then stripping the solution by pressure reduction without additional heating (45). In 1924, Williamson and Mathews (46) studied the rate of absorption of CO_2 in potassium carbonate solution and found that increasing the absorption temperature from 25°C (77°F) to 75°C (167°F) greatly increased the rate of absorption. The work of the U.S. Bureau of Mines, however, constitutes a major contribution, in that it resulted in the development of an economical commercial process. A patent covering one aspect of this work has been issued to Benson and Field in Great Britain (47). This patent describes the use of potassium carbonate solution as an absorbent at temperatures near its atmospheric boiling point and its regeneration by flashing and steam stripping.

As a result of the high absorber-temperature, the steam which other regenerative processes require to heat the solution to stripping temperature is not required in the hot potassium carbonate system. In addition, the need for heat-exchange equipment between the absorber and stripper is eliminated. The high temperature also increases the solubility of potassium bicarbonate, thus permitting operation with a highly concentrated solution.

Process Description

As can be seen from the flow diagram, Figure 5-17, the process is extremely simple. In the split-stream process shown, a portion of the lean solution from the regenerator is cooled and fed into the top of the absorber while the major portion is added hot at a point some distance below the top. This simple modification improves the purity of the product gas by decreasing the equilibrium vapor pressure of CO_2 over the portion of solution last contacted by the gas. A somewhat more complex scheme termed two-stage, has also been used for applications in which more complete CO_2 removal is required. In this modification, the main solution-stream is withdrawn from the stripping column at a point above the reboiler so that only a portion of the solution passes down through the bottom of the stripping column to the reboiler. Since this portion of the solution is regenerated by the total steam supply to the stripping column, it is thoroughly regenerated and is capable of reducing the CO_2 content of the gas to a low value. The main solution-stream is fed into the mid-point of the absorber, while the more completely regenerated portion is fed at the top.

In the flow scheme employed for removal of hydrogen sulfide and carbon dioxide from natural gas, the entire stream of lean solution is fed to the top of the absorber.

Several modifications of the three basic flow patterns have been used, aimed primarily at greater heat economy. One such modification consists of two-stage regeneration with flash cooling by use of ejectors, as shown in Figure 5-19 (48). The Benfield LOHEAT process represents a further development in the direction of maximum utilization of internal heat sources and, consequently, energy economy. As described by Baker and McCrea (81) usable heat recovered from gas and liquid streams may reduce outside energy requirements by as much as 60 percent and, in some cases, result in export of low pressure steam.

Another modification described by Benson and Parrish (78) is the "Hi Pure" process which reportedly is capable of producing a treated gas containing less than one part per million of H_2S and less than 50 parts per million of CO_2. This process uses two independent but compatible circulating solutions in series to achieve high purity combined with high efficiency. The gas is first contacted with normal hot potassium carbonate followed by contact with a second solution of somewhat different composition. The hot potassium carbonate serves to remove the bulk of the acid gases while final purification is achieved with the second solution. The two solutions are regenerated separately in two sections of a regenerator with the stripping steam leaving the lower section of the regenerator being re-used in the upper section. The combined heat required for the two solutions is generally lower than that for a conventional hot carbonate system. Although the capital cost of a Hi Pure unit is somewhat higher than that of a normal Benfield unit, the savings in heat energy are sufficiently great—on the order of 20 percent—to make this process quite attractive.

Basic Data

A large amount of comprehensive physical data on the potassium carbonate-potassium bicabonate-carbon dioxide-water and potassium carbonate-potassium bicarbonate-potassium bisulfide-carbon dioxide-hydrogen sulfide-water systems is available in the literature (42, 43, 49, 50, 51, 52). Some typical data are given below; however, for complete information, the reader is referred to the original sources.

The effect of temperature and percentage conversion to bicarbonate on the solubility of the salts in the system, potassium carbonate bicarbonate, has been determined by Benson, Field, and Jimeson (42), and their data are presented in Figure 5-20, together with data from other literature sources (53-55). Lines on the chart represent the conditions under which crystals of potassium bicarbonate begin to precipitate for varying potassium carbonate concentrations. At 240°F, for example, the 60 percent solution can be con-

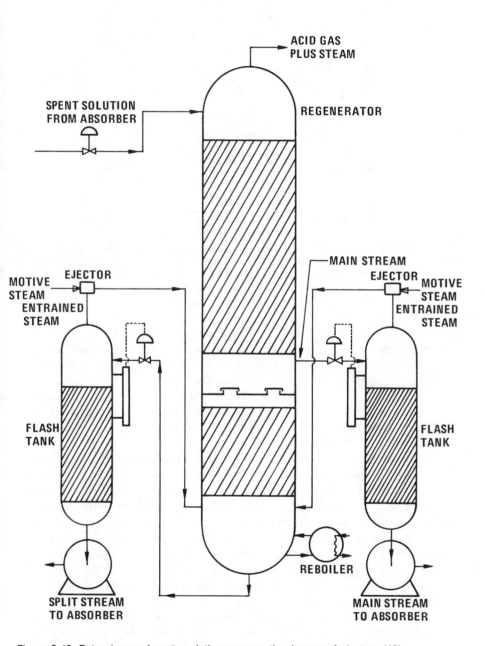

Figure 5-19. Potassium carbonate solution regeneration by use of ejectors (48).

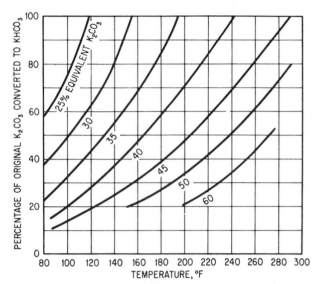

Figure 5-20. Effect of temperature and percentage conversion to bicarbonate on solubility of K₂CO₃ plus KHCO₃. Lines represent solubility limits for given salt concentrations (measured as equivalents of K₂CO₃). *Forty, Fifty, and Sixty percent lines based on data of Benson, Field, and Jimeson (42); other lines based on data of Perry (53), Seidell (54), and Hill (55)*

verted to only about 30 percent bicarbonate without the formation of a precipitate. A 50 percent solution can be 50 percent converted; and a 40 percent solution can theoretically be converted 100 percent. On the basis of these data, it is concluded that a 40 percent equivalent concentration of potassium carbonate is about the maximum that can be used for the treating operation without precipitation occurring, and a 30 percent solution is considered to be a reasonable design value for most applications.

If cooling of the solution occurs at any point in the system, even 30 percent potassium carbonate may be too great a concentration. On the basis of commercial-plant experience with units treating natural gas, Buck and Leitch (56) recommend 30 percent potassium carbonate equivalent as a maximum concentration. They found no appreciable effect on the absorptive capacity of the solution when its concentration was reduced to as low as 20 percent.

The equilibrium vapor pressure of CO_2 over a solution containing the equivalent of 20 percent potassium carbonate as a function of conversion to bicarbonate, based on the data of Tosh et al. (49), is presented in Figure 5-21. These authors, who investigated vapor-liquid equilibria for 20, 30, and 40 percent equivalent potassium carbonate solutions, found that the CO_2 equilibrium vapor pressure remains practically the same for the range of 20 to 30 percent. This in effect confirms the observation of Buck and Leitch (56) for commercial operations.

The experimental CO_2 vapor pressure data were used by Tosh et al. (49) as the basis for calculating the equilibrium constant, K, for the three solution concentrations according to the expression:

$$K = (KHCO_3)^2/(K_2CO_3)P_{CO_2} \qquad (5\text{-}12)$$

with $KHCO_3$ and K_2CO_3 expressed in gram moles per liter and P_{CO_2} in

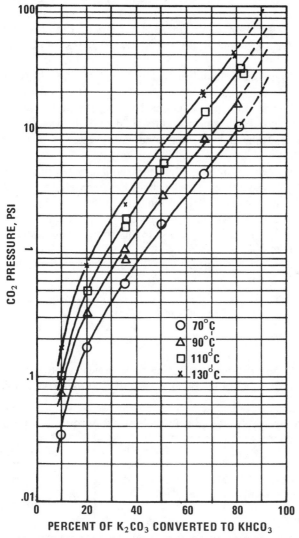

Figure 5-21. Equilibrium vapor pressure of CO_2 over 20 percent equivalent potassium carbonate solution (49).

mmHg. K was found to be constant at a given temperature for each degree of conversion for the 20 and 30 percent potassium carbonate solutions. From the values of K, the equilibrium vapor pressure can be calculated for any conversion within the given range of solution concentrations. Table 5-8 shows average K values (arithmetic mean of all experimental data points) for 20 and 30 percent solutions (49).

The equilibrium vapor pressure of water over a solution containing the equivalent of 20 percent potassium carbonate as a function of conversion to bicarbonate is shown in Figure 5-22. This chart is also based on data obtained by Tosh et al. (49). Here, again, there is not much difference in the vapor pressure of water for 20 and 30 percent equivalent potassium carbonate concentrations.

Additonal vapor-liquid equilibrium data based on published information and experimental work have been reported by Bocard and Mayland (52). Other physical data on the potassium carbonate-potassium bicarbonate carbon dioxide system are shown in Figures 5-23 to 5-26. The data of Bocard and Mayland (52) have been converted to a series of nomographs by Mapstone (57).

The potassium carbonate-potassium bicarbonate-potassium bisulfide carbon dioxide-hydrogen sulfide-water system has been studied extensively by Tosh et al. (50) and Field et al. (58) of the U.S. Bureau of Mines. It was found that in order to use this system in a continuous gas treating operation, it is necessary that the gas to be treated contains some carbon dioxide, besides hydrogen sulfide.

If hydrogen sulfide were the only acid gas to be absorbed, the following reactions would occur:

$$K_2CO_3 + H_2S \rightleftharpoons KHCO_3 + KHS \tag{5-13}$$

$$2KHCO_3 \rightleftharpoons CO_2 + H_2O + K_2CO_3 \tag{5-14}$$

Since with each absorption-regeneration cycle some carbon dioxide would be lost from the system, all of the potassium in the solution would eventually be

Table 5-8. Average Values of K for 20 and 30 percent K_2CO_2 Solutions		
Temperature °C	K, 20% Solution	K, 30% Solution
70	0.042	0.058
90	0.022	0.030
110	0.013	0.017
130	0.0086	0.011
Source. Data of Tosh et al. (49).		

converted to potassium bisulfide which was found to be essentially non-regenerable according to the following reaction (50):

$$2KHS \rightleftharpoons K_2S + H_2S \tag{5-15}$$

Equilibrium pressures of hydrogen sulfide, carbon dioxide, and water over solutions of 30 percent equivalent potassium carbonate are shown in Figures 5-27 to 5-33. The term "equivalent potassium carbonate" means that all potassium is assumed to be present as the carbonate.

(text continued on page 225)

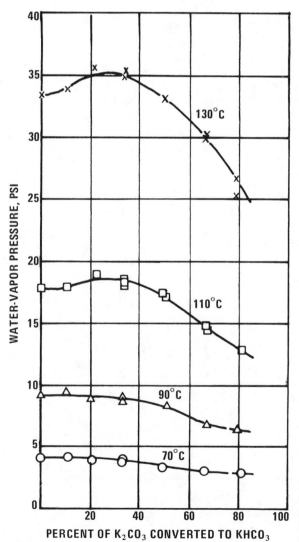

Figure 5-22. Equilibrium vapor pressure of water vapor over 20 percent equivalent potassium carbonate solution (49).

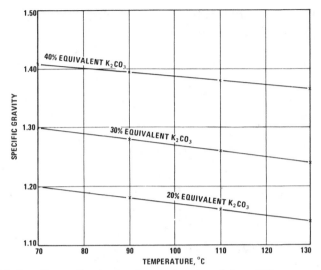

Figure 5-23. Specific gravity of potassium carbonate solutions (51).

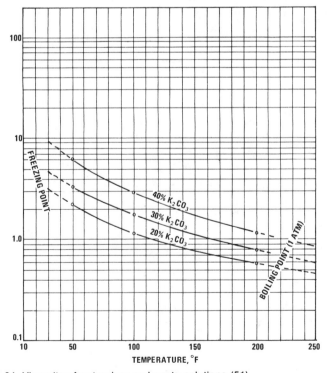

Figure 5-24. Viscosity of potassium carbonate solutions (51).

Figure 5-25. Heat capacity of potassium
carbonate solutions.

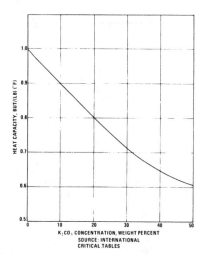

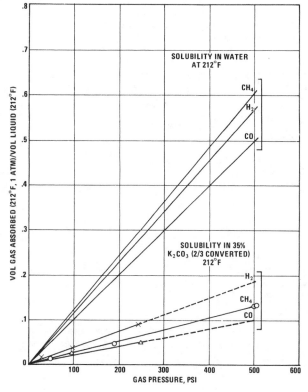

Figure 5-26. Solubility
of CH_4, CO and H_2 in
35 percent potassium
carbonate solution and
water (58).

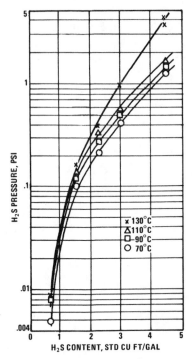

Figure 5-27. Equilibrium vapor pressure of H_2S over 30 percent potassium carbonate solution (50).

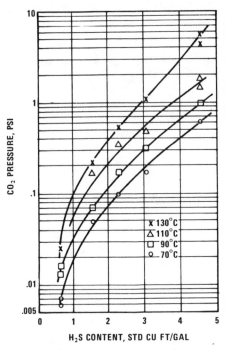

Figure 5-28. Equilibrium vapor pressure of CO_2 over 30 percent potassium carbonate solution containing H_2S (50).

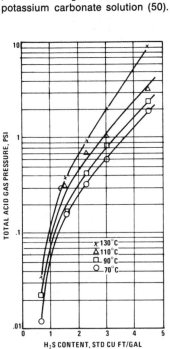

Figure 5-29. Equilibrium vapor pressure of H_2S and CO_2 over 30 percent potassium carbonate solutions (50).

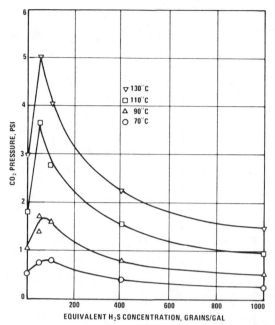

Figure 5-30. Equilibrium vapor pressure of CO_2 over 30 percent potassium carbonate, one third converted to $KHCO_3$ + KHS (50).

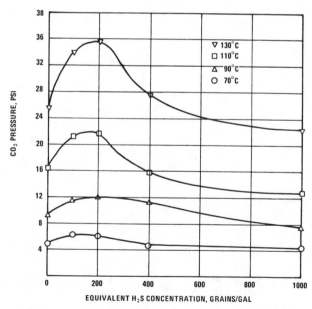

Figure 5-31. Equilibrium vapor pressure of CO_2 over 30 percent potassium carbonate solution, two thirds converted to $KHCO_3$ + KHS (50).

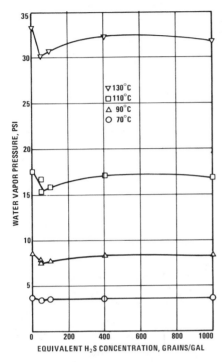

Figure 5-32. Equilibrium vapor pressure of water vapor over 30 percent potassium carbonate solution, one third converted to $KHCO_3$ + KHS (50).

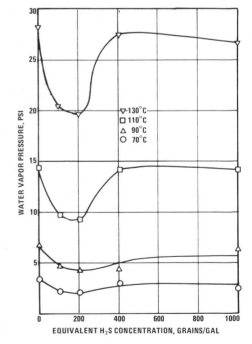

Figure 5-33. Equilibrium vapor pressure of water vapor over 30 percent potassium carbonate solution, two thirds converted to $KHCO_3$ + KHS (50).

Values of the equilibrium constant:

$$K_1 = (P_{H_2S}) (KHCO_3)/(P_{CO2}) (P_{H_2O}) (KHS) \tag{5-16}$$

are given by Tosh et al. (50) for a 30 percent equivalent solution, two-thirds converted to potassium bicarbonate and potassium bisulfide and are shown in Table 5-9. Estimated heats of reaction of hydrogen sulfide with 30 or 40 per cent potassium carbonate solutions of 11 and 22 B.t.u. per cu. ft. of hydrogen sulfide (at 0°C and 760 mm of mercury) have been reported by Tosh et al. (50).

Design and Operating Data

As a result of the comprehensive pilot-plant studies of the U.S. Bureau of Mines (42, 43, 58), adequate data are available for the design of hot-potassium carbonate-process plants. A limited amount of commercial-plant operating data has also been published (56). Three series of pilot-plant tests on the hot potassium carbonate process have been reported. In the first (42), the absorber consisted of a 4-in. diameter, schedule 80 steel pipe, packed with 0.5-in. raschig rings to a depth of about 9 ft. In the second (43), the absorber consisted of a 6-in. schedule 80 pipe, packed to a height of 30 ft with 0.5-in. porcelain raschig rings. The regenerator for the first series consisted of a 6-in. diameter, schedule 40 steel pipe, packed to a depth of 4.75 ft with 0.5-in. raschig rings, and, for the second series, an 8-in. diameter, schedule 40 pipe was used, packed to a height of 25 ft with the 0.5-in. rings. Heat losses from the pilot plants were minimized by adequate insulation and use of steam tracing.

While the first two pilot-plant studies were concerned primarily with carbon dioxide removal, the third study was conducted to investigate the suitability of the hot potassium carbonate process for the removal of

Table 5-9. Values of K_1 for 30 percent Equivalent K_2CO_3, Two-Thirds Converted to $KHCO_3$ and KHS		
Equivalent H_2S Concentration, Grains/gallon	K_1 110°C	K_1 130°C
100	0.12	0.39
200	0.12	0.41
400	0.09	0.39
1,000	0.09	0.37

hydrogen sulfide and carbon dioxide (58). The equipment used was arranged for single stage split-stream, or two-stage split-stream operation. The absorber consisted of a 6-in. diameter, schedule 80 pipe and the regenerator of an 8-in. diameter, schedule 40 pipe. An additional regenerator consisting of a 10-ft packed section of 6-in. diameter pipe was used in the two-stage operation. The absorber contained two sections of packing: a lower section packed to a height of 24 ft 8 in. with ½-in. porcelain raschig rings and an upper section packed to a height of 3 ft 10 in. with ¼-in. porcelain raschig rings. This upper section was later increased in length to 10 ft. The split stream was fed to the top of the upper section while the main stream entered the absorber above the lower packed section. The main regenerator was packed to a height of 25 ft with ½-in. raschig rings.

Commercial-plant operating data which have been reported were obtained in a relatively small unit, treating natural gas containing about 7½ percent CO_2. The plant was designed to remove 50 percent of the carbon dioxide from 8.5 MMscf/day of feed gas; it employs perforated trays in both absorber and stripping columns.

Absorption. Absorption coefficients are presented for pilot-plant runs with the 4-in. absorber (42), and, although a complete variable study was not made, certain trends are apparent. Fortunately, solution and gas concentrations, temperatures, and liquid/gas ratios are in the range typically encountered in commercial-column design so that the data are quite useful.

All of the overall gas coefficients, tabulated by Benson et al. (42) for runs in the 4-in. column with hot 40 percent K_2CO_3, are presented in Figure 5-34 as $K_Ga/L^{2/3}$ versus partial pressure of CO_2. The plot is intended to show the effect of CO_2 partial pressure on the absorption coefficient; $L^{2/3}$ is introduced with K_Ga to minimize the spread of points due to different liquid-flow rates. As can be seen, the points appear to fall on two straight lines on the semi-logarithmic coordinates. The first portion of the curve, which extends from 0.0 to 0.7 atm CO_2 partial pressure, has a slope which indicates $K_Ga/L^{2/3}$ to be proportional to e^{-26p}, while the second portion of the curve, which extends from a CO_2 partial pressure of 0.7 atm to the maximum value tested (5 atm), has a much smaller negative slope. For comparison purposes, a curve of data from Shneerson and Leibush (59), on the absorption of CO_2 in 5 M monoethanolamine in a very small packed column, is also included (see Chapter 2). Over the partial-pressure range covered (0.0 to 0.5 atm), the points are very well correlated by a line plotting the equation

$$K_Ga \propto e^{-3.5p} \qquad (5\text{-}17)$$

which is surprisingly close to the relationship for absorption in hot potassium carbonate solution over the same range.

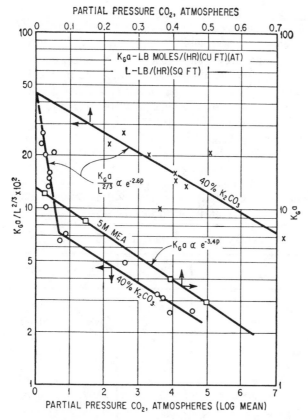

Figure 5-34. Effect of CO_2 partial pressure on absorption coefficients for hot potassium carbonate and monoethanolamine solutions. *Potassium carbonate data from Benson, Field, and Jimeson (42), for 4-in. diameter column packed with 9 ft of ½-in. ceramic rings; monoethanolamine data from Shneerson and Leibush (59), for 1-in. column packed with 5- to 6-mm glass rings.*

The pronounced effect of the partial pressure of CO_2 on K_Ga is the result of the chemical reaction in the liquid phase and illustrates the difficulty of using K_Ga to correlate absorption with chemical reaction. The occurrence of a reaction causes an increase in the liquid film coefficient over that which would be observed with physical absorption alone. The degree of liquid film coefficient enhancement is a complex function of concentrations, reaction rates, and diffusivities in the liquid phase. Discussions of enhancement factors for various generalized types of chemical reactions are given in several texts (60, 82, 83, 84).

A proposed methodology for the design of isothermal packed towers covering the specific case of CO_2 absorption in a hot aqueous solution of potassium carbonate is presented by Joshi et al. (85). They compare the pilot plant data provided by Benson et al. (42) with predicted values, and conclude that equations based on a bimolecular irreversible reaction provide a satisfactory approximation of the rate enhancement due to chemical reaction in this case.

Typical results from the pilot-plant test reported by Field et al. (58) for simultaneous absorption of hydrogen sulfide and carbon dioxide are shown in Table 5-10. The absorption pressure was 300 psig and the regeneration pressure slightly above atmospheric. No attempt was made to remove hydrogen sulfide selectively.

Table 5-10. Simultaneous Removal of Carbon Dioxide and Hydrogen Sulfide				
(Absorption Pressure—300 psig)				
Run:	115A	126A	131C	136A
Type of Flow:	Split Stream	2-Stage	Single Stream	2-Stage
Feed Gas: Rate, ft³/Hr *	1,262	1,214	1,185	583
CO_2, vol. %	11.5	11.35	11.0	11.9
H_2S, vol. %	2.65	0.51	0.88	10.0
Treated Gas:				
CO_2, vol. %	0.6	0.3	0.8	0.07
H_2S, vol. %	0.111	0.002	0.035	0.056
Solution:				
Rate, gal/hr.				
Main stream	34	33	42	40.5
Split stream	17	9	—	12
K_2CO_3, %	34	38	35	20
Temp., °C				
Main stream	114	109	109	102
Split stream	95	91	—	74
Capacity ft³/gal				
CO_2	2.6	3.2	2.9	1.3
H_2S	0.565	0.152	0.24	1.15
Regeneration Efficiency:				
ft³ acid gas/lb stream	6.8	5.7	4.7	5.0
Percent Removal:				
CO_2	94.6	97.4	93.7	99.6
H_2S	95.5	99.7	96.4	99.6

Source: Data of Field et al. (58).
*At O°C and 760 mm Hg, dry

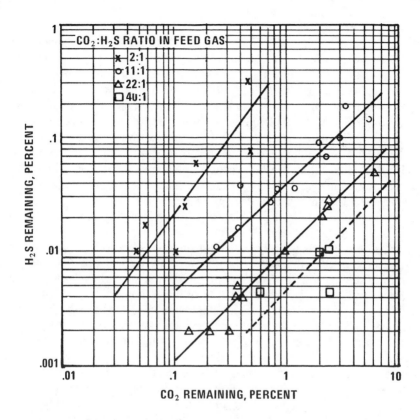

Figure 5-35. Relationship of H₂S and CO₂ in treated gas (58).

The relationship of the degree of hydrogen sulfide removal and that of carbon dioxide removal, based on pilot-plant test results with hydrogen sulfide concentrations in the feed gas ranging from 0.2 to 5 percent, is shown in Figure 5-35. Again, no attempt was made to remove hydrogen sulfide selectively (58).

The pilot-plant tests indicate that when properly designed, the hot potassium carbonate process should be suitable for the production of pipeline specification gas.

Since hydrogen sulfide is absorbed much more rapidly in potassium carbonate solutions than carbon dioxide, it would be expected that the process could be made at least partially selective for hydrogen sulfide. Lowering the absorption temperature would decrease the rate of carbon dioxide absorption without effect on the hydrogen sulfide absorption rate, thus resulting in increased selective. It is claimed that selective hydrogen sulfide removal can be obtained by special design (44). Rib et al. (86) report on the operation of a two-stage Benfield system used in a demonstration plant for selectively removing sulfur compounds from coal gas. The total acid-gas content of the gas varied from 6.4 to 17.1 percent CO_2, 2,600 to 3,200 ppmv H_2S and 106 to 216 ppmv COS. In order to obtain high selectivity the unit was operated at 85°C (185°F) and the effects of lean and semi-lean solution flow, and reboiler steam rates were studied for different feed gas compositions. The results of this work showed that under the operating conditions investigated, over 90 percent of the H_2S and about 40 percent of the COS could be removed, with about 70 percent of the CO_2 retained in the product gas.

Contaminants present in gas streams in relatively small concentrations such as carbonyl sulfide, carbon disulfide, mercaptans, thiophene, hydrogen cyanide, ammonia, and sulfur dioxide are removed by hot potassium carbonate solution to various degrees. Parrish and Neilson (79) report laboratory and commercial scale data obtained in purifying natural gases and gas from a pressurized Lurgi gassifier.

Carbonyl sulfide is hydrolyzed almost quantitatively to hydrogen sulfide and carbon dioxide and better than 99 percent removal can be achieved. Carbon disulfide is hydrolyzed in two steps, first to carbonyl sulfide and hydrogen sulfide, followed by hydrolysis of the carbonyl sulfide to hydrogen sulfide and carbon dioxide. The two-step reaction is slower than the simple hydrolysis of carbonyl sulfide with the result that only about 75 to 85 percent carbon disulfide removal is obtained under normal operating conditions.

Mercaptans which are slightly acidic react with the hot potassium carbonate solutions forming mercaptides. Since the acidity of mercaptans decreases with their molecular weight, less complete removal of higher molecular weight mercaptans is obtained than of the lower mercaptans, i.e. methyl and ethyl mercaptan. However, the lower molecular weight mercaptans normally constitute the bulk of those present in gas streams. Removal efficiencies of up to 92 percent have been reported with a Hi Pure flow pattern. Thiophene does not react chemically with potassium carbonate, and any removal of this compound could only be attributed to physical solubility in the carbonate solution.

Ammonia is readily absorbed by potassium carbonate solutions. Sulfur dioxide and hydrogen cyanide, both acidic compounds, are also readily absorbed by hot carbonate solutions. After absorption, these compounds react further forming a variety of compounds such as sulfates, thiosulfates, thiocyanates, polysulfides, and elemental sulfur which accumulate in the solution.

Typical operating data, which have been reported for a commercial absorber, are presented in Table 5-11. Contact in this absorber is by means of perforated trays. Unfortunately, the number of trays is not specified so that tray efficiencies cannot be calculated from the data. However, since this has been shown to be a liquid-film controlled system, tray efficiencies would be expected to be quite low. Data reported from other installations have indicated tray efficiencies (Murphree vapor) to be on the order of 5 percent for absorption of CO_2 in hot potassium carbonate solutions.

Table 5-11. Operating Data for Commercial Perforated-Tray Absorber (56)

Absorber Data	1	2	3	4	5
Feed-gas rate, MMSCF/day	1.13	2.60	4.32	5.80	8.56
Lean-solution rate, gpm	95.0	95.9	95.9	84.2	117.5
Gas composition, per cent CO_2:					
Inlet	7.2	7.8	7.7	7.6	7.7
Outlet	0.4	2.0	3.0	2.7	4.0
Lean-solution characteristics:					
Specific gravity	1.23	1.27	1.26	1.205	1.275
Equivalent K_2CO_3 concentration, wt. per cent by		28.9	28.0	30.0	27.7
Free K_2CO_3 concentration, per cent by wt		18.5	16.8		16.7
$KHCO_3$ concentration (as equivalents of K_2CO_3), per cent by wt		10.4	11.2		11.0
CO_2 concentration in solution, cu ft/gal		3.0	3.3		3.2
Rich-solution characteristics:					
$KHCO_3$ concentration (as equivalent K_2CO_3), per cent by wt		13.8	16.7		17.5
CO_2 concentration in solution, cu ft/gal		4.0	4.8		5.1
Per cent of K_2CO_3 reacted		47.6	59.5		63.1
Per cent of inlet CO_2 absorbed	95.0	68.3	62.7	68.6	48.5

Regeneration. As shown in Figure 5-21, the CO_2 concentration in the solution decreases rapidly with decreased partial pressure of CO_2 over it. Because of this, solution regeneration can be carried out most effectively at a very low pressure. Benson et al. (43) employed 2 to 10 psig and found the 2-lb regeneration pressure more favorable. Because of the strong dependence of CO_2 concentration in the solution on partial pressure, an appreciable portion of the regeneration is effected when the pressure is reduced over the rich solution leaving the absorber. From ⅓ to ⅔ of the absorbed carbon dioxide is released during this flashing operation.

The heat of reaction of carbon dioxide with the hot potassium carbonate solution is relatively small (about 32 Btu/cu ft of CO_2) (42), and, theoretically, this heat would not need to be supplied in the regeneration step if only a simple flash were used and cooling of the solution were allowed to occur. Since a simple flash does not give adequate regeneration, it is necessary to provide stripping vapor in the regeneration column, and this is done by evaporating water in the reboiler. Essentially all of the heat added goes to the vaporization of water, and this heat must be removed from the system in the overhead condenser of the regenerator. A small amount of additional heat must, of course, be added to make up for heat losses from vessels and lines and to provide latent heat of vaporization for any water evaporated into the gas stream in the absorber.

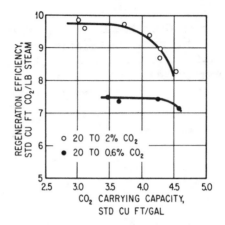

Figure 5-36. Effect of product gas purity and CO₂-carrying capacity of solution on regeneration efficiency (split stream at 300 psig and 12,000 SCF/hr). *Data of Benson, Field, and Haynes* (43)

The effect on regeneration efficiency of altering the operation to give increased solution-carrying capacity is illustrated   (Figure 5-36) for two product-gas purity conditions (43). This figure is based on operations of the U.S. Bureau of Mines pilot plant, in which the solution-capacity changes were effected by varying the degree of stripping of the lean solution. As can be seen, attempting to increase the solution-carrying capacity to more than about 4 cu ft CO_2/gal resulted in a decreased efficiency for both product-purity conditions. Comparison of the two curves shows that operating to produce a 0.6 percent CO_2 product gas instead of a 2 percent CO_2 product results in a steam-consumption increase of about 30 percent. The tests illustrated were made with split-stream flow at an absorption pressure of 300 psig and with a feed gas containing 20 percent CO_2 and 80 percent nitrogen.

Since operation of the hot potassium carbonate process is based primarily on the difference in the solubility of CO_2 at high partial pressure (absorber conditions) and that at low partial pressure (stripper conditions), it would be expected to be more efficient with increased CO_2 partial pressure in the feed gas. The magnitude of this effect for the pilot-plant operations is shown in Figure 5-37, which includes curves for two outlet-gas compositions.

Guidelines for the selection of the most advantageous flow system and equipment, with respect to feed-gas composition, treated gas purity required, and heat consumption, have been presented by Benson and Field (48). A convenient chart, developed by these authors, is shown in Figure 5-38. A sample of a design calculation for a hot-potassium-carbonate plant has been given by Maddox and Burns (61).

Operating Problems. In the initial tests made by the U.S. Bureau of Mines, severe corrosion of carbon steel was encountered, especially where the conversion to bicarbonate was high or where carbon dioxide and steam were released by pressure reduction (62). Potassium dichromate was found to be an effective corrosion inhibitor, and 0.2 percent was used in the solution for subsequent CO_2-absorption tests. This material is recommedned as a corrosion inhibitor only for plants handling sulfur-free gas. Concentrations of 1,000 to 3,000 ppm have been found satisfactory for commercial installations.

Since H_2S and, possibly, other sulfur compounds rapidly reduce the chromate ion, the use of this inhibitor is uneconomical in the presence of appreciable quantities of such impurities. In addition to increasing operating costs by destroying the inhibitor, the reduction reaction results in the formation of insoluble precipitates which cause erosion of equipment and other operating difficulties. Fortunately, H_2S itself appears to inhibit CO_2 corrosion somewhat so that carbon steel can still be used for most of the equipment in plants handling gas containing both CO_2 and H_2S. Some success has also been noted with other corrosion inhibitors such as the organic film-forming types which are not destroyed by H_2S.

Stainless steels (types 304 and 316) are recommended for the portions of the plant subject to particularly corrosive conditions, e.g., the reboiler-tube bundle, and for critical items of equipment, e.g., pressure let-down valves and

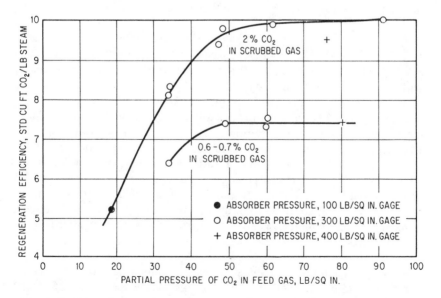

Figure 5-37. Effect of partial pressure of CO_2 in feed gas on regeneration efficiency (split-stream operation). *Data of Benson, Field, and Haynes* (43)

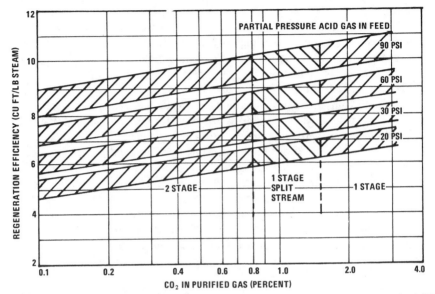

Figure 5-38. Guide for selecting process scheme for hot potassium carbonate plant (48).

pumps. In the Bureau of Mines pilot plant, both absorber and stripper were packed with porcelain rings, and this material is apparently satisfactory, although laboratory tests (7) indicate that some dissolution of porcelain rings may occur.

As stated above, a 40 percent solution has been found to be too concentrated for commercial use in some cases because of its tendence to form a slurry of bicarbonate crystals if the solution is cooled at any point in the circuit. In the presence of bicarbonate slurry, carbon steel cases and impellers of pumps were found to last only a few hours. Reduction of the solution concentration to below 30 percent eliminates this severe erosion; however, stainless-steel (type 316) impellers, case rings, and throat bushings are recommended in pumps handling the K_2CO_3 solution as a further safety measure (56).

Column flooding and dumping have also been experienced in hot potassium carbonate process plants; however, it appears that this problem can be solved by the use of additives such as silicone foam inhibitors. According to Armstrong and Palo (63), the additive greatly increases the column efficiency of both bubble-cap and perforated-tray absorbers by improving the bubbling action.

Activated Solutions. As pointed out earlier in this chapter (3, 4) addition of activators or promoters to the solution which accelerate the rate of absorption of carbon dioxide should result in appreciable improvement of the process with regard to investment and operating costs, as well as better product

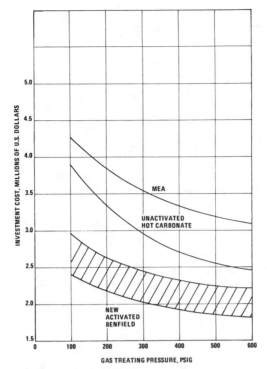

Figure 5-39. Comparison of capital investment for CO_2 removal by MEA, hot carbonate, and new activated Benfield process. Basis: gas flow — 5 MMSCF/H.; % CO_2 in — 20; % CO_2 out — 0.1. *Data of Benfield Corporation* (44)

quality. The "New Activated Benfield Process," utilizing an undisclosed activator, is claimed to be capable of producing high-purity treated gas (less than 500 ppm of CO_2) with capital and operating costs substantially below those of the process using unactivated solutions (see Figure 5-39) (44). A large number of plants using this process is presently in operation. The activator is reportedly inexpensive and stable and introduces no operating problems.

As discussed in Chapter 2, sterically hindered amines are reported to be very effective promoters. Sartori and Savage (87) compared absorption rates and vapor-liquid equilibria of CO_2 in unpromoted hot potassium carbonate solutions with solutions promoted with diethanolamine (DEA) and solutions promoted with sterically hindered amines. Both DEA and sterically hindered amines were found to be very effective in increasing the rate of CO_2 absorption. However, the equilibrium loading obtained with the DEA promoted solution was not significantly higher than that of the unpromoted solution for a given partial pressure of CO_2. The sterically hindered amines, on the other hand, showed appreciably higher equilibrium loadings than the DEA promoted solutions, an effect which became more pronounced as CO_2 partial pressure was increased.

Economics. Cost data on the hot potassium carbonate process published since its first disclosure (64,65,66,67) are of limited value because of substantial changes in engineering and construction costs during the past years and, more importantly, because of the rapid development of technology. While the process was originally thought to be most economical for bulk removal of carbon dioxide and, in some cases, hydrogen sulfide, with a necessary final purification step, using, for example, an ethanolamine solution, recent developments have shown that product gas of high purity can be obtained economically with hot carbonate solutions alone. The economics of the process are primarily determined by the overall heat utilization in each specific case, and generalizations are not particularly useful.

Data for capital investment and operating costs supplied by the Benfield Corporation (44) are shown in Figures 5-39 and 5-40. These data apply primarily to ammonia synthesis gas and crude hydrogen. The cost data for the new activated Benfield process are shown as a band rather than a line because this process is claimed to be quite flexible, permitting substantial savings under the proper circumstances (68).

The trend of the curves for MEA and the unactivated solutions indicates that the economic advantage of the hot potassium carbonate process diminishes with decreasing pressure and, by inference, with decreasing carbon dioxide content (partial pressure) of the feed gas.

The Catacarb Process

The Catacarb process, which was disclosed by Eickmeyer (69, 70) and is licensed by Eickmeyer and Associates of Prairie Village, Kansas, is a modification of the hot potassium carbonate process. The solution used in the Catacarb process contains a potassium salt and an undisclosed additive which catalyzes absorption and desorption of acid gases, especially of carbon dioxide. The additive, which also incorporates a corrosion inhibiting component, is claimed to be nontoxic and nonpoisonous to reformer and methanation catalyst (71). Although primarily used for carbon dioxide removal from crude hydrogen and ammonia synthesis gas, the process is also suitable for hydrogen sulfide and carbon dioxide removal from natural gases, yielding a product gas containing less than 1/4 grain of H_2S per 100 cu ft (72). If required, the carbon dioxide content of hydrogen or ammonia synthesis gas can be lowered to 300 ppm (71). In March, 1971, 36 Catacarb plants, treating more than 3,000 MMscf per day of gas, were reported to be in operation around the world.

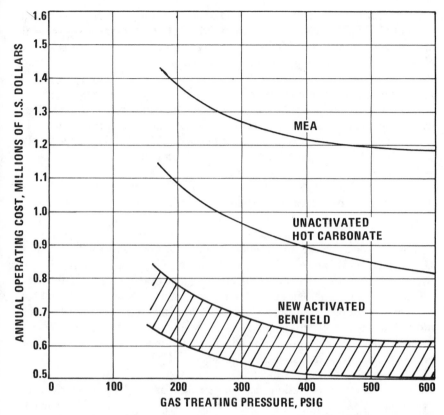

Figure 5-40. Comparison of operating costs for CO_2 removal by MEA, hot carbonate, and new activated Benfield process. Basis: same as on Figure 5-39. Heat — $0.50/MMBTU; power — $0.008/KWH; cooling water — $0.015/thousand gallons; 330 operating days/year. *Data of Benfield Corporation* (44)

The flow scheme of a Catacarb plant is identical to that used for the hot potassium carbonate process. When high-purity product gas is required, a split-stream cycle, with cooling of the stream flowing to the top of the absorber, and two- (or three-) stage regeneration are used. Carbon steel is a satisfactory material of construction for the entire plant.

The principal advantages of the Catacarb process are higher treated gas purity than obtainable with uncatalyzed hot potassium carbonate solutions, appreciably lower steam consumption than required for hot potassium carbonate or monoethanolamine, and smaller equipment, resulting in 20 to 30 percent savings in capital investment when compared to the conventional monoethanolamine process (71).

Comparative operating data for two plants using solutions with and without Catacarb catalyst are shown in Table 5-12 (70). Plant A operated at an absorption pressure of 360 psig with single stage, uncooled absorption. In plant B split-stream absorption at 300 psig with cooling of the smaller liquid stream was used.

These data show that addition of the catalyst resulted in increased capacity and appreciably lower steam consumption for the same carbon dioxide concentration in the treated gas in runs 1 and 2 in plant A. In runs 3 and 4 in plant A and runs 5 and 6 in plant B, use of the catalyzed solution appreciably improved the treated gas purity and the heat economy.

The Giammarco-Vetrocoke Process

The use of sodium and potassium arsenite solutions for the absorption of carbon dioxide at elevated temperatures and pressures has been disclosed by Giammarco (73) and is being licensed by Giammarco-Vetrocoke of Marghera, Italy. A precursor of this process, also developed by Giammarco, utilized relatively small amounts of an organic activator in conventional hot carbonate solutions; however, this process was soon replaced by the much more efficient process using arsenites. The Giammarco-Vetrocoke process for hydrogen sulfide removal, which is based on the use of aqueous solutions of alkali arsenite and arsenate, is described in Chapter 9.

Addition of essentially stoichiometric proportions of arsenic trioxide to aqueous sodium or potassium carbonate solutions results in a marked increase in the rate of absorption and desorption of carbon dioxide, as compared with conventional carbonate solutions. Figure 5-41 illustrates this phenomenon by comparing, qualitatively, the rate of absorption of carbon dioxide at 1 atm partial pressure and room temperature in 40 percent

Table 5-12. Comparison of Hot Potassium Carbonate and Catacarb Process Performance						
Run no.	1	2	3	4	5	6
Plant	A	A	A	A	B	B
Catalyst content, %	0.0	6.8	0.0	7.0	0.0	5.0
Gas rate, MSCF/Hr	405.0	509.0	464.0	472.0	1,193.0	1,193.0
Inlet gas CO_2, %	23.4	22.8	22.9	23.0	15.2	14.8
Outlet gas CO_2, %	1.0	1.1	2.3	0.6	1.5	0.6
Liquid circulation, gpm	860.0	982.0	810.0	815.0	1,360.0	1,220.0
Steam, lb/MCF CO_2	169.0	121.0	169.0	149.0	218.0	167.0

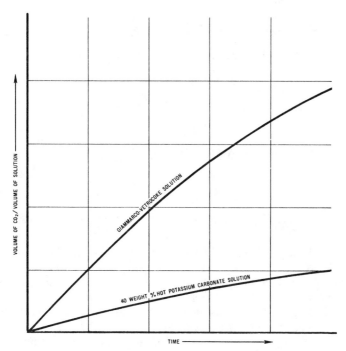

Figure 5-41. Comparative rate of absorption of CO_2 in 40 percent potassium carbonate and typical Giammarco-Vetrocoke solution (74).

potassium carbonate and in a typical solution used in the Giammarco-Vetrocoke process (74). The effects of the more rapid absorption and desorption are appreciable savings in regeneration heat, reduction in equipment size and production of treated gas of higher purity than is possible with ordinary hot carbonate solutions.

The process, which is suitable for treating synthesis gas, hydrogen, as well as natural gas is in wide use world-wide, with several hundred plants in operation. However, it has found practically no acceptance in the United States, probably because of the toxicity of the arsenite solution, which may require special precautions in handling. One large installation, treating high pressure natural gas, has been operating successfully in the United States (75, 76).

Basic Chemistry

The chemical reactions occurring during absorption and desorption of carbon dioxide can be symbolized by the following equations:

Absorption

$$6 CO_2 + 2K_3AsO_3 + 3H_2O \rightleftharpoons 6KHCO_3 + As_2O_3 \qquad (5\text{-}18)$$

$$CO_2 + K_2CO_3 + H_2O \rightleftharpoons 2KHCO_3 \qquad (5\text{-}19)$$

Desorption

$$6KHCO_3 + As_2O_3 \rightleftharpoons 2K_3AsO_3 + 6CO_2 + H_2O \qquad (5\text{-}20)$$

$$2KHCO_3 \rightleftharpoons K_2CO_3 + CO_2 + H_2O \qquad (5\text{-}21)$$

It is claimed that the addition of arsenic trioxide not only increases the rate of carbon dioxide absorption but also the carrying capacity of the solution. Jenett (76) reports that an arsenite solution of 30 percent equivalent potassium carbonate content, regenerated with 0.45 lb of steam per gallon has about 25 percent more carrying capacity for carbon dioxide than a potassium carbonate solution of the same equivalent concentration. The carbon dioxide content of the gas treated with the arsenite solution is substantially lower than that of the gas treated with the carbonate solution. These observations indicate that arsenic trioxide increases the rate of hydration of carbon dioxide to carbonic acid in the absorption step and also the shift of pH toward the acid side in the regeneration step, resulting in more complete expulsion of the absorbed carbon dioxide. The net effect of these two phenomena is higher solution capacity and lower carbon dioxide content of the purified gas.

Process Description

Typical flow diagrams for the Giammarco-Vetrocoke process are shown in Figures 5-42 and 5-43. The flow scheme shown in Figure 5-42 illustrates the version of the process in which steam is used for solution regeneration. This arrangement is used primarily for carbon dioxide removal from synthesis gases or crude hydrogen when the gas to be treated is hot and inexpensive steam is available. The flow is quite conventional; the hot gas is washed countercurrently with the solution in the absorber which usually is a packed column, although trays can also be used. The rich solution flows first to a flash drum where a portion of the carbon dioxide is removed by pressure reduction. The partially stripped solution is then heated by exchange with the lean solution before entering the regenerator where it is stripped of its remaining carbon dioxide by steam rising in the column. The regenerator is also a packed column provided with a reboiler for supply of heat. The carbon dioxide leaving the top of the regenerator is cooled; the condensed water is

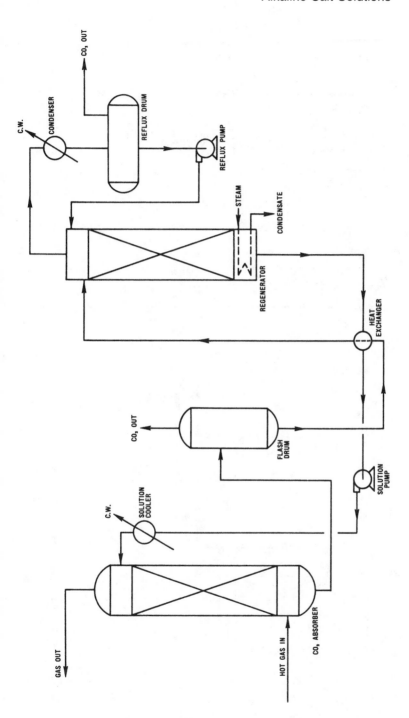

Figure 5-42. Typical flow diagram of Giammarco–Vetrocoke CO₂ removal process with steam regeneration.

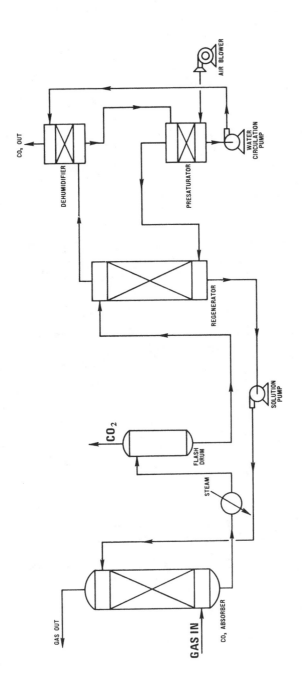

Figure 5-43. Typical flow diagram of Giammarco-Vetrocoke CO₂ removal process with air regeneration.

collected in a reflux drum and recycled to the regenerator while the cold carbon dioxide is vented to the atmosphere.

The regenerated solution leaving the bottom of the regenerator is first cooled by heat exchange against the partially stripped solution and then further cooled in a cooler before being returned to the absorber. The second cooling step is optional, depending on the gas purity required.

Air regeneration of the solution is illustrated in Figure 5-43. The rich solution leaving the absorber is heated by steam before entering the flash drum, where part of the carbon dioxide is flashed off. The partially stripped solution then flows to the top of the regenerator where final stripping is obtained by countercurrent contact with a stream of preheated and water saturated air. The mixture of carbon dioxide and air leaving the top of the regenerator is washed with cold water in the dehumidifier in order to recover heat and condense water contained in the gas stream. The cooled mixture of air and carbon dioxide is vented to the atmosphere, while the water flows from the dehumidifier to the gas presaturator. In this manner maximum heat economy is achieved.

Process Operation

Operating data for a large plant treating natural gas at high pressure have been reported by Jenett (76). The feed gas to the unit contained a small amount of hydrogen sulfide which was removed by an arsenate-arsenite solution (see Chapter 9) prior to carbon dioxide removal. The potassium arsenite solution was regenerated by air stripping. Operating data for this plant are shown in Table 5-13

Table 5-13. Operating Data for Giammarco-Vetrocoke Plant for Natural-Gas Treating (76)	
Number of absorbers	3
Number of regenerators	6
Inlet gas, MMSCF/day	180
CO_2, %	28
H_2S, gr/100 SCF	2.5
Outlet gas, MMSCF/day	130
CO_2, %	2.0
H_2S, gr/100 SCF	<0.25
Solution circulation, gpm	8,000
Steam, lb/hr	65,000
Power, kw	1,400
Fuel, MMSCF/day	3.8
Absorber Pressure, psig	1,000

Note: CO_2 removal was followed by dehydration with triethylene glycol.

The entire plant is constructed from carbon steel, and no evidence of corrosion has been reported. Also, no discharge of toxic material to the atmosphere with the regeneration air has been observed. Economic comparisons of the Giammarco-Vetrocoke process with the hot potassium carbonate and ethanolamine processes have been presented by Riesenfeld and Mullowney (74) and by Jenett (76). These comparisons show a definite economic advantage of the Giammarco-Vetrocoke process over the other two processes with respect to both capital investment and operating costs. However, the comparisons do not take into account the more recently developed activated potassium carbonate solutions and design improvements in the ethanolamine process and, therefore, are of limited value.

References

1. Danckwertz, P.V., and Sharma, M.M. 1966. *Chem. Eng.* (Oct.):CE 244-280.
2. Garner, F.H.; Long, R.; and Pennell, A. 1958. *J. Appl. Chem.* (May 8):325-336.
3. Riou, P., and Cartier, P. 1927. *Compt. Rend.* 184:325-326.
4. Riou, P., and Cartier, P. *Compt. Rend.* 186:1727-1729.
5. Quinn, E.L., and Jones, C.L. 1936. *Carbon Dioxide.* American Chemical Society Monograph Series. New York: Reinhold Publishing Corporation.
6. Reich, G.T. 1931. *Chem. & Met. Eng.* 38:136-145.
7. The Fluor Corporation, Ltd. Unpublished data.
8. Harte, C.R.; Baker, E.M.; and Purcell, H.H. 1933. *Ind. Eng. Chem.* 25:528.
9. Seidell, A. 1940. *Solubilities of Inorganic and Metal Organic Compounds,* 3rd ed. Princeton, N.J.: D. Van Nostrand Company, Inc.
10. Sherwood, T.K., and Pigford, R.L. 1952. *Absorption and Extraction,* 2nd ed. New York: McGraw-Hill Book Company, Inc., pp. 358-364.
11. Furnas, C.C., and Bellinger, F. 1938. *Trans. Am. Inst. Chem. Engrs.* 34:251.
12. Tepe, J.B., and Dodge, B.F. 1943. *Trans. Am. Inst. Chem. Engrs.* 39:255.
13. Spector, N.A., and Dodge, B.F. 1946. *Trans. Am. Inst. Chem. Engrs.* 42:827.
14. Sperr, F.W., Jr. 1921. *Am. Gas Assoc., Proc.* 3:282-364.
15. Gollmar, H.A. 1945. *Chemistry of Coal Utilization,* vol. 2. Edited by H.H. Lowry. New York: John Wiley & Sons, Inc., p. 984.
16. Sperr, F.W., Jr., and Hall, R.E. 1925. U.S. Patent 1,533,733.
17. Powell, A.R. 1941. U.S. Patent 2,242,323.
18. Gollmar, H.A. 1945. *Chemistry of Coal Utilization,* vol.2. Edited by H.H. Lowry. New York: John Wiley & Sons, Inc., p. 985.
19. Gollmar, H.A. 1949. U.S. Patent 2,464,805.
20. Gollmar, H.A., and Hodge, W.W. 1956. *Proc. Ind. Waste Conf., 10th Conf.* 89(Mar.):35-48.
21. Kurtz, J.K. 1955. *Problems and Control of Air Pollution.* New York: Reinhold Publishing Corporation, p. 215.
22. Litvinenko, M.S. 1952. *J. Appl. Chem. (U.S.S.R.)* 25:775-794.
23. Kastens, M.L., and Barraclough, R. 1951. *Ind. Eng. Chem.* 43(Sept.):1882-1892.
24. Smith, G.A. 1953. *Gas World* (Coking Section) (Oct. 3):65-69.
25. Koppers Company, Inc. 1951. *Koppers Vacuum Carbonate Process for Sulfur Recovery.*

26. Rosenstein, L., and Kramer, G.A. 1934. U.S. Patent 1,945,163.
27. Rosebaugh, T.W. 1938. *Proc. Am. Petrol. Inst.* (Sect. 3) 19M(May):47-52.
28. Wainwright, H.W., Egleson, G.C.; Brock, C.M.; Fisher, J.; and Sands, A.E., 1953. *Ind. Eng. Chem.* 45(June):1378.
29. Liedholm, G.E. (Shell Development Company). 1958. Personal communication, (Feb. 17).
30. Zahnstecher, L.W. 1950. *Heat Eng.* 25(Apr.):61.
31. Shreve, R.N. 1945. *The Chemical Process Industries,* 1st ed. New York: McGraw-Hill Book Company, Inc., p. 106.
32. Powell, A.R. 1938. *The Science of Petroleum,* vol. 3. Edited by A.E. Dunstan. New York: Oxford University Press, p. 1804.
33. Bähr, H., and Mengdehl, H. 1935. U.S. Patent 1,990,217.
34. Reed, R.M., and Updegraff, N.C. 1950. *Ind. Eng. Chem.* 42(Nov.):2269.
35. Bähr, H. 1938. *Proc. Am. Petrol. Inst.* (Sect. 3). 19M(May):37.
36. Pasternak, R. 1962. *Brennstoff-Chem.* 3(43):65-67.
37. Leuhddemann, R.; Noddes, G.; and Schwarz, H.G. 1959. *Oil Gas J.* 57(Aug. 3):100-104.
38. Anon. 1957. *Sulphur* (special number):21-23.
39. Pasternak, R. 1963. *Brennstoff-Chem.* 4(44):105-110.
40. Farbenindustrie, I.G.; G.m.b.H. 1935. French Patent 787,782.
41. Peck, E.B. 1938. *Proc. Am. Petrol. Inst.* (Sect. 3). 19M:51.
42. Benson, H.E.; Field, J.H.; and Jimeson, R.M. 1954. *Chem. Eng. Progr.* 50:356.
43. Benson, H.E.; Field, J.H; and Haynes, W.P. 1956. *Chem. Eng. Progr.* 52:433.
44. The Benfield Corporation. 1971. *The Way to Low Cost Scrubbing of CO_2 and H_2S from Industrial Gases.*
45. Behrens, E.A., and Behrens, J. 1904. German Patent 162,655.
46. Williamson, R.V., and Mathews, J.H. 1924. *Ind. Eng. Chem.* 16:1157-1161.
47. Benson, H.E., and Field, J.H. 1955. British Patent 725,000.
48. Benson, H.E., and Field, J.H. 1960 *Petrol. Refiner* 39(Apr.):127-132.
49. Tosh, J.S.; Field, J.F.; Benson, H.E.; and Haynes, W.P. 1959. *U.S. Bur. Mines, Rept. Invest., No. 5484.*
50. Tosh, J.S.; Field, J.H.; Benson, H.E.; and Anderson, R.B. 1960. *U.S. Bur. Mines, Rept. Invest., No. 5622.*
51. Solvay Process Division, Allied Chemical Corporation. 1961. *The Hot Potassium Carbonate Process for Acid Gas Absorption.* Tech./Ser. Rept. No. 6.61.
52. Bocard, J.P., and Mayland, B.J. 1962. *Hydro. Process.* 41(Apr.):128-132.
53. Perry, J.H. (ed.). 1950. *Chemical Engineers' Handbook,* 3rd ed. New York: McGraw-Hill Book Company, Inc., p. 198.
54. Seidell, A. 1940. *Solubilities of Inorganic and Metal Organic Compounds,* vol. 1. Princeton, N.J.: Van Nostrand Company, Inc., p. 727.
55. Hill, A.E., and Hill, D.G. 1927. *J. Am. Chem. Soc.* 49(Apr. 7):968.
56. Buck, B.O., and Leitch, A.R.S. 1958. *Oil Gas J.* 56(Sept. 22):99-103.
57. Mapstone, G.E. 1966. *Hydro. Process.* 45(Mar.):145-148.
58. Field, J.H.; Johnson, E.G.; Benson, H.E.; and Tosh, J.S. 1960. *U.S. Bur. Mines, Rep. Invest., No. 5660.*
59. Shneerson, A.L., and Leibush, A.G. 1946. *J. Appl. Chem. (U.S.S.R.)* 19(9):869-80.
60. Sherwood, T.K., and Pigford, R.L. 1952. *Absorption and Extraction.* 2nd ed. New York: McGraw-Hill Book Company, Inc., p. 337.
61. Maddox, R.N., and Burns, M.D. 1967. *Oil Gas J.* 65(Nov. 3):122-131.
62. Bienstock, D., and Field, J.H. 1961. *Corrosion* 17(July):337t-339t.
63. Palo, R.O., and Armstrong, J.B. 1958. *Petrol. Refiner* 37(Dec.):123-128.

64. Mullowney, J.F. 1957. *Petrol. Refiner* 36(Dec.):149.
65. Eickmeyer, A.G. 1958. *Chem. Eng.* 65(Aug.):113.
66. Katell, S., and Faber, J.H. 1960. *Petrol. Refiner* 39(Mar.):187-190.
67. Benson, H.E. 1961. *Petrol. Refiner* 40(Apr.):107-108.
68. The Benfield Corporation. 1971. Private communication.
69. Anon. 1961. *Chem. Week* (Jan. 28).
70. Eickmeyer, A.G. 1962. *Chem. Eng. Progr.* 58(Apr.):89-91.
71. Morse, R.J. 1968. *Oil Gas J.* 66(Apr. 22).
72. Eickmeyer, A.G. 1971. *Oil Gas J.* 69(Aug. 9):74-75.
73. Italian patents 5J8, 145, 564, 203, 545, 908, 563, 853, 535, 177, 563, 854.
74. Riesenfeld, F.C., and Mullowney, J.F. 1959. *Oil Gas J.* 57(May 11):86-91.
75. Anon. 1960. *Chem. Eng.* (Sept. 19):166-169.
76. Jenett, E. 1962. *Oil Gas J.* 60(Apr. 30):60-72.
77. Massey, M.Y., and Dunlap, R.W. 1975. *Journal of the Air Pollution Control Association* 25(10):1019-1027.
78. Benson, H.E., and Parrish, R.W. 1974. *Hydro. Process.* 53(4):81-82.
79. Parrish, R.W., and Neilson, H.B. 1974. "Synthesis Gas Purification Including Removal of Trace Contaminants by the BENFIELD Process," Paper presented at the 167th National Meeting of the American Chemical Society, Division of Ind. & Eng. Chem., Los Angeles, Calif. (March, 1974).
80. Clayman, M.A., and Clark, J.R. 1980. "Low Energy Natural Gas Purification Using Benfield Processes," paper presented at 1980 Gas Conditioning Conference, University of Oklahoma.
81. Baker, R.L., and McCrea, D. H. 1981. "The Benfield LOHEAT Process: An Improved HPC Absorption Process," paper presented at AIChE 1981 Spring National Meeting, Houston, Texas, April 5-9.
82. Astarita, G. 1967. *Mass Transfer with Chemical Reaction,* Elsevier, Amsterdam.
83. Danckwerts, P. V. 1970. *Gas-Liquid Reactions,* McGraw-Hill Book Company, New York.
84. Astarita, G., Savage, D.W., and Bisio, A. 1983. *Gas Treating with Chemical Solvents,* John Wiley & Sons, New York.
85. Joshi, S. V., Astarita, G., and Savage, D. W. 1981. *Transport with Chemical Reactions,* AIChE Symposium Series No. 202, Vol 77, p. 63.
86. Rib, D.M., Kimura, S.G., and Smith, D.P. 1982. "Performance of a Coal Gas

6

Water as an Absorbent for Gas Impurities

The principal advantage of water as an absorbent for gas impurities is, of course, its ready availability at a low cost; this factor alone is sufficient to make the use of water worth considering for the removal of gas impurities which are reasonably soluble in it. Water is particularly applicable to the treatment of large volumes of low-pressure exhaust gas for the prevention of air pollution because solvent losses are difficult to avoid in such installations. Organic solvents, in general, have sufficient vapor pressure to cause the occurrence of appreciable losses by vaporization into the purified gas stream. Practically any chemical absorbent other than water requires a tight system and, unless a salable reaction product is formed, a regenerative cycle. Water, on the other hand, can be used in simple scrubbing units with less concern over leakage and frequently on a once-through basis with the rich solution being discarded.

Water may also be applicable to the washing of high-pressure gases where the solubility of an impurity (such as CO_2), which is only sparingly soluble at low pressure, is brought up to an economically high level by the high partial pressure.

Although a detailed discussion of particulate removal from gases is beyond the scope of this text, it should be noted that water scrubbing is frequently used for this purpose. In many cases, a water wash is used to remove both particles and vapor phase impurities. Water contact stages are also frequently used to quench high-temperature gases and in some instances, the quench step serves four distinct purposes, i.e., temperature reduction, condensation of liquids from the vapor phase, absorption of soluble impurities, and removal of particulate material. The quenching of the product from coal gasification units with water is an example of such a multipurpose operation.

Table 6-1. Selection Guide for Wet Scrubbers (Based on Data of Hanf and MacDonald (47))

Type of Wet Scrubber	Gas Phase Impurities		Entrained Liquids		Particles			
	High Solubility	Low Solubility	Mists <10μ	Drops >10μ	Fumes <1μ	Dusts 1-5μ	Dusts over 5μ Low Loading	Dusts over 5μ High Loading
Countercurrent Packed Tower	E	E	F	E	NR	NR	E	NR
Cocurrent Packed Tower	G	F	F	E	NR	NR	E	NR
Crossflow Packed Scrubber	E	F	F	E	NR	NR	E	NR
Parallel Flow Washer	NR	NR	P	E	NR	NR	NR	NR
Wet Cyclone	F	NR	P	E	NR	NR	G	G
Venturi	F	NR	E	E	E	E	E	E
Spray Tower	F	NR	NR	G	NR	NR	E	F
Eductor Venturi (Jet)	F	NR	E	E	G	E	E	E

Code for collection efficiencies: E = 95-99%, G = 85-95%, F = 75-85%, P = 50-75%, NR = Not Recommended

Soluble components absorbed by the water in such processes include ammonia, hydrogen cyanide, phenol, and hydrochloric acid. Heavy hydrocarbons are condensed as a separate liquid phase, and particles of ash and unreacted coal are captured by the liquids.

A generalized selection guide for wet scrubbers covering both absorption and particulate collection functions is given in Table 6-1, which is based on a review of the subject by Hanf and MacDonald (47).

Gas-phase impurities which are absorbed commercially in water-scrubbing operations include ammonia, sulfur dioxide, carbon dioxide, hydrogen fluoride, silicon tetrafluoride, hydrogen chloride, and chlorine. The absorption of ammonia (and other nitrogen bases) from gases has little importance as a gas-purification operation except as it pertains to coke-oven gas and other gas streams in which hydrogen sulfide and carbon dioxide are also present. Since the processes developed for removing ammonia from such gases with water are closely interrelated with those for removing the acidic components, they are discussed together in Chapters 4 and 10. Sulfur dioxide absorption in water is the basis for a process used commercially for removing sulfur dioxide from power-plant flue gases (the Battersea process). However, in this application, an alkaline water is used (from the Thames River), and additional lime is added to maintain the alkalinity. The process is, therefore, described along with other SO_2 absorption systems in Chapter 7.

The impurities which are specifically considered in this chapter are carbon dioxide, hydrogen sulfide, hydrogen fluoride, silicon tetrafluoride, hydrogen chloride, and chlorine. All of these form acids when they are in aqueous solution, and the prevention of corrosion is a common problem. They are by no means equally corrosive, however, and differ also in the nature of the absorption processes. Carbon dioxide, hydrogen sulfide, and chlorine are relatively insoluble in water and their absorption rates are determined largely by the liquid-film resistance. Hydrogen fluoride, silicon tetrafluoride, and hydrochloric acid, on the other hand, are very soluble in water, and their absorption is generally found to be gas-film controlled.

When any of the above-mentioned impurities are absorbed from a gas stream, small amounts of the primary gaseous components are also absorbed. This effect is particularly noticeable in high-pressure operations and can materially affect the economics of the water-wash process for such cases. Water-solubility data for components which typically make up the major portion of gases to be purified by the water-wash process are presented in Table 6-2.

Carbon Dioxide Absorption in Water

The absorption of carbon dioxide in water has been of considerable industrial importance for the purification of certain high-pressure gas streams, particularly in ammonia synthesis. However, the process has generally been

Table 6-2. Solubility of Various Gases in Water*

Gas	H, at Temperature, °C					
	0	10	20	30	40	50
Carbon monoxide, CO	3.52×10^4	4.42×10^4	5.36×10^4	6.20×10^4	6.96×10^4	7.61×10^4
Hydrogen, H$_2$	5.79×10^4	6.36×10^4	6.83×10^4	7.29×10^4	7.51×10^4	7.65×10^4
Methane, CH$_4$	2.24×10^4	2.97×10^4	3.78×10^4	4.49×10^4	5.20×10^4	5.77×10^4
Nitrogen, N$_2$	5.29×10^4	6.68×10^4	8.04×10^4	9.24×10^4	10.4×10^4	11.3×10^4
Oxygen, O$_2$	2.55×10^4	3.27×10^4	4.01×10^4	4.75×10^4	5.35×10^4	5.88×10^4

Source: International Critical Tables (1)
*Values given are for H in Equation $P = Hx$, where x = mole fraction of solute in the liquid phase and P = partial pressure of solute in the gas phase, atmosphere.

replaced by more efficient systems which employ either chemical or physical solvents of higher capacity. Such systems are discussed in Chapters 2, 3, 5, and 14.

A flow diagram of a simple water-absorption operation is shown in Figure 6-1, and a photograph of the absorption train of a large plant is shown in Figure 6-2. In its simplest form, the plant consists of nothing more than an absorption tower operating at elevated pressures, a flash chamber where CO_2 is disengaged from the water after pressure reduction, and a pump to return the water to the top of the absorber. In the flow diagram shown, a power-recovery turbine has been added to reclaim some of the power available from the pressure-reduction of the liquid and subsequent expansion of the absorbed gas, and a degasifying tower provides more complete removal of CO_2 from the water than could be obtained with a simple flash. In an arrangement of this type, it is possible to operate the flash chamber at an intermediate pressure and obtain from it a gas sufficiently rich in combustible components to be used as a low-Btu fuel gas. Alternatively, the gas from the intermediate pressure flash tank may be recompressed and returned to the absorber inlet.

In general, the process is limited to gas streams containing carbon dioxide at a partial pressure greater than about 50 psi in order to ensure an economically useful carbon dioxide capacity of the solvent water. This in effect limits the process to absorption pressures over about 200 psig since ammonia-synthesis gas streams typically contain about 25 percent CO_2.

The use of water as an absorbent offers the following advantages as compared, for example, to monoethanolamine solutions:

1. Simple plant design (no heat exchangers or reboilers)
2. No heat load
3. Inexpensive solvent

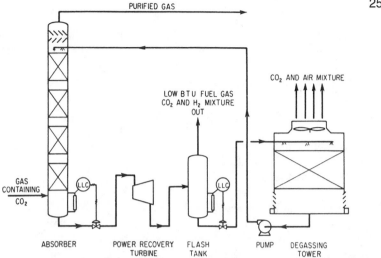

Figure 6-1. Simplified flow diagram of process for absorption of carbon dioxide from gas streams with water.

Figure 6-2. Photograph of absorption train of large CO_2-removal plant employing water-wash process. *E.I. DuPont de Nemours and Company*

4. Solvent not reactive with COS, O_2, and other possible trace constituents
5. No nitrogenous solvent vapors introduced into the gas stream

The principal disadvantages of the water process are

1. Substantial loss of hydrogen or other valuable constituents of the gas stream
2. Very high pumping load
3. Poor CO_2-removal efficiency
4. Impure by-product CO_2

Since ammonia is frequently converted to urea at the plant site by reacting it with carbon dioxide, the need for CO_2 of high purity often precludes the use of the water-wash process. The relative economics of the two processes depend to a large extent upon the costs assigned to thermal energy and power. If waste heat is available from other process steps, processes using chemical solvents are generally more economical.

Basic Data

The solubility of carbon dioxide in water, considered as a function of pressure and temperature, is presented in Figures 6-3 and 6-4, which are based upon the compilation of Dodds et al. (2). Figure 6-3, which covers pressures from 1 to 700 atm, represents smooth curves of the data from numerous investigators. Figure 6-4, which covers primarily the region of sub-atmospheric pressures, is based entirely on the data of Morgan and Maass (3) and the experimental points are therefore shown.

Design and Operation

Packed-Tower Design

The absorption of carbon dioxide in water has been shown to be almost entirely liquid-film controlled—presumably because of the relatively low solubility of carbon dioxide. Considerable research has, therefore, been conducted on the CO_2-H_2O system in connection with both absorption and desorption to determine the liquid-film resistance to mass transfer when various packings are used. Some of the data obtained are directly applicable to the design of commercial installations for carbon dioxide absorption and desorption.

The work of Cooper, Christl, and Peery (4) is particularly valuable because they employed commercial-size packing (2- by 2- by $\frac{1}{16}$-in. steel raschig rings) and covered the range of very high liquid-flow rates commonly encountered in practice where substantially complete CO_2 removal is required. These investigators found that H_{OL} values for liquid-flow rates greater than about 20,000 lb/(hr)(sq ft) were considerably higher than those indicated by previous data obtained at lower liquid-flow rates. Cooper, Christl, and Peery suggest that this apparent discrepancy may result from the high ratio of liquid- to gas-velocities in a column operating in this manner. Under these

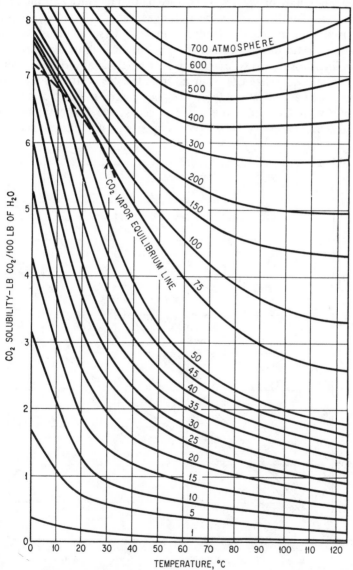

Figure 6-3. Solubility of carbon dioxide in water at pressures of 1 atm and greater. *From Dodds, Stutzman, and Sollami (2)*

conditions, when the ratio of average linear liquid- and gas-velocities is greater than unity, the slower gas stream may be carried downward by the liquid, thus tending to destroy the true countercurrent action and thereby increasing the observed H_{OL}. The results of this investigation are summarized in Figures 6-5 and 6-6.

In Figure 6-5, lines of equal mG_M/L_M are plotted in which m is the slope of the equilibrium line, y_e/x, G_M is the molar gas velocity, lb moles/(hr)(sq ft), and L_M is the molar liquid velocity, lb moles/(hr)(sq ft). This parameter represents the ratio of the slope of the equilibrium line to the slope of the operating line. It is normally less than 1.0 for practical absorption towers, in which essentially complete CO_2 removal from the gas is effected with excess water and may be as high as 2.0 for stripping towers, in which almost complete removal of CO_2 from the water is necessary. The data, therefore, indicate that for towers utilizing 2- by 2- by $\frac{1}{16}$-in., steel raschig rings, H_{OL}

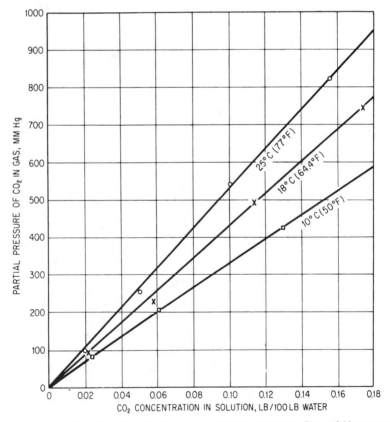

Figure 6-4. Solubility of carbon dioxide in water at low pressure. *Data of Morgan and Maass* (3)

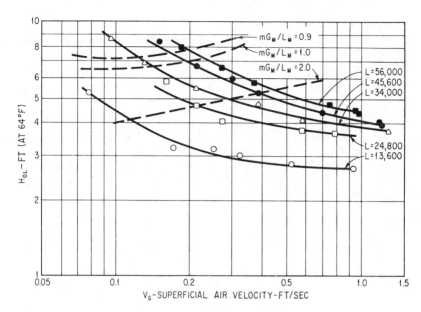

Figure 6-5. Effect of superficial gas velocity on height of transfer units for carbon dioxide absorption in water. *Data of Cooper, Christl, and Peery (4) for 2- by 2- by ¹⁄₁₆-in. steel raschig rings*

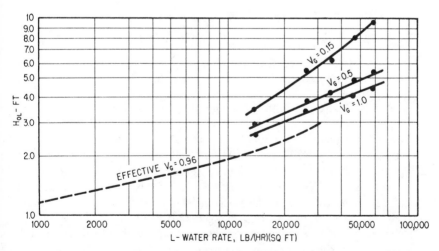

Figure 6-6. Effect of superficial liquid-flow rate on height of transfer units for CO_2 absorption in water. *Data of   Cooper, Christl, and Peery (4) Solid lines are data for 2- by 2- by ¹⁄₁₆-in. steel raschig rings. Dotted line based on data of Sherwood and Holloway (5) for 2-in. ceramic raschig rings*

values greater than 7 ft may be expected in the absorber and values from 4 to 5½ ft in the stripper. Data of Sherwood and Holloway (5) for 2-in. ceramic rings have been included in Figure 6-6 (which is a cross plot of Figure 6-5) for purposes of comparison. The effective V_G (superficial gas velocity) indicated for the ceramic-ring curve represents a value corrected for the difference in free volume between ceramic and steel rings.

The data of Sherwood and Holloway (5) are of general value in estimating H_L (or $k_L a$) for the lower liquid-flow rate region. Although these data were largely obtained in experiments involving the desorption of oxygen, sufficient data were also obtained with carbon dioxide to indicate that the resulting correlation (Equation 6-1), or the corresponding form (Equation 6-2), applies equally well to this gas.

$$\frac{k_L a}{D_L} = \alpha \left(\frac{L}{\mu_L}\right)^{1-n} \left(\frac{\mu_L}{\rho_L D_L}\right)^{1-s} \tag{6-1}$$

or

$$H_L = \frac{1}{\alpha} \left(\frac{L}{\mu_L}\right)^{n} \left(\frac{\mu_L}{\rho_L D_L}\right)^{s} \tag{6-2}$$

Where $k_L a$ = volume coefficient of mass transfer for liquid phase, lb moles/(hr)(cu ft)(lb mole/cu ft)

H_L = height of an individual liquid-film transfer unit, ft

L = liquid mass velocity, lb/(hr)(sq ft)

μ_L = viscosity of liquid, lb mass/(ft)(hr)

ρ_L = density of liquid, lb/cu ft

D_L = diffusion coefficient for liquid phase, sq ft/hr (approximately 6.6×10^{-5} sq ft/hr for CO_2 at 18°C)

α, n, and s = constants with the values given in Table 6-3

The preceding equations (Equations 6-1 and 6-2) are intended to provide a means for calculating the liquid-film coefficient k_L and height of transfer unit H_L for the general case. If the gas film also presents an appreciable resistance, an overall coefficient (or height) must be calculated by means of the conventional equations for adding the effects of both films. In the case of CO_2 absorption, the gas-film resistance can generally be assumed to be of negligible importance and values calculated by Equations 6-1 and 6-2 can be used directly as approximate overall values for $K_L a$ and H_{OL}.

Rixon (6) has presented absorption data from a small commercial plant designed to produce relatively pure hydrogen from fermentation gas containing approximately 9.4 percent CO_2, 50.2 percent CO, 34.6 percent H_2, 1.4

Table 6-3. Values of Constants α, *n* and *s*

Packing	α	*n*	*s*
2-in. rings.............	80	0.22	0.50
1.5-in. rings...........	90	0.22	0.50
1.0-in. rings...........	100	0.22	0.50
0.5-in. rings...........	280	0.35	0.50
⅜-in. rings.............	550	0.46	0.50
1.5-in. saddles.........	160	0.28	0.50
1.0-in. saddles.........	170	0.28	0.50
0.5-in. saddles.........	150	0.28	0.50
3-in. tile..............	110	0.28	0.50

percent CH_4, and 4.4 percent N_2 (by volume). In addition, he has presented extensive absorption and desorption data obtained during a pilot-plant study which preceded construction of the commercial unit. Results are given for both the absorption and desorption stages of the pilot plant and for the absorption step only of the commercial unit. His results, which are correlated in terms of K_La, are in general agreement with those of Cooper, Christl, and Peery (4) in showing a decrease in the absorption coefficient with decreased gas rate at the high liquid rates used in the absorber. No effect of gas rate was observed in the stripper. The value of K_La was found to increase with increased liquid rate in both the stripper and the absorber.

The effect of gas rate on K_La as observed by Rixon is illustrated in Figure 6-7, which presents results of tests in both the pilot and full-scale absorbers. The pilot-plant absorption coefficients in this figure represent values calculated for small sections of the tower. Unexpectedly high values were observed for the lowest section, presumably because of the effect of liquid spray below the packing, and these values are, therefore, not included in the plot. Pilot-plant desorber data are correlated in Figure 6-8 to show the effect of liquid-flow rate. Operating data for typical runs on both plants are presented in Table 6-4, which also gives physical dimensions for the equipment tested.

Absorption in Plate Columns

Plate-efficiency data for the carbon dioxide-water system have been presented by Walter and Sherwood (7). For the case of a single plate, 18 in. in diameter with seven 4-in. caps, Murphree plate efficiencies from 1.8 to 2.6 percent were observed with operating temperatures averaging 10 to 12°C (50 to 54°F) and liquid/gas flow-ratios from 2.2 to 16 moles liquid per mole gas. These data are of interest in illustrating the very low efficiencies obtainable

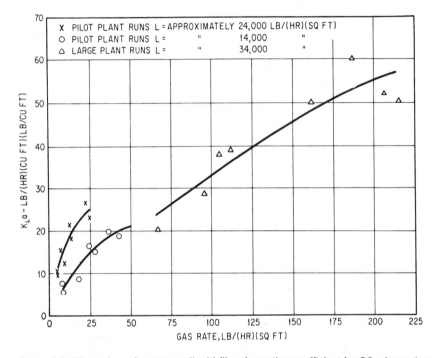

Figure 6-7. Effect of gas-flow rate on liquid-film absorption coefficient for CO_2 absorption in water. Pilot-plant absorption at 11.9 atm pressure with 1-in. stoneware rings; commercial plant, 21.4 atm pressure with 1½-in. stoneware rings. *Data of Rixon* (6)

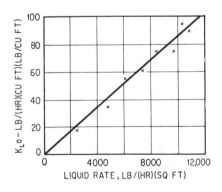

Figure 6-8. Effect of liquid-flow rate on desorption of CO_2 from water with air in pilot-plant tower packed with 1-in. stoneware rings. *Data of Rixon* (6)

Table 6-4. Typical Test Data for Absorption and Desorption of CO₂ from Water in Packed Towers

Run Variables	Pilot-Plant Absorption, Run No.			Pilot-Plant Desorption, Run No.			Commercial-Plant Absorption, Run No.		
	2	4	6	2	25	10	1	6	11
Physical dimensions:									
Column ID, in.	9¾			15			27		
Packed height, ft.	25.5 (25 ft., 5½ in.)			15			18 (two 9-ft sections)		
Packing, stoneware rings	1 × 1 in.			1 × 1-in.			1½ × 1½-in.		
Water rate, lb/(hr)(sq ft)	24,030	14,450	17,400	10,390	7,340	2,445	33,200	33,200	22,400
Water temp., °F	59.0	63.5	54.5	ambient	ambient	ambient	84.2	84.2	78.8
Inlet gas rate, lb/(hr)(sq ft)	134.0	116.0	120.0	136.0	96.0	332.0	76.2	181.0	231.5
Absolute pressure, atm	11.9	11.9	11.9	1.0	1.0	1.0	21.4	21.4	21.4
CO_2 in inlet gas, % by vol	61.2	60.6	59.1	neg.	neg.	neg.	9.63	10.12	9.20
CO_2 in exit gas, % by vol	0.49	4.97	3.9	3.7*	1.80*	0.17*	0.35	0.63	2.14
CO_2 in inlet water, lb/cu ft	3.1×10^{-4}	3.1×10^{-4}	$3.1 \times .0^{-4}$	4.4×10^{-4}	3.1×10^{-4}	4.4×10^{-4}	1.25×10^{-4}	1.25×10^{-4}	1.37×10^{-4}
CO_2 in outlet water, lb/cu ft	0.337	0.471	0.404	not available			0.0288	0.0709	0.0993
K_La, lb/(hr)(cu ft)(lb/cu ft)	Not calculated because of spray-zone effect			95.0	61.0	17.3	20.1	50.0	35.2

Source: Data of Rixon (6)
*Percent CO₂ in air at point 9 ft above bottom of packing.

with equipment of this type; however, conventional bubble-cap columns are not normally utilized for the absorption of carbon dioxide in water because they do not lend themselves to the very high ratios of liquid- to gas-rate employed. Several modified tray designs have been developed, however. which appear to offer promise. In one of these (8) each plate consists of a series of parallel troughs. The liquid which fills the troughs overflows through the spaces between them to form curtains of falling liquid in the space below the plate. The liquid falls into the troughs of the plate below entraining gas bubbles in the liquid. The primary gas flow is horizontal between each plate in a direction parallel to the liquid curtains and upward through spaces at alternate ends of adjacent plates.

In another somewhat similar design, a level of water several inches deep is maintained on each tray, and the bottom of each tray is perforated so that water jets flow downward to the tray below, entraining gas bubbles in the liquid on that tray. As in the previous design, the gas flows horizontally in the space between adjacent trays, then up through an opening at one end of the upper tray and back in the opposite direction in the next tray space. Plates of this type are sometimes referred to as "shed" or "shower" trays.

The Kittel plate has also been proposed for this service, and data on the results of replacing raschig rings with Kittel plates have been presented by Pollard (9). The tower tested had a diameter of 6 ft 6 in. and a packed height of 36 ft. The packing was replaced by 42 polygon Kittel plates in place of the rings. A comparison of performance before and after the change is presented in Table 6-5.

It is interesting to note that if equilibrium were obtained in the above installation, the outlet water would contain 6.73 scf of CO_2 per cubic foot of water, indicating a theoretical minimum water consumption of 41 cu ft of water per 1,000 scf of gas. The two cases required 80 and 35 percent more water, respectively, than the theoretical minimum.

Table 6-5. Comparison of Kittel Plates and Raschig Rings for Carbon Dioxide Absorption in Water

Absorber Variables	Raschig Rings (2½-in. diam.)	Kittel Plates (42 polygon plates)
Gas rate, SCF/hr	480,000	480,000
Carbon dioxide in feed, %	29.0	29.2
Carbon dioxide in product gas, %	1.8	1.8
Pressure, atm	27	27
Water rate, cu ft/hr	35,300	27,000
Water temperature, °F	63	63
Water consumption, cu ft/1,000 cu ft gas	73.5	56.2

Source: Data of Pollard (9)

Operating Data

Typical flow rates and gas analyses have been presented by Yeandle and Klein (10) who discuss various ammonia-synthesis gas-purification processes. According to these authors, a typical water-absorption unit would handle 5,-000 scf/min of synthesis gas at 250 to 275 psig with a flow of 4,000 gpm of water which has been deaerated and cooled. With a conventional packed absorber, the gas leaving the top of the column would contain 0.5 to 0.8 percent carbon dioxide. Analyses of two typical synthesis-gas feed streams and streams of gas released from the corresponding low-pressure flash operation are presented in Table 6-6.

Although no detailed operating data are presented with the above analyses, it is apparent that the gas released from the water in the reformed natural-gas plant does not represent all of the CO_2, as this high a hydrogen loss could not be tolerated with the total CO_2 stream. Presumably this analysis represents gas flashed at an intermediate pressure of 15 to 40 psig. In normal practice, 3 to 5 percent of the hydrogen is lost with the CO_2 in water-scrubbing operations on ammonia-synthesis gas.

More complete data are presented in Table 6-7 for the water-wash section of a large ammonia plant (11). Data are presented for two periods representing summer and winter conditions. It will be noted that considerably less water is required in the winter because of the lower ambient temperature, and this results in a reduced loss of hydrogen with the flashed gas. In general, the

Table 6-6. Typical Gas Analyses for Water-wash CO_2 Absorption Process

Component Gas	Content of Gas Feed to Absorber, %		Content of Gas Released from Flash Step, %	
	Case A*	Case B†	Case A*	Case B†
Hydrogen................	51.0	60.0	6.0	31.0
Nitrogen................	16.5	20.0	0.0	8.5
Carbon dioxide...........	28.0	16.0	92.3	58.5
Carbon monoxide..........	3.5	3.5	1.5	2.0
Methane + argon.........	0.5	0.5	0.0	0.0
Hydrogen sulfide..........	0.5	0.0	0.2	0.0
Total..................	100.0	100.0	100.0	100.0

Source: Data of Yeandle and Klein (10)
*Case A: Gas feed to absorber; coke-oven gas after shift conversion of carbon monoxide.
†Case B: Gas feed to absorber; synthesis gas from steam-natural gas reforming after shift conversion of carbon monoxide.

temperature of the gas entering the absorber is 15 to 30°F higher than that of the feed water. The water at this plant contains 96 ppm dissolved solids and 119 ppm total solids and has a pH of 7.2. The absorber is 10 ft ID by 81 ft high and is packed with raschig rings in three sections as follows:

1. Top 19 ft, 3-in. aluminum rings
2. Middle 17 ft, 3-in. ceramic rings
3. Bottom 17 ft, 2-in. ceramic rings

Table 6-7. Operating Data on Removal of CO₂ from Synthesis Gas by Water Scrubbing

Scrubber Design Variables	July	December
Temperatures, °F:		
Water entering	80	52
Gas leaving	85	55
Pressures, psig:		
Gas entering	254	260
Gas leaving	248	256
Flow rates:		
Feed gas SCF/min	10,550	12,150
Product gas (calculated) SCF/min	8,350	9,800
Water feed, gpm	8,250	5,275
NH_3 production in synthesis section, lb/min	162	189
Gas composition, %:		
Feed gas:		
CO_2	16.7	16.2
CO	2.1	2.8
H_2	59.8	60.1
CH_4	0.3	0.2
N_2 (including argon)	21.1	20.7
Product gas:		
CO_2	0.81	0.72
CO	2.70	2.87
Flashed gas:		
CO_2	75.2	79.5
O_2	0.0	0.0
CO	0.7	0.8
H_2	18.0	15.1
CH_4	0.3	0.1
N_2 (including argon)	5.8	4.5

Source: Data courtesy of Tennessee Valley Authority (11)

Water scrubbing has recently been proposed for the removal of CO_2 from methane produced by anaerobic digestion. A pilot unit for this type of process was first put into operation in 1978 at a wastewater treatment plant in Modesto, California. (49) The process, which is called the Binax system, uses an absorption tower operating in the range of 100 to 500 psig and an atmospheric pressure air-blown regeneration tower. About 90 percent of the methane is recovered from feed gas to the absorber, which contains 30 to 50 percent carbon dioxide and trace amounts of hydrogen sulfide. The product gas is 98 percent methane. Plans to construct a Binax plant to produce approximately 650,000 scfd of pipeline quality gas from digester gas containing about 35 percent CO_2 at Baltimore's Back River Wastewater Treatment Plant have been reported. (50) Water scrubbing is attractive for this type of application because of its relatively low capital cost in small sizes, simplicity of operation and maintenance, and use of a readily available nonhazardous absorbent.

Power Recovery

The use of power-recovery equipment of some type is desirable in water-absorption plants because of the large amount of power required for pumping. Turbines of either the impulse (Pelton) or reaction (Francis) types have been employed. Pelton-wheel turbines generally permit recovery of 50 to 60 percent of the energy in the high-pressure water stream; Francis turbines can provide an energy recovery up to about 80 percent. Since water-pump efficiency is generally on the order of about 80 percent, the power-recovery equipment can, at best, supply approximately 65 percent of the required input energy to the pumps.

An unusual power-recovery system has been employed at DuPont's Belle, West Virginia, ammonia plant (12). Water from the scrubbers, which operate at 30 atm pressure, is forced to the top of a 600-ft mountain located adjacent to the plant. Carbon dioxide is flashed from the water at atmospheric pressure at the top of the mountain, and final stripping is accomplished at this location by air blowing. The elevated position of the water after final stripping provides a high suction-pressure for the pumps which feed the water back into the absorbers and, therefore, decreases the horsepower requirement. About 60 percent of the energy in the high-pressure water is recovered in this manner—the principal remaining power requirement is used to overcome friction in the long lines up and down the mountain.

Corrosion

Unlike the amine- and alkali-salt-process solutions or essentially anhydrous organic solvents which are generally basic or neutral, water becomes quite acid when appreciable quantities of carbon dioxide are absorbed. As would be expected, this results in corrosion problems in water-wash CO_2-removal plants; however, a compensating factor is that, unlike the other processes, the water system operates at or near ambient temperatures throughout the cycle. The low temperature level is, of course, a favorable factor with regard to corrosion, and the absence of heat exchangers reduces the amount of corrosible metal exposed.

The corrosion of steel by carbon dioxide and other dissolved gases has been studied on a laboratory scale by Watkins and Kincheloe (13) and Watkins and Wright (14). These investigators found that the presence of oxygen greatly accelerates the rate of corrosion by carbon dioxide, while hydrogen sulfide in small quantities inhibits carbon dioxide corrosion. Comparative data taken from their corrosion curves are presented in Table 6-8.

The tests were conducted in a dynamic test apparatus with water containing dissolved gas flowing at a constant rate past the specimens. Although such tests are of value in illustrating comparative effects, extreme caution snould be used in extrapolating to plant-equipment design. The authors found, in fact, that the mere substitution of a different lot of mild steel changed the corrosion rate by a factor of three.

In commercial operations, corrosion is minimized by the addition of inhibitors such as potassium dichromate to the water, the use of stainless steel in areas of high turbulence, and the application of protective coatings to the interior of the absorber and other vessels. Wooden cooling towers are adaptable to the stripping operation where countercurrent contact with air is required. Conventional water treatment to control algae may also be required

Table 6-8. Corrosion of Mild Steel by Carbon Dioxide and Other Gases in Water*

Gas Concentration, ppm		Corrosion of Mild Steel, mils/year	
O_2	H_2S	A: CO_2 conc, 200 ppm	B: CO_2 conc, 600 ppm
8.8	0	28	60
4.3	0	18	44
1.6	0	12	34
0.4	0	17	27
<0.5	35	6	6
<0.5	150	15	16
<0.5	400	17	21

Source: Data of Watkins and Kincheloe (13) and Watkins and Wright (14)
*Temperature 80°F, exposure time 72 hr.

in installations where the water is exposed to light. Foaming is not usually a problem; however, when it does occur, due to the presence of oil or other impurities in the water, conventional foam-inhibitors of the high-molecular-weight alcohol (e.g., ocenol) or silicone types have been found to be effective.

Hydrogen Sulfide Removal by Absorption in Water

Although hydrogen sulfide is appreciably more soluble in water than carbon dioxide, the use of water for removing this impurity from gas streams is not a commercially important process. Several attempts were made to commercialize the process, especially for treating natural gases of very high hydrogen sulfide content. The fact that no heat is required for acid-gas desorption, and, consequently, no heat exchange equipment is necessary, suggested the possibility of substantial savings in capital and operating costs over the conventionally used ethanolamine processes. However, the savings resulting from the reduced energy requirements are not as significant as expected because large amounts of steam are generated in the conversion of hydrogen sulfide to elemental sulfur in Claus type units which are necessary adjuncts of desulfurization plants, especially when large amounts of hydrogen sulfide are involved. In many cases sufficient steam is produced to drive the rotating equipment and satisfy the heat requirements for solution regeneration in an amine plant. Furthermore, the relatively high hydrocarbon content of the acid gas evolved from the water leads not only to appreciable losses of product gas but also to a considerable increase in the size of the Claus unit, besides operational difficulties due to cracking of hydrocarbons on the Claus catalyst. Lastly, design improvements made in ethanolamine units, particularly those using the S.N.P.A.-DEA process, have resulted in substantial reduction of energy requirements. These considerations and the fact that water washing alone cannot produce pipeline quality gas have eliminated the process from large-scale industrial use.

In spite of the above considerations, the absorption of H_2S in water may be economical in special cases, and it has been demonstrated to be a technologically feasible operation (15, 16). It should also be noted that, in some instances, small quantities of H_2S are removed simultaneously with carbon dioxide in water-wash plants designed primarily for carbon dioxide absorption.

Pilot-plant experiments were conducted in Canada (15), and a rather large commercial water wash installation, treating about 60 million cu ft per day of natural gas containing 15 percent hydrogen sulfide and 10 percent carbon dioxide at a pressure of 1,100 psig was operated in Lacq, France, by Societe Nationale des Petroles d'Aquitaine (S.N.P.A.) (16). This unit was intended to remove the bulk of the hydrogen sulfide and carbon dioxide from the sour gas prior to final purification with diethanolamine. However, the plant has been converted to the S.N.P.A.-DEA process (see Chapter 2), and, at present, no commercial units using water washing for acid-gas removal from high-pressure natural gas are known to be in operation. The solubility of hydrogen sulfide in water at moderate pressures has been determined by Wright and Maass (17) and the behavior of the hydrogen sulfide-water system evaluated in more detail by Selleck et al. (18) up to a pressure of 5,000 psig. Fortunately, the data from both of these investigations indicate that Henry's law holds reasonably well for the system at conditions which would normally be encountered in gas-purification operations. Equilibrium gas and liquid compositions can readily be calculated from Henry's law coefficients such as those presented in Table 6-9.

Experimentally determined vapor-liquid equilibrium data for the system hydrogen sulfide-carbon dioxide-methane-water at pressures ranging from atmospheric to 1,014 psia and temperatures from 85 to 115°F have been reported by Froning, Jacoby, and Richards (19). These authors found that the equilibrium constants K can be represented by the following equations:

$$K_{\text{Methane}} = 306,000/P + 2.19t + 3,910 \ t/P - 145.0 \ AG - 121.6R$$

$$\tag{6-3}$$

$$K_{\text{CO}_2} = -3,500/P + 0.12t + 360.0 \ t/P + 8.30 \ AG - 5,825R/P$$

$$\tag{6-4}$$

$$K_{\text{H}_2\text{S}} = 4.53 - 1,087/P + 110.0 \ t/P + 4.65 \ AG \tag{6-5}$$

where K = mole fraction in gas phase/mole fraction in
water phase
P = system pressure, psia
t = system temperature, °F
AG = mole fraction $CO_2 + H_2S$ in gas phase
R = mole fraction, H_2S/AG

Comparison of calculated K values using the above equations, with those reported in the literature for the binary systems methane-water, carbon dioxide-water and hydrogen sulfide-water shows quite good agreement.

In its simplest form, the water-wash-process plant consists of nothing more than an absorber, flash vessel, and recycle pump. As justified by the economics, additional equipment may be added to reduce hydrocarbon losses, recover energy from the high-pressure water, or control operating temperature. A process-flow diagram of a pilot plant used to study water absorption for natural-gas purification is shown in Figure 6-9. The pilot-plant study was made by the Shell Oil Company to investigate the feasibility of absorbing H_2S and CO_2 from the gas of a Canadian field which analyzed about 45 percent acid gas. Results of the study indicated that 80 to 90 percent of the

Table 6-9. Solubility of H₂S in Water

Pres- sure, atm	H at Various Temperatures, °C						
	5	10	20	30	40	50	60
1	3.12×10^2	3.64×10^2	4.78×10^2	6.04×10^2	7.35×10^2	8.65×10^2	9.81×10^2
2	3.19×10^2	3.69×10^2	4.80×10^2	6.06×10^2	7.39×10^2	8.77×10^2	10.02×10^2
3	3.26×10^2	3.72×10^2	4.83×10^2	6.09×10^2	7.42×10^2	8.83×10^2	10.11×10^2

Source: Based on data of Wright and Maass (17)
*Values given are for H in Equation $P = Hx$, where x = mole fraction of solute in the liquid phase and P = partial pressure of solute in the gas phase, atmospheres.

H₂S (initial concentration 35 percent) and about 60 percent of the CO₂ (initial concentration about 10 percent) could be removed by water washing at 500 to 700 psig. The pilot unit purified about 300,000 scf/day of gas with a water circulation of about 50 gpm (20, 21). A detailed analysis of the feed gas processed in the pilot plant is presented in Table 6-10.

An extensive study of corrosion was made in Shell's pilot-plant investigation inasmuch as this was expected to be one of the major operating problems. The results of the corrosion study have been presented in detail by Bradley and Dunne (21). It was found that carbon steel in the absorber and rich-solution line was corroded rapidly at first (up to 100 mils per year (mpy) during the first 2 days), but corrosion slowed down appreciably (to 10 to 30 mpy after about 10 days) as a result of the formation of a heavy protective iron sulfide scale. The protective film was found to form more slowly on steel samples in the flash vessel. Hydrogen blistering and sulfide-corrosion cracking were found to be potential problems requiring the use of resistant materials and stress relieving. A corrosion inhibitor of the water-dispersible amine type was found to reduce all forms of corrosion markedly when used in a concentration of 0.1 percent in the absorption water. In general, the data indicate that stress-relieved carbon steel could be used for vessels and major piping with special alloys, e.g., type 304, type 316, or K-Monel, preferable for critical components such as pump impellers, Bourdon tubes, thermometer wells, orifice plates, and relief-valve springs.

The commercial unit described by Barbouteau and Galaud used a more complex flow scheme than is shown on Figure 6-9. The sour gas was first contacted with water in a column, provided with bubble-cap trays, at about 1,100 psig and from there passed to the diethanolamine unit for final acid-gas removal. The water leaving the high-pressure absorption column passed through a power recovery turbine to a flash drum maintained at 220 psig

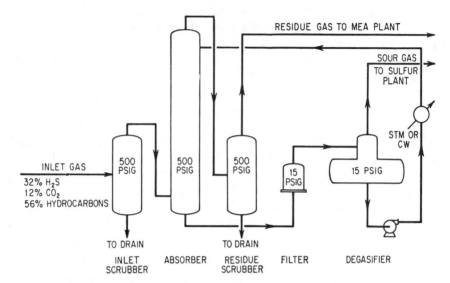

Figure 6-9. Simplified flow diagram of a pilot plant used to study H₂S removal from natural gas by water absorption. *Data of Bradley and Dunne* (21)

where most of the hydrocarbon gas dissolved in the water, together with substantial amounts of hydrogen sulfide and carbon dioxide were flashed off. The flash gas was washed with water in a secondary absorption tower and then sent to the fuel gas system.

The effluent water from the first flash drum together with the wash water from the secondary absorber was flashed, through another power recovery turbine, to essentially atmospheric pressure and stripped in a regenerating column. To facilitate expelling of acid gases, a small amount of deoxygenated flue gas was injected at the bottom of the regeneration column. The acid gases were fed to a Claus type sulfur plant.

Corrosion studies conducted in the unit gave similar results as those reported by Bradley and Dunne (21).

Absorption of Fluorides

The removal of fluorine compounds from industrial exhaust gases is of increasing importance as an air-pollution-control measure. Fluorides may be emitted from many processes and actual vegetation damage or cattle fluorosis have reportedly resulted from the operation of plants producing aluminum, phosphate fertilizer, iron, enamel, and bricks. The two operations of major importance from the standpoint of quantity of fluorine compounds liberated are phosphate-rock treatment and aluminum-metal reduction.

The chemical reactions which lead to the evolution of gaseous fluorine compounds in the above operations are not completely understood; however,

Table 6-10. Composition of Gas Feed to Water-wash Hydrogen Sulfide Absorption Pilot Plant

Component	Content, %
Hydrogen sulfide	35.00
Carbon dioxide	10.12
Nitrogen	1.31
Methane	52.25
Ethane	0.69
Propane	0.16
Butanes	0.10
Pentanes	0.07
Hexane plus	0.30

Source: Data of Bradley and Dunne (21)

the resulting compounds are usually silicon tetrafluoride, SiF_4, and hydrogen fluoride, HF. Particulate matter is also usually present in the exhaust-gas streams and this may contain other nonvolatile fluorine compounds. Available data on the mechanism by which fluorides are liberated have been reviewed by Semrau (22). He concludes that the principal mechanism in high-temperature operation is pyrohydrolysis resulting in the formation of HF, while the formation of SiF_4 is limited to low-temperature operations in which fluosilicates or other fluoride compounds are decomposed with acids. In practice, aluminum-plant exhaust gases are usually found to contain primarily HF with some particulate fluorides such as NaF and AlF_3; gases from the acidulation of phosphate rocks contain primarily SiF_4. Other phosphate-rock processing operations which involve heat, e.g., those using nodulizing kilns and calcium phosphate furnaces, evolve primarily hydrogen fluoride.

Fortunately, both hydrogen fluoride and silicon tetrafluoride are very soluble in water, and most commercial installations for the control of fluoride emission make use of this fact. Since large volumes of low-pressure gas must be handled, the design of exhaust-gas contacting equipment is governed to a large extent by the requirement of a minimum pressure drop. Low investment and operating costs are also important with equipment of this type because the resulting acid solutions do not usually represent an economically recoverable by-product. A further factor influencing the design of fluoride scrubbers is the presence of solid particulate matter in the gas stream and the formation of solids by reactions occurring in the scrubbing liquid. As a result of the considerations discussed above, fluoride-absorption units usually are based on water-spray systems or relatively open grid packing. The effluent solution may be recycled to build up the acid concentration, treated with lime slurry to precipitate the fluoride ion, or discarded without further treatment.

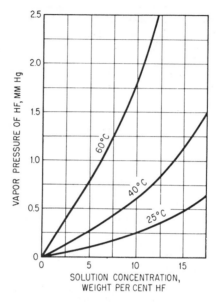

Figure 6-10. Vapor pressure of hydrofluoric acid over dilute aqueous solutions. *Data of Brosheer, Lenfesty, and Elmore* (23)

Basic Data

The vapor pressure of hydrogen fluoride over aqueous solutions is shown in Figure 6-10, which is based on the data of Brosheer et al. (23). As can be seen, HF is quite soluble in water at relatively low temperatures. A gas containing 130 ppm (by volume) of HF, for example, representing a vapor pressure of 0.1 mm Hg, could theoretically produce a scrubber-effluent solution containing almost 5 percent HF. In commercial installations, however, a very large excess of water is usually used and the discharged solutions seldom contain over 0.1 percent HF. When the data of Figure 6-10 are plotted as the logarithm of the partial pressure versus the reciprocal of the absolute temperature, essentially straight lines are obtained, the slope of which indicates that the heat of vaporization is in the range of 10.8 to 11.2 Cal/g mole (19,800 Btu/lb mole) for solution concentrations from 2 to 30 percent HF.

Although hydrogen fluoride in solution is toxic and very corrosive to most materials, it is actually a relatively weak acid. It has an ionization constant of 7.4×10^{-4} at 25°C; this makes it only slightly stronger than acetic acid.

When silicon tetrafluoride is absorbed by water, it reacts to form fluosilicic acid. The mechanism of the absorption process has been studied by Whynes (24), who suggests that the reaction probably occurs in steps, as represented by Equations 6-6 and 6-7:

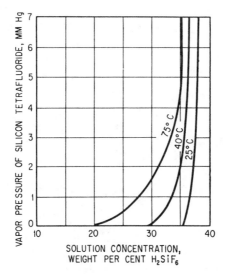

Figure 6-11. Vapor pressure of silicon tetrafluoride over aqueous solutions of fluosilicic acid. *Data of Whynes* (24)

$$SiF_4 + 2H_2O \leftrightharpoons SiO_2 + 4HF \qquad (6\text{-}6)$$

$$2HF + SiF_4 \leftrightharpoons H_2SiF_6 \qquad (6\text{-}7)$$

The simple fluosilicic acid probably reacts with additional SiF_4 or SiO_2 to form a more complex form of this compound. The second reaction (Equation 6-7) is reversible to the point that solutions of fluosilicic acid exert a definite vapor pressure of HF and SiF_4.

Whynes has presented data on the concentration of fluosilicic acid in equilibrium with SiF_4 vapors; these are reproduced in Figure 6-11. The curves indicate that an exhaust gas containing 0.3 percent silicon tetrafluoride can be made to form a solution containing about 32 percent fluosilicic acid at temperatures below 70°C; in fact, this indication has been confirmed in a plant wash-tower by continuously recirculating the solution for several hours.

Tail gas scrubbers for phosphate operations typically utilize very dilute aqueous solutions. Data on the vapor pressure of HF and SiF_4 over solutions containing less than 2.5 percent H_2SiF_6 are given in Figures 6-12 and 6-13 which are based on Russian data (51) as presented by Hansen and Danos. (52)

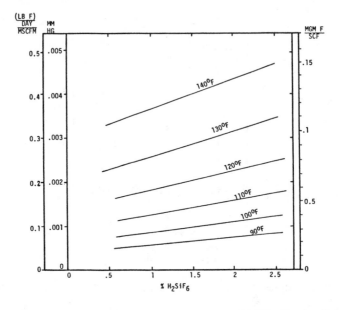

Figure 6-12. Vapor Pressure of HF Over Dilute Solutions of H_2SiF_6. Russian Data (51) as Presented by Hansen and Danos (53).

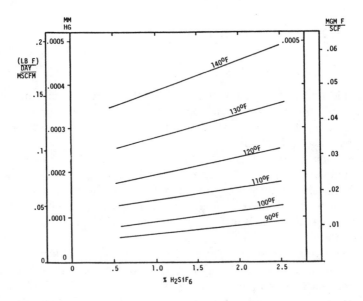

Figure 6-13. Vapor Pressure of SiF_4 Over Dilute Solutions of H_2SiF_6. Russian Data (51) as Presented by Hansen and Danos (53).

Table 6-11. pH values of Fluosilicic Acid Solutions

Concentration of H_2SiF_6, % by wt	pH
1.0	1.4
0.1	2.2
0.01	3.0
0.001	3.8

Fluosilicic acid solutions exhibit very low pH values, as indicated by the data in Table 6-11 presented by Sherwin (25) for industrial acid samples.

Absorber Design

As both hydrogen fluoride and silicon tetrafluoride are very soluble in water, the gas-film resistance would be expected to be the controlling factor in their absorption. This has generally been verified by experimental evidence although Whynes (24) found the SiF_4 absorption to be complicated by a tendency to form mist particles in the gas; there is also a tendency of the silica (produced by reaction with water) to form a solid film on the outside of the water droplets, thus hindering absorption.

Absorbers which have been used or proposed for the removal of fluoride vapors from gas streams with water include the following general types:

1. Spray chambers (vertical and horizontal)
2. Counterflow packed towers (typically low-pressure drop packing)
3. Venturi scrubbers, in which high gas velocities cause atomization of the water
4. Ejectors, in which high-velocity water jets are used to scrub and exhaust the gas
5. Cross-flow packed scrubbers

A wide variety of spray-chamber designs has been employed, from tall cylindrical towers, such as illustrated in Figure 6-14, to horizontal vessels with cross- or cocurrent flow of water and gas. In typical spray chambers, water is introduced at several points from nozzles operating with water pressures of 15 to 60 psig.

Gas velocities are generally low, in the range of 4 to 5 ft per second, to avoid excessive entrainment of liquid droplets. This results in the requirement for large-diameter vessels; however, it also results in very low gas-side pressure drop, typically less than 3 in. of water.

Packed towers are generally more efficient than spray chambers but suffer from the disadvantages of a somewhat higher pressure drop and a greater

Figure 6-14. Photograph of commercial spray-tower installation for HF removal from alumi-num-plant exhaust gases. *Fluor Products Company.*

tendency to clog. The most successful packings are of the grid type which pre-sent less pressure drop to the gas than conventional raschig rings and berl saddles. Grid and mesh type packings are usually employed, although wood and open slat-type plastic packing are also used for this type of service (see Figure 6-15).

Venturi scrubbers are very efficient in providing atomization of the water and good contact between the gaseous and liquid phases; however, they re-quire a relatively high gas-side pressure drop and, consequently, a good deal of power. This type of scrubber is probably in more widespread use for re-moving particulate matter than for absorbing soluble vapors. A schematic diagram of a venturi scrubber is shown in Figure 6-16.

Ejectors offer the advantage of simplicity in that they serve as exhausters for the gas as well as contactors for absorption of the soluble vapors. Their principle drawback is a relatively large power requirement because of the comparative inefficiency of ejectors, when used for moving gas, as compared to well-designed blowers. The power is expended in pumping the liquid. An ejector absorption unit is shown in Figure 6-17.

Crossflow scrubbers are normally designed with one or more relatively shallow panels of packing slightly inclined toward the incoming gas. An open packing such as woven plastic mesh has reportedly given good results and can be cleaned readily with a high-pressure water hose (53). During operation, the packing must be uniformly wetted, and this is usually accomplished by the use of low-pressure spray nozzles aimed at the face of the bed.

Wet cells composed of beds of wetted saran fibers have been proposed for fluoride absorption and tested on a pilot scale (26). However, no commercial installations are known.

Because of the basically different operating principles of the various types of contacting equipment employed, they cannot logically be compared on the basis of conventional mass-transfer coefficents or heights of transfer units In fact, the volume coefficient of absorption $K_G a$ has little value as a mea $_\jmath$ of correlating spray equipment because of the area for mass transfer a \ ries with liquid rate, nozzle design, liquid pressure, distance from nozzle, and other factors. Peformance of different fluoride-vapor absorbers may, however, be compared on the basis of the number of transfer units in a given

Figure 6-15. Photograph of an open slat-type injection-molded plastic packing proposed for exhaust-gas scrubbing. *Fluor Corp.*

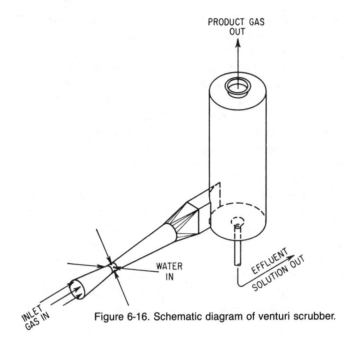

Figure 6-16. Schematic diagram of venturi scrubber.

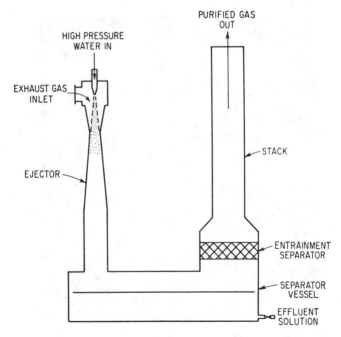

Figure 6-17. Diagrammatic sketch of ejector absorption unit.

piece of equipment, or more simply, on the basis of percentage removal efficiency.

An interesting approach to the problem of correlating data from equipment of this type has been suggested by Lunde (27), who attempted to relate the number of transfer units to the total power requirements of full-scale equipment.

For the case of a dilute gas stream, the number of transfer units based on an overall gas-phase driving force is defined as follows:

$$N_{OG} = \int_{y_2}^{y_1} \frac{dy}{y - y_e} = \frac{K_G a P h}{G_M} \tag{6-8}$$

where N_{OG} = number of transfer units (overall gas)

y = concentration of vapor in gas stream, mole fraction

y_e = vapor concentration in equilibrium with liquid, mole fraction

$K_G a$ = overall mass transfer coefficient based on the gas, lb moles/(hr)(cu ft)(atm)

P = total pressure, atm

h = tower height, ft

G_M = gas-flow rate, lb moles/(hr)(sq ft)

In HF and SiF_4 absorption units, where a large excess of water is used, the concentration of acid in the solution is very low and y_e can be neglected. The equation is thus simplified to

$$N_{OG} = \int_{y_2}^{y_1} \frac{dy}{y} = \ln \frac{y_1}{y_2} \tag{6-9}$$

Since the absorption efficiency (E) is directly related to y_1 and y_2 [$E = (y_1 - y_2)/y_1 \times 100$] the number of transfer units can be calculated directly from the following equation:

$$N_{OG} = \ln \left(\frac{1}{1 - E/100} \right) \tag{6-10}$$

In accordance with this equation, an absorption efficiency of 95 percent requires three transfer units, while 99 percent efficiency requires about five units.

Lunde suggests that the number of transfer units for a packed tower is primarily dependent upon the height of the tower and is only slightly affected by the power introduced in the gaseous or liquid phase in accordance with the following general relationship:

$$N_{OG}::\frac{P_L{}^{0.1-0.5}h^{0.6-1.0}}{P_G{}^{0.1}}$$

(6-11)

where P_L = power introduced in the liquid
P_G = power introduced in the gas
h = tower height

With spray towers, on the other hand, the proposed equation is

$$N_{OG}::\frac{P_L{}^{\sim 1}}{P_G{}^{0.1}}$$

(6-12)

That is, the number of transfer units is expected to be approximately proportional to the power introduced in spraying the liquid. This power would, of course, increase with an increase in either the quantity of water or the pressure drop across the nozzles.

No equation is proposed for venturi or jet scrubbers; however, it should be noted that, with the former, most of the power must be supplied with the gas stream, which atomizes the liquid introduced; in the latter type, all of the power is introduced with the water, which provides atomization and moves the gas stream.

Data presented by Lunde for the absorption of HF in several types of equipment are reproduced in Table 6-12 and plotted in Figure 6-18. He concludes that for a given requirement of HF- or SiF_4-removal efficiency, grid-packed towers require the least power and venturi scrubbers the most, with spray towers in between. His analysis of the data of Berly et al. (26) and First and Warren (28), for the absorption of HF in wetted beds of fiber (wet cells), indicates that the power consumption of this type of equipment is intermediate between those of the spray and venturi scrubbers. Data on the absorption of SiF_4 in spray, packed, and jet scrubbers, presented by Pettit (29, 30) and Sherwin (31), were found to show no correlation with power introduced, although the spray-scrubber points fall reasonably well on the line for spray scrubbers handling HF, as shown in Figure 6-18.

Sherwin (31) also presents data relating to the effect of nozzle design, number of nozzles, and water-flow rate on the performance of spray towers for SiF_4 absorption. The three types of nozzles tried, which included one solid-cone and two hollow-cone spray-pattern designs, did not appear to give appreciably different performances, nor did any improvement result from the substitution of as many as nine nozzles for one centrally located one. As

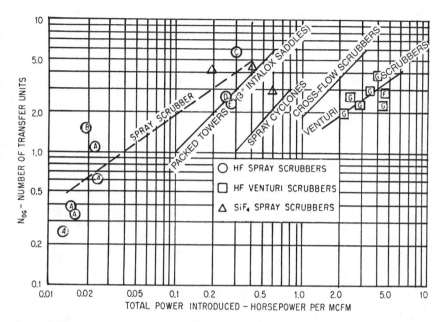

Figure 6-18. Correlation of number of transfer units with total power consumed by HF and SiF₄ scrubbers. Letters in HF spray and venturi scrubber points refer to installations listed in Table 6-12. Spray and venturi scrubber data from Lunde (27), other data from Hansen and Danos (52).

would be expected, however, the water-flow rate was found to have a very strong effect. Using the average performance values for all nozzles, N_{OG} has been calculated for several water-flow rates from Sherwin's study, and the results are shown in Table 6-13.

Data on which the table is based were obtained in the first of six towers of a commercial plant. Gas and liquid flowed cocurrently downward through the unit, which was 7 ft square by 28 ft high. Approximately 10,000 cfm of gas containing about 0.6 percent SiF₄ entered the unit at a temperature of about 60°C (140°F). Sufficient data are not available to permit plotting the points on Figure 6-18; however, if power input is assumed to be proportional to water-flow rate, the data would form a line closely paralleling that plotted for spray scrubbers.

Sherwin's data on the relative efficiencies of each of the six spray towers operating in series are also of interest. The six units were arranged for downward gas flow in the first, then alternately upward and downward flow, with the gas leaving the top of the last section. Water was fed to the system through 24 spray nozzles (four per tower), at a pressure of 8 psig and a rate of 3.8 gpm per section. Each unit was 7 ft square by 28 ft high. Pertinent results of this study are tabulated in Table 6-14.

Table 6-12. Hydrogen Fluoride Absorption

Installation	Type of Absorbing Equipment	Absorbent	Gas Rate, lb/(hr)(sq ft)	Liquid Rate, lb/(hr)(sq ft)	Horsepower Expended per 1,000 cu ft gas/min.			$K_G a$ lb mole/(hr)(cu ft)(atm)	N_{OG}
					Gas	Liquid	Total		
A	Crossflow spray	Water	2,110	72	~0.0067	0.0089	~0.016	~11	~0.33
A	Crossflow spray	Water	1,880	72	~0.0056	0.0098	~0.015	~12	~0.38
A	Crossflow spray	Water	2,080	103	~0.0067	0.0067	~0.013	~12	~0.25
A	Crossflow spray	Water	1,830	84	~0.0067	0.017	~0.024	~15	~0.62
A	Crossflow spray	Water	1,400	92	~0.0061	0.017	~0.023	~25	~1.09
B	Crossflow spray	Lime water	2,050	105	~0.006	0.013	~0.019	~35	~1.50
C	Counterflow spray	Water	2,000	800	~0.2	0.10	~0.3	9	~5.85
D	Parallel flow spray	Lime water	13,800	3,800	~0.23	0.017	~0.25	51	2.58
E	Counterflow spray	Water	2,000	380	0.24	0.02	0.26	~4.3	2.5
F	Venturi	Water	76,000*	42,000*	4.7	0.071	4.8	.....	2.9
G	Venturi	Water	~70,000*	~40,000–65,000*	2.1	0.074	2.2	.....	2.0
G	Venturi	Water	~70,000*	~40,000–65,000*	2.4	0.11	2.5	.....	2.7
G	Venturi	Water	~70,000*	~40,000–65,000*	2.9	0.071	3.0	.....	2.3
G	Venturi	Water	~70,000*	~40,000–65,000*	3.5	0.095	3.6	.....	3.0
G	Venturi	Water	~70,000*	~40,000–65,000*	4.0	0.12	4.1	.....	3.9
G	Venturi	Water	~70,000*	~40,000–65,000*	4.5	0.13	4.6	.....	2.3

Source: Data of Lunde (27)   *Based on throat cross section.

Table 6-13. Effect of Water-flow Rate on SiF$_4$ Absorption in Cocurrent Spray Tower		
Water-Flow Rate, gpm	Approx. Absorption Efficiency, %	N_{OG}
3	30	0.36
9	45	0.60
15	60	0.92
21	72	1.27
27	81	1.66

The high apparent efficiency of the final stage is believed due to the presence of some packing which was installed in this section for the purpose of removing silica particles from the gas. It is interesting to note the small difference between countercurrent and cocurrent operation in the various sections. This is due to the fact that solution concentration has little effect on the absorption rate in the ranges encountered. In a separate experiment to evaluate this factor, solution from the first two towers was recirculated continuously over the first tower to build up the fluosilicic acid concentration. The fluoride-removal efficiency was found to be relatively constant (60 to 80 percent) for solution concentrations from 0 to 15 percent fluosilicic acid and then to drop rapidly to about 35 percent as a concentration of 25 percent acid was approached.

A detailed compilation of engineering design methods for wet scrubbers is presented by Calvert et al. (48). The study includes a sample calculation for the design of a water scrubber to remove SiF$_4$ from gases evolved from a phosphoric acid manufacturing operation. The calculation is based on processing gas from the digestion of 2,000 T/day of phosphate rock in sulfuric acid to produce 600 T/day P$_2$O$_5$ equivalent. Approximately 400 lb/hr of SiF$_4$ are emitted from the reactor and enter the scrubber in approximately 220,000 ft^3/hr of gas saturated with water at 144°F. In order to obtain a removal efficiency of 99.8 percent, it is concluded that two scrubber stages are required. The first stage is simply a vertical duct approximately 3 ft diam. by 33 ft high, with cocurrent downward flow of gas and water. Approximately 80 gpm of water are sprayed in at the top of the duct. The second stage is a 3 ft diam. by 12 ft high cyclonic spray scrubber employing 22 nozzles, each operating with 6 gpm of water and a water pressure of 60 psig. An efficiency of 91.8 percent fluoride removal is predicted for the first stage and over 99.8 percent for the overall system. Detailed discussions of the design of wet scrubbers for phosphate plants are also presented by Hansen and Danos (52, 53).

Table 6-14. Relative Absorption Efficiencies of Spray Towers in Series of SiF₄ Absorption.

Section No.	Direction of Gas Flow	Fluorine Absorbed, % of Total	Fluorine Entering Section, % of Total	Efficiency of Section, %	K_Ga, lb moles/ (hr)(cu ft)(atm)	N_{OG}
1	Down	30	100	30	0.31	0.36
2	Up	25	70	36	0.38	0.45
3	Down	16	45	36	0.38	0.45
4	Up	13	29	45	0.51	0.60
5	Down	7	16	44	0.49	0.58
6	Up	8	9	89	1.87	2.19
Over-all performance		99	100	99		4.63

Source: Data of Sherwin (31)

Operating Data

In the production of superphosphate fertilizer from phosphate rock, the rock is normally pulverized, mixed with sulfuric acid, and discharged into a den where the reaction between rock and acid proceeds. The fresh superphosphate is then removed from the den by an elevator and conveyed to storage for curing. Exhaust gases containing fluorine compounds are drawn from the mixing and den operations and, in some instances, from the elevator and other units.

Typical Florida pebble phosphate rock contains about 3.6 percent fluoride expressed as elemental fluorine, and approximately 32 percent of this is released during the acidulation process (30). Almost all of the fluoride vapors are evolved in the mixer and den although a slight evolution of vapor occurs during subsequent handling and storage operations. The composition of mixer gases and total flue gases from a plant handling Morocco rock is presented in Table 6-15.

The fluorine evolved from this rock corresponds to approximately 1 percent of the weight of the rock or 25 percent of the fluorine originally present.

Comparative data on three types of scrubbers used in superphosphate plants are presented by Pettit (29, 30). Although no conclusions are drawn in the study, the data, which are summarized in Table 6-16, indicate that the horizontal scrubber offered the highest efficiency with the lowest water-flow rate. The high efficiency of this unit probably resulted from the use of high-pressure nozzles and the long tortuous path which the gas stream was forced to follow.

Table 6-15. Composition of Exhaust Gases from Continuous Superphosphate Acidulation Process

Exhaust Conditions	Mixer Gases	Flue Gas (mixer plus den)
Gas temperature, °C............	78 (172° F)	60 (140°F)
Gas quantity, lb moles/hr......	146.8	792.4
Analysis, mole per cent:		
Air......................	34.5	78.9
CO_2......................	16.1	3.0
SiF_4......................	2.0	0.6
H_2O......................	47.4	17.5

Source: Data of Sherwin (31)

Table 6-16. Comparative Performance Data on Superphosphate Plant Exhaust-gas Scrubbers

Design Variables	Vertical-Spray Towers*	Ejector Unit†		Horizontal-Spray Scrubber‡
		Sprays	Ejector	
Water-flow rate, gpm.................	292	60	300	60
Water pressure, psig.................	32	30	60	62
No. of nozzles (or ejectors)............	12	24	1	80
Exhaust-gas rate (max), cfm...........	13,000	7,500		14,000
Plant size, tons superphosphate/hr......	15	15		40
Water-use rate, gpm/1,000 cfm gas......	24.3	48		4.3
Apparent average efficiency of fluoride removal, %......................	95	98		99

Source: Data of Pettit (29, 30)

*Two cylindrical vertical towers 42 in. in diameter by 40 ft high and one tower 42 in. in diameter by 20 ft high. Gases pass down through first tower, up through second, and down through third. First two towers are fitted with six spray nozzles, each pointed against the gas flow. Last tower is dry.
†Ejector has gas-suction and discharge connections 30 in. in diameter and a 3-in. water connection. Nozzle opening is 30 cm. Unit discharges into covered sump containing 20 spray nozzles mounted on top. Four spray nozzles are also mounted in base of stack. To prevent entrainment, wooden baffles and a thin bed of 1-in. rachig rings are installed in the stack.
‡Horizontal scrubber is 33 ft 3¼ in. long, 4 ft 9¼ in. wide, and 11 ft 10 in. high. Unit is divided into eleven compartments so that gas goes up through first, then down through second compartment, and continues alternately up and down throughout the unit. The first 10 compartments contain 8 spray nozzles each; the last is dry.

Data on 13 scrubbers handling superphosphate-den gas have been presented by Sherwin (31). Ten of these are more or less conventional spray-tower systems, one is a packed tower, and one is a jet-exhauster system. The spray towers show values for K_Ga ranging from 0.62 to 2.65; the packed tower, a K_Ga of 3.7, and the jet exhauster, a K_Ga of 15.6. The volume for the jet exhauster is based on the volume of the "tower" which would enclose the vertical venturi pipe from the jet level to the level of the liquid in the tank below. The system obviously gives a very high volume-coefficient of performances; however, power consumption was reported high and overall fluorine-removal efficiency for two units in series was not as high as that of the majority of the spray installations. A portion of the data on these units is summarized in Table 6-17.

It will be noticed that the two-stage spray tower (installation 4) gives slightly better performance than the six-stage spray tower. The two major reasons for this appear to be (a) the lower gas velocity which allows the mist formed to settle out and (b) the appreciably higher water-circulation rate. Silica-deposition problems generally favor the use of a spray tower for this service over the more compact packed tower.

Table 6-17. Summary of Data for Typical Commercial Superphosphate Den-gas Scrubbers

Scrubber Variables	Spray Units		Packed Tower	Jet System
	A	B		
Installation number......................	1	4	12	13
Stages in series.........................	6	2	3	2
Size of dens, tons superphosphate/hr........	28	9	6.5	25
SiF_4 in feed gas, % by volume............	0.29	0.54	0.17	0.88
SiF_4 in exit gas, ppm by volume...........	38	50	110	200
Gas flow, cfm/(ton superphosphate)(hr).....	250	134	462	120
Liquid flow, gpm/(ton superphosphate)(hr)..	0.75	1.42	....	12.0
Gas velocity, ft/sec......................	24	0.77	6.3	
Fluoride-removal efficiency, %.............	98.6	99.1	93.1	97.8
Effective height, ft (height × no. of passes)..	170.0	35.0	45	49
Tower volume, cu ft/(ton superphosphate) (hr).....................................	293.0	101.0	56	4.8
Liquid circulation rate gal/1,000 cu ft of gas.	3.0	10.7	10	100
Gas contact time, sec....................	70	46	7.1	0.2
K_Ga..................................	0.616	1.04	3.68	15.6
N_{OG}.................................	3.8	4.7	2.7	3.8

Source: Data of Sherwin (31)

Hansen and Danos report on experience with a large (18 ft × 8½ ft × 46 ft) crossflow scrubber in a phosphoric acid plant. (52) The scrubber consisted of a spray chamber followed by multiple packed beds of plastic woven mesh. With regard to the spray chamber section of the scrubber, they conclude that a spray nozzle pressure over 60 psig is required to attain 80 percent fluoride removal efficiency (1.5 transfer units); the amount of spray chamber water should be about 20 to 30 gpm/1000 acfm; and full cone spray nozzles directed countercurrent to the gas flow are preferred. The plastic woven mesh may be irrigated with low-pressure co-current sprays; however, the nozzles should be mounted so that they are equidistant from the packing face and should be designed so that, when partially plugged, they do not create a single jet of water that can wear holes in the woven mesh.

Data on a commercial installation for removing HF and other fluoride compounds from the exhaust gases of a nodulizing kiln have been reported in considerable detail (32) and are reproduced in Table 6-18.

Materials of Construction

As would be expected with chemicals as corrosive as hydrofluoric and fluosilicic acids, the selection of suitable construction materials is an important design consideration. As a general principle, it can be stated that unprotected carbon steel is not suitable for aqueous solutions of these acids at any concentration and, therefore, corrosion-resistant alloys, organic materials (wood or polymers), concrete, or brick must be employed.

Exhaust gas containing hydrofluoric acid is normally produced at an elevated temperature, and carbon-steel ducts may be employed to convey the gas to the purification unit provided its temperature is well above the dew point. It is usually necessary to precool the gas before it enters the actual absorption equipment in order that the latter can be constructed from, or lined with, wood or other organic materials. The gas can be most readily cooled by the use of water sprays within the ducts, and the spray section of the conveyor duct should be constructed of stainless steel or other material which can stand both high temperatures and corrosive liquids. After cooling, the gas and dilute aqueous hydrofluoric acid can be handled in equipment constructed of wood or lined with resistant plastics such as polyvinyl chloride, polyethylene, Kel F, and neoprene. For large aluminum-plant scrubbers such as shown in Figure 6-14, the hydrofluoric acid-absorption towers are constructed of clear heart-redwood staves with internal piping of polyvinyl chloride and brass, stainless-steel, or Monel nozzles.

Ejectors for hydrofluoric acid service may be constructed of cast iron or steel and lined with a protective film such as neoprene or Kel F, or they may be fabricated entirely from a hydrofluoric acid-resistant material. Asplit "F,"

Table 6-18. Operating Data on Water-spray Tower Treating Nodulizing-kiln Exhaust Gas

Design Variables	
Tower diameter, in..........................	18
Tower height, ft...........................	80
No. of spray injection-points...............	6
Gas inlet temperature, °C...................	300
Gas exit temperature, °C....................	72
Gas inlet rate, cfm........................	52,000
Gas exit rate, cfm.........................	26,500
Rate of inlet HF, lb F_2/day...............	4,000
Rate of exit HF, lb F_2/day................	97
Fluorine removed (as HF), %..............	97.6
Rate of inlet NaF as dust, lb F_2/day.......	340
Rate of exit NaF as dust, lb F_2/day........	14
Removal of F_2 in dust, %.................	95.9
Water flow, gpm..........................	700
Hydrated lime consumption, lb/day.........	24,000

Source: Data from *Air Pollution Handbook* (32)

a modified phenolic resin fortified with an inert carbon filler, has proved to be satisfactory for this service (33).

In silicon tetrafluoride operations, the gas is usually available at moderate temperature (from phosphate-rock acidulation); however, relatively high fluoride concentrations may be developed. Depending on whether recirculation of fluosilicic acid solution is practiced, the fluosilicic acid concentration in the liquid may range from a fraction of 1 percent to over 10 percent. Because of the wide range of operating conditions and performance requirements of fertilizer-plant exhaust-gas purifying systems, a considerable variety of construction materials have been utilized.

Experience in Great Britain (31) indicates that good-quality engineering brick without frogs, bonded with a latex hydraulic-type cement, is the most satisfactory type of massive construction material for silicon tetrafluoride-absorption towers. In the United States, the towers are more commonly constructed of wood, with or without a protective organic coating. Tower basins and sumps are usually constructed of conventional portland-cement concrete. This material is apparently protected from severe attack by the precipitation of silica and other compounds in the pores as a result of the initial reaction between fluosilicic acid and constituents of the cement.

Monel metal and high chromium-molybdenum-nickel stainless steel are generally the most resistant materials of construction for equipment in contact with fluosilicic acid. However, they are very expensive, and their use is generally reserved for precision parts in very severe service, such as impellers and nozzles, particularly where recirculation is practiced. Common brass has proved quite satisfactory for exterior piping and, in some cases, for spray nozzles. Lead is not satisfactory for handling fluosilicic acid solutions primarily because of the erosion caused by silica particles.

In addition to causing erosion of solution-handling equipment, silica particles, which precipitate from both gaseous and liquid phases, complicate the design of lines and vessels. The silica coats the inside of ducts and columns, plugs packings, and settles out of the solution stream wherever its velocity is reduced. The problem is best handled by designing the equipment so that all surfaces in the absorber are flushed with adequate quantities of water and the resulting solution is kept moving as rapidly as possible in lines and collection basins. In spite of such precautions, it is common practice to flush out silicon tetrafluoride-absorption systems with high-pressure water once every 2 or 3 days and to physically scrape deposited silica from the walls and ducts several times a year.

Disposal of Absorbed Fluoride

In both HF- and SiF$_4$-absorption operations, the product water may be too acid and toxic for direct disposal to sewers. It is, therefore, common practice to neutralize the effluent with limestone (or lime) in a separate tank. The fluoride ion is precipitated as calcium fluoride, and where other components such as silica and iron are present, these may also be precipitated as a result of the pH change. The solids can be readily separated by settling or filtration and disposed of separately, although occasionally the neutralized mixture, which is relatively innocuous, can be disposed of as a dilute slurry.

Efforts have been made in some installations to recover the fluorine evolved in phosphate-rock processing as cryolite or other marketable products. The quantity of fluorine evolved in such operations is very large—approximating the total consumption of fluorine in the United States during 1956 (34)—so that steps to recover it would appear warranted. A process developed by the Tennessee Valley Authority (TVA) and described by Farr is claimed to accomplish this (34). In the TVA process, the water used to scrub nodulizing-kiln exhaust gas is maintained at a pH of 5 to 6 by the continuous addition of ammonia, and the resultant rich liquor is treated to precipitate impurities and yield a valuable NH$_4$F solution. The absorber solution is recirculated to bring its fluorine content up to about 35 g/liter. The rich solution is then treated with sufficient ammonia to raise the pH to 9, thus precipitating iron, silica, and part of the phosphorus. The precipitate is filtered off, and the solution is then used to make cryolite (by adding sodium sulfate and alum at a pH of 6) or aluminum fluoride.

Additional studies conducted by TVA on the removal of fluorine from aqueous scrubber effluents and the production of useful fluorine compounds have been reported by Tarbutton (35) et al. and by Barber and Farr (36). The latter study describes cryolite recovery from effluents from phosphorus furnaces.

Hydrogen Chloride Absorption

The absorption of hydrogen chloride in water (or in dilute HCl) is a very important operation in the manufacture of commercial hydrochloric acid. It is much less important as a gas-purification process because hydrogen chloride is seldom present in industrial gas streams other than those associated with its manufacture. Where HCl vapor is present as an impurity in gas streams, it can readily be removed by washing with water. The only complicating factors are the extreme corrosiveness of the resultant solution and the problem of its disposal.

Vapor-pressure data for the hydrogen chloride-water systems are presented in Table 6-19. As can be seen, the vapor pressure of hydrogen chloride over dilute aqueous solutions is extremely low although it increases appreciably with increased temperature. The heat of solution is considerable, however (240 Btu/lb of 35 percent hydrochloric acid produced at room temperature), and if it is desired to effect very complete removal of hydrogen chloride from a concentrated gas stream or to produce a solution of maximum concentration, heat removal is necessary. This may be accomplished by using cooled absorbers or by recycling the acid through a cooler and back to the absorption unit.

Because of the very high solubility of HCl in water and the rapidity with which the reaction with water occurs, the absorption is completely gas-film controlled. With concentrated gas streams, in which the low concentration of inert gas permits rapid diffusion through the gas stream, absorption is extremely rapid and very simple devices such as externally cooled, wetted-wall columns may be used. With less concentrated gases, the absorption rate is reduced somewhat, and conventional ceramic-packed towers may be employed.

Although the absorption of concentrated HCl vapors in the manufacture of hydrochloric acid is beyond the scope of this book, the removal of traces of HCl from the tail gas of such a process is a gas-purification problem. This type of operation is particularly important in the manufacture of HCl as a by-product of hydrocarbon-chlorination reactions, because the hydrogen chloride is associated with appreciable volumes of inert gas in the effluent from such processes. An excellent review of hydrochloric acid manufacture is presented in the *Encyclopedia of Chemical Technology* (38). A standard-design tail-gas absorber, as described by this text, is filled with stoneware packing and takes liquid loadings of 1 gal/(min)(sq ft) or higher and gas

Table 6-19. Partial Pressure of Hydrogen Chloride Gas over Hydrochloric Acid Solutions in Water (Variation with Temperature)

Conc. of HCl, lb per 100 lb H₂O	Partial pressure HCl, mm Hg, at			
	0°C	20°C	50°C	110°C
78.6	510			
66.7	130	399		
56.3	29.0	105.5	535	
47.0	5.7	23.5	141	
38.9	1.0	4.90	35.7	760
31.6	0.175	1.00	8.9	253
25.0	0.0316	0.25	2.12	83
19.05	0.0056	0.0428	0.55	28
13.64	0.00099	0.0088	0.136	9.3
8.70	0.000118	0.00178	0.0344	3.10
4.17	0.000018	0.00024	0.0064	0.93
2.04	....	0.000044	0.00140	0.280

Source: Data from Perry (37)

velocities from 1 to 3 ft/sec. A weak acid containing 20 percent or less HCl is produced in the tower, and the HCl concentration in the gas vented is reduced to the range of 0.1 to 0.3 percent. The towers are usually operated under suction with a fan located in the vent-gas stream.

Experimental data on the operation of a complete hydrogen chloride-absorption pilot plant, including both the primary cooler-absorber and the tails tower, have been presented by Coull, Bishop, and Gaylord (39). The unit was constructed of Karbate (trademark of National Carbon Company for impervious carbon and graphite products) and consisted of a vertical water-jacketed tube through which the HCl gas and liquid flowed cocurrently for production of strong acid, plus a packed column through which unabsorbed tail gas was passed countercurrent to a stream of water. A schematic diagram of the unit together with data from one typical run are presented in Figure 6-19.

The cooler-absorber contained a Karbate tube, ⅞-in. ID by 1¼-in. OD and 108 in. long, which served as a cooled, wetted-wall column, and the tails tower consisted of a 4-in. ID Karbate tube packed to a depth of 4 ft 5 in. with ½-in. carbon raschig rings. Losses of hydrogen chloride in the vent gas were found to be negligible, and, in general, the absorption of hydrogen chloride in the tails tower appeared to take place in the lower portion, as indicated by the height of the hot zone. The run used to illustrate the schematic diagram of Figure 6-19 was not typical of this regard, however, as the hot zone extended 40 in. up from the bottom of the 5-ft high tails tower.

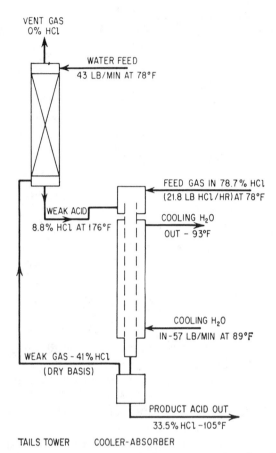

Figure 6-19. Schematic diagram of hydrochloric acid—absorption pilot plant with operating data from a typical run. From *Coull, Bishop, and Gaylord* (39)

A similar installation was employed by Dobratz et al. (40), who evaluated three different absorption tubes in the cooler-absorber (1½-in. ID Karbate, 1-in. ID tantalum, and 0.88-in. ID stainless steel) and investigated the production of 36 to 40 percent hydrochloric acid solutions. These authors present complete data, including heat-transfer and absorption-coefficient correlation for the cooler-absorber but give no data on the performance of their 4-in. by 4-ft packed tails tower.

Kantyka and Hincklieff (41) have shown that a single-tower adiabatic absorber is capable of recovering a 28 percent hydrochloric acid of commercial quality from by-product gas produced during batch chlorination of organic compounds.

An evaluation of wet-fiber filters for use in the absorption of hydrogen chloride vapors by water in connection with an unusual gas-purification problem was presented by First and Warren (28). The hydrogen chloride-containing gas in this case was produced as the result of an operation used to make very pure silica by the combustion of silicon tetrachloride. The gas contained a white fume of silicon dioxide and some unburned silicon tetrachloride as well as the hydrogen chloride. The gas absorber consisted of two stages of wetted, glass-fiber mats plus a third-stage dry mat which served as a mist eliminator. The mats were constructed of curly glass fibers of 50-μ diameter. The wet stages were each 4 in. thick, and the dry pad was 2 in. thick. Gas entered at a temperature of 350°F and was precooled by a spray nozzle in the inlet duct. Absorbent was sprayed on the fiber mats at a rate of 3.8 gpm/sq ft for a gas-flow rate of 216 scfm/sq ft. Pressure drop through the unit was found to be 4 in. of water gauge.

With water as the absorbent, the HCl content of the gas was reduced from 3.15 mg/cu m to 0.0025 mg/cu m, a removal efficiency of 99.9 percent. Other tests made, using a 5 percent aqueous solution of sodium carbonate as the absorbent, revealed essentially the same removal efficiency. Although the apparatus proved very effective for HCl removal, little or no decrease in the quantity of fine silica fume was effected by the system.

Hydrochloric acid is extremely corrosive to most metals, and great care must be exercised in the choice of construction materials. Special alloys such as Durichlor, the Chlorimets, and the Hastelloys are suitable for hydrochloric acid solutions under some conditions. Pure tantalum is inert to hydrochloric acid at all concentrations and at temperatures up to 350°F. Among the suitable nonmetals are acidproof brick, chemical stoneware, chemical porcelain, glass, glass-lined steel, rubber (natural and synthetic for low-temperature service), plastics (vinyl chloride, polyethylene, polystyrene, filled phenolics, and fluorocarbons), and various forms of carbon. Carbon and graphite have found extensive use in the handling of wet or dry hydrogen chloride at temperatures up to about 750°F. Karbate, a resin-impregnated carbon or graphite material, is widely used for heat-transfer and absorption equipment.

Chlorine Absorption in Water

The need for the removal of chlorine from gas streams occurs most frequently as a gas-purification problem in connection with the manufacture, liquefaction, transfer, and storage of elemental chlorine. The problem also occurs to some extent as a result of magnesium chloride electrolysis, hydrocarbon chlorinations, and other chlorine-producing or -consuming operations.

The principal source of chlorine-containing gas in caustic-chlorine plants is the liquefaction step where noncondensables are vented from chlorine condensers as "sniff" gas containing 30 to 40 percent chlorine by weight. Dilute gas may be collected at other points in the operation; this also requires purification before it can be vented to the atmosphere. A number of processes have been developed to recover the chlorine from the vent-gas streams, including its use for the manufacture of bleach. Where the demand for bleach does not justify this operation, a regenerative recovery system is necessary, and one of the simplest of these involves absorption in water. The absorption of chlorine gas in water is also an important step in the manufacture of certain types of wood pulp. In this application, the process is intended primarily to provide a source of concentrated bleaching solution; however, design data which have been obtained for the absorption step are equally applicable to gas-purification or chlorine-recovery operations.

A schematic diagram of the water-absorption process for recovering chlorine from caustic-chlorine process "sniff" gas developed by the Hooker Electrochemical Company (42) is shown in Figure 6-20. In this process chlorine-containing noncondensable gas from the liquefaction stage of chlorine manufacture is scrubbed countercurrently with water in a packed absorber. The resulting chlorine-free gas can be vented to the atmosphere, and the aqueous chlorine solution is used to cool impure chlorine from the electrolytic cell by direct contact in a second packed tower. Chlorine remaining in the water after contact with the hot-cell gas is removed by steam stripping in the lower portion of the same tower, and the stripped water is discarded.

Solubility Data

Solubility data for the chlorine-water system have been published by Whitney and Vivian (43). These authors point out that the two-phase system may be visualized as consisting of chlorine gas at such a partial pressure that it is in equilibrium with molecular chlorine in the solution—a relationship which can be assumed to follow Henry's law. In addition, the dissolved molecular chlorine will be in equilibrium with hypochlorous acid and hydrogen and chloride ions in the solution in accordance with the following reaction:

$$Cl_{2(aq)} + H_2O \rightleftharpoons HOCl + H^+ + Cl^- \tag{6-13}$$

For the case of chlorine in pure water, the total dissolved chlorine can be expressed as

$$C = \frac{p}{H'} + \left(\frac{K_e p}{H'}\right)^{\frac{1}{3}} \tag{6-14}$$

where C = total concentration of chlorine in water, lb moles/cu ft

p = partial pressure of chlorine vapor over the solution, atm

K_e = Equilibrium constant for reaction of dissolved chlorine and water given above

H' = Henry's law coefficient for equilibrium between gaseous chlorine and dissolved but unreacted chlorine in water (atm)(cu ft)/lb mole

Values for the equilibrium and Henry's law constants given by Vivian and Whitney (44) are presented in Table 6-20.

Absorption Coefficients

An extensive investigation of the absorption of chlorine by water in packed columns was made by Vivian and Whitney (44), who confirmed earlier findings (45, 46) that the absorption coefficients for this system are much lower than would be predicted from liquid-film correlations. Vivian and Whitney proposed an explanation of the low coefficients based on the relative rates of

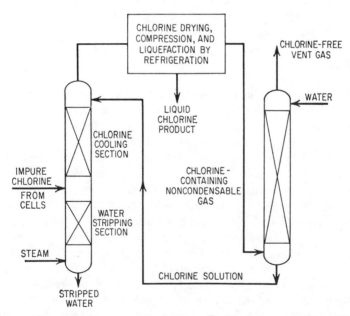

Figure 6-20. Simplified diagram of water-absorption process for recovering chlorine from waste gases of electrolytic caustic chlorine plants. *Hooker Electrochemical Company process* (42).

Table 6-20. Henry's Law and Reaction Equilibrium Constants for Chlorine in Water		
Temperature, °F	H', (atm)(cu ft)/lb mole	K_e, (lb moles/cu ft)^2
50	141	7.10×10^{-7}
59	171	8.55×10^{-7}
68	213	10.7×10^{-7}
77	256	12.8×10^{-7}
Source: Data of Vivian and Whitney (44)		

hydrolysis and of diffusion and suggested the use of a pseudocoefficient using an unhydrolyzed-chlorine driving force for correlation of absorption data for this system.

The absorption coefficient $K_L a$ (normal) was found to increase with liquid rate in accordance with the relationship $K_L a \propto L^{0.6}$. This exponent is somewhat lower than the value observed by Sherwood and Holloway (5) for other sparingly soluble gases. Gas-flow rate was found to have no appreciable effect on the value of the coefficient over a tenfold range of gas rates—further confirming the liquid-film-controlled nature of the absorption.

The normal absorption coefficient was found to increase with increased temperature. The data plotted as a straight line on semilog coordinates indicated a relationship of the form:

$$K_L a = me^{nt} \tag{6-15}$$

With t in °F, n was found to have the value 0.0115. As an alternate correlation, Vivian and Whitney noted that the absorption coefficient is approximately proportional to the sixth power of the absolute temperature over the relatively narrow range studied (35 to 90°F).

Vivian and Whitney obtained data in two towers (4-in. and 14-in. ID). However, for convenience of operation, the major portion of the runs was made in the 4-in. column. The results for both towers are correlated for a single typical temperature in Figure 6-21. The 4-in. tower was packed to a height of 2 ft with 1-in. tile raschig rings while the 14-in. tower was packed to a height of 4 ft with similar rings. As can be seen, the towers show different performance characteristics when calculated as normal liquid-film coefficients, but this difference disappears when pseudocoefficients are used. The pseudocoefficient values are observed to be closer to predicted liquid-film coefficients (based on the oxygen-water system) at high liquid rates, while the normal coefficients are closer to predicted values at very low liquid rates. This is explained on the basis of the relatively slow chemical reaction between chlorine and water. At low liquid rates (and, therefore, low absorption rates) the reaction has time to proceed almost to equilibrium in the liquid film and a conventional coefficient applies. At high liquid rates, on the other hand, chlorine molecules are absorbed faster than they can react with water in the film zone so that an appreciable portion of the dissolved chlorine may penetrate into the body of the liquid in the molecular state. If this occurs, the driving force which is available to produce diffusion is the difference in molecular-chlorine concentration between the interface and the body of the liquid rather than the difference in total chlorine concentration.

The use of pseudocoefficients as proposed by Vivian and Whitney represents a procedure by which chlorine absorbers can be designed on the

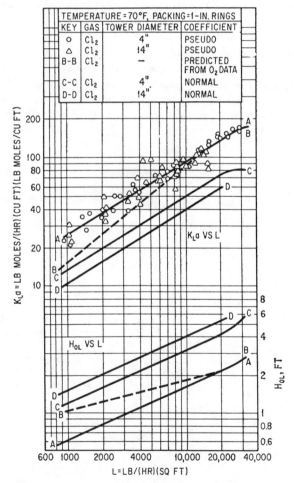

Figure 6-21. Effect of liquid-flow rate on $K_L a$ and H_{OL} for chlorine absorption in water. *Data of Vivian and Whitney* (44)

basis of liquid-film coefficients obtained with other systems. However, if actual chlorine-absorption data, such as those presented by Vivian and Whitney, are available for the desired packing, the normal liquid-film coefficients can probably be employed in the design of commercial equipment with reasonable assurance.

Materials of Construction

The chlorine-water system is both an acid and a strong oxidizing agent and, therefore, constitutes an extremely corrosive environment. Materials normally resistant to oxidizing conditions, such as stainless steels, may be attacked by aqueous Cl_2 solutions because the oxidation products formed are soluble in water and cannot protect the underlying metal. Carbon steel is rapidly attacked. However, rubber is resistant to the solutions at moderate temperatures so that rubber-lined steel equipment can be employed. Rubber-lined absorbers are used, for example, in the Hooker "sniff"-gas chlorine-recovery process previously described (42). Ceramics and glass are unaffected by chlorine solutions so that glass pipe and stoneware packing are excellent construction materials for this system. Among the metals, the best resistance is shown by nickel, silver, tantalum, Chlorimet 3, and Durichlor.

References

1. *International Critical Tables,* vol. 4. New York: McGraw-Hill Book Company, Inc.
2. Dodds, W.S.; Stutzman, L.F.; and Sollami, B.J. 1956. *Ind. Eng. Chem.* (Data Series) 1(1):94.
3. Morgan, O.M., and Maass, O. 1931. *Can. J. Research* 5:162.
4. Cooper, C.M.; Christl, R.J.; and Peery, L.C. 1941. *Trans. Am. Inst. Chem. Engrs.* 37:979.
5. Sherwood, T.K., and Holloway, F.A.L. 1940. *Trans. Am. Inst. Chem. Engrs.* 36(Feb.25):21.
6. Rixon, F.F. 1948. *Trans. Inst. Chem. Engrs. (London)* 26:119-130.
7. Walter, J.F., and Sherwood, T.K. 1941. *Ind. Eng. Chem.* 33:493.
8. Cooper, C.M. 1945. U.S. Patent 2,398,345.
9. Pollard, B. 1957. *Trans. Inst. Chem. Engrs. (London)* 35:69-75.
10. Yeandle, W.W., and Klein, G.F. 1952. *Chem. Eng. Progr.* 48(July):349-352.
11. Walthall, J.H. (Tennessee Valley Authority). 1958. Personal communication, (Oct. 29).
12. Anon. 1930. *Ind. Eng. Chem.* 22:433.
13. Watkins, J.W., and Kineheloe, G.W. 1958. *Corrosion* 14(7):55-58.
14. Watkins, J.W., and Wright, J. 1953. *Petrol. Engr.* 25(Nov.):B-50-57.
15. Burnham, J.G. 1959. The Refining Eng. (Feb.):C-15-C-19.
16. Barbouteau, L., and Galaud, R. 1964. *Gas Purification Processes.* Edited by G. Nonhebel, London: George Newnes Ltd., Chap. 7.
17. Wright, R.H., and Maass, O. 1932. *Can. J. Research* 6:94.
18. Selleck, F.T.; Carmichael, L.T.; and Sage, B.H. 1952. *Ind. Eng. Chem.* 44(Sept.):2219.

19. Froning, H.R.; Jacoby, R.H.; and Richards, W.L. *Hydro. Process.* 43(Apr.):125-130.

20. Anon. 1955. *Chem. Engr.* 62(Feb.):110.

21. Bradley, B.W., and Dunne, N.R. 1957. *Corrosion* 13(Apr.):238t.

22. Semrau, K.T. 1956. *Emission of Fluorides from Industrial Processes.* Paper presented at American Chemical Society Air Pollution Symposium, Atlantic City, N.J. (Sept. 17-19).

23. Brosheer, J.C.; Lenfesty, F.A.; and Elmore, K.L. 1947. *Ind. Eng. Chem.* 39(Mar.):423.

24. Whynes, A.L. 1956. *Trans. Inst. Chem. Engrs. (London)* 34:118.

25. Sherwin, K.A. 1955. *Chem. & Ind. (London)* 41:1274-1281.

26. Berly, E.M.; First, M.W.; and Silverman, L. 1954. *Ind. Eng. Chem.* 46:1769-1777.

27. Lunde, K.E. 1958. *Ind. Eng. Chem.* 50(Mar.):293.

28. First, M., and Warren, R. 1956. *J. Air Pollution Control Assoc.*, 6:32-34.

29. Pettit, A.B. 1951. *Chem. Eng.* 58(8):250-252.

30. Pettit, A.B. 1951. *Air Pollution and Smoke Prevention Assoc. Am. Proc.* 44:98.

31. Sherwin, K.A. 1954. *Trans. Inst. Chem. Engrs. (London)* 32(Suppl. 1):S129-140.

32. Magill, P.L.; Holden, F.R.; and Ackley, C. 1956. *Air Pollution Handbook.* New York: McGraw-Hill Book Company, Inc., p. 96.

33. Brown, C.R., and Tomlinson, R.W. 1952. *Air Pollution and Smoke Prevention Assoc. Am. Proc.* 45:69-74.

34. Anon. 1957. *Chem. Eng. News* 35(Part 3/Sept. 23):81.

35. Tarbutton, G.; Farr, T.D.; Jones, T.M.; and Lewis, H.T., Jr. 1958. *Ind. Eng. Chem.* 50(10):1525-1528.

36. Barber, J.C., and Farr, T.D. 1970. *Chem. Eng. Progr.* 66(Nov.):56-62.

37. Perry, J.H. (ed.) 1950. *Chemical Engineers' Handbook.* 3rd ed. New York: McGraw-Hill Book Company, Inc., p. 675.

38. Kirk, R.E., and Othmer, D.F. 1951. *Encyclopedia of Chemical Technology.* New York: Interscience Publishers, Inc.

39. Coull, J.C.; Bishop, H.; and Gaylord, W.M. *Chem. Engr. Progr.* 45(Aug.):525-531.

40. Dobratz, C.J.; Moore, R.J.; Barnard, R.D.; and Meyer, R.H. 1953. *Chem. Engr. Progr.* 49(Nov.): 611.

41. Kantyka, T.A., and Hincklieff, H.R. 1954. *Trans. Inst. Chem. Engrs. (London)* 32:236-243.

42. Anon. 1957. *Chem. Eng.* 64(Part 1/June):154.

43. Whitney, R.P., and Vivian, J.E. 1941. *Ind. Eng. Chem.* 33:741.

44. Vivian, J.E., and Whitney, R.P. 1947. *Chem. Eng. Progr.* 43(Dec.):691.

45. Adams, S.W., and Edmonds, R.G. 1937. *Ind. Eng. Chem.* 29:447.

46. Whitney, R.P., and Vivian, J.E. 1940. *Paper Trade J.* 110(20):29.

47. Hanf, E.W. and MacDonald, J.W. 1975. *Chem. Eng. Progr.* 71,3(March):48-52.

48. Calvert, S.; Goldshmid, J.; Leith, D.; and Mehta, D. 1972. *Wet Scrubber System Study, Vol. 1, Scrubber Handbook.* EPA-R2-72-118a (August, 1972) Distributed by NTIS as PB-213016.

49. Henrich, R. A. 1981. Biocycle, May/June, p. 27.

50. Henrich, R. A., and Ross, B. 1983. "Landfill and Digester Gas Purification by Water Extraction and Case Study of Commercial System, Baltimore's Back River Wastewater Treatment Plant," presented at the IGT Symposium on Energy from Biomass and Waste, Disney World, Florida, January 27.

51. Illanionev, W. V. et al., Zhumal Prikladnoi Khimll. 1963. *36.* No. 2, (February) p. 237.

52. Hansen, A. O. and Danos, R. J. 1982. *Chem. Eng. Prog. 78,* No. 3, (March) p. 40.

53. Hansen, A. O. and Danos, R. J. 1980. "The Design and Selection of Scrubbers for Granulation Plants," presented at the Central Florida Section, AIChE Annual meeting, Clearwater, Florida, May 24.

7

Sulfur Dioxide Removal

The problem of sulfur dioxide removal from stack gases has probably been the subject of more research than any other gas purification operation. This research had very little commercial impact until the 1970s when an explosive growth occurred in the number of flue gas desulfurization (FGD) units installed in the United States and Japan. Unfortunately, this rapid growth is more attributable to regulatory pressures than to research breakthroughs. The FGD systems, in general, continue to be costly adjuncts to industrial operations which would otherwise release excessive amounts of SO_2 to the atmosphere. Their success is measured in terms of minimizing cost and operating problems rather than making a profit from recovered sulfur values.

Requirements for SO_2 Removal

One of the first precedents for establishing limits on sulfur dioxide discharge in terms of ground level concentration was set in connection with the operation of smelters in the Salt Lake district of Utah in 1920 and resulted in the imposition of a regime which limited the sulfur dioxide concentration to 1 ppm (for an hourly average) at the level of vegetation during the growing season (1, 2).

The Trail, Canada, smelter of the Consolidated Smelting and Mining Company of Canada, Ltd., was the subject of a prolonged international investigation which resulted in the establishment of an operating regime setting the maximum permissible discharge in terms of tons per hour under certain weather conditions and restricting the emission in terms of ground concentration and duration (3). The first known instance where sulfur dioxide removal was a legal requirement for the operation of a large power plant was at the

Battersea Station of the London Power Company constructed in 1929 (4). More recently, blanket restrictions have been placed upon industry in many areas limiting the quantities and concentrations of SO_2 which can be emitted. In the United States, the Air Quality Act of 1967 (Public Law 90-148, November 1967) required the Secretary of Health, Education, and Welfare to establish air quality criteria for pollutants, such as SO_2, to publish and report on the technology that can be used to control these pollutants, and to select air quality regions. It also authorized an effort to "model" the pollution levels in the atmosphere of all metropolitan areas. This latter effort was not particularly successful, and the Clean Air Act Amendments of 1970 (Public Law 91-604, December 31, 1970) dictated a revised strategy with emphasis on procedures to obtain standards to be met in each federally designated region by 1975 (5).

Table 7-1 shows the Environmental Protection Agency (EPA) ambient air quality standards promulgated April 30, 1971. The primary standards were designed to protect public health, while the secondary standards were prescribed to protect the public welfare. To meet the primary standards, the 1970 Clean Air Act Amendments required each state to adopt (and submit to the administrator) a State Implementation Plan (SIP) to provide for meeting the standard as soon as possible, but no later than 3 years after approval of the SIP. In addition, the act called for New Source Performance Standards (NSPS) which require all new industrial sources to install the best demonstrated pollution control technology regardless of location. New Source Performance Standards established by the EPA for a number of industrial source categories are provided in the Code of Federal Regulations (40 CFR, Part 60). Table 7-2 summarizes the NSPS for the principal sources

Table 7-1. National Sulfur Oxides Air Quality Standards (5)

Standard	Concentration $\mu g/m^3$	ppm	Description
Primary	80	0.03	Annual Arithmetic mean.
	365	0.14	24 hr maximum, not to be exceeded more than once per year.
Secondary	60	0.02	Annual Arithmetic mean.
	260	0.1	24 hr maximum, not to be exceeded more than once per year.
	1300	0.5	3 hr maximum, not to be exceeded more than once per year.

of sulfur dioxide and also includes the proposed NSPS for sulfur recovery plants reported in the Federal Register (41FR43866) October 4, 1976 (6). The Clean Air Act Amendments of 1977 (Public Law 95-95) introduced further changes into the law including a modification of the definition of "Standard of Performance" to include not only emission limitations but also the achievement of a percentage reduction in emissions. This reduction may be attained by pretreating the fuel, but the use of naturally low-sulfur fuels is no longer in itself satisfactory, even if the established emission limitations are met. This section does not authorize EPA to require the use of any particular

Table 7-2. Federal New Source Performance Standards for Sulfur Oxides (6)

Source Category	Affected Facilities	Maximum Emissions
Fossil-Fueled Steam Generators	Coal- and Oil-Fired Boilers	Solid Fuel: 2.2 g $SO_2/10^6$ cal (1.2 lb $SO_2/10^6$ Btu)* Liquid Fuel: 1.4 g $SO_2/10^6$ cal (0.8 lb $SO_2/10^6$ Btu)
Sulfuric Acid Plants	Process Equipment	2 Kg $SO_2/10^6$ cal (4 lb SO_2/ton H_2SO_4) 0.074 kg acid mist/mton H_2SO_4 (0.15 lb acid mist/ton H_2SO_4)
Petroleum Refineries	Refinery Process Equipment including waste-heat boilers and fuel gas combustors	Fuel gas max H_2S: 230 mg/dry Std M^3 (0.10 grain/dry Std ft^3)
Primary Copper Smelters	Roaster, Smelting Furnace Copper Converter	0.065% SO_2 by vol
Primary Zinc Smelters	Roaster, Sintering Machine	0.065% SO_2 by vol
Primary Lead Smelters	Sintering Machine, Dross Reverberatory Furnace, Electric Smelting Furnace and Converter	0.065% SO_2 by vol
Petroleum-Refinery Sulfur Recovery Plants	Claus Plants	0.025% SO_2 by vol (with incineration) or 0.030% by vol reduced sulfur compounds and 0.0010% by vol H_2S (reduction only)

*Under the 1977 Clean Air Act Amendments, EPA has proposed that 90% removal of sulfur input be required with a floor of 0.2 lb $SO_2/10^6$ and a maximum emission of 1.2 lb $SO_2/10^6$ Btu.

system of emission reduction, but the standards may be established so that stack gas scrubbing is a necessity in many cases.

Specific standards were set for utility boilers of 250 million Btu/h or more under a rule adopted in 1979. The restrictions range from 70 percent removal for low-sulfur (1 percent) western coal to 90 percent removal for high-sulfur (3.5 percent) eastern coal.

The effect of implementation of the Clean Air Amendments on the amount of sulfur dioxide released to the atmosphere is quite dramatic. It has been estimated (7) that the total emissions of SO_2 from all sources in the United States will decrease from 33 to 28 million tons per year from 1975 through 1990. This assumes that the 1975 State Implementation Plan enforcement levels are continued through 1990, and 90 percent availability is realized for all SO_2 recovery systems on new units subject to New Source Performance Standards. It also assumes "business as usual" coal consumption trends with no major coal substitution or energy conservation measures instituted. Without the requirement for flue gas desulfurization, it is estimated that the amount of SO_2 released would increase by a factor of three during the same time period.

Flue gases from combustion processes normally contain less than 0.5 percent sulfur dioxide. The relationship between sulfur content of the fuel and sulfur dioxide content of the resulting flue gas is shown in Table 7-3. This table gives the sulfur dioxide content of combustion gases from several typical fuels. The values given are based upon combustion with approximately 15 percent excess air.

Stack gas from smelters handling sulfur ores, on the other hand, may range up to as high as 8 percent sulfur dioxide. The economics of recovering sulfur values from such gases can be much more favorable. Of course, the problems of discharging such gases without sulfur dioxide recovery are also much more acute.

Many smelting operations, which produce relatively high concentrations of sulfur dioxide in the gaseous product, feed the gas stream directly into a more or less conventional sulfuric acid plant. The design and operation of acid plants of this type are not discussed in this text, as they are considered to represent a separation and chemical manufacturing operation, not a gas purification process. The removal of unconverted sulfur dioxide from the acid plant tail gas, on the other hand, constitutes a gas purification problem.

Sulfur Trioxide Formation

Most combustion gases, which contain sulfur dioxide, also contain a small but significant amount of sulfur trioxide (or its reaction product with water, sulfuric acid). This component is of considerable importance because of its highly corrosive nature, its effect on the chemistry of many sulfur dioxide recovery processes, and its suspected critical role in air pollution problems.

The amount of sulfur trioxide formed during combustion is primarily a function of air/fuel ratio, fuel composition, temperature, time at temperature, and the presence or absence of a catalyst. The equilibrium concentrations of the principal sulfur species in the combustion gas from a typical fuel oil at several air/fuel ratios have been calculated by Pebler (8). The results show that in excess air mixtures at equilibrium, SO_2 is the most stable compound above 1000°K; SO_3 is the predominant sulfur compound between 900° and 600°K; while on further cooling, H_2SO_4 gains dominance over SO_3.

The sulfuric acid condenses as a liquid phase below 400°K. Fortunately, the equilibrium conditions are not attained in conventional combustion processes. However, the presence of catalytically active material, such as vanadium in oil and iron pyrites in coal can increase SO_3 formation. In the absence of actual analytical data for specific cases, a rough estimate of the SO_3 concentration, expected in combustion gases from oil and coal, may be obtained from Table 7-4. Data compiled from several other sources by Pierce (9) are presented.

The sulfuric acid dew point of combustion gases, which is the key parameter with regard to stack corrosion, has been studied by a number of investigators. The available data have been reviewed and correlated by Pierce, who presented the result in graphical form. Selected points from his correlations are given in Table 7-5. Used in combination, Tables 7-4 and 7-5 show, for example, that typical combustion gases from 2 percent sulfur coal should not be cooled below 285°F in equipment which can be corroded by dilute sulfuric acid (e.g., carbon steel).

Process Categories and Economics

A great many processes have been proposed for removing sulfur dioxide from gas streams, however, relatively few have attained commercial status. Although the primary emphasis of this text is on industrial processes, some discussions are presented relative to processes in the development stage which appear to have the potential for future commercialization. In addition, discussions include several processes which are no longer considered viable, but once represented major developmental efforts or commercial operations and

Table 7-3. Sulfur Dioxide Concentration in Combustion Flue Gases*	
Fuel	SO₂ in Flue Gas,%
Coal, 4% sulfur..........................	.... 0.35
Fuel oil, 2% sulfur......................	.... 0.12
Fuel oil, 5% sulfur......................	.... 0.31
Refinery acid sludge, 40% sulfuric acid..	.... 2.0
*Fifteen percent excess air.	

Table 7-4. Estimate of Sulfur Trioxide in Combustion Gases (9)

Fuel	Excess Air %	Oxygen in Gas, %	Sulfur Trioxide Expected in Gas, ppm With Fuel Sulfur Content of:					
			0.5%	1.0%	2.0%	3.0%	4.0%	5.0%
Oil	5	1	2	3	3	4	5	6
Oil	11	2	6	7	8	10	12	14
Oil	17	3	10	13	15	19	22	25
Oil	25	4	12	15	18	22	26	30
Coal	25	4	3-7	7-14	14-28	20-40	27-54	33-66

Table 7-5. Sulfuric Acid Dewpoint of Typical Combustion Gases (9)

Fuel	Water Vapor in Gas %	Acid Dewpoint °C (°F) for Sulfur Trioxide Concentrations of:					
		1 ppm	5 ppm	10 ppm	25 ppm	50 ppm	100 ppm
Oil	10	116 (241)	131 (268)	135 (275)	144 (290)	150 (302)	157 (315)
Coal	6	110 (230)	125 (257)	130 (266)	139 (281)	145 (293)	152 (306)

can, therefore, provide valuable background data pertaining to the development of improved new processes.

It is common practice to categorize FGD methods as nonregenerable or regenerable depending upon the disposition of the spent sorbent. However, this breakdown does not provide a logical basis for discussing the technologies involved because the identical SO_2 removal step may be used with both nonregenerable and regenerable processes. Accordingly, FGD processes have been categorized solely by the initial SO_2 removal step. This categorization is used as a general basis of organization for the balance of the chapter. A detailed categorization of FGD processes, in this manner, is given in Figure 7-1. Not all process types indicated in the figure are represented by commercial processes. In fact, a list of U.S. power plant FGD systems operational, under construction, or planned as of June 1983, includes only nine different processes (see Table 7-6). The number of operating plants increased from 29 to 114 between 1977 and 1983 while the number of processes employed increased by only one during this period (11, 133). Japan had almost 1,000 FGD plants operational in mid-1977 utilizing about 15 basically different types of processes (10); however, the growth of FGD capacity in Japan was slow after 1977 (37 plants were built in the 1978-83 time period), and

Table 7-6. U.S. Power Plant Flue Gas Desulfurization Systems
as of June 1983 (133)

| | Number of Systems | | |
	Operational	Under Construction	Planned
Limestone Scrubbing (Nonregenerative)	52	14	11
Lime Scrubbing (Nonregenerative)	35	5	2
Dry Lime	5	7	4
Dry Sodium Carbonate	1	—	—
Sodium Carbonate Scrubbing (Nonregenerative)	6	—	—
Sodium Sulfite Scrubbing (Wellman-Lord, Regenerative)	7	—	—
Magnesia Scrubbing (Regenerative)	3	—	—
Sodium Carbonate (ACP, Regenerative)	1	—	—
Sodium Carbonate (Double Alkali, Nonregenerative)	4	1	1
Total Number of Plants	114	27	18

virtually no plants involving processes other than lime/limestone were built. (134) A comprehensive evaluation and status report covering 189 flue gas desulfurization processes and 24 subsystems with regard to their applicability to power plants has been published by the Electric Power Research Institute (EPRI) (135).

Several attempts have been made to compare the economics of the various processes. Such studies are of questionable value for comparing proposed new processes with proven systems because of the large uncertainties in cost estimates for processes still in the developmental stage. In addition, the results are strongly affected by the assumed bases, including scope of items included in capital costs; unit costs (and credits) assumed for calculating operating costs, assumed on-stream time; and the time frame of the estimate. Nevertheless, some idea of the overall economics of flue gas desulfurization can be gained from Table 7-7 which provides data on nine different FGD processes compiled from several sources. (12, 13, 14, 15, 16, 136) Reisdorf et al. (136) evaluated 17 processes and concluded that the Chiyoda thorough-bred 121 process is the most economical of the proven techniques for high-sulfur coal applications. The injection of trona into the flue gas was deter-mined to be the most economical process for low-sulfur plants, although its level of development is lower than that of the lime spray dryer process which was a close second.

In subsequent sections of this chapter, specific sulfur dioxide removal processes are described. The order generally follows that of Figure 7-1, and is based upon the agent used for initially contacting and removing SO_2 from the gas stream.

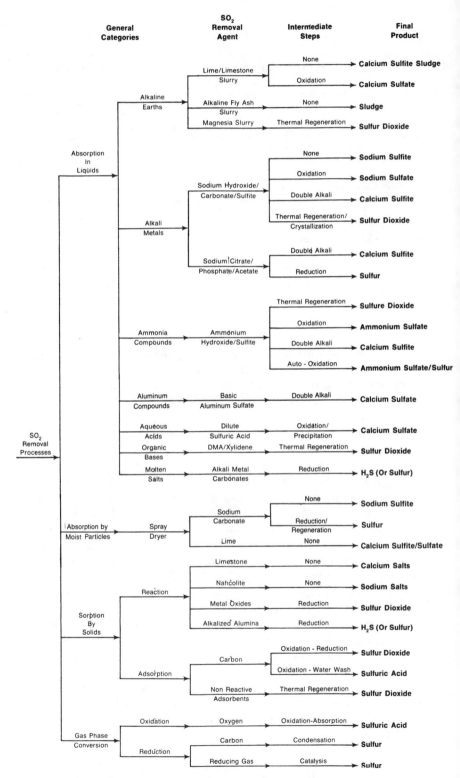

Figure 7-1. Categorization of sulfur dioxide removal processes.

Alkaline Earth Processes

Lime/Limestone Process

The lime/limestone process is currently the predominant process for power plant flue gas scrubbing and has been the subject of numerous studies and publications. Two particularly valuable documents are the "Limestone FGD Scrubbers: Users Handbook" (137) and the "Lime FGD Systems Data Book." (138) The gas is contacted with an aqueous slurry of lime or limestone. Sulfur dioxide in the gas reacts with the slurry to form calcium sulfite (and sulfate). These compounds are collected as a relatively inert sludge for disposal, and the purified gas is discharged to the atmosphere (after passing through a mist eliminator and a reheater). A schematic flow diagram is shown in Figure 7-2. For purposes of this discussion, the use of slurries of alkaline fly ash or special reactive lime-based sorbents will be considered as modifications of the basic lime/limestone process.

The process is believed to have originated with Eschellman (17) who, in 1909, patented a method of purifying burner gases using milk of lime (a slurry of lime in water). The history of the development of the process through the early British plants in the 1930s to large, modern installations has recently been traced by Marten (18).

The earliest commercial applications of the lime/limestone process were in London, England. The first unit at the Battersea Power Station was put into operation in 1931. This was followed by improved units at Bankside, Swansea, and Fulham. The initial process was primarily a once-through water wash, using the natural alkalinity of the Thames River. However, a small amount of additional reactant in the form of chalk slurry was added to the water before it entered the absorber column. A flow diagram of the Battersea process is shown in Figure 7-3, and a photograph of one of the absorbers of the Bankside plant during construction is shown in Figure 7-4. The process used at the Fulham Power Plant, which was installed in 1938, was called the cyclic lime process because water was recycled in the process after removal of calcium sulfate sludge. This process is basically the same as the modern lime/limestone systems.

A modification of the process which was installed on several United States power plants in the early 1970s involves the injection of limestone into the furnace followed by the scrubbing of flue gas with a lime slurry (19, 20). This process appeared to offer the advantage of in-situ calcination of limestone to lime and 2-stage contacting (dry plus wet). Early operations, however, encountered numerous operational problems. A simple lime/limestone wet scrubber technique (Figure 7-2) has been generally preferred for more recent installations.

Another early process "improvement" which has not been favored in recent plants is the concept of combining fly ash particulate removal with the

Table 7-7. Comparative Economics of Processes for Desulfurizing Power Plant Flue Gas

Process	Capital Costs, $/Kw						Operating Costs, mils/kWh					
Reference Date Basis	A '76-'78	B '75-'78	C '77-'80	D 1973	E 1975	F 1982	A 1978	B 1978	C 1980	D 1973	E 1975	F 1982
Lime	53.9	61.1	—	—	—	165	4.57	4.15	—	—	—	20
Limestone	62.6	68.4	97	69.8	82	175	4.39	4.01	4.03	3.43	3.65	18
Double-Alkali	81.9	—	101	—	—	150	5.29	—	4.19	—	—	17
Magnesium Oxide	74.9	71.7	—	—	—	270	4.20	3.23	—	—	—	19
Sodium Sulfite (Wellman-Lord)	82.5	84.4	—	75.6	—	275	4.79	3.04	—	3.98	—	26
Catalytic Oxidation (Cat-Ox)	—	109.6	—	99.0	—	—	—	4.99	—	4.37	—	—
Citrate	—	—	150	—	—	—	—	—	6.65	—	—	—
Ammonia (ACPI-IFP)	—	—	—	—	103	—	—	—	—	—	3.43	—
Chiyado Thoroughbred	121	—	—	—	—	140	—	—	—	—	—	14

Notes:
1. All except F are 500 MW power plants burning 3.5% sulfur coal, with 90% recovery from the flue gas. A, B, D, and E studies are for new plants; C is a new FGD system retrofitted to an existing boiler, F is a 1000 MW plant (2-500 MW units) burning 4% sulfur coal.
2. Operating costs include normal operating expenses, capital charges, and credits for byproducts where applicable.
Data sources: A - PEDCO Environmental (12), B - McGlamery et al (13), C - Torstrick et al (14),
    D - M.W. Kellogg Co. (15), E - Jimeson & Maddocks (16), F - Reisdorf et al (136)

SO$_2$ removal scrubber. This scheme offers a very large potential for cost savings by eliminating the need for an electrostatic precipitator or baghouse. However, it also introduces several problems. First, it eliminates the clean dry fan location ahead of the wet scrubber. Operation of the fan after the scrubber creates the potential for corrosion and deposition problems. Second, the chemical and physical operating characteristics of the wet scrubber system are affected by the addition of the fly ash particulates. And more importantly for many utilities, this arrangement eliminates the possibility of

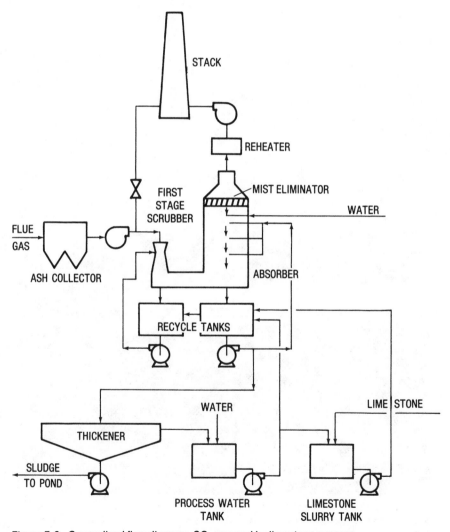

Figure 7-2. Generalized flow diagram, SO$_2$ removal by limestone process.

310

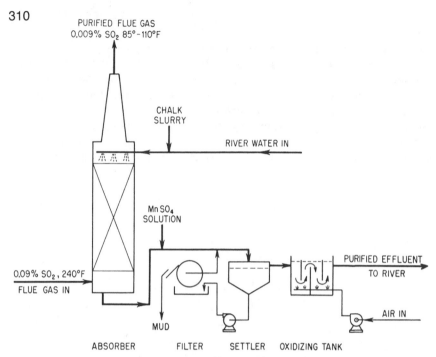

Figure 7-3. Flow diagram of the Battersea process for removal of sulfur dioxide from exhaust gas.

Figure 7-4. View of upper section of absorption chamber at Bankside power plant during construction operations. The wooden-slat packing is visible. *Central Electricity Authority, London, England.*

bypassing the SO_2 scrubber system without also bypassing particulate removal. A further advantage of a separate dry ash removal step is that the dry ash can then be mixed with the wet sludge to produce a product for disposal with a higher total solids content than that produced from a single wet system. Nevertheless, the economies associated with elimination of the electrostatic precipitator or other dry fly ash removal systems are substantial, and the combined system may regain favor as the above problems are solved (21).

Process Description

The basic requirements of the lime/limestone process are a scrubber for contacting the gas with the absorbent slurry, a holding tank for the slurry where the chemical reactions can proceed, a separator to remove spent solids from the liquid absorbent, and a system to feed fresh lime or limestone into the circuit. Refinements needed for most installations include a particulate removal device to preclean the feed gas, a saturator to cool and add water to the feed gas before it enters the scrubber, a mist eliminator following the scrubber, and a reheater to warm the product gas before it is exhausted to the atmosphere. Pumps, blowers, and controls are, of course, necessary accessories to the process. Many of the major system components are visible in the photograph (Figure 7-5) of a scale model of a large limestone process FGD plant.

Key items which must be evaluated in the design of lime/limestone systems include: absorbent selection (lime versus limestone), absorbent cycle design (slurry concentration, recycle rate), absorber type, mist eliminator design and operation, reheater design, sludge handling technique, and materials of construction. These items are discussed in subsequent sections.

Basic Chemistry

When sulfur dioxide dissolves in water, a portion of it ionizes according to the following equations:

$$SO_{2(g)} \rightleftarrows SO_{2(aq)} \tag{7-1}$$

$$SO_{2(aq)} + H_2O \rightleftarrows H^+ + HSO_3^- \tag{7-2}$$

$$HSO_3^- \rightleftarrows H^+ + SO_3^= \tag{7-3}$$

The solubility of sulfur dioxide in pure water in equilibrium with pure gases is given in Figure 7-6 which is based on the data of Parkison (22). As indicated by the equations, the amount of sulfur dioxide absorbed by an aqueous system can be increased by reducing the hydrogen ion concentration

Figure 7-5. Photograph of a scale model of the limestone process FGD plant on Unit 2 of Alabama Electric Cooperative's Tombigbee Station. Components from left to right are: Inlet ductwork; I.D. fans; common recycle tank, spray tower absorbers, one behind the other; slurry preparation tank; limestone silo; and limestone ball mill (in enclosure). *Peabody Process Systems, Inc.*

or by removing HSO_3^- or $SO_3^=$. The addition of calcium oxide or carbonate to the system accomplishes both of these actions. In the presence of lime and limestone, the following reactions occur:

lime dissolution,

$$Ca(OH)_{2(s)} \rightleftarrows Ca(OH)_{2(aq)} \tag{7-4}$$

$$Ca(OH)_{2(aq)} \rightleftarrows Ca^{++} + 2OH^- \tag{7-5}$$

limestone dissolution,

$$CaCO_{3(s)} \rightleftarrows CaCO_{3(aq)} \tag{7-6}$$

$$CaCO_{3(aq)} \rightleftarrows Ca^{++} + CO_3^= \tag{7-7}$$

reaction with dissolved SO_2,

$$Ca^{++} + SO_3^= \rightleftarrows CaSO_{3(aq)} \tag{7-8}$$

$$CaSO_{3(aq)} + \tfrac{1}{2}H_2O \rightleftarrows CaSO_3 \cdot \tfrac{1}{2}H_2O_{(s)} \tag{7-9}$$

oxidation,

$$HSO_3^- + \tfrac{1}{2}O_2 \rightleftarrows SO_4^= + H^+ \tag{7-10}$$

$$Ca^{++} + SO_4^= \rightleftarrows CaSO_{4(aq)} \tag{7-11}$$

$$CaSO_{4(aq)} + 2H_2O \rightleftarrows CaSO_4 \cdot 2H_2O_{(s)} \tag{7-12}$$

coprecipitation,

$$Ca^{++} + (1-x)SO_3^= + xSO_4^= + \tfrac{1}{2}H_2O \rightleftarrows$$
$$Ca(SO_3)_{1-x}(SO_4) \cdot \tfrac{1}{2}H_2O_{(s)} \tag{7-13}$$

liberation of CO_2 (from limestone),

$$CO_3^= + H^+ \rightleftarrows HCO_3^- \tag{7-14}$$

$$HCO_3^- + H^+ \rightleftarrows H_2CO_{3(aq)} \tag{7-15}$$

$$H_2CO_{3(aq)} \rightleftarrows CO_{2(g)} + H_2O \tag{7-16}$$

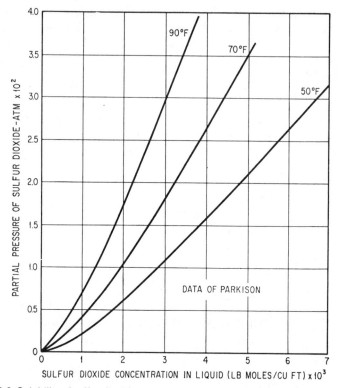

Figure 7-6. Solubility of sulfur dioxide in pure water. *Data of Parkison* (22)

Equilibrium conditions for Reactions 7-2 and 7-3 (sulfite/bisulfite distribution) and for Reactions 7-14 and 7-15 (carbonate/bicarbonate distribution) are defined by the curves of Figure 7-7 (23).

The individual steps involved in the removal of SO_2 from gas streams by the lime/limestone process may be summarized as follows:

1. Transfer of SO_2 to the gas/liquid interface
2. Solution of SO_2 in water at the interface
3. Ionization of dissolved SO_2 (Note: hydrolysis of dissolved SO_2 to sulfurous acid is often included as a step, but there is no strong evidence of its existence.) (24) (25)
4. Transfer of HSO_3^-, H^+, and $SO_3^=$ ions from the interface into the liquid interior
5. Dissolving of $Ca(OH)_2$ or $CaCO_3$
6. Ionization of dissolved calcium salts to form Ca^{++}
7. Reactions of Ca^{++} with $SO_3^=$ to form $CaSO_3$
8. Precipitation of $CaSO_3 \cdot \frac{1}{2}H_2O$

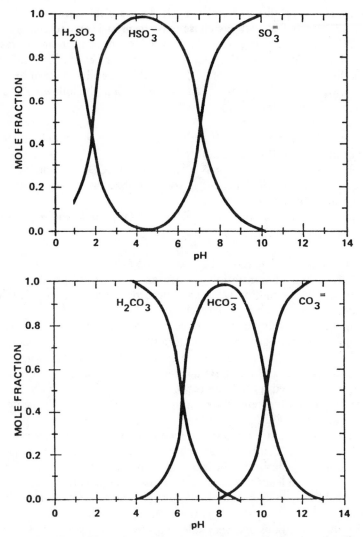

Figure 7-7. Bisulfite-sulfite and bicarbonate-carbonate distributions as a function of pH.

9. Oxidation of sulfite to sulfate
10. Reaction of Ca^{++} with $SO_4^=$ to form $CaSO_4$
11. Precipitation of $CaSO_4 \cdot 2H_2O$
12. Co-precipitation of $Ca(SO_3)_{1-x}(SO_4)_x \cdot \frac{1}{2}H_2O$

Research indicates that the rate-controlling mechanisms for SO_2 absorption are step 1, gas phase mass transfer; step 4, liquid phase mass transfer; and step 5, dissolving of $CaCO_3$. In the ideal case, steps 4 and 5 become sufficiently fast that gas phase mass transfer is controlling. There is evidence that this case is approached at low SO_2 concentrations in the gas and relatively high solution pH. High SO_2 removal efficiencies are obtained under these conditions. The efficiency of SO_2 removal is generally lower at high SO_2 concentrations in the gas (greater than 3000 ppm), and the following chain of events has been postulated (25):

1. Excess dissolved SO_2 causes a drop in pH of the slurry and creates an oversupply of sulfite (or bisulfite) ions.
2. The locally high pH around a dissolving $CaCO_3$ particle decreases $CaSO_3$ solubility and causes it to crystallize on the surface of the limestone particle.
3. Therefore, there is "blinding" of the $CaCO_3$ particle and inhibition of further dissolution and reaction.

Thus, it can be concluded that the absorption of SO_2 by a limestone slurry from relatively high-concentration gas streams is controlled primarily by the rate of limestone dissolution (25). The above problem does not occur with lime because of its much higher reactivity.

A comprehensive study of the chemistry of SO_2 absorption via limestone was conducted by the Radian Corporation (26). In the Radian model, SO_2, CO_2, O_2, and NO_x in the gas phase are considered to be in equilibrium with components in the liquid. The system is represented by 28 chemical reactions involving 50 condensed species. Bechtel Corporation utilized the Radian data in the development of a computer program for calculating chemical compositions in a plant system (23).

The equilibrium data do not tell the whole story, however, because non-equilibrium effects contribute strongly to observed process operating problems and characteristics. The absorption of sulfur trioxide or oxidation of sulfite in solution leads to the formation of sulfate (equation 7-10 and step 9). This would eventually be expected to lead to the precipitation of gypsum scale (equation 7-12 and step 11). However, it has been found that if the recycling slurry contains seed crystals and the concentration of $CaSO_4$ in solution is less than ~1.35 times the saturation value, precipitation will occur on the seed crystals rather than on the equipment.

An alternative approach for the prevention of calcium sulfate scale deposition is to operate with a solution less than saturated with respect to $CaSO_4$. This can be accomplished by continuously removing $CaSO_4$ from solution as a co-precipitate with calcium sulfite at a rate equal to that at which it is formed by oxidation. Up to ~12 to 15 mol percent sulfate can be incorporated into the calcium sulfite crystals as a solid solution (26). Therefore, if oxidation is less than ~15 percent of the sulfite formed, absorber operation free of gypsum deposition is possible. This mode of operation has been observed at the EPA test facilities at Shawnee and Research Triangle Park, North Carolina (27) (28). According to Borgwardt (29), lime is superior to limestone with respect to ease of development of unsaturated conditions because of the higher precipitation rates and lower oxidation rates that are characteristic of the lime system.

The process chemistry is strongly affected by the presence of magnesium and chloride ions. The nature of this effect is described in detail by Head in the third progress report of the EPA alkali scrubbing test facility: advanced program (23). He points out that any cation (such as Mg^{++} or Na^+) which forms a soluble sulfite or sulfate will increase the concentration of alkaline species and, therefore, increase the rate of SO_2 removal. Alkaline species are defined as those ions that are a base to aqueous SO_2. Chloride ions are acidic with respect to SO_2 and can cancel otherwise effective cations. $MgCl_2$ is, therefore, not effective in improving SO_2 absorption, and the effective Mg^{++} concentration is defined by the following simple equation:

$$2(Mg^{++})_{effective} = 2(Mg^{++})_{total} - (Cl^-) \qquad (7\text{-}17)$$

The effect of high chloride concentrations in the scrubbing liquor is of increasing importance because of the trend toward tightly closed water loops in FGD systems. The chloride may originate from the coal (and be introduced as HCl in the flue gas) or from the makeup water. Extensive experimental work has been performed on the effects of chloride on SO_2 absorption (139, 140, 141). In general, it has been found that high concentrations of chloride reduce SO_2 absorption efficiency. Downs et al. (139) found that the negative effect could be suppressed by increasing L/G, increasing holdup of the liquor, or decreasing limestone particle size. Laser et al. (140) found a strong cation effect. Calcium chloride decreased both pH and SO_2 absorption efficiency; magnesium chloride actually increased efficiency at Cl^- concentrations below 40,000 ppm, while no significant effects were observed for sodium chloride at Cl^- concentrations less than 50,000 ppm. Chan and Rochelle (142) developed an integrated scrubber model to predict SO_2 removal as a function of NaCl and $CaCl_2$ concentration, which accurately simulated the available test data. According to this model, Cl^- reduces SO_2 removal by reducing the apparent equilibrium constant for SO_2 hydrolysis, and Ca^{++} reduces SO_2 removal by reducing the concentrations of HCO_3^- and $SO_3^=$ in solution.

Selection of Absorbent

The choice between lime or limestone is largely an economical one, as both reagents can normally accomplish the required SO_2 removal efficiency. The use of a fly ash slurry, on the other hand, requires that the coal being burned have an ash with a relatively high concentration of alkaline metal oxides (e.g., CaO or MgO) and a relatively low sulfur content. Generally, the coals best suited to alkaline fly ash scrubbing are the western lignites and sub-bituminous coals (6).

Limestone is much lower in cost than lime, and because of this, is generally the preferred absorbent for plants which are located close to usable limestone deposits. However, limestone has several drawbacks which cause lime to be preferred for some applications:

1. Limestone must be finely pulverized to react at a reasonable rate; good results have been obtained with 80 to 90 percent of the particles passing through a 200-mesh screen.
2. Transportation costs are higher for limestone because of its greater weight per mole and because of its poorer stoichiometry (typically, 1.3 to 1.5 moles limestone per mole SO_2 in the feed gas versus 1.1 moles lime).
3. Scrubbing plant equipment is larger because of the increased holdup time required for limestone reaction and the greater amount of absorbent used.

Limestones may be characterized as 1) high-calcium limestones containing at least 95 percent $CaCO_3$; 2) high-magnesium limestones which are predominately $CaCO_3$ but contain over 5 percent $MgCO_3$; and 3) dolomitic limestones containing $CaCO_3$ and $MgCO_3$ in approximately equal molar concentrations. The first of these categories is the best for use as uncalcined limestone, the second is less effective, and the third is generally not suitable. $MgCO_3$ is relatively unreactive and can also render some of the calcium unreactive in the form of $MgCO_3 \cdot CaCO_3$ crystals. However, when limestone containing $MgCO_3$ is calcined, the magnesium content is converted to a soluble reactive form. As previously discussed, the presence of magnesium ions in the scrubbing solution greatly improves process operation. If the limestone is used without calcining, it is frequently desirable to add a small proportion of a soluble magnesium compound to the circulating slurry. Calcined dolomite or magnesium sulfate are typical additives.

Several patents relating to the use of magnesium oxide as an additive to lime scrubbing systems have been obtained by the Dravo Corporation, Pittsburgh, Pennsylvania (30, 31, 32). Their proprietary system is offered as the Thiosorbic Flue Gas Desulfurization Process. Pilot plant and commercial experience with the process are described by Selmeczi and Stewart (33) who point out that the use of a magnesium oxide additive increases SO_2 removal efficiency, prevents scale deposition in the scrubber, and permits clarifier overflow to be used successfully for washing the mist eliminators. The Pullman Kellogg Company also offers a system incorporating a magnesium compound additive based on a patent (U.S. 3,883,639) which covers the use of a soluble sulfate (such as $MgSO_4$) in concentrations of approximately 3 to 27 percent. It is claimed that the improved absorption efficiency obtained with the modified calcium-based (lime or limestone) system permits the use of a low contact-time absorber such as the Weir Horizontal Spray Contactor (34).

Lime is much more reactive than limestone. As a result, the pH of the solution can be adjusted quickly by changing the rate of lime addition. With limestone, it is necessary to maintain a lower pH in the scrubbing slurry to force the absorbent into solution. The resulting mixture is buffered by the $CaCO_3$ so that rapid pH shifts cannot occur. A final point in favor of limestone relates to the mechanical problems of materials handling. Limestone is a relatively inert material which can be shipped and stored in the open. Lime, however, can absorb water from the air to form calcium hydroxide. This results in appreciable heat release and caking of the product. Lime is, therefore, shipped and stored in sealed containers, a requirement which can have a significant impact on materials handling costs.

The addition of buffering agents such as carboxylic acids has been found to enhance the rate of SO_2 absorption by lime/limestone slurries. Chang and Rochelle conclude that adipic, sulfopropionic, sulfosuccinic, and β-hydroxypropionic acids are the most promising; (143) however, adipic acid has been studied most extensively (144, 145, 146). Lower cost mixtures of organic dibasic acids (DBA) obtained as byproducts have also been evaluated (147, 148).

The extent of improvement in SO_2 removal efficiency brought about by the addition of adipic acid and DBA mixtures to a limestone scrubber is illustrated by Figure 7-8 from Chang and Dempsey (147). The chart is based on data obtained in a three-stage turbulent contact absorber (TCA) pilot plant with a flue gas capacity of about 7.5 m^3/min (~ 0.1 MW).

The absorber used 1-1/2-in. diameter nitrile foam spheres arranged in three beds each 4 in. in depth (static). During all of the tests, the liquid/gas ratio was maintained at 8 l/m^3 (60 gal/mcf) with a liquid flow rate of 1.1 l/sec (17 gal/min). The flue gas contained 2800-ppm SO_2 and 6–8% oxygen. The following additives were tested:

1. Pure adipic acid.
2. Badische acid water, byproduct of a cyclohexanone plant containing about 11 percent adipic acid, 9 percent hydroxycaproic acid, and 15 percent other organic acids in water.
3. Badische caustic water, byproduct of a cyclohexanone plant containing the sodium salts of valeric, caproic, hydroxycaproic, succinic, adipic, and other acids.
4. Monsanto DBA, byproduct of adipic acid manufacture containing primarily glutaric, succinic, and adipic acids.

5. Du Pont DBA, byproduct of adipic acid manufacture containing primarily glutaric, succinic, and adipic acids.

The results showed all of the acids to enhance SO_2 removal significantly and to be roughly equal in effectiveness when their concentrations are expressed in terms of milliequivalents of organic acid per liter (meq/l). For pure adipic acid, one meq/l is equivalent to 73 ppm.

The test also showed that the DBAs performed in a similar manner to adipic acid with regard to limestone utilization. At a concentration of 13.7 meq/l and a pH of 5.0, for example, the additives caused limestone utilization to increase from about 85 percent to about 95 percent in the pilot system. Economic evaluations have been reported for converting several commercial FGD installations to organic acid enhanced limestone systems (148). The results generally showed the use of DBA to be the most attractive of the various options considered.

Commercial tests of DBA have shown significant performance gains, with SO_2 removal efficiency increasing markedly with increased DBA concentrations. For example, during one test with the scrubbing liquid maintained at a pH of 5.4, the scrubber outlet gas SO_2 concentration decreased from about 400 ppm to < 100 ppm when the DBA concentration was increased from 400- to 900-ppm adipic acid equivalent (149).

The main drawback of organic acid addition is the cost of makeup. During the commercial-scale DBA test described above, DBA consumption averaged about 20 lbs per ton of SO_2 absorbed.

Absorber System Design

The key factors affecting performance of the absorber system are the reactant used (lime or limestone), liquid/gas ratio, gas velocity, slurry pH, SO_2 concentration in the gas, absorber design, solids concentration in the slurry, concentration of other components such as magnesium and chloride ions, and degree of oxidation of the slurry. The stoichiometric ratio (moles of lime or limestone added per mol of SO_2 absorbed) is often considered to be an important design parameter, but it becomes a dependent variable if pH of the entering slurry and the other design factors are fixed. Typical values for the above factors are given in Table 7-8.

The effects of individual parameters on SO_2 removal have been evaluated by Bechtel at the EPA Alkali Scrubbing Test Facility located at TVA's Shawnee Power Station, Padukah, Kentucky (23). Two types of absorbers were tested at the facility, a venturi/spray tower system and a Turbulent Contact Absorber (TCA) System. Each system is capable of treating approximately 10 MW equivalent (up to 35,000 acfm at 300°F) of flue gas containing 1,500 to 4,500 ppm SO_2. The results of these tests and the conclusions are summa-

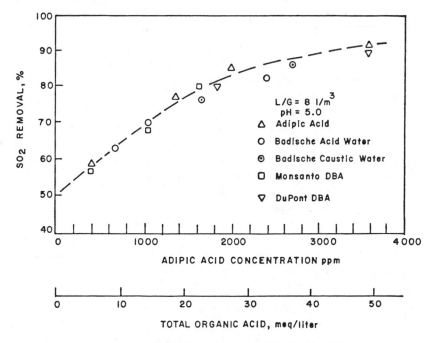

Figure 7-8. Effect of Adipic Acid and DBA Mixtures on SO_2 Removal Efficiency in a Turbulent Contact Absorber (TCA); from Chang and Dempsey (147).

Table 7-8. Typical Operating Conditions for SO_2 Absorption by Lime and Limestone Slurries in a Spray Tower [Based on Shawnee Test Facility Operations (23)]

Absorbent	Lime	Limestone
SO_2 in Feed Gas, ppm	4000	4000
Solids in Slurry, %	12	12
Gas Velocity, ft/sec	8	8
L/G, gal/MCF	60	60
pH of Feed Slurry	7.5	5.6
Stoichiometric Ratio	1.1	1.3
Magnesium Concentration (effective ppm)	3000	3000
Chloride Concentration (ppm)	6000	6000
SO_2 Removal, %	95	75

rized in the nomographs of Figures 7-9 through 7-12. Briefly, the following effects were noted:

1. With both lime and limestone, percent SO_2 removal in the venturi and spray tower decreased slightly with increased gas rate and constant liquid rate. In the TCA, sulfur removal increased slightly or did not change with increased gas velocity.
2. Increasing the liquid flow rate increased SO_2 removal in all systems.
3. Increasing the pH of the feed slurry from 5 to 6 for limestone and from 6 to 9 for lime strongly increased SO_2 removal in all systems.
4. Increasing the effective magnesium ion concentration increased the SO_2 removal. (Effective magnesium ion concentration is defined as the concentration in excess of that required to neutralize chloride ions.)
5. Increasing the chloride ion concentration increased SO_2 removal for all cases except lime scrubbing in the TCA (at constant pH).
6. Increasing SO_2 concentration generally decreased percent removal. However, no quantitative effect was noted for the case of lime scrubbing in the spray tower.

The observed beneficial effect of increasing the chloride ion concentration during these tests was the result of operating at constant pH. If no other changes are made, the addition of Cl^- will lower the pH of the solution. If the system is controlled to operate at constant pH, this results in the addition of more lime or limestone. The increase in SO_2 absorption attributed to chloride is, therefore, actually the result of increasing stoichiometry.

A computer model has been developed by Bechtel National, Inc. and TVA based on data obtained from the Shawnee test program. (150) This model consists of two programs that produce design and economic data for full-scale limestone or lime, wet scrubber FGD systems.

Numerous scrubber designs have been proposed for the lime/limestone process. The principal requirements are low pressure drop, freedom from plugging problems, adequate contact efficiency to meet the SO_2 removal requirements, and low cost. Special designs are required if the scrubber is intended to remove both fly ash and SO_2.

The leading scrubber types are the spray tower, venturi, static packed bed, mobile packed bed, sieve tray, and rod tray. If fly ash and SO_2 are to be removed simultaneously, a two (or more) stage contactor may be necessary. Typical combinations are a venturi followed by a spray tower, two venturis in series, and a multi-stage mobile bed contactor.

One scrubber type consists of beds of marbles. Problems were encountered with early units and several of these have since been converted to spray towers (35). A related type, the Turbulent Contact Absorber (TCA) originally used hollow spheres similar to ping pong balls. These did not wear well. More recent designs are based on a foam-rubber ball with solid skin (36). In general,

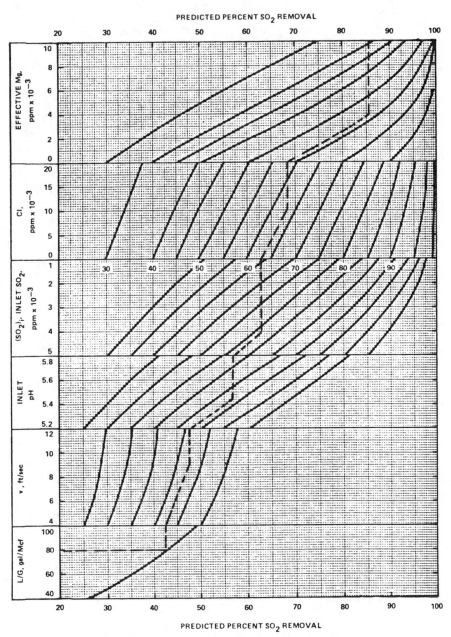

Figure 7-9. Nomograph for percent SO₂ removal by limestone scrubbing in a spray tower.
Data of Head (23)

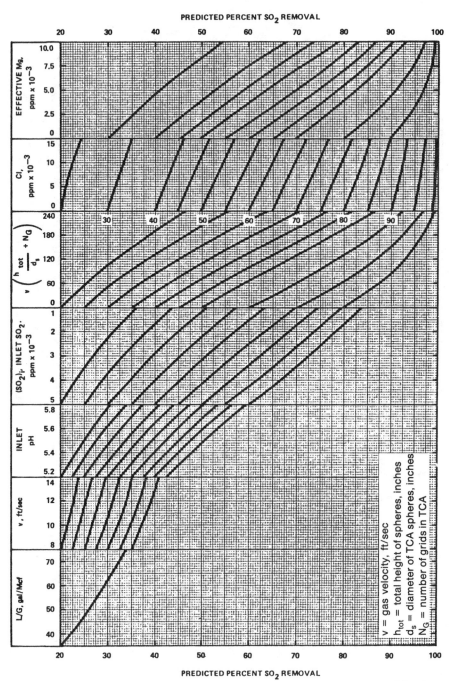

Figure 7-10. Nomograph for percent SO$_2$ removal by limestone scrubbing in a TCA. *Data of Head* (23)

PREDICTED PERCENT SO$_2$ REMOVAL

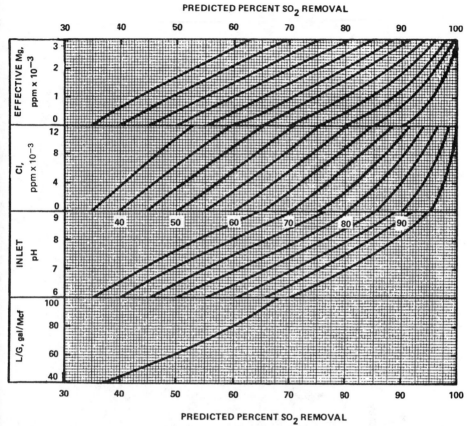

Figure 7-11. Nomograph for percent SO$_2$ removal by lime scrubbing in a spray tower. *Data of Head (23)*

the trend appears to be away from complex contactors with internal components which may offer sites for deposition of solids in the form of scale, unused reactants, and uncollected fly ash. Most major system suppliers now offer spray tower designs. Although spray towers offer design simplicity, they require high liquid to gas ratios (between 60 and 100 gpm per 1,000 cfm) in order to provide a sufficiently high absorption efficiency for SO$_2$. Most of the spray contactors are vertical columns, however, a horizontal design described by Weir (38) appears quite promising from the standpoint of gas pressure drop.

Simplified drawings of a venturi/spray combination and a TCA unit are shown in Figures 7-13 and 7-14. The illustrated units are approximately to scale and depict the relatively small pilot plant equipment at the Shawnee Power Station (23).

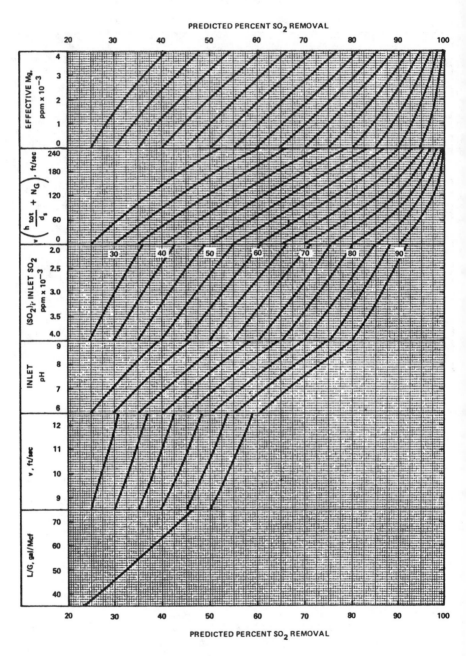

Figure 7-12. Nomograph for percent SO₂ removal by lime scrubbing in a TCA. *Data of Head* (23)

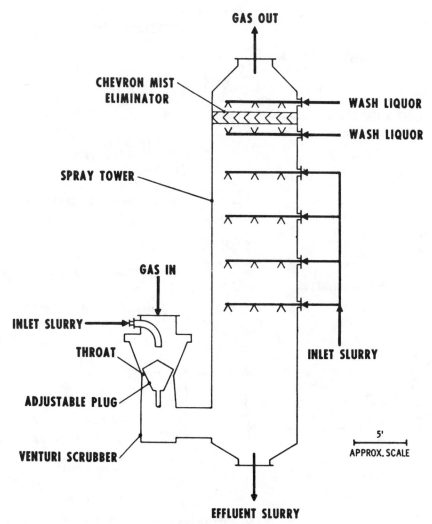

Figure 7-13. Schematic of Venturi Scrubber and Spray Tower (Shawnee Test Facility).

Mist Elimination

The elimination of fine droplets of entrained slurry from the gas leaving the absorber has proven to be a difficult problem, primarily because of the tendency of the recovered liquid to deposit solid material on the mist eliminator surfaces. Techniques which have been developed to minimize the problem include frequent washing of the mist eliminator surfaces, the use of wash-water trays to prewash the gas before it encounters the mist

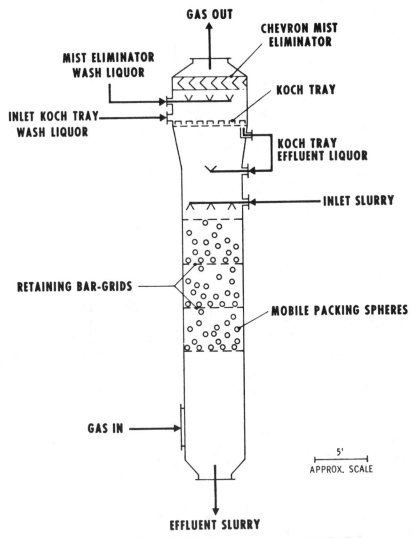

GAS OUT

CHEVRON MIST ELIMINATOR

MIST ELIMINATOR WASH LIQUOR

KOCH TRAY

INLET KOCH TRAY WASH LIQUOR

KOCH TRAY EFFLUENT LIQUOR

INLET SLURRY

RETAINING BAR-GRIDS

MOBILE PACKING SPHERES

GAS IN

5'
APPROX. SCALE

EFFLUENT SLURRY

Figure 7-14. Schematic of Three-Bed TCA Scrubber (Shawnee Test Facility).

eliminators, and the adjustment of absorber cycle stoichiometry to minimize the carry-over of unreacted absorbent. A number of specific design and construction trends were noted by Laseke and Devitt (37) as follows:

1. Chevron designs (continuous vaned construction) predominate over baffle designs (discontinuous slats).
2. Fiberglass reinforced plastic is preferred over stainless steel (316-L) and other types of plastics.

3. The horizontal configuration (vertical gas flow) is preferred over the vertical configuration.
4. Two-stage designs are selected more frequently than single stage.
5. Perforated plates and other precollection devices are installed as integral parts of the mist eliminator systems.
6. Wash-water trays are frequently used as part of the mist eliminator system.
7. Three-pass designs are preferred over two-pass designs except where prior stages or precollectors are incorporated.
8. Wash systems which use intermittent high velocity sprays are preferred to continuous wash systems.

The use of intermittent high velocity sprays minimizes water usage and as a result, the water required as makeup to the absorber cycle is generally sufficient for operation of this type of mist eliminator system. When necessary, clarifier overflow liquor is used in conjunction with fresh water for washing the mist eliminators.

Stack Gas Reheat

Lime/limestone FGD plants generally employ some type of product gas reheat system to avoid condensation and consequent corrosion of downstream equipment, to suppress the formation of a visible plume, and to improve rise and dispersion of the stack gas. Flue gas reheat of 50°F (28°C) is typical. The two most commonly used reheater designs are 1) in-line heat exchange employing steam, and 2) direct combustion of gas or oil, and injection of the hot combustion products into the main stack gas stream after it passes through the mist eliminator. A third type uses a steam/air heat exchanger with the hot air injected into the flue gas before it enters the stack. This technique eliminates the possibility of entrained droplets fouling the heat exchanger surfaces, but is relatively expensive. Perhaps the simplest approach is to bypass part of the flue gas around the absorber. This technique, however, can only be employed when the product gas purity requirements are met by scrubbing only part of the flue gas.

Slurry Processing and Sludge Disposal

Slurry from the scrubbing units flows into one or more reaction or hold tanks which perform the following functions:

1. Dissolution of the added lime or limestone
2. Reaction to form and precipitate calcium sulfite and co-precipitate calcium sulfite and sulfate
3. Precipitation of calcium sulfate

Residence time in the hold tanks is generally established by the need to precipitate calcium sulfate and thereby prevent the formation of sulfate scale. However, it has also been observed that increased residence time generally increases limestone utilization. At a pH of 5.6, for example, tests at Shawnee (23) indicated a limestone utilization of about 80 percent with 6 minutes of hold tank residence time versus about 85 percent with 12- to 20-minutes time. The same program also produced data indicating that three tanks in series perform better than a single tank with the same residence time. Commercial plants operate with hold tanks providing residence times in the range of 10 to 40 minutes (based on recycle rate).

In many limestone process plants, air is bubbled into the hold tank to provide forced oxidation of dissolved sulfite to sulfate as a means of scale control. The calcium sulfate so formed is precipitated in the hold tank as gypsum, provided sufficient time and seed crystals are available. This reduces the amount of dissolved sulfite returned to the absorber, and thus minimizes the possibility for sulfite oxidation and sulfate scale deposition to occur on equipment surfaces. Forced oxidation is not required for scale control with lime systems (39) since sulfate scaling is controlled either by the naturally occurring seed crystals or by co-precipitation of sulfate with sulfite as described previously in the Process Chemistry section.

A stream of slurry is continuously removed from the hold tanks and processed to concentrate the suspended solid particles which consist primarily of calcium sulfite, calcium sulfate, and unreacted absorbent. The solids are normally removed by gravity settling in a thickener forming a thick sludge containing 35 percent to 45 percent solids which can be pumped to storage ponds or further treated to improve handling and ultimate disposal.

A number of proprietary processes have been developed to "fix" or stabilize sludge so that it can be used as landfill or disposed of without containment. These processes generally include the addition of agents which bind the soluble constituents of the sludge to minimize leaching and harden the material to improve stability at filled sites.

The waste sludge from lime or limestone scrubbing operations is primarily calcium sulfite with lesser amounts of calcium sulfate unless an operation is included to ensure a high level of oxidation. The calcium sulfite tends to form a dilute sludge which presents a difficult disposal problem. The material has a high viscosity at low shear rates which causes plugging, while its thixotropic property makes land disposal somewhat complicated (40). In addition, its high water content increases both water consumption of the FGD process and the volume of sludge requiring disposal. As a result, techniques have been developed for improving the properties of lime/limestone process sludges.

The addition of fly ash to the sludge is one approach which helps to resolve many of the problems associated with its physical properties. Another approach is to convert calcium sulfite in the sludge to calcium sulfate. Tests have shown this to eliminate the thixotropic properties of the sludge and permit the generation of filter cakes of very low water content. Scanning electron microscope photographs show that the oxidation process converts the crystal structure from flat plate-like $CaSO_3 \cdot \frac{1}{2} H_2O$ particles to dense monoclinic $CaSO_4 \cdot 2 H_2O$ (gypsum) crystals (41).

Some oxidation occurs in conjunction with normal flue gas scrubbing operations due to the absorption of oxygen from the flue gas. However, the complete oxidation of calcium sulfite to calcium sulfate normally requires the introduction of a step which is optimized for oxygen absorption. Because of the low solubility of oxygen in water, the absorption is aided by an extended liquid residence time. Typically, the operation is carried out in an agitated tank with air sparging.

Test data from a pilot plant program which simulated the oxidation of slurries circulated in the quencher (flooded disc scrubber) and main absorber circuits of a limestone FGD plant have been reported by Goodwin (41). The results show that complete oxidation to calcium sulfate can be attained in the quencher sump tank by the use of air sparging. Specific results were reported as:

1. Efficient oxidation will not take place with high residual limestone ($CaCO_3$) concentrations; i.e., it is necessary to operate with high limestone utilization if oxidation is desired.
2. Oxidation rate increases with increased air sparge rate. With a slurry depth of 10 feet and a pH of 4.0, the degree of oxidation varies from 40 percent at an air rate equivalent to 1.5 times the amount required by stoichiometry to 96 percent at an air stoichiometry of 2.9.
3. Oxidation rate is strongly affected by pH. Conditions which would produce 90 percent conversion at a pH of 4.0 would oxidize only about 50 percent of the calcium sulfite at a pH of 5.0.
4. Oxidation rate is dependent upon slurry depth. To attain a conversion of 90 percent at 10-foot depth, an amount of air equivalent to 2.5 times stoichiometric was required while the attainment of equivalent conversion at 9-foot depth required the use of 5.0 times the stoichiometric air rate.
5. The oxidized sludge can be dewatered more readily than unoxidized sludge. Filter leaf tests produced filter cakes with about 55 percent solids content from sludge containing 55 percent sulfite versus over 80 percent solids content from sludges containing less than 5 percent sulfite.

Air sparging is apparently not required to oxidize calcium sulfite when alkaline fly ash is used as the primary source of alkali. The large commercial

FGD system at the Colstrip Power Station in Montana has consistently produced an effluent approaching 100 percent oxidation (42). This is accomplished by maintaining a low pH (less than 5.6), operating with a high level of suspended solids in solution (12 percent to 15 percent by weight), and providing a long residence time for slurry in a stirred tank external to the scrubber (8 to 10 hours, based on bleed rate). The Colstrip sludge has been observed to increase in pH from about 5, as discharged from the absorber loop, to about 8, after 10 to 20 hours in the settling pond. This is believed to contribute to the self-hardening characteristic (fixation) that slowly occurs with the Colstrip sludge.

Materials of Construction

The combination of sulfur dioxide, oxygen, and water is highly corrosive to carbon steel, and as a result, special alloys, coatings, and nonmetallic materials are widely used in FGD systems. The lime/limestone process has the additional problem of erosion, particularly where the recirculated slurry contacts solid surfaces at high velocity (e.g., pump impellers and nozzles).

The preferred material of construction for scrubber vessels, inlet and outlet ducts, mist eliminator vanes, and other surfaces contacting the flue gas and absorbent slurry is 316L stainless steel. This material has proven to be appreciably better than 304 or 304L stainless steel and all low-alloy ferritic steels. The data of Crow and Horsman (43) indicate that 316L is not as resistant for general use in the system as Hastelloy C-276, Inconel 625, Carpenter 20Cb-3, or Incoloy 825. These alloys are appreciably more expensive than 316L but may be justified for components which encounter particularly severe conditions such as the inlet gas saturator or prescrubber.

Rubber coatings (particularly, neoprene) have proven to be quite effective against both corrosion and erosion. Pumps, fans, tanks, and pipes can all be rubber-coated. The key to satisfactory performance appears to be the method of application. Scrupulous care in preparing the base metal surface is required to assure good adhesion.

Erosion of 316L nozzles can be excessive, and harder alloys are recommended for these components. Stellite Alloy 6 nozzles gave good service at the Shawnee Test Facility (43).

The performance of plastic linings on carbon steel equipment has been somewhat unpredictable. Both successes and failure have been reported. The failures have generally been blamed on improper application to the metal surface. It appears that a number of plastic coatings, including proprietary fiberglass reinforced (FRP) materials are capable of giving excellent service in hold tanks, washwater tanks, and other moderately severe environments. Solid FRP pipes, tanks, and other components are also satisfactory for many

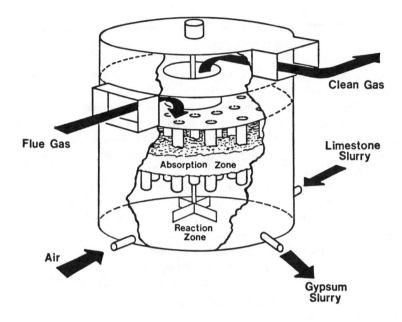

Figure 7-15. Jet Bubbling Reactor (JBR); from Kaneda et al. (151).

secondary applications. FRP has been used extensively as the construction material for mist eliminators with considerable success.

Chiyoda Thoroughbred 121 System

This process differs from conventional lime/limestone systems in that the absorbent is maintained at a pH between 4 and 5 compared with 5.5–6 for other units, and a single reactor is used to remove SO_2, oxidize the reaction products, and form calcium sulfate. Reportedly, one installation has been operating since 1982 in Japan, and in 1983 five other units were in the planning or construction stages. (150)

A key feature of the process is the special reactor design which is illustrated in Figure 7-15. The unit is called a jet bubbling reactor (JBR). This design, which provides a large inventory of absorbent in the reaction zone, eliminates the need for liquid recycle. The low pH favors the oxidation of sulfite to sulfate so complete conversion of the solid product to gypsum is feasible.

In the process as operated by the Mitsubishi Petrochemical Company at their Yokkaichi, Japan, complex flue gas from a boiler burning 3-4 percent sulfur oil is treated to remove 97 percent of the SO_2. (151, 152) Flue gas from the air preheater is first blown into a precooler where a fine spray of recirculated water humidifies and cools the gas while removing particulates and other impurities (such as chlorides).

The humidified gas flows into the JBR where it bubbles through a shallow zone of absorbent. Sulfur dioxide is absorbed, oxidized, and reacted with calcium ions in solution to precipitate calcium sulfate and form a slurry of gypsum. Makeup limestone is added as 20 percent slurry, air is blown into the bottom reactor zones of the JBR to enhance the oxidation reactions, and the product gypsum is continuously withdrawn as a slurry containing about 15 percent $CaSO_4 \cdot 2H_2O$.

The gypsum slurry is dewatered in a solid bowl decanter centrifuge. The byproduct gypsum, which is essentially dry, is taken to storage, and the mother liquid is recycled. A bleed stream of the precooler water is continuously circulated through a thickener to concentrate captured fly ash. The fly ash slurry is mixed with a bleed stream of mother liquor from the centrifuge, neutralized with limestone slurry, and filtered. The resulting solids are discarded. A portion of the filtrate is also discarded to maintain a water balance and dispose of soluble salts while the main stream is recycled to the absorption circuit.

According to Kaneda et al., (151) the plant has given excellent performance. Operation has been smooth and trouble-free. SO_2 removal efficiencies have been in the 97-99 percent range with inlet SO_2 concentrations from 1000-2000 ppm. Limestone utilization has been greater than 99 percent, and the gypsum, which is sold to cement and wallboard manufacturers, typically contains about 99.2 wt. percent $CaSO_4 \cdot 2H_2O$. In a cost study covering eight throwaway FGD process operations on high-sulfur coal, the Chiyoda thoroughbred 121 process (and the DOWA process) showed the lowest cost. (136) The total levelized busbar costs for these two processes were found to be 14 mil/kWH compared to 18 mil/kWH for a conventional limestone plant (December 1982 dollars).

Magnesium Oxide Process

Regenerable magnesium oxide processes were developed in the late 1960's by the Grillo Company of Hamborn, Germany; the Chemico Construction Company in the United States (18); and by Showa Denko in Japan. A commercial-size demonstration unit was installed on a 150-MW oil-fired steam-generating boiler at the Mystic Power Station of the Boston Edison Company. The unit was started up in 1972 and operated intermittently until June 1974. It was designed by Chemico and incorporated a single-stage Venturi module. Numerous operating problems were encountered. However, the plant successfully demonstrated that the basic concepts were sound. Ninety percent SO_2 removal could be accomplished, magnesia could be regenerated and recycled, and high-quality sulfuric acid could be recovered from the SO_2 removed (44).

Philadelphia Electric Company (PECo) and United Engineers and Constructors initiated a study in 1971 which led to construction of a demonstration unit at PECo's Eddystone Unit 1 by 1975 and culminated in the installation and startup of three commercial systems in 1982. These were retrofitted to Eddystone Units 1 and 2 (335 and 355 MW respectively) and Cromby Unit 1 (160 MW). Emission compliance tests conducted shortly after startup showed that all units met or exceeded state requirements for SO_2 and particulate control. (153) According to Ando (10), a process developed by Chemico-Mitsui is operating in Japan on an oil-fired boiler. The plant provides 90 percent SO_2 removal efficiency with 100 percent operability.

Process Description

Magnesium oxide is not as suitable for a throwaway process as calcium oxide because magnesium sulfate is quite soluble and magnesium sulfite is several times as soluble as the equivalent calcium compound. Furthermore, magnesium compounds are generally more expensive than similar calcium materials. However, the principal product of the absorption of SO_2 by magnesium oxide, magnesium sulfite, decomposes at relatively modest temperatures. This characteristic makes it suitable for a regenerative cycle using a calcination step to release the absorbed SO_2.

Simplified flow diagrams representing the magnesium oxide process are shown in Figures 7-16 and 7-17. These flow diagrams are based on the initial Eddystone No. 1 installation. An aqueous slurry containing magnesium oxide is used as the absorbent in a scrubbing step similar to lime slurry scrubbing and requiring the same type of equipment. Magnesium oxide is converted to magnesium sulfite and sulfate which are removed from the solution and dried. The dried magnesium sulfite is regenerated by calcining at about 1600°F in the presence of carbon (which is required to reduce magnesium sulfate). The calcining operation produces magnesium oxide which is returned to the absorption system and a sulfur-dioxide rich gas which can be fed to a sulfuric acid plant or reduced to elemental sulfur.

Process Chemistry

The basic chemistry of the magnesium oxide process is discussed in detail by McGlammery, et al. (45). The principal reactions are as follows:

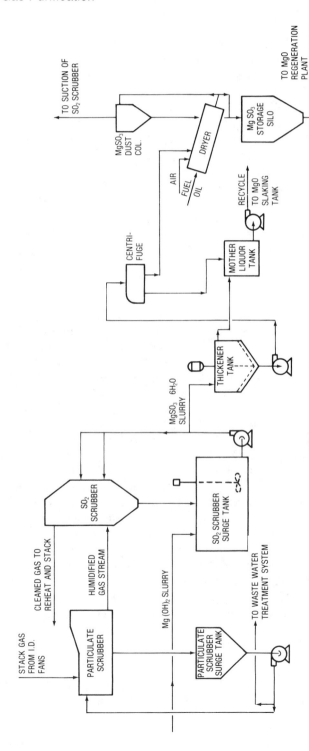

Figure 7-16. Simplified flow diagram of magnesium oxide process, stack gas scrubbing section.

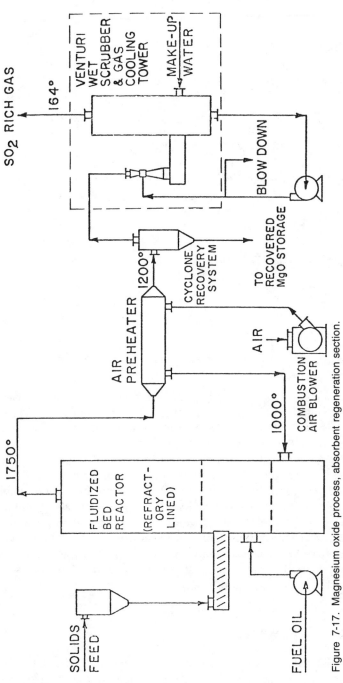

Figure 7-17. Magnesium oxide process, absorbent regeneration section.

Slaking:

$$MgO_{(s)} + H_2O = Mg(OH)_{2(aq)} \qquad (7\text{-}18)$$

absorption:

$$Mg(OH)_{2(aq)} + SO_{2(aq)} = MgSO_{3(aq)} + H_2O \qquad (7\text{-}19)$$

$$MgSO_{3(aq)} + H_2O + SO_{2(aq)} = Mg(HSO_3)_{2(aq)} \qquad (7\text{-}20)$$

$$Mg(HSO_3)_{2(aq)} + Mg\,(OH)_{2(aq)} + 4H_2O = 2MgSO_3\,3H_2O_{(s)} \qquad (7\text{-}21)$$

oxidation:

$$MgSO_{3(aq)} + \tfrac{1}{2}O_2 = MgSO_{4(aq)} \qquad (7\text{-}22)$$

regeneration:

$$MgSO_3 = MgO + SO_2 \qquad (7\text{-}23)$$

$$MgSO_4 + \tfrac{1}{2}C = MgO + SO_2 + \tfrac{1}{2}CO_2 \qquad (7\text{-}24)$$

The flue gas is contacted with a recycle stream containing magnesium sulfite and sulfate in solution and particles of magnesium sulfite and hydroxide in suspension. SO_2 reacts with the dissolved sulfite to form bisulfite in accordance with equation 7-20. Equilibrium vapor pressure data for this reaction are given in Table 7-9 which is based on the data of Pinaev (46). In the scrubber surge tank, a slurry of slaked magnesium oxide (MgO and $Mg(OH)_2$) is added to the scrubber liquor at a rate sufficient to maintain the pH at about 6.3, converting the Mg $(HSO_3)_2$ to relatively insoluble $MgSO_3$. Equation 7-21 shows this reaction leading to the precipitation of magnesium sulfite trihydrate. Actual operating experience at the Eddystone Plant indicated that the magnesium sulfite precipitates initially as the hexahydrate in the surge tank but transforms to the more stable trihydrate during the period it is in the thickener and centrifuge (47).

Design and Operation

The Eddystone Unit 1 plant employs three venturi rod-type absorbers while the Eddystone Unit 2 plant uses two spray towers (plus one spare).

Table 7-9. Vapor Pressure of Sulfur Dioxide over Solutions Containing Magnesium Sulfate and Magnesium Sulfite-Bisulfite
(Active Mg refers to magnesium associated with sulfite-bisulfite ions.)

Temp. °C	MgSO$_4$ Concentration g/liter	Active Mg g/liter	SO$_2$ g/liter	Mole ratio SO$_2$/Mg	P$_{SO2}$ mm Hg
		Solution Composition			
30	54.75	2.01	7.04	1.31	0.030
30	52.00	2.16	11.0	1.49	0.041
30	51.67	4.22	18.2	1.62	0.052
30	51.30	5.53	25.0	1.69	0.073
60	51.75	3.77	11.9	1.18	0.060
60	53.65	3.48	12.7	1.37	0.069
60	50.20	6.32	23.6	1.40	0.111
60	51.75	8.12	31.6	1.46	0.141
30	106.50	4.15	18.1	1.63	0.073
60	105.50	6.23	23.6	1.38	0.133

Source: Data of Pinaev (46)

Both plants handle about 1×10^6 acfm gas and are designed to remove 92 percent of the incoming SO$_2$. The Cromby Unit 1 plant utilizes two spray towers (plus one spare) for 416,000 acfm gas and is designed to remove 95 percent of the SO$_2$. A comparison of actual vs design performance for one Cromby absorber is given in Table 7-10.

All of the magnesium oxide plants, including the one in Japan, encountered numerous operating problems. According to Ando (10), the problems at the Chemico-Mitsui plant were mainly in the regeneration steps. The problems were solved, and the plant operates quite satisfactorily with almost 100 percent operability.

**Table 7-10.
Comparison of Actual and Design Performance for Magnesium Oxide Scrubbing at PECo Cromby Unit 1 Plant (153)**

Parameter	Actual	Design
Gas Flow, m^3/s (acfm)	104 (220,000)	98.2 (208,100)
L/G, l/m^3 (gal/1000 acf)	5.4–6.7 (40–50)	6.7 (50)
Pressure Drop, Pa (in W.C.)	500–623 (2.0–2.5)	1245 (5.0)
pH	6.7–6.9	6.8
% Solids	15–20	15
Crystal Form	Trihydrate	Trihydrate
SO$_2$ Removal, %	96–98	95

Operating problems encountered in the U.S. plants included pipe and pump corrosion; plugging of systems for feeding MgO powder, slaking MgO and feeding MgO slurry; caking in the $MgSO_3$ dryer; dust production in the dryer; solids handling malfunctions; regenerator product gas filter plugging; and high MgO losses.

It should be noted that many of the operating problems were satisfactorily resolved at the plants where they were encountered. None are considered to represent fundamental barriers to the successful application of the magnesium oxide process for SO_2 removal.

Alkali Metal Processes

Many processes have been developed which are based on SO_2 removal by absorption in an aqueous solution of a soluble alkali metal compound. Sodium compounds are preferred over potassium or the other alkali metals strictly on the basis of cost. In its simplest form, the process consists of contacting the gas with a solution of sodium carbonate (or sodium hydroxide) to form sodium sulfite, followed by disposal of the spent absorbent solution as waste or as a raw material for some other industrial process. More complicated forms have been proposed to reduce the costs of active absorbent make-up and spent absorbent disposal. These processes incorporate a variety of steps to regenerate the absorbent and produce a readily disposable (or salable) by-product.

Nonregenerable, Alkali Metal-Based Processes

The nonregenerable, alkali metal-based process is particularly applicable to situations involving relatively small quantities of SO_2, where the advantage of a simple system with low capital cost outweighs the operating cost penalties associated with continuous chemical usage and disposal.

The process is also applicable to special situations where the resulting sodium sulfite has a significant market value. This is the case in Japan where an estimated 335 such plants exist, producing sodium sulfite or sodium sulfate (10). About 80 percent of the Japanese plants produce sodium sulfite for paper mills and the rest oxidize the sulfite by air-bubbling to sulfate, which is either used in the glass industry or purged in waste water. Operating data on a typical plant employing NaOH as the absorbent and producing Na_2SO_3 as a byproduct are given by Ando (10). The plant treats 190,000 Nm³/hr (112,000 scfm) of flue gas containing 1,400 ppm SO_2 in a packed tower and produces an outlet gas containing only 6 ppm SO_2. An L/G of 1.2 1/Nm³ (~9 gal/1,000 scf) is used with a liquor pH of 6.5.

In a modification developed by the Environmental and Energy Systems Division of Rockwell International, the solution is evaporated to dryness during the absorption step (by use of a spray dryer) so that the resulting product is in a more convenient form for disposal. This and a regenerative version of the process are discussed in the "Spray Dryer Processes" section.

A number of small non-regenerative sodium base absorption units are also in use in the U.S. for industrial boilers and other applications. Three large-scale systems were placed in service on coal-fired utility boilers in 1974 and 1976. The units were designed and installed by Combustion Equipment Associates in association with A.D. Little, Inc. (11) at the Reid Gardner Power Station of the Nevada Power Company. This type of system was economically attractive at the Reid Gardner Station because of a combination of factors including the availability of low-cost sodium carbonate ore (trona), the use of relatively low-sulfur coal (0.5 to 1.0 percent sulfur), and the plant location in a warm, arid zone where the effluent can be evaporated in a pond without causing pollution problems.

In each of the three Reid Gardner units approximately 473,000 acfm of gas at 350°F is processed. It first goes through a venturi scrubber where it is cooled to about 119°F, then through a scrubber containing a single-sieve tray, and finally through a radial mist eliminator before flowing to the stack. Heated air added to the gas after it exits the absorber reheats the gas. About 5,000 gpm of recycled solution is pumped to the venturi, and about 900 gpm to the absorber. Makeup of alkali and water is provided by 53 gpm (maximum) of concentrated trona solution and 129 gpm of water from the ash pond. Sulfur dioxide removal is reported to be 85 percent.

In order to avoid the cost and disposal problems of once-through processes employing alkali metal compounds, a considerable amount of research and development effort has been expended on techniques for regenerating this type of absorbent. Proposed processes employ precipitation of insoluble compounds (double alkali, zinc oxide), thermal decomposition (Wellman-Lord), low-temperature reduction of sulfite (citrate process, potassium formate process), high-temperature reduction (aqueous carbonate process), and electrolysis (Stone & Webster/Ionics Process). Brief descriptions of these processes are provided in subsequent sections.

Double Alkali Process

In the double alkali (or dual alkali) process for flue gas desulfurization, the gas is contacted with a solution of soluble alkali, such as sodium sulfite or hydroxide, which absorbs the SO_2. The resulting solution is reacted with a second alkaline material (normally lime or limestone) to precipitate the absorbed SO_2 as insoluble calcium sulfite and regenerate the absorbent solution. Several alkali combinations are possible, however, this discussion is limited to the sodium/calcium case.

Process Description

The overall effects of the double alkali process are identical to those of the lime/limestone slurry process—SO_2 is removed from the gas, lime or limestone is consumed, and a calcium sulfite or sulfate sludge is produced. The intermediate steps, however, are quite different and result in a complete separation of the SO_2 absorption and sludge precipitation reactions. This approach permits the gas to be contacted with a clear solution of highly soluble salts, thereby minimizing scaling, plugging, and erosion problems in the absorbent circuit. The use of a clear reactive solution instead of a slurry also offers the potential for a higher SO_2 absorption rate because the SO_2 removal reaction is not limited by the rate of dissolution of solid particles.

The double alkali process was first described in a patent issued in 1918 to Howard & Stantial (48). In the proposed process, a 2.5 percent solution of sodium hydroxide or sodium carbonate was used as the first alkali, followed by a lime solution as the second. Interest in the process was revived in the 1960s and early 1970s as a result of the serious operational problems then being encountered with early lime/limestone slurry plants. General Motors initiated pilot plant work on the process in 1969 which culminated in the construction of a commercial plant at GM's Chevrolet Parma Plant near Cleveland, Ohio. This plant, which operates with a dilute absorbent solution, was started in 1974 (49). A somewhat different version of the process employing a more concentrated absorbent solution was pioneered by the FMC Corporation (50) who placed a unit into operation at their Modesto, California plant in December, 1971. Concentrated mode processes were also developed in Japan by Kureha Chemical Industry Company, Limited, working with Kawasaki Heavy Industries, Limited, and by Showa Denko KK jointly with Ebara Manufacturing Company (51). Both of the Japanese processes feature the use of limestone for regeneration, sulfuric acid for sulfate removal, and a separate oxidizer to convert precipitated calcium sulfite to gypsum.

The double alkali process has not been as widely accepted as lime/limestone slurry scrubbing. However, a significant number of commercial units have been installed in Japan and the United States. Ando lists 47 indirect lime/limestone flue gas desulfurization plants operational in Japan at the end of 1977 (mostly sodium/calcium double alkali systems). He concludes that such processes are about equal to direct lime-limestone processes with regard to SO_2 removal efficiency, power consumption, and operability (10). In the U.S., six relatively small double alkali plants were operational and two large power plant installations were under construction in mid-1977 (11, 51).

The process appears to be gaining in favor for power plant applications. By mid-1983, six large double alkali systems had been sold, all for Midwest power plants burning high-sulfur coals. Four of the plants were in operation during 1983, and data on these plants are given by Glancy et al. (154) All consistently met their SO_2 removal performance criteria, and availabilities were found to be generally higher than for direct limestone scrubbing. Two independent cost studies published in 1983 indicate that the cost of owning and operating a limestone double alkali system is less than that of conventional limestone scrubbing for relatively high-sulfur fuels. (136, 155) Figure 7-18 shows the scrubber section of a large double alkali plant during installation.

A generalized flow sheet for the double alkali process is shown in Figure 7-19. The gas is contacted with a clear solution containing sodium sulfite, sodium sulfate, sodium bisulfite, and in some cases, sodium hydroxide or carbonate. Sodium sulfite is the principal reactive component and is converted to bisulfite by the absorbed SO_2. A sidestream of the recycling absorbent solution is removed and treated with lime (or limestone) which reacts with the

Figure 7-18. Venturi Scrubber System being installed in a double alkali FGD plant. The unit shown handles 80,000 CFM of stack gas from the combustion of coal containing 3.2 percent sulfur. *FMC Corporation*

bisulfite solution to form insoluble calcium sulfite and soluble sodium sulfite (and hydroxide). The insoluble precipitate is removed by settling in a thickener followed by filtration of the sludge and washing of the filter cake. The clear liquor from the settling and filtration steps is returned to the absorber. A small amount of sodium ion makeup is required. This is typically added as sodium carbonate solution to the thickener or absorber circuit.

Two distinct types of sodium/calcium-based double alkali processes have evolved, generally designated as "concentrated" and "dilute" mode. The distinction is based on the concentration of active alkali, NaOH, Na_2CO_3, $NaHCO_3$, Na_2SO_3, and $NaHSO_3$, in the circulating solution. (Salts such as Na_2SO_4 and NaCl, which are not directly involved in the SO_2 absorption reactions, are considered to be inactive.) A system is operating in the concentrated mode when the concentration of dissolved sulfite is so high that calcium is precipitated as calcium sulfite rather than sulfate (gypsum). This occurs at an active Na^+ concentration greater than about 0.15 M and is caused by the very low solubility of calcium sulfite relative to calcium sulfate. High concentrations of sodium sulfite cause calcium to be precipitated so completely that the solubility product for the more soluble calcium sulfate cannot be exceeded. Although $CaSO_4 \cdot 2H_2O$ will not normally precipitate from a

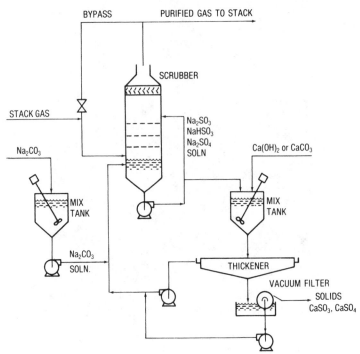

Figure 7-19. Generalized flow diagram for double alkali flue gas desulfurization process.

concentrated mode operation, some calcium sulfate can co-precipitate with calcium sulfite. In the dilute mode, either $CaSO_4 \cdot 2H_2O$ or a mixture of $CaSO_4 \cdot 2H_2O$ and $CaSO_3 \cdot \frac{1}{2}H_2O$ will precipitate from solution when calcium hydroxide is added. This provides a convenient method of purging sulfate from the system and, as a result, dilute mode of operation is favored when the scrubber operation is such that the considerable amount of oxidation to sulfate occurs. However, where it is applicable, the concentrated mode operation has the advantage of a lower absorbent circulation rate with attendant lower operating costs.

Basic Chemistry

The chemistry of the Double Alkali Process has been described in considerable depth by Kaplan (51, 52). The primary reactions involved are as follows:

absorption:

$$2NaOH + SO_2 = Na_2SO_3 + H_2O \tag{7-25}$$

$$Na_2CO_3 + SO_2 = Na_2SO_3 + CO_2 \tag{7-26}$$

$$Na_2SO_3 + SO_2 + H_2O = 2NaHSO_3 \tag{7-27}$$

regeneration with lime:

$$Ca(OH)_2 + 2NaHSO_3 = Na_2SO_3 + CaSO_3 \cdot \frac{1}{2}H_2O_{(s)} + \frac{3}{2}H_2O \tag{7-28}$$

$$Ca(OH)_2 + Na_2SO_3 + \frac{1}{2}H_2O = 2NaOH + CaSO_3 \cdot \frac{1}{2}H_2O_{(s)} \tag{7-29}$$

regeneration with limestone:

$$CaCO_3 + 2NaHSO_3 = Na_2SO_3 + CaSO_3 \cdot \frac{1}{2}H_2O_{(s)} + \frac{1}{2}H_2O + CO_2 \tag{7-30}$$

oxidation:

$$HSO_3^- + \frac{1}{2}O_2 = SO_4^= + H^+ \tag{7-31}$$

$$SO_3^= + \frac{1}{2}O_2 = SO_4^= \tag{7-32}$$

sulfate removal:

$$Na_2SO_4 + Ca(OH)_2 + 2H_2O = 2NaOH + CaSO_4 \cdot 2H_2O_{(s)} \tag{7-33}$$

$$Na_2SO_4 + 2CaSO_3 \cdot \tfrac{1}{2}H_2O + H_2SO_4 + 3H_2O$$
$$= 2NaHSO_3 + 2CaSO_4 \cdot 2H_2O_{(s)} \tag{7-34}$$

softening:

$$Ca^{++} + Na_2CO_3 = 2Na^+ + CaCO_{3(s)} \tag{7-35}$$

$$Ca^{++} + CO_2 + H_2O = 2H^+ + CaCO_{3(s)} \tag{7-36}$$

$$Ca^{++} + Na_2SO_3 + \tfrac{1}{2}H_2O = 2Na^+ + CaSO_3 \cdot \tfrac{1}{2}H_2O_{(s)} \tag{7-37}$$

The absorption reactions are relatively straightforward. The principal reaction 7-27 is reversible. This fact is utilized in the Wellman Lord Process described in a subsequent section. Equilibrium data relating to this reaction are given in Figure 7-20.

The process configuration is controlled to a considerable extent by the regeneration reactions, which are complicated by the presence of both sulfite and sulfate. Sodium sulfate can be regenerated by Reaction 7-33 only when the concentration of $SO_3^=$ is quite low because this ion forms a calcium compound which is much less soluble than gypsum. In addition, the OH^- concentration must be low and the $SO_4^=$ relatively high. In actual practice, General Motor's process is maintained at 0.1 M OH^- and 0.5 M $SO_4^=$ (49). Because of the difficulty of causticizing Na_2SO_4 by Reaction 7-33, excess lime is required. This causes sufficient calcium ions to be present in the regeneration solution (approximately 800 ppm) to produce scaling in the scrubber. As a result, dilute mode process designs typically employ a softening step in which sodium carbonate and carbon dioxide are added to precipitate calcium by Reactions 7-35 and 7-36.

A number of techniques have been developed to remove sulfate from systems employing solutions which are too concentrated to permit the precipitation of gypsum. As previously pointed out, some calcium sulfate will coprecipitate with calcium sulfite. In some cases, this plus normal losses with solution entrained in the filter cake is sufficient to take care of oxidation. A technique which is used in full-scale double alkali systems operating in Japan makes use of Reaction 7-34 and involves the addition of sulfuric acid to a small slip stream of the circulating absorbent to which is added calcium sulfite filter cake. The process works because the resulting drop in pH converts insoluble $CaSO_3$ to more soluble $CaHSO_3$, thus increasing the Ca^{++} concentration so that the solubility product for calcium sulfate is exceeded. Work is also reportedly underway on an electrolytic process for removing sulfate ions (51).

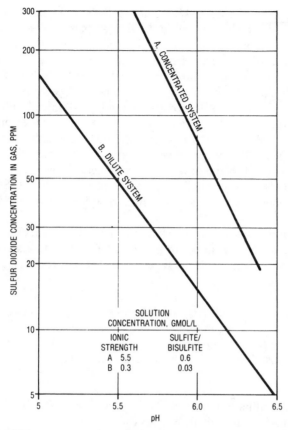

Figure 7-20. Equilibrium concentration of SO_2 in gas over sodium sulfite/bisulfite/sulfate solutions as a function of pH at 1 atm pressure and 130°F. Concentrated system, FMC laboratory data, dilute system-calculated. *Data of Legatski et al.* (50)

Design and Operation

Detailed design and operating data have been made available on a prototype double alkali FGD System (53, 54). The plant is one of three prototype systems testing different FGD processes located at the Scholtz Electric Generating Station of Gulf Power Company near Chattahoochee, Florida. The double alkali process unit was provided by Combustion Equipment Associates, Inc./Arthur D. Little, Inc. (CEA/ADL). The design basis for the double alkali prototype system is given in Table 7-11. The gas-scrubbing system consists of a variable throat plumb bob type venturi scrubber followed by an absorber designed to operate as either a tray tower or a spray contactor. The venturi scrubber and tower are equipped with separate sumps and mist eliminators (Chevron type) so that they can be tested independently.

Performance of the CEA/ADL plant at the Scholtz Station is summarized in Figure 7-21 which is based on the data of Rush and Edwards (54). The curves represent a large number of individual runs conducted with active alkali concentrations ranging from 0.15 to 0.4 M. Over a 15-month operating period, the average sulfur dioxide removal efficiency was 95.5 percent with the venturi and tray tower in operation, and 90.7 percent with the venturi alone. Efficiencies over 95 percent were readily attainable with liquid-to-gas ratios (L/G) on the order of 25 gallons per 1000 ACF in the venturi (primarily for dust removal) and 5 to 7 gallons per 1000 ACF in the tray tower. These results are generally consistent with reported data from other concentrated mode double alkali plants. FMC, for example, utilizes a scrubbing solution with a pH of approximately 6.5 and typically observes SO_2 collection efficiencies in excess of 90 percent with a relatively low pressure drop scrubber (50).

On the basis of the Scholtz Station test results, Rush and Edwards (54) concluded that the overall performance of a properly designed and operated double alkali system should be superior to that of direct lime and limestone systems because:

1. The system is highly resistant to upset and the potential for scaling is eliminated except in extreme upset conditions.
2. The handling of slurries in the absorption section is eliminated.
3. The most important control parameter, pH, has a wide acceptable range of operation.

The principal limitation which they observed relates to the inability of concentrated mode double alkali systems to reject large amounts of sulfate. This characteristic results in a lower limit of sulfur content in fuel which can be burned in a typical plant using the process. For fuels containing less than one percent sulfur, a concentrated mode system cannot be operated at the excess air levels typical of pulverized coal boilers without an intentional purge of sodium sulfate. Above two percent sulfur, the operation is excellent. In the one to two percent range, successful operation is highly dependent on combustion air rate and other factors. Another potential problem with double alkali systems results from the presence of soluble sodium salts in the calcium sulfite/sulfate sludge. This not only results in an operating cost for makeup sodium compounds, but also raises questions with regard to leaching of soluble salts from sludge disposal sites. The problem can be minimized by sludge fixation or the use of linings under ponds used for disposal of filter cake material.

Materials of construction for double alkali process plants are quite similar to those used for lime/limestone systems—stainless steel (316L) venturis and scrubbers, rubber lined pumps and slurry lines, Hastelloy G tubes for direct steam tube gas reheat, and fiberglass reinforced plastic or plastic-lined carbon steel tanks.

Table 7-11. Design Basis for CEA/ADL Prototype Double Alkali System

Flue Gas, Inlet:	
Flow rate, acfm	75,000
Temperature, °F	275
O_2 concentration, % dry basis	6.5 (max)
Particulate loading, gr/scf dry	0.02 (from precipitator)
SO_2 concentration, ppm (dry)	1800-3800
Design Performance:	
SO_2 removal, %	90 (min)
Maximum SO_2 removal rate, lb/hr	1530
Particulates, gr/scf (dry)	0.02

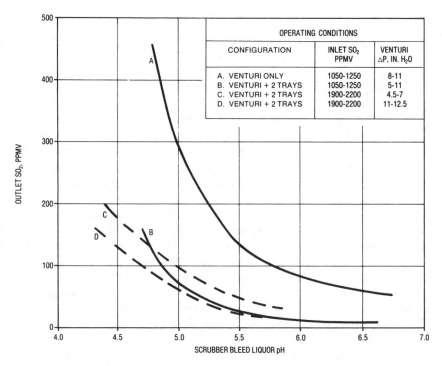

Figure 7-21. Performance of double alkali absorber system at Scholtz Electric Generating Station.

Zinc Oxide Process

The zinc oxide process was developed by H.F. Johnstone and A.D. Singh at the Engineering Experimental Station of the University of Illinois (55) as a portion of a research project sponsored by the Commonwealth Edison Company and several other utility companies. The process has not been used commercially; however, a considerable amount of pilot-plant work has been done on it, and features of the process design have been worked out in considerable detail. The process is illustrated in Figure 7-22. The flue gases are contacted with a solution of sodium sulfite and bisulfite, and sulfur dioxide is absorbed, thus causing an increase in bisulfite content. The solution is next passed into a clarifier, in which particulate matter removed from the gas steam is separated, and finally into a mixer in which it is treated with zinc oxide. At this point, the original ratio of sulfite to bisulfite is restored, and zinc sulfite is precipitated in accordance with the following reactions:

$$ZnO + NaHSO_3 + 2\tfrac{1}{2}H_2O \rightarrow ZnSO_3 \cdot 2\tfrac{1}{2}H_2O + NaOH \qquad (7\text{-}38)$$

$$NaOH + NaHSO_3 \rightarrow Na_2SO_3 + H_2O \qquad (7\text{-}39)$$

After agitation to promote crystal growth, the precipitate is removed by settling and filtration, and the filter cake is dried and calcined. Calcining of the zinc sulfite results in a gas containing 70 percent water and 30 percent sulfur dioxide, which may be cooled, dried, and compressed to produce a nearly pure liquid sulfur dioxide as the final product. Zinc oxide obtained in the calciner is recycled to the process.

As in most processes for recovery of sulfur dioxide from flue gas, oxidation of sulfur dioxide to sulfate introduces a complication. In this case, the sulfate is removed as calcium sulfate which is formed by treatment with lime.

Lime is added to a clarified side stream of the solution; this results in the precipitation of insoluble calcium sulfite to form a slurry which is added to the main solution-stream leaving the gas washer. The resulting thin slurry is passed into a clarifier. The calcium sulfite and any fly ash which may have been picked up are then removed as slurry. This slurry is acidified by contacting it with a portion of the product sulfur dioxide. Acidification results in conversion of the calcium sulfite to the more soluble bisulfite form and reaction of the dissolved calcium ions with any sulfate present in the solution to form calcium sulfate, which is relatively insoluble under these conditions. Precipitated calcium sulfate and undissolved ash are removed together in a small filter. The resulting desulfated solution containing dissolved calcium bisulfite is then treated with lime to form the slurry which is recycled to the process.

As the zinc oxide process has not been applied commercially since the presentation of complete data in 1940, and recent economic evaluations have

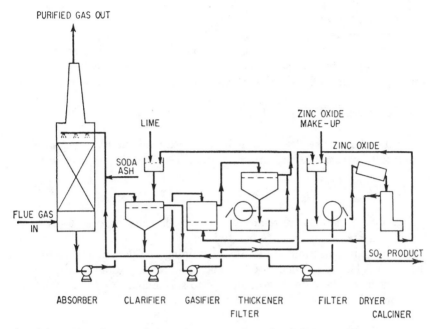

Figure 7-22. Flow diagram of zinc oxide process for sulfur dioxide recovery.

not shown it to be economically competitive with other processes, no detailed design data are given. However, it should be noted that this process has been developed very thoroughly with regard to chemical-engineering design data. Much of the work may be useful in connection with other systems particularly the studies of tower packings (56) and the use of wet cyclone scrubbers (57).

Alkali Metal Sulfite-Bisulfite (Wellman-Lord) Process

The Wellman-Lord process was initially developed to use a potassium sulfite-bisulfite cycle. In this version, the process makes use of the decreased solubility of potassium pyrosulfite at reduced temperatures to provide a means of concentrating the SO_2 absorbent. More recently the process has been modified to use the less expensive sodium salts. Regeneration of the spent sodium absorbent is accomplished by evaporating the solution and precipitating sodium sulfite crystals.

Process Description

In the potassium salt cycle, the gas is first scrubbed with water or sulfuric acid solution to remove particulates and sulfur trioxide. It is then contacted with a potassium sulfite solution which removes SO_2 by the reaction:

$$K_2SO_3 + SO_3 + H_2O \rightarrow 2KHSO_3 \qquad (7\text{-}40)$$

A bleed stream of the absorbent liquid is cooled to about 40°F to convert the bisulfate to pyrosulfite and cause crystallization of this less soluble form:

$$2KHSO_3 \rightarrow K_2S_2O_5 + H_2O \qquad (7\text{-}41)$$

The $K_2S_2O_5$ crystals are removed, mixed with water to form a slurry, and fed into a steam stripper. Separation of the $K_2S_2O_5$ as a solid provides maximum concentration of the SO_2-containing compound. In the stripper, the slurry is heated to about 250°F, which causes the crystals to dissolve and form bisulfite. The latter is decomposed to sulfite and SO_2 by the following reaction:

$$2KHSO_3 \rightarrow K_2SO_3 + H_2O + SO_2 \qquad (7\text{-}42)$$

Under optimum conditions, each pound of SO_2 produced requires 4 to 4.5 lb of steam. The vapor leaving the top of the stripper is a saturated mixture of steam and SO_2. Most of the steam is condensed on cooling the mixture to about 110°F (58). The condensate, which is a saturated solution of SO_2 in water, is returned to the stripper. The SO_2 may be fed to an acid plant, marketed as liquid SO_2, or reduced to elemental sulfur.

Initial pilot tests were conducted at Tampa Electric Company's Gannon Station and indicated a 90 percent plus removal of SO_2. Another unit rated at 56,500 cfm was installed at the Crane Station of the Baltimore Gas & Electric Co. (59). It is reported that this pilot unit was not fully successful (60). A significant problem in the original process was loss of expensive potassium salts. As a result, the process was modified to use sodium sulfite as the active absorbent. Such a cycle was incorporated into a commercial unit installed on an Olin Corporation sulfuric acid plant in Paulsboro, New Jersey (61).

The chemistry of the sodium cycle is extremely simple and may be represented by the following reaction for both absorption and regeneration:

$$Na_2SO_3 + H_2O + SO_2 = 2\,NaHSO_3 \qquad (7\text{-}43)$$

Following successful operation of the Paulsboro Plant, the Wellman-Lord Process has been applied commercially in a considerable number of installations in Japan and the U.S. As of the end of 1977, nineteen such plants were operating in Japan—thirteen Wellman-Mitsubishi Kakoki and six Sumitomo-Wellman (10)—while in the U.S. one large plant was in operation and two more were under construction (11).

A simplified flow diagram of the Wellman-Lord Process as applied to utility installations is shown in Figure 7-23. This diagram and the process description is based on a paper by E.E. Bailey of Davy Powergas, Inc. (62). Gas from the power plant at a temperature of 250° to 300°F first enters a gas-saturation prescrubber unit. This unit serves to remove fly ash and reduce the gas temperature to the 120° to 130°F range. An acidic water-fly ash slurry is recirculated through the prescrubber and a bleed stream is withdrawn continuously to remove fly ash to the disposal pond. A mist eliminator is provided in the lower section of the absorber tower to minimize carry-over of fly ash slurry into the main absorbing circuit. After flowing through the mist eliminator, the flue gas passes into the sulfur dioxide absorption section of the tower where it is contacted with sodium sulfite/bisulfite solution. The active component in the solution is primarily in the form of sulfite when it enters the top of the column and becomes progressively richer in bisulfites as it proceeds downward. Typically the column will contain three to five sieve or valve trays.

In order to provide the necessary amount of liquid to hydraulically load each tray and maintain countercurrent operation in the overall tower, it is necessary to recirculate liquid over each tray by using an external pump. This detail is not shown in the flow diagram. The product gas passes through a conventional mist eliminator and reheater (50°F reheat is typical) and is then vented to the stack. The rich solution from the bottom of the absorption section is pumped to a storage tank which provides feed to the regeneration portion of the plant.

Regeneration of the sodium bisulfite-rich solution is accomplished in a forced-circulation vacuum evaporator (single or double effect). The increased temperature and steam stripping vapor cause decomposition of the bisulfite to sulfite which crystallizes from solution to form a slurry. Steam and sulfur dioxide are carried overhead to a series of condensers. In large plants, the first condensor is the heat exchanger of a second-effect evaporator. The SO_2 and steam are ultimately cooled to as low a temperature as possible to reduce the load on the vacuum pump and provide a relatively pure SO_2 to the next operation. Normally, the product gas will contain approximately 85 percent SO_2. The regenerated absorbent (slurry) leaving the evaporators is combined with steam stripped condensate from the partial condensors. The resulting solution is pumped to the absorbent solution storage tank for use in the absorber.

In the absorbent cycle of the Wellman-Lord Process, as in all alkali-based SO_2 removal processes, some oxidation to sulfate occurs due to oxygen in the gas stream. In addition, at the temperature of regeneration, disproportionation is possible by the reaction:

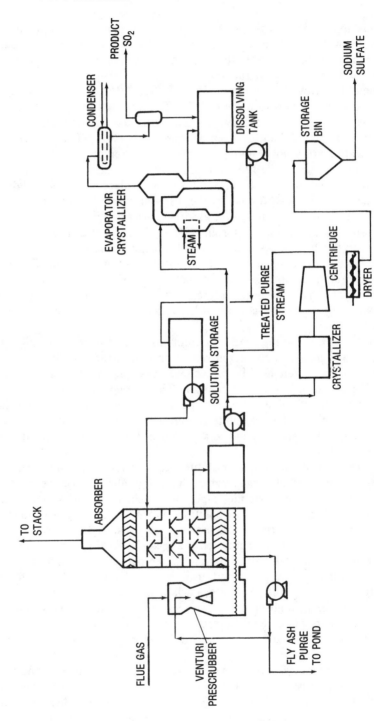

Figure 7-23. Simplified flow diagram of the Wellman-Lord SO₂ removal process.

$$2NaHSO_3 + 2Na_2SO_3 = 2Na_2SO_4 + Na_2S_2O_3 + H_2O \qquad (7\text{-}44)$$

Both of these mechanisms result in the formation of inactive salts which must be purged from the recirculating absorbent. In the Wellman-Lord Process, this is accomplished by continuously removing a small slip stream of the absorbent solution for disposal or processing. One processing technique consists of a fractional freeze crystallization operation which produces solids containing approximately 70 percent sodium sulfate and 30 percent sodium sulfite from a typical absorber solution containing 7.1 percent sulfate, 5.7 percent sulfite, and 21 percent bisulfite (63). Although the solid effluent produced by this operation represents a considerable reduction in quantity compared to a liquid purge stream, it can represent a significant disposal and sodium makeup problem. As a result, techniques are being evaluated to minimize oxidation in the main process and to reduce and recycle oxidized sulfur compounds which are formed so that no purge is required.

Design and Operation

Link and Ponder (64) have reported on the initial operation of the first large Wellman-Lord plant installed on a coal-fired utility power station in the U.S. The FGD plant was retrofitted to Northern Indiana Public Service Company's 115 MW pulverized coal-fired Unit 11 at the Dean H. Mitchell Station in Gary, Indiana. In addition to the Wellman-Lord SO_2 removal system, the plant includes an Allied Company process for converting the SO_2 to elemental sulfur. The acceptance test on the entire FGD system was completed on September 14, 1977, with the following results:

1. SO_2 efficiency exceeded the required removal of 90 percent of the SO_2 when the boiler was firing a nominal 3 percent sulfur coal and producing 320,000 acfm of flue gas at 300°F. Actual removal averaged 91 percent.
2. Operating costs for electricity, steam, and natural gas averaged $43/hr for the test period, 77 percent of the maximum allowable consumption specified. (The costs are based on electricity at 7 mil/kWh, steam at 50¢/1,000 lb, and natural gas at 55¢/10^6 Btu.)
3. Sodium carbonate addition rate averaged 6.2 tons/day, 94 percent of the maximum specified.
4. The FGD plant produced approximately 25 tons of molten sulfur per day. Sulfur purity was approximately 99.9 percent versus 99.5 percent specified.
5. The plant consistently operated within the established particle emission limitation of 0.1 lb/10^6 Btu.

Although some minor mechanical problems were encountered, general operation of the plant was considered to be extremely satisfactory. It was accepted by the utility on September 16, 1977.

Materials of Construction

The materials of construction used in four Wellman-Lord process plants were reviewed by Bailey and Heinz (65). The following generalized observations are based on this review: absorber vessels constructed of either carbon steel lined with flake glass polyester or 316-L stainless steel hold up quite well. The preferred materials for absorber internals are 316-L stainless steel supports, fiberglass reinforced polyester (FRP) liquid distributors, polypropylene packing, and polypropylene mist eliminators.

Piping for the absorbing solutions may be FRP, but an internal resin-rich layer of polyester with Dynel is recommended for improved wear resistance. Stainless steel (316-L) is also generally suitable for piping and is the alloy of choice for pumps.

Storage tanks for absorbent solution may be of FRP or flake glass polyester lined carbon steel. In the evaporation section of the plant, 316-L stainless steel is required almost exclusively because of the high temperatures involved and the presence of highly corrosive SO_2 vapors. Due to its susceptibility to accelerated attack by chloride ion, conventional stainless steel is not recommended for installations where high chloride concentrations are encountered. In one such plant, Incoloy 825 was successfully used for the evaporator and Durimet 20 for wetted parts of pumps instead of 316-L stainless steel.

Citrate Process

This process was developed specifically for the removal of sulfur dioxide from smelter gases by the Salt Lake City Metallurgy Research Center of the U.S. Bureau of Mines (71). The absorbent used is an aqueous solution containing approximately 190g of citric acid and 80g of sodium carbonate per liter, and is capable of absorbing 10 to 20g of sulfur dioxide per liter.

Smelter gases containing 1 to 3 percent of sulfur dioxide are first freed of particulate matter, then cooled to about 120°F, and subsequently contacted countercurrently with the citrate solution in an absorption tower. The loaded solution is reacted with hydrogen sulfide in a stirred, closed vessel, and elemental sulfur is precipitated. The sulfur slurry flows to a thickener, and the thickener underflow is centrifuged to separate the sulfur from regenerated citrate solution which, together with the thickener overflow, is returned to the absorption tower. The sulfur cake is heated in an autoclave. Liquid sulfur is separated from the residual citrate solution, which is also returned to the system.

Two-thirds of the molten sulfur product is converted to hydrogen sulfide to be used for solution regeneration.

The process was operated (with the exception of the conversion of elemental sulfur to hydrogen sulfide) in a pilot plant processing 400 cu. ft. min. of reverberatory furnace gas, located at the San Manuel, Arizona, smelter of Magma Copper Company. Operating data collected over a period of several months indicate sulfur dioxide removal efficiencies exceeding 90 percent. Subsequent to the smelter plant tests, pilot scale operations were conducted on stack gas discharge from a coal fired industrial boiler simulating a utility application. The results of the process on both smelter and boiler plant stack gases were considered sufficiently promising to warrant construction of a full-scale demonstration plant. A detailed description of the plant, which is located at the G.F. Weaton Electric Generating Station of Saint Joe Minerals Corporation, has been presented by Madenburg and Kurey (72). The system is designed to treat 156,000 scfm of flue gas from a 120 MWe power station, reducing the SO_2 concentration from 2,000 to 200 ppm (90 percent efficiency). These authors also give cost data for a large (500 MWe) plant firing coal containing 2.5 percent sulfur. Capital costs for such a plant are estimated at \$76/kW (1977 dollars) while operating costs are 2.07 mils/kWhr (1978 dollars).

Electrolytic Regeneration
(Stone & Webster/Ionics Process)

Stone & Webster Engineering Corporation and Ionics, Incorporated joined forces to carry out development of this process for removing sulfur dioxide from power-plant flue gases (73). The process is based on the absorption of sulfur dioxide in an aqueous solution of sodium hydroxide to form sodium sulfite and bisulfite. These compounds are reacted with sulfuric acid to release sulfur dioxide, which is evolved as a pure gas, and form sodium sulfate in solution. The key to the process is in the reconversion of the inert sodium sulfate to an active absorbent for sulfur dioxide. This is accomplished in an electrolytic cell which generates both sodium hydroxide and sulfuric acid for reuse in the process. The principal chemical reactions involved in the process are shown below:

$$2NaOH + CO_2 \rightarrow Na_2CO_3 + H_2O \qquad (7\text{-}45)$$

$$Na_2CO_3 + SO_2 \rightarrow Na_2SO_3 + CO_2 \qquad (7\text{-}46)$$

$$Na_2SO_3 + \tfrac{1}{2}O_2 \rightarrow Na_2SO_4 \qquad (7\text{-}47)$$

$$Na_2SO_3 + SO_2 + H_2O \rightarrow 2NaHSO_3 \qquad (7\text{-}48)$$

$$Na_2SO_3 + H_2SO_4 \rightarrow Na_2SO_4 + SO_2 + H_2O \qquad (7\text{-}49)$$

$$2NaHSO_3 + H_2SO_4 \rightarrow Na_2SO_4 + 2SO_2 + 2H_2O \tag{7-50}$$

$$Na_2SO_4 + 3H_2O \xrightarrow{\text{electrolysis}} 2NaOH + H_2SO_4 + H_2 + \tfrac{1}{2}O_2 \tag{7-51}$$

Reactions 7-45 through 7-48 occur in the absorber. It is desirable that Reaction 7-48 be maximized in order to capture as much SO_2 as possible per unit of regenerated sodium hydroxide. Reaction 7-47 is an undesirable but unavoidable side reaction. However, oxidation of sulfite to sulfate is not as serious in this process as in many other aqueous systems since it does not interfere with the process chemistry or result in a loss of absorbent. Reactions 7-49 and 7-50 represent the SO_2 release step. Both reactions result in the formation of sulfur dioxide gas and sodium sulfate in solution. The final reaction, 7-51, depicts the overall result of electrolysis. Sodium hydroxide and hydrogen are produced at the cathode, while sulfuric acid and oxygen are produced at the anode. The sodium hydroxide is recycled to the absorber, and the sulfuric acid is used to liberate SO_2 from the rich solution. The sulfur dioxide thus represents the only net product of the process.

The process has been tested in a 2000-acfm pilot plant at Wisconsin Electric Power Company's Valley Station in Milwaukee (74). The test program showed the process to be technically feasible and demonstrated that process reliability can be designed into the system as required by the power generation industry. The average SO_2 removal was 85 to 95 percent with inlet SO_2 concentrations from about 1000 to 3600 ppm. The effluent gas contained 200 to 300 ppm SO_2.

Oxidation of SO_2 in the absorber varied from 7 to 25 percent. As a result of the oxidation, it is necessary to have two types of electrolytic cells. A three-compartment (A type) cell is the basic design which converts sodium sulfate into caustic soda and an impure sulfuric acid solution suitable for reacting with sodium bisulfate in the process. The second type of cell, a four-compartment (B type) unit, produces substantially pure, approximately 10 percent sulfuric acid. This acid can be withdrawn from the system for sale or disposal and represents the net production of sulfate ion by oxidation. The amount of sulfate formed in the absorber determines the number of B-type cells required. In the pilot plant, 36 percent of the electrolyzer cells were of the four-compartment B-type.

The majority of operating problems experienced at the site were mechanical in nature. After they were resolved, the process had an overall operation availability of greater than 90 percent. One early problem was found to be caused by the presence of precipitate-forming impurities in the feed to the electrolizer cells. This was resolved by installation of an adequate solution cleanup system. The pilot plant program included the preliminary design of a 75-MW prototype. However, no commercial units have been placed in operation to date.

Sulfur Dioxide—Recovery Processes Employing Ammonia

A number of processes based upon the absorption of sulfur dioxide in aqueous solutions of ammonia have been proposed, and several have been developed to commercial or advanced pilot-plant operations. The processes differ primarily in the method of removing the sulfur dioxide from the ammonia-containing solution. Techniques used include steam or inert-gas stripping, oxidation to sulfate, reduction to elemental sulfur, and displacement by a stronger acid.

The possibility of using an aqueous solution of ammonia to absorb sulfur dioxide was considered as early as 1883 (75), and the use of countercurrent stagewise washing was disclosed by Hansen in 1929 (76). This latter patent also described the use of sulfuric (or other strong acid) to release the absorbed sulfur dioxide. The use of a cyclic ammonia system to concentrate sulfur dioxide, which is later reduced to elemental sulfur with hot carbon, was disclosed in 1934 by Gleason and Loonam (77) in a patent assigned to Guggenheim Bros. Pilot-scale development work on the Guggenheim process was conducted by the American Smelting and Refining Company at its Garfield, Utah plant (78). However, this work was terminated without construction of a commercial plant.

H.F. Johnstone of the University of Illinois made important contributions to the early development of the ammonia process, particularly with regard to systems employing heat regeneration. Patents covering certain aspects of the operation were obtained as a result of this work (79, 80).

Commercialization of the ammonia process was pioneered by the Consolidated Mining & Smelting Company, Ltd. (Cominco), which operated a 3-ton/day sulfur-producing pilot unit at their Trail plant in 1934 and placed a 40-ton/day commercial plant in operation in 1936 (81). The sulfur dioxide recovered in these early units was reduced to elemental sulfur. Later changes in the market picture made it more economical to use the concentrated sulfur dioxide streams as feed to sulfuric acid plants. Sulfur dioxide-absorption processes using both heat and acid neutralization were developed at Trail. The present operations use only the neutralization process.

More recently, processes based on the absorption of SO_2 in ammonia solutions have been commercialized in Japan and Russia, while development work on advanced concepts has been conducted in France and the United States (82).

Basic Data

Vapor Pressure of Sulfur Dioxide and Ammonia

Equilibrium partial vapor pressures over solutions of the ammonia-sulfur dioxide-water system have been reported by Johnstone (83). His data cover temperatures from 35° to 90°C as well as concentrations in the range likely to be encountered in a cyclic process in which the solution is regenerated by distillation. Johnstone proposed the following equations to predict the partial pressure of sulfur dioxide and ammonia over aqueous solutions:

$$p_{SO_2} = \frac{M\ (2S - C)^2}{(C - S)} \tag{7-52}$$

$$p_{NH_3} = \frac{N\ C(C - S)}{(2S - C)} \tag{7-53}$$

where C = concentration of ammonia, moles/100 moles H_2O
S = concentration of sulfur dioxide, moles/100 moles H_2O,
and M and N are empirical constants which vary with temperature according to the equations:

$$\log M = 5.865 - 2,369/T \quad (T = °K) \tag{7-54}$$

$$\log N = 13.680 - 4,987/T \tag{7-55}$$

In ordinary operations on waste gases, oxidation occurs to form sulfate ions which tie up some of the ammonia as ammonium sulfate. If this occurs, Equations 7-52 and 7-53 become:

$$p_{SO_2} = \frac{M\ (2S - C + nA)^2}{C - S - nA} \tag{7-56}$$

$$p_{NH_3} = \frac{N\ (C)(C - S - nA)}{2S - C + nA} \tag{7-57}$$

where A = concentration of sulfuric acid (or other strong acid)
n = valence of acid ion (2 for sulfate)

Comparison of experimental data with values calculated from the equations shows good agreement except near the bisulfite ratio. At this ratio, S approaches C, and small analytical errors are greatly magnified. Figures 7-24, 7-25, and 7-26 present vapor-pressure data for typical solutions in graphical form.

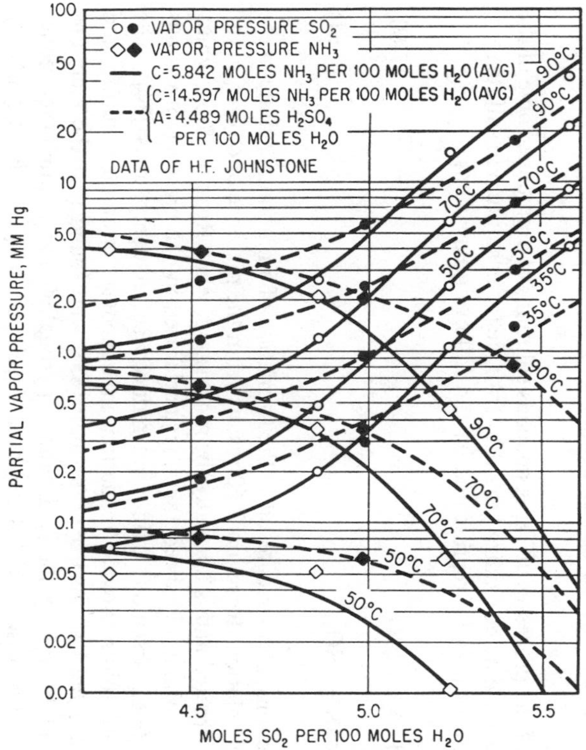

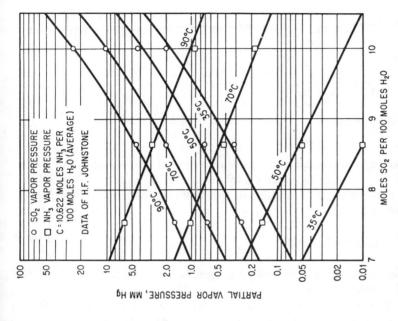

Figure 7-25. Vapor pressures of sulfur dioxide and ammonia over a solution containing 10.622 moles ammonia/100 moles water. *Data of Johnstone (83)*

Figure 7-24. Vapor pressures of sulfur dioxide and ammonia over a solution containing 5.842 moles ammonia/100 moles water and over a solution containing an equivalent concentration of free ammonia plus dissolved ammonium sulfate. *Data of Johnstone (83)*

The vapor pressure of water over the sulfur dioxide-ammonia-water system was found to follow Raoult's law quite well and can be estimated from the relationship

$$p_{H_2O} = p_W\left(\frac{100}{100 + C + S}\right)$$

(7-58)

where p_W = vapor pressure of pure water at the same temperature

In the same discussion, pH measurements of the solution are also presented by Johnstone (83), and the following empirical equation is stated to reproduce the observed data within 0.1 pH unit over the range of concentrations studied.

$$pH = -4.62(S/C) + 9.2$$

(7-59)

The equation cannot be extrapolated entirely to the bisulfite ratio (at which the pH is approximately 4.1).

Heats of absorption for both sulfur dioxide and ammonia in the dilute aqueous solutions considered have been estimated from slopes of vapor-pressure curves on the log P versus $1/T$ scale. The values obtained for sulfur dioxide vary from -9500 to $-11,500$ cal/mole, while those of ammonia vary from $-19,400$ to $-22,900$ cal/mole.

Oxidation of Absorbed Sulfur Dioxide

In all of the processes utilizing aqueous solutions of ammonia, some of the sulfur dioxide may be oxidized to sulfur trioxide (or in solution to sulfate ion). The reactions may be represented by the following equations:

$$2SO_2 + O_2 \rightleftarrows 2SO_3$$

(7-60)

$$3SO_2 \rightleftarrows S + 2SO_3$$

(7-61)

Both reactions, of course, result in the formation of ammonium sulfate in the absorbing solution. Reaction 7-60 results from the presence of oxygen in the gas being treated and is normally the most rapid, while Reaction 7-61, which can occur without added reactants, is accelerated by the presence of various catalysts (e.g., arsenious oxide and reduced sulfur compounds [84]). The problem of sulfur dioxide oxidation in aqueous ammonium sulfite and bisulfite has been studied by Wartman (85), who found that the reactions could be inhibited by gallic acid, tannin, pyrogallol, and certain other reducing agents. Processes which have been developed, however, have generally

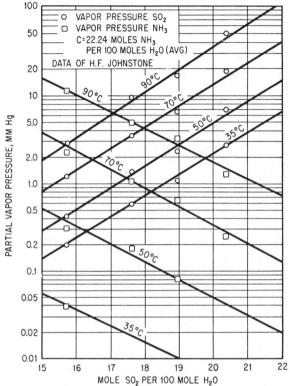

Figure 7-26. Vapor pressures of sulfur dioxide and ammonia over a solution containing 22.24 moles ammonia/100 moles water. *Data of Johnstone (83)*

been designed to accept or enhance the oxidation which occurs rather than inhibit it.

Heat-Regenerative Process

The heat-regenerative process for recovering sulfur dioxide from gas streams with ammonia solutions was developed on a pilot scale by the American Smelting and Refining Company, but first commercialized by the Consolidated Mining & Smelting Company, Ltd., as their "exorption" process. This operation was subsequently converted to regeneration with sulfuric acid, and no commercial installations of the process are now known. The process is based upon the reversible reaction:

$$2NH_4HSO_3 \rightleftharpoons (NH_4)_2SO_3 + SO_2 + H_2O \qquad (7\text{-}62)$$

The two principal problems of this process are oxidation of sulfur dioxide to form ammonium sulfate and loss of ammonia by vaporization. The oxidation problem can be alleviated by acidifying a portion of the circulating absorbent to release sulfur dioxide and produce ammonium sulfate solution or by carefully controlling the ammonium sulfate concentration in the circulating stream so that the required amount can be removed by a crystallization step. At Trail, the first operation was used because ammonia-acidification units were available elsewhere in the plant. Control of the ammonia-vapor-loss problem requires maintenance of minimum temperatures in the absorbers and careful adjustment of solution concentrations.

Design of Distillation Unit

The design of a distillation system for regenerating dilute ammonia solutions containing absorbed sulfur dioxide is complicated by the fact that compound formation takes place in the liquid phase, although all three components are volatile. A method for calculating the number of theoretical plates and quantity of steam required for the cases of distillation by direct steam addition and indirect heating has been developed by Johnstone and Keyes (86). The method is based upon a tray-by-tray calculation of the equilibrium composition at each theoretical contact starting with the desired lean-solution composition. Results of such calculations for regeneration with direct steam addition and indirect heating of the regenerator are shown in Figure 7-27. The plot is based upon use of a solution containing 22.27 moles ammonia per 100 moles water which, according to the authors, is approximately the highest concentration that can be used in this system. It will be noted that the most efficient steam utilization is obtained with direct steam addition over most of the range of solution compositions. The use of 10 rather than 5 theoretical trays produces only a small decrease in the quantity of steam required. More than 10 trays yield no significant improvement. The feed-solution composition is a function of absorber conditions and can be estimated on the basis of the equilibrium vapor-pressure composition curves of Figures 7-24 to 7-26. As shown by the equilibrium curves, a reduction in the sulfur dioxide concentration of the solution during stripping will result in a decrease in sulfur dioxide concentration and an increase in ammonia concentration in the vapor. Obviously, a point will be reached where no further enrichment of the vapor with regard to sulfur dioxide occurs. Johnstone and Keyes (86) point out that this limiting sulfur dioxide concentration lies somewhat below the point where the sulfur dioxide and ammonia have the same vapor pressures.

Cominco Sulfur Dioxide-Recovery Process

The process developed at the smelter of the Consolidated Mining & Smelting Company, Ltd., at Trail, Canada, to absorb sulfur dioxide from exhaust-gas streams produced by their metallurgical operations and sulfuric acid plant is known as the Cominco sulfur dioxide-recovery process. The process is based upon the absorption of the sulfur dioxide in an aqueous solution of ammonium sulfite and the liberation of absorbed sulfur dioxide by the addition of sulfuric acid to the solution, forming ammonium sulfate as a by-product. The process has also been applied to acid-plant tail gases by the Olin-Mathieson Chemical Corporation at their Pasadena, Texas plant. Olin-Mathieson has acquired the rights to license this process in the United States. A flow sheet of the process as employed by Olin-Mathieson is shown in Figure 7-28.

Absorption Step

Operating data on four absorption systems of plants utilizing the Cominco Process are presented in Table 7-12. Observed sulfur dioxide removal efficiencies vary from 85 to 97 percent. The degree of sulfur dioxide removal which is attainable in a system of this type is obviously dependent upon a large number of variables. Chief among these are the following:

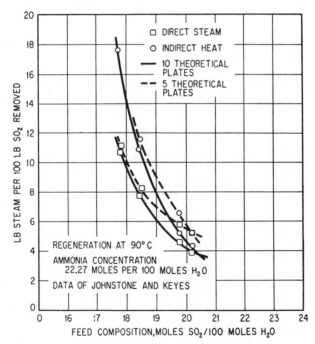

Figure 7-27. Effect of feed composition and number of plates on the quantity of steam required for regenerating a relatively concentrated ammonia solution. *Data of Johnstone and Keyes (86)*

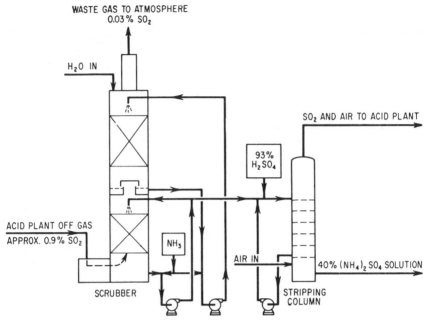

WASTE GAS TO ATMOSPHERE
0.03% SO₂

H₂O IN

SO₂ AND AIR TO ACID PLANT

93%
H₂SO₄

ACID PLANT OFF GAS
APPROX. 0.9% SO₂

NH₃

AIR IN

40% (NH₄)₂SO₄ SOLUTION

SCRUBBER

STRIPPING
COLUMN

Figure 7-28. Flow diagram of Cominco sulfur dioxide-recovery process.

Table 7-12. Cominco Process Absorber Operating Data

Plant Factors	Trail Lead-Sinter-Plant* Gas	Trail Zinc Roaster Plant Gas	Trail Acid-Plant Gas	Olin-Mathieson Acid-Plant Gas
Gas volume, SCF/min.....	150,000	20,000(avg.)	50,000–60,000	
Feed gas, % SO₂..........	0.75	5.5	1.0	0.9
Tail gas, % SO₂..........	0.10	<0.2	0.08	0.03
SO₂ in rich sol., g/liter.....	500	400–550		
Gas velocity (superficial), ft/sec.................	4.0	1.7	2.9	
No. of stages in series......	3	4	1	2
Packing height per stage, ft.	17	17	25	
Approx. circulation rate, gpm:				
Stage 1‡................	1,200–1,500	450	1,000	
Stage 2................	1,200–1,500	450	....	900†
Stage 3................	600– 800	450		
Stage 4................	....	450		
SO₂ removal efficiency, %..	85	97	92	97

*Per unit.
†For stages 1 and 2 combined.
‡Gas feed to stage 1.

1. Height (and type) of packing in each stage
2. Number of stages
3. Solution-circulation rate in each stage
4. Gas-flow rate
5. Solution composition (with respect to both ammonia and sulfur dioxide) in each stage
6. Temperature

Wood-slat packing is used in all of the absorbers described in Table 7-12. Packing of the lead-sinter plant and zinc-roaster plant gas-absorbers is described (87) as 2- by 6-in. boards on edge, 2 in. apart, with each layer arranged at right angles to the one below it. At intervals, 2- by 8-in. boards are used in place of 2- by 6-in. boards so that the alternate layers are about 2 in. apart, permitting lateral flow of gas.

Aqua ammonia of approximately 30 percent concentration is used as makeup in the Trail absorbers. Where several stages are used, the fresh ammonia additive is divided so that a portion goes to the circulating stream of each tower to maintain the proper pH for optimum absorption with minimum ammonia loss. The pH of the solution in the various absorption units ranges from about 4.1 to 5.4. The low figure represents the richest solution with regard to sulfur dioxide; this circulates in the first stage of the zinc-roaster system and contacts gas containing 5.5 percent sulfur dioxide. The pH of solutions in the last stages (with respect to gas) of the Trail absorption systems are approximately 5.1 for the lead-sinter plant unit, 5.2 for the zinc-roaster plant unit, and 5.4 for the acid-plant retreatment tower (single stage).

Temperatures of absorption must be kept as low as possible to minimize ammonia loss and maintain a favorable equilibrium for absorption of sulfur dioxide. Heat of reaction is removed from the sulfur dioxide-absorption units handling smelter gases by passing the circulating streams of solution through aluminum-tube coolers so that the final gas-contact temperature is no higher than about 35°C (95°F). Temperature control is greatly simplified for the absorbers handling acid-plant tail gas, as this gas stream is so dry that evaporation of water to saturate it provides ample cooling if the gas does not contain more than about 1 percent sulfur dioxide. A heat balance for a typical case is presented by Burgess (88) and is based upon the following overall heat of reaction for the absorption of sulfur dioxide in a circulating solution to which 28 percent ammonia solution is added:

$$SO_{2(g)} + NH_{3(28\% \, aq)} + H_2O_{(l)} \rightleftharpoons NH_4HSO_{3(aq)}$$
$$\Delta H = -42,750 \text{ Btu/lb mole} \tag{7-63}$$

Simple calculations show that for a 1 percent gas, heat generated by Reaction 7-63 is approximately equal to the heat required for the evaporation of water into vapor at 25°C (77°F), assuming that equilibrium is attained with regard to water, and the vapor pressure of water over the solution is about 80 percent that of pure water.

Burgess (88) also points out the importance of feeding a clean gas into the absorber to minimize ammonia losses. The presence of an acid fog in the gas stream from the H_2SO_4 plants can cause formation of an ammonium sulfate aerosol which is not recovered by the scrubber solution and can result in a tenfold increase in ammonia losses. Too high a pH in the absorbing solution can also cause a fogging condition due to the formation of ammonium sulfite in the gas stream.

As noted above, the pH of the solution in the Cominco operation ranges from 4.1 to 5.4 with the lowest value for the most concentrated gas stream. Hein et al. (89) found that with a gas containing about 0.3 percent sulfur dioxide, essentially no absorption took place when the pH was 5.6 or less. Their work was conducted using a 2-ft ID pilot-plant scrubber packed with 2-in. ceramic rings. As would be expected, increasing the pH gave increased SO_2 recovery but also increased ammonia loss. To attain 80 percent recovery with 8 ft of packed height, for example, a pH of about 6.4 was required, and an ammonia loss of about 5 percent occurred. This was found to be recoverable, however, by introducing a second-stage absorber.

Stripping Operation

At the Trail plant, a quantity of solution equivalent to both the ammonia added and the sulfur dioxide absorbed is continuously withdrawn from the base of the absorbers and pumped to the acidifiers. These are steel tanks lined with acidproof brick, 8 ft in diameter and 10 ft high. Only one is used at a time. In this vessel, 93 percent sulfuric acid is added to convert the ammonia to ammonium sulfate and to free the sulfur dioxide. After neutralization, the solution is still saturated with sulfur dioxide, and it is necessary to remove the last traces by stripping with steam or air. A tall, packed tower is used for the purpose and serves to reduce the sulfur dioxide content of the solution to below 0.5 g/liter. The air-sulfur dioxide mixture from the top of the stripper contains about 30 percent sulfur dioxide and is used along with the pure sulfur dioxide from the neutralizer as feed to the acid plants.

At Olin-Mathieson's Pasadena, Texas plant, 93 percent sulfuric acid is added to the solution withdrawn from the scrubber in a Karbate mixing tee. The mixture passes into a lead-lined, steel bubble-plate column through which air is blown. Acid-feed rate to the stripper is regulated by a controller operating on the pH of the neutralized solution. The air-sulfur dioxide mixture is forced into the drying tower of the sulfuric acid plant, and the ammonium sulfate solution is processed in an adjacent fertilizer plant.

Ammonia—Ammonium Bisulfate (ABS) Process

Development work on this process was conducted by TVA at its Colbert Power Plant (82, 90). The concept is closely related to the Cominco process in that the SO_2 is absorbed in an ammonium sulfite solution and then liberated by acidulation of the solution. It differs, however, in the technique employed for acidulation.

Spent solution from the SO_2 absorber, containing ammonium bisulfite and ammonium sulfite, is reacted with ammonium bisulfate according to the following equations:

$$NH_4HSO_3 + NH_4HSO_4 = (NH_4)_2SO_4 + H_2O + SO_2 \qquad (7\text{-}64)$$

$$(NH_4)_2SO_3 + 2NH_4HSO_4 = 2(NH_4)_2SO_4 + H_2O + SO_2 \qquad (7\text{-}65)$$

The resulting liquor is stripped with air or steam to remove the SO_2 and is then fed to a crystallizer for the production of ammonium sulfate crystals. The crystals are then decomposed by heating to approximately 700°F. The decomposition reaction produces ammonium bisulfate for acidulation and ammonia for recycle to the process in accordance with the following reaction:

$$(NH_4)_2SO_4 \xrightarrow{700°F} NH_3 + NH_4HSO_4 \qquad (7\text{-}66)$$

In a commercial application of the process, the concentrated SO_2 stream produced by the acidulation and stripping operations would typically be fed to a sulfuric acid plant or sulfur production unit.

Catalytic/IFP/CEC Ammonia Scrubbing Process

A major problem which has been encountered with ammonia scrubbing systems is the appearance of a characteristic "blue plume." The plume is caused by the precipitation of ammonium salts from the vapor phase as extremely small solid particles. A research program was carried out by Catalytic, Inc., and its parent company, Air Products & Chemicals, Inc., to develop an understanding of the mechanism of blue plume formation and techniques for avoiding it (91, 92). This work led to the establishment of "fumeless" design criteria which were subsequently patented (93). The essence of the Air Products/Catalytic fume avoidance technology as presented by Quackenbush, et al. (91), is shown in Figure 7-29. The data upon which the figure is based indicate that any combination of vapor pressures and temperatures representable as a point below the lower curve is not conducive to fume formation, while any point above the upper curve represents condi-

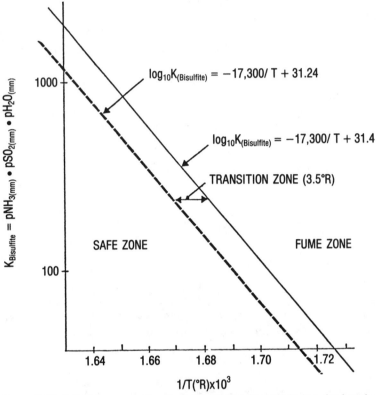

Figure 7-29. Criteria proposed by Air Products and Chemicals, Inc., for fumeless operation of SO_2 absorbers utilizing ammoniacal solutions. *Data of Quackenbush et al.* (91)

tions which are likely to result in the generation of a visible plume. The region
between the curves represents a transition zone and provides a margin of
safety when the lower curve is used for design purposes.

The proposed fume avoidance correlation is based on the precipitation of
ammonium bisulfite by the following reaction:

$$NH_{3(vap)} + SO_{2(vap)} + H_2O_{(vap)} = NH_4HSO_{3(solid)} \qquad (7\text{-}67)$$

The presence of chlorides in the gas phase can also cause the formation of a
plume (NH_4Cl). In order to avoid this occurrence, Quackenbush, et al.
recommend that inlet chlorides be kept below 10 ppm by efficient aqueous
scrubbing of the gas before it enters the SO_2 absorber (91). They also point
out the importance of operating the entire absorption system within the safe
zone defined by the curves. For example, the direct contact of ammoniacal

solution with hot dry gas or the presence of localized hot or cold spots within the absorber must be avoided. It is also necessary to remove ammonia from the product gas to prevent the formation of fumes outside the scrubber. This can best be accomplished by contacting the gas with slightly acidic water followed by an efficient mist elimination step prior to discharging the product gas into the atmosphere.

Catalytic, Inc. has proposed a complete flue gas purification system which combines their ammonia scrubber technology with the liquid Claus sulfur recovery system developed by Institut Francais du Petrole (IFP) and, if desired, the NO_x removal technology of Chisso Engineering Company (CEC). The IFP process has reportedly operated successfully in a 30 MW scale test installation in France, and the CEC process has been piloted in Japan (91). The CEC process requires that catalysts (EDTA and ferrous ion) be present in the scrubbing solution. These promote the absorption of NO, forming an adduct. The additives also cause the SO_2 to form a dithionate ($(NH_4)_2S_2O_2$) so a decomposition step for this compound is needed in addition to the standard ammonium sulfite process steps. The NO absorption reaction actually produces additional ammonia by the overall reaction:

$$2NO + 5SO_2 + 8NH_3 + 8H_2O = 5(NH_4)_2SO_4 \qquad (7\text{-}68)$$

In the combined process, ammonium sulfate formed by oxidation of sulfite is concentrated to a slurry, then decomposed at 600°F to 700°F in a step similar to that used in the ABS process. The liberated SO_2 is reduced in an H_2S generating unit which can utilize reducing gas from a coal gasifier. The generation of H_2S is controlled to produce a two-to-one ratio of H_2S to SO_2 for feed to the IFP liquid Claus reactor. The product of this unit is molten sulfur and ammonia. The ammonia is condensed from the Claus plant tail gas, concentrated, and recycled to the absorber (63).

Aqueous Aluminum Sulfate Processes

Dowa Dual Alkali Process

The Dowa process uses a solution of basic aluminum sulfate to absorb SO_2, air injection to oxidize sulfite to sulfate, and limestone to precipitate the resulting excess sulfate in the form of gypsum.

The process was developed by the Dowa Mining Company of Japan in the early 1980's, and, by 1983, they had ten commercial systems operating on a variety of smelters, sulfuric acid plants, and one oil-fired boiler. (156) The process has been demonstrated in the United States at TVA's Shawnee Steam Plant. (157)

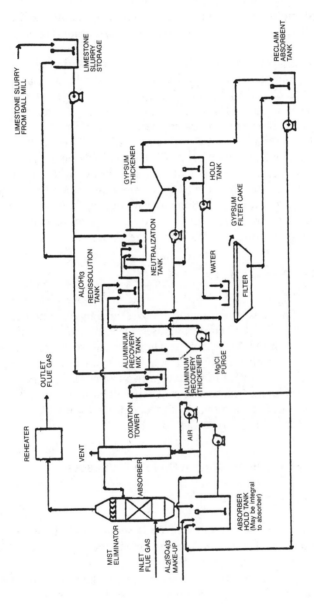

Figure 7-30. Dowa Process Flow Diagram.

A simplified flow diagram of the Dowa process is shown in Figure 7-30. A solution of basic aluminum sulfate is used to absorb SO_2 from the gas in a packed contactor. The resulting rich liquor is then pumped through an oxidation tower where air is injected to achieve essentially 100 percent conversion of sulfite to sulfate. Most of the liquor is recycled to the absorber to provide a sufficiently high L/G ratio for efficient absorption. A slip stream is continuously removed and passed through an external neutralization loop where it is first used to redissolve precipitated aluminum hydroxide then neutralized with limestone to regenerate basic aluminum sulfate solution and precipitate gypsum. The gypsum is removed from the slurry by conventional settling and filtration techniques and the clear solution is returned to the absorption loop. The basic chemical reactions of the process can be identified as follows:

Absorption

$$Al_2(SO_4)_3 \cdot Al_2O_3 + 3SO_2 \rightarrow Al_2(SO_4)_3 \cdot Al_2(SO_3)_3 \qquad (7\text{-}69)$$

Oxidation

$$Al_2(SO_3)_3 \cdot Al_2(SO_4)_3 + 3/2O_2 \rightarrow Al_2(SO_4)_3 \cdot Al_2(SO_4)_3 \qquad (7\text{-}70)$$

Neutralization

$$Al_2(SO_4)_3 \cdot Al_2(SO_4)_3 + 3CaCO_3 + 2H_2O \rightarrow \qquad (7\text{-}71)$$
$$Al_2(SO_4)_3 \cdot Al_2O_3 + 2CaSO_4 \cdot 2H_2O + CO_2$$

Aluminum Hydroxide Precipitation

$$Al_2(SO_4)_3 + 3CaCO_3 + 2H_2O \rightarrow 2Al(OH)_3 + 3CaSO_4 \cdot 2H_2O + 3CO_2 \qquad (7\text{-}72)$$

The pH of the process solution is maintained in the range of 3.0 to 3.5. Because of the low pH and the complete oxidation realized, all of the limestone reacts to precipitate gypsum. Unlike conventional lime/limestone or dual alkali processes, there is no calcium sulfite produced. The low pH also affects the rate of absorption of SO_2 and an efficient contactor must be used to obtain high levels of SO_2 removal.

The removal of soluble chloride and magnesium salts is handled by treating a small bleed stream of the process liquor to precipitate the valuable aluminum compounds before purging the clean solution from the system. As shown on the flow sheet, the bleed stream is removed from the regenerated process liquor and contacted with limestone. An excess of limestone is used to raise the pH and precipitate $Al(OH)_3$ together with gypsum. The resulting mixture is separated in a thickener to produce the clean solution purge and a concentrated slurry of aluminum hydroxide, gypsum, and unreacted limestone. This slurry is added to the main process solution loop in the dissolution tank. In this tank, the aluminum hydroxide is dissolved by the acidic solution from the oxidation tower and the primary neutralization reactions are initiated.

The Dowa process is claimed to offer several advantages in comparison with both conventional lime/limestone and sodium dual alkali systems (156). These include:

1. 100 percent limestone utilization. A secondary advantage is the use of low-cost limestone rather than lime.
2. Scale-free operation. Calcium sulfate and sulfite concentrations are well below saturation levels.
3. No slurry problem in the absorber. Erosion and plugging can be problems in lime/limestone systems.
4. Tolerance to load swings. This is due to the high buffering capacity of the solution.
5. Stable gypsum product. Gypsum crystals settle and filter more readily than sulfite/sulfate mixtures.
6. No requirement to limit oxidation. This is a requirement in sodium dual alkali systems where the sulfate concentration must be kept relatively low for efficient regeneration.

Although some system operating problems were encountered in the Shawnee tests, the test program was generally successful in demonstrating the basic operability and reliability of the process over a range of operating conditions representative of coal-fired utility boilers. The economics of the process, as evaluated by Reisdorf et al., (136) were found to be very favorable when compared to the more conventional (and more thoroughly developed) processes for desulfurizing the flue gas from a power plant burning high-sulfur coal.

Dilute Acid Processes

Chiyoda Thoroughbred 101 Process

This process represents another alternative to the double alkali process. Sulfur dioxide is absorbed in dilute sulfuric acid, oxidized to sulfate by air blowing, then precipitated as gypsum by the addition of limestone. The oxidation rate is increased by the use of iron as a catalyst in the circulating acid and is also enhanced by the low pH of the solution. The process, which was developed by Chiyoda Chemical Engineering and Construction Company,

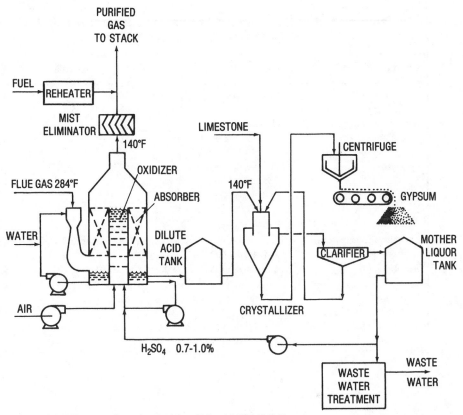

Figure 7-31. Process flow sheet of the Chiyoda (CT-101) Process.

Limited, of Yokohama, has been used quite extensively in Japan. Fourteen plants were reportedly in operation at the end of 1977 (10).

A flowsheet of the Chiyoda Thoroughbred 101 (CT-101) process is shown in Figure 7-31. This flowsheet is based on the unit treating one half of the flue gas from a 500-MW boiler at the Toyama-Shinko Power Plant of the Hokuriku Electric Power Company, Limited. Detailed operating data have been made available for this plant by Tamaki (94). Major process equipment items are listed in Table 7-13.

The chemistry of the process is defined by the following equations:

Absorption and Oxidation:

$$2SO_2 + O_2 + 2H_2O = 2H_2SO_4 \tag{7-73}$$

Table 7-13. Key Components in the Chiyoda Process Unit at Toyama—Shinko Power Plant		
Component and Quantity	Description	Material
Prescrubber (2)	Venturi type	Titanium upper section 316L lower section
Absorber/Oxidizer (1)	69-ft dia. x 79-ft ht	
Absorber section	Packed annulus; dia. 69 and 31 ft x 30-ft ht	316L
Oxidizer section	Flooded perforated tray column 30-ft dia. x 30-ft ht	316L
Crystallizer (1)	Cylindrical, inner circulation type, 17,000 cu ft, 22 kVA	316L
Clarifier (1)	Cylindrical type, 10,600 cu ft	Rubber lined
Absorbent recycle pump (3; two plus one spare)	Centrifugal, 48,400 gpm 66-ft head; 800 kVA	Rubber lining and 316L
Air blower (2)	Two stage turbo fan; 4,400 scfm, 28 psig, 300 kVA	Carbon steel
Centrifuge (3)	Basket type, automatic 1.7 T/H cake ea. 37 kVA	316L
Mist Eliminator (1)	Impingement type, two stage 95% mist removal efficiency	Rubber lining Polypropylene plates

$$2FeSO_4 + SO_2 + O_2 = Fe_2(SO_4)_3 \qquad (7\text{-}74)$$

$$Fe_2(SO_4)_3 + SO_2 + 2H_2O = FeSO_4 + 2H_2SO_4 \qquad (7\text{-}75)$$

Crystallization:

$$H_2SO_4 + CaCO_3 + H_2O = CaSO_4 \, 2H_2O + CO_2 \qquad (7\text{-}76)$$

When the Toyama-Shinko boiler burns 1 percent sulfur fuel oil, the flue gas contains approximately 450-ppm SO_2 and 2.5 to 4 percent oxygen. The gas passes through an electrostatic precipitator which reduces its particulate concentration to about 0.012 gr/scf and is then fed to the FGD unit at a rate of 467,000 scfm. It flows first through a venturi prescrubber which cools it from 284 to 140°F, then through a packed absorber where it contacts dilute sulfuric acid, and finally through a mist eliminator before being reheated and released to the stack.

The plant has obtained a 90 percent desulfurization efficiency, using 2.3 percent sulfuric acid concentration and a liquid-to-gas ratio of 210 gal/1000 scf (97,000 gpm). The solution is oxidized by contact with air in a flooded perforated tray column located in the center of the absorber. Spent dilute sulfuric acid is pumped from the absorber sump to the bottom of the oxidizer and flows upward through this unit cocurrently with 1900 scfm of air. The oxidized absorbent overflows from the oxidizer to a distributor system, then percolates down through the absorber which is packed with 3-in. Tellerettes.

A portion of the recirculating H_2SO_4 stream is continuously withdrawn to a crystallizer where the acid is neutralized by the addition of limestone to a concentration of 0.7 to 1.0 percent. Approximately 30.6 T/day of limestone are required when processing gas from 1 percent sulfur fuel oil. The product of the crystallizer is a slurry of gypsum crystals in dilute acid which is sent to a basket-type centrifuge. Gypsum cake from the centrifuge is in the form of a relatively dry powder containing less than 10 percent free water. This material is widely used in Japan for the manufacture of wallboard and as a retarder for Portland cement.

The liquid from the centrifuge flows into a clarifier from which settled particles are returned to the crystallizer and clarified liquor (containing 0.7 to 1.0 percent sulfuric acid) is returned to the absorber/oxidizer circuit. The flow rate of the return stream is approximately 500 gpm. A small amount of the dilute acid is continuously purged from the system to prevent build-up of chlorides which could cause corrosion of stainless steel equipment. A limit of 200-ppm chlorine has been specified for the plant. The catalyst, iron sulfate, is not a major cost item. It is estimated that 1,700 lb/day would be required for a CT-101 plant designed for 90 percent SO_2 removal from flue gas of a 250-MW boiler burning 3 percent sulfur fuel oil. The same plant would need about 116 T/day of limestone (94).

A prototype CT-101 process plant was recently built and tested at the Sholtz Electric Generating Station of Gulf Power Company. The object of this program was to establish the applicability of the process to a coal-fired steam generator. Detailed operating data for a 25-month test period from 1975 to 1977 have been presented by Rush and Edwards (54). Although several mechanical problems were encountered, they conclude that the overall performance of a properly designed and operated CT-101 system should be superior to that of direct lime and limestone systems because of its resistance to upsets, freedom from scaling, elimination of slurry handling in the absorption section, and wide latitude with regard to pH or concentration control.

Possible drawbacks of the process are the high corrosivity of the absorbent, the need for very high liquid-to-gas ratios brought about by the low solubility of SO_2 in acid solutions, and the requirement for careful water management to avoid chloride build-up and pollution problems.

Sulfur Dioxide—Recovery Processes Employing Aromatic Amines

Processes for the recovery of sulfur dioxide based on absorption in aromatic amines, particularly xylidine* and dimethylaniline, have been applied commercially for purifying smelter fumes, however, no applications on gases containing less than about 3.5 percent sulfur dioxide are known. The use of aromatic amines to absorb sulfur dioxide was disclosed in 1932 in British Patent 371,888 which specifically claimed aniline and its homologues. The first commercially successful process of this type was developed in Germany before World War II (Sulphidine process) (95, 96), and used xylidine. Dimethylaniline was first used commercially in a sulfur dioxide-absorption plant at the Falconbridge Nickel Company plant in Kristiansand, Norway (97). Later, the American Smelting and Refining Company (ASARCO) developed a novel flow system based upon dimethylaniline (98, 99) and, in 1947, installed a 20-ton/day sulfur dioxide plant at their Selby, California smelter. The novel features of the ASARCO process are based upon the flow pattern rather than the solvent. Under some circumstances, it could be advantageous to use xylidine rather than dimethylaniline in the proposed flow cycle.

Basic Data

Properties of the three principal aromatic amines which have been proposed for sulfur dioxide absorption are presented in Table 7-14. As can be seen, dimethylaniline boils at a somewhat lower temperature than either xylidine or toluidine and has a correspondingly higher vapor pressure under the conditions in the absorber. Because of this, losses of dimethylaniline by vaporization (or chemical costs to recover it from the gases) may be higher than those of the other amines.

However, xylidine (which is apparently preferable to toluidine) also has disadvantages. Its sulfate is only sparingly soluble in water so that if oxidation of sulfur dioxide occurs in the solution or if sulfur trioxide is present in the gas stream, precautions must be taken to prevent the formation of crystals and subsequent plugging of equipment. The solubility of xylidine sulfite in the solvent is also not as high as would be desirable. At 20°C, for example, crystallization occurs when the concentration of sulfur dioxide reaches 108 g/liter (100). Because of this, xylidine is normally used in a mixture with water. Xylidine sulfite is quite soluble in water so that crystallization is avoided, and when sufficient sulfur dioxide has been absorbed, the xylidine-water mixture becomes a single phase.

---

* A mixture of the xylidines is used.

Table 7-14. Properties of Aromatic Amines*

Property	Dimethylaniline	Xylidine (ortho, para, and meta mixture)	Toluidine (ortho)
Formula....................	$C_6H_5N(CH_3)_2$	$(CH_3)_2C_6H_3NH_2$	$CH_3C_6H_4NH_2$
Molecular wt.................	121.18	121.18	107.15
Boiling pt., °C...............	193	212–223	200
Solubility in water............	Very slight	Very slight	Slight
Vapor pressure at 20°C, mm Hg..	0.35 mm	0.20 mm	
Temperature at which vapor pressure = 1 mm, °C........	29.5	50	44
Specific gravity, 20°/4°........	0.956	0.97–0.99	1.0
Cost per lb, tank-car lots (FOB works), $.................	0.35	0.38	0.26

*Commercial grades.

A comparison of the sulfur dioxide-carrying capacities of dimethylaniline and various xylidine-water mixtures is presented in Figure 7-32. Data for pure xylidine follow closely the curve for the 2:1 mixture. It will be noted that at low SO_2 concentrations, all of the xylidine-water mixtures have appreciably higher capacities than dimethylaniline, while at high sulfur dioxide concentrations, the anhydrous dimethylaniline appears to have the advantage. Solubility data for xylidine are based upon the work of Pastnikov and Astasheva (100). These authors found that at 40°C a maximum solubility of sulfur dioxide in xylidine-water mixtures occurred at a ratio of seven parts xylidine to one part water (by volume). The solubility of sulfur dioxide at this ratio, about 455 g/liter of mixture (in equilibrium with pure sulfur dioxide), corresponds to the compound

$$C_6H_3 \underset{\diagdown (CH_3)_2}{\overset{\diagup NH_2}{}} \cdot H_2SO_3$$

The melting point of this compound was found to be 53°C.

The effects of temperature and xylidine/water ratio on the solubility of sulfur dioxide in xylidine-water mixtures are shown in Figure 7-33. This figure is based upon pure sulfur dioxide gas. According to Roesner (96), the heat of reaction for the absorption of sulfur dioxide in xylidine is 4.7 Cal/g mole sulfur dioxide absorbed (132 Btu/lb sulfur dioxide).

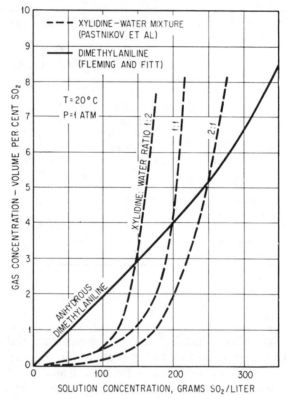

Figure 7-32. Solubility of sulfur dioxide in anhydrous dimethylaniline and various xylidine-water mixtures. *Data of Pastnikov and Astasheva (100) and Fleming and Fit (97)*

Sulphidine Process

The Sulphidine process was developed in Europe by the Gesellschaft für Chemische Industrie in Basel and the Metallgesellschaft, A.G., of Frankfurt (101, 102). The absorbent used in this process is a mixture of xylidine and water (approximately 1:1). The mixture fed into the top of the absorber consists of two layers, but during the absorption of sulfur dioxide, water-soluble xylidine sulfite is formed. The liquid from the bottom of the absorber therefore consists of an aqueous solution of xylidine sulfite. This is stripped by heating to drive off the sulfur dioxide. Sodium carbonate is added to the solution to convert any xylidine sulfate formed to sodium sulfate. The gas feed to the sulfur dioxide-recovery plant is first cleaned in electrostatic precipitators, then passed through two packed absorbers in series where it is contacted with the xylidine-water absorbent. Xylidine vapors are recovered

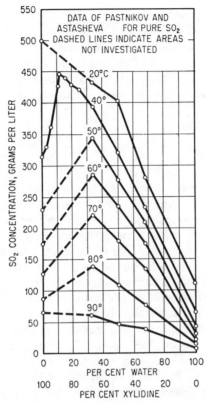

Figure 7-33. Effect of temperature and xylidine/water ratio on solubility of pure SO_2 gas at 1 atm pressure in xylidine-water mixtures. *Data of Pastnikov and Astasheva (100)*

from the gas stream by washing with dilute sulfuric acid before it is vented to the atmosphere. The tail gas from the process contains 0.05 to 0.1 percent sulfur dioxide. The SO_2-rich absorbent, carrying 130 to 180 g/liter, is pumped to the top of a raschig ring-packed stripping column in which the sulfur dioxide is removed by heating. The reboiler is heated indirectly by steam, and a temperature of 95° to 100°C is attained. Vapors from the stripping column are first passed through a cooler where water and xylidine vapors are condensed, then through a water-wash column to reduce further the xylidine content of the sulfur dioxide. Water from this column and from the condenser, which is saturated with sulfur dioxide and contains some xylidine, is returned to the stripping column.

The stripped xylidine-water mixture from the bottom of the stripping column is passed to a separator in which excess water is removed to purge the system of Na_2SO_4. Xylidine and water in the proper proportions are pumped through a cooler to the top of the second absorber. An aqueous solution of sodium carbonate is added periodically to the circulating liquid stream in the second absorber. The added sodium carbonate is converted by the free sulfur dioxide to sodium sulfite and carbon dioxide. The latter passes out of the column with the waste gas. The sodium sulfite combines with sulfate ions, which may be formed by oxidation, and the resulting sodium sulfate is removed from the system with the waste-water stream.

Asarco Process

Process Description

The process developed by the American Smelting and Refining Company for the absorption of sulfur dioxide from smelter gases represents an improvement over the Sulphidine process with regard to reagent loss, steam consumption, and labor requirements. Although the process is reported to be applicable to either dimethylaniline or xylidine, dimethylaniline has been used in all commercial installations of the process. The principal advantage of dimethylaniline is that it does not require water in admixture to dissolve the sulfur dioxide compound formed. In addition, as shown by Figure 7-32, at high concentration of sulfur dioxide in the feed gas, dimethylaniline can absorb larger quantities of sulfur dioxide than xylidine. Note, however, that with low gas-concentrations, the use of xylidine may have an economic advantage. The American Smelting and Refining Company has used the dimethylaniline process in their own plant at Selby, California, and has licensed its use by the Tennessee Copper Company in Copperhill, Tennessee. A third plant is operated by the Falconbridge Nickel Company at Kristiansand, Norway. Figure 7-34 is a photograph of the Selby unit.

A flow diagram of the process is presented in Figure 7-35. Sulfur dioxide containing flue gas from the ore-roasting process is first thoroughly cleaned and cooled, then contacted with anhydrous dimethylaniline in the lower portion of the absorbing tower. Gas from this section of the absorber, which contains dimethylaniline vapor and a small percentage of sulfur dioxide, next passes through several trays where it is contacted with a dilute soda solution. Sulfur dioxide is absorbed by the soda, converting this to sodium sulfite or bisulfite which is used at a subsequent stage in the process. The gas, which is now essentially sulfur dioxide-free but still contains vaporized dimethylaniline, is finally contacted with dilute sulfuric acid in the upper portion of the tower. This effectively removes dimethylaniline by forming the sulfate of this compound. The finally cleaned gas escapes to the atmosphere.

Rich dimethylaniline solution from the bottom of the absorber is heated by indirect heat exchange with hot, lean dimethylaniline and is then fed near the top of the stripping column where it is stripped of its sulfur dioxide content by heat and steam-stripping vapor. The resulting hot, lean dimethylaniline (with

Figure 7-34. View of ASARCO-process plant for recovery of sulfur dioxide from smelter exhaust gas. *American Smelting and Refining Company*

condensed steam) is passed through the exchanger, cooled further, and pumped to a separator from which lean dimethylaniline is withdrawn as liquid feed to the absorber.

A portion of the heat and stripping vapor for the dimethylaniline-stripping operation is provided by the steam stripping of dimethylaniline from its water solution in the lower section of the stripper. Feed to this section of the column is a mixture of dimethylaniline sulfate solution from the top section of the absorber, sodium sulfite (or bisulfite) solution from the central section of the absorber, and water separated from the lean dimethylaniline stream. The dimethylaniline sulfate and sodium sulfite react to form sodium sulfate, which remains in the water discharged from the stripper, and dimethylaniline sulfite, which is dissociated by the heat into dimethylaniline and sulfur dioxide. The latter two substances are carried out with the steam into the dimethylaniline-stripper section.

Sulfur dioxide liberated in the stripping column is passed through a rectifier section where it is contacted with water to absorb dimethylaniline

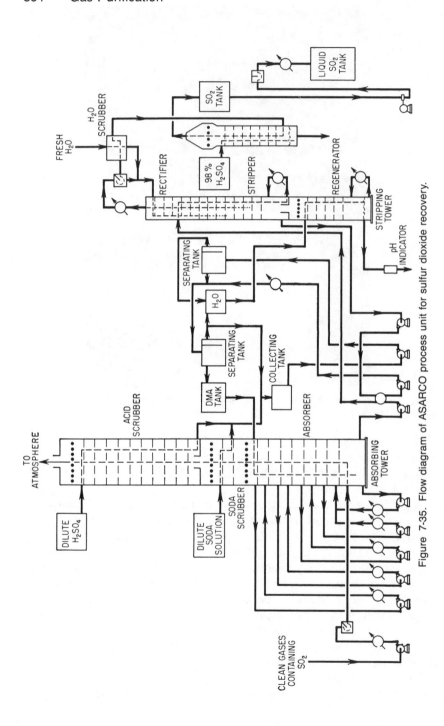

Figure 7-35. Flow diagram of ASARCO process unit for sulfur dioxide recovery.

vapors (as dimethylaniline sulfite). It is then cooled and washed with a stream of fresh water to eliminate dimethylaniline completely. The cold, purified sulfur dioxide is finally dried by countercurrent washing with 98 percent sulfuric acid, compressed, and condensed as product liquid sulfur dioxide.

Design and Operating Data

Fleming and Fitt (78) have presented design and operating data for the Selby plant which are summarized in Table 7-15. The sulfur dioxide absorption section of the absorber contains eight bubble-cap trays and five water coolers. It is stated that the sulfur dioxide content of the dimethylaniline coming off the bottom tray is about 60 percent of the equilibrium value with a tail gas containing less than 0.10 percent sulfur dioxide (97) (the tail gas is stated to contain 0.05 to 0.06 percent). These data would indicate that the rich dimethylaniline solution carries about 140 g/liter of sulfur dioxide and that an average vapor-phase tray-efficiency somewhat greater than 50 percent is obtained (this assumes zero vapor pressure of sulfur dioxide over the lean solution).

The second unit of the absorption column, in which sodium carbonate is converted to sodium sulfite, contains two trays; the top section, in which dimethylaniline is absorbed in sulfuric acid, contains nine trays. Efficient absorption in the final stage is very important to prevent losses of dimethylaniline and to minimize consumption of acid. The nine trays are sufficient to provide almost 100 percent recovery of dimethylaniline with essentially complete saturation of the acid. It is stated that the acid from the top tray contains less than one-twentieth of the saturation quantity of dimethylaniline.

Table 7-15. Data from Operations of Selby, California Sulfur Dioxide Plant During 1949

Plant capacity (design), tons/day	20
Feed gas, avg. SO_2, % by vol	5.0
Recovery of SO_2, %	99.0
Dimethylaniline consumed, lb/ton SO_2 produced	1.1
Sodium carbonate consumed, lb/ton SO_2 produced	35.5
Sulfuric acid consumed, lb/ton SO_2 produced	40.0
Steam required, tons/ton SO_2 produced	1.2
Power, kwhr/ton SO_2 (incl. compression of SO_2)	145
Cooling water at 65°F., gpm	300
Operating labor:	
Foreman, day shift only	1
Shift operators	3
Utility man, day shift only	1

The stripping column is also divided into three sections: the regenerator, where dimethylaniline is stripped from the neutralized acid-wash stream, which contains 7 bubble-cap trays; the stripping section, where sulfur dioxide is released from the rich dimethylaniline stream, which contains 14 bubble-cap trays; and the rectifying section, where dimethylaniline is removed from the sulfur dioxide stream, which contains 5 trays.

Materials of Construction

Some of the materials of construction are presented in a flowsheet describing the Selby plant (103) and a Lead Industries Association bulletin describing the Tennessee Copper Company plant (104). They are listed below.

Absorbing tower:

1. Shell—lead cage-type construction of 12-lb chemical-sheet lead supported within a steel framework of columns and right angles
2. Bubble-cap trays—¾-in. thick, 8½ percent antimonial lead supported on steel spiders
3. Bubble-caps—4-in. diameter caps of antimonial lead

Stripping tower:

1. Shell—¼-in. lead lining in ¼-in. steel shell
2. Bubble-cap trays and caps—antimonial lead trays supported and bonded by steel bars

Feed-gas blower: 316 stainless steel.
Feed-gas cooler: 316 stainless steel.
Dimethylaniline pump: 316 stainless steel.
Rich-lean dimethylaniline heat exchanger: concentric extruded lead pipe
Dimethylaniline coolers: 6 percent antimonial lead pipe, 1-in. ID, in helical coils (water sprayed over outer surfaces).
Tanks for weak sulfuric acid, dimethylaniline supply and separation, solution collectors, etc.: lead-lined steel or lead cage-type construction.

Sulfur Dioxide Removal by Absorption in Molten Salts

The only process in this category that has received significant research and development attention is the Molten Carbonate Process in which the absorbent is a melt made up of a mixture of alkali metal carbonates (105, 106). Although operation of a complete integrated system has not been accomplished, the process is of interest because of the unique technology involved. Its principal advantages are the ability to treat the stack gas at an elevated temperature and the production of a valuable sulfur product.

The basic flow diagram for the molten carbonate process is shown in Figure 7-36. The process operates as a closed cycle in which a molten eutectic mixture of sodium, lithium, and potassium carbonates is circulated to react with the sulfur oxides in the flue gas. Properties of the carbonate melt are listed in Table 7-16. The flue gas is directed from the boiler at an appropriate temperature ($\sim 800°F$), its fly ash is removed, and it is contacted with the molten salt mixture in a scrubber. Removal of SO_2 occurs without a significant change in flue-gas temperature, and the hot purified gas is then directed back to the boiler to continue its original path. The balance of the process is concerned with the regeneration of the molten salt stream and recovery of sulfur.

The principal chemical reactions occurring in the process are listed below:

$$SO_2 + M_2CO_3 \rightarrow M_2SO_3 + CO_2 \tag{7-77}$$

$$SO_3 + M_2CO_3 \rightarrow M_2SO_4 + CO_2 \tag{7-78}$$

$$4M_2SO_3 \rightarrow 3M_2SO_4 + M_2S \tag{7-79}$$

$$M_2SO_4 + 2C \rightarrow M_2S + 2CO_2 \tag{7-80}$$

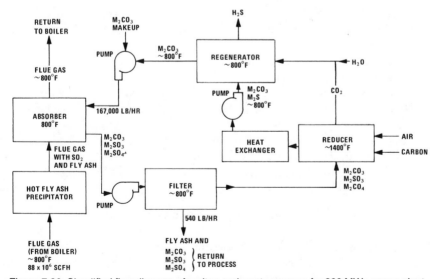

Figure 7-36. Simplified flow diagram of molten carbonate process for 800 MWe power plant.

Table 7-16. Properties of Carbonate Melt	
Composition	
Li_2CO_3	32 wt %
Na_2CO_3	33 wt %
K_2CO_3	35 wt %
Melting point	747°F
Viscosity at 800°F	12 cp
Density at 800°F	125 lb/ft³
Heat capacity at 800°F	0.40 Btu/lb °F
Surface tension at 800°F	220 dynes/cm

$$M_2S + CO_2 + H_2O \rightarrow M_2CO_3 + H_2S \qquad (7\text{-}81)$$

In these reactions, M stands for the mixture of lithium, sodium, and potassium cations. Reaction 7-77 is the principal SO_2 absorption mechanism. Reaction 7-78 also occurs in the absorber and results in the removal of SO_3 from the flue gas. Reaction 7-79 is a disproportionation that occurs rapidly at system temperatures and results in the formation of additional M_2SO_4 as well as M_2S. The reduction step, which is illustrated by Reaction 7-80, involves the conversion of M_2SO_4 in the melt to the sulfide form. Although carbon is shown as the reducing agent, carbon monoxide or hydrogen can also be used. A relatively high temperature ($\sim 1400°F$) is required to make the reduction proceed at a reasonably rapid rate. This step is analogous to Kraft paper industry reduction furnace operations.

In the regeneration step, Reaction 7-81, the sulfides are converted to carbonates with the release of hydrogen sulfide. This is accomplished by contacting the melt with a mixture of carbon dioxide and water vapor at a temperature of about 800°F. The reaction is quite exothermic, and a means must be provided for rejecting heat from the regeneration column. The exit gas from the regeneration column is expected to contain about 30 percent H_2S and be suitable for feeding directly into a Claus type sulfur plant.

Spray Dryer Processes

The use of spray dryers for SO_2 removal has experienced a remarkable growth. The first U.S. contract for a spray dryer absorber was awarded in 1977 and by late 1983 seventeen spray dryers had been sold to electric utilities and 21 units, of smaller capacity, had been sold for industrial applications. (158)

In spray dryer processes, sulfur dioxide is removed from the flue gas by contact with an atomized spray of reactive absorbent such as sodium carbonate solution or lime slurry. The sulfur dioxide reacts with the absorbent while the thermal energy of the flue gas vaporizes the water in the droplets to produce a fine powder of spent absorbent. The dry product, consisting of sulfite and sulfate salts, unreacted absorbent, and fly ash, is collected in a fabric filter or electrostatic precipitator.

Two types of spray dryer processes have attained commercial status. These employ sodium carbonate solution and lime slurry, respectively in nonregenerative (throw-away) process configurations. The lime process is by far the most widely used; however, the sodium carbonate system, exemplified by the Aqueous Carbonate Process (ACP) was developed first.

Aqueous Carbonate Process

In this process, which was developed by the Environmental and Energy Systems Division of Rockwell International, the SO_2 is removed by passing the flue gases through a modified spray dryer where efficient contact with a fine mist of an aqueous sodium carbonate is achieved. The SO_2 reacts with the sodium carbonate (Na_2CO_3) to form sodium sulfite (Na_2SO_3), some of which is further oxidized to sodium sulfate (Na_2SO_4).

The thermal energy of the flue gases (at around 300°F) is sufficient to vaporize the water in the spray dryer completely without saturating the gas. Thus, the absorbent reaction products are carried along with the gas as a dry powder, and reheat is not required to avoid saturated plumes. This powder is then separated from the gas for either disposal or further processing.

Pilot tests of the modified spray dryer absorption unit have been described (66) which indicate that 90 percent removal of incoming SO_2 can be realized with liquid to gas ratios of less than 0.4 gal/1000 scf. The $NaCO_3$ absorbent solution was generally maintained as dilute as possible in the tests consistent with desired SO_2 removal and the generation of a dry product. Absorbent utilization exceeded 80 percent in a single pass through the dryer.

Two versions of the Aqueous Carbonate Process (ACP) have been developed: 1) an open-loop configuration in which the dry spent absorbent is simply removed from the system for disposal and fresh alkali is continuously fed to the spray dryer, and 2) a closed-loop process in which the spend absorbent is regenerated and reused. The open-loop system has been selected for installation on the 410 MW Coyote Station at Beulah, North Dakota (67). The plant will use a two-stage process for simultaneous SO_2 removal and dust control. As indicated by the flow diagram of Figure 7-37, the process is extremely simple. The second stage unit for the Coyote Station is a fabric filter which has been shown to be more effective for the collection of the fine particulate product material than a cyclone or electrostatic precipitator. The fil-

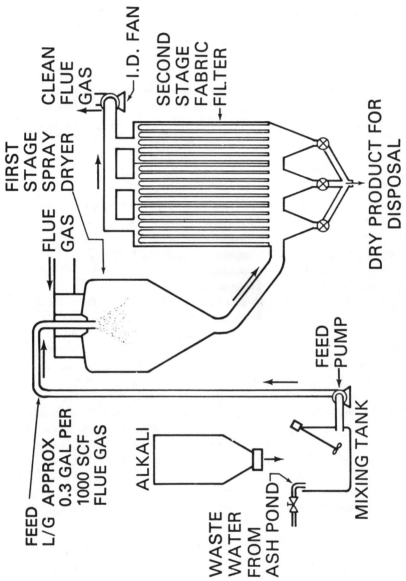

FIRST
STAGE
SPRAY
DRYER

CLEAN
FLUE
GAS

I.D. FAN

SECOND
STAGE
FABRIC
FILTER

FLUE
GAS

FEED
L/G APPROX
0.3 GAL PER
1000 SCF
FLUE GAS

ALKALI

WASTE
WATER
FROM
ASH POND

FEED
PUMP

MIXING TANK

DRY PRODUCT FOR
DISPOSAL

Figure 7-37. Flow sheet of Aqueous Carbonate Process—Open Loop, Two Stage System.

ter also provides a second gas-solids contact zone after the spray dryer so that additional absorption and reaction of SO_2 with the alkaline material can occur.

The two-stage open-loop process was tested by the Environmental and Energy Systems Division of Rockwell International (developer of the ACP system) and Wheelabrator-Frye, Inc. (supplier of the fabric filter) in a joint program conducted at the Leland Olds Station of the Basin Electric Power Corporation in Stanton, North Dakota, during 1977. In addition to sodium carbonate, lime, fly ash, and a fly ash/lime mixture were tested as SO_2 reactants. The lime- and fly-ash-containing absorbents were fed to the spray dryer as slurries. Data from the test program have been presented by Estcourt, et al. (68). Typical results are given in Table 7-17 for the two-stage open-loop system utilizing a sodium carbonate solution and a lime slurry.

As would be expected on the basis of chemical reactivity, the lime slurry is appreciably less efficient with regard to both SO_2 removal and reagent utilization. However, it is not necessarily less economical. With low sulfur coals, emission requirements can be met with a relatively low SO_2 removal efficiency, and in such cases, a large excess of lime is not required. Since lime typically costs less than soda ash on a weight basis, and its equivalent weight is much lower, lime can be economically competitive with soda ash even when much higher stoichiometric feed ratios are required. With soda ash at $75 per ton and lime at $53 per ton (delivered), it has been shown that absorbent costs will be lower for lime when the coal contains about 1.3 percent sulfur or less and lower for sodium carbonate at any sulfur concentration above that value (68).

The closed-loop ACP system using sodium carbonate as the absorbent was selected for a 100 MW "second generation" FGD demonstation plant under a program sponsored by the Empire State Electric Energy Research Corporation (ESEERCO) and the U.S. Environmental Protection Agency (EPA) (69, 70). A simplified flow diagram of the process used in the demonstration plant is shown in Figure 7-38. In this application the spent absorbent particles from the spray dryer are collected in cyclones, with final removal of the remaining particles in an electrostatic precipitator. The design emission rate to the stack is 0.01 grain/scf or less. The gas will be at least 50°F above its dew point at the stack inlet so reheat is not required.

The dry spent absorbent is mixed with carbon (petroleum coke, or coal) and fed into a refractory-lined reducer vessel which contains a pool of molten sodium carbonate and sodium sulfide at a temperature of about 1800°F. Air is injected to oxidize part of the carbon to CO and CO_2 in order to provide the heat needed by the endothermic reduction reactions and maintain the overall system at the reaction temperature.

The reduced molten salt mixture containing typically 62 percent Na_2S, 8 percent Na_2SO_4, 25 percent Na_2CO_3, and 5 percent unreacted carbon and ash

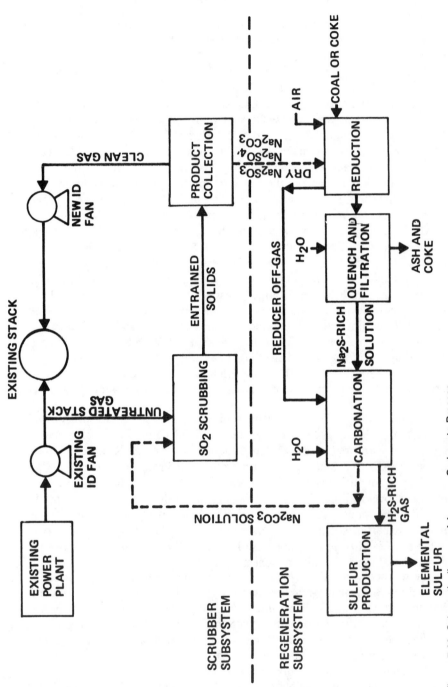

Figure 7-38. Schematic diagram of Aqueous Carbonate Process.

Table 7-17. Performance of Spray Dryer—Fabric Filter System for SO₂ Absorption

Stoichiometric Ratio	SO₂ Removal, %			Utilization Efficiency, %		
	Spray Dryer	Fabric Filter	Total	Spray Dryer	Fabric Filter	Total
A: Tests with Sodium Carbonate Solution						
0.5	40	8	48	80	16	96
1.0	82	10	92	82	10	92
1.5	86	12	98	57	8	65
B: Tests with Lime Slurry						
0.66	35	18	53	53	28	81
0.94	50	19	69	53	21	74
1.21	50	25	75	41	21	62

Notes:
1. SO₂ concentration in feed gas 800-2800 ppm
2. Fabric filter temperature approximately 200°F
3. Fabric filter performance values are based on feed to the spray dryer, not on feed to the filter

is continuously discharged from the reducer vessel and quenched in an aqueous slurry. Soluble constituents of the melt are dissolved in the aqueous medium which is then filtered to remove unreacted carbon and ash. The clear liquor is reacted with carbon dioxide gas in a series of sieve tray columns to produce, ultimately, a solution of sodium carbonate and a gas stream containing H_2S and CO_2. This gas is fed to a conventional Claus plant where the H_2S is converted to elemental sulfur. The sodium carbonate solution is recycled to the spray dryer as the active absorbent for SO_2.

Lime Slurry Spray Dryer Process

Almost all of the spray dryer FGD systems installed for utility power plants and industrial applications use lime slurry as the absorbent. (158, 159) Although lime is not as reactive as sodium carbonate, it is usually preferred because of its lower cost and, equally important, because the spent absorbent can be disposed of more readily than soluble sodium salts. Advantages of spray dryer systems over wet scrubbers include dry waste production, lower capital cost, and lower operating cost for fuels of low sulfur content. Their chief disadvantage is a higher absorbent cost due to the requirement for a relatively high stoichiometric ratio (CaO/SO_2) and the need to use lime (or sodium carbonate) instead of limestone.

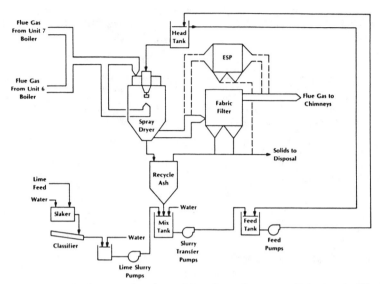

Figure 7-39. Flow Diagram of Lime Slurry Spray Dryer Process as Tested at the Riverside Station of Northern States Power Company (160).

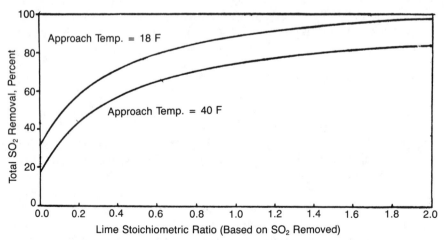

Figure 7-40. SO_2 Removal by the Lime Slurry Spray Dryer Process as a Function of Stoichiometric Ratio and Approach Temperature (160).

A simplified diagram for a lime slurry spray dryer system is shown in Figure 7-39. This diagram represents the 100 MW Demonstration Plant on Units 6 and 7 of the Riverside Station of Northern States Power Company. The plant was used for extensive testing of the effects of operating variables on a large 46 foot diameter spray dryer absorber. (160)

Spray dryer FGD plants consist of four major subsystems: absorbent preparation, absorption and drying, solids collection, and solids disposal. Because of the quantity of lime required it is not economical to purchase lime in the hydrated form and pebble lime is normally slaked on site. Several types of slakers have been tested including ball mills, paste, and detention slakers. Ball mill slakers have been used for most large utility applications because they pulverize uncalcined limestone and other solids which the simpler slakers reject as waste material. The water used for slaking is critical. Fresh water is generally required; however cooling tower blowdown or other impure water supply may be used for dilution. (159)

Key factors affecting SO_2 absorption efficiency are the dryer design (e.g., atomizer performance, gas residence time); the sorbent stoichiometry; and the approach to saturation. The general effects of the latter two variables are indicated by Figure 7-40 which is based on tests at the Riverside plant. (160)

Slurry atomization is generally accomplished by the use of rotary atomizers or two-fluid nozzles. Rotary atomizers appear to be preferred for large installations and nozzles for small units; however, this is not a universal rule. The rotary atomizers have the advantage of high capacity per unit, uniformly fine drop formation, and a lower power requirement. Two-fluid nozzles using air or steam to provide the atomizing energy, are simpler and easier to maintain. The degree of atomization and chamber design must be coordinated to assure that the droplets are fine enough to provide adequate surface for SO_2 removal, moist enough during most of their flight to be effective absorbents, and sufficiently dry when they strike the walls to flow as a powder without adhering to the walls or to each other. Most lime spray dryer absorbers have a flue gas residence time of 10 to 12 seconds and operate with an approach to saturation in the range of about 20 to 50°F at the dryer outlet. (158)

As indicated by Figure 7-40, reducing the approach to saturation can result in a significant improvement in SO_2 absorption. Unfortunately this technique is limited because a margin of safety is required to prevent water condensation in downstream equipment. With given gas inlet conditions (rate, temperature, and humidity) the approach to saturation in the spray dryer is established by the amount of water fed with the lime slurry. For typical operating conditions this generally results in liquid-to-gas ratios ranging from 0.2 to 0.3 gal/MSCF.

Absorption efficiency can also be improved by increasing the stoichiometry. This can be done by raising the concentration of lime in the feed slurry.

Another approach is to recycle a portion of the spent absorbent using solids which have dropped out in the spray dryer or have been collected in the baghouse. Recycle can increase SO_2 removal efficiency and/or sorbent utilization and also increase the utilization of alkalinity in the fly ash.

The particulate removal equipment represents a key component in spray dryer FGD systems. Not only does it prevent the release of particulates into the atmosphere, it also can provide considerable SO_2 removal capability. This feature is the subject of a U.S. Patent (161). Both a fabric filter and an electrostatic precipitator (ESP) were used in the Riverside Station tests (160). The fabric filter typically averaged about 15 percent efficiency for SO_2 removal while the ESP averaged only 6 percent. Essentially all commercial installations built to date have employed fabric filters.

The solid waste removed from the spray dryer and particulate control system is a dry flowable powder containing calcium sulfite, calcium sulfate, fly ash, and unreacted absorbent. It is normally conveyed pneumatically to a silo for storage prior to disposal and is ultimately disposed of in landfill operations. Water is often added for dust control. This causes pozzolanic reactions to occur resulting in a final waste product of low permeability and desirable landfill characteristics (159).

Sulfur Dioxide Removal by Dry Sorption with Reaction

A considerable amount of developmental effort has been aimed at dry absorption processes primarily because of their apparent advantage over aqueous absorption systems of permitting stack-gas treatment at an elevated temperature. Processes involving chemical reaction with the absorbent require either a very low-cost material (such as limestone) or a regenerative cycle. The limestone dry-injection process has reached the most advanced state of development although its SO_2 removal efficiency is limited. The regenerable dry processes have generally suffered from excessive losses of the absorbent due to attrition and high investment costs due to the complexities of the regeneration system. However, development work on such processes is still underway, and some of the early results show considerable promise.

The oxides of 48 metals were screened by the Tracor Co. in a project conducted for the U.S. National Air Pollution Control Administration to determine which were best suited for the removal of sulfur oxides from flue gases by chemical reaction (107). The screening was accomplished by consideration of the thermodynamic requirements for efficient sulfur oxide removal and product regeneration. Sixteen potential sorbents were selected as a result of this screening process. These were the oxides of titanium, zirconium, hafnium, vanadium, chromium, iron, cobalt, nickel, copper, zinc, aluminum, tin, bismuth, cerium, thorium, and uranium.

The 16 metal oxides were further screened on the basis of their rate of reaction with SO_2 in a flue-gas atmosphere. The oxides were prepared in a kinetically active form by calcining a salt which decomposed to the oxide at a relatively low temperature. The rate data were collected, using an isothermal gravimetric technique whereby weight gain of SO_2 was recorded as a function of time. The oxides of copper, chromium, iron, nickel, cobalt, and cerium were found to have economically feasible reaction rates with SO_2. After further evaluation of such factors as sorption reaction stoichiometry, formation of product layers which affect the reaction rate, and SO_3 partial pressure over the sorption product, two materials, copper and iron oxides, were selected as most promising. Finally, preliminary design and economic studies were made for a process employing the oxides with a fluidized bed gas-solid contactor for both sorption and thermal regeneration steps. Copper oxide (CuO) and iron oxide (Fe_2O_3) were found to have promise as potential sorbents for an economically feasible SO_2 removal process. Copper oxide processes proposed by others are described in a subsequent section.

Injection of Alkaline Solids

The concept of injecting dry powered alkaline solids into the boiler firebox (or into the flue gas), then collecting the reacted material together with fly ash, offers the promise of an extremely simple SO_2 removal technique. As a result, a considerable amount of research and development work has been conducted on this approach. However, no commercial applications are known.

In the limestone (or dolomite) injection process, finely ground limestone is added to the coal or injected into the furnace. A review of this process has been presented by Falkenberg and Slack (108), based on work done by the Tennessee Valley Authority (TVA). The work included a conceptual design and cost study, basic research on limestone reactivity, and full-scale testing of the dry limestone injection process on a 150-Mw unit. The full-scale test program began in May, 1970, on a Babcock and Wilcox boiler at TVA's Shawnee Station near Pudacah, Kentucky. Initial tests resulted in relatively low sulfur dioxide removal efficiencies (below 35 percent) (109).

The problems of injecting limestone for dry absorption are similar to those encountered in the previously mentioned lime injection with wet scrubbing process. However, lime reactivity and distribution problems are even more critical in the dry process because all of the SO_2 absorption must occur during the very short period that the hot gases and reactive lime are in contact. The calcination reaction proceeds best at temperatures near 2000°F. As a result, the point of injection must be carefully selected. Temperatures in the range of 3000°F occur near the bottom of typical boiler furnaces so dead burning can occur if the material is injected at too low a point. Injection directly with the fuel has resulted in low removal efficiencies presumably because of the excessive temperature encountered by the limestone.

Reaction between sulfur dioxide and the calcined limestone particles occurs primarily in the temperature range from about 2000° to 1000°F. The efficiency of removal is affected by numerous operating factors, chief among them being the quantity of limestone used. In tests conducted by Combustion Engineering (110), only about 20 percent of the SO_2 was removed from hot flue gas when a stoichiometric amount of raw dolomite was injected. Similarly, the Wisconsin Electric Co. observed very low removal efficiencies with slightly below stoichiometric quantities of injected limestone but obtained 40 to 50 percent reduction in SO_2 with 75 percent excess limestone (111).

Because of the low efficiencies observed with lime, attempts were made to find a more reactive additive. In mid-1974, tests were conducted at the Nucla Station of Colorado-Ute Electric Association involving the injection of dry nahcolite powder into the flue gas stream and onto the fabric filter bags being used for particulate removal. Nahcolite, a naturally occurring sodium bicarbonate ore, was selected because of its availability in large deposits and its high reactivity with SO_2. The tests indicated a 69 percent SO_2 removal efficiency with 56 percent utilization of the injected alkali (68).

In late 1976, a relatively large-scale test was conducted at the Leland Olds plant of the Basin Electric Power Cooperative. In this test, two modules, each containing six full-size fabric filter bags (11½-in. diameter × 30 ft long) were arranged to receive approximately 3,000 acfm of flue gas. Typical results were 83 percent removal efficiency at 77 percent utilization, and 90 percent removal efficiency at 60 percent utilization (68).

Although the results were considered promising, uncertainties with regard to the availability of low-cost nahcolite led to further modification of the technical approach and test program at the Leland Olds plant. The principal modification consisted of installing a spray dryer ahead of the fabric filter and the injection of the reactive material as a fine spray of liquid droplets. The resulting two-stage system is discussed in the previous section entitled "Aqueous Carbonate Process."

A commercial-scale (22-MW) demonstration of the dry injection process was conducted at the Public Service Company of Colorado Cameo Unit 1 under Electric Power Research Institute sponsorship (162). Both nahcolite and trona were used as reagents for the project and found to perform satisfactorily. Pure sodium carbonate was also tried and found to provide very limited SO_2 removal (nominally 10 percent). It was concluded that nahcolite and trona are effective because they thermally decompose to release CO_2 and H_2O, which greatly enhances porosity and reactive surface area of the residual sodium carbonate. The decomposition of trona occurs at significant rates above 200°F while nahcolite requires temperatures above about 275°F for decomposition.

Following the commercial test, an economic study of the process was made (163). It was generally concluded that capital costs would be much lower for a dry injection system than for a spray dryer system, while operating costs would be somewhat higher due primarily to the high cost of reagent.

Levelized busbar costs for the combined FGD/particulate removal plant were estimated to be 8.6 mills/kWh for trona injection and 9.6 mills/kWh for nahcolite injection compared to 10.8 mills/kWh for a spray dryer process. These are 1983 costs based on two 500-MW units burning 0.48 percent sulfur coal and providing 70–75 percent SO_2 removal.

Alkalized Alumina

The alkalized alumina process was developed by the U.S. Bureau of Mines and carried through the pilot-scale testing phase (112, 113). It has not been applied commercially. The process uses dawsonite, sodium aluminate, $[NaAl(CO_3)(OH)_2]$, as the absorbent. The material is activated at 1200°F to form a high surface area, high porosity, dry solid which removes sulfur dioxide from flue gas at temperatures between 300° and 650°F. The process is similar to the dry limestone injection process in that an alkaline additive is used to contact the sulfur dioxide containing flue gas. Process conditions are quite different, however, and the process differs from the basic limestone injection process in that the absorbent is regenerated.

A simplified flow diagram of the process is shown in Figure 7-41. Flue gases are ducted from a point upstream of the air preheater through a dust collector and into the reaction column at about 625°F. In this column, the up-flowing flue gas is contacted by a free or baffled-fall stream of absorbent particles. (A modified version of the process, developed by the Central Electricity Generating Board of Great Britain, employs fluidized beds for the absorbent-gas contacting step [114].) Gas from the reactor then passes through the air preheater and to the stack. The spent absorbent particles, containing sodium sulfate as the principal reaction product, are heated to about 1200°F and enter the regenerator where they are reduced. A reducing gas, typically producer gas (primarily H_2, CO, and CO_2), is required at a rate of about 2 percent of the primary flue-gas rate. Effluent gas from the regenerator containing 10 to 25 percent H_2S is passed to a Claus unit for sulfur recovery while the regenerated absorbent is recycled to the flue-gas contactor for reuse.

The alkalized alumina process offers the attractions of producing a very desirable by-product, i.e., elemental sulfur, and releasing the stack gases at a high enough temperature to maintain buoyancy. Its two principal problems appear to be process complexity and absorbent attrition. Both can lead to unfavorable economics. A completely satisfactory answer to the problem of excessive attrition has not yet been reported.

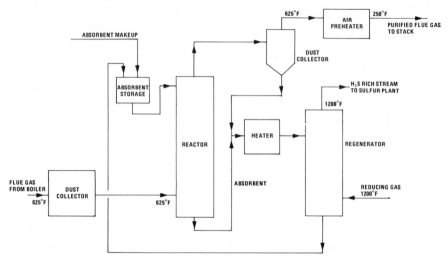

Figure 7-41. Flow diagram of alkalized alumina process.

Copper Oxide Processes

Laboratory-scale work on a copper oxide process for SO_2 removal has been conducted by the U.S. Bureau of Mines (115). The effort was aimed at developing a dry, regenerable absorbent for SO_2 which would not have the problems of alkalized alumina, i.e., physical degradation, excessive consumption of reducing gas, and a high temperature difference between absorption and regeneration steps. An absorbent which appeared to meet these requirements was prepared by the impregnation of copper oxide into porous alumina supports.

The following chemical reactions are important in copper oxide, SO_2 removal processes:

$$CuO + SO_2 + \frac{1}{2}O_2 \rightarrow CuSO_4 \qquad (7\text{-}82)$$

$$CuSO_4 \rightarrow CuO + SO_3 \qquad (7\text{-}83)$$

$$CuSO_4 + 2H_2 \rightarrow Cu + SO_2 + 2H_2O \qquad (7\text{-}84)$$

$$CuSO_4 + \frac{1}{2}CH_4 \rightarrow Cu + SO_2 + \frac{1}{2}CO_2 + H_2O \qquad (7\text{-}85)$$

Reaction 7-82 goes essentially to completion at temperatures below about 450°C. At higher temperatures, $CuSO_4$ decomposes by either the reverse reaction of 7-82 or Reaction 7-83. At about 700°C regeneration should be

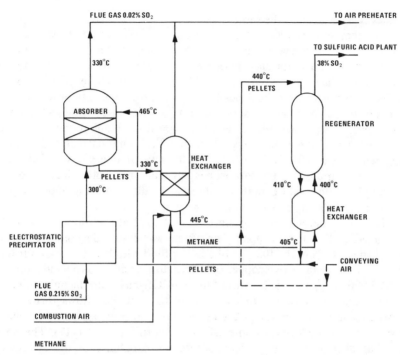

Figure 7-42. Conceptual process flow diagram of SO$_2$ removal with copper oxide as proposed by Bureau of Mines (115).

complete enough to allow recovery of evolved sulfur oxides and recycle of CuO. Regeneration by the use of reducing gases as in Reaction 7-84 and 7-85 can be accomplished at much lower temperatures. The Bureau of Mines work indicated methane to be preferable to hydrogen for regeneration because of the tendency for reduction to proceed all the way to sulfide with hydrogen at low temperature. A conceptual design of a process was proposed as shown in the flow diagram of Figure 7-42. Although no pilot or full-scale plant operations of this process have been conducted to date, it appears to offer considerable promise. The major drawbacks are the requirements for a large expensive reactor which results from a low rate of absorption of SO$_2$ on the copper oxide-alumina pellets, and the high consumption of reducing gas.

Work on a copper oxide process has also been conducted by Shell in the Netherlands (116). This process, which has been named the Shell Flue Gas Desulfurization (SFGD) Process, also uses CuO on an alumina support. It is unique, however, in that it uses fixed beds, and the absorption and regeneration steps are carried out in the same vessel. Two units are used to provide continuous operation. One is used for gas purification and the other undergoes regeneration at the same time. Both steps are accomplished at about the same temperature (400°C). Regeneration is accomplished by the use of a reducing gas such as hydrogen, carbon monoxide, or methane and results in the production of a sulfur dioxide rich gas (Reactions 7-84 and 7-85). Hydrogen and CO are preferred as reducing agents for the Shell process because of problems of coke deposition with hydrocarbons. In order to avoid plugging of the fixed bed by soot and ash particles in the flue gas, a novel reactor system was developed in which the gases are made to pass alongside large surfaces of absorbent mass rather than through a particle bed. A pilot plant with a capacity of 21,000 to 35,000 scfh of flue gas was built and operated to demonstrate the parallel passage reactor and other features of the process. The pilot plant operated quite successfully, removing about 90 percent of the SO_2 from the gas passing through it.

A large commercial unit was started up in 1973 at the Showa Yokkaichi Sekiyo (SYS) Refinery in Japan. The system was designed to remove 90 percent of the SO_2 from the flue gas of a boiler fired with heavy high-sulfur fuel oil. The plant contains two copper-oxide reactors each sized to handle the entire 125,000 Nm^3/h (77,500 scfm) of flue gas. Off-gas from the unit undergoing regeneration is cooled to condense water and then flows through an absorber-stripper which damps the cyclic flow and provides a near constant flow of concentrated SO_2 to an existing Claus sulfur plant (117). The SYS Unit has also been used to demonstrate the capability of the system to remove NO_x simultaneously with SO_2 (63). A 60 to 70 percent reduction in NO_x concentration was obtained by adding NH_3 to the flue gas entering the SFGD System. The ammonia reduces NO to N_2. Carryover of unreacted ammonia with the flue gas was reported to be 2 ppm maximum.

In order to determine the applicability of the process to flue gas from coal combustion, a small scale (0.6 MW) unit was built and placed into operation by UOP, Inc. at the Tampa Electric Company (TECO), Big Bend Station in North Ruskin, Florida. The pilot plant was started up in 1974 and operated for approximately 2 years. During this time, six runs were made involving over 13,000 acceptance and regeneration cycles with the same acceptor loading. The test program demonstrated:

1. The commercial acceptor material has good stability, achieving 90 percent SO_2 removal across a 4 m bed at a space velocity of 5,000/h after 13,000 cycles.
2. The reactor design is tolerant of high fly ash loadings. Techniques were developed to clean fouled internals in situ.
3. Metal oxides in the fly ash and halogen compounds in the flue gas do not interfere with the redox cycle.
4. Mechanical components including flapper valves, the sequence controller, and the reactor vessel gave good performance.

The availability of design information on the SFGD process is excellent. However, little commercial experience is available to date. The principal drawback of the process is its high requirement for reducing gas, particularly if the end product is elemental sulfur. Consumption figures reportedly show 6.2 moles of H_2 per mole of SO_2 removed (63).

The process has a potential advantage over conventional SO_2 removal processes in that it can be operated to simultaneously remove NO_x. This is accomplished by adding ammonia to the flue gas ahead of the copper oxide bed and using the bed material as a combination acceptor for SO_2 and catalyst for the $NO_x + NH_3$ reaction. An independent evaluation based on pilot-scale test results indicated that the process is capable of reducing NO_x and SO_2 emissions by 90 percent when applied to a high-sulfur coal-fired boiler, but its economics are less favorable than a separate NO_x catalytic reduction process combined with a conventional FGD system for the specific case studied (164).

Sulfur Dioxide Recovery by Adsorption

Basic Data

The removal of sulfur dioxide from gas streams by adsorption on an activated solid represents an attractive approach because of the ease of regeneration compared to chemical reaction absorption. However, adsorbents generally have a lower capacity than solid absorbents, and relatively large contactors are required. Although research work had been conducted on SO_2 removal based on zeolite (118, 119), silica gel (120), and ion exchange resins (121, 122), the only adsorption processes which are of commercial significance at this time involve the use of carbon. A comparison of the equilibrium capacities of the above four adsorbent types is given by Figure 7-43.

When sulfur dioxide is adsorbed on activated carbon in the presence of excess oxygen, the carbon acts as a catalyst for oxidation of SO_2 to SO_3. It has been found that the reaction takes place at an impractically slow rate in the absence of water (123). In the presence of water, the reaction becomes

$$SO_2 + \tfrac{1}{2}O_2 + H_2O \xrightarrow{\text{activated carbon}} H_2SO_4 \qquad (7\text{-}86)$$

Regeneration can be accomplished by washing the adsorbent with water to produce a dilute solution of sulfuric acid or heating to reduce the sulfuric acid to SO_2, which can then be converted to concentrated sulfuric acid or sulfur. The reduction reaction can be represented by the following reaction:

$$2H_2SO_4 + C \rightarrow 2SO_2 + 2H_2O + CO_2 \qquad (7\text{-}87)$$

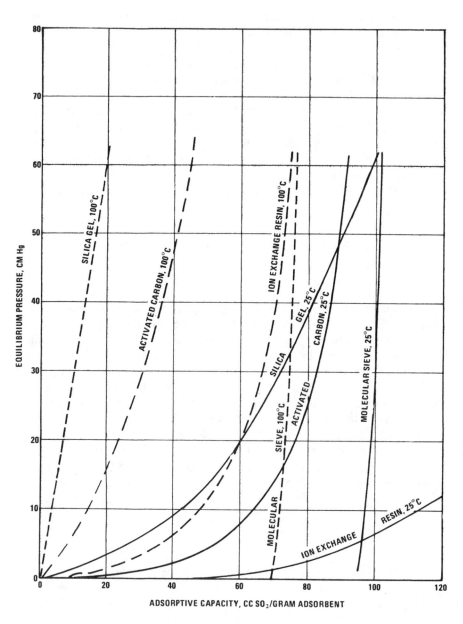

Figure 7-43. Adsorptive capacity for SO₂ of silica gel, activated carbon, ion exchange resin (Rohm & Haas Co. IRA-400) and molecular sieve (Linde Co. Type 5-A). *Data of Cole and Shulman (121)*

Although the water-wash process is simplest, it has the drawback of producing a dilute sulfuric acid product which is difficult to store, ship, or market. The thermal regeneration process, on the other hand, has the drawback of consuming part of the adsorbent to accomplish the reduction. As a result, processes which employ thermal regeneration require a very low-cost form of activated carbon, such as coke.

Carbon Adsorption with Water Wash Regeneration

This process is typified by the Sulfacid process developed by Lurgi Gesellschaft für Chemie and Hüttenwesen m.b.H., Frankfurt, Germany, and the activated carbon process developed by Hitachi Mfg. Co. Ltd, Tokyo. The two versions of the process differ primarily in the method of flue-gas contacting. In the Lurgi Sulfacid process, the impure gas is first contacted by weak sulfuric acid from the adsorption step. This cools the gas and also concentrates the sulfuric acid product somewhat. The cooled gas then passes through a fixed bed adsorber containing the activated carbon. Water is sprayed in intermittently without interrupting the gas flow to remove sulfuric acid which has formed in the pores of the carbon. An acid strength of only about 7 percent H_2SO_4 is attainable in the liquid effluent from the adsorber; however, this is upgraded in the gas cooler to as high as 15 percent depending upon the temperature of the entering gas. A sulfur dioxide removal efficiency of over 90 percent is reported (124). The process has been used to treat emissions from a sulfuric acid plant (100,000 cfm) and from a coal fired unit equivalent to a 2,000-kw power plant (125).

The Hitachi process, which has been described in considerable detail by Tamura (126), uses a somewhat more complex gas-contacting arrangement. A schematic flow diagram of the process as employed in a 55-Mw pilot plant is shown in Figure 7-44. The plant processes a portion of the stack gas from a 350-Mw boiler. About 1450 scfm of gas is removed from the boiler flue duct after the preheater, cleaned of dust, passed through the adsorption beds, and returned to the boiler stack-gas line. At any time in the cycle, four of the carbon beds are operated in parallel for SO_2 adsorption while the remaining one is being washed and regenerated. Each unit operates on stack gas (drying the bed and adsorbing SO_2) for 48 hours and is then washed for 12 hours making a total cycle time of 60 hours. Six washing tanks are employed, each containing a different concentration of H_2SO_4. During the washing step, the tower is washed with acid from the six tanks in sequence, starting with the most concentrated acid and ending with fresh water. A stream of the highest concentration acid, 20 percent H_2SO_4, is continuously removed and fed to the submerged combustion concentrator. The final product is 65 percent H_2SO_4, which is used primarily for phosphate fertilizer manufacture. Although minor problems were encountered during initial operations, the general perfor-

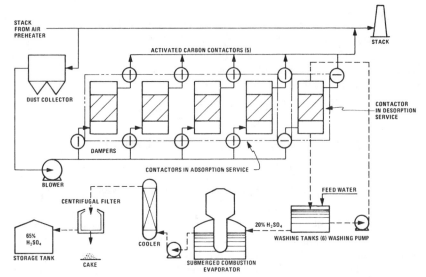

Figure 7-44. Schematic flow diagram of Hitachi activated carbon process—55 Mw pilot plant.

mance of the plant was considered to be very successful; a mean SO_2 removal efficiency of 80 percent was observed after 3000 hours of operation.

One plant in Japan has operated for 5 years without appreciable problems (as of the end of 1977). Carbon consumption has been very low—about 2 percent per year. The dilute sulfuric acid product (17 percent) from this plant is reacted with limestone to produce a saleable gypsum (10).

A continuous process has been proposed by Joyce et al. (123), however, only laboratory-scale tests were conducted to develop process data. In the proposed process, a fixed adsorption bed is used, with wash water continuously sprayed on the top surface and dilute sulfuric acid removed from the bottom of the adsorption vessel. The laboratory data indicated that a maximum H_2SO_4 concentration of 12 to 15 percent could be obtained from typical flue-gas streams. Controlling variables in the adsorption operation were found to be the O_2/SO_3 mole ratio and gas-contact time. The results indicate, for example, that with 60-sec contact time, increasing the O_2/SO_2 ratio from 1.8 to 5.3 increased the conversion from about 70 percent to almost 100 percent. With the O_2/SO_2 ratio fixed at 1.8, increasing the contact time from 20 to 80 sec caused the conversion efficiency to go from less than 40 percent to over 80 percent.

The carbon adsorption process with water-wash regeneration appears to be quite simple technically and economically attractive for cases where the dilute acid produced is of local value. It is of particular interest for cleaning the tail gas from contact sulfuric acid plants. In such an installation, the weak acid produced could be used instead of makeup water in the acid plant, increasing the total sulfuric acid production while abating air pollution by purifying the tail gas.

Carbon Adsorption with Thermal Regeneration

This type of system has been developed into a commercial process by the German companies, Reinluft GmbH and Chemiebau Dr. A. Zieren GmbH, and is marketed as the Reinluft (Clean Air) Process (127, 128). Work on another version has been carried out by Bergbau-Forschung GmbH (Research Center of the German Coal Mining Industry) (129). The Bergbau-Forschung process differs from the Reinluft process in its use of a special activated char rather than low-temperature coke and its use of hot sand instead of hot gas to provide heat for regeneration. Both processes employ moving bed adsorbers and make use of the same chemical reactions in the cycle (Reactions 7-86 and 7-87).

A flow diagram of Chemiebau's design of a Reinluft plant is shown in Figure 7-45. Flue gas, at a temperature in the range of about 200° to 320°F, enters the bottom of the adsorption column in which it contacts downflowing streams of granular adsorbent (coke). The size of individual coke particles is between 0.1 and 1.3 in. The purified flue gas leaves the top of the adsorber at approximately its original temperature. Ash entrained in the entering flue gas is retained by coke in the lower portion of the adsorber and has little effect on its activity. Coke containing adsorbed sulfuric acid is removed from the bot-

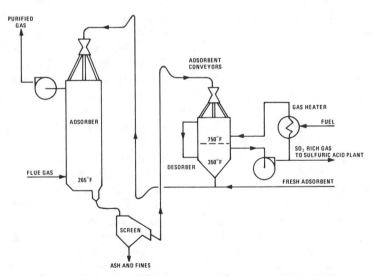

Figure 7-45. Schematic diagram of Reinluft process-Chemiebau design (128).

tom of the adsorption column, passed over a screen which removes fine particles of coke and ash, and conveyed to the desorber.

The regeneration step consists of reducing adsorbed H_2SO_4 to SO_2 (consuming some of the coke in the process) and stripping the resulting SO_2, H_2O, and CO_2 from the bed. This is accomplished by passing the spent coke downward in the top section of the desorber countercurrent to a hot gas stream. In this part of the desorber, the high-temperature gas strips off the reaction products and is cooled by the downflowing coke, which is heated to about 720°F in the process. The cooled gas is withdrawn from the top of the desorber and readmitted to the bottom of the lower section. In this section it again undergoes heat exchanger with the coke, the upflowing gas being heated and the coke being cooled. The regenerated and cooled coke is then conveyed to the top of the adsorber with added fresh material to make up for the amount lost by reaction or attrition. The gas is removed from the top of the lower section of the desorber, a portion of it is withdrawn from the system for conversion of SO_2 to sulfur or sulfuric acid, and the balance is passed through the heater and recycled to the desorber.

The key to the economic potential of the Reinluft process is the low-cost adsorbent used. Instead of employing activated charcoal as makeup adsorbent, the Reinluft process uses low-temperature coke. The relatively inexpensive material is made by carbonizing geologically young fuels such as peat, lignite, and, in some cases, coal at temperatures below 1300°F. The coke is not activated before use but becomes activated by the adsorption-regeneration cycle. After 3 to 10 cycles, the coke activity reaches a maximum comparable to gas-adsorption charcoal.

Four Reinluft process plants or pilot plants have been constructed and operated. Although a number of operating problems were encountered, it is reported that the plants adequately demonstrated the basic processes, and a new design has been developed to overcome the observed shortcomings. The principal operating problems were fires in the coke beds and corrosion of equipment by sulfuric acid. The new design (Figure 7-45) avoids these problems by completely separating the adsorber and desorber portions, maintaining constant (and low) coke temperature at critical locations and improving the coke-feed and discharge systems to provide uniform distribution and movement of coke particles.

The Bergbau-Forschung process has been tested in two prototype systems: a 35-MW unit at Lunen, West Germany, and a 20-MW (equivalent) installation at Gulf Power Company's Scholtz Electric Generating Station (54). The latter installation includes a section to reduce sulfur dioxide to elemental sulfur by contact with crushed anthracite coal. The reduction process was developed by Foster Wheeler Energy Corporation and is called the RESOX process. The combined process used at the Scholtz Station is identified as the FW/BF process. Neither the mechanical reliability nor the process operability of the Scholtz Station plant was considered acceptable. The major problems were hot spots in the adsorber, poor reliability of the char/sand separator, and plugging of the RESOX system condenser with a mixture of sulfur and carbon particles. Bergbau reportedly had better success operating the adsorption and regeneration sections of the process at Lunen. However, the Lunen plant incorporates a modified Claus process for conversion of sulfur dioxide to sulfur instead of the newer RESOX process. Because of the potential advantages of using coal directly to reduce SO_2, development work is being continued on the RESOX process, and tests are planned for this operation at the Lunen plant (54).

An activated carbon process which uses fluidized beds in both the adsorption and regeneration steps is under development by Westvaco (63). Five shallow fluidized beds are specified for an adsorber designed to provide 90 percent removal of SO_2. It is claimed that 99 percent SO_2 removal can be achieved with only about a 20 percent increase in carbon bed depth (and pressure drop). Adsorbed SO_2 is catalytically oxidized to sulfuric acid on the carbon and the acid-laden carbon is regenerated by a multistep reduction process employing hydrogen. The following reactions form the basis for the regeneration system:

Sulfur Generator/Acid Converter:

$$H_2SO_4 + 3H_2S = 4S + 4H_2O \tag{7-88}$$

H_2S Generator/Sulfur Stripper:

$$H_2 + S = H_2S \tag{7-89}$$

Westvaco has operated an integrated pilot plant which utilized the process to purify a 330 cfm slip stream from an oil-fired boiler. The unit operated for 350 hours with one short period of downtime due to sulfur plugging. The process requires relatively large vessels (due to a design gas velocity of about 4 ft/sec) and has a somewhat higher pressure drop than most wet scrubbers. Hydrogen consumption is reported to be 3.3 to 3.9 moles H_2 per mole SO_2 (63).

Sulfur Dioxide Recovery by Catalytic Oxidation

The removal of SO_2 from dilute flue-gas streams by catalytic oxidation represents an adaptation of the contact catalytic process used in the manufacture of sulfuric acid. Sulfur dioxide is oxidized to sulfur trioxide by reaction with oxygen in the flue gas in the presence of a vanadium pentoxide type catalyst. The resulting SO_3 combines with water vapor in the flue gas to form sulfuric acid which is condensed by cooling and separated from the gas stream. Some of the special problems encountered in adapting the well-known contact sulfuric acid process to flue-gas treating are (a) preventing plugging or poisoning of the catalyst bed by impurities in the gas, (b) maintaining the catalyst at an optimum conversion temperature (about 900°F), and (c) avoiding corrosion by the acid produced. A considerable amount of effort has gone into resolving these problems and developing commercial processes for purifying flue gas by catalytic oxidation.

Cat-Ox Process

The system which appears to be most advanced at this time is the Cat-Ox process developed by the Monsanto Company (130, 131). This process was originally tested in a small pilot plant built and operated under a joint effort by Pennsylvania Electric Company, Air Preheater Company, Research-Cottrell Corporation, and Monsanto Company. More recently a larger prototype unit was built and operated by Monsanto and Metropolitan Electric Company. The unit was designed to treat 24,000 scfm or approximately 6 percent of the total flue gas from a 250-Mw unit of Metropolitan Edison's Portland, Pennsylvania, generating station. A flow diagram of the process is shown in Figure 7-46.

Flue gas is taken from the boiler at about 950°F and passed through a high-temperature electrostatic precipitator designed to remove over 99.5 percent of the fly ash from the gas stream. The hot gas then passes through a bed of catalyst in the converter where oxidation of SO_2 to SO_3 occurs. The gas is then cooled by passing first through a tubular economizer and then a regeneration type air preheater. In order to avoid serious corrosion at this point, it is necessary to limit cooling to about 450°F, which is above the dew point of sulfuric acid in the gas stream. The gas then passes through a packed absorber in which it is contacted with a cool stream of sulfuric acid. The gas is cooled to about 225°F, and the resulting hot acid is recycled through a shell and tube heat exchanger in which the heat is transferred to cooling water. Excess acid is drained off as product and further cooled to 110°F for storage.

Gas from the absorber contains some sulfuric acid mist formed during the cooling step plus a small amount of liquid entrained from the absorption tower. Removal of this mist is the key step in the process. In the Monsanto Cat-Ox prototype plant, the mist is removed by a fiber-packed cartridge type eliminator which produces an effluent gas containing less than 1.0 mg 100 percent H_2SO_4/scf. This amounts to 10 ppm of acid as mist. The vapor pressure of sulfuric acid at 225°F contributes another 11 ppm so the total quantity of sulfuric acid in the exit stack is about 20 ppm. This is less than the amount of sulfuric acid in the flue gas as it comes from the boiler. The sulfur

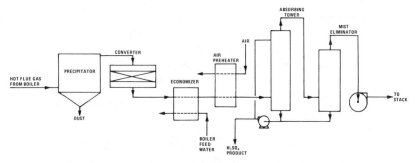

Figure 7-46. Flow diagram of catalytic oxidation process of SO_2 removal.

dioxide content of the product gas is only about 10 percent of its original value.

Operation of the prototype plant at Metropolitan Edison's Portland Station provided several years of operating experience on the process. Unfortunately, the high-temperature precipitation and special heat-recovery system made the arrangement tested impractical for all but new stations. As a result, Monsanto developed a "back fit" cycle in which the stack gas is reheated to the 800° to 900°F (60). This permits existing low temperature precipitators to serve the SO_2 conversion plant.

A 103-MW commercial demonstration plant was installed on a coal-fired boiler at Illinois Power Company's Wood River Power Station, Unit 4 (132). Since this was a retrofit application, it was necessary to include equipment for reheating the flue gas to 850-900°F before feeding it to the sulfuric acid converter. After start-up in September, 1972, natural gas was used as fuel for the in-line heaters to reheat the feed gas. The plant operated successfully for 444 hours.

Natural gas became unavailable in October, 1972, and it was necessary to substitute oil as fuel for the heaters. Following modification of the burners, an additional period of satisfactory operations was experienced, bringing the total operating time to 602 hours by July, 1973. It became apparent, however, that the in-line combustion of oil would cause rapid deterioration of the catalyst. As a result, a major plant modification was undertaken to change to an "external" oil-fired reheat system.

During subsequent attempts to start-up and operate the plant, a number of additional mechanical and structural problems developed. According to a recent report, the unit would require refurbishing before additional operations could be conducted (63). As a result, the current commercialization status of the Cat-Ox process is in doubt, although the basic technology appears to be sound and relatively well established.

References

1. Katz, M., and Cole, R.J. 1950. *Ind. Eng. Chem.* 42(11):2258-2269.
2. Swain, R.E. 1921. *Chem. & Met. Eng.* 24:463-465.
3. Dean, R.S., and Swain, R.E. 1944. *U.S. Bur. Mines, Bull. No. 453.*
4. Hewson, G.W.; Pearce, S.L.; Pollitt, A.; and Rees, R.L. 1933. *Soc. Chem. Ind. (London), Chem. Eng. Group, Proc.* 15:67.
5. Engdahl, R.B. 1973. *J. of the Air Pollution Control Assoc.*, 23(May):364-375.
6. Kaplan, N., and Maxwell, M.A. 1977. *Chem. Eng.* 84(Oct. 17):127-135.
7. Bauman, R.D., and Crenshaw, J.O. 1977. *Proc. 12th Intersociety Energy Conversion Engineering Conference,* Washington, D.C., Aug. 28-Sept. 2. Paper 779009:629.
8. Pebler, A.R. 1974. *Combustion* 46(Aug.):21-23.
9. Pierce, R.R. 1977. *Chem. Eng.* 84(April 11):125-128.
10. Ando, J. 1977. "Status of SO_2 and NO_x Removal Systems in Japan," presented at the EPA Symposium on Flue Gas Desulfurization, Hollywood, Florida (Nov. 8-11, 1977).
11. Pedco Environmental Specialists, Inc., "Summary Report—Flue Gas Desulfurization Systems—June-July, 1977," prepared for U.S. EPA.
12. Anon, 1977. *Chem. Eng.* 84(May 23):101-103. (Cost data from Pedco Environmental Specialists, Inc., report to EPA, June 1976.)
13. McGlammery, C.G.; Faucett, H.L.; Torstrick, R.L.; and Henson, L.J. 1976. *Proc. Symp. on Flue Gas Desulfurization,* New Orleans, March 1976, Vol. I, EPA-600/2-76-136a (May 1976):79-99.
14. Torstrick, R.L.; Henson, L.J.; and Tomlinson, S.V. 1977. "Economic Evaluation Techniques, Results, and Computer Modeling for Flue Gas Desulfurization," presented at the EPA Symposium on Flue Gas Desulfurization, Hollywood, Florida (Nov. 8-11, 1977).
15. The M.W. Kellogg Company, 1975. "Comparison of Flue Gas Desulfurization, Coal Liquefaction, and Coal Gasification for Use at Coal Fired Power Plants," EPA-450/3-75-047, (April 1975).
16. Jimeson, R.M., and Maddocks, R.R. 1976. *Chem. Eng. Progr.* 72(Aug.):80-88.
17. Eschellman, G. 1909. U.S. Patent 900,500.
18. Anon, 1977. *J. of the Air Pollution Control Assoc.* 27(Oct.):948-961, condensed from "The Status of Flue Gas Desulfurization Applications in the United States, A Technological Assessment," Federal Power Commission (Historical Research by J.C. Marten).
19. Miller, D.M. 1976. *Proc. Symp. on Flue Gas Desulfurization* New Orleans, March 1976, Vol. 1, EPA-600/2-76-136a (May 1976):373-385.
20. Jonakin, J., and McLaughlin, J.F. 1969. *Proc. American Power Conference* 31:543-552.
21. McIlvaine, R.W. 1976. "Flue Gas Desulfurization—Controversial Aspects Examined," presented at the American Power Conference, Chicago, Illinois, April 20-22, reprinted in *Combustion,* (Oct. 1976):33-36.
22. Parkison, R.V. 1956. *Tappi* 39(May):517-519.
23. Head, H.N. 1977. "EPA Alkali Scrubbing Test Facility: Advanced Program, Third Progress Report," EPA-600/7-77-105 (Sept. 1977).
24. Jolly, W.J. 1966. *The Chemistry of the Non-Metals* Prentice Hall, Englewood Cliffs, New Jersey:67.
25. Nannen, L.W.; West, R.E.; and Kreith, F. 1974. *J. of the Air Pollution Control Assoc.* 24(Jan.):29-39.

26. Radian Corporation, 1976. "Experimental and Theoretical Studies of Solid Solution Formation in Lime and Limestone SO$_2$ Scrubbers—Vol. 1," Final Report, EPA-600/2-76-273a (Oct. 1976).

27. Bechtel Corporation, 1976. "EPA Alkali Scrubbing Test Facility: Advanced Program Second Progress Report," EPA-600/7-76-008 (Sept. 1976).

28. Ottmers, D., Jr.; Phillips, J.; Burklin, C.; Corbett, W.; Phillips, N.; and Shelton, C. 1974. "A Theoretical and Experimental Study of Lime/Limestone Wet Scrubbing Process," EPA-650/2-75-006 (Dec. 1974).

29. Borgwardt, R.H. 1975. *Combustion* 47, 37-42(Oct.).

30. Selmeczi, J.G. 1975. U.S. Patent 3,914,378.

31. Selmeczi, J.G. 1975. U.S. Patent 3,919,393.

32. Selmeczi, J.G. 1975. U.S. Patent 3,919,394.

33. Selmeczi, J.G., and Stewart, D.A. 1978. *Chem. Eng. Progr.* 74(Feb.):41-45.

34. Raymond, W.J., and Sliger, A.G. 1978. *Chem. Eng. Progr.* 74(Feb.):75-80.

35. Gogineni, M.R. 1975. *Combustion* 47(Oct):9-15.

36. Sorenson, P.; Takvoryan, N.E.; and Jaworowski, R.J. 1976. *Proc. Symp. on Flue Gas Desulfurization,* New Orleans, March, 1976, Vol. 1, EPA-600/2-76-136a (May 1976):999.

37. Laseke, B.A., and Devitt, T.W. 1977. "Status of Flue Gas Desulfurization Systems in the United States," presented at EPA Symposium on Flue Gas Desulfurization, Hollywood, Florida (Nov. 8-11).

38. Weir, A., Jr.; Papay, L.T.; Jones, D.G.; Johnson, J.M.; and Martin, W.C. 1976. *Proc. Symp. on Flue Gas Desulfurization,* New Orleans, March, 1976, Vol. 1, EPA-600/2-76-136a (May 1976):325.

39. Gogineni, M.R., and Maurin, P.G. 1975. *Combustion* 47(Oct.):9-15.

40. Rossoff, J., and Rossi, R.C. 1974. "Disposal of Biproducts from Non-Regenerable Flue Gas Desulfurization Systems: Initial Report," EPA-650/2-74-037a (May 1974).

41. Goodwin, R.W. 1978. *J. of the Air Pollution Control Assoc.* 28(Jan.):35-39.

42. Grimm, C.; Abrams, J.Z.; Leffmann, W.W.; Raben, I.A.; and LaMantia, C. 1978. *Chem. Eng. Progr.* 74(Feb.):51-57.

43. Crow, E.L., and Horsman, H.R. 1976. "Third TVA interim Report of Corrosion Studies: EPA Alkali Scrubbing Test Facility," March, 1976, Appendix L of EPA-600/7-76-008 (Sept. 1976).

44. Koehler, G.R., and Dober, E.J. 1974. *Proc. Symp. on Flue Gas Desulfurization,* Atlanta, Georgia, Nov., 1974, Vol. II, EPA 650/2-74-126-6 (Dec. 1974):673-708.

45. McGlammery, G.G. et al. 1973. "Sulfur Oxide Removal from Power Plant Stack Gas—Magnesia Scrubbing—Regeneration and Production of Sulfuric Acid," EPA-R-2-73-244 (May).

46. Pinaev, V.A. 1963. "SO$_2$ Pressure over Magnesium Sulfite-Bisulfite-Sulfate Solution," *Journal of Appl. Chem.* (USSR), 36(Oct.):2049.

47. Gille, J.A., and Mackenzie, J.S. 1977. "Philadelphia Electric's Experience with Magnesium Oxide Scrubbing," presented at the EPA Symposium on Flue Gas Desulfurization, Hollywood, Florida (Nov. 8-11, 1977).

48. Howard, H., and Stantial, F.G. 1918. U.S. Patent 1,271,899 (July 9).

49. Dingo, T.T. 1974. *Proc. Symp. on Flue Gas Desulfurization,* Atlanta, Georgia, Nov., 1974, Vol. I, EPA-650/2-74-126a (Dec. 1974):519-538.

50. Legatski, L.K.; Johnson, K.E.; and Lee, L.Y. 1976. *Proc. Symp. on Flue Gas Desulfurization,* New Orleans, March, 1976, Vol. I, EPA-600/2-76-136a (May 1976):471-502.

51. Kaplan, N. 1976. *Proc. Symp. on Flue Gas Desulfurization*, New Orleans, March, 1976, Vol. I, EPA-600/2-76-136a (May 1976):387-422.

52. Kaplan, N. 1974. *Proc. Symp. on Flue Gas Desulfurization*, Atlanta, Georgia, Nov., 1974, Vol. I, EPA-650/2-74-126a (Dec. 1974):445-515.

53. LaMantai, C.R.; and Lunt, R.R.; Rush, R.E.; Frank, T.M.; and Kaplan, N. 1976. "Operating Experience—CEA/ADL Dual Alkali Prototype System at Gulf Power/Southern Services, Inc.," *Proc. Symp. on Flue Gas Desulfurization*, New Orleans, March, 1976, Vol. I, EPA-600/2-76-136a (May 1976):423-470.

54. Rush, R.E., and Edwards, R.A. 1977. "Operating Experience with Three 20 MW Prototype Flue Gas Desulfurization Processes at Gulf Power Company's Scholtz Electric Generating Station," presented at EPA Flue Gas Desulfurization Symposium, Hollywood, Florida (Nov. 8-11).

55. Johnstone, H.F., and Singh, A.D. 1940. *Univ. Illinois Bull. Eng. Expt. Sta. Bull. No. 324* (Dec. 31).

56. Johnstone, H.F., and Singh, A.D. 1937. *Ind. Eng. Chem.* 29(Mar.):286.

57. Johnstone, H.F., and Kleinschmidt, R.V. 1938. *Trans. Am. Inst. Chem. Engrs.* 34(Apr. 25):181.

58. Chemical Construction Corp. 1970. *Engineering Analysis of Emissions Control Technology for Sulfuric Acid Manufacturing Process*, Vol. 1. PB 190 393.

59. Besner, D. 1970. *Control of the Emissions of Sulfur Oxide*. Austin, Texas: Ralph McElroy Co., p. 102.

60. Farthing, J.G. 1971. *Electrical World* (May 15):34-39.

61. Martinez, J.L.; Earl, C.B.; and Craig, T.L. 1971. *The Wellman-Lord SO₂ Recovery Process—A Review of Industrial Operation*. Paper presented at the Environmental Quality Conference for the Extractive Industries of the Am. Inst. Mining, Metallurgical, and Petroleum Engineers, Inc., Washington, D.C. (June 7-9).

62. Bailey, E.E. 1974. *Proc. Symp. on Flue Gas Desulfurization*, Atlanta, Georgia, Nov. 1974, Vol. II, EPA 650/2-74-126-6 (Dec. 1974):745-760.

63. Radian Corporation, 1977. "Evaluation of Regenerable Flue Gas Desulfurization Procedures," Vol. I, EPRI FP-272 (Jan. 1977).

64. Link, F.W., and Ponder, W.H. 1977. "Status Report on the Wellman Lord/Allied Chemical Flue Gas Desulfurization Plant at Northern Indiana Public Service Company's Dean H. Mitchell Station," presented at EPA Flue Gas Desulfurization Symposium, Hollywood, Florida (Nov. 8-11).

65. Bailey, E.E., and Heinz, R.W. 1975. *Chem. Eng. Progr.* 71(3):64-68.

66. Gehri, D.C., and Gylfe, J.D. 1973. *The Atomics International Aqueous Carbonate Process for SO₂ Removal—Process Description and Pilot Test Results*. Paper No. 73-306. Presented at the APCA Meeting in Chicago, Illinois.

67. Botts, W.V.; Fockler, R.B.; and Phelan, J.H. 1978. "Dry Scrubber Systems," presented at American Public Power Association Workshop, San Francisco, California (Feb. 28-Mar. 2).

68. Estcourt, V.F.; Grutle, R.O.M.; Gehri, D.C.; and Peters, H.J. 1978. "Tests of a Two-Stage Combined Dry Scrubber/SO₂ Absorber Using Sodium or Calcium," presented at the 40th Annual Meeting, American Power Conference, Chicago, Illinois (Apr. 26).

69. Aldrich, R.G., and Oldenkamp, R.D. 1977. "A 100 MW Second Generation SO₂ Removal Plant for New York State Utilities," presented at the 34th Annual Meeting, American Power Conference (Apr. 18).

70. Binns, D., and Aldrich, R.G. 1977. "Design of the 100 MW Atomics International Aqueous Carbonate Process Regeneration FGD Demonstration Plant," presented at the EPA Symposium on Flue Gas Desulfurization, Hollywood, Florida (Nov. 8-11).

71. Rosenbaum, J.B.; George, D.R.; Crocker, L.; Nissen, W.J.; May, S.L.; and Beard, H.R. Paper presented at AIME Environmental Quality Conf., Washington, D.C. (June 7-9).
72. Madenburg, R.S., and Kurey, R.A. 1977. "Citrate Process Demonstration Plant—A Progress Report," presented at the EPA Symposium on Flue Gas Desulfurization, Hollywood, Florida (Nov. 8-11).
73. Humphries, J.J., Jr., and McRae, W.A. 1970. *Proc. Am. Power Conf.* 32:663.
74. Meliere, K.A.; Gartside, R.J.; McRae, W.A.; and Seamans, T.F. 1974. *Proc. Symp. on Flue Gas Desulfurization,* Atlanta, Georgia, Nov. 1974, EPA-650/2-74-126b (Dec. 1974):1109-1126.
75. Ramsay, W. 1883. British Patent 1247.
76. Hansen, C. 1929. U.S. Patent 1,740,342.
77. Gleason, G.H., and Loonam, A.C. 1934. U.S. Patent 1,972,883 (to Guggenheim Bros.).
78. Fleming, E.P., and Fitt, T.C. 1950. *Ind. Eng. Chem.* 42(11):2249-2253.
79. Johnstone, H.F. 1937. U.S. Patent 2,082,006.
80. Johnstone, H.F. 1938. U.S. Patent 2,134,481.
81. King, R.A. 1950. *Ind. Eng. Chem.* 42(11):2241-2248.
82. Breed, C.E., and Hollinden, G.A. 1974. *Proc. Symp. on Flue Gas Desulfurization,* Atlanta, Georgia, Nov. 1974, Vol. II, EPA-650/2-74-126b (Dec. 74):1069-1108.
83. Johnstone, H.F. 1935. *Ind. Eng. Chem.* 27:587.
84. Howat, D.D. 1940. *Chem. Age (London)* 43(Nov. 30):249.
85. Wartman, F.C. 1937. *U.S. Bur. Mines, Rept. Invest. No. 3339.*
86. Johnstone, H.F., and Keyes, D.B. 1935. *Ind. Eng. Chem.* 27(June):659.
87. Ontario Research Foundation. 1947. *The Removal of Sulfur Gases from Smelter Fumes.* Province of Ontario: Department of Mines.
88. Burgess, W.D. 1956. *Chem. in Can.* (June):116-120.
89. Hein, L.B.; Phillips, A.B.; and Young, R.D. 1955. *Problems and Control of Air Pollution.* Edited by F.S. Mallette. New York: Reinhold Publishing Corporation, p. 55.
90. Tennessee Valley Authority, 1974. "Pilot Plant Study of an Ammonia Absorption—Ammonium Bisulfate Regeneration Process, Topical Report Phases I and II," EPA-650/2-74-049a (June).
91. Quackenbush, V.C.; Polek, J.R.; and Agarwal, D. 1977. "Ammonia Scrubbing Pilot Activity at Calvert City," presented at EPA Symposium on Flue Gas Desulfurization, Hollywood, Florida (Nov. 8-11).
92. Ennis, C.E. 1977. "SO$_2$ Removal with Ammonia; A Fresh Perspective," Second Pacific Chemical Engineering Congress (PACHEC 1977), Denver, Colorado.
93. Spector, M.L., and Brian, P.L.T. 1974. U.S. Patent 3,843,789.
94. Tamaki, A. 1975. *Chem. Eng. Progr.* 77(5):55-58.
95. Weidmann, H., and Roesner, G. 1935. *Metallges., Periodic Rev.* 11.
96. Roesner, G. 1937. *Metall u. Erz* 34:5.
97. Fleming, E.P., and Fitt, T.C. 1950. *Ind. Eng. Chem.* 42(11):2253-2258.
98. Fleming, E.P., and Fitt, T.C. 1946. U.S. Patent 2,399,013.
99. Fleming, E.P., and Fitt, T.C. 1946. U.S. Patent 2,295,587.
100. Pastnikov, V.F., and Astasheva, A.A. 1940. *J. Chem. Ind. (U.S.S.R.)* 17(3):14-19.
101. Weidmann, H., and Roesner, G. 1936. *Ind. Eng. Chem. (News Ed.)* 14(Mar. 20):105.
102. Weidmann, H., and Roesner, G. 1936. *Metallges., Periodic Rev.* 11(Feb.):7-13.
103. Anon. 1953. *Chem. Eng.* 60(Apr.):274-277.
104. Lead Industries Assoc. 1952. *Lead Chemical Construction Bulletin No. 6* (Dec. 5).

105. Oldenkamp, R.D., and Margolin, E.D. 1969. *Chem. Eng. Prog.* 65(Nov.):73.
106. Katz, B., and Oldenkamp, R.D. 1969. Paper presented at the ASME Winter Annual Meeting, Los Angeles, Calif. (Nov. 11-20). ASME Publ. 69-WA/APC-6.
107. Thomas, A.D.; Davis D.L.; Parsons, T.; Schroeder, G.D.; and DeBerry, D. 1969. *Applicability of Metal Oxides to the Development of New Processes for Removing SO₂ from Flue Gases.* PB 185 562 (July 31).
108. Falkenberg, H.L., and Slack, A.V. 1969. *Chem. Eng. Progr.* 65(Dec.):61.
109. Slack, A.V., and Harrington, R.E. 1970. *Removal of Sulfur Dioxide from Power Plant Stack Gas: Status of Limestone Process.* Paper presented at the Second International Clean Air Congress of the International Union of Air Pollution Prevention Association, Washington, D.C. (Dec. 6-11).
110. Plumley, A.L.; Whiddon, O.D.; Shukto, F.W.; and Jonakin, J. 1967. *Proc. Am. Power Conf.* 29.
111. Pollock, W.A.; Tomany, J.P.; and Frieling, G. 1967. *Mech. Eng.* (Aug.):21.
112. Bienstock, D.; Field, J.H.; and Myers, J.G. 1964. *J. Eng. for Power, Trans. ASME,* Series A, 86:457-464.
113. Bienstock, D.; Field, J.H.; and Myers, J.G. 1967. *Process Development in Removing Sulfur Dioxide from Hot Flue Gases.* U.S. Dept. Int., Bur. Mines, R.I. 7021 (July).
114. Newell, J.E. 1969. *Chem. Eng. Progr.* 65(Aug.):62.
115. McCrea, D.H.; Forney, A.J.; and Meyers, J.G. 1970. *J. Air Pollution Control Assoc.* 20(Dec.):819-824.
116. Doutzenberg, F.M.; Naber, J.E.; and van Ginneken, A.J.J. 1971. *The Shell Flue Gas Desulfurization Process.* Paper presented at AIChE 68th National Meeting, Houston, Texas (Feb. 28-Mar. 4).
117. Arneson, A.D.; Nooy, F.M.; and Pohlenz, J.B. 1977. "The Shell FGD Process, Pilot Plant Experience at Tampa Electric," presented at the EPA Symposium on Flue Gas Desulfurization, Hollywood, Florida, (Nov. 8-11).
118. Tamboli, J.K., and Sand, L.B. 1970. *SO₂ Sorption-Properties of Molecular Sieve Zeolites.* Paper presented at the Second International Clean Air Congress of the International Union of Air Pollution Prevention Association, Washington, D.C. (Dec. 6-11).
119. Martin, D.A., and Bently, F.E. 1962; *U.S. Bur. Mines RI 6321.*
120. McGavack, J., and Patrick, N.A. 1920. *J. Am. Chem. Soc.* 42:946.
121. Cole, R., and Shulman, H.L. 1960. *Ind. Eng. Chem.* 52(Oct.):859.
122. Glowiak, B., and Gostomczyk, A. 1970. *Sulfur Dioxide Sorption on Anion Exchangers.* Paper presented at Second International Clean Air Congress of the International Union of Air Pollution Prevention Association, Washington, D.C. (Dec. 6-11).
123. Joyce, R.J.; Lynch, R.T.; Sutt, R.F.; and Tobias, G.J. 1970. *Effective Recovery of Dilute SO₂.* Paper presented at Third Joint Meeting of AIChE and Instituto Mexicani de Ingenieros Quimicos, Denver, Colo. (Aug. 30-Sept. 2).
124. Dennis, R., and Bernstein, R.H. 1968. *Engineering Study of Removal of Sulfur Oxides from Stack Gas.* Report prepared by GCA Corp. for Am. Petrol. Inst. (GCA-TR-68-15-G).
125. Maurin, P.G., and Jonakin, J. 1970. *Chem. Eng.* 77(Deskbook Issue):173-180.
126. Tamura, Z. 1970. *Desulfurization Method from Stack Gas by Activated Carbon.* Paper presented at Second International Clean Air Congress of the International Union of Air Pollution Prevention Association, Washington, D.C. (Dec. 6-11).
127. Furkert, H. 1970. *Proc. Am. Power Conf.* 32:673.

128. Anon. 1967. *Chem. Eng.* (Oct.):94.

129. Juntzen, H.; Knoblauch, K.; and Peters, W. 1970. *SO₂ Removal from Flue Gases by Special Carbon.* Paper presented at Second International Clean Air Congress of the International Union of Air Pollution Prevention Association, Washington, D.C. (Dec. 6-11).

130. Stites, J.G., Jr., and Miller, J.G. 1969. *Proc. Am. Power Conf.* 31:553.

131. Stites, J.G., Jr.; Horlacher, W.R.; Bachofer, J.L., Jr.; and Bartman, J.S. 1969 *Chem. Eng. Progr.* 65(Oct.):74.

132. Jamgochian, E.M., and Miller, W.E. 1974. *Proc. Symp. on Flue Gas Desulfurization,* Atlanta, Georgia, Nov. 1974, Vol. II, EPA-650/2-74-126b (Dec. 1974):762-806.

133. Laseke, B. A. Jr.; Melia, M. T.; and Kaplan, N. 1983. "Trends in Commercial Applications of FGD," presented at the EPA/EPRI Symposium on Flue Gas Desulfurization, New Orleans, LA, November 1-4.

134. Ando, J. 1983. "Status of SO₂ and NO₂ Removal in Japan," presented at the EPA/EPRI Symposium on Flue Gas Desulfurization, New Orleans, LA, November 1-4.

135. Radian Corp. 1984. "The Evaluation and Status of Flue Gas Desulfurization Systems," EPRI CS-3222, January.

136. Reisdorf, J. B.; Keeth, R. J.; Miranda, J. E.; Scheck, R. W.; and Morasky, T. M. 1983. "Economic Evaluation of FGD Systems," presented at the EPA/EPRI Symposium on Flue Gas Desulfurization, New Orleans, LA, November 1-4.

137. Smith, E. O.; Swenson, D. O.; Morgan, W. E.; Meadows, M. L.; Cannell, A. L.; and Gustke, J. M. 1983. *Lime FGD Systems Data Book,* 2nd Ed., EPRI-CS-2781; DE830 07786, January.

138. Henzel, D. S.; Laseke, B. A.; Smith, E. O.; and Swenson, D. O. 1981. Limestone FGD Scrubbers: Users Handbook, EPA-600/8-81-017, April.

139. Downs, W.; Johnson, D. W.; Aldred, R. W.; Tonty, L. V.; Robards, R. F.; and Runyan, R. A. 1983. "Influence of Chlorides on the Performance of Flue Gas Desulfurization," presented at the EPA/EPRI Symposium on Flue Gas Desulfurization, New Orleans, LA, November 1-4.

140. Laslo, D.; Chang, J. C. S.; and Mobley, J. D. 1983. "Pilot Plant Tests on the Effects of Dissolved Salts on Lime/Limestone FGD Chemistry," presented at the EPA/EPRI Symposium on Flue Gas Desulfurization, New Orleans, LA, November 1-4.

141. Rader, P. C.; Borsare, D. C.; and Frabotta, D. 1982. "Process Design of Lime/Limestone FGD Systems for High Chlorides," presented at Coal Technology '82, Houston, Texas, December 7-9.

142. Chan, P. K. and Rochell, G. T. 1983. "Modeling of SO₂ Removal by Limestone Slurry Scrubbing: Effects of Chlorides," presented at the EPA/EPRI Symposium on Flue Gas Desulfurization, New Orleans, LA, November 1-4.

143. Chang, C-S and Rochell, G. T. 1981. "Transport with Chemical Reactions," AIChE Symposium Series 202, Vol 77, p. 78.

144. Head, H. N.; Wang, S. C.; Rabb, D. T.; Borgwardt, R. H.; Williams, J. E.; and Maxwell, M. A. 1979. "Recent Results from EPA's Lime/Limestone Scrubbing Programs—Adipic Acid as a Scrubber Additive," Proceedings: Symposium on Flue Gas Desulfurization, Las Vegas, Nevada, EPA-600/7-79-167, July.

145. Wasag, T.; Galka, J.; and Fraczak, M. 1975. Air Conservation, Vol 9, *3,* 16.

146. Colley, J. D.; Hargrove, O. W.; and Mobley, J. D. 1983. "Results of Industrial and Utility Boiler Full-Scale Demonstration of Adipic Acid Addition to Limestone Scrubbers," Proceedings: Symposium on Flue Gas Desulfuration, Hollywood, Florida, May 1982, EPRI CS-2897, March 1983.

147. Chang, J. C. S.; and Dempsey, J. H. 1983. "Pilot Plant Evaluation of By-product Dibasic Acids as Buffer Additives for Limestone Flue Gas Desulfurization Systems," Proceedings: Symposium on Flue Gas Desulfurization, Hollywood, Florida, May 1982, EPRI CS-2897 March 1983.

148. Dickerman, J. C., and Mobley, J. D. 1983. "Technical/Economic Feasibility Studies for Full Scale Application of Organic Acid Technology for Limestone FGD Systems" presented at the EPA/EPRI Symposium on Flue Gas Desulfurization, New Orleans, LA, November 1-4.

149. Hicks, N. D., and Fraley, D. M. 1983. "Commercial Application Experience with Organic Acid Addition at City Utilities of Springfield," Proceedings: Symposium on Flue Gas Desulfurization, Hollywood, Florida, May 1982, EPRI CA-2897, March.

150. Anders, W. L.; and Torstrick, R. L. 1981. "Computerized Shawnee Lime/Limestone Scrubbing Model Users Manual," EPA-600/8-81-008 (NTIS PB-178963) U.S. EPA.

151. Classen, D. D. 1983. "Status of the Chiyoda Thoroughbred 121 Flue Gas Desulfurization Process," Proceedings of Seminar on FGD sponsored by Canadian Electrical Association, Ottowa, Canada, September 19-20.

152. Kaneda, S.; Nichimura, M.; Wakui, H.; Kuwahara, I.; and Classen, D. D. 1983. "Operating Experience with the Chiyoda Thoroughbred 121 FGD Systems," presented at the EPA/EPRI Flue Gas Desulfurization Symposium, New Orleans, LA, November 1-4.

153. MacKenzie, J.; Bove, H.; and Bitsko, R. 1983. "Operating Experience with the United/PECo Magnesium Oxide Flue Gas Desulfurization Process," Proceedings of Seminar on FGD sponsored by Canadian Electrical Association, Ottowa, Canada, September 19-20.

154. Glancy, D. L.; Grant, R. J.; Legatski, L. K.; Wilhelm, J. H.; Wrobel, B. A. 1983. "Utility Double Alkali Operating Experience," presented at EPA/EPRI Symposium on Flue Gas Desulfurization, New Orleans, LA, November 1-4.

155. Hollinden, G. A.; Stephenson, C. D.; and Stensland, J. G. 1983. "An Economic Evaluation of Limestone Double Alkali Flue Gas Desulfurization Systems," presented at the EPA/EPRI Symposium on Flue Gas Desulfurization, New Orleans, LA, November 1-4.

156. Nolan, P. S.; and Seaward, D. O. 1983. "The Dowa Process Dual-Alkali Flue Gas Desulfurization with a Gypsum Product," Proceedings of Seminar on Flue Gas Desulfurization sponsored by the Canadian Electrical Association, Ottowa, Ontario, Canada, September.

157. Hollinden, G.; Runyan, R.; Newton, S.; Garrison, F.; Pfeffer, S.; and Smith, D. 1983. "Results of the Dowa Technology Tests at the Shawnee Scrubber Facility," Proceedings of the Symposium on Flue Gas Desulfurization, Hollywood Beach, Florida, May 1982, EPRI CS-2897, March 1983.

158. Palazzolo, M. A.; Kelly, M. E.; and Brno, T. G. 1983. "Current Status of Dry SO_2 Control Systems," presented at the EPA/EPRI Symposium on Flue Gas Desulfurization, New Orleans, LA, November 1-4.

159. Liegois, W. A. 1983. "Status of Spray Dry Flue Gas Desulfurization," Conference Proceedings, "Effective Use of Lime for Flue Gas Desulfurization," sponsored by National Lime Association, Denver, Colorado, September 27-28.

160. Gutslke, J. M.; Morgan, W. E.; and Wolf, S. H. 1983. "Overview and Evaluation of Two Years of Operation and Testing of the Riverside Spray Dryer System," presented at the EPA/EPRI Symposium on Flue Gas Desulfurization, New Orleans, LA, November 1-4.

161. Gehri, D. C.; Adams, R. L.; and Phelan, H. 1980. U.S. Patent 4,197,278, April 4.

162. Muzio, L. J.; Sonnichsen, T. W. 1982. "Demonstration of SO_2 Removal on a 22-MW Coal-Fired Utility—Boiler by Dry Injection of Nahcolite," Final Report, EPRI RP-1682—2, April.

163. Naulty, D. J.; Scheck, R. W.; McDowell, D. A.; and Hopper, R. G. 1983. "Economics of Dry FGD by Sorbent Injection," presented at 76th Annual Meeting of the Air Pollution Control Association, Atlanta, Georgia, June 19–24.

164. Burke, J. M. 1983. "Shell NO_x/SO_2 Flue Gas Treatment Process: Independent Evaluation," EPA-600/7-82-064 (PB 83-144816), March.

8

Dry Oxidation Processes for Removal of Sulfur Compounds

Dry processes for the removal of hydrogen sulfide and other sulfur compounds such as mercaptans, carbonyl sulfide, carbon disulfide, and thiophene from gases by oxidation can be classified into the following two categories:

1. Oxidation to sulfur
2. Oxidation to oxides of sulfur

These processes are selective for sulfur compounds and other oxidizable agents and have no effect on gas impurities which do not undergo oxidation under the operating conditions. Although these methods are used to some extent in the purification of natural gas, especially in cases where the sulfur content of the gas is relatively low and where extremely high purities are required, their primary application is in the desulfurization of manufactured gas, coke-oven gas, carburetted water gas, and synthesis gas. Since these gases are used to a much larger extent in Europe than in the United States, it is understandable that the technology of these processes was brought to its highest development in European countries.

Oxidation to Sulfur

Chemically the removal of sulfur compounds from gases by oxidation is very simple and can be symbolized (for the case of hydrogen sulfide) by the following equations:

$$H_2S + \frac{1}{2}O_2 = H_2O + S \tag{8-1}$$

$$H_2S + \tfrac{3}{2}O_2 = H_2O + SO_2 \qquad (8\text{-}2)$$

$$SO_2 + \tfrac{1}{2}O_2 = SO_3 \qquad (8\text{-}3)$$

$$2H_2S + SO_2 = 2H_2O + 2S \qquad (8\text{-}4)$$

However, the rates of these reactions are too low at ordinary temperatures to be useful in a practical desulfurization process. Sufficiently high reaction rates are obtainable by operating at fairly high temperatures, preferably in the presence of catalysts, or by use of intermediary oxygen carriers which react readily with the sulfur compounds at ordinary temperatures. The oxygen carriers are either used dry or are incorporated into absorptive liquid systems. Processes employing regenerative liquid systems will be discussed in Chapter 9.

Iron Oxide Process

The iron oxide gas-purification process is one of the oldest methods used for the removal of objectionable sulfur compounds from industrial gases. Around the middle of the nineteenth century, the iron oxide process was introduced in England to replace a wet purification process utilizing calcium hydroxide as the active agent. The process is still used on a large scale for the treatment of coal gases. However, recently developed wet purification processes which are capable of producing purified gases of equally high purity as that obtainable with the iron oxide process are gradually replacing oxide box purifiers. The liquid processes (see Chapter 9) have substantial economic advantages as they require less space and eliminate the high labor cost associated with the handling of the iron oxide.

The first installations utilized the simplest form of the dry-box process. In this form of the process, hydrogen sulfide is removed completely by reaction with hydrated ferric oxide, resulting in the formation of ferric sulfide. Upon exposure to atmospheric oxygen, ferric sulfide is oxidized to elemental sulfur and ferric oxide, which is subsequently used to react with additional hydrogen sulfide. This cycle can be repeated several times until the elemental sulfur covers most of the surface of the oxide and fills most of the interstices between the oxide particles, thus causing loss of activity of the purifying material and excessive pressure drop through the bed. After the sulfur has been removed, the oxide may be reused in the process.

Later, it was determined that a much simpler and more economical process results when the iron oxide is revivified *in situ*, by addition of small amounts of air or oxygen to the gas at the inlet of the purification plant. Another method of *in situ* revivification consists of circulation of oxygen-containing gas after the oxide has been fouled with oxygen-free gas. The primary advantages of this process are large savings in labor cost for loading and unloading the purifiers and the realization of higher sulfur content of the oxide before final removal of the material from the purifiers is necessary.

The iron oxide purification process finds its widest application in the treatment of gases where complete removal of hydrogen sulfide is essential. Another consideration favoring use of the process is a low permissible pressure drop in the purification step. Finally, some manufactured and synthesis gases contain impurities which react irreversibly with the chemicals used in certain liquid purification processes and therefore make the use of such methods uneconomical.

In Europe it is common practice to treat gases containing up to 1,000 grains $H_2S/100$ scf in oxide purifiers. It has been stated that the iron oxide process is economical for the treatment of high-pressure natural gas of similar hydrogen sulfide content, provided the total daily sulfur production does not exceed 8 to 10 tons (1). However, in some cases, hydrogen sulfide absorption in a regenerable liquid, followed by sulfur recovery in a Claus plant, is more economical well below the 8 to 10 tons/day level. In cases where the gas contains large amounts of carbon dioxide, besides hydrogen sulfide, making sulfur recovery in a Claus plant difficult, a liquid oxidative process may be indicated for daily sulfur productions of one to 20 tons. Perry (2) recommends the iron oxide process for desulfurization of high pressure gases containing less than 20 grains $H_2S/100$ scf.

Disposal of sulfur, after complete loading of the oxide, is usually effected by burning and subsequent utilization of the sulfur dioxide in the manufacture of sulfuric acid. In some instances the sulfur is extracted with selective solvents such as carbon disulfide, perchlorethylene, or toluene, and subsequently recovered by crystallization from the solvent or by removal of the solvent by distillation.

Basic Chemistry

The basic chemistry of the process can be represented by the following equations:

$$2Fe_2O_3 + 6H_2S = 2Fe_2S_3 + 6H_2O \tag{8-5}$$

$$2Fe_2S_3 + 3O_2 = 2Fe_2O_3 + 6S \tag{8-6}$$

Combining Equations 8-5 and 8-6,

$$6H_2S + 3O_2 \rightarrow 6H_2O + 6S \tag{8-7}$$

It should be understood that these equations represent a simplified mechanism and that, depending on the operating conditions, a large number of other reactions may occur. Variables principally influencing the reaction mechanism are temperature, moisture content, and pH of the purifying material.

There are several forms of ferric oxide known, but only two, namely α $Fe_2O_3 \cdot H_2O$ and γ $Fe_2O_3 \cdot H_2O$, are useful as purifying materials (3,4). These two forms react readily with hydrogen sulfide, and the resulting ferric sulfide is easily reoxidized to an active form of ferric oxide. The cycle proceeds most satisfactorily at moderate temperatures, approximately 100°F, and in an alkaline environment. At temperatures above 120°F and in neutral or acid environment ferric sulfide loses its water of crystallization and changes into a mixture of FeS_2 and Fe_8S_9. At pH 8.0 or higher, decomposition does not take place below 190°F. These sulfides are not readily reconverted to hydrated ferric oxide but oxidize slowly to ferrous sulfate and polysulfides, neither of which are useful for H_2S removal. Excellent discussions on the properties of purification materials have been presented by Bengt Smith (5) and Ward (4).

The iron oxides used for gas-purifying materials are classified as unmixed and mixed oxides. Unmixed oxides consist essentially of pure hydrated ferric oxide containing water and, in some cases, fibrous materials such as those occurring in natural iron oxide ores. Mixed oxides, on the other hand, are prepared artificially by supporting finely divided iron oxide on materials of large surface and loose texture.

Unmixed oxides are prepared chemically from iron ore or metallic iron and contain approximately 75 percent ferric oxide and 10 percent water. Another source of unmixed oxides is the residue from the purification of bauxite which contains approximately 25 to 50 percent ferric oxide and 10 to 50 percent water. This material is sometimes sold under the trade name "Lux." In Europe, the most widely used materials for dry-box purification are natural bog ores, especially those found in Denmark and Holland. These ores consist of a highly active form of hydrated ferric oxide mixed with fibrous and peaty material containing approximately 45 percent water. In order to obtain the proper pH, alkaline substances such as sodium hydroxide or sodium carbonate are usually added to the natural bog ores. A very active purification material which is used in recently developed continuous processes is obtained by granulating unmixed oxides. Granule sizes of approximately ¼ to ⅓ in diameter with a porosity of 60 percent are recommended. It is important that the granulated materials have sufficient strength to withstand handling in the process of loading and unloading of purification plants.

The mixed oxides, or sponges, are prepared by supporting the finely divided oxide on media such as wood shavings and granulated or crushed slags; their common property is large surface area. The ferric oxide used in these

Table 8-1. Composition of Purifying Materials		
Characteristic	Bog Ore	Oxide Supported on Wood Shavings
Moisture, %................	45–55	38–42
Loss on ignition (dry basis), % .	25–35	36–41
Fe_2O_3 (dry basis), %..........	45–55	46–48
Bulk density, lb/cu ft....	.....	19–21

materials may be obtained by air oxidation of iron borings in the presence of water and lime. The advantage of mixed oxides is that the bulk density, the iron oxide content, the moisture, and the pH of the materials can be controlled better than in unmixed oxides. In addition, mixed oxides have less tendency to cake, and free passage of the gas and higher final sulfur loading can be achieved.

As a result of a comprehensive study Moignard (6,7) recommends supported iron oxides of a coarse and open texture, low bulk-density (on the order of 20 to 25 lb/cu ft, dry) and the highest possible moisture content, 30 to 50 percent by weight, consistent with free gas flow. Oxides containing less than 17 percent and more than 55 percent water do not function properly. An optimum pH range of 8 to 8.5 is also recommended. Desirable compositions of unmixed and mixed oxides as reported by Hopton (14) and Moignard (6) respectively are shown in Table 8-1. The physical properties of the purifying materials are equal in importance to the chemical reactivity.

The physical mechanism of hydrogen sulfide absorption by iron oxide was investigated thoroughly by Avery (8) and by Dent and Moignard (9). The studies showed that under proper conditions of temperature, moisture content, and pH, the sulfur formed on the oxide particle is continuously displaced by iron oxide, which migrates from the particle center to the surface, thus exposing fresh oxide for further reaction with hydrogen sulfide and oxygen.

Migration is prevented by dehydration and subsequent hardening of the oxide and, also, by excessive localized deposition of sulfur. Close moisture control and even distribution of gas flow through the purifying mass is therefore of greatest importance to ensure proper functioning of the process.

The two most important criteria for the selection of purifying materials are capacity and activity. Theoretically 1 lb of ferric oxide absorbs 0.64 lb of hydrogen sulfide when completely transformed to iron sulfide. In operations without oxygen addition, this value is not attainable although capacities of as high as 0.56 lb sulfur/lb ferric oxide can be achieved (1,2). In general, it is assumed that 50 percent of the theoretical sulfiding capacity is attainable in the first cycle of operation and that the capacity decreases progressively with additional cycles. When revivification is carried out simultaneously with the

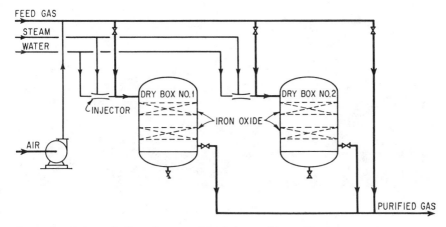

Figure 8-1. Schematic flow diagram of basic iron oxide purification process.

sulfiding reaction, by means of oxygen present in the gas, considerably higher capacities can be obtained. It is not unusual to reach final sulfur contents of 2.5 lb sulfur/lb iron oxide (1). Actual capacities of purifying materials are determined empirically by laboratory tests. On the basis of actual, plant-operating experience, it has been determined that maximum utilization of the iron oxide can be attained by removing it at least once from the purifying vessel before complete fouling is reached.

The activity of the purifying materials is measured by the rate at which hydrogen sulfide is absorbed by the oxide. A large number of variables such as moisture content, pH, texture, etc., influence the activity, and empirical methods are used for its determination. Methods commonly used by the gas industry in the United States to determine oxide capacity and activity are described by Seil (10).

Process Description

A simplified flow diagram of the basic, iron oxide purification process is shown in Figure 8-1. There are several modifications of the process presently in use which can be separated into the following categories:

1. Conventional-box purifiers
2. Deep-box purifiers
3. Tower purifiers and tower boxes
4. Continuous processes
5. High-pressure processes

Conventional-Box Purifiers. Purifiers of this type, representing the classical process, consist of large rectangular boxes constructed of cast iron,

Figure 8-2. Conventional dry-box purification plant for treatment of 12 million cu ft of gas per day. Section in front consists of six boxes 35 by 30 by 10 ft arranged in two lines of three boxes each. Section in rear consists of six boxes 30 by 30 by 10 ft also arranged in two lines of three boxes each. *Robert Dempster & Sons, Ltd.*

steel, or concrete with removable plates at the top. These boxes may be placed at ground level, partly sunk into the ground, or mounted on legs. In some instances a floor is put on the top of the boxes. One of the main considerations in the design of such installations includes the area occupied by the boxes themselves as well as the additional area necessary for handling partly spent oxide. It is obvious that considerable saving in ground space can be made by placing the purifying boxes on legs. Several publications describing detailed mechanical design of all types of purifiers are available in the literature (4, 11-14).

A purifying plant consists of one or more series of four to six boxes with the piping arranged in such a manner that the inlet gas can enter each box first, thus allowing changes in the order of the boxes. Usually provisions are made for upward or downward flow of the gas or admission into the center of the box. Each box contains one or several layers of oxide supported on wooden or steel grids (15). Although the depth of the layers varies considerably, it is common practice to use a thickness of 2 ft per layer. A photograph of a large installation of this type is shown in Figure 8-2.

Deep Boxes. Deep boxes are similar in design to ordinary box-purifiers except that the depth of the oxide layers is considerably greater than in the conventional equipment. In general, an oxide depth of 4 to 6 ft is used; this depth has the obvious advantage of greater capacity for a given plant area. Because of the higher pressure drop through the deeper beds, these boxes are frequently operated on the split-flow principle by admitting the gas between two layers, one-half passing upward and the other half downward. Deep-box purifiers are especially useful in purification plants in which infrequent dis-

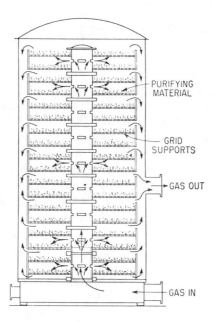

Figure 8-3. Thyssen-Lenze tower.

charging is desired. They are also used for high-pressure operations where pressure drop is not a critical factor.

Tower Purifiers. The Thyssen-Lenze towers constitute an improvement over the conventional purification-boxes primarily because of better space utilization and reduced labor cost. A schematic diagram of a purification tower is shown in Figure 8-3.

The tower consists of a carbon-steel shell and contains up to 14 super-imposed removable baskets, each holding two layers of oxide supported on wooden grids. The baskets are constructed in such a manner that a central gas-inlet tube is formed when they are placed in the tower. Each section of the tube is perforated with rectangular openings through which the gas enters the space between the two oxide layers of each basket. After contact with the oxide, the gas flows into the annular space between the baskets and the outer shell and from there to the gas outlet.

Tower installations usually consist of four to six operating towers and two additional structures of similar dimensions: one for storage of baskets of spent oxide and the other for storing fresh oxide.

A modification of the Thyssen-Lenze tower was introduced by Kloenne. Towers based on this design consist of cylindrical or rectangular containers with approximately the same height as diameter (35 to 40 ft). The purifying material is arranged on 14 to 18 removable trays in layers, each about 18 in. deep. The operation of Kloenne towers is quite similar to that of Thyssen-Lenze towers. Mechanical design features of tower purifiers have been given by Guntermann and Schnürer (16).

Tower boxes, which are still another modification of tower purifiers, consist of rectangular tall vessels, built with common walls and containing baskets with single layers of oxide. In these installations the gas flows in parallel and in the same direction through all the baskets. The design and operation of tower box-purifiers has been described by Bairstow (17).

The advantages of tower purifiers lie primarily, as stated above, in the reduction of the ground area required for the installation, as compared to that needed for conventional boxes, and in simplification of oxide handling. The baskets are loaded and unloaded by use of traveling cranes, devices which permit appreciable reduction in the labor force necessary for oxide handling.

Continuous Processes. In an attempt to reduce the size of dry purification plants and to obtain more efficient utilization of the oxide, work on continuous iron oxide processes was undertaken in Europe after World War II. A pilot plant operated by the North Thames Gas Board has been described by Hollings (18). A schematic diagram of the purifying equipment is shown in Figure 8-4.

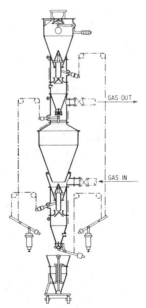

Figure 8-4. Schematic diagram of pilot plant for continuous iron oxide purification process (18).

The installation consists of two towers, each 2 ft in diameter, containing a packed section 8 ft 6 in. high. The purifying material, consisting of pellets made from copperas, 'is fed intermittently to the top of each tower and, after being partly spent, is withdrawn from the bottom of the towers. Because of the physical characteristics of this material, free flow by gravity is obtained.

Another continuous process which was developed in Germany and commercialized under the name "Gastechnik" purification process is described by Moore (19) and Sexauer (20). It is stated that over 60 plants using this process were operating in Germany in 1956, and a considerable number of installations are presently in operation in England (21,22). This process utilizes a series of cylindrical towers, each of which is filled with a continuous deep bed of specially pelleted oxide. The pellets used are roughly spherical, $\frac{1}{2}$ to $\frac{3}{4}$ in. in diameter, and have a bulk density of 38 lb/cu ft at a moisture content of 10 percent. They consist of a mixture of specially activated iron oxide, sawdust, cement, and lime. The pellets flow freely by gravity and are charged and discharged from the vessels, through gas locks, without interruption of the gas flow.

A typical installation consists of three towers connected in series as shown diagrammatically in Figure 8-5. The operating data shown in Figure 8-5 are representative for the treatment of a typical coal gas. Each tower is normally 9 ft 6 in. in diameter and 40 ft high and contains the pellets in a continuous bed. A plant consisting of three such towers has a daily treating capacity of 2 million cu ft of gas, containing 450 to 750 grains of $H_2S/100$ cu ft, when operating at essentially atmospheric pressure. Multiples of this basic unit are used to treat larger volumes of gas (21). The operation of this process is similar to that described by Hollings, in that addition and withdrawal of the oxide pellets are carried out intermittently. The gas flow may be countercurrent or cocurrent to the flow of pellets. Countercurrent flow is used when gases containing relatively small amounts of H_2S (60 to 250 grains/100 cu ft) and some oxygen are treated. In such cases the iron oxide is partially revivified *in situ* and completely oxidized by exposure to atmospheric oxygen after discharging from the towers. Cocurrent flow in the first two towers and countercurrent flow in the last tower are used for the treatment of oxygen-free gases containing H_2S in concentrations up to 1,000 grains/100 cu ft. In such installations the pellets are continuously oxidized *in situ* by injection of air at a point about 10 ft below the top of the bed. It is claimed that this method of operation results in complete revivification of the pellets in the towers, thus eliminating external regeneration (21).

The flow of the pellets can be followed on the scheme shown in Figure 8-5. Revivification between towers is, of course, not necessary if complete oxidation is accomplished *in situ*. The iron oxide pellets withdrawn from the first tower are extracted with perchlorethylene for removal of sulfur and recycled to the second tower. It will be noted from the data shown in Figure 8-5 that

430   Gas Purification

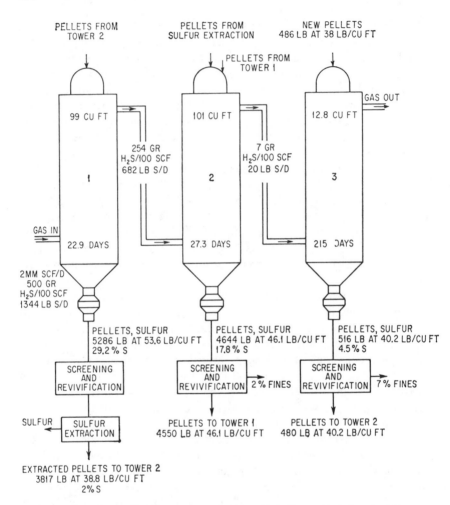

Figure 8-5. Flow chart of Gastechnik purification plant. (Representative operating conditions for treatment of coal gas; pellet weights adjusted for sulfur content and 10 percent moisture.) *Robert Dempster & Sons, Ltd.*

about 80 percent of the pellets in the system are extracted material and that new pellets are added only to replace screening losses from the various stages. The fines removed from the system are suitable for use in conventional iron oxide boxes. For optimum operation of the process, the inlet-gas temperature should be 40° to 80°F, and the gas should not contain more than 0.4 grain/100 cu ft of tar and dust (21). A photograph of a typical installation with two lines, each treating 2 million cu ft of coal gas, is shown in Figure 8-6.

The primary advantage of this process is a large saving in oxide-handling cost, because of the relative ease of charging and discharging the oxide

Figure 8-6. Gastechnik-type treating plant. *Robert Dempster & Sons, Ltd.*

pellets. The main disadvantage is that a final sulfur loading of only 30 percent by weight can be achieved; in many cases this is considered uneconomical.

High-Pressure Purifiers. A rather recent development in dry-box purification is the adaptation of the process to the removal of hydrogen sulfide from gases at high pressure. Several plants, ranging in capacity from 6 to 20 million cu ft gas/day, are being operated in Germany at pressures of 100 to 200 psia (5). A plant utilizing this process in the United States has been described by Turner (23). In this installation 15 MMscf of natural gas, containing 10 grains $H_2S/100$ scf, are treated daily at 325 psig in two series of four cylindrical vessels, each containing an iron oxide bed of 10-ft depth. No air is added to the gas, and the fouled oxide is revivified by exposure to air outside the vessels. After 10 to 12 foulings the material is discarded. Because of the relatively high permissible pressure drop through the beds, considerably higher space velocities (and consequently lower oxide volumes) can be used in installations of this type than in conventional purification plants.

Another high-pressure plant located in Canada, treating 7 MMscf/day of natural gas containing 23 grains $H_2S/100$ scf, one grain of mercaptans/100 scf, and 4.2 percent CO_2 at 1,000 psig has been described by Duckworth and Geddes (24). In this plant periodic revivification by circulation of a mixture of gas and air is being practiced. The high-pressure process has become well established in the United States and Canada, and it is customary to use groups of one to four cylindrical vessels containing 10 ft deep beds of wood shavings impregnated with iron oxide (iron sponge). These installations are claimed to be competitive with liquid-purification systems for treating gas streams of low hydrogen sulfide concentration and small daily sulfur production (1, 2, 24). As in other iron oxide purification systems, the process is most efficient when the gas contains enough oxygen to effect continuous revivification of the oxide. However, alternate fouling by use of oxygen-free gas and regeneration by circulating air through the exhausted beds can also be carried out in a satisfactory manner (1). A high-pressure installation is shown in Figure 8-7.

A bench-scale study of hydrogen sulfide removal from synthesis gases with iron oxide impregnated wood chips has been reported by Jolenson, Field, Decker, and Jimeson of the U.S. Bureau of Mines (25). The purpose of this study was to investigate the suitability of the process for sufficiently complete hydrogen sulfide removal in order to protect sulfur sensitive catalyst in subsequent processing steps. The targets for hydrogen sulfide content of the treated gas were 0.02 and 0.1 grain/100 scf, and the effects of pressure, temperature, and space velocity on the efficiency of the process were investigated.

The results of this work show that the required gas purities can be achieved with an inlet gas containing about 10 grains of hydrogen sulfide per 100 scf. Increased pressure, over the range of atmospheric to 400 psig increased the allowable gas throughput linearly. Increasing the operating temperature from about 80° to 212°F resulted in decreased removal efficiency. At a pressure of 400 psig, space velocities of as high as 500 cu ft/(hr) (cu ft) were found to be feasible. The authors state that saturation of the iron oxide with hydrogen sulfide approaching the stoimetric value can probably be obtained in commercial operation using a bed of about 10-ft depth.

Regeneration of the iron oxide bed with air in a limited number of cycles was found to reestablish the activity of the bed.

Design of Purifiers

Determination of Bed-Size. Process-design methods for iron oxide purifiers are based largely on empirical rules. In the United States the most

Figure 8-7. Iron oxide gas-purification plant operating at high pressure. *Canadian Western Natural Gas Co., Ltd.*

commonly used method for sizing iron oxide boxes was proposed by the Steere Engineering Company and is expressed by the following formula (26):

$$A = \frac{GS}{3,000(D + C)} \tag{8-8}$$

$$G = \frac{3,000(D + C)A}{S} \tag{8-9}$$

where G = maximum amount of gas to be purified, in scf/hr
S = correction factor for hydrogen sulfide content of inlet gas
D = total depth of oxide, ft. through which the gas passes consecutively in the purifier set. In boxes with split flow where half the gas volume passes through each layer, the area exposed to the gas is twice the cross-sectional area of the box, while the depth D is the depth of one layer of oxide
A = cross-sectional area through which the gas passes on its way through any one box, in series, of a set
C = factor: 4 for 2 boxes, 8 for 3 boxes, and 10 for 4 boxes, respectively
3,000 = constant

The values of S are tabulated below:

Grains $H_2S/100$ scf of Unpurified Gas	Factor
1,000 or more	720
900	700
800	675
700	640
600	600
500	560
400	525
300	500
200 or less	480

The values for the factor C are based on the number of boxes in series in actual use, disregarding boxes taken off the line for changing or revivifying spent oxide. This factor is also based on the assumption that the flow will be reversed during purification.

European practices in oxide-purifier design differ somewhat from American practices and are most commonly based on space velocity through one box, as described by Clayton, Williams, and Avery (27), known as the R ratio, which is expressed by the following equation:

$$R = \text{cu ft of gas per hr/cu ft of oxide in one box} \qquad (8\text{-}10)$$

The values for R vary over a considerable range, from 15—in very conservative designs—to as high as 100 in certain tower installations. Obviously, since R is calculated on the basis of one box only, a higher value can be used if the oxide is divided among a larger number of boxes. In general, an R ratio of 20 to 50 is considered good design for systems operating at essentially atmospheric pressure and with revivification *in situ*.

Considerably higher space velocities are attainable in high-pressure processes where a high pressure-drop through the beds is permissible. The actual space velocity in the installation described by Turner (23), calculated at 325 psig (see Table 8-4), is 37.4 cu ft/(hr) (cu ft of oxide), which corresponds to an R ratio of 112. It is claimed that this is conservative and that space velocities twice as high can be used at high pressures with gases containing up to 50 grains $H_2S/100$ scf, provided air is added to the gas. Taylor (1) recommends a maximum space velocity of 240 cu ft/(hr) (cu ft of iron oxide sponge) if the gas contains sufficient oxygen to effect revivification *in situ*. The gas volume is calculated at flowing conditions. In the absence of oxygen and with high hydrogen sulfide content, space velocities on the order of 90 cu ft/(hr) (cu ft of oxide) under the same operating conditions appear to be satisfactory.

Table 8-2. Hydrogen Sulfide and Oxygen Content of Coal Gas Throughout Purifiers		
Location	Hydrogen Sulfide	Oxygen, %
Entering first box	550*	1.2
Entering second box	25*	0.9
Entering third box	0.5*	0.8
Entering fourth box	0.17†	0.7
Leaving fourth box	0.02†	0.7

* In grains per 100 SCF.
† In ppm.

Perry (2) presents a simplified design method for high-pressure treaters applicable to gases containing a maximum of 30 grains H_2S/100 scf, without continuous oxygen addition. The method, which is based on the use of iron oxide sponge containing 15 lbs of iron oxide per bushel of sponge, a maximum space velocity of 60 cu ft/(hr) (cu ft of iron oxide sponge), and a maximum linear velocity of 10 ft/min, is consistent with that recommended by Taylor (1) and other authors.

The size and number of boxes will depend on the desired frequency of changes of iron oxide. The rapidity of charging and discharging and expected periodic overloads should also be considered. A box life of 60 days is considered minimum; however, in some installations exposure periods of 6 months or more are used. In general, not more than six boxes are used and most installations operate with four boxes.

Treating Efficiency. Performance of a set of four purifiers described by Hollings and Hutchison (28) is shown in Table 8-2. The boxes have a cross-sectional area of 0.5 sq ft/1,000 cu ft of gas per day and a bed-depth of 4 ft, corresponding to a rather conservative space velocity of 20 cu ft of gas/ (hr) (cu ft of oxide) (box).

It is seen that 95 percent of the hydrogen sulfide is removed in the first box, a somewhat higher efficiency than that attainable in most commercial installations. A hydrogen sulfide removal of 80 percent in the first box is considered satisfactory, and in most cases a concentration of 0.1 grain/100 scf at the outlet of the purifier set satisfies even the most stringent purity requirements. The data of Hollings and Hutchison were obtained with oxygen admission to the gas prior to its entry into the first box.

Brown, Porter, and Thompson (29) report a study on the effect of frequency of bed rotation on treating efficiency. Laboratory experiments and tests on a large installation consisting of 12 working towers (each 22 ft 3 in. in diameter and 61 ft high) designed to purify 24 MMscf per day of coal gas containing 550 to 650 grains $H_2S/100$ scf indicated increased efficiency with more frequent rotation. A change in rotation frequency from 24 to 4 hr increased the margin of safety of the purification process significantly.

Pressure Drop. A major factor in the design of iron oxide dry beds is the selection of the proper particle-size of iron oxide, which is related to the pressure drop through the beds. The pressure drop through porous masses is usually related to the Reynolds number by means of a friction coefficient and is proportional to this coefficient times the square of the velocity. The coefficient is inversely proportional to velocity at low flow-rates and independent of velocity at high rates. The pressure drop is, therefore, proportional to a power of velocity which varies from one to two and depends upon the regime of flow. Speers (30) reported that material of a particle-size less than $\frac{1}{40}$ in. causes excessively high pressure-drop in oxide purifiers. This observation was confirmed by Pyke and Monoghan (31), who developed the following relation for mixtures of iron oxide and sawdust containing 15 percent of water.

$$p = 3.14d^{-0.61} \tag{8-11}$$

where p = pressure drop, in. water/ft of depth of oxide bed at a gas rate of 1 cu ft/(min) (sq ft of box cross section)
d = mean particle-size, thousandths of an inch

These authors also found that the pressure drop is directly proportional to the gas rate, indicating laminar flow. Since it is customary to operate oxide purifiers at a gas-flow rate of 1.4 fpm, based on the empty box, and to use four boxes in series with bed-depths of 4 ft, Equation 8-11 would indicate a pressure drop of 10 in. of water for material of 1/40-in. diameter.

Considerably higher pressure drop is permissible for a purifier operating at high pressure. Perry (2) recommends, for operation at 800 to 1,000 psig, allowable pressure drops of 1 to 2 psi/ft of bed, with linear gas velocities of 5 to 10 ft/min.

Heat Effects. The reaction of hydrogen sulfide with ferric oxide and the subsequent reaction of ferric sulfide with oxygen are highly exo-thermic. The overall reaction is shown in simplified form in the following equation:

$$H_2S + \frac{1}{2}O_2 \rightarrow H_2O + S \qquad -\Delta H = 94{,}500 \text{ Btu/lb mole } H_2S \tag{8-12}$$

The heat of reaction is distributed so that approximately 10 percent is generated in the sulfiding reaction and 90 percent in the revivification. Thompson (32) reports the heat release for the sulfiding and oxidation reactions as 0.087 Btu/grain H_2S and 0.326 Btu/grain H_2S, respectively, giving a

Table 8-3. Thermal Effects in Tower-Box Purifiers

Order of Box in Stream	Gas Temperature, °F		Water Loss from Oxide		Condensate from Box, lb/day
	Inlet of Box	Leaving Oxide Bed	lb/day	% Oxide	
1	55.4	96.8	6,550	1.9	0
2	93.2	102.2	3,250	1.0	0
3	98.6	102.2	1,850	0.6	600
4	98.6	98.6	750	0.2	740
5	98.6	93.2	0	0	3,330

SOURCE: Courtesy of Butterworth's Scientific Publications (London).

total heat release of 0.413 Btu/grain H_2S. If water is condensed, 0.081 Btu/grain H_2S has to be added. If a specific heat of 3 Btu/(°F) (100 cu ft) is taken for coal gas saturated with water vapor and an approximate specific heat of 0.30 Btu/(°F) (lb) for the iron oxide, it is seen that a considerable temperature rise occurs in the oxide beds when gas containing a high concentration of hydrogen sulfide is treated. In view of the effects of temperature and moisture on the activity of the iron oxide, it is evident that temperature control is of great importance. The simplest way to control temperature in the purifiers is to provide sufficient external surface for dissipation of the heat by radiation and convection.

An experimental study on the disposal of the heat of reaction in a purifier installation has been conducted by Moignard (6). Certain empirical rules have been developed, and it appears that a rate of sulfur deposition of 15 grains/ (sq ft of cross-sectional area of the bed) (min) is the optimum. In practice this varies from as low as 2 grains to as high as 60 grains. Conventional boxes are usually large enough to provide for adequate external heat dissipation, so that overheating is avoided. However, in compact tower installations, appreciable overheating may occur, especially if ambient temperatures are fairly high.

An estimation of the thermal effects in an installation consisting of six tower boxes, treating 7.5 million cu ft/day of coal gas containing 1,000 grains H_2S/100 scf, was presented by Hopton (14) and is shown in Table 8-3.

Each tower box has a cross section 28 ft 9½ in. by 16 ft 3 in. and a height of 56 ft and contains 14 trays of 338-sq ft cross-sectional area with a bed-depth of 2 ft 3 in. per tray. This corresponds to a space velocity of 6.1 cu ft gas/(hr) (cu ft of oxide), with five towers in operation or an R of 30.5. For the purpose of the thermal calculations, an inlet temperature of 55°F (saturated) and an ambient temperature of 39°F were assumed. The hydrogen sulfide—removal efficiency was assumed to be 90 percent in the first and second boxes, with complete removal in the third box. The oxidation reaction was distributed, with 50 percent in the first box, 25 percent in the second, 15 percent in the third, and 10 percent in the fourth. Oxide with a moisture content of 30 percent and a bulk density of 50 lb/cu ft was used. The assumption has also been made that condensate is not absorbed by the oxide, which is generally the case.

Typical design-and operating-conditions of plants of different types are shown in Table 8-4.

Operation of Purifiers

Operating difficulties encountered in iron oxide purifers are almost entirely due to loss of activity because of hardening and contamination of the purify-

Table 8-4. Typical Operating Conditions of Iron Oxide Purifiers

Variable Condition	Type of Purifier				
	Conventional Boxes (33)	Deep Boxes (7)	High-Pressure (23)	Tower Purifiers (34)	Con-tinuous (19)
Gas volume treated, MMCF/day..........	6.0	4.3	15.0	24.0	2.0
Hydrogen sulfide content, grains/100 SCF	1,000	740	10	500–950	1,000
Pressure, psig...........	low	40*	325	low	
Number of units in series (boxes or towers)......	5	6	4†	6§	3
Cross-sectional area per unit, sq ft.............	960	1,200	24	776	71
Number of layers per unit.	4	1	1	28	1
Depth of layer, ft........	2.25; 1.50	4	10	1.4	40
Temperature, °F:					
In...................	60	73	.....	85	
Out..................	70	93	.....	100	
Space velocity, cu ft/(hr) (cu ft)..............	7.15	6.66	37.4‡	5.38	9.4
R ratio................	35.6	40	112‡	32	28

* Inches of water.
† Two series of 4 units, 3 units in operation in each series.
‡ At 325 psig.
§ Two series of 6 towers.

ing material. These difficulties are especially prevalent in installations where gases containing high concentrations of hydrogen sulfide are treated. The studies of Moignard have resulted in considerable clarification of the physical and chemical phenomena involved in the process and consequently in certain recommendations to alleviate operating difficulties.

Contamination and subsequent plugging of the iron oxide beds can be prevented by maximum removal of tar from the gas prior to purification. Tar fog is eliminated from the gas by cooling to as low a temperature as possible and subsequent passage through electrostatic precipitators. In some cases special coolers located immediately before the precipitator may be justified. These measures are usually adequate to reduce the tar content of the gas sufficiently so that the spent oxide contains no more than 1 to 1.5 percent of tar. However, in special cases where very low tar-content of the oxide is required, washing of the gas with oil before purification may be necessary.

Hydrogen cyanide reacts irreversibly with iron oxide causing a loss of purifying material. It is therefore desirable to remove hydrogen cyanide from the gas prior to its treatment with iron oxide, especially if hydrogen cyanide is present in relatively high concentrations. In many cases it is practical to place an exhausted box ahead of the purifying plant. Hydrogen cyanide is removed from the gas by reaction with iron sulfide, which is converted to Prussian blue.

Optimum treating-efficiency is obtained by selection of a purifying material of appropriate but not excessive activity and by proper moisture and temperature control in the purifier. The activity of the oxide may be controlled by using a mixture of fresh and partly spent material. The pH of the iron oxide can be maintained at 8.0 to 10.0 by admixture of basic materials and, in many plants, by the presence of small amounts of ammonia contained in the gas. Ammonia concentrations of 2 to 5 grains/100 scf are adequate to ensure the desired alkalinity of the oxide. High concentrations of carbon dioxide in the gas may decrease the efficiency of the purifying material. This difficulty is probably due to the formation of iron bicarbonate and can be prevented by injection of ammonia into the gas.

Since the sulfiding and revivifying reactions occur primarily in the first box, it is necessary to "rotate" boxes in order to maintain proper distribution of sulfur and moisture on the oxide. The system usually employed is backward rotation in which a box occupies successively positions nearer to the gas outlet end.

Gas Inlet						Gas Outlet
	1	2	3	4,		
	4	1	2	3,		
	3	4	1	2,		
	2	3	4	1,		
	1	2	3	4,		
			etc.			

If rotation is carried out correctly, the oxide is almost completely regenerated by the time it reaches the last position and, therefore, is most effective for removing final traces of hydrogen sulfide. In addition to rotation of the boxes, reversal of the gas flow through the boxes should be practiced at regular intervals. Moignard recommends that rotation of the boxes and reversal of flow be practiced once every 24 hr, although larger intervals are permissible, especially in plants where gas of low hydrogen sulfide concentration is treated. As stated earlier, Brown, Porter, and Thompson recommend more frequent rotation, on the order of 8 or 4 hr (29).

Since the oxidation reaction of ferric sulfide releases considerably more heat than the sulfiding reaction but proceeds with less speed, the temperature in the beds can be somewhat controlled by feeding oxygen to the second box rather than to the first. The optimum temperature range for the process lies between 60° and 100°F. An excess of oxygen over the stoichiometric quantity required is necessary for proper functioning of the process. On the basis of experience, it is recommended that the rate of admission be controlled so that the purified gas contains from 0.4 to 0.7 percent oxygen.

Although it is most practical to operate the process with continuous addition of a small amount of air to the gas ahead of the purification boxes, this is at times impossible because the presence of residual oxygen in the treated gas is undesirable. In such cases the spent oxide can be revivified *in situ* by recirculating gas to which oxygen is added until the ferric sulfide is reconverted to the oxide. Because of the strongly exothermic nature of the reaction, the rate of oxygen addition has to be controlled so that the temperature of the bed does not rise above 120°F. Taylor (1) recommends a gas-flow rate of 0.3 to 0.6 cu ft/(min) per cubic foot of bed and a slow increase of the oxygen concentration to a maximum value of 8 percent. Depending on the degree of fouling of the bed, 18 to 36 hr may be required for complete revivification. Operating data with this method of revivification are reported by DiRienzo (35) and Vandaveer (36).

Overheating of the oxide and loss of water can sometimes be avoided by supersaturation of the inlet gas with water and by introduction of water to the beds through spray systems.

Free gas flow through the oxide beds and even distribution of sulfur deposition are maintained by the highest possible throughput and the shortest possible exposure time compatible with economic operation. Although in some installations the oxide is left in the box until total exhaustion, it is recommended that total fouling be reached in at least two exposures. It has been found that higher final sulfur loading of the oxide can be obtained by removing the partly spent oxide from the purifier box at least once during its lifetime and restoring its original physical properties by disintegration and humidification. In plants operating at high pressure and treating natural gas, hydrate formation in the purifiers may be a problem. Duckworth and Geddes report that this difficulty can be eliminated by maintaining the feed-gas temperature above the hydrate-formation temperature of the gas (24).

Because of the complexity of variables involved in dry-box purification, no rigid rules can be given for optimum operation of such plants, and these conditions will have to be determined for each installation. This will require close observation of plant performance, keeping of extensive records, and the performance of a considerable number of analyses. However, it is felt that with the data presently available a set of operating conditions can be developed which will satisfy practically all the conditions prevailing in a commercial purification plant.

Sulfur Recovery from Spent Oxides

It is common practice in European installations to recover sulfur from spent purifying materials. Unmixed oxides containing no combustible materials other than sulfur can be burned, and the sulfur dioxide can be fed directly to a sulfuric acid plant. Mixed oxides are usually extracted with solvents from which the sulfur is recovered by crystallization or distillation. Carbon disulfide is the solvent used in most commercial installations. The oxide is extracted at atmospheric temperature in several countercurrent stages, and the carbon disulfide is subsequently removed from the sulfur by distillation. After the solvent has been removed, the temperature is raised sufficiently to melt the sulfur, which is then handled as a liquid. In order to yield pure, yellow sulfur, the oxide has to be freed from tar, which dissolves in carbon disulfide together with the sulfur. Experimental work on the production of pure sulfur from spent oxide has been reported by Dent and Moignard (9).

A process using perchloroethylene as the solvent has been developed in Germany for the extraction of iron oxide pellets (19). The production of pure, yellow sulfur is not attempted by this method.

The oxides remaining after sulfur extraction are reusable as purification materials and in some cases possess higher activities than the original materials.

In the United States it is usually considered uneconomical to extract sulfur from spent oxide, especially if the materials are used for purifying gases of low sulfur-content. However, the spent oxides are sometimes used for agricultural purposes.

Removal of Mercaptans, Hydrogen Cyanide, and Nitric Oxide

Mercaptans are removed by iron oxide in a manner analogous to hydrogen sulfide removal. Iron mercaptides are formed which, upon regeneration of the iron oxide with air, are converted to disulfides. Zapffe (37) describes an installation designed to treat 21 MMscf of natural gas free of hydrogen sulfide, but containing sufficient amounts of mercaptans to result in unacceptable mercaptan content of the liquid products extracted from the gas.

Hydrogen cyanide is removed in the iron oxide process, especially if revivification *in situ* is not practiced, or if oxygen is admitted to the second box. The rather complicated reaction mechanism involved has been described by Clayton, Williams, and Avery (27). The reaction products known as "blue" are sometimes extracted from the spent oxide and used in the manufacture of printing ink. In some instances it has been found profitable to operate plants in such a manner that a concentration of 20 percent of "blue" is obtained in the first box. Under such conditions, the first box is operated beyond its capacity for hydrogen sulfide removal and serves solely as a hydrogen cyanide absorber.

Nitric oxide forms a stable complex with ferric oxide under conditions normally present in dry-box purifiers. The nitric oxide is not removed from the spent oxide by oxidation at normal temperatures but can be stripped from granular oxide or sponge by air blowing at 1500°F.

Activated-Carbon Process

This process, which was developed by I.G. Farbenindustrie during the twenties, takes advantage of the catalytic action of activated carbon in promoting the oxidation of H_2S to elemental sulfur at ordinary temperatures. The sulfur deposited on the activated carbon is recovered by extraction with an appropriate solvent, ammonium sulfide, and the carbon is reused until attrition of the carbon particles becomes excessive. The activated-carbon process has the distinct advantage that very pure sulfur is obtained in a relatively simple operation. Its main drawback lies in the fact that the carbon is deactivated rather rapidly by deposition of tar and polymeric materials on the particle surfaces and that complete removal of all such materials from the gas prior to treatment is necessary.

Because of this limitation, the process has not been applied on a large commercial scale, but it is being used for treating large quantities of blue watergas. Although the process was originally designed for H_2S removal only, it is claimed (38) that by selection of the proper carbon, both H_2S and organic sulfur compounds can be removed in two successive treating steps.

Process Description

A diagrammatic flow scheme of the process as described by Engelhardt (39) and Kronacher (40) is shown in Figure 8-8. The sour gas, from which tar and ammonia have been removed in a previous purification step, is passed to carbon-bed 1, after addition of air and a small amount of ammonia. In order to ensure complete reaction, it is customary to add approximately 50 percent more air than is stoichiometrically required. The ammonia, which increases the rate of oxidation quite appreciably, is added in the proportion of 5 volumes of ammonia to 100 volumes of H_2S (41). When the bed is saturated,

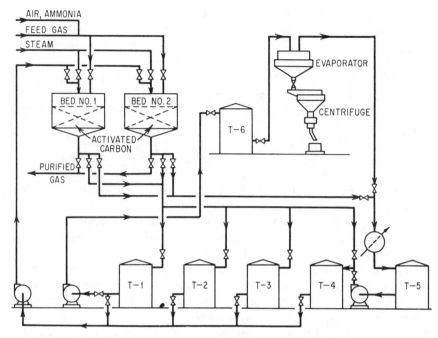

Figure 8-8. Schematic flow diagram of activated-carbon process for hydrogen sulfide removal.

as evidenced by the appearance of small amounts of H_2S in the treated gas, the gas flow is switched to bed 2, and bed 1 is regenerated.

Regeneration is carried out by extraction of the sulfur in several successive stages with a 15 percent aqueous ammonium sulfide solution, followed by steaming of the bed for the removal of residual ammonium sulfide. Solution is first pumped into the saturated bed from tank T-1 until the carbon layer is completely covered with liquid. A few minutes are allowed for dissolving the sulfur, and the solution is then drained back into T-1. This treatment is repeated with the solutions from tanks T-2, T-3, and T-4 so that the last solution contacts essentially sulfur-free carbon. The carbon, which contains practically sulfur-free ammonium sulfide solution, is now treated with saturated steam at 212°F and is then ready for further service. The vapors from the steam treatment, which contain ammonia, H_2S, and water, are condensed in a spray condenser, and the condensate is accumulated in tank T-5, whence it is pumped to solution-tank T-4.

After the extraction process has been repeated several times, the solution in tank T-1 becomes saturated; this is indicated by a sulfur content of about 1.7 to 2.5 lb/gal of liquid. The saturated solution is pumped from T-1 to another tank, T-6, from which it flows by gravity to the evaporator. Here the solution is heated by addition of steam and polysulfides are decomposed. The overhead vapors, containing H_2S, ammonia, and water are condensed in the condenser, and solid sulfur with some water removed from the bottom of the evaporator. The water, which is separated from the sulfur in a centrifuge, is sprayed into the condenser. The sulfur obtained is granular and contains 1 to 2 percent moisture. By operating at pressures of 1.5 to 2 atm and correspondingly higher temperatures, liquid sulfur can be withdrawn from the bottom of the evaporator.

Several modifications of the basic scheme have been described since the process was first placed in operation. Francis (41) states that the sulfur is extracted best by spraying the saturated carbon-bed with several batches of 15 percent ammonium sulfide solution, each batch containing decreasing amounts of sulfur. It is stated specifically that the bed should never be flooded with the solution. Spichal (38) proposes a process in which H_2S removal and extraction of the sulfur are carried out at different locations. In this process, the carbon is placed in a removable circular container, as shown diagrammatically in Figure 8-9. The gas is fed into the center of the container and flows horizontally through the bed. After saturation, the container is removed from the purification plant and a regenerated container is inserted. Regeneration is carried out in a separate plant by immersing the carbon-bed first in hot water and then in several batches of ammonium sulfide solution. The main advantages claimed for this process are reduction of the ground area required and the decrease of down time caused by regeneration. In additon, the construction of the carbon containers supposedly prevents ac-

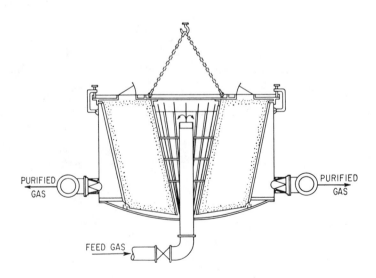

PURIFIED GAS

PURIFIED GAS

FEED GAS

Figure 8-9. Activated-carbon container used in process of Spichal (38).

cumulation of fines and scum on the carbon surface during extraction; this accumulation appears to be a problem in the conventional process.

Design and Operation

Since the rate of reaction of H_2S and oxygen in the presence of activated carbon is appreciably faster than that of H_2S and iron oxide, considerably shorter contact times can be used. Space velocities of 350 to 400 volumes of gas per hour per volume of carbon are customary, as compared to 20 to 40 volumes per hour per volume used in conventional iron oxide boxes. A typical activated-carbon purifier, as used in Germany, consists of a cylindrical carbon-steel vessel, 13 ft in diameter, in which the carbon is placed on a horizontal grid to a depth of approximately 4 ft. Such a unit is capable of processing approximately 200,000 cu ft of gas/hr with a pressure drop of about 25 in. of water column (38). The activated carbon normally used in Germany retains approximately 25 to 35 lb of sulfur/cu ft at saturation. If, therefore, water-gas containing 200 grains H_2S/100 cu ft is treated, a service life of approximately 2 weeks between regenerations is obtained.

In order to accomplish complete H_2S removal, the temperature of the bed has to be maintained below 140°F. Because of the high heat of reaction, gases containing more than 400 grains H_2S/100 cu ft cannot be treated satisfactorily. However, satisfactory operation can be achieved by recycling a portion of the purified gas or by cooling the bed. The rate of oxidation is influenced favorably by the presence of water vapor in the gas, and it is therefore advantageous to saturate the gas with water before it contacts the activated carbon. As already mentioned, small amounts of ammonia increase the catalytic activity of the carbon considerably.

The activated carbon used in the I.G. Farbenindustrie process for H_2S removal is prepared from low temperature, brown-coal coke which is ground to a particle size of 1 to 4 mm in diameter. The carbon is activated by heating with combustion gases and steam at approximately 1500°F for several hours. The resulting product has a bulk density of about 25 lb/cu ft and is capable of absorbing sulfur to the extent of 100 to 150 percent of its weight. A good grade of activated carbon will withstand 20 to 30 cycles of saturation and regeneration.

Spichal (38) describes a two-stage process for the removal of both H_2S and organic sulfur compounds. Activated carbon, derived from brown-coal coke in the manner described above, is used for H_2S removal in the first stage, while a carbon prepared from anthracite coal by the same activation method is utilized in the second stage for complete removal of organic sulfur compounds such as carbonyl sulfide, carbon disulfide, and thiophene. Complete absence of H_2S and hydrocarbons is necessary for the proper functioning of the organic-sulfur-removal step. Spichal suggests that if ammonia and oxygen are added to the water-saturated gas in concentrations somewhat in excess of the stoichiometric proportions, carbonyl sulfide is converted quantitatively to ammonium sulfate and thiourea, carbon disulfide to ammonium sulfate and ammonium thiosulfate, and thiophene to thiourea. Operating conditions include space velocities of 350 to 400 vol/(hr)(vol) and temperatures ranging from 80° to 100°F. The activated carbon is said to be capable of absorbing 10 to 12 percent of its weight of organic sulfur compounds. Regeneration is accomplished by first extracting the saturated carbon with steam condensate at 175°F; this is followed by treatment with steam super-heated to 750°F at a pressure of 0.5 atm.

In view of the known difficulty of removing organic sulfur compounds, especially thiophene, by oxidation even at high temperatures and in the presence of very active catalysts, the reaction mechanism proposed by Spichal is somewhat questionable. Formation of thiourea from carbonyl sulfide, thiophene, and ammonia is highly improbable, considering the low partial-pressure of the reactants and the low operating temperatures. Removal of organic sulfur compounds is most probably due to adsorption. This is indicated by the relatively low capacity of the carbon for organic sulfur—as compared to H_2S—and the necessity of using superheated steam under vacuum to strip the adsorbed compounds from the absorbent.

The Sulfreen Process

The Sulfreen Process was developed jointly by Lurgi Gesellschaft für Wärme und Chemotechnik of West Germany and Société National des Petroles d'Aquitaine (now Société National Elf Aquitaine) of France with the specific purpose of reducing residual sulfur compounds in the tail gases of Claus type sulfur recovery plants. The process is, in essence, an extension of the Claus process discussed later in this chapter. The reaction between hydrogen sulfide and sulfur dioxide is carried out at lower temperatures at which the conversion to elemental sulfur is more complete than at the temperatures normally used in the Claus process. While in the Claus process the reaction is carried out at temperatures above the dew point of sulfur, the operating temperatures used in the Sulfreen process are below the sulfur dew point and the sulfur formed is adsorbed on a solid catalyst.

The catalyst originally used in the process was a specially prepared activated carbon. This catalyst, although highly efficient, requires high temperatures to vaporize the adsorbed sulfur during regeneration. In recent years, an alumina catalyst similar to that used in Claus units was developed, and practically all Sulfreen plants are now operating with alumina catalyst. Results from an extensive investigation of the performance of alumina catalysts under conditions representative of the Sulfreen process, i.e. at temperatures below the sulfur dew point, have been reported by Pearson (79). The study showed that the catalyst retains its activity even when loaded up to

50 percent by weight with condensed sulfur. However, sulfation of the catalyst leads to rapid deactivation. Use of alumina permits regeneration at relatively low temperatures, about 570°F, which not only reduces fuel consumption but also makes it possible to construct the entire plant of carbon steel. In older plants utilizing activated carbon catalyst, stainless steel reactors were required to withstand corrosion at the high regeneration temperatures.

Although the Sulfreen process is cyclic, alternating between adsorption and desorption of the sulfur formed, the use of several reactors in adsorbing, desorbing and cooling service permits continuous operation.

Process Description

A flow diagram of the Sulfreen process is shown in Figure 8-10. In this arrangement (42, 43) six reactors are used, four in parallel for adsorption, while one is desorbed and one is cooled. This flow sheet is representative of the original form of the process using activated carbon catalyst and high regeneration temperatures. In more modern plants where alumina catalyst is

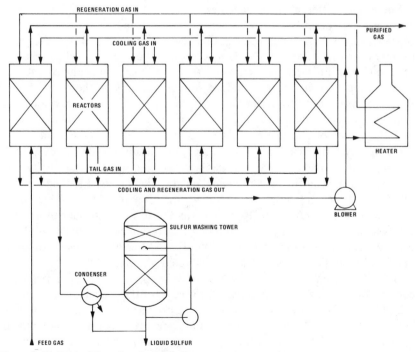

Figure 8-10. Typical flow diagram of Sulfreen process.

used (80, 81, 82), the number of reactors is smaller and the sulfur wash tower is replaced with a sulfur condenser.

The Claus plant tail gas, containing typically about 1.0 volume percent hydrogen sulfide, 0.5 volume percent sulfur dioxide, and some sulfur vapor, passes through four reactors in parallel at 260° to 300°F, and about 75 to 85 percent of the hydrogen sulfide and sulfur dioxide are converted to elemental sulfur.

When one reactor is saturated with sulfur, it is taken off the line and a regenerated reactor is placed on line. The sulfur adsorbed by the carbon is desorbed by a stream of recirculated hot inert gas. The hot regeneration gas is first cooled in a sulfur condenser where 50 psig steam is produced and liquid sulfur is condensed. The gas then passes to a tower where it is further cooled by washing with a stream of liquid sulfur and where additional sulfur is condensed.

The effluent gas from the sulfur wash tower is divided into two streams, one of which flows to a heater and from there to the reactor being desorbed. The other stream is used to cool the reactor after desorption is completed.

The pressure in the regeneration system is maintained constant by either venting or adding inert gas. The various steps of the process cycle are automatic, according to a fixed program.

Somewhat different flow schemes, used in plants containing alumina catalyst, are described by Cameron (80), Grancher et al (81), and Nougayrede (82). An important design feature of such plants is that when regeneration of a catalyst bed is completed, as indicated by a bed temperature of about 570°F, a small stream of reducing gas is introduced into the regeneration loop in order to reduce sulfates formed on the catalyst. This technique has proved to be very effective in preserving high catalyst activity over long periods of time.

Performance of alumina catalyst in Sulfreen units is discussed in a study presented by Grancher (89). The problem of sulfation of the catalyst due to the presence of oxygen in the feed gas was solved quite successfully by addition of hydrogen sulfide as a reducing agent to the gas used for regeneration of saturated reactors. Figure 8-11 shows this effect on the efficiency of regenerated catalyst. Another approach was the use of a special proprietary catalyst named AM catalyst which reportedly removes oxygen and is also an effective catalyst for the Claus reaction. The AM catalyst consists of an undisclosed metallic salt supported on activated alumina (89).

In 1976, sixteen Sulfreen units were either in operation or in the design and construction stages. Operating data for several of these plants were reported (42, 80, 81, 82). It is claimed that with a properly designed and operated plant, better than 99 percent overall sulfur recovery is attainable with the combination of the Claus-Sulfreen processes. As with a Claus unit, optimum conversion in a Sulfreen unit depends on maintaining the stoichiometric ratio

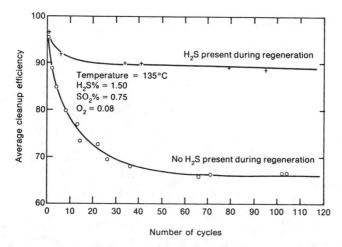

Figure 8-11. Effect of H_2S in Regeneration Gas on Cleanup Efficiency of Alumina Catalyst (89). Courtesy *Hydrocarbon Processing,* September 1978.

of two mols of hydrogen sulfide per mol of sulfur dioxide in the Claus tail gas, which requires close control of the Claus plant operation.

Utility consumption of six Sulfreen units ranging in capacity from 0.32 to 7.3 MMscf/hour of tail gas from Claus units producing from 90 to 2,100 long tons of sulfur per day are shown in Table 8-5 (82). A photograph of a Sulfreen unit at Lacq, France is shown in Figure 8-12.

The AMOCO Cold Bed Adsorption (CBA) Process

This process, which was developed by AMOCO Canada Petroleum Company, Ltd., was described by Goddin, Hunt, and Palm (83) and by Nobles,

Table 8-5. Utility Requirements for Sulfreen Units Treating Claus Tail Gases (82)

Plant	A	B	C	D	E	F
Catalyst	Carbon	Carbon	Alumina	Alumina	Alumina	Alumina
Tail Gas, MMscf/hr	4.6	7.3	7.3	7.0	6.7	0.32
Claus Capacity, LTD Sulfur	1,200	2,000	2,100	1,800	1,600	90
Electricity, h.p.	1,000	1,700	1,500	1,500	1,450	40
Steam Production, lbs/hr	12,000	24,000	20,000	23,000	23,000	850
Fuel Gas, Mscf/hr	25	40	27	33	33	1.25

Figure 8-12. Sulfreen plant, treating tail gases from a 1,000 long tons per day, Claus type sulfur plant at Lacq (France). *Lurgi Gesellschaft fur Warme and Chemotechnik and Societe Nationale des Petroles d' Aquitaine*

Palm, and Knudtson (84). At present, one plant is in operation in conjunction with a large Claus type sulfur recovery plant located at East Crossfield in Alberta, Canada. In principle, the CBA process is quite similar to the Sulfreen process, as the Claus reaction is also carried out at sufficiently low temperatures to cause condensation of sulfur on the catalyst. However, unlike the Sulfreen process where a closed loop of inert gas is used to desorb the adsorbed sulfur, a hot slip stream from the Claus unit is used for this purpose.

As a result, the only items required to integrate a CBA unit into an existing Claus unit are two additional reactors and sulfur condensers.

The flow of gases through a CBA process unit during the regeneration and cooling periods is shown in Figures 8-13 and 8-14. Following the flow in Figure 8-13, it is seen that CBA Reactor No. 1 is being regenerated, while CBA Reactor No. 2 is adsorbing. Gas from Claus Reactor No. 2 is cooled to 260°F in condenser No. 3 and sent to CBA Reactor No. 2. The effluent from

CBA Reactor No. 2 is sufficiently low in sulfur content that it can be sent directly to the incinerator. Regeneration of CBA Reactor No. 1 is effected by use of a slip stream withdrawn from the feed to Claus Reactor No. 1 at 450° to 550°F. Heating of the catalyst bed being regenerated is accomplished not only by the sensible heat of the feed gas, but also by the heat of the Claus reaction. As the temperature of the catalyst bed rises, adsorbed sulfur is gradually vaporized. Regeneration is continued until a temperature plateau of 600° to 700°F is reached which corresponds to the outlet temperature of Claus Reactor No. 1. The regeneration effluent is returned to the inlet of Claus Reactor No. 1 after passing through condenser No. 4.

When regeneration of CBA Reactor No. 1 is completed, the valves are switched to start cooling the hot reactor as shown in Figure 8-14. During the cooling period, tail gas from Condenser No. 3 is diverted to CBA Reactor No. 1 and then to CBA Reactor No. 2 via condenser No. 4. When cooling is completed, the effluent from CBA Reactor No. 1 is sent to incineration and regeneration of CBA Reactor No. 2 is started.

The catalyst used in the CBA reactors is alumina, the same as that used in the Claus reactors. Nobles et al (84) report conversion efficiencies of better than 99 percent for the combined Claus-CBA plants. A discussion of catalyst performance in low temperature Claus processes has been presented by Pearson (79).

The MCRC Sulfur Recovery Process

This process, which was described by Heigold and Berkeley (97) and is licensed by Delta Engineering Corporation, is quite similar to the Amoco CBA process just described. The process uses a Claus type reaction furnace followed, after cooling of the gas and condensation of sulfur, by three or four catalytic converters. One or two of the converters operate at temperatures below the dew point of sulfur, and periodic removal of condensed sulfur is required. Regeneration is made continuous by switching converters to different positions in the system. Sulfur recovery levels of 98.5 to 99 percent are reported for three converters and 99 to 99.5 percent for four converters (97).

The Claus Process

The Claus process is not a gas-purification process in the true sense of the word, as its principal objective is recovery of sulfur from pure gaseous hydrogen sulfide or from acid gas streams containing hydrogen sulfide in high concentrations. Typical streams of this type are the acid gases stripped from

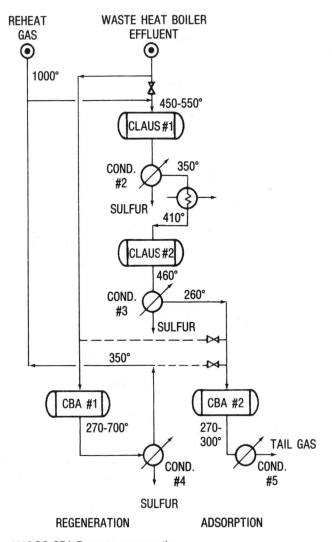

Figure 8-13. AMOCO-CBA Process, regeneration.

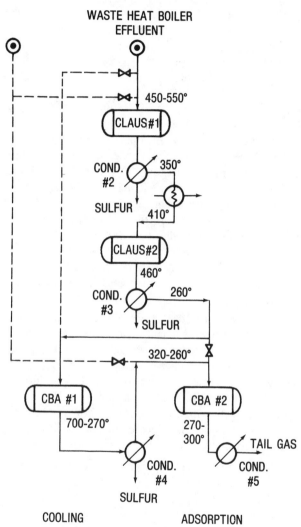

Figure 8-14 AMOCO-CBA Process, cooling.

regenerable liquids, e.g., alkaline solutions or physical solvents used for the purification of sour gases. The effluent gases from the Claus plant are without value and are vented to the atmosphere, usually after incineration of residual sulfur compounds to sulfur dioxide. In view of air pollution control regulations existing in most industrialized countries which prohibit discharge of large amounts of sulfur compounds to the atmosphere, Claus plants are often mandatory adjuncts of gas-desulfurization installations, and, consequently, the Claus process is of considerable significance within the general scope of gas-purification technology. Furthermore, the Claus process yields sulfur of extremely good quality and thus is a source of a valuable basic chemical.

Considering growing concern with air pollution, sulfur recovery in Claus type units is increasing to the point where units which normally would not be considered economical are being installed, strictly for the purpose of air pollution control. In addition, the recovery efficiency of Claus type plants is continuously improved by better design methods. Unfortunately, complete conversion of hydrogen sulfide to elemental sulfur is precluded by the equilibrium relationships of the chemical reactions upon which the process is based. As a result of this limitation, the process is, in many instances, not adequate to reduce emission of sulfur compounds to the atmosphere to the level required by air pollution control authorities and has to be supplemented with processes specifically designed to remove residual sulfur compounds from the Claus plant tail gases. Such processes are discussed elsewhere in this text.

Since the disclosure of the process by Claus in 1883, it has undergone several modifications, most of which are, however, only of historical interest. As presently used, the various processes on the market are similar in their basic concept and differ only in the design and arrangement of the equipment.

The literature describing the theoretical as well as design and operational aspects of the Claus process is quite voluminous (44-59). In view of this extensive coverage, the following discussion will be directed primarily toward the design and operation of high-efficiency plants giving maximum sulfur recovery and minimum emission of sulfurous air pollutants to the atmosphere.

Basic Data

The basic chemical reactions occurring in the Claus process are represented by Equations 8-1, 8-2, and 8-4, with Reactions 8-1 and 8-2 taking place in the thermal stage (reaction furnace) and Reaction 8-4 in the catalytic

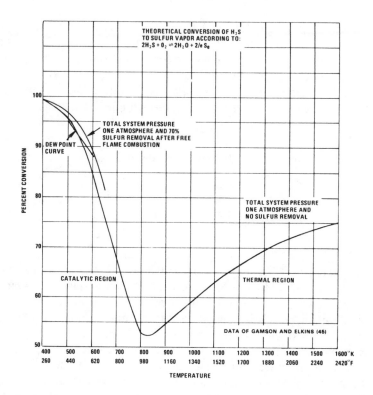

Figure 8-15. Theoretical equilbrium conversion of hydrogen sulfide to sulfur vapor (45).

stage (catalytic converters). The thermodynamics and kinetics of the reactions have been investigated by Gamson and Elkins (45), and theoretical conversion as a function of temperature and pressure was computed. These data are shown in Figure 8-15. The equilibria are complicated by the presence of various species of sulfur in the sulfur vapor at different temperatures. Grancher (89) reports that significantly higher conversions can be obtained in the catalytic stages, up to 10 percent more than those predicted by the data of Gamson and Elkin.

It can be seen from Figure 8-15 that the process can be separated into two stages, i.e., a thermal stage, above about 1000°F, and a catalytic stage, usually between 700°F and a temperature somewhat above the sulfur dew point of the gas mixture. The lower this temperature, the more complete conversion can be attained. It is, therefore, advantageous to have several catalytic stages with condensation of the sulfur formed after each stage.

The heat of reaction for the combination of Reactions 8-2 and 8-4, i.e.,

$$3H_2S + \frac{3}{2}O_2 = 3H_2O + 3S \qquad (8\text{-}13)$$

is 262,000 to 311,000 Btu (45), with 223,000 to 248,000 Btu for Reaction 8-2 and 38,000 to 63,000 Btu for Reaction 8-4 (45). Since Equation 8-4 represents the catalytic stage, it is seen that the temperature rise in the catalyst beds is relatively small, permitting operation at low temperatures and, consequently, the attainment of high conversion. Since the reactions are reversible and water is a reaction product, removal of water between catalytic stages would be a means to increase conversion. However, attempts to do this have so far failed, primarily because of the corrosiveness of the aqueous condensate and plugging of the equipment with solid sulfur. Thus, the presence of water vapor in the reaction gases throughout the plant imposes a definite limitation on the degree of conversion.

A further limitation on conversion is the occurrence of a number of side reactions, due to the presence of carbon dioxide and light hydrocarbons in the feed gas, resulting in the formation of carbonyl sulfide and carbon disulfide in the thermal stage of the process. These compounds are quite stable and may pass unchanged through the catalytic converters unless provisions are made for their conversion to hydrogen sulfide and carbon dioxide. Formation of carbonyl sulfide and carbon disulfide is a significant consideration in the design and operation of high-efficiency Claus plants as the sulfur associated with these compounds may amount to an appreciable percentage of the sulfur in the plant feed gas.

Various reaction mechanisms have been proposed for the formation of carbonyl sulfide and carbon disulfide (44, 45, 54, 57, 60, 89) and for their conversion to hydrogen sulfide, elemental sulfur, and carbon dioxide. The limited plant data available indicate that carbonyl sulfide is formed primarily from hydrogen sulfide and carbon dioxide following reasonably well the equilibrium relationships prevailing at reaction furnace temperatures (see Chapter 13). However, carbon disulfide has been found in much higher concentrations than would be indicated by chemical equilibria, and an additional mechanism, probably involving hydrocarbons and sulfur vapor, is surmised. This is particularly true for the partial combustion process, where the entire feed gas is passed through the reaction furnace, and substantial amounts of sulfur are formed in the flame.

From a practical standpoint, the exact mechanism of side reaction products formation is less important than their conversion to hydrogen sulfide and sulfur to prevent emission to the atmosphere. It is generally assumed that carbonyl sulfide and carbon disulfide undergo hydrolysis (see Chapter 13) at temperatures on the order of 700°F, in the presence of an aluminum oxide type catalyst at sufficiently rapid rates to attain thermochemical equilibrium. Thus, by maintaining sufficiently high temperatures in the first catalytic converter and by use of an active catalyst, substantial conversion of carbonyl sulfide and carbon disulfide can be obtained. These conclusions have been substantiated by the work of Grancher (89) who investigated formation and hydrolysis of COS and CS_2 in large Claus units at Lacq, France.

The catalyst used in the Claus process is either natural bauxite or alumina which may be shaped into pellets or balls. For high-efficiency plants, an alumina catalyst of high activity is usually preferred. Besides activity, resistance to attrition and to relatively high temperatures during regeneration are important catalyst properties. Furthermore, the shape of the catalyst must be such that no excessive pressure drop is incurred at space velocities ranging from 1,000 to 2,000 volumes of gas per volume of catalyst per hour.

Process Description

There are two basic forms of the process which may be termed the partial combustion and the split-stream processes. In the partial combustion process, which is shown diagrammatically in Figure 8-16, the entire acid-gas stream and the stoichiometric amount of air to burn one-third of the hydrogen sulfide to sulfur dioxide is fed through a burner to the reaction furnace. At the temperatures prevailing in the furnace, typically 2000° to 3000°F, a substantial amount of elemental sulfur is formed (see Figure 8-14), which is condensed after cooling the gases first in a waste heat boiler and subsequently in a sulfur condenser. While high-pressure steam can be generated in the waste heat boiler, it is preferable to produce low-pressure steam (25 psig) in the sulfur condenser in order to cool the reaction gases to obtain maximum sulfur condensation. The reaction gases leaving the sulfur condenser are reheated and flow through the first catalytic converter where additional sulfur is produced by the reaction of hydrogen sulfide with sulfur dioxide. Reheating the gas is necessary to maintain the temperature of the reaction gas above the sulfur dew point as it passes through the catalytic converter because condensation of

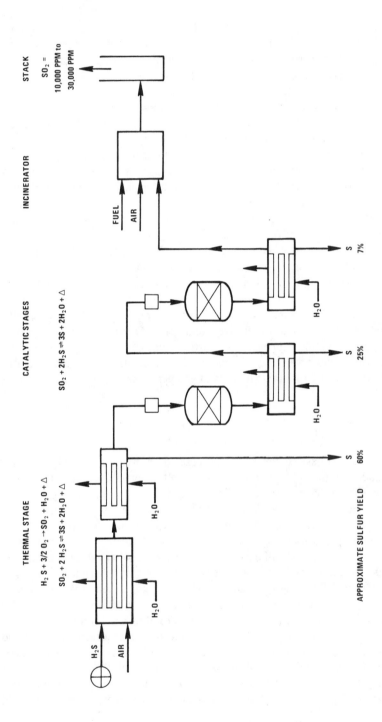

Figure 8-16. Typical flow diagram of two-stage Claus process plant.

sulfur leads to rapid catalyst deactivation. The gases leaving the first catalytic converter are again cooled, and sulfur is condensed. The process of reheating, catalytically reacting, and sulfur condensing may be repeated, in one, two, or even three additional catalytic stages. As conversion progresses through the catalytic stages and more and more sulfur is removed from the gas mixture, the sulfur dew point of the reaction gases is lowered, permitting operation at progressively lower temperatures, thus improving conversion (see Figure 8-15). After leaving the last sulfur condenser, the exhaust gases, which still contain appreciable amounts of sulfur compounds and a small amount of sulfur vapor, are either incinerated, in order to convert all sulfur compounds to sulfur dioxide, before venting to the atmosphere, or further treated, in a separate process, for removal of residual sulfur values.

The partial combustion process is used for gas streams of high hydrogen sulfide content (above 50 percent). Depending on the hydrogen sulfide concentration in the acid gas, conversion efficiencies of 94 to 95 percent can be attained with two catalytic stages and of 96 to 97 percent with three catalytic stages. A fourth catalytic converter is normally not economical as it increases conversion by less than 1 percent. It should be noted that the conversion efficiencies indicated above do not take into account sulfur losses caused by carbonyl sulfide and carbon disulfide formation and that the overall conversions must be adjusted downward by the amount of such losses.

As stated earlier, carbonyl sulfide and carbon disulfide hydrolyze fairly readily at temperatures around 700°F in the presence of Claus catalyst. It is therefore advantageous to design the first catalytic converter for operation at such a temperature level if high conversion efficiency is required. However, this results in inefficient operation of the first converter with respect to the Claus reaction, and installation of a third converter may be desirable to compensate for this loss in efficiency. If, because of air pollution control requirements, an additional processing step is needed to minimize sulfur emission to the atmosphere, it is usually economical to use only two catalytic converters, and remove residual sulfur compounds in the tail gas treating unit.

The split-stream process is used for acid-gas streams containing hydrogen sulfide in such low concentrations that stable combustion could not be sustained if the entire gas stream were fed to the reaction furnace. In this process one third of the acid gas is fed to the reaction furnace, and all the contained hydrogen sulfide is burned to sulfur dioxide with the stoichiometric amount of air. The hot gases are cooled in a waste heat boiler and then combined with the remaining two thirds of the acid gas before entering the first catalytic conversion stage. The rest of the process is identical with the partial combustion process.

Since in the split-stream process all hydrogen sulfide contained in the portion of the acid gas fed to the furnace is burned to sulfur dioxide, no carbonyl sulfide and carbon disulfide should be formed; consequently, it should not be necessary to operate the first catalytic converter at high temperatures to hydrolyze these compounds.

Combination of the partial combustion and split-stream processes may prove to be advantageous for the processing of acid-gas streams containing hydrogen sulfide concentrations on the order of 35 to 50 percent. In such cases varying amounts of acid gas may be bypassed around the furnace. Special techniques, including pre-heating of the air and acid gas streams, and generation of sulfur dioxide by burning of recycled sulfur are required to process acid gases of very low hydrogen sulfide content, on the order of less than 20 percent. A review of a number of such designs has been presented by Beavon and Leeper (85). Figures 8-17 and 8-18 show photographs of a large and small Claus plant.

Design and Operation

Process design procedures have been presented by Valdes (50, 51) and by Opekar and Goar (54). Since sufficient thermodynamic and kinetic data are available in the literature, these methods are straightforward, and design optimization by the use of computer techniques is possible. One such method has been proposed by Boas and Andrade (78). Another method for predicting Claus products by use of a modified Gibbs free energy minimization technique has been presented by Maadah and Maddox (90).

Several mechanical designs and arrangements of the major equipment have been reported in the literature. The reaction furnace normally used combines the burners, the combustion chamber, and the waste heat boiler in one integral vessel. Methods for the design of firetube reaction furnaces have been reported by Valdes (52). A different design consisting of a separate combustion furnace and waste heat boiler has been described by Sawyer, et al. (44). However, this design is not widely used in modern Claus plants.

Heating of the reaction gases may be effected by several methods (49, 50, 51, 57), i.e., bypassing of hot gases from the reaction furnace, auxiliary in-line acid-gas burners, gas-to-gas heat exchangers, and indirect fuel-gas fired heaters. The first two methods result in slightly lower conversion as some of the acid gas is bypassed around one or more catalytic converters. Gas-to-gas heat exchangers and fuel-gas fired indirect heaters are more expensive than hot gas bypasses and in-line burners, but their use may be justified if very high conversion is required. The catalyst beds may be arranged in horizontal or vertical vessels, with more than one bed located in one vessel. In very large installations, care has to be taken to assure even gas distribution and to avoid channeling.

It is customary to install mist eliminating devices after the last sulfur condenser in order to minimize entrainment of sulfur droplets into the incinerator. Installation of mist eliminators after each sulfur condenser is of

Figure 8-17. Claus-type sulfur plant producing 875 long tons of sulfur per day. *The Ralph M. Parsons Company*

value as catalyst deactivation by entrained sulfur may be a problem. Effective mist elimination devices are vessels packed with Raschig rings or wire mesh pads.

The mechanical design of small plants is often substantially different from that of large installations as small units lend themselves to compact packaging. Designs of this type of unit with capacities of less than 50 long tons of sulfur per day have been presented by Grekel, Kunkel, and McGalliard (49).

The most important variable in the operation of Claus plants is the ratio of hydrogen sulfide to sulfur dioxide in the reaction gases entering the catalytic converters. Maximum conversion requires that this ratio be maintained constant at the stoichiometric proportion of 2 moles of hydrogen sulfide to 1 mole of sulfur dioxide. Appreciable deviation from the stoichiometric ratio leads to

Figure 8-18. Claus-type sulfur plant producing 7 long tons of sulfur per day. *The Ralph M. Parsons Company*

drastic reduction in conversion efficiency (51). The proper ratio is maintained by control of the air flow to the reaction furnace which is easily done by automatic air to acid-gas ratio flow control. However, this method is only successful if the hydrogen sulfide content of the acid gas is constant, as it does not compensate for the variations in the actual amount of hydrogen sulfide flowing to the reaction furnace. Several methods based on controlling the air flow by continuous analysis of the ratio of hydrogen sulfide to sulfur dioxide in the plant tail gas have been developed. One such method has been described by Carmassi and Zwilling (61) and Grancher (89) and Taggart (91). Several analytical instruments based on vapor chromatography and ultraviolet absorption which can be used in the control circuit are commercially available.

A serious operating problem in Claus plants is deactivation of the catalyst by deposition of carbonaceous materials and, in some cases, of sulfate. Acid gases usually contain small amounts of hydrocarbons, especially if the sour gas from which the acid gases have been removed is relatively rich in high molecular weight aliphatic and aromatic hydrocarbons which are somewhat soluble in the liquids used in the desulfurization unit. When acid gases of high hydrogen sulfide content are processed, the temperature in the reaction furnace is usually high enough to result in complete combustion of all hydrocarbons to carbon dioxide and water, and no catalyst deactivation is experienced.

However, at the low combustion temperatures occurring in plants processing gases containing less than about 40 percent hydrogen sulfide, hydrocarbons crack and the products from the incomplete combustion are carried into the catalytic reactors, gradually deteriorating the catalyst. The catalyst can be regenerated by air oxidation of the carbonaceous deposits. Care must be taken during regeneration not to exceed a temperature of about 1000°F in order to avoid changes in the catalyst structure. A good catalyst can be regenerated several times although the activity decreases somewhat with each regeneration.

Equally serious as deposition of carbonaceous materials on the catalyst is the gradual accumulation of sulfate which destroys the capability of the catalyst to hydrolyze carbonyl sulfide and carbon disulfide. This problem has been investigated extensively (86, 87, 88, 89, 92, 93, 94) and catalyst formulations have been prepared which are quite resistant to deactivation by sulfation. Another operating problem is condensation of sulfur on the catalyst resulting in rapid deactivation. This can be avoided by maintaining the temperature in the catalytic converters above the sulfur dew point of the gas mixture. Should sulfur condense on the catalyst, raising of the gas temperature is usually sufficient to vaporize the sulfur and to reestablish catalyst activity.

Special techniques have to be used for the processing of gas streams containing appreciable amounts of ammonia, such as effluents from sour water strippers. The ammonia must be destroyed in the reaction furnace in order to avoid deposition of ammonium salts on the catalyst beds. This is usually achieved by burning the ammonia containing stream in a separate zone under conditions which assure complete combustion of the ammonia without formation of nitrogen oxides (95, 96).

The liquid sulfur produced in Claus units contains dissolved hydrogen sulfide which, upon cooling during storage, is released into the gas space above the liquid surface. Accumulation of H_2S, especially in unvented vessels, can lead to a lethal concentration and may easily exceed the lower explosive limit of H_2S. Sulfur is normally produced at temperatures between 140 to 170°C (284 to 338°F) but upon storage and during transportation the temperature can drop to as low as 125°C (258°F). At the higher temperature the H_2S is present as dissolved H_2S and as polysulfide, H_2S_x. The polysulfide becomes unstable at lower temperature and slowly dissociates into H_2S and sulfur, causing the emission of H_2S into the vapor space.

In order to prevent hazardous H_2S concentrations, the sulfur must be degassed and the H_2S content reduced from several hundred ppm to 10 to 15 ppm. Several commercial methods, with or without addition of catalysts, are described by Watson et al. (98) and Lagas (99). Basically, degassing is effected by vigorous agitation of the liquid sulfur, by circulating or spraying in a collecting pit for several hours, or by stripping with gas in a column. Several catalysts, for example ammonia (98), have been found to be effective in accelerating the operation (99).

Miscellaneous Processes

A modification of the Claus process operating at elevated pressure has been disclosed by Kerr et al. (100). This process, named Richards Sulfur Recovery Process (RSRP), was developed jointly by Alberta Energy Company Ltd. and Hudsons Bay Oil and Gas Company Ltd. The process was tested in the laboratory and a conceptual design of a commercial plant proposed. Operation is conducted at pressures ranging from 70 to 300 psig with the catalyst immersed in circulating liquid sulfur which acts as a cooling medium. An interesting feature of the process is that the combustion furnace of a Claus plant is replaced with a catalytic oxidizer where a portion of the H_2S in the feed gas is partially oxidized with cool liquid sulfur being sprayed over the catalyst bed to maintain the temperature at 700 to 800°F. Conversion of H_2S to sulfur of better than 99 percent is claimed.

Another modification of the Claus process, operating at much lower temperatures and reportedly usable for H_2S removal from hydrocarbon gases, was developed through the pilot-plant stage in the United States by the Jefferson Lake Sulfur Company (62). In the process, the SO_2 necessary for the reaction is obtained by burning sulfur in an external burner. The gas to be treated is preheated to moderately high temperatures and then contacted with the SO_2 in several chambers containing a special catalyst. The elemental sulfur formed is removed from the gas by condensation. In order to maintain high efficiency of the catalyst, periodic regeneration is required. It is claimed that hydrocarbons pass unaffected through the catalyst beds. This process can also be used for acid-gas streams of such low hydrogen sulfide content that combustion could not be sustained even with only one third of the gas stream.

A process also based on the reaction of H_2S with SO_2 is described by Audas (63). In this process, which was developed through the pilot-plant stage and is covered by British Patent 653,317, SO_2 is added to the gas in slight excess over the amount theoretically required to react with the H_2S. The gas is then passed through a bed of alumina chips or granules from which it emerges free from sulfur compounds. Depending on the H_2S content of the gas, the operating temperature ranges from 85 to 190°F. Elemental sulfur, water, and excess SO_2 are retained on the alumina. A portion of the alumina is continuously withdrawn from the bottom of the bed and regenerated alumina is added to the top of the purification vessel. The spent alumina is regenerated by recycling combustion gases at a temperature of 850°F. Sulfur and SO_2 are recovered from these gases. By adjustment of the oxygen content of the circulating gas, the sulfur can be completely converted to SO_2. The advantages claimed for this process are (a) drastic reductions in plant volume and ground area, if compared to an equivalent iron oxide dry-box installation; (b) recovery of relatively pure sulfur and SO_2; and (c) production of extremely dry gas.

Direct catalytic oxidation of hydrogen sulfide to elemental sulfur in the presence of gaseous paraffinic hydrocarbons or other gases has been reported by Greckel (47). Although pilot-plant work indicated good conversions and no effect on the hydrocarbons, the process has so far not been successfully commercialized.

A process using synthetic zeolites for recovery of sulfur from sour gases under pressure has been described by Haines et al. (64). In this process which has been tested in a pilot plant but which has not been commercialized, hydrogen sulfide is adsorbed on the zeolite which is then regenerated with sulfur dioxide containing gas at high temperatures. The elemental sulfur formed is condensed, and residual hydrogen sulfide and sulfur dioxide are vented to atmosphere.

Oxidation to Oxides of Sulfur

Although the iron oxide process is capable of satisfying the most stringent H_2S-content requirements of domestically used gases, its shortcomings, i.e., the poor quality of the sulfur produced and its ineffectiveness for organic-sulfur removal, led to the development of dry processes in which H_2S and organic sulfur compounds are catalytically converted to oxides of sulfur which are subsequently removed with aqueous solvents and converted to pure sulfates and elemental sulfur. The most important of these processes are

1. The "Katasulf" process in which H_2S and a portion of the organic sulfur are oxidized catalytically to SO_2
2. The North Thames Gas Board process in which organic sulfur compounds are catalytically converted to SO_2

3. The Iron-Soda process in which organic sulfur compounds are converted to SO_2 and SO_3

As mentioned above, iron oxide does not react with organic sulfur compounds such as carbonyl sulfide, carbon disulfide, mercaptans, and thiophene at ordinary operating temperatures. All of these compounds are present in gases manufactured from sulfur-bearing materials in concentrations varying from a few grains to 50 grains/100 cu ft. Since the concentration of organic sulfur compounds in manufactured gases is always appreciably lower than that of hydrogen sulfide and, also, because of the less objectionable odor and lower toxicity of these compounds, the removal of organic sulfur is usually of no particular concern if the gas is used strictly for domestic purposes. Practically all legal restrictions for the sulfur content of gas refer to hydrogen sulfide, and, in general, no limits are set for organic-sulfur content.

With the advent of catalytic processes for the manufacture of synthetic fuels, ammonia, alcohols, etc., from hydrogen, nitrogen, and carbon monoxide, removal of all sulfur compounds from synthesis gas became of prime importance. Catalysts used in the synthesis processes are sensitive to sulfur poisoning; this results in rapid decrease of activity and, in some cases, the necessity to replace expensive catalysts after short periods of operation. Total sulfur concentrations on the order of 0.1 grain/100 cu ft or, in some instances even as low as 0.01 grain/100 cu ft, are required to ensure the proper functioning and a satisfactory life of synthesis catalysts.

Considerable work on organic-sulfur removal from many kinds of gases has been conducted during the past 40 years in England, Germany, and the United States, and several processes were developed on the laboratory, pilot-plant, and commercial scale. These processes can be divided into the following categories:

1. Catalytic oxidation of the organic sulfur compounds to SO_2, which is removed in aqueous solvents.
2. Catalytic oxidation to SO_3, which reacts with an alkaline constituent of the catalyst and accumulates as sulfate in the catalyst bed.
3. Catalytic conversion to H_2S followed by removal of the H_2S.
4. Catalytic conversion to H_2S, which is retained on the catalyst as sulfide and periodically removed by oxidation to SO_2.

The latter two categories will be discussed in Chapter 13. Fairly complete conversion of carbonyl sulfide, carbon disulfide, and mercaptans appears feasible by all of these techniques, although several catalytic contacts are usually required. None of the existing catalytic processes is capable of complete thiophene conversion. At present, several commercial organic-sulfur-removal plants utilizing catalytic conversion are operating in Europe and the United States and producing gas of varying degrees of purity.

The Katasulf Process

The Katasulf process was developed in the early 1930s in Germany by Hans Bähr (65). The primary purpose was to develop an economical method by which the hydrogen sulfide and ammonia, present in coke-oven gases, could be removed in one purification step and converted to relatively pure ammonium sulfate and elemental sulfur. Several combination processes for the simultaneous removal of hydrogen sulfide and ammonia had been developed, but none of these processes were accepted widely, primarily because of their inability to cope in a satisfactory manner with the difficulties arising from the presence of tar and hydrogen cyanide in coke-oven gases. Another shortcoming of these processes was the fact that they did not provide proper means of compensating for variations in the hydrogen sulfide and ammonia contents of the gas to be treated. It is claimed that the Katasulf process, which is based on the catalytic oxidation of H_2S to SO_2 followed by removal of SO_2 with an ammonium sulfite-bisulfite solution, overcomes both of these difficulties.

Process Description

The basic form of the Katasulf process involves a relatively simple flow scheme. The gas is first passed through an electrostatic filter for tar removal and then preheated to 750°F by heat exchange with the gas leaving the catalytic converter. After addition of air, the gas is admitted to the catalyst chamber where the H_2S is converted to SO_2. Because of the exothermic nature of the oxidation reaction, the gas temperature increases considerably during conversion. The effluent from the catalytic converter is partially cooled by heat exchange with the incoming gas and then flows to the bottom of an absorption tower, where it is washed countercurrently with an aqueous ammonium sulfite-bisulfite solution which absorbs both the SO_2 and ammonia. The last traces of impurities are removed in a final water wash.

This form of the process is suitable for the treatment of gases containing H_2S and ammonia at a maximum weight ratio of 1.5:1 because higher ratios result in solutions containing excessive amounts of bisulfite. The high vapor-pressure of SO_2 over such solutions causes incomplete SO_2 removal from the gas and may lead to serious corrosion of carbon-steel equipment.

Gases containing large amounts of H_2S are treated by using the following modifications of the process; these are shown schematically in Figure 8-19.

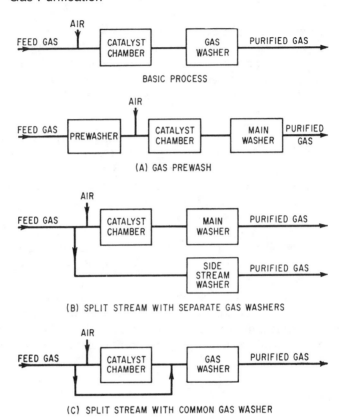

Figure 8-19. Various operating schemes of Katasulf process.

1. The inlet gas is washed with a slip stream of sulfite-bisulfite solution before it enters the catalyst chamber. Ammonia, and a portion of the hydrogen sulfide, are removed, and ammonium thiosulfate is formed in the solution. The remaining hydrogen sulfide is converted to sulfur dioxide in the catalyst chamber and absorbed in the main wash tower. Depending on the SO_2 content of the gas leaving the catalyst chamber, ammonium bisulfite is formed in the solution, which in turn determines the quantity of hydrogen sulfide absorbed in the prewash. Because of this, the solution composition automatically adjusts to handle variations in feed-gas composition.

2. The inlet-gas stream is split into two portions: the main stream passes directly through the catalyst bed and the sulfite-bisulfite wash; the side stream is bypassed around the catalyst chamber and contacted with the sulfite-bisulfite solution in a separate wash column.

3. A variation of this scheme consists of bypassing the catalyst chamber with part of the gas and recombining the two streams ahead of the wash column.

The basic process and the process using a prewash produce a gas diluted with nitrogen, depending on the volume of air added before the catalytic step. The split-stream process, if operated with two wash columns, produces one stream containing the same percentage of nitrogen as the conventional process and another stream free of nitrogen. The split-stream process with a common wash column produces gas containing less nitrogen than would be obtained in the conventional column, depending on the volume of the side stream. A distinct advantage of the split-stream process is the fact that smaller catalyst chambers are required.

If hydrogen cyanide is present in the gas, it is desirable to saturate the gas with water so that the hydrogen cyanide is hydrolyzed to ammonia and carbon monoxide in the catalyst chamber.

The circulating ammonium sulfite-bisulfite solution is maintained at a salt concentration of 50 to 60 percent by continuous withdrawal of a side stream. This stream is acidified with sulfuric acid, and the ammonium sulfite, bisulfite, and thiosulfate are converted to ammonium sulfate and elemental sulfur by heating under pressure at 290°F. The ammonium sulfate and sulfur obtained are of good commercial quality. Ammonium nitrate or phosphate can also be produced by using nitric or phosphoric acid instead of sulfuric acid. A schematic flow diagram of the process is shown in Figure 8-20.

Design and Operation

The Katasulf process is based on the somewhat surprising discovery that H_2S reacts with oxygen in the presence of hydrogen at approximately 750°F, forming SO_2 and water according to Equation 8-2.

In the presence of certain catalysts and at high space-velocities (66-68) practically complete conversion of H_2S to SO_2 is obtained. At the same time no measurable amounts of hydrogen react with the oxygen.

Since the reaction between hydrogen sulfide and oxygen is highly exothermic, considerable heat is evolved, resulting in an appreciable increase of the gas temperature. The temperature of the gas rises approximately 20°F/45 grains H_2S/100 cu ft, and the heat generated is used to preheat the inlet gas by heat exchange. If a gas containing 225 grains H_2S/100 cu ft is processed, no addition of external heat is necessary.

Catalysts used in the Katasulf process are activated carbon, bauxite, or, preferably, combinations of two metals. One of these metals, e.g., iron, nickel, and copper, combines with the hydrogen sulfide to form metal sulfide and the other metal, e.g., tungsten, vanadium, or chromium, serves as an oxygen carrier. The combined oxygen reacts with sulfur, resulting in the formation of sulfur dioxide. The addition of lead to the catalyst improves its efficiency, and smaller amounts of oxygen-carrier metal can be used. The catalysts are used as alloys and may be formed into wire or screens. The oxides of these metals, supported on suitable carriers, are equally effective.

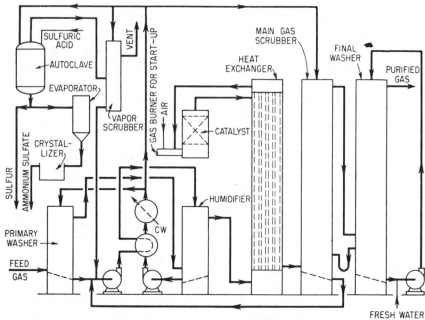

Figure 8-20. Schematic flow diagram of Katasulf process.

The Katasulf process was first tested in a pilot plant treating 500,000 cu ft of gas/day and subsequently commercialized in plants ranging in capacity from 6 million to 27 million cu ft of gas/day. The design of the commercial plants was conventional except that a prewash column ahead of the catalyst chamber was used. Operating results, from a commercial plant processing 6 million cu ft of coke-oven gas/day at Hüls, Germany, are presented in Table 8-6 (65). As can be seen, the purified gas contains less than 1 grain $H_2S/100$ cu ft, and the organic-sulfur content of the gas is reduced by approximately 50 percent. The ammonium sulfate produced in this installation is of good quality, i.e., of light color and fairly large crystal size.

A larger commercial unit, processing 27 million cu ft/day, was designed with three catalytic sections, each treating 9 million cu ft, and a common absorption plant. In order to ensure maximum removal of hydrogen cyanide, the gas was humidified to a water dew point of 100°F between the prewash and the catalyst chamber. Although no detailed operating data are available, it is reported that the catalytic sections of this installation operated at 50 percent above capacity.

The North Thames Gas Board Process for Organic-Sulfur Removal

This process can be considered as an outgrowth of the Carpenter-Evans process which will be discussed in Chapter 13, as the same catalytic material,

Table 8-6. Operating Data of Katasulf Process

Design Variables	In	Out
Substance to be removed; grains/100 SCF		
H₂S....................................	450	0.4–0.8
NH₃....................................	250	0.4–0.8
HCN....................................	50	25*
HCN....................................	50	2.5–5†
Org. S..................................	23	12
Oxygen, %.............................	0.3	0.15
Nitrogen, %............................	3–5	8–10
Gas volume, MMSCF/day...............	6.0	6.5
Heating value, Btu/cu ft...............	580	550

* Dew point, 77°F.
† Dew point, 100°F.

nickel subsulfide, is used. However, the process is basically different from that disclosed by Carpenter and Evans because conversion of organic sulfur compounds is effected by oxidation rather than by reduction. In addition, the catalyst is of a different physical form, lower temperatures are used, and no external preheating is necessary. The process which has been in commercial use at the Harrow Station of the North Thames Gas Board since 1938 has been described by Griffith (69), Plant and Newling (70), and others (71) and is covered by several patents assigned to the North Thames Gas Board (72).

Process Description

The flow scheme of the process as practiced at Harrow is illustrated schematically in the diagram shown in Figure 8-21. After purification in dry boxes the pressure of the gas is raised to 20 to 30 in. water. A small amount of air is added to the gas, which flows through the shell of the heat exchanger and from there to three of four catalyst chambers. Leaving the catalyst chambers, the gas flows through the tubes of the heat exchanger, imparting a portion of its heat to the incoming gas, and from there to the bottom of the absorption tower where it is contacted countercurrently with a dilute solution of sodium carbonate. The absorption tower also serves as a gas cooler. The effluent gas from the absorption tower passes through a conventional dry-box purifier and from there to its ultimate use. The solution is recycled after cooling in a forced-draft cooling tower. In order to maintain proper alkalinity and concentration of salts in the solution, sodium carbonate is added periodically, and a portion of the solution is withdrawn from the system. The catalyst is regenerated in a separate unit by controlled combustion with oxygen-

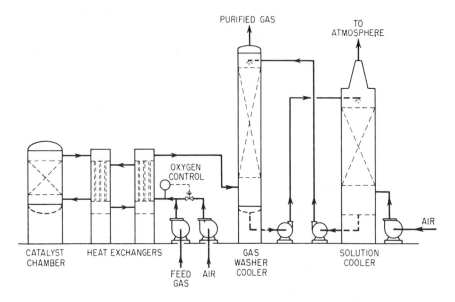

Figure 8-21. Schematic flow diagram of North Thames Gas Board Process for organic sulfur removal.

containing gas. The sulfur oxides evolved are removed in a cast iron washer-cooler by water washing.

Design and Operation

The catalyst consists of nickel subsulfide supported on ¼ by ¼-in. china pellets fired at 1650°F. The pellets are impregnated by boiling in a solution of nickel sulfate, followed by drying at 750°F and sulfiding with hydrogen sulfide contained in coal gas. This method of preparation gives a catalyst of large surface area, containing approximately 8 percent nickel subsulfide.

The kinetics of the oxidation of carbon disulfide and carbonyl sulfide on nickel subsulfides was investigated by Crawley and Griffith (73) who found that carbon disulfide is chemisorbed and oxidized by impact with oxygen, while carbonyl sulfide reacts by collision with chemisorbed oxygen.

The following two reactions are also catalyzed by nickel subsulfide:

$$2H_2 + O_2 = 2H_2O \tag{8-14}$$

$$SO_2 + 3H_2 = H_2S + 2H_2O \tag{8-15}$$

The first of these reactions is strongly exothermic and provides heat for internal temperature-control of the process. At normal operating temperatures, the rate of reaction is proportional to the oxygen concentration of the gas. The oxidation of organic sulfur compounds occurs in spite of the presence of high concentrations of hydrogen which would tend to reduce sulfur dioxide. This reaction (Equation 8-15) only takes place to an appreciable extent at very low oxygen concentrations.

The catalyst is fouled by deposition of gummy and carbonaceous material, the rate depending on the operating temperature. Fouling of the catalyst inhibits the reaction of hydrogen sulfide with oxygen, and when the carbon content of the catalyst reaches approximately 6 percent at operating temperatures of about 700°F, or 2 percent at an operating temperature of 480°F, regeneration is necessary. At the higher temperatures the deposits are carbonized materials which are less objectionable. It is therefore desirable to operate the process at the highest possible temperature. Fouling is caused by picrates, acetylene, diolefins, and cyclopentadiene, while hydrogen cyanide and nitric oxide are not harmful. The catalyst is regenerated by combustion of the deposits with air. The gummy materials begin to react at 470°F, but higher temperatures are necessary for removal of the carbonized materials. The reaction is highly exothermic, and the temperature should not be allowed to exceed 1050°F. During regeneration, the nickel subsulfide reacts with oxygen, forming a mixture of nickel oxide and nickel sulfate which is resulfided by the hydrogen sulfide contained in the gas. If the regeneration temperature reaches 1100°F, the nickel starts reacting with the silica of the china clay, and at 1800°F approximately 10 percent of the nickel is converted to nickel silicate, which is not active as a conversion catalyst.

Thiophene is not decomposed over the nickel subsulfide catalyst although free nickel catalyzes oxidation. However, the formation of free nickel from the catalyst is inhibited by certain compounds present in coal gas.

The heat exchanger used at Harrow is a two-pass shell-and-tube unit, with each pass containing 178 steel tubes 12 ft long, 1.09-in. ID, and on 1.56-in. triangular pitch, located in a cylindrical shell 35.25 in. ID. The gas flow is parallel to the tubes. Heat-exchanger designs projected for future plants of larger capacity than the Harrow installation provide for larger surfaces with lower gas exit-temperatures (70).

The catalyst vessels are 4 ft 6 in. high and 3 ft in diameter; they are provided with conical grids with 45° slope leading to a 2-in. discharge pipe extending through the gas-inlet connection. The vessels are charged with catalyst at a point located in the top cover. Plant and Newling (70) propose an improved design consisting of a single vessel subdivided into compartments and providing for horizontal gas flow. To start up the plant, an air burner is used to heat the catalyst to the operating temperature. Preheating of the catalyst normally requires 5 to 6 hr; after this time the reaction is self-supporting. Im-

Table 8-7. Typical Operating Data of North Thames Gas Board Process

Operating Variables	
Catalytic Step:	
Gas volume, MSCF/day	1,500
Total sulfur in, grains/100 SCF	26–31.5
Oxygen in, %	1.2
Total sulfur out, grains/100 SCF	7.9–10.2
Thiophene out, grains/100 SCF	6–8
Space velocity through catalyst, SCF/(hr)(cu ft of catalyst)	1,000
Temperatures, °F:	
Gas to catalyst	430
Maximum in catalyst	695
Gas to heat exchanger	670
Treated gas to wash tower	295
Wash operation:	
Wash solution circulation, gal/100 SCF	30
pH	7.5
Catalyst regeneration:	
Gas-flow rate, MSCF/hr	10
Oxygen in, %	2
Temperatures, °F:	
Gas in	800
Gas out	1050 (max.)
Duration of regeneration, hr	6
Catalyst life, days	90–120

proved heat-exchanger and catalyst-chamber designs are expected to result in a 20 to 30 percent reduction in oxygen requirements.

The washer-cooler is of conventional design and consists of a vessel 30 ft high and 3 ft in diameter, packed to a height of 18 ft 2 in. with wooden slats. The wash solution enters the vessel at 65°F.

Typical operating conditions of the Harrow installation are shown in Table 8-7 (70).

The exact temperature conditions during the catalytic step depend on the age of the catalyst. Inlet temperatures range from 430° to 570°F, with the optimum probably being nearer to 570°F because of the formation of gummy materials at lower temperatures. Maximum temperatures in the catalyst on the order of 700°F are desirable because no uncarbonized gums can exist at these temperatures.

The scrubbing solution is maintained at a concentration of 1.8 percent sodium carbonate by three daily additions of sodium carbonate and by purging 300 gal solution/million cu ft gas.

Because of side reactions occurring in the catalytic contact, the organic sulfur compounds are decomposed to approximately 15 to 20 percent hydrogen sulfide and 2 to 3 percent sulfur trioxide, the balance being sulfur dioxide.

The residual sulfur in the purified gas consists primarily of thiophene with approximately 1.5 to 2 grains/100 cu ft of other sulfur compounds.

The catalyst is regenerated in a separate vessel. Each batch of 7 cu ft of catalyst is placed in a drum which when inverted serves as an inlet hopper to the kiln. The kiln is refractory-lined and insulated and is provided with a double discharge-cone of special sheet steel. The 7 cu ft of fouled catalyst are placed on a bed of 3½ cu ft of oxidized catalyst left in the kiln from the previous regeneration, and inert gas containing 2 percent oxygen is fed to the kiln at the rate of 10,000 cu ft/hr [30 to 40 cu ft/(hr)(sq ft)]. The temperature in the regeneration vessel is maintained between 800° and a maximum of 1050°F. The oxygen is completely consumed, and the oxidation is monitored by a carbon dioxide indicator. The effluent gas is scrubbed with water in a small packed column, and a portion of the cooled gas is recycled after addition of air by means of a venturi tube.

Small Sulfur-Removing Plants

Small plants utilizing the catalytic process and ranging in capacity from 500 to 12,000 cu ft/hr are being used in locations where relatively small quantities of sulfur-free gas are required. The operation of these units is essentially the same as that of the Harrow plant, except that it is sometimes more economical to preheat the gas externally than to provide heat exchangers.

The Soda-Iron Process for Organic-Sulfur Removal

One of the most widely used processes for the removal of organic sulfur compounds from synthesis gas was disclosed in Germany in 1934 and is known as the "soda-iron process." This process can be considered as a further development of the classical iron oxide process. Its basis is the oxidation of organic sulfur compounds to oxides of sulfur, primarily SO_3, at elevated temperatures over a catalyst consisting of hydrated iron oxide and sodium carbonate. The oxides of sulfur react with the sodium carbonate and are retained on the catalytic material as sodium sulfate. The oxygen necessary for the oxidation of the organic sulfur compounds is supplied by the addition of a small amount of air ahead of the catalytic conversion chambers. The soda-iron process has been used successfully in many German synthetic-fuel plants producing gas with sufficiently low organic sulfur content to be usable in catalytic-synthesis processes.

Process Description

The process as practiced by Ruhrchemie A.G. at Oberhausen-Holten was described by Sands, Wainwright, and Schmidt (74) and is shown diagrammatically in Figure 8-22. The installation consists of a heater, a heat ex-

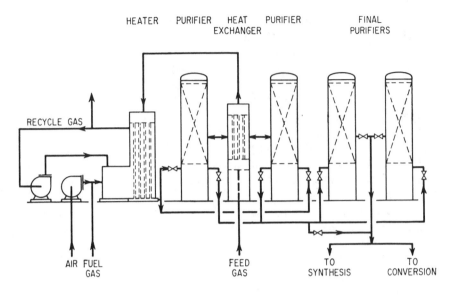

HEATER   PURIFIER   HEAT   PURIFIER   FINAL
                      EXCHANGER   PURIFIERS

RECYCLE GAS

AIR FUEL   FEED   TO   TO
     GAS   GAS   SYNTHESIS   CONVERSION

Figure 8-22. Schematic flow diagram of soda-iron process for organic sulfur removal.

changer, two purification towers in series, and two final purification towers in parallel. At Oberhausen-Holten the gas is purified in five parallel units, each consisting of two purification towers in series, followed by one unit of two final-purification towers in parallel through which the entire gas stream is passed.

The gas is first heated in the heat exchanger by exchange with the outlet gas from the first tower. It is then further heated in an external heater and from there flows through the first purification tower, the heat exchanger, and the second tower to the final purifiers. The first purification tower, containing partially exhausted catalyst, operates at a somewhat higher temperature than the second tower which contains fresh catalyst. The temperature in the final purification towers is again somewhat lower than that in the second purification tower.

Design and Operation

The catalyst used consists of a mixture of natural iron oxide ore with sodium carbonate. Commercial catalysts used in Germany contain small quantities of impurities consisting of oxides of aluminum, titanium, calcium, and silica. The concentration of sodium carbonate generally employed in German installations is on the order of 30 percent, although catalyst compositions containing 70 percent sodium carbonate are also used. In the German process the catalyst is used until approximately 90 percent of the alkaline material is converted to sodium sulfate; it is then discarded. The catalyst appears to be quite effective for the conversion of carbonyl sulfide, carbon disulfide, and mercaptans but does not accomplish anywhere near complete conversion of thiophene.

Although two kinds of purification towers were used at Oberhausen-Holten, it appears that the so-called screen-type tower is preferred. A diagrammatic view of this tower is shown in Figure 8-23. In this vessel, which is approximately 35 ft high and 14 ft in diameter, 77 tons of catalyst are contained in an annular space between two screens with diameters of approximately 6.2 ft and 12.8 ft respectively. The screens are placed in the tower in six sections stacked on top of each other. In the older (tub-type) purifier 66 tons of catalyst were contained in tubs which were also stacked on top of each other. The considerably easier loading and unloading of the screens as compared to the tubs made the use of screens definitely preferable.

The gas is admitted to the tower through one of two nozzles, located at the bottom and the mid-point of the tower respectively, flows through the center of the tower, through the catalyst, and is collected in the annular space between the outer screen and the brick-lined vessel wall, whence it is discharged through the other nozzle. When the position of a tower in series is changed, the gas flow is reversed and the inlet becomes the outlet. In order to discharge catalyst from a tower, the screens are removed through the top of the vessel; the catalyst falls through the center of the tower and is withdrawn through a discharge opening in the tower bottom.

Typical operating conditions for the Oberhausen-Holten plant are shown in Table 8-8 (74).

The space-velocities reported are somewhat lower than those reported elsewhere. Key (75) reports that 10 grains organic sulfur/100 cu ft can be removed from water gas to leave a residue of 0.05 to 0.01 grain/100 cu ft at space-velocities of 200 volumes per volume per hour, and temperatures ranging from 325° to 400°F with a catalyst containing 70 percent sodium carbonate. Experiments conducted by the U.S. Bureau of Mines (76) show that, by operating at a temperature of 480°F, a pressure of 300 psig, and a space-velocity of 450 vol/(vol)(hr), 25 grains of carbonyl sulfide per 100 cu ft could be reduced to less than 0.1 grain per 100 cubic feet in a simulated synthesis gas.

The temperatures in the purification towers are interdependent and have to be increased as the exhaustion of the catalyst progresses. When the first purification tower has reached a temperature of 500°F, the material is discharged and the second tower is placed in the first position, while the first tower is recharged with fresh material and placed in the second position. An average catalyst life of 40 to 50 days in each position can be expected.

While good results were obtained by Ruhrchemie A.G. in the purification of water gas made from coke, much poorer purification was obtained at the

478

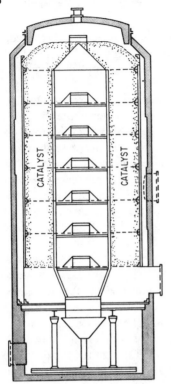

Figure 8-23. Screen-type purification tower used in soda-iron process.

Table 8-8. Operating Data of Soda-iron Plant at Ruhrchemie, Oberhausen-Holten

Gas-flow rate, MSCF/hr...............	550–630
Space-velocity, vol./vol.h*..............	81–92
Temperatures, °F	
First tower.....................	390–500
Second tower.....................	325–400
Final purifiers...................	300–360
Pressure, in. H_2O.....................	12
Organic sulfur, grains/100 SCF:	
In...............................	18–20
After first tower..................	5–10
After second tower................	0.05–0.5
After final purifier...............	0.05
Hydrogen sulfide, grains/100 SCF:	
In...............................	0.5–0.7
After first tower..................	0.05–0.1
After second tower................	0.05
After final purifier...............	0.05
Oxygen in, mole %....................	0.2

* Based on two purification towers.

Wintershall A.G. plant in Lutzgendorf, where gas derived from coal, containing 15 to 45 grains organic sulfur/100 cu ft, could only be purified to an organic sulfur content of 0.5 to 1.0 grain/100 cu ft. The unsatisfactory performance of the soda-iron process in this case can probably be ascribed to the presence of dust and other impurities in the coal gas and, also, to the relatively high thiophene content of the gas. The difficulty in this plant was remedied by placing a bed of activated carbon ahead of the soda-iron catalyst.

The Appleby-Frodingham Process

In this process, which has been described by Reeve of the Appleby-Frodingham Steel Company (77), hydrogen sulfide and organic sulfur compounds are removed from coke-oven gas by absorption on iron oxide in a continuous fluidized system at a temperature of approximately 660°F. The fouled iron oxide which contains about 10 percent by weight of sulfur (as iron sulfide) is regenerated with air at 1110° to 1470°F and recycled to the absorption stage. The sulfur dioxide leaving the regenerator is used for the manufacture of sulfuric acid. A pilot plant treating a daily volume of 2.5 million cu ft of gas containing 600 grains H_2S and 20 grains organic sulfur/100 cu ft is described. Total sulfur-removal efficiencies of 80 percent and 98 percent are obtained by use of a one- and two-stage absorber, respectively. The main advantages claimed for the process are low labor cost and excellent heat economy because two thirds of the heat required is obtained by heat exchange between the inlet and outlet gas and the remaining third is provided by the hot regenerated iron oxide.

References

1. Taylor, D.K. 1956. *Oil Gas J.* 54(79):125; (81):260; (83):133; (84):147.
2. Perry, C.R. 1970. *A New Look at Iron Sponge Treatment of Sour Gas.* Paper presented at Gas Conditioning Conference, University of Oklahoma (Mar. 31-Apr. 1).
3. Griffith, R.H. 1954. *Gas J.* 280:254-256.
4. Ward, E.R. 1964. *Gas Purification Processes.* Edited by G. Nonhebel. London: Georges Newnes Limited, Chap. 8.
5. Smith, B. 1957. "Dry methods for removing hydrogen sulfide from gases." *Trans. Chalmers Univ. Technol., Gothenburg,* No. 184.
6. Moignard, L.A. 1952. "The performance of oxide purifiers." *Gas Council (Gt. Brit.) Research Commun. GC7.*
7. Moignard, L.A. 1955. *Gas World* 142:816-818.
8. Avery, H.B. 1939. *Chem. & Ind. (London)* 58:171.
9. Dent, F.J. and Moignard, L.A. 1950. "The purification of town gas by means of iron oxide." *Gas Research Board, Commun.* GRB 52.
10. Seil, G.E. 1943. *Dry Box Purification of Gas.* New York: American Gas Association, Inc.

11. Morgan, J.J. 1931. *Textbook of American Gas Practices*, 2nd ed. Maplewood, N.Y.: J.J. Morgan.

12. Pearson, G.C. 1929. *Gas J.* 186:797-807.

13. Duwel, G. 1953. *Gas-u. Wasserfach* 94(23):684.

14. Hopton, G.V. 1956. *Chemical Engineering Practice*. Edited by H.N. Crewer and T. Davies. New York: Academic Press, Inc.

15. Taylor, H.F. 1950. *Gas J.* 264:421.

16. Guntermann, W., and Schnurer, F. 1957. *Gas-u. Wasserfach* 100(25):643-649.

17. Bairstow, R. 1957. *Gas J.* 289:243.

18. Hollings, H. 1952. "Progress in gas purification." *Inst. Gas Engrs., Commun. 407.*

19. Moore, D.B. 1956. *Gas World* 143:153-155.

20. Sexauer, W. 1959. *Gas-u. Wasserfach* 100(27):687-690.

21. Robert Dempster & Sons Ltd., Elland, Yorkshire, England. 1958. Private Communication (Apr. 22).

22. Powdrill, J. 1959. *Operation of Liquid and Gastechnik Purification Plants at Cardiff*. The Institution of Gas Engineers.

23. Turner, C.F. 1944. *Oil Gas J.* (Annual Pipeline Number), 42(Sept.23):191, 192, 195, 199.

24. Duckworth, G.L., and Geddes, J.H. 1965. *Oil Gas J.* 63(Sept. 13):94-96.

25. Jolenson, G.E.; Field, J.H.; Decker, W.A.; and Jimeson, R.M. 1962. *U.S. Bur. Mines, Rept. Invest. 6023.*

26. The Pacific Coast Gas Association. 1934. *Gas Engineers' Handbook*. New York: McGraw-Hill Book Company, Inc.

27. Clayton, R.H.; Williams, H.E.; and Avery, H.B. 1931. *Oxide of Iron Purification of Coal Gas*. Miles Platting, England: The Manchester Oxide Company, Ltd.

28. Hollings, H., and Hutchison, W.K. *J. Inst. Fuel* 8:360.

29. Brown, J.M.; Porter, S.C. and Thompson, R.J.S. *The Influence of Frequency of Rotation on the Behavior of Oxide Purifiers*. The Institution of Gas Engineers. Publ. 535.

30. Speers, J.A. 1936. *Gas J.* 214:172.

31. Pyke, H.G., and Monoghan, H.F. 1947. *Australian Nat. Gas Bull.* 11:20.

32. Thompson, R.J.S. 1960. *Gas J. (British)* 301:(Jan.)146-157.

33. Hayston, W., and Wright, D.G. 1955. *Gas J.* 284:727, 731-33.

34. Warr, R.G., and Pickering, E.T. 1952. "Oxide purification in towers." *Inst. Gas Engrs., Commun. 406.*

35. DiRiezo, J.B. 1948. *Am. Gas Assoc., Proc.* 30:622.

36. Vandaveer, F.E. 1951. *Am. Gas Assoc., Proc.* 33:514.

37. Zapffe, F. 1963. *Oil Gas J.* 61(Aug. 19):103-104.

38. Spichal, W. 1953. *Gas-u. Wasserfach* 94(23):679-684.

39. Engelhardt, A. 1928. *Gas-u. Wasserfach* 71(13):290.

40. Kronacher, H.K. 1931. *Gas Age-Record* 68(2):37.

41. Francis, W. 1951. *Engineering* 172(Aug. 10):180.

42. Guyot, G., and Martin, J.E. 1971. *The Sulfreen Process*. Paper presented at the Canadian Natural Gas Processors Association Meeting, Edmonton (June).

43. Lurgi Gesellschaft für Wärme und Chemotechnik. 1971. Private communication.

44. Sawyer, F.G.; Hader, R.N.; Herndon, L.K.; and Morningstar, E. 1950. *Ind. Eng. Chem.* 42(10):1939-1950.

45. Gamson, B.W., and Elkins, R.H. 1953. *Chem. Eng. Progr.* 49(4):203-215.

46. Blohm, C.L. 1952. *Petrol. Engr.* (April):C-68-C72.

47. Grekel, H. 1959. *Oil Gas J.* 57(July 20):76-79.
48. Graff, R.A. 1960. *Oil Gas J.* 58(Oct. 17):118-121.
49. Grekel, H.; Kunkel, L.V.; and McGalliard, R. 1965. *Chem. Eng. Progr.* 61(9):70-73.
50. Valdes, A.R. 1964. *Hydroc. Process.* 43(3):104-108.
51. Valdes, A.R. 1964. *Hydroc. Process.* 43(4):122-124.
52. Valdes, A.R. 1965. *Hydroc. Process.* 44(5):223-229.
53. Giusti, P.G. 1965. *Oil Gas J.* 63(Feb. 22):99-103.
54. Opekar, P.C., and Goar, B.G. 1966. *Hydroc. Process.*45(6):181-185.
55. Grekel, H., and Kilmer, J.W. *Oil Gas J.* 66(44):88-101(Oct. 28, 1968).
56. Chute, A.E. 1968. *Trans. AIME Soc. Mining Eng.* 241(Sept.):314-318.
57. Peter, S., and Woy, H. 1969. *Chemie Inginieur Technik* 41(1,2):1-6.
58. Beavon, D.K., and Cameron, D.J. 1970. *Gas Process. (Can.)* 62(May/June):16-21.
59. Estep, J.W.; McBride, G.T.; and West, J.R. 1962. *Advances in Petroleum Chemistry and Refining,* vol. 6. New York: Interscience Publishers.
60. Beavon, D.K., and Vaell, R.P. 1971. *Prevention of Air Pollution by Sulfur Plants.* Paper presented at Eighth Annual Technical Meeting, Southern California Section, AIChE (Apr. 20).
61. Carmassi, M.J., and Zwilling, J.P. 1967. *Hydroc. Process.* 46(Apr.):117.
62. Anon. 1951. *Oil Gas J.* 50(4):59.
63. Audas, F.G. 1951. *Coke and Gas* 13:229.
64. Haines, H.W.; Van Wielingen, G.A.; and Palmer, G.H. 1961. *Petrol. Refiner* 40(Apr.):123-126.
65. Bahr, H. 1938. *Chem. Fabrik* 11(1/2):10.
66. 1930. German Patent 510,488.
67. 1933. German Patent 576,137.
68. 1936. German Patent 634,427.
69. Griffith, R.H. 1952. *Ind. Eng. Chem.* 44(5):1011.
70. Plant, J.H.G., and Newling, W.B.S. 1948. "The catalytic removal of organic sulfur compounds from coal gas." *Inst. Gas Engrs, Commun. 344.*
71. Anon. 1953. *Gas. J.* 276:461-462.
72. British Patents 489,398; 528,848; 529,711; 535,482; 554,691; 556,169; 597,501; 601,320; 619,585.
73. Crawley, B., and Griffith, R.H. 1938. *J. Chem. Soc.* 720.
74. Sands, A.E.; Wainwright, H.W.; and Schmidt, L.D. 1948. *Ind. Eng. Chem* 40(4):607.
75. Key, A., and Eastwood, A.H. 1945. Forty-fifth Report of the Joint Research Committee of the Gas Research Board and the University of Leeds, Discussion and Reply, p.37.
76. Sands, A.E.; Wainwright, H.W.; and Egelson, G.C. 1950. "Organic sulfur removal in synthesis gas, occurrence, determination and removal." *U.S. Bur. Mines Rept. Invest. No. 4699* (July).
77. Reeve, L. 1958. *J. Inst. Fuel* 31:319.
78. Boas, A.H., and Andrade, R.C. 1971. *Hydroc. Process.* 50(Mar.):81-84.
79. Pearson, M.J. 1976. *Energy Processing/Canada* (July-August):38-42.
80. Cameron, L.C. 1974. *Oil Gas J.* 72(25):110-118.
81. Grancher, P., and Nougayrede, J. 1976. *Pollution Atmospherique* No. 70 (April/June):129-134.
82. Nougayrede, J. 1976. *Sulphur* No. 127(Nov./Dec.):37-41.

83. Goddin, G.S.; Hunt, E.B.; and Palm, J.W. 1974. *Hydrocarbon Processing* 53(10):122-124.

84. Nobles, J.E.; Palm, J.W.; and Knudtson, D.K. 1977. "Design and Operation of First AMOCO CBA Unit," paper presented at Gas Conditioning Conference, Norman, Oklahoma (March 7-9).

85. Beavon, D.K., and Leeper, J.E. 1977. *Sulphur* No. 133(Nov./Dec.):35-39.

86. Dalla Lana, J.G. 1973. *Gas Processing/Canada* (March-April):20-25.

87. Pearson, M.J. 1973. *Hydrocarbon Processing* 52(2):81-85.

88. Norman, W.S. 1976. *Oil Gas J.* 74(46):50-60.

89. Grancher, P. 1978. *Hydrocarbon Processing* 57:7, July (Part 1) and 57:9, September (Part 2).

90. Maadah, A. G., and Maddox, R. N. 1978. *Hydrocarbon Processing* 57:(8), August p. 143.

91. Taggart, G. W. 1980. *Hydrocarbon Processing* 59:(4), April p. 133.

92. Pearson, M. J. 1978. *Hydrocarbon Processing* 57:(4) April p. 99.

93. Pearson, M. J. 1981. *Hydrocarbon Processing* 60:(4) April p. 131.

94. Dupin, T. and Voirin, R. 1982. *Hydrocarbon Processing* 61:(11), Nov. p. 189.

95. Wiley, S. 1980. *Hydrocarbon Processing* 59:(4), April p. 127.

96. Beavon, D. K. 1977. U.S. Patent 4,038,036.

97. Heigold, R. E., and Berkeley, D. E. 1983. "The MCRC Sub-Dewpoint Sulfur Recovery Process," paper presented at 1983 Gas Conditioning Conference, University of Oklahoma.

98. Watson, E. A., Hartley, D., and Ledford, T. H. 1981. *Hydrocarbon Processing* 60:(5), May p. 102.

99. Lagas, J. A. 1982. *Hydrocarbon Processing* 61:(10), October p. 85.

100. Kerr, R. K., Jagodzinzki, R. F., and Dillon, J. 1982. "The RSRP: A New Sulfur Recovery Process," paper presented at the 1982 Gas Conditioning Conference, University of Oklahoma, March 8-10 1982.

9

Liquid Phase Oxidation Processes for Hydrogen Sulfide Removal

The disadvantages inherent in the dry purification processes caused a search for more efficient methods for hydrogen sulfide removal from industrial gases as early as the beginning of this century. The primary objectives of this work were to reduce the large ground-space requirements and high labor costs of the purification plants and to increase the value of the sulfur produced. Quite naturally the technology turned to purification methods employing liquids in regenerative cycles and capable of yielding pure elemental sulfur. One important goal was the development of processes which would permit the utilization of the sulfur contained in coal gases for absorption of ammonia and subsequent production of ammonium sulfate. Such a process, which ideally would effect simultaneous removal of hydrogen sulfide and ammonia, would eliminate the need to purchase sulfuric acid for ammonia removal and thus result in an appreciable improvement of the economics of coal-gas purification. Most early processes were used only for limited periods of time or were designed for specific installations and therefore have little significance from the point of view of general utility.

On the other hand, several more recently developed desulfurization processes based on oxidation of hydrogen sulfide in liquid media have become firmly established in the gas industry, and the number of operating installations is growing steadily. Processes of this type are primarily applicable to the treatment of gases containing relatively low concentrations of hydrogen sulfide in the presence of substantial concentrations of carbon dioxide. The high ratio of carbon dioxide to hydrogen sulfide in the acid-gas portion of the gas to be treated would make processing of the acid gas in a Claus plant rather difficult if such gases were treated in a process using an absorption-stripping cycle (e.g., ethanolamines). The selectivity of the liquid oxidation processes for hydrogen sulfide overcomes this difficulty.

However, a serious drawback of such processes is the fact that the considerable reaction heat generated by the oxidation of hydrogen sulfide (see Chapter 8) has to be dissipated at a low-temperature level and cannot be used, as in the Claus process, for generation of valuable steam. A further drawback of these processes is the relatively low capacity of the solutions for hydrogen sulfide, requiring large liquid circulation rates and large facilities for handling the precipitated sulfur. In general, plants of this type are not economical if the amount of sulfur to.be removed from the gas exceeds about 10 tons per day. Processes of general interest will be discussed in some detail in this chapter, while others will be mentioned briefly, mainly for historical reasons.

Oxidation of hydrogen sulfide to elemental sulfur is achieved by oxygen carriers dissolved or suspended in aqueous or nonaqueous liquid media. The liquid processes are therefore based on the same chemical mechanism as the dry oxidation processes described in the preceding chapter. The following compounds are used as oxygen carriers, either in suspensions or solutions:

1. Polythionates
2. Iron oxide
3. Thioarsenates
4. Iron-cyanide complexes
5. Organic compounds (alone or in conjunction with heavy metal salts)
6. Sulfur dioxide
7. Potassium permanganate and sodium or potassium dichromate

The first six of these agents are used in regenerative systems while permanganate and dichromate solutions are employed in nonregenerative processes.

Polythionate Solutions

The early development of liquid oxidation processes using polythionate solutions for the removal of hydrogen sulfide from gases derived from coal is to a large extent identified with the work of Feld in Germany. Feld started his studies before the outbreak of World War I, and his principal aim was to devise a process by which hydrogen sulfide and ammonia could be removed simultaneously from coal gas and subsequently converted to ammonium sulfate and elemental sulfur. The chemistry of this process can be illustrated schematically by the following overall equation showing the oxidation of ammonium sulfide to ammonium sulfate.

$$(NH_4)_2S + 2O_2 = (NH_4)_2SO_4 \qquad (9\text{-}1)$$

The cyclic process which he proposed involves the absorption of hydrogen sulfide and ammonia in aqueous solutions of ammonium tri- and tetrathionate which are thus converted to ammonium thiosulfate and sulfur. The spent solution is regenerated by addition of sulfur dioxide which reacts with ammonium thiosulfate forming again tri- and tetrathionate. The regenerated solution is recycled for further absorption of hydrogen sulfide and ammonia. When the solution contains 30 to 45 percent thiosulfate, it is subjected to a final treatment with SO_2 and heated, and the polythionates are converted to ammonium sulfate, SO_2, and elemental sulfur. Unconverted thiosulfate remaining in the solution reacts with polythionate and is decomposed into ammonium sulfate and elemental sulfur.

A great deal of difficulty was encountered with this process primarily because of the numerous chemical reactions involved. Proper functioning was not only dependent on the concentration of H_2S and NH_3 (two moles of ammonia per mole of hydrogen sulfide) but also on the maintenance of close temperature control because of the complicated solubility relationships of the various salts present in the system. Feld also experimented with solutions containing zinc and iron polythionates, and, more recently, Terres and coworkers described a process using manganese polythionates and manganese sulfate (1,2). The complicated chemistry of the various polythionate processes has been described in some detail by Terres (1).

Other processes based on the original work of Feld are the Gluud combination process and the Koppers C.A.S. process (3). In the Gluud combination process the hydrogen cyanide-free gas is washed with a solution containing a mixture of thio- compounds and iron sulfate. The spent solution is first saturated with sulfur dioxide and then aerated in a tall tower where the hydrosulfides and the sulfites are converted to thiosulfates; these are then used for further absorption of hydrogen sulfide. During the cyclic operation thiosulfates accumulate, and a portion of the solution is withdrawn continuously for conversion of the thiosulfates to ammonium sulfate and elemental sulfur.

In the Koppers C.A.S. process—symbolizing cyanogen, ammonia, sulfur—the hydrogen cyanide is first removed with a recirculating solution containing ammonia and elemental sulfur, with the formation of ammonium thiocyanate. Hydrogen sulfide and ammonia are then removed from the gas by contact with a solution containing ammonium polythionate, ammonium sulfite, ammonium thiosulfate and some iron compounds. This solution is circulated in three towers operated in series, and ammonia is injected into the solution at various points of the system to obtain complete H_2S removal. The following scheme is used to regenerate the solution and to produce ammonium sulfate and elemental sulfur.

A portion of the solution is withdrawn from all four towers and divided into three parts. One part is regenerated by aeration during which iron sulfide and ammonium hydrosulfide oxidize to form elemental sulfur. The second part is treated with sulfur dioxide, and the sulfur formed is recovered by filtration. The third part is also treated with sulfur dioxide, having first been decanted from the solids. The resulting liquid is then combined with the filtrate from the second part of the solution and heated under pressure at 350°F. In this step the thiocyanate and the other thio- compounds are converted to ammonium sulfate and elemental sulfur.

Many variations of these processes were developed in Europe, but most of them did not prove to be practical in commercial operation. An excellent review of these processes is presented by Hill (4). In spite of this considerable effort, extended over many years, development of a satisfactory commercial polythionate process was not achieved.

Iron Oxide Suspensions

A logical step in the development of processes employing liquids in regenerative cycles was the utilization of the reaction between iron oxide and hydrogen sulfide followed by conversion of iron sulfide to iron oxide and elemental sulfur. Several processes using iron oxide suspended in alkaline aqueous solutions were developed in Europe and the United States, beginning with the work of Burkheiser shortly before the first world war. During the 1920s the Ferrox process was introduced by the Koppers Company of Pittsburgh, Pennsylvania, and an almost identical process was disclosed by Gluud. More recently a modification of the Ferrox process was developed in England and is known as the Manchester process. At present the Burkheiser process is not used commercially although its possible utility in a novel scheme of coal-gas purification has been considered by Terres (1). Most Ferrox plants have been replaced by installations using more efficient processes, although a few plants of this type are still operating in the United States. The Gluud process is used to some extent on the continent. The Manchester process, which for a while enjoyed some popularity in Great Britain, is being gradually replaced by the more efficient Stretford process, discussed later in this chapter.

Basic Chemistry

The chemistry of all these processes is based on the reaction of H_2S with an alkaline compound, either sodium carbonate or ammonia, followed by the reaction of the hydrosulfide with iron oxide. Regeneration is effected by converting the iron sulfide to elemental sulfur and iron oxide by aeration. This portion of the cycle involves essentially the same reactions as those occurring in dry-box purifiers. The following equations represent the reaction mechanism:

$$H_2S + Na_2CO_3 = NaHS + NaHCO_3 \tag{9-2}$$

$$Fe_2O_3 \cdot 3H_2O + 3NaHS + 3NaHCO_3 = Fe_2S_3 \cdot 3H_2O \qquad (9\text{-}3)$$
$$+ 3Na_2CO_3 + 3H_2O$$

$$2Fe_2S_3 \cdot 3H_2O + 3O_2 = 2Fe_2O_3 \cdot 3H_2O + 6S \tag{9-4}$$

Besides the main reactions, several side reactions (mostly leading to the formation of undesirable sulfur compounds) occur in the process, depending on the operating conditions and the composition of the gas to be treated. Usually a certain amount of thiosulfate formation is inevitable. In some cases it may even be desirable to operate the processes in such a manner that hydrogen sulfide is quantitatively converted to thiosulfate according to the following equations:

$$2NaHS + 2O_2 = Na_2S_2O_3 + H_2O \tag{9-5}$$

$$Na_2S + 1\tfrac{1}{2}O_2 + S = Na_2S_2O_3 \tag{9-6}$$

A further side reaction in which sulfur is also converted to an undesirable product is caused by the absorption of hydrogen cyanide in the alkaline material. The hydrogen cyanide is first converted to sodium cyanide and then oxidized by elemental sulfur to thiocyanate:

$$HCN + Na_2CO_3 = NaCN + NaHCO_3 \tag{9-7}$$

$$NaCN + S = NaSCN \tag{9-8}$$

Normally only a small portion of the hydrogen cyanide reacts in this manner, and, therefore, this reaction is not a major source of sulfur loss. The majority of the absorbed hydrogen cyanide is stripped from the solution by the regeneration air.

The presence of hydrogen cyanide in the gas to be purified leads to still another side reaction which may have considerable influence on the operation of the process. It was observed (5) that very noticeable color changes occur in the solution when gas containing relatively large amounts of hydrogen cyanide (approximately 10 percent of the hydrogen sulfide) are treated. The oxidized solution displays a blue coloration indicating the presence of ferric-ferrocyanide complexes, such as Prussian blue, while the fouled solution is pale yellow. In addition, it was noticed that, while the reaction between hydrogen sulfide and iron oxide is quite slow, the presence of blue complexes results in rapid conversion of hydrogen sulfide to elemental sulfur. Although the exact chemical nature of the blue and yellow compounds is not known, it is hypothesized that the reactions involve oxidation of H_2S to elemental sulfur by conversion of the ferric-ferrocyanide complex to ferrous ferrocyanide. In the regeneration step the ferric-ferrocyanide complex is reestablished. The reactions occurring in the cycle can be represented schematically by the following equations:

$$2H_2S + Fe_4[Fe(CN)_6]_3 + 2Na_2CO_3 \qquad (9\text{-}9)$$
$$= 2Fe_2Fe(CN)_6 + Na_4Fe(CN)_6 + 2H_2O + 2CO_2 + 2S$$

$$2Fe_2Fe(CN)_6 + Na_4Fe(CN)_6 + O_2 + 2H_2CO_3 \qquad (9\text{-}10)$$
$$= Fe_4[Fe(CN)_6]_3 + 2Na_2CO_3 + 2H_2O$$

It is quite probable that under these conditions hydrogen sulfide does not react at all with iron oxide and that these reactions are the only ones occurring. The iron, which in most cases is added to the solution as soluble iron sulfate, serves only to replenish the iron-cyanide compounds lost with the sulfur.

The Burkheiser Process

This process, which was developed in Germany at approximately the same time as Feld conducted his work, is described in some detail by Terres (1). The difference between the two processes is that in the Feld process H_2S and ammonia are absorbed simultaneously while in the Burkheiser process they are removed in two consecutive steps. However, only the H_2S-removal portion of the process was successful and applied on a commercial scale.

Hydrogen sulfide and hydrogen cyanide are absorbed in an aqueous solution containing ammonia, iron oxide, and elemental sulfur. The spent solution leaving the absorber is introduced into a "sulfur dissolving" vessel where the suspended free sulfur is converted to ammonium polysulfide by the action of gaseous ammonia and hydrogen sulfide. Subsequently, the iron sulfide is removed from the solution by filtration and regenerated by contact with atmospheric oxygen. The iron oxide and elemental sulfur formed in this operation are resuspended in an aqueous solution of ammonia and recycled for further absorption of hydrogen sulfide. The filtrate, which contains ammonium polysulfide, ammonium cyanide, ammonium thiocyanate, and ammonia, is heated to approximately 200°F, and the ammonium polysulfide is decomposed into ammonia, hydrogen sulfide, and elemental sulfur. The gaseous ammonia and hydrogen sulfide are absorbed by the spent solution in the "sulfur dissolving" vessel and used to convert elemental sulfur to ammonium polysulfide as indicated above. The free sulfur is separated from the filtrate and the residual solution, containing cyanides and thiocyanates, is treated with a suspension of calcium hydroxide. The precipitated calcium cyanide and calcium thiocyanate are filtered and added to the coal used in gas manufacturing. During gasification the cyanogen compounds are converted to hydrogen sulfide and ammonia. The advantage of this rather complicated process is that the only end products of gas purification are elemental sulfur and ammonia.

The Ferrox Process

Although the process proposed by Burkheiser was not generally accepted, additional processes using suspensions of metal oxides in alkaline solutions were developed independently in Europe and the United States. Besides iron, nickel was considered to be an active agent for hydrogen sulfide removal. However, nickel forms soluble salts with hydrogen cyanide from which it cannot be regenerated for further use. Because of this and its relatively high price, nickel was never used on a large scale. The perfection of flotation techniques for recovering elemental sulfur after oxidation was an important step in commercializing these processes.

The Ferrox process was disclosed by Sperr of the Koppers Company, Pittsburgh, Pennsylvania (5, 6) and subsequently used on a fairly extensive scale. Although most Ferrox plants have been replaced by installations using other processes, a few are still in operation.

The Ferrox process is a marked improvement over dry-box purification because the plants occupy only a fraction of the ground area necessary for dry boxes treating equivalent volumes of gas. In addition, the labor cost is reduced appreciably, and the initial installation cost is somewhat lower than that for dry-box purifiers. The principal disadvantage of the process is the fact that complete removal of H_2S cannot be obtained as readily and regularly as by the use of dry boxes. Operating costs of a typical Ferrox installation may be estimated from Table 9-1 (5).

The Ferrox process may also be considered an improvement over the Seaboard process, discussed earlier, because more complete hydrogen sulfide elimination is obtained, while only small amounts of carbon dioxide are removed at the same time.

Process Description

A schematic flow diagram of the Ferrox process is shown in Figure 9-1. The solution normally containing 3.0 percent sodium carbonate and 0.5 per-

Table 9-1. Operating Requirements of Ferrox Process

Type of gas	Coal gas
Plant capacity, MMSCF/day	10.0
H₂S removed, grains/100 SCF	400
HCN removed, grains/100 SCF	40
Labor, hr/day	32
Power, hp-hr/day	2,550
Sodium carbonate, lb/day	3,500
Iron compound, lb/day	2,800

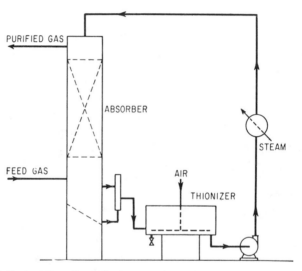

Figure 9-1. Typical flow diagram of Ferrox process.

cent ferric hydroxide is pumped to the top of the absorber where it is counter-currently contacted with the gas fed into the bottom of the vessel. The hydrogen sulfide-containing solution flows from the bottom of the absorber to the thionizer, or regenerator, where elemental sulfur is formed by contact of the solution with air. The sulfur accumulates on the liquid surface as a froth, enters the slurry tank, and is pumped from there to a filter where excess liquid is removed. The regenerated solution is pumped from the thionizer through a heater to the absorber, thus completing the cycle. The liquid obtained in the filter is usually discarded, thus providing a means for continuously purging the system of undesirable salts.

Design and Operation

The absorber used in Ferrox installations consists of two portions, a lower portion, or saturator, and an upper portion, the absorber proper. The saturator contains a continuous liquid phase, several feet high, through which the raw gas is passed before it enters the upper section. The function, of the saturator is to provide sufficient contact time to complete the reaction between sodium hydrosulfide and ferric oxide before regeneration of the solution. If essentially complete reaction is achieved, thiosulfate formation in the regenerator is kept at a minimum. The upper part of the absorber contains sprays and wooden hurdles similar to those used in the Seaboard process and usually has a total height of 60 ft. (5).

The thionizer consists of long shallow tanks, each containing several compartments arranged so that the solution may be transferred from one compartment to another. The compartments contain directional baffles which ensure proper flow of the solution along the total length of the tanks. The depth of the liquid in the thionizers is approximately 42 in. Air is admitted to the bottom of the thionizer and divided into fine bubbles by passage through cloth-covered tubes. Typical cloth tubes are about 5 in. in diameter by 10 ft in length and are mounted on steel pipes which are manifolded above the thionizer tank (5).

The liquid is circulated at such a rate that a two- to threefold excess of ferric hydroxide over the stoichiometric quantity necessary for the complete reaction with hydrogen sulfide is present. Gollmar (3) states that the process can be operated with less than the stoichiometric concentration of iron oxide and interprets the function of the iron as a catalytic oxygen carrier. Data from several plants indicate that operation with an excess of iron oxide over the stoichiometric properties is commonly practiced. This excess seems to be required for complete removal of hydrogen sulfide and, also, to minimize thiosulfate formation in the thionizer.

The air requirements for oxidation of ferric sulfide are stated to be 300 cu ft/1,000 cu ft of gas; this would be equivalent to a ratio of about 10 moles of oxygen to 1 mole of hydrogen sulfide for a gas containing 400 grains hydrogen sulfide per 100 cu ft, assuming complete removal (5). The air requirements depend to a large extent on the proper oxygen utilization in the thionizer. Since theoretically only ½ mole of oxygen is required per mole of H_2S, it is evident that improved design of thionizers should permit the use of appreciably smaller quantities of air.

The efficiency of hydrogen sulfide removal by this process varies from 85 to practically 100 percent. Sperr (5) reports that hydrogen sulfide concentrations in the treated gas, sufficiently low to satisfy the U.S. Bureau of Standards lead acetate test, can be achieved in a single absorber. However, when high-purity gas is required, two-stage absorption is recommended. Depending on the hydrogen cyanide content of the gas and the raie of thiosulfate formation, 70 to 80 percent of the hydrogen sulfide can be recovered as elemental sulfur.

The solids obtained in the filters contain from 30 to 50 percent elemental sulfur, approximately 50 percent moisture, and 10 to 20 percent salts, mostly entrapped ferric hydroxide and sodium carbonate. Because of this loss of sodium carbonate and ferric hydroxide, these chemicals have to be added continuously to the solution. This affects the economics of the process materially, and a point may be reached where the quantity of recoverable sulfur is so small that regeneration of the solution is uneconomical. In the United States there is no market for the sulfur obtained from a Ferrox plant, a fact which should be considered in deciding whether to operate a plant on a regenerative or nonregenerative basis.

One of the major drawbacks of the Ferrox process is the corrosiveness of the treating solution which causes fairly rapid destruction of carbon-steel equipment. The use of alloys is uneconomical in most installations, but lining of the major vessels with rubber and the possible use of coated wooden tanks for thionizers may be considered.

A modification of the Ferrox process has been described by Gard (7) and Bailey (8) of Union Oil Company of California. The design and operation of this process are appreciably different from the original process described by Sperr (5), and, therefore, the operating requirements shown in Table 9-1 are not applicable. Union Oil Company uses the Ferrox process to purify three natural gas streams containing hydrogen sulfide within the range of 40 to 100 grains per 100 cu ft and carbon dioxide between 4 and 26 percent. Operating pressures are 80 to 160 psi. Although under normal operating conditions essentially all of the hydrogen sulfide is removed from the feed gas, the Ferrox unit is followed by dry-box purifiers in order to ensure continuous production of pipeline quality gas. The plant consists of six contactors, 12 regeneration troughs, a liquid surge pit, and the necessary auxiliary equipment. The contactors vary in diameter from 82 to 108 in. and in height from 20 to 24 ft. For corrosion control the contactors are gunite lined, and the gas inlet spargers are made from stainless steel. Plastic pipe is used for the lines connecting the contactors and the regeneration troughs. The regeneration troughs are made of redwood and are about 5 ft deep, 24 ft long and 2½ ft wide. Air diffusers are located at the bottom of the troughs.

The sour gas enters the bottom of the contactors through a sparger and bubbles through a column of liquid, the level of which is maintained at approximately 5 ft below the the top of the contactor. Fresh solution enters the contactor with the feed gas. Spent solution is withdrawn through a draw-off tray and sent to the regeneration troughs where it is aerated, and the elemental sulfur formed is skimmed from the liquid. The regenerated liquid is then returned to the contactors. A photograph of this installation is shown in Figure 9-2.

Figure 9-2. Ferrox plant used for removal of hydrogen sulfide from natural gas. *Union Oil Company of California*

The Gluud Process

This process was developed in Germany independently of the American Ferrox process (9). The chemical reactions in this process are the same as those of the Ferrox process with the exception that a dilute solution of ammonium carbonate is used instead of a sodium carbonate solution. The principal difference between the two processes is that Gluud employs tall regenerators which permit much more efficient utilization of oxygen. This, however, is accompanied by the necessity to pump air against a higher pressure head. While 300 cu ft air/1,000 cu ft gas are required in a typical Ferrox installation, the equivalent plant using the Gluud process is said to require only 30 cu ft air/1,000 cu ft gas for solution regeneration (9). Assuming again a typical coal gas, containing 400 grains hydrogen sulfide per 100 cu ft, this would be equivalent to about one mole of oxygen per mole of hydrogen sulfide. A schematic flow diagram of the Gluud process is shown in Figure 9-3.

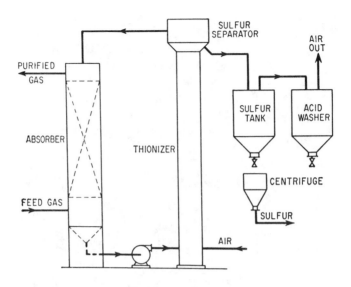

Figure 9-3. Typical flow diagram of Gluud process.

The Manchester Process

A modification of the Ferrox process was developed in England at the Rochdale Works of the Manchester Corporation Gas Department and is known as the Manchester process. This process, which is covered by British Patents 550,272 and 611,917, has been used in several installations primarily in British gas works. In one installation, the Manchester process is used to remove 250 ppm of hydrogen sulfide from the exhaust air from a viscose cellulose manufacturing plant. In this installation, which has a capacity of more than 80 million cu ft air/day, the hydrogen sulfide content of the air is reduced to approximately 5 ppm (10).

Process Description

The principal difference between the Manchester and Ferrox processes is the use of multistage treatment with fresh solution fed to each washing stage in the Manchester process, whereas a single contact is used in the Ferrox process. In order to ensure completion of the reaction between hydrogen sulfide and iron oxide the Manchester process provides for a separate delay vessel between the absorbers and the regenerators. The regenerators are tall vessels providing for relatively long contact times between the solution and the air, which is introduced by means of several rotary or turbo diffusers.

Design and Operation

The design and operation of a typical Manchester plant located at Linacre, Liverpool, is described in detail by Townsend (11). In this plant, which processes about 3 MMscf gas/day, the gas passes consecutively through six cylindrical absorption towers 7 ft 6 in. in diameter by 25 ft in overall height. The towers are packed with wooden boards, and the liquid is distributed in the first two and last two towers by rotating distributors and in the middle two towers by serrated troughs. Fresh solution is pumped into each tower and collected in a flooded manifold whence it flows into the delay tanks. The delay tanks are of sufficient size to provide for the residence time of 7 to 10 min which appears to be necessary to drive the reaction between sulfide ion and iron oxide to completion. From the delay tanks, the solution flows to the oxidizers which are 10 ft in diameter by 25 ft in overall height. The effective height of the oxidizers is approximately 16 ft because of the arrangement of

Figure 9-4. Manchester process plant, treating 12 million SCF/day of coal gas. Absorbers shown in foreground. *Southwestern Gas Board, England, and W.C. Holmes & Co., Ltd.*

the diffusers in the bottom of the vessels. The elemental sulfur liberated in the regenerators is removed as a froth from the top of the vessels and collected in sludge-receiving tanks. From there, the sulfur is processed further and finally recovered in 90 to 95 percent pure form, the contaminants being primarily iron oxide, sodium carbonate, some thiosulfate, and water. Figure 9-4 shows a photograph of a Manchester plant based on this design. In a modified design (12) rectangular vessels are used in a somewhat different arrangement than that described by Townsend. A photograph of this type of plant is shown in Figure 9-5.

Typical operating data of two plants, each treating about 3 million cu ft gas daily, are shown in Table 9-2.

Horizontal spray absorbers may be used for removing relatively small amounts of H_2S from large volumes of air by means of iron oxide suspensions (13). An absorber of this type which is used to reduce the H_2S content of 7.5 million cu ft/hr of exhaust air from a viscous rayon plant from

Figure 9-5. Manchester process plant treating 3 million SCF/day of coal gas. Rectangular absorbers and delay tanks shown. *R. & J. Dempster, Ltd.*

Table 9-2. Operating Data of Manchester Process		
Design variables	Plant A (11)	Plant B (12)
Gas-flow rate, MSCF/hr	125	140
Liquid-flow rate, gpm	2,400	1,700
Air-flow rate (to oxidizers), MSCF/hr	54	21
Temperatures, °F:		
Inlet gas	.....	80
Outlet gas	.....	86
H_2S, grains/100 SCF:		
In	738	705
After 1st stage	328	190
After 2d stage	102	90
After 3d stage	17	15
After 4th stage	1	trace
After 5th stage	0.11	nil
After 6th stage	0.04	nil
Suspended iron in sol., % Fe_2O_3	0.046	0.033
Soluble iron in sol., % Fe	0.16	
Alkalinity, N	0.23	0.3
Thiosulfate, N	0.18	
"Blue," %	0.11	
Reagents used, lb/MMSCF		
Copperas	344	
Sodium carbonate	207	

300 to 13 ppm is shown in Figure 9-6. The absorber is 55 ft long, 30 ft wide, 15 ft high, and semicircular in cross section. The solution (133 gpm) is sprayed upwards at a 45° angle from 140 orifice jets placed at ground level along each side of the vessel. The main advantages of this type of absorber as compared to vertical, packed vessels are lower construction cost, lower power consumption for solution pumping, and much lower pressure drop (0.3 in. of water for the installation shown).

Thioarsenate Solutions

The Thylox Process

The Thylox process was disclosed by Gollmar and Jacobsen (14, 15, 16) in the late twenties and commercialized by the Koppers Company, Inc. This process was used extensively for many years, especially for the purification of coke-oven and other manufactured gases. A partial list of Thylox plants operating in 1945 shows a daily gas volume of 266 million cu ft, resulting in the production of about 60 long tons of sulfur. In recent years, especially

Figure 9-6. Horizontal spray absorber for removal of hydrogen sulfide from exhaust air
Courtaulds, Ltd.

since the advent of large-scale use of natural gas and the development o
more efficient processes for desulfurization of natural and manufactured gas·
es, the Thylox process has lost a great deal of its prominence.

The early application of the Thylox process was aimed at about 80 to 90
percent hydrogen sulfide removal, but essentially total hydrogen sulfide
removal could be achieved through subsequent process improvements. One
variation of the process, the so-called "modified Thylox process," is
applicable in cases where complete purification of gases containing small
amounts of hydrogen sulfide is required.

A neutral or slightly alkaline solution, containing sodium or ammonium
thioarsenate as the active ingredient, is used in the Thylox process. Hydrogen
sulfide is converted to elemental sulfur of sufficiently high purity to be usable
in agriculture (17). Since the sulfur contains less than 0.5 percent arsenic, it
can also be used as a raw material for the manufacture of various chemicals.

The Thylox process offers some economic advantages over processes
previously discussed in this chapter. The consumption of alkali due to
thiosulfate formation is reduced markedly, and the sulfur is produced in a
much more valuable form. Operating costs for a plant treating 5 million cu
ft/day of refinery gas containing 1,000 grains hydrogen sulfide/100 cu ft may
be estimated from Table 9-3 (18).

Process Description

A basic flow diagram of the Thylox process is shown in Figure 9-7. The gas
enters at the bottom of the absorber and is washed countercurrently with the
solution entering at the top of the vessel. Essentially all of the hydrogen sul-

Table 9-3. Operating Requirements of Thylox Process

Type of gas......................	Refinery or natural
Plant capacity, MMSCF/day...........	5.0
H₂S content, grains/100 SCF...........	1,000
Sulfur removal, %....................	98
Gas pressure, psig....................	60
Labor (operating and maintenance), hr/day..........	14
Power, kw-hr/day.............................	1,200
Steam, lb/day................................	15,000
Sodium carbonate, lb/day.......................	600
Arsenic trioxide, lb/day........................	150
Credit: sulfur recovered, tons/day................	3

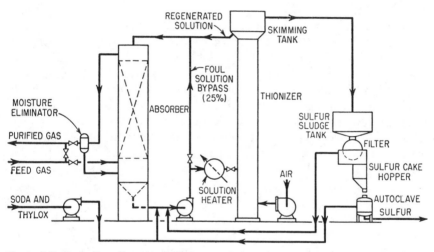

Figure 9-7. Basic flow diagram of Thylox process.

fide and hydrogen cyanide are removed in this operation. The foul solution is pumped from the bottom of the absorber through a heater, where it is heated to approximately 110°F, into the bottom of the thionizer where it flows upwards and is regenerated by a cocurrent stream of air. The air not only releases sulfur but also acts as a flotation agent for the sulfur which collects at the surface of the solution as a froth. The liquid level in the thionizer is maintained below a weir over which the sulfur froth flows to the sulfur-sludge tank. The regenerated liquid returns by gravity flow to the top of the absorber. The sulfur sludge leaves the sludge tank and is passed through the filters, whence the sulfur cake is further processed for final product preparation. The sulfur can be recovered as a wet paste, a dry powder, or as cast crude sulfur. Pure sulfur can be obtained by distillation of the crude product. Various means for sulfur recovery are discussed extensively by Gollmar (3). The filtrate is returned to the foul-solution stream, or it can be partly or totally discarded, thus serving as a system purge. Normally a small portion of the solution is continuously withdrawn from the system to prevent the accumulation of thiosulfates and thiocyanates. The arsenic may be reclaimed from this liquid by adding acid and filtering the arsenic sulfide formed. The arsenic sulfide is dissolved in sodium carbonate solution and returned to the system. A photograph of a Thylox plant is shown in Figure 9-8.

Various modifications of the basic process have been developed, including two-stage absorption for more complete hydrogen sulfide removal and bypassing a portion of the foul solution (approximately 25 percent) around the thionizer. In the two-stage absorption process, freshly prepared Thylox solution is used in the second absorber. A plant using two-stage absorption is described by Powell (19).

The so-called "modified Thylox process" (20) is nonregenerative and is recommended for installations where gas of very high purity is required, and the quantities of hydrogen sulfide to be removed are quite small. A flow diagram of this process is shown in Figure 9-9. Fresh Thylox solution is prepared in a mixing tank and pumped to the top of the absorber where it contacts the hydrogen sulfide-containing gas. The fouled solution is withdrawn from the bottom of the absorber and enters a neutralizing tank where acid is added to precipitate arsenic sulfide. After the sludge settles, the clear water is discarded, and the sludge is usually treated for recovery of the arsenic.

Basic Chemistry

The chemistry of the Thylox process is described in detail by Gollmar (21). The actual reaction mechanism taking place during the various phases of the process is probably quite complicated because of the possible existence of a large variety of ionic species. The principal reactions involve replacement of one atom of oxygen by one atom of sulfur in the thioarsenate molecule during absorption, and the reverse during regeneration. The reactions can be symbolized by the following equations:

$$Na_4As_2S_5O_2 + H_2S = Na_4As_2S_6O + H_2O \qquad (9\text{-}11)$$
$$\text{(Absorption)}$$

$$Na_4As_2S_6O + \tfrac{1}{2}O_2 = Na_4As_2S_5O_2 + S \qquad (9\text{-}12)$$
$$\text{(Regeneration)}$$

Figure 9-8. Thylox gas-purification plant for removing hydrogen sulfide and producing sulfur. *Koppers Company*

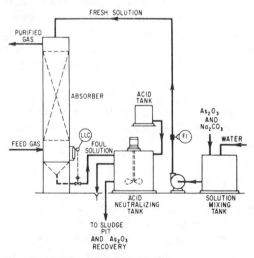

Figure 9-9. Flow diagram of the modified Thylox process.

These two reactions take place at a rapid rate and are undoubtedly the main reactions occurring under most operating conditions. In cases where gas containing very high concentrations of hydrogen sulfide is treated, or when long contact-times are provided, other, much slower reactions (Equations 9-13 and 9-14) may also take place to some extent.

$$Na_4As_2S_6O + H_2S = Na_4As_2S_7 + H_2O \tag{9-13}$$
$$\text{(Absorption)}$$

$$Na_4As_2S_7 + \tfrac{1}{2}O_2 = Na_4As_2S_6O + S \tag{9-14}$$
$$\text{(Regeneration)}$$

The solution is prepared by dissolving arsenic trioxide and sodium carbonate, in the proportion of 2 moles of sodium carbonate to 1 mole of arsenic trioxide, in water. The resulting solution contains sodium carbonate and bicarbonate, sodium arsenite, and arsenious acid which, upon alternate reactions with hydrogen sulfide and oxygen, are converted to sodium thioarsenate, $Na_4As_2S_5O_2$. During oxidation of the arsenite the pH is lowered and carbon dioxide is gradually expelled. When the solution is completely oxidized, it contains practically no carbon dioxide. In preparing the solution, it is important to maintain a ratio of a least 2 atoms of sodium to 1 atom of arsenic. If insufficient sodium is present, the pH decreases below 7.5 and arsenic sulfide precipitates. The process operates best within a pH range of 7.5 to 8.0. Ammonia can be substituted for the sodium carbonate without changing the characteristics of the process.

Part of the sulfur is converted to thiosulfate, although the rate of formation is appreciably lower in the essentially neutral Thylox solution than in more alkaline solutions used in other processes. Hydrogen cyanide, which is absorbed in the absorber, reacts readily with the sulfur formed in the thionizer to yield sodium thiocyanate. Because of these side reactions, the active thioarsenate has to be replenished continuously by addition of arsenic oxide and sodium carbonate.

The presence of relatively large amounts of carbon dioxide—8 percent and more—in the gas does not interfere with the removal of hydrogen sulfide. The Thylox solution is not sufficiently alkaline to absorb carbon dioxide to an appreciable extent.

Design and Operation

The absorber is a steel tank packed with wooden hurdles; it is 90 ft high and up to 20 ft in diameter, depending on the gas throughput. Since most Thylox plants operate at essentially atmospheric pressure, large vessels are required. The thionizer, which operates under slight pressure, is usually an empty vessel of much smaller diameter and a height of 120 ft.

A somewhat different thionizer design is reported by Powell (19). Thionizers of this type are 66 ft high and have an outer shell 12 ft 6 in. in diameter. Inside of this shell is a concentric inner shell, 6 ft in diameter and slightly higher than the outer shell. The annular space between the two shells is divided into two equal compartments by means of two vertical partitions, and each compartment is divided into three sections by radially arranged vertical baffles. The solution flows upward through one compartment, then downward through the center and upward again through the second compartment. Air is admitted only to the compartments in the annular space and the air flow is guided by the vertical baffles.

Thylox solution is somewhat corrosive; therefore, pumps and valves should have stainless-steel trim, and stainless-steel clad tube sheets, and stainless-steel tubes should be used in solution heaters. Other equipment is essentially all carbon steel, with corrosion allowance for the foul solution.

The circulation rate of the treating solution is such that considerable excess of arsenic oxide, over the stoichiometric quantity required to react with the hydrogen sulfide, is maintained. This is necessary because of the incomplete regeneration obtained in the thionizers. In general, 4 to 5 moles of thioarsenate (measured as As_2O_3) per mole of hydrogen sulfide are circulated through the absorber. If the solution is properly reactivated, the vapor pressure of hydrogen sulfide over the solution is extremely small, and it is possible to produce a purified gas containing 0.2 to 0.3 grain hydrogen sulfide per 100 cu ft gas. In most installations, however, there is no attempt to reach such low concentrations in a single-stage absorption process.

Reactivation of the solution is also carried out with an excess of oxygen over the stoichiometric quantity required. The rate of air flow reported in the literature varies considerably from plant to plant, but it appears that about 5 moles of oxygen per mole of hydrogen sulfide are required for proper solution reactivation.

The considerable heat of reaction generated in the oxidation of hydrogen sulfide to elemental sulfur is dissipated by evaporation of water in the thionizer. As long as solutions of low arsenic concentration are used and, consequently, large volumes of liquid are circulated, no appreciable temperature increase of the solution is noticed.

Operating results from three Thylox plants were reported by Powell (19), Denig (22), Farquhar (23), and McBride (24). The plant described by McBride employs a solution containing ammonium thioarsenate instead of sodium thioarsenate. Operating results from three of these installations are presented in Table 9-4.

From these data it can be seen that approximately 95 percent hydrogen sulfide removal can be obtained with a single absorber, while essentially total hydrogen sulfide removal is possible in a two-stage absorption system.

Foxwell and Grounds (25) report operating data of two Thylox plants treating coal gas and blue water-gas respectively. Ammonium thioarsenate

Table 9-4. Operating Data of Thylox Plants			
	Plant*		
Design variables	A (23)	B (19)	C (24)
Gas-flow rate, MSCF/hr....................	210	900	2,200
Solution-flow rate, gpm:			
Primary solution........................	280	1,350†	5,000‡
Secondary solution......................	....	850	
Air-flow rate, MSCF/hr....................	16.4	34	12
Steam rate, lb/hr.........................	1,100		
Temperature, °F:			
Inlet gas..............................	73	....	75
Outlet gas............................	83	....	100
Solution to absorber....................	97	95	
Composition:			
H_2S content, grains/100 SCF:			
Inlet gas..............................	316	61	160
Outlet gas:			
Primary absorber....................	9	3.5	3
Secondary absorber..................	....	0.25	
HCN content, inlet gas, grains/100 SCF.........	12		
As_2O_3 content, %:			
Primary solution.......................	0.7	0.3	
Secondary solution.....................	....	0.8	
Sulfur recovered, %......................	64.0		

*Plant B was operating at less than 50% of rated capacity when the operating data were obtained.
†315 gpm bypassed around thionizer.
‡25% bypassed around thionizer.

solutions are used in a single-stage absorption process. In the plant treating coal gas, the hydrogen sulfide is reduced from 580 grains/100 cu ft to 8 grains/100 cu ft or 98.6 percent. In the other installation, where blue water-gas is treated, 99.99 percent hydrogen sulfide removal—from 410 grains to 0.06 grain/100 cu ft—is achieved.

A process used in Russia, which is practically identical with the Thylox process, is described in some detail by Jegorov, Dimitriev, and Sikov (26). Besides a rather detailed description of process equipment, certain design criteria and a sample calculation of a typical plant for the treatment of coke-oven gas are presented.

Absorbers used in Russian installations are cylindrical vessels packed with wood slats about ½ in. thick and 4 to 5 in. high. The slats are arranged in horizontal decks in which individual slats are spaced about ½ in. apart. Linear gas velocities of 2 to 4 ft/sec and pressure drops of 2 to 3 in. of water are used for the design of typical absorbers operating at essentially atmospheric pressure. An absorption coefficient, based on the logarithmic mean driving force in the gas, with the numerical value of 2.5 to 3.5 g H_2S/(sq m)(hr)(g H_2S/cu m) [approximately 0.08 to 0.12 lb H_2S/(sq ft)(hr)(atm)] is employed. In accordance with European practice, the absorption coefficient K_G is used alone in conjunction with packing area rather than the volume coefficient $K_G a$ which is more commonly employed in the United States. Use of this coefficient results in the calculation of the packing surface-area which then must be multiplied by the area per unit volume to get the required absorber volume. The treating solution, which contains 8 g/liter of As_2O_3 and 10 g/liter of Na_2CO_3, is circulated at a rate equivalent to a 50 percent excess over the stoichiometric quantity required.

The regenerator described by Jegorov et al. consists of an empty tower, 100 to 120 ft high, containing several metal screens for redistribution of the air. The volume of the vessel is based on a solution residence time of 40 to 50 min and the diameter on an air velocity of 500 to 800 cu ft/(hr)(sq ft) of tower cross section. The air-flow rate is equivalent to a ratio of 2.5 moles of oxygen per mole of hydrogen sulfide.

An interesting part of the Russian process is a two-step method for the complete recovery of arsenic from solution waste-streams. In the first step, which is similar to the recovery method used in the Thylox process, the solution is heated to 70°C (158°F), and arsenic sulfide is precipitated by the addition of 75 percent sulfuric acid. The precipitate is separated from the liquid by filtration, dissolved in aqueous sodium carbonate, and returned to the circulating solution-stream. The clear liquid is passed to the second step where it is made alkaline with sodium carbonate solution and treated with a solution of ferric sulfate. In this operation the small amount of arsenic remaining in the solution after the first step is precipitated as ferric arsenite and arsenate. The precipitate is removed by filtration, and the filtrate, which contains about 10 to 20 ppm of arsenic, is either discarded or processed for recovery of thiosulfate.

Wooden tanks lined with acid-resistant materials are used in both steps of the arsenic-recovery operation. Each tank is sized for a solution residence-time of 4 hr and provided with a mechanical agitator.

The Giammarco-Vetrocoke Process

This process, which also uses aqueous sodium or potassium carbonate solutions containing arsenic compounds for the absorption of hydrogen sulfide, was first disclosed by Giammarco (27) and later described in some detail in the technical literature (28-31). Like many other processes of this type, the Giammarco-Vetrocoke process is primarily used for the desulfurization of coke-oven and synthesis gases; however, at least one large installation utilizing the process for selective removal of small quantities of hydrogen sulfide from natural gas is in operation in the United States (29,30). Although the presence of arsenic salts in the treating solution suggests a similarity with the Thylox process, it is claimed that the chemistry and operating characteristics of the two processes are fundamentally different. The process is stated to be capable of producing purified gas containing less than 1 ppm of hydrogen sulfide even when operated at absorption temperatures up to 300°F and in the presence of substantial concentrations of carbon dioxide in the gas to be treated. Furthermore, it is claimed that by proper choice of the operating conditions, formation of side-reaction products, primarily thiosulfate, can be kept to a minimum. Finally, the treating solution reportedly is noncorrosive toward carbon steel, thus obviating the use of alloys in any portion of the plant in contact with the solution.

Basic Chemistry

The chemistry of the process is probably quite complex, but the overall reaction mechanism of the absorption-regeneration cycle can be represented by the following equations:

$$KH_2AsO_3 + 3H_2S = KH_2AsS_3 + 3H_2O \qquad (9\text{-}15)$$

$$KH_2AsS_3 + 3KH_2AsO_4 = 3KH_2AsO_3S + KH_2AsO_3 \qquad (9\text{-}16)$$

$$3KH_2AsO_3S = 3KH_2AsO_3 + 3S \qquad (9\text{-}17)$$

$$3KH_2AsO_3 + 1\tfrac{1}{2}O_2 = 3KH_2AsO_4 \qquad (9\text{-}18)$$

The overall reaction which can be written as shown in Equation 9-19 is oxidation of hydrogen sulfide to elemental sulfur, achieved by exchange of one oxygen atom for one sulfur atom in the pentavalent arsenic compound.

$$3H_2S + 3/2\,O_2 = 3S + 3H_2O \qquad (9\text{-}19)$$

The first step of the process, absorption of H_2S, as represented by Equation 9-15, is rapid, and the rate of absorption is favored by an excess of arsenite. The equilibrium vapor pressure of H_2S over the solution is very low so that it is possible to produce treated gas of very high purity even at elevated absorption temperatures.

In the second step thioarsenite reacts with arsenate forming monothioarsenate (Equation 9-16). This is claimed to be the most critical step in the process as the monothioarsenate is stable toward oxygen, and, thus, formation-of thiosulfate is largely prevented. The stoichiometry indicates that the presence of 1 mole of pentavalent arsenic is required for each mole of hydrogen sulfide absorbed. However, in order to drive the reaction to completion, an excess of arsenate is recommended. This reaction is relatively slow, requiring substantial periods of time, but the reaction rate can be increased by an excess of arsenate and by increasing the temperature.

In the third step represented by Equation 9-17, monothioarsenate is decomposed into arsenite and elemental sulfur. This is achieved by lowering pH of the solution, either in a separate operation or simultaneously with the last step, oxidation of arsenite to arsenate. In the first case the solution is treated with a stream of carbon dioxide, usually at somewhat elevated pressure, in order to convert essentially all of the carbonate to bicarbonate, resulting in sufficient lowering of pH to precipitate elemental sulfur. This procedure is used when treating solutions of relatively high pH are utilized as is common for treatment at elevated temperatures. With solutions of rather low pH, the increase in acidity resulting from the formation of arsenate during oxidation leads to the precipitation of elemental sulfur.

The final step of the process shown in Equation 9-18 is reoxidation of trivalent to pentavalent arsenic, usually by contact with air. The rate of the oxidation reaction is quite slow but may be markedly increased by addition of certain catalysts.

The treating solutions used in the Giammarco-Vetrocoke process vary over a considerable range of concentrations. Jenet (29) reports sodium or potassium carbonate concentrations ranging from 0.5 to 15 percent, presumably with corresponding concentrations of arsenic compounds.

Process Description

A typical flow diagram of the Giammarco-Vetrocoke process is shown in Figure 9-10. The diagram depicts the basic form of the process where acidification of the solution and oxidation of trivalent to pentavalent arsenic take place simultaneously in the regenerator or oxidation vessel. In the other version where carbon dioxide is used for solution acidification, a separate vessel is located between the "digester" and the regenerator in which the solution is contacted with carbon dioxide.

The gas enters the bottom of the absorber where it is contacted countercurrently with the treating solution. Essentially all of the H_2S contained in the feed gas and some of the other impurities, such as hydrogen cyanide, are removed in this operation. The rich solution leaves the bottom of the absorber and flows to a surge vessel called "digester," where the relatively slow conversion of thioarsenite to monothioarsenate is completed. From the digester the solution flows to the oxidizer where trivalent arsenic is oxidized to pentavalent arsenic by contact with air and from there is recycled to the

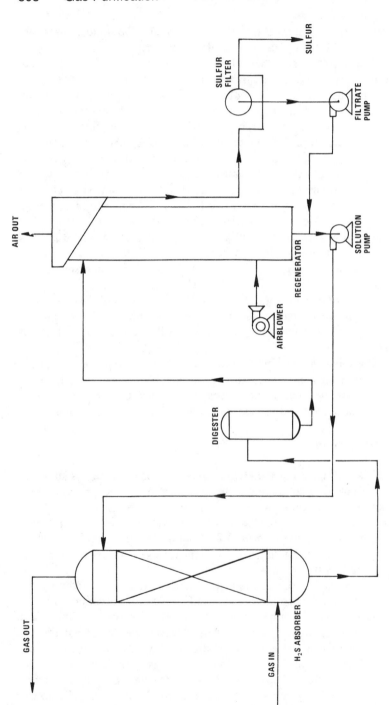

Figure 9-10. Typical flow diagram of Giammarco-Vertrocoke process.

absorber. Elemental sulfur is formed in the oxidizer and removed from the solution by flotation. A froth containing about 10 percent sulfur and 90 percent solution is withdrawn from the top of the oxidizer and further treated either in a rotary vacuum filter or in a centrifuge. The filter cake, usually containing about 50 percent solids, is washed with fresh water which is then discarded. The filtrate is recycled to the process.

After washing, the sulfur contains about 0.3 percent arsenic (as arsenite and thioarsenate) on a dry basis (31). If required the sulfur can be further purified by solvent extraction. In general, especially in cases where small amounts of sulfur are produced in the process, the sulfur paste is disposed of as such. However, in some cases it is economical to further process the paste in an autoclave and to produce liquid or solid sulfur of relatively high quality.

In spite of the slow rate of undesirable side reactions, thiocyanates (if the feed gas contains hydrogen cyanide) and sulfates gradually build up in the solution. These compounds are removed by treatment in an autoclave for destruction of thiocyanates and by concentration and precipitation of sulfate (31).

Design and Operation

The absorber may be any efficient liquid-gas contacting device such as a packed tower or a column provided with multiple spray nozzles. Since it is claimed that no solids precipitate during absorption, no special provisions are needed to prevent plugging. The digester may be located in the bottom of the absorber or may be a separate vessel of sufficient capacity to allow the reaction to go to completion. The acidifier, if required, may be a stirred vessel or a column provided with baffle trays.

Empty towers of sufficient height to permit effective sulfur flotation are used as oxidizers. Vacuum filters and centrifuges are satisfactory although in small installations filters appear to be more economical. Typical operating data for the Giammarco-Vetrocoke process reported by Jenet (29) are shown in Table 9-5. A comparison of operating characteristics of the Giammarco-Vetrocoke process with those of other liquid oxidation processes is given in Table 9-6 (28).

Iron Cyanide Solutions and Suspensions

Several processes utilizing iron-cyanide complexes as oxidation agents were developed in Europe shortly before or during World War II. All of these processes are identical with respect to the basic chemistry involved in the absorption-regeneration cycle. Trivalent iron is reduced to divalent iron in the absorption step, and the divalent iron is reoxidized during regeneration. Hydrogen sulfide is oxidized to elemental sulfur, which is recovered in a

Table 9-5. Typical Operating Data for Giammarco-Vetrocoke Process

Alkalı	Na or K
Alkali Concentration, wt.%	0.5 - 15
Solution capacity, cu.ft. H_2S/Gal.	0.15 - 2.8
Air requirement, cu.ft./1,000 grains H_2S	15 - 20
Maximum absorption Temperature, °F	300
H_2S in treated gas, ppm	0.01 - 1.0

Table 9-6. Process Comparison

Process	Giammarco-Vetrocoke	Thylox	Ferrox	Manchester
Pressure, psig	Atm.	Atm.	Atm.	Atm.
Inlet gas temp., °F	100-300	100	100	100
H_2S in, grains/100 SCF	300-500	300-500	300-500	500-1,000
H_2S out, grains/100 SCF	<0.1*	10†	5	0.25‡
Solution capacity, grains/gal.	600	40	70	10

*One-stage absorption.
†One-stage absorption; 0.25 grain/100 SCF claimed for two-stage absorption.
‡Six-stage absorption with fresh solution to each stage.

manner similar to that used in the Thylox process. The reactions are symbolized in the following equations.

$$2Fe^{+++} + H_2S = 2Fe^{++} + S + 2H^+ \qquad (9\text{-}20)$$
$$\text{(Absorption)}$$

$$2Fe^{++} + \tfrac{1}{2}O_2 + H_2O = 2Fe^{+++} + 2OH^- \qquad (9\text{-}21)$$
$$\text{(Regeneration)}$$

The differences between the processes lie in the type of iron-cyanide complexes used and the method of regeneration. In two processes, that of the Gesellschaft für Kohlentechnik and the Fischer process, alkaline aqueous solutions of potassium ferricyanide and ferrocyanide are used, and regeneration is carried out by contact with air and electrolysis, respectively. The other two processes of this category, the Staatsmijnen-Otto and the Autopurification processes, employ suspensions of complex ferric-ferrocyanide compounds in alkaline solutions which are regenerated by air contact. The latter two processes are essentially identical, although they were developed independently in the Netherlands and in England, respectively. The main advantages claimed for these processes are (a) complete removal of hydrogen sulfide and (b) the production of relatively pure sulfur. An additional advantage may be the fact that hydrogen cyanide, which is present in most coal gases, is utilized in the process. In many cases sufficient hydrogen cyanide can be obtained from the gas, and the requirements for chemicals consist solely of the occasional addition of iron sulfate.

The iron-cyanide processes have not been employed on a large industrial scale, and only a few publications describing their operation have appeared in the literature. The Fischer (32) process has been described by Mueller (33) and Thau (34). The Staatsmijnen-Otto process has been discussed in detail, especially with respect to its complex chemistry, by Pieters and van Krevelen (35). A study on the operation of a plant using the Autopurification process has been presented by Craggs and Arnold (36).

The Fischer Process

This process, which was developed and commercialized at the gas works at Hamburg, Germany, utilizes an aqueous solution containing about 20 percent potassium ferrocyanide and 6 percent potassium bicarbonate. The solution is subjected to electrolysis, which converts a portion of the ferrocyanide to ferricyanide and an equivalent amount of the bicarbonate to carbonate, releasing at the same time a proportionate volume of hydrogen. The presence of both ferricyanide and carbonate enables the solution to absorb H_2S rapidly and to convert it immediately to elemental sulfur. The reactions involved can be expressed by the following equations:

$$2K_4Fe(CN)_6 + 2KHCO_3 = 2K_3Fe(CN)_6 + 2K_2CO_3 + H_2 \qquad (9\text{-}22)$$
$$\text{(Electrolysis)}$$

$$2K_3Fe(CN)_6 + 2K_2CO_3 + H_2S = 2K_4Fe(CN)_6 + 2KHCO_3 + S$$
$$\text{(Absorption)} \qquad (9\text{-}23)$$

The overall reaction can also be written differently, indicating the decomposition of hydrogen sulfide to hydrogen and sulfur.

$$H_2S = H_2 + S \qquad (9\text{-}24)$$

In practice, the process as described by Mueller (33) and Thau (34) operates as follows. The crude gas is contacted with the regenerated solution in a centrifugal contactor, and the H_2S is converted to sulfur of very small particle size. A centrifugal contactor was chosen because of plugging difficulties experienced with equipment of conventional design. The spent solution is pumped to a settling tank and from there flows to a filter press in which the sulfur is removed. If a large settling tank is used, only a portion of the solution has to be filtered with the rest of the liquid being decanted. The combined sulfur-free liquids are pumped to a specially designed battery where regeneration is achieved by electrolysis. The regenerated solution is then recycled to the contactor.

Operating data of the plant at the Hamburg gasworks show that the regenerated solution contains about 2 to 5 percent potassium ferricyanide. Approximately 2,000 gal/hr are circulated to treat an hourly volume of about 100,000 cu ft of water-gas containing 175 grains $H_2S/100$ cu ft. Electric power consumption of about 1.8 kwhr/lb of sulfur is rather high (twice the theoretical required) but may be justified in view of the good quality of the sulfur produced.

The Staatsmijnen-Otto and Autopurification Processes

Process Description

In these processes, which are practically identical, hydrogen sulfide is removed—primarily from coal gases which also contain ammonia and hydrogen cyanide—by contact with an ammoniacal solution containing suspended ferric-ferrocyanide complexes, usually referred to as "iron blue." The solution also contains ammonium salts which are necessary to stabilize the cyanide complexes and, after long periods of use, thiocyanate and thiosulfate. Regeneration of the spent solution is carried out by compressed air in a tall aerating tower. Because of the exothermic nature of the oxidation, reaction heat is liberated in the regenerator and causes an increase in the solution temperature. After leaving the regenerator, the oxidized solution flows through a cooler and from there is returned to the absorber.

Elemental sulfur separates from the solution during regeneration and is collected as a froth at the top of the regenerator. The sulfur froth, which contains 70 percent water, some ammonia, and a small amount of "blue," is filtered and washed for removal of the bulk of the impurities. It is then heated with water in an autoclave at a temperature above its melting point and recovered as fairly pure molten sulfur. An extremely pure product may be obtained by further heating of the sulfur at about 600°F—at which temperature organic impurities are decomposed—followed by distillation.

The chemistry of the process is rather complicated, primarily because of the complex behavior of the "blue" and the many side reactions which may occur, depending on the gas composition and operating conditions. A very extensive study of process variables was made by Pieters and van Krevelen (35). Basically, during absorption, hydrogen sulfide reacts with ammonia to form ammonium hydrosulfide. In the regeneration portion of the process, the hydrosulfide is oxidized to elemental sulfur by reduction of the "blue," which acts as an oxygen carrier. The reduced "blue" is then reoxidized. It is obvious that this interpretation is an oversimplification of the mechanism actually occurring, especially if it is kept in mind that the solution contains dissolved iron-cyanide salts which can react directly with the hydrogen sulfide.

Process Operation

It is claimed that H_2S is removed quantitatively and for the greater part converted to elemental sulfur. However, a certain amount of thiosulfate formation does take place, especially at high pH and low concentrations of "blue." In some instances it may even be desirable to operate the process in such a way that the H_2S is completely converted to thiosulfate.

Hydrogen cyanide contained in the gas is absorbed quantitatively and converted to ammonium thiocyanate. In order to avoid losses of sulfur caused by this reaction, the hydrogen cyanide may be removed from the gas, before it enters the desulfurization plant, and converted to alkali ferrocyanide. Since the process requires continuous ferrocyanide makeup, to compensate for losses of "blue" with the sulfur, this procedure is economically advantageous. The hydrogen cyanide may be absorbed by a solution of alkali carbonate in a cast-iron vessel filled with iron filings. If the absorption temperature is about 200°F, HCN reacts rapidly with the iron and H_2S and CO_2 are not absorbed. A portion of the solution is withdrawn from the vessel at regular intervals, and the ferrocyanide is salted out by adding alkali carbonate. The crystals are removed and the remaining solution is returned to the absorber.

Typical operating data of the Staatsmijnen-Otto process are presented in Table 9-7. These data were obtained at a plant operated by the Société Carbochimique at Tertre, Belgium; this consists of two parallel installations, each composed of a contactor and regenerator, and one final-purification installation, also containing a contactor and regenerator.

Operating experience with a small plant utilizing the Autopurification process has been reported by Craggs and Arnold (36).

The plant is located at Billingham, England, and processes an hourly volume of 6,000 cu ft of coke-oven gas containing about 400 grains $H_2S/100$ cu ft. The composition of the treating solution is shown in Table 9-8.

Operational tests conducted over a period of 4 years resulted in the conclusion that although the plant was capable of producing gas containing less than 1ppm of H_2S, the operation was erratic, with occasional H_2S contents of as high as 200 ppm in the treated gas. Laboratory experiments conducted in parallel with the plant tests revealed a fairly good correlation between the ammonia content of the solution and the H_2S content of the outlet gas. It was found that the treating efficiency of the process is much more sensitive to the ammonia content than to the total iron and cyanide contents of the solution.

Table 9-7. Typical Operating Data of Staatsmijnen-Otto Process (35)	
Gas flow rate, MSCF/hr	1,400
Solution flow rate, gpm	3,500–5,300
Air flow rate, MSCF/hr	56
Cooling-water flow rate, gpm	1,750
Gas contents, grains/100 SCF:	
H_2S in inlet gas	170–210
H_2S after first contactor	0.5
H_2S after second contactor	0
HCN in inlet gas	15–26
HCN after first contactor	0
NH_3 in inlet gas	200–250
Solution composition, g/liter:	
Total solids	300–400
"Blue"	1.0 (min)
Chemical consumption:	
Ferrous sulfate, lb/day	330
Sulfur recovery, %	30–60

Table 9-8. Composition of Plant Solution	
Constituent	g/liter
Iron (total)	0.15–2.25
CN^-	0.25–1.34
Free NH_3	0.60–5.00
Dissolved salts	3.50–15.60
Free S	0.20–0.50
Total suspended solids	1.20–6.00
pH	8.1–9.0

As a result of these studies, a solution containing 2.0 g/liter of total iron, 2.8 g/liter of cyanide, and 4.0 g/liter of ammonia was specified for optimum performance.

Another shortcoming of the process was the large amount of air required for solution regeneration. The plant is equipped with a tall regenerating tower, and an air volume equal to about 15 percent of the gas volume treated was required for adequate regeneration. Pilot-plant studies of the regeneration step indicated that more efficient regeneration could be obtained by relatively shallow tanks operated in parallel, instead of the tall aeration tower, and distribution of the air by means of impeller aerators.

A comparative cost analysis indicated that, at the time of the study, the Autopurification process was not competitive with dry-box purification.

Iron Complex Solutions

Since iron has proved to be an excellent agent for removing hydrogen sulfide from gas streams, attempts have been continuing to devise methods for the use of iron in solutions without the disadvantages inherent in handling suspensions, as in the Ferrox and Manchester processes, and toxic materials as those utilized in processes based on iron cyanide complexes. Several processes have been developed based on the use of chelating agents to hold the iron in solution and prevent precipitation. In these processes, the iron acts as a catalyst for the oxidation of absorbed hydrogen sulfide to elemental sulfur. The overall reaction is symbolized by Equations 9-20 and 9-21.

Basic Data

Macák et al. (69) conducted a survey of chelating agents and selected ethylene-diamino-tetraacetic acid (EDTA) for pilot and demonstration plant testing. This compound was found to be very stable under conditions anticipated for commercial plants. In the absence of oxygen, the free chelating agent is stable to about 250°C. This temperature is reduced to about 170°C if oxygen is present and is further reduced to about 120°C when the chelating agent is in the form of the iron complex.

For H_2S absorption systems, it is necessary that the iron chelate scrubbing solution be maintained at a pH in the range of about 7.0 to 9.0. This can be accomplished by the addition of buffering salts or ammonia. If the alkalinity is too high (above about 10), precipitation of the metal hydroxide can occur. On the other hand, some alkalinity is required to promote the initial absorption of H_2S to yield HS^-. The reaction of HS^- with the EDTA chelate of iron can be depicted as follows:

$$2\,(FE^{3+}\ EDTA^{4-})^- + HS^- + H_2O$$
$$= 2\,(Fe^{2+}\ EDTA^{4-})^{2-} + S + H_3O^+ \qquad (9\text{-}25)$$

The elemental sulfur is precipitated as very fine particles. Elevated temperature operation (about 50°C) was found to improve the filterability of the precipitated sulfur and also to increase the rate of the reactions (69).

The various commercial processes differ somewhat in the method of removing elemental sulfur from the circulating liquid. Small quantities may be removed by filtering the process fluid and discarding the filter cake as waste. Larger quantities require a more complex treatment. The process typically involves an initial settling or flotation step to produce a sulfur concentrate

(5–10 percent solids) followed by filtration or centrifugation with a water wash to produce a relatively clean dry cake (∿30 percent water). When liquid "Claus plant type" sulfur is desired, the wet sulfur is heated with low-pressure steam in an autoclave to produce a lower phase of pure liquid sulfur and an upper phase of dilute salt solution which may be discarded or returned to the scrubber circuit.

Reactivation of the solution is accomplished by contact with oxygen which causes the following reaction to occur:

$$4 \ (Fe^{2+} \ EDTA^{4-})^{2-} + 4 \ H_3O^+ + O_2 = 4 \ (Fe^{3+} \ EDTA^{4-})^- \\ + 6 \ H_2O \quad (9\text{-}26)$$

If the H_2S is to be removed from an oxygen-containing gas, the oxidation reaction will occur in the absorber and a separate regeneration step is not required. More commonly, the H_2S is present in an oxygen-free gas stream such as natural gas and it is necessary to continuously remove solution from the scrubber and oxidize it with air in a separate vessel. The air oxidation step is very slow due primarily to the low solubility of oxygen gas, and adequate liquid residence time must be provided for complete conversion of ferrous iron to the ferric state. Between 30 and 60 min is recommended for a completely reduced complex (69).

In pilot-scale tests of the EDTA-based process (69), 90 to 99 percent removal of H_2S was achieved from a gas-containing 20–70 g/m³ H_2S; 70–90% CO_2; and up to 0.1 percent O_2. A venturi scrubber was used for the tests with a scrubbing solution containing 0.08–0.12 mol/l EDTA, 0.07–0.10 mol/l Fe, and 30–70 g/l $NaHCO_3$.

In addition to the desired reactions which result solely in the conversion of H_2S to elemental sulfur, some side reactions are inevitable. The most significant is the oxidation of a portion of the dissolved H_2S to thiosulfate by the reaction:

$$2 \ HS^- + 2O_2 = S_2O_3 = + H_2O \quad (9\text{-}27)$$

This and other side reactions are favored by high temperature which results in a practical upper temperature limit for the process of about 50°C. Thiosulfate gradually builds up in the solution (as sodium or ammonium thiosulfate). The presence of these salts does not materially affect process operation at moderate concentration (less than about 30 percent) but steps are required in commercial plants to prevent excessive buildup.

Sulfint Process

The Sulfint process is licensed by Integral Engineering of Vienna, Austria, and has been described in considerable detail (70). A flow diagram is shown

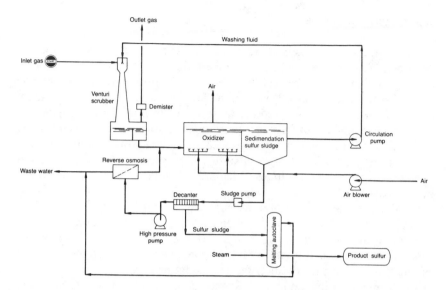

Figure 9-11. Schematic Flow Diagram of Sulfint Process (70). Courtesy *Hydrocarbon Processing,* March 1982.

in Figure 9-11. The process is based on the use of EDTA as the chelating agent for iron, a venturi scrubber for H_2S absorption, and a separate air-blown oxidizer. An interesting feature of the process is the use of reverse osmosis to separate water with dissolved salts from the chelate and avoid buildup of salts in the scrubbing solution. The chelate complex cannot pass through the membrane and is returned to the main solution circuit as a concentrated solution.

Design data for a Sulfint plant are given in Table 9-9.

Table 9-9. Sulfint Plant Data (70)	
Feed gas quantity, Nm^3/h	40,000
H_2S in feed gas, g/Nm^3	20
H_2S removal efficiency, %	99.9
Sulfur production, Tons/D	18
Feed gas temperature, °C	40
Process temperature, °C	20–40
Utilities	
Electricity, kW approx.	1500
Steam, tons/h	1.2
Waste water, m^3/h	50

The Cataban Process

This process which has been developed by Rhodia Inc. of New York was described in some detail by Meuly and Ruff (56, 57). The basis of the process is the oxidation of hydrogen sulfide to elemental sulfur by reduction of ferric ion to ferrous ion, followed by oxidation of the ferrous to ferric ion by contact with air. The reactions involved are those shown in Equations (9-20) and (9-21). The elemental sulfur formed is present in the solution as a fine crystalline suspension and can be removed by mechanical means such as decantation, filtration, or separation as molten sulfur. Mercaptans contained in the gas stream are oxidized to disulfides which are insoluble in the solution and can be removed by skimming.

Cataban, the catalytic agent, is a clear orange-brown aqueous solution containing two to four percent ferric ion in chelated form. It is reportedly stable over a pH range from 1.0 to 11.0, and over a temperature range from below room temperature to at least 260°F (57). The reaction between the hydrogen sulfide and ferric ion is extremely fast. Reoxidation of ferrous ion is also quite fast but limited by the amount of oxygen dissolved in the solution. Since oxygen is only sparingly soluble in aqueous solutions, rapid oxidation requires highly efficient air-liquid contact.

Complete removal of hydrogen sulfide requires at least the stoichiometric equivalent of the Cataban reagent, or 14 pounds of Cataban per cubic foot of hydrogen sulfide. For a 5 percent Cataban solution this is equivalent to a circulation rate of 40 gallons per minute in order to remove 1,000 ppm of hydrogen sulfide from 1,000 cfm of gas. However, in practice about twice this amount is used.

The effectiveness of the process is considerably influenced by the pH of the solution. Hydrogen sulfide removal is incomplete at pH below 7.0 but no significant adverse effect is noted within the pH range of 7.0 to 10.0. Above pH 11.5, iron hydroxide is precipitated and the solution becomes ineffective.

Air requirements for solution regeneration depend on the effectiveness of air-liquid contact. However, a four fold excess over the theoretical quantity of oxygen is recommended (57).

The process is not particularly temperature sensitive. Data reported by Meuly (57) indicate only minor effects over a temperature range of 80°C to 250°C.

Gas impurities such as CO_2 and CO are not removed by the Cataban solution, but their presence is not injurious to the Cataban reagent. Hydrogen cyanide will react with the Cataban reagent forming ferrocyanide complexes. However, since such complexes are in themselves capable of removing hydrogen sulfide, their presence does not deactivate the Cataban solution. It is reported (57) that at concentrations of up to 100 to 200 ppm in the gas to be treated, hydrogen cyanide has little effect on the Cataban system.

The Cataban process is primarily useful for the removal of small amounts of hydrogen sulfide from exhaust gas streams, when recovery of sulfur is not the prime objective. The process is quite simple, requiring essentially two contact vessels for absorption and regeneration, and auxiliary equipment such as pumps, an air blower, and sulfur filter. If the gas to be purified contains oxygen, the reduction-oxidation reaction takes place simultaneously in one vessel.

The Konox Process

The Konox process, which was developed in Japan (58), is also based on reduction-oxidation of an iron complex, resulting in the formation of elemental sulfur from the absorbed hydrogen sulfide. The active iron complex is present as a ferrate, Na_2FeO_4 which is reduced by reaction with hydrogen sulfide to $NaFeO_2$ which, in turn, is re-oxidized to the ferrate state by blowing with air.

The solution is reported to have a high capacity for hydrogen sulfide resulting in liquid circulation rates several times smaller than those required in other liquid oxidation processes. Side reactions, such as formation of thiosulfate and thiocyanate are minimized because, as a consequence of the rapid reaction between hydrogen sulfide and the iron complex, essentially no hydrosulfide or sulfide ions exist in the solution. Practically all of the hydrogen cyanide co-absorbed with the hydrogen sulfide is stripped from the solution during regeneration with air. It is claimed that utility and chemical requirements compare favorably with those of competing processes.

Organic Compounds

In the search for liquid gas-purification processes to replace the expensive and cumbersome iron oxide boxes, the possibility of using organic compounds as catalysts or oxygen carriers for the oxidation of hydrogen sulfide to elemental sulfur was recognized as early as 1921 when British Patent 168,504 was granted to Evans and Stanier for a process using certain organic dye stuffs. Later, Davis, Mills, and Ryden (37) developed a process utilizing methylene blue dissolved in aniline as the active agent. However, this process had serious drawbacks and was abandoned after fairly extensive pilot-plant studies. The processes currently in commercial operation employ moderately alkaline aqueous solutions containing water-soluble salts of quinone-type compounds which are capable of rapidly converting hydrosulfide ions to elemental sulfur by shifting from the oxidized to the reduced form. The compounds are readily reoxidized by contact with air, thus permitting their use in a cyclic process. In some processes the organic constituent of the solution is the sole carrier of oxygen, while in others, specifically the Stretford process, heavy metal ions, kept in solution by chelating agents, are used to accelerate the process of oxygen transfer.

These processes have several advantages over other liquid oxidation processes and, consequently, have found wide application in the treatment of many types of gas streams. The solutions are considered nontoxic and their disposal constitutes no pollution problem. Formation of undesirable side-reaction products is slow, and techniques are available to reclaim valuable reagents from contaminated solutions. Finally, the sulfur produced is of quite good quality and is usable for sulfuric acid manufacture or for agricultural purposes.

The Perox Process

The operation of this process, which was developed in Germany after World War II, consists of absorption of hydrogen sulfide in an aqueous ammonia solution containing 0.3 gram per liter of an organic oxidation catalyst, usually hydroquinone, followed by oxidation of ammonium hydrosulfide to elemental sulfur by contact with air. The process is being used successfully for the purification of coal gas, and it is reported that three commercial units with a combined daily capacity of approximately 30 million cu ft were operating in Germany in 1956 (38).

The flow scheme used in the process is quite simple and similar to that of other oxidation processes. The crude gas which contains hydrogen sulfide, hydrogen cyanide, and ammonia is first passed through a cooler in which the temperature and ammonia content are adjusted by direct contact with water. From there the gas flows to the contactor in which it is washed counter-currently with the Perox solution and practically all of the H_2S and HCN and a portion of the ammonia are removed. The spent solution is regenerated in the oxidizer by contact with compressed air and returned to the contactor. Elemental sulfur which separates as a froth at the top of the oxidizer is filtered and further processed by conventional methods. The filtrate is returned to the solution circuit.

The water balance of the system is maintained by cooling the inlet gas to a temperature somewhat below that of the circulating solution. In this manner the water produced in the oxidation of hydrogen sulfide to elemental sulfur is continuously carried away by the purified gas. Excessive accumulation of side-reaction products, such as thiosulfates and thiocyanates, in the solution is prevented by losses naturally occurring in the process. Since these solution losses also entail continuous loss of catalyst, periodic additions of the catalytic compound are required.

Operating results from two Perox plants reported in the literature (39-41) are shown in Table 9-10.

Table 9-10. Typical Operating Data of Perox Process		
Design variables	Plant	
	A (39)	B (40, 41)
Gas-flow rate, MSCF/hr	300	519
Solution-flow rate, gpm	1,300	2,650
Air-flow rate, MSCF/hr	12	31.8
Solution temp., °F	72	71
Inlet gas, grains/100 SCF:		
H₂S	240	399
NH₃	230	372
HCN	45	54
Outlet gas, grains/100 SCF:		
H₂S	0.5	0.06
NH₃	150	295
HCN	2.5	1.8
pH of solution	8.78	8.83
Solution loss, gal/day	390	550
Catalyst loss, lb/day	2.0	9.9

The Lo-Cat Process

The Lo-Cat process was developed by Air Resources Inc., Palatine, Illinois specifically for removing small quantities of hydrogen sulfide—in the parts per million range—from exhaust air in order to eliminate an odor problem and thus comply with air pollution control regulations (59, 70). The process is also based on the reduction-oxidation cycle forming elemental sulfur from hydrogen sulfide. The active agent is a water soluble organo-metallic catalyst designated ARI-300. All constituents of the catalyst reportedly are safe and nontoxic, and their spillage or discharge into sewers will not cause undue pollution problems (59).

It is claimed that when used for removal of small amounts of hydrogen sulfide from large exhaust gas streams, the Lo-Cat process is the most economical process presently available (59).

The Stretford Process

The Stretford process, developed jointly by the North Western Gas Board (now North West Gas) and the Clayton Aniline Company, Ltd., was intended initially for the removal of hydrogen sulfide from coal gas. However, the process proved to be equally suitable for desulfurization of a variety of other gas streams, such as refinery gases, synthesis gas, and natural gas, as well as of hydrocarbon liquids. Another rather important application of the Stretford process is its use as the hydrogen sulfide removal step of the Beavon Sulfur Removal Process (see Chapter 13).

As originally conceived and described by Nicklin and Brunner (42), the process utilized an aqueous solution containing sodium carbonate and bicarbonate in the proportion of about 1:3 (resulting in a pH range of 8.5 to 9.5) and sodium salts of the 2,6 and 2,7 isomers of anthraquinone disulfonic acid (ADA). The postulated reaction mechanism involves five steps.

1. Absorption of hydrogen sulfide in alkali
2. Reduction of ADA by addition of hydrosulfide ion to a carbonyl group
3. Liberation of elemental sulfur from reduced ADA by interaction with oxygen dissolved in the solution
4. Reoxidation of the reduced ADA (by air)
5. Reoxygenation of the alkaline solution providing dissolved oxygen for Step 3 of the process

Although this form of the process was tested successfully in commercial installations, it was soon found that certain inherent features imposed serious limitations on its economic operation. Because the process depended on dissolved oxygen for the conversion of hydrosulfide ion to elemental sulfur, a maximum sulfide loading of only about 40 ppm could be achieved without formation of thiosulfate. This resulted in very large liquid circulation rates and considerable power consumption. Furthermore, the formation of elemental sulfur was slow, requiring large reaction tanks and large liquid inventories. Finally, in order to obtain satisfactory rates of hydrogen sulfide absorption when treating gas streams containing appreciable amounts of carbon dioxide, partial decarbonation of the solution was required before recycle to the absorber.

In order to improve the economics of the process, a number of compounds were tested as possible additives, with the principal purpose of increasing the solution capacity for hydrogen sulfide and the rate of conversion of hydrosulfide to elemental sulfur. Among the compounds tested, alkali vanadates proved to be outstanding. It was found that hydrosulfide is reduced quite rapidly by vanadate to elemental sulfur, with a simultaneous valance change of vanadium from five to four. By introducing vanadate as the oxidant, it was no longer necessary to rely on oxygen dissolved in the solution, thus permitting substantially higher solution loadings. At present, sulfide loadings of 500 ppm are common, and many plants have been designed with loadings of as high as 1000 ppm (46). Because of the fast rate of reaction, it is possible to convert hydrosulfide completely to sulfur in relatively short periods of time, thus reducing the size of the reaction tanks.

Although vanadate reacts readily with hydrosulfide to produce sulfur, a solution containing vanadate alone cannot be regenerated by blowing with air. However, in the presence of ADA, complete oxidation of reduced vanadate is achieved, and the reduced ADA is readily reoxidized by contact with air. The vanadate-ADA system works at a lower pH than the original Stretford system without loss of washing efficiency, obviating any necessity for decarbonation of the treating solution. Furthermore, the complete conversion of hydrosulfide to sulfur prior to solution regeneration and operation of the process at relatively low pH results in a minimum of thiosulfate formation.

The process has been described by Nicklin and Holland (43), Thompson and Nicklin (44), Ellwood (45), the staff of the North Western Gas Board (46), and Ludberg (47). The application of the Stretford process to the purification of natural gas has been described by Nicklin, Riesenfeld, and Vaell (60). A modification of the process, called the Holmes-Stretford process, developed by Peabody-Holmes of Huddersfield, England has been described by Moyes and Wilkinson (61), and by Vasan (62). The paper by Moyes and Wilkinson gives a particularly good description of this process. The improvements claimed include certain proprietary equipment designs and, most importantly, a system for handling waste streams, eliminating all liquid effluents from the process. This system, which is based on reductive incineration of fixed salts formed in the solution, is discussed below. Improvements developed recently by British Gas Corporation, specifically operation of the Stretford process at elevated pressure, treatment of waste streams, and prevention of bacterial growth in the solution were reported by Wilson and Newell (72).

Basic Data

The chemistry of the process can be represented by the following idealized equations:

$$Na_2CO_3 + H_2S = NaHS + NaHCO_3 \tag{9-28}$$

$$4\ NaVO_3 + 2\ NaHS + H_2O = Na_2V_4O_9 + 2S + 4\ NaOH \tag{9-29}$$

$$Na_2V_4O_9 + 2\ NaOH + H_2O + 2\ ADA \tag{9-30}$$

$$= 4\ NaVO_3 + 2\ ADA\ (reduced)$$

$$2\ ADA\ (reduced) + O_2 = 2\ ADA + H_2O \tag{9-31}$$

Equation 9-28 represents absorption of hydrogen sulfide in the solution. Although the rate of absorption is favored by high pH, the rate of conversion of the absorbed hydrogen sulfide to elemental sulfur is adversely affected by pH values above 9.5. The process is therefore best operated within a pH range of 8.5 to 9.5.

Conversion of hydrogen sulfide to elemental sulfur is represented by Equation 9-29. This reaction is quite rapid and is essentially a function of the vanadate concentration in the solution as shown in Figure 9-12 (43). According to Thompson and Nicklin (44), the reaction is of second order and follows the equation:

$$t = 1/K(a - b) \times \log_e b \ (a - x)/a \ (b - x) \tag{9-32}$$

where t = time in minutes
   K = reaction rate constant,
      $1/[(\text{moles}/1) \ (\text{hr})]$
   a = initial concentration of vanadium in moles/1
   b = initial hydrosulfide concentration in moles/1
   x = moles/1 hydrosulfide reacted

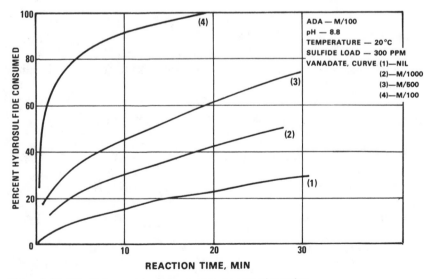

Figure 9-12. Effect of vanadate concentration on rate of reaction.

Numerical values for the rate constant as a function of pH are shown in Figure 9-13. A general correlation for estimating the time required for conversion is given in Figure 9-14. The values shown in Figure 9-14 are based on a reaction rate constant of 6,000 (pH about 8.5) and a vanadate concentration of 0.01 mole per liter. Such values are common in commercial use. For other values of the rate constant and molar concentrations, the reaction time is obtained by dividing the reaction time given on the graph by the value of the rate constant multiplied by the molar concentration (44). Equation 9-29 shows that 2 moles of vanadate are required for each mole of hydrogen sulfide. In practice an excess of vanadate is used in order to avoid overloading the solution with sulfide and subsequent formation of thiosulfate during solution regeneration.

Reduced vanadate is oxidized by ADA according to Equation 9-30. Since ADA acts as the effective oxidant for vanadate, its concentration must be sufficient to satisfy the overall equation:

$$2H_2S + O_2 = 2H_2O + 2S \tag{9-33}$$

Reoxidation of ADA by contact with air is fairly rapid. However, the rate of oxidation can be appreciably accelerated by the presence of small amounts of iron salts kept in solution by a chelating agent. In general, a concentration of 50 to 100 ppm of iron combined with about 2,700 ppm of ethylene diamine tetraacetic acid (EDTA) is satisfactory (44).

Besides the principal reactions, a number of side reactions occur in the process. It has been reported that it is possible under some conditions for the solution to absorb more hydrogen sulfide than the vanadate can oxidize, causing the vanadium to precipitate as a black complex vanadium-oxygen-sulfur compound. This compound can only be restored to a soluble vanadate by continuous air blowing over long periods of time. However, the presence of sodium tartrate in the solution largely prevents the formation of this compound (43).

The most serious side reactions are those involving hydrogen cyanide present in the feed gas and conversion of hydrosulfide to thiosulfate.

Hydrogen cyanide is absorbed by the alkaline solution and partially expelled during regeneration by air blowing. A portion of the cyanide reacts with elemental sulfur forming thiocyanate which accumulates in the solution and eventually may have to be discarded.

Formation of thiosulfate is dependent on the conversion of hydrosulfide to sulfur prior to contact with oxygen, pH of the solution, and operating temperature. The effects of temperature and pH are shown in Figures 9-15 and 9-16 (43). It is claimed that in a properly operated plant thiosulfate formation can be controlled at less than 1 percent of the sulfur in the feed gas (46).

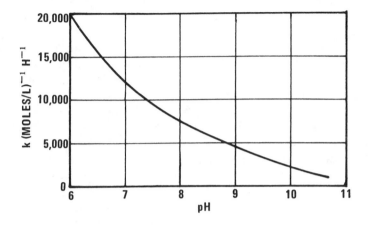

Figure 9-13. Rate of reaction versus pH for conversion of hydrosulfide to sulfur.

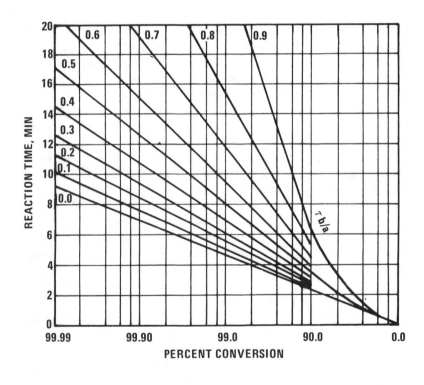

Figure 9-14. Time of reaction versus percent conversion of hydrosulfide to sulfur.

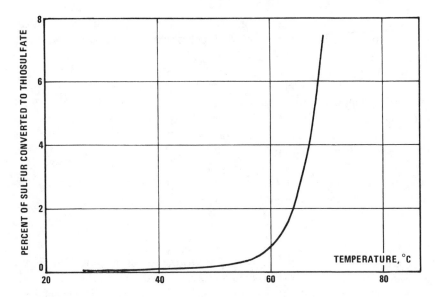

Figure 9-15. Effect of temperature on thiosulfate formation.

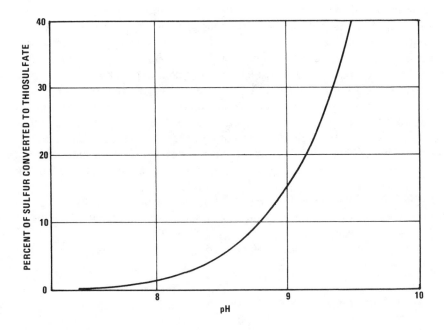

Figure 9-16. Effect of pH on thiosulfate formation.

Carbonyl sulfide and carbon disulfide are not absorbed by the Stretford solution to a significant degree. Methyl mercaptan is largely removed, probably by oxidation and conversion to disulfide.

Carbon dioxide is partially absorbed by the alkaline solution, resulting in the formation of bicarbonate and consequent lowering of pH. When gases containing high concentrations of carbon dioxide are treated, the absorption efficiency of the solution may be sufficiently lowered to require an appreciable increase in the absorber height (43). However, pilot plant studies conducted at total pressures up to 40 bar, CO_2 partial pressures ranging from 0.05 to 1.5 atmosphere, and a wide range of gas and liquid velocities indicated that the effects of total pressure and CO_2 partial pressure are much less than had been observed at atmospheric operation pressure (72).

Process Description

A schematic flow diagram of the Stretford process is shown in Figure 9-17. The raw gas is contacted countercurrently with the solution in the absorber where practically all hydrogen sulfide is removed. The treated gas contains less than 1 ppm of hydrogen sulfide. The solution flows from the absorber to a reaction tank where the conversion of hydrosulfide to elemental sulfur is completed. The reaction tank may be the bottom of the absorber or a separate vessel. From the reaction tank the solution flows to the oxidizer where it is regenerated by intimate contact with air, usually in cocurrent flow. In the oxidizer sulfur is separated from the solution by flotation and removed at the top as a froth containing about 10 percent solids. The relatively sulfur free regenerated solution is recycled to the absorber.

The sulfur froth is collected in a tank and subsequently further processed in filters or centrifuges to separate the solution remaining in the froth. In general, it is necessary to wash the sulfur cake with water to recover chemicals contained in the solution and to produce relatively pure sulfur. For reasons of water balance in the system, wash water and water produced by the reaction has to be evaporated either with the gas or in evaporators, depending on the quantity of water involved.

The sulfur cake which contains about 50 to 60 percent solids may be further processed by melting in an autoclave. In this manner high grade liquid or solid sulfur is produced.

A photograph of a Stretford plant treating coke-oven gas is shown in Figure 9-18.

Design and Operation

The absorber may be any efficient gas-liquid contacting device. Most absorbers presently in operation are packed with wood slats or with metal

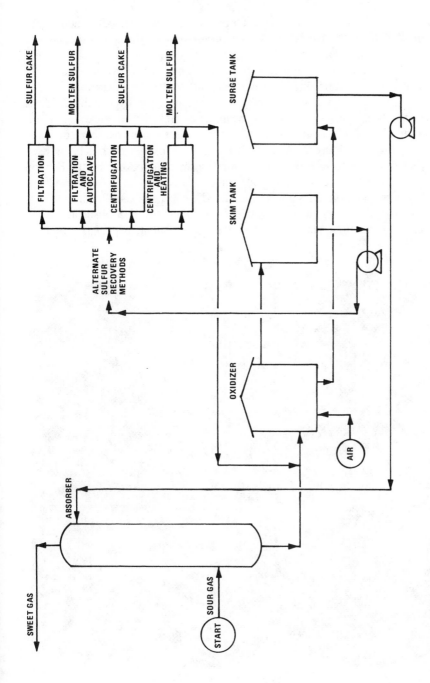

Figure 9-17. Typical flow diagram of Stretford process.

splash plates 2 in. wide on 6-in. centers. Large plastic rings or saddles are claimed to be equally effective. One problem which arises, especially if gases containing relatively high concentrations of hydrogen sulfide (above 1 percent) are treated, is plugging of the packing with elemental sulfur. In some installations this difficulty has been alleviated by contacting the gas first in a spray column where about 80 percent of the hydrogen sulfide is removed. Final purification is subsequently achieved in a packed column. Another type of absorption device is a venturi scrubber-absorber combination as shown in Figure 9-21 (Unisulf Process). This approach is being used very successfully for Stretford units working in conjunction with the Beavon Sulfur Removal Process for the treatment of Claus plant tail gases (see Chapter 13). However, pilot plant tests conducted by British Gas Corporation (72) indicated that venturi absorbers are less efficient for H_2S absorption than packed towers. Pilot plant operation also showed that, especially for high

Figure 9-18. Stretford plant for treatment of coke-oven gas. *North Western Gas Board, Manchester, England*

pressure operation, co-current absorption may be more economical than and as efficient as countercurrent operation (72).

The reaction tank is an empty vessel sized sufficiently large to allow complete conversion of hydrosulfide to elemental sulfur (see Figure 9-14). Normally a liquid residence time of 10 to 20 minutes is required.

Several types of oxidizers have been employed (44). The simplest and most commonly used form is a cylindrical vertical tank containing at the bottom a device for effective air distribution. The liquid and air flow cocurrently upward, and the sulfur froth is removed by overflowing a weir at the top of the vessel. The solution, which is essentially sulfur free, is withdrawn at a point located below the top of the oxidizer. Oxygen utilization of 15 to 20 percent is achieved.

Another type of oxidizer is similar to surface aerators used in waste-water treatment. Such machines are claimed to be particularly suitable for solutions with low sulfide loadings (44).

Yet another type of oxidizer, with reportedly very high oxygen utilization (44), consists of coils through which a mixture of air (or oxygen) and solution is pumped at very high Reynolds numbers. These devices are recommended for plants operating at high pressure and particularly where oxygen is available at elevated pressures.

Stretford plants can be constructed entirely of carbon steel, with inert linings, e.g., cold cured epoxy resins, for oxidizers and sulfur froth tanks. Stainless steel linings are recommended for solution and sulfur slurry pumps. However, care should be taken to avoid sulfur deposits on unprotected metal surfaces. The Stretford process operates best within a temperature range of 70° to 110°F. There are no limitations on absorption pressure.

Operating data on several commercial units ranging in capacity from 240,000 cu ft per day to 28 million cu ft per day have been reported by Nicklin and Holland (43), Ludberg (47), Penderleith (64), Carter et al. (65), and Donovan et al. (69). Design characteristics of a plant treating 28 million cu ft of coke-oven gas are shown in Table 9-11 (47).

One rather serious operating problem is associated with the presence of hydrogen cyanide in the feed gas to the Stretford unit. As pointed out earlier in this text, in the discussion of the Vacuum Carbonate process and other processes using alkaline salt solutions for hydrogen sulfide removal, hydrogen cyanide is largely co-absorbed with other acid gases. In alkaline solutions not containing oxidants, the hydrogen cyanide is stripped during regeneration; however, if the solution comes in contact with air thiocyanate is formed slowly and gradually builds up in the liquid. In oxidative processes this phenomenon is aggravated, as the absorbed hydrogen cyanide is largely converted to thiocyanate during re-oxidation of the reduced oxidant during regeneration. High concentrations of thiocyanate reduce the effectiveness of the treating solution and continuous discharge of a side stream and addition of fresh chemicals is required to maintain solution strength. Because of the

Table 9-11. Design Data of Stretford Plant (47)	
Type of gas	Coke oven
H₂S in, grains/100 SCF	275
H₂S out, grains/100 SCF	<1
Gas volume, cu ft/day	28,000,000
Absorber (2 units in parallel):	
Height	97 ft
Diameter	12 ft
Packing type	Wooden grids
Packing height	4 layers each 11 ft, 7 in.
Reaction tank:	Bottom of absorber, 24 ft high
Solution pumps:	3 units (one stand-by), 90,000 igph*
Oxidizer (2 units in parallel):	
Height	24 ft
Diameter	20 ft
Solution depth	20 ft
Air blower (2 units)	60,000 cu ft/hr
Filter	Rotary vacuum

*Imperial gallons per hour.

toxicity and high chemical oxygen demand of the waste stream, its disposal could create a serious pollution problem. To protect the Stretford solution from compounds formed by HCN, i.e., thiocyanate, it is desirable to remove HCN ahead of the Stretford absorber. However, existing methods are only capable of removing about 90 percent of the HCN and some thiocyanate will inevitably be formed in the solution (64, 73).

Besides thiocyanate, which is only a problem when the gas to be treated contains hydrogen cyanide, thiosulfate and sulfate are formed to some degree in the operation of the Stretford and other oxidation processes in the liquid phase. Although these compounds are much less objectionable from a pollution standpoint than thiocyanates, they can only be allowed to accumulate in the solution to a limited extent, about 30 percent, before adverse effects are encountered. Consequently, a portion of the solution has to be discarded continuously, or periodically, which again may cause a pollution problem, but also results in the loss of valuable chemicals.

A review of methods to cope with the cyanide problem in coke-oven gas desulfurization has been presented by Massey and Dunlop (63). One approach is pretreatment of the gas prior to its entry into the desulfurization unit, either by washing with water or with a polysulfide solution containing suspended sulfur. In the first case, the absorbed hydrogen cyanide is stripped from the water with a gas stream which is subsequently incinerated, and the water is recirculated. In the second case, the hydrogen cyanide is converted to thiocyanate which accumulates in the solution which eventually contains large concentrations of thiocyanate, polysulfide, hydrogen sulfide, sulfur, and alkali. Here again disposal of spent solution is a very serious pollution problem. Detailed operating experience with polysulfide scrubbers for hydrogen cyanide removal has been reported by Penderleith (64), Carter et al. (65), and Donovan et al. (69).

Two hydrogen cyanide removal units, used by Dofasco Inc. at Hamilton, Ontario, Canada have been described by Donovan and Laroche (73). In one unit, operating in conjunction with a fixed salts reductive incinerator (developed by Peabody-Holmes) (61, 64, 65) the gas is contacted countercurrently with a solution containing 1.5 to 2.0 percent sodium carbonate, 1.0 to 1.5 percent sulfur, and 0.5 to 1.0 percent polysulfides, in a vertical wash tower at a pressure of about 500 psig. The HCN absorbed reacts irreversibly, forming sodium thiocyanate. The salts in the solution are allowed to build up to a total solids concentration of 25 percent by controlling the purge stream and the feed of fresh solution. The purge stream is collected in a tank and subsequently processed in the Fixed Salts Recovery plant, together with the purge stream from the Stretford unit.

The other HCN removal unit uses an ammonia based-aqueous solvent in packed towers ahead of the ammonia absorbers in the coke oven gas purification train. Alkalinity is provided by absorption of ammonia and 11 to 14 grams per liter of finely divided sulfur are added to the solution. Hydrogen cyanate is converted to ammonium thiosulfate and, because of the presence of oxygen in the gas, ammonium thiocyanate is formed as a byproduct. The concentration of ammonium thiocyanate is maintained at 200 grams per liter in the effluent. Thiosulfate content will be about 5 percent by weight in the recirculating solution.

The effluent spent solution is further processed for recovery of chemicals in a Zimpro wet air oxidation process.

The Fixed Salts Recovery plant, where the soluble salts, such as thiosulfate and thiocyanate, are converted to usable chemicals, utilizes a reductive incineration process, as shown in Figure 9-19. The spent solution is pumped from the collection tank into a concentration zone, where the solids content is increased to 40 percent by evaporating water into the hot combustion gases. The concentrated solution is then sprayed into the substoichiometric reducing atmosphere, generated by a multi nozzle burner, using coke oven gas. Evaporation of any remaining water is effected by the 1500 to 1800°C combustion products. This evaporation lowers the reaction zone temperature to 900°C. Sodium thiosulfate, thiocyanate, and sulfate are converted to sodium sulfide and carbonate. ADA, citrate, and elemental sulfur contained in the effluent from the Stretford unit are converted to sodium carbonate and a number of gaseous products, primarily hydrogen sulfide.

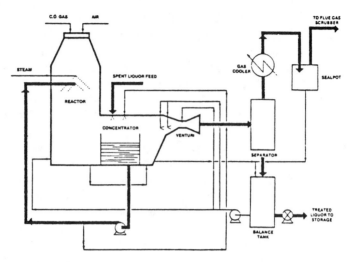

Figure 9-19. Fixed Salt Recovery Unit (73).

After leaving the reaction zone, the hot gases, containing finely divided solids, as well as gaseous reaction products, are sprayed with spent and recirculated solution, resulting in a solids concentration in the solution of 40 percent and cooling of the gases to 150 to 200°C. The cooled incineration products then enter a venturi scrubber where the solids are removed from the gas with recovered solution. Gas and solution leaving the venturi scrubber are separated, the liquid is returned to the balance tank and the gases pass to a condenser, where they are sufficiently cooled to maintain water balance in the system. The final step for the gas stream is removal of the acid gases in the flue gas scrubber, which is a small tower using Stretford solution, and subsequent venting to the atmosphere.

Reported operating results are somewhat disappointing as conversion rates of 65 percent for NaSCN, 65 percent for $Na_2S_2O_3$, and 55 percent for Na_2SO_4 were attained. The low conversion rates were attributed to low residence time and low reaction temperature in the reaction zone.

The Zimpro wet air oxidation process used in Dofasco's ammonia based HCN removal system is also described by Donovan and Laroche (69). This process, which is in wide use for the treatment of industrial and municipal waste waters, is based on the principle that oxidation of waste material will occur when the aqueous waste stream is contacted with oxygen and heated to an elevated temperature at a corresponding high pressure. Ammonium thiocyanate is converted to ammonium sulfate and CO_2, ammonium thiosulfate is converted to ammonium sulfate and sulfuric acid, elemental

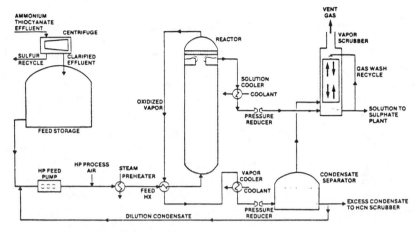

Figure 9-20. Zimpro Flow Diagram (73).

sulfur is converted to sulfuric acid, and organics are converted to CO_2 and water. The product solution from oxidation contains 35 percent ammonium sulfate and 1.5 to 2.0 percent sulfuric acid. Gaseous products consisting primarily of water vapor, oxygen, carbon dioxide and carbon monoxide are vented to the atmosphere. A flow diagram of the Zimpro unit is shown in Figure 9-20.

Besides these two processes, a process developed by Nittetsu Chemical Engineering Ltd. of Japan is also available for disposal of waste streams from Stretford units and from pretreaters for HCN removal (66).

A process for handling purge streams containing side reaction products; i.e., sodium sulfate and thiosulfate, was developed by the British Gas Corporation through the pilot plant stage (72). However, this process is only applicable in plants treating gas containing no HCN or very little HCN. Sodium thiosulfate is converted to sodium sulfate and sulfur, by acidification with concentrated sulfuric acid at high temperature. By sparging the acid into the lower part of the acidification vessel the SO_2 produced in the initial reaction is used in further conversion of the thiosulfate according to the following reactions

$$Na_2S_2O_3 + H_2SO_4 \rightarrow Na_2SO_4 + SO_2 + H_2O + S^o \qquad (9\text{-}34)$$

$$2Na_2S_2O_3 + SO_2 \rightarrow 2Na_2SO_4 + 3S^o \qquad (9\text{-}35)$$

$$3Na_2S_2O_3 + H_2SO_4 \rightarrow 3Na_2SO_4 + H_2O + 4S^o \qquad (9\text{-}36)$$

At present the process is conceived as a batch operation. Approximately 90 percent of the thiosulfate is converted in about 3 hours when a slight excess of acid is used. Sodium carbonate contained in the solution is, of course, neutralized but ADA is not destroyed.

The sulfur produced is very coarse and granular and can be separated by known techniques. After separation of the sulfur, sodium sulfate decahydrate is crystallized at low temperature to prevent destruction of ADA. If desired, the crystals can be further processed to anhydrous sodium sulfate.

One problem encountered in Stretford plants treating gas streams other than coke-oven gases, is contamination of the system by bacterial growth. Bacteria in the Stretford solution were first discovered in 1978 in a solution sample from a petroleum refinery in Illinois. The plant had serious operating problems, specifically, large amounts of uncontrollable light foam, rather than normal sulfur froth, in the reactivator, decrease in pH from 9.6 to 7 and excessive consumption of alkali. Subsequently it was found, in this and other plants, that the sulfur produced was sticky and adhered to all surfaces, leading to plugging problems. Finally, some corrosion of carbon steel equipment was found (72).

The British Gas Corporation instituted an in depth investigation to elucidate this phenomenon. Several strains of bacteria were identified in every solution from plants, except from those treating coke oven gas. This work is discussed in great detail by Wilson and Newell (72). To combat the problem, a number of biocides were developed which proved to be effective when added to the Stretford solution in concentrations of about 1,000 ppm or, in some cases, less. The disadvantage of this approach is that the biocides are quite expensive which, of course, affects the operating costs adversely. Wilson and Newell point out that each Stretford plant behaves somewhat differently and that specific solutions must be found for each case after thorough investigation.

The Unisulf Process

This process, which was developed by the Union Oil Company of California, is quite similar to the Stretford process, as can be seen in the flow diagram shown in Figure 9-21. The process, as described by Fenton and Gowdy (74) and by Hass, Fenton, Gowdy and Bingham (75) is a homogeneous catalytic process for oxidizing H_2S to sulfur, with minimum chemical consumption and without liquid waste effluent. It is suitable for the treatment of gas streams rich in H_2S, but its main application is expected to be in recovering more than 99.9 percent of H_2S from gas streams containing less than 10 mole percent H_2S. Purities of less than 10 ppmv of H_2S in the treated gas can be obtained.

The agent employed in the Unisulf process is an aqueous solution of unique composition containing typically the following components:

Sodium carbonate and bicarbonate
Vanadium complex
Thiocyanate ions
Carboxylated complexing agent
A water soluble aromatic compound
Water

The principal difference from the Stretford process is the absence of quinone compounds in the solution. Instead, carboxylated complexing agents, for example, sodium 1-hydroxybenzene 4 sulfonate and sodium 8-hydroxyquinoline-5-sulfonate are used (74). It is claimed (75) that there is no chemical consumption, except for entrainment of solution with sulfur filter cake. Under some conditions, at the start of the run, very slight formation of thiosulfate and sulfate may occur, but upon reaching steady state operation no further buildup of these ions takes place.

The Takahax Process

The Takahax process, which was developed by Tokyo Gas Company, Ltd., and described by Hasebe (48), utilizes naphthoquinone compounds as the oxygen carrier. The preferred solutions contain salts of 1, 4-naphthoquinone 2-sulfonic acid dissolved in alkaline aqueous media within a pH range of 8 to 9.

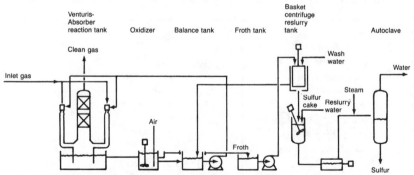

Figure 9-21. Schematic Flow Diagram of Unisulf Process (74). Courtesy *Hydrocarbon Processing,* April 1982

It is reported that the process is capable of producing treated gas containing no detectable amounts of hydrogen sulfide even when the raw gas contains substantial quantities of carbon dioxide.

Typical operating conditions for coke-oven gas desulfurization, reported by Hasebe (48), are shown in Table 9-12.

The flow scheme of the Takahax process is quite similar to that of other processes of this type discussed in this chapter. However, it is claimed that, in the Takahax process, oxidation of hydrosulfide to elemental sulfur occurs in the absorber instantaneously, and no delay tank is required to complete the reaction prior to solution regeneration with air. One other difference with respect to other processes is that the precipitated sulfur is of very fine grain size and not amenable to flotation. The technique used is continuous recirculation of a sulfur slurry of relatively high solids content and removal of sulfur from a slip stream in a filter press.

Improvements made in the process since its early use are described by Keizumi et al. (67). One reportedly important feature is the present use of ammonia as the alkaline component of the solution instead of the previously used sodium carbonate. It is claimed that this resulted in an appreciable improvement of the process economics. Waste effluent from the process, containing ammonium thiosulfate, sulfate and thiocyanate is incinerated and the sulfur values are converted to sulfur dioxide, which is further processed to sulfuric acid, ammonium sulfate, and gypsum. The combustion is carried out in such a manner that a minimum of nitrogen oxides is formed. About a hundred units are in operation in Japan, primarily in gas works, steel plants, and

Table 9-12. Typical Operating Conditions of Takahax Process

Gas composition:	
H_2S, volume %	0.4
CO_2, volume %	5-10
Solution composition:	
Na_2CO_3	40 grams/liter
Catalyst	.0015-0.002 gram mol/liter
Volume ratio gas—liquid:	22
Volume ratio air—liquid:	1.9
Chemical consumption:	
Na_2CO_3	0.4 lb/lb of recovered sulfur
Quinone	0.225 gram mol/lb of recovered sulfur

chemical plants. The process is licensed in the United States by Ford, Baken and Davis of Dallas.

The Fumaks Process

This process, developed by Sumitomo Metals Industries Ltd. of Japan (76), utilizes picric acid as the oxygen carrier in a Redox system similar to those described for the preceding processes. Alkalinity is provided by ammonia contained in coke oven gas to be treated. Waste products are incinerated and converted to sulfuric acid and gypsum. Hydrogen sulfide removal efficiency, reported for eleven units operating on coke oven gas between 1972 and 1977, ranges from 90 to 99.5 percent.

Sulfur Dioxide

The processes falling within this category are, in essence, extensions of the Claus process discussed in Chapter 8. The reaction involved is the oxidation of hydrogen sulfide to elemental sulfur and water using sulfur dioxide as the oxidant, according to Equation 9-37.

$$2H_2S + SO_2 = 3S + H_2O \tag{9-37}$$

Since equilibrium conversion of hydrogen sulfide to sulfur is favored by low temperatures, it is advantageous to conduct the reaction at the lowest temperature compatible with reasonable reaction rates. This objective can be achieved by reacting hydrogen sulfide and sulfur dioxide at temperatures below the sulfur dew point in a liquid medium and in the presence of a catalyst. Depending on the operating temperature, solid or liquid sulfur is produced.

The Townsend Process

This process, which has been disclosed by Townsend and Reid (49, 50), was proposed as a method for high-pressure natural gas desulfurization and production of elemental sulfur in one operation, thus combining the conventional process of first absorbing hydrogen sulfide and, if present, carbon dioxide in an aqueous alkaline solution (e.g., ethanolamine), followed by processing the stripped acid gases in a Claus type sulfur plant. However, the process is claimed to be equally applicable to the treatment of acid-gas streams, such as the effluents from ethanolamine plant regenerators. In this application the Townsend process would be a substitute for the Claus unit.

A schematic flow diagram of the Townsend process, as applied to high-pressure natural gas treating, is shown in Figure 9-22. The sour natural gas enters the base of the reactor column at atmospheric temperature and is contacted countercurrently with a concentrated stream of di- or triethylene glycol (typically 98 percent glycol—2 percent water) containing dissolved sulfur dioxide. The reaction of hydrogen sulfide with sulfur dioxide is very rapid as the water present acts as a catalyst. The amount of glycol circulated should be sufficiently large to result in a water concentration not exceeding 5 percent by weight after absorption of the water contained in the feed gas and the water formed in the reaction.

The treated gas leaving the reactor column is washed in a high-pressure column with concentrated glycol to remove sulfur dioxide carried by the gas out of the reactor. The effluent gas from this column is claimed to be within pipeline specifications with respect to hydrogen sulfide and to be free of sulfur dioxide.

A slurry of elemental sulfur in glycol flows from the bottom of the reactor to a settling tank where the mixture is heated to 250° to 275°F. Excess sulfur dioxide is stripped from the solution and returned to the reactor. Liquid sul-

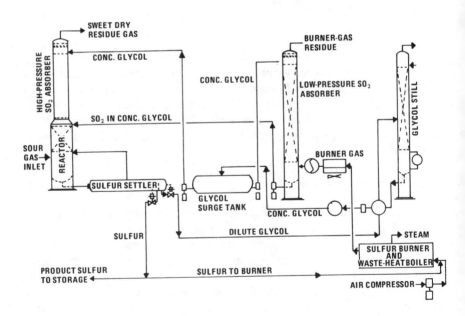

Figure 9-22. Typical flow diagram of Townsend process for high pressure natural gas treating.

fur is withdrawn from the bottom of the settling tank. A portion of the liquid sulfur flows to a sulfur burner where it is burned to supply sulfur dioxide required in the process.

The dilute glycol flows from the settling tank to the glycol still where water is removed and the glycol is reconcentrated. The concentrated glycol then flows to a surge tank where it is split into two streams. One stream flows to the top of the high-pressure sulfur dioxide absorber to remove sulfur dioxide from the purified gas. The other stream passes to a low-pressure column where it absorbs sulfur dioxide from the sulfur burner flue gases and is then recycled to the top of the reactor column.

Application of the process to the treatment of acid-gas streams is shown schematically in Figure 9-23. In this version the sulfur dioxide is supplied by burning one-third of the acid gas in the same manner as in the Claus process using split flow (see Chapter 8). The glycol reactor is the equivalent of the catalytic converters of the Claus plant.

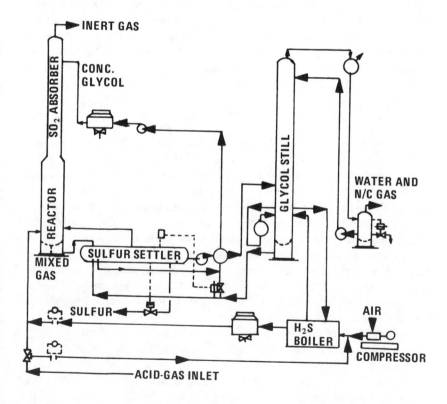

Figure 9-23. Flow diagram of Townsend process for acid gas treating.

The IFP Process

A process, somewhat similar to the Townsend process, has been disclosed by Renault (51) and Barthel et al. (52) of Institut Francais du Petrole (IFP). In this process, which has been developed specifically for the removal of hydrogen sulfide and sulfur dioxide from Claus unit tail gases, the conversion reaction (Equation 9-37) takes place in an essentially anhydrous liquid medium containing a catalyst at temperatures above the melting point of sulfur but below the sulfur dew point of the gas mixture to be treated. The solvent first proposed (51) was tributyl orthophosphate containing an alkaline substance as the catalyst. However, it is understood that other undisclosed solvents e.g. polyglycols and catalysts are also useful in the process, as long as they satisfy the requirements of good thermal and chemical stability, low vapor pressure, low solubility of sulfur in the solvent and of the solvent in sulfur, low cost, and ready availability.

Since the last edition of this text was published, the process has become widely accepted and some additional information has been published. A recent article by members of the Institute Francais du Petrole (68) states that in 1976 twenty-seven plants were either in operation or in various stages of design and construction. It is claimed that among all the processes capable of reducing sulfur compounds in Claus plant tail gases to 1,000 ppm of sulfur dioxide after incineration, the IFP process requires the lowest capital investment and the lowest expenditure of energy and labor.

Process Description and Operation

The flow scheme of the IFP process, shown in Figure 9-24, is extremely simple. The gas, containing hydrogen sulfide and sulfur dioxide, is contacted countercurrently with the solvent in a packed tower at about 260°F. The liquid sulfur formed separates readily by gravity and is withdrawn from the bottom of the tower. Since the solvent is only slightly soluble in liquid sulfur, no further purification of the sulfur product is required. The sulfur-free solvent is recycled to the top of the packed tower, after injection of condensate to maintain the solvent temperature between 260° and 280°F. The condensate and the water formed in the reaction are evaporated and carried out of the tower by the purified gas. Proper operation depends primarily on the ratio of hydrogen sulfide to sulfur dioxide and on the total content of hydrogen sulfide and sulfur dioxide in the feed gas.

For maximum conversion, it is necessary to maintain the ratio of hydrogen sulfide to sulfur dioxide in the feed gas within ±5 percent of the stoichiometric proportion of two to one. This requires reliable control instruments, such as in-line gas chromatographs or ultraviolet spectrophotometers in the Claus unit to assure constant composition of the Claus unit tail gas.

The effect of total sulfur content (in the form of hydrogen sulfide and sulfur dioxide) in the feed gas on conversion is shown in Table 9-13 (52). It should be noted that sulfur compounds such as carbonyl sulfide and carbon disulfide which may be present in Claus unit tail gases in appreciable concentrations are unaffected by the solvent. The term "conversion" therefore applies only to hydrogen sulfide and sulfur dioxide.

The IFP process may be combined with a Claus unit in a variety of ways as shown in Tables 9-14 and 9-15. Since operation of the process is based on physical solubility (i.e., partial pressure) of hydrogen sulfide and sulfur dioxide in the solvent, the solvent flow rates and tower dimensions are essentially the same for a relatively wide range of hydrogen sulfide and sulfur dioxide concentrations in the feed gas. This enables the process to compensate for upsets in the Claus unit.

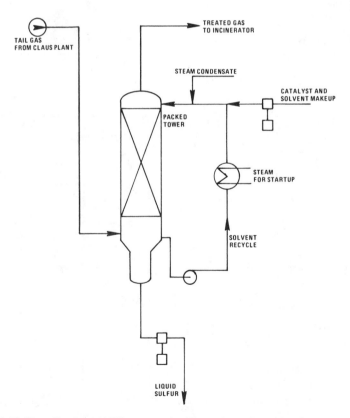

Figure 9-24. Flow diagram of I.F.P. process for Claus plant tail gas treating.

Table 9-13. Effect of Sulfur on Conversion	
Volume % H_2S + SO_2	%Conversion (H_2S + SO_2)
0.4 - 0.8	80
0.8 - 1.5	90
1.5	95

Table 9-14. Combination of Claus-I.F.P. Processes (52)

Scheme	H_2S + SO_2 Conversion—%		
	Claus unit	I.F.P. unit	Overall
Furnace + 3 converters + I.F.P. unit	96.8	80	99.30
Furnace + 2 converters + I.F.P. unit	94.4	90	99.44
Furnace + 1 converter + I.F.P. unit	86.1	95	99.36

Table 9-15. Application of I.F.P. Process for Treating Claus Unit Tail Gas (52).

	Effluent from		
	1st Converter	2nd Converter	3rd Converter
Gas composition, mol% H_2S	1.84	0.59	0.34
SO_2	0.74	0.29	0.17
S	1.26	0.14	0.13
H_2O	28.58	29.96	30.25
N_2, CO_2, misc.	67.94	69.02	69.11
Temperature, °F	260	260	260
Pressure, psig	0.5	0.5	0.5
H_2S + SO_2 conversion, %	95	90	80
Gas to incinerator, H_2S + SO_2-ppm	1,100	900	1,000

The Wiewiorowski Process

In this process, which was disclosed by Wiewiorowski of Freeport Sulfur Company (53, 54), the reaction between hydrogen sulfide and sulfur dioxide takes place in molten sulfur within the temperature range of 240° to 320°F (low-viscosity range of liquid sulfur). A basic nitrogen compound, such as ammonia or an amine, in concentrations of 1 to 5,000 ppm is added as the catalyst.

The operation of the process is analogous to that of the IFP process, except that only one liquid phase, i.e., molten sulfur, is present. Any efficient gas-liquid contacting device (54) such as a packed tower or a turboreactor, is claimed to be suitable for carrying out the reaction. As in the IFP process, it is important to maintain the ratio of hydrogen sulfide to sulfur dioxide close to the stoichiometric proportion of two to one in order to achieve maximum conversion to elemental sulfur. It is stated that the process will operate with feed gases containing from 1 to 67 percent hydrogen sulfide and from 0.5 to 34 percent sulfur dioxide and that essentially complete conversion to elemental sulfur can be obtained (53). At present, the process is not known to be in commercial operation.

Other Oxidation Processes

Permanganate and Dichromate Solutions

Buffered aqueous solutions of potassium permanganate and sodium or potassium dichromate are used to remove completely traces of hydrogen sulfide from industrial gases. Processes employing such solutions are non-regenerative and, because of the high cost of the chemicals used, are only economical when very small amounts of hydrogen sulfide are present in the gas. Permanganate solutions are used quite extensively for the final purification of carbon dioxide in the manufacture of dry ice.

A schematic flow diagram of the process is shown in Figure 9-25. The gas is contacted with the solution in two packed towers operating in series. The solution, which in the case of the permanganate process contains about 4.0 percent potassium permanganate and 1.0 percent sodium carbonate, is circulated until approximately 75 percent of the permanganate in either tower is converted to manganese dioxide. At that point the spent solution is discarded and fresh solution is pumped into the tower. This can be done without interruption of the gas flow if sufficient active permanganate is available in each contact stage to ensure complete H_2S removal. Dichromate solutions usually contain 5 to 10 percent potassium dichromate, zinc sulfate, and

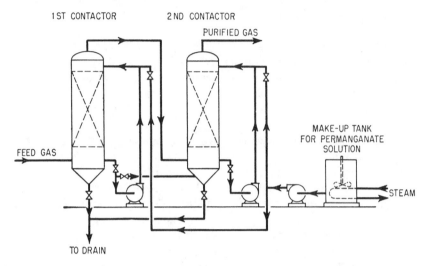

Figure 9-25. Flow diagram of permanganate and dichromate processes.

borax. In addition to hydrogen sulfide, traces of organic compounds such as amines are also removed quantitatively in these processes.

The Lacey-Keller Process

The Lacey-Keller process reportedly has been practiced in two commercial installations (55). Based on the rather incomplete description in the literature, the flow scheme may be visualized as follows. The feed gas is contacted in a column with a solution of undisclosed composition, and hydrogen sulfide and mercaptans are oxidized to elemental sulfur, which precipitates and is held in the solution as a colloidal suspension. The fluid flows from the bottom of the contactor to a bank of flotation cells where the sulfur is flocculated. A thick slurry accumulates on the surface of the liquid and from there flows to a vacuum filter. The filtrate is combined with the clear effluent liquid from the flotation cells and regenerated in electrochemical cells before being returned to the contactor.

Most of the equipment has to be constructed of stainless steel or coated with baked-on epoxy resins. The process is claimed to be suitable for the treatment of gas streams containing a few grains of hydrogen sulfide per 100 cu ft. It does not appear to be economical if the total quantity of sulfur produced exceeds 1 ton per day.

EIC Copper Sulfate Process

This process, which has been designated Cuprosol by its developers, the EIC Corporation, is specifically aimed at removing hydrogen sulfide from geothermal steam at relatively high temperatures and pressures. The process has been described by Brown and Dyer (77) as consisting of the following four key steps:

1. Scrubbing. In this step the geothermal steam is contacted with a solution of copper sulfate. Hydrogen sulfide and ammonia are absorbed to form a precipitate of copper sulfide and a solution of ammonium sulfate and sulfuric acid. Other impurities in the raw steam are also removed by the scrubbing operation.
2. Liquid-Solids Separation. A slip stream of the circulating scrubbing liquid is separated into a more concentrated slurry of copper sulfide solids and a purge stream of clear solution.
3. Regeneration. The slurry of copper sulfide solids is contacted with air to oxidize the precipitate to soluble copper sulfate. The resulting solution is recycled to the scrubber.
4. Copper Recovery. The purge stream of clear liquid from Step 2 is treated to recover dissolved copper for recycle to the process.

Features of the process are described in a 1980 patent to Harvey and Makrides (78), and the results of demonstration tests are presented in a U.S. Department of Energy report (79).

In the demonstration testing a stream of approximately 100,000 lb/h of steam was treated in a 6-ft-diameter by 30-ft-high scrubber containing a single sieve tray plus mesh-type demister pads. Hydrogen sulfide removal efficiency averaged 97 percent during the 6-month test campaign and occasionally exceeded 99 percent. Approximately 80 percent of the ammonia in the feed stream was removed. The other process steps, including regeneration of copper sulfide, were also demonstrated.

Hydrogen sulfide release to the atmosphere at geothermal power plants is currently minimized by adding an iron salt to the circulating condensate water and using the Stretford process on vent gases from the condenser. The EIC process is claimed to offer the advantages of treating a single low volume gas stream (high-pressure steam); purifying steam before it enters the turbine (and thus assuring that steam which may be vented from the turbine will be clean); removing other harmful impurities before they enter the turbine; and eliminating the need for operating the condensate system as a slurry of suspended iron compound particles in water (80).

References

1. Terres, E. 1953. *Gas-u. Wasserfach* 94(9):260.
2. Terres, E.; Buscher, H.; and Matroff, G. 1954. *Brennstoff-Chem.* 35(9/10):144.
3. Gollmar, H.A. 1945. *Chemistry of Coal Utilization.* Edited by H.H. Lowry. New York: John Wiley & Sons, Inc., Chap. 26.
4. Hill, W.H. 1945. *Chemistry of Coal Utilization.* Edited by H.H. Lowry. New York: John Wiley & Sons, Inc., Chap. 27.
5. Sperr, F.W. 1926. *Gas Age-Record* 58:73.
6. Sperr, F.W. 1932. U.S. Patent 1,841,419.
7. Gard, C.D. 1948. *Calif. Oil World* 41(Dec.):3.
8. Bailey, E.J. 1966. *Proc. Western Gas Processors and Oil Refiners Assoc.* (Oct. 6-7).
9. Gluud, W., and Schoenfelder, R. 1927. *Chem. & Met. Eng.* 34(12):742.
10. Roberts, C.B., and Farrar, H.T. 1956. "The treatment of gaseous and liquid effluents attendant in producing viscose cellulose film." *Roy. Soc. Promotion Health J.* 76:36-44.
11. Townsend, L.G. 1953. "Operation of the Manchester liquid purification plant at Linacre, Liverpool." *Inst. Gas Engrs., Commun. 429.*
12. R. & J. Dempster, Ltd. (Manchester, England). 1957. Private communication (July 12).
13. Anon. 1957. *Chem. Engr.* 64:141.
14. Gollmar, H.A. 1929. U.S. Patent 1,719,177.
15. Gollmar, H.A. 1929. U.S. Patent 1,719,762.
16. Jacobson, D.L. 1929. U.S. Patent 1,719,180.
17. Sauchelli, V. 1933. *Ind. Eng. Chem.,* 25:363.
18. Dunstan, A.E. (ed.). 1938. *The Science of Petroleum.* New York: Oxford University Press, p. 1807.
19. Powell, A.R. 1936. *Chem. & Met. Eng.* 43(6):307.
20. Koppers Company, Inc. *Form D-1, E-7700.*
21. Gollmar, H.A. 1934. *Ind. Eng. Chem.* 26(2):130.
22. Denig, F. 1933. *Gas Age-Record* 71:593.
23. Farquhar, N.G. 1944. *Chem. & Met. Eng.* 51(7):94.
24. McBride, R.S. 1933. *Chem. & Met. Eng.* 40(8):399.
25. Foxwell, G.E., and Grounds, A. 1939. *Chem. & Ind. (London)* 58(8):163.
26. Jegorov, N.N.; Dimitriev, M.M.; and Sikov, D.D. 1954. *Desolforazione dei gas.* Milan: Editore Ulrico Hoepli.
27. Giammarco, G. 1955. Italian Patent 537,564; 1956. Italian Patent 560,161; 1957. Italian Patent 565,320.
28. Riesenfeld, F.C., and Mullowney, J.F. 1959. *Petrol. Refiner* 38(May):161.
29. Jenett, E. 1962. *Oil & Gas J.* 60(Apr. 30):72.
30. Anon. 1960. *Chem. Eng.* 67(19):166-169.
31. Anon. 1960. *Sulfur* (Special Issue):35-38.
32. Fischer, F. 1932. U.S. Patent 1,891,974.
33. Mueller, H. 1931. *Gas-u. Wasserfach* 74(28):653.
34. Thau, A. 1932. *Gas World* 97:144.
35. Pieters, A.J., and van Krevelen, D.W. 1946. *The Wet Purification of Coal Gas and Similar Gases by the Staatsmijnen-Otto Process.* Amsterdam: Elsevier Publishing Company.
36. Craggs, H.C., and Arnold, M.H.M. 1947. *Chem. & Ind. (London)* 59:571.
37. Davis, J.E.; Mills, J.E.; and Ryden, C. 1958. *Gas Council (Gr. Brit.), Research Commun. GC 45.*

38. Reinhardt, K. 1956. *Energietechnik* 6(10):454.
39. Pippig, H. 1953. *Gas-u. Wasserfach* 94:62.
40. Anon. 1957. *Coke and Gas* 19:412.
41. Brommer, H., and Luhr, W. 1956. *Stahl u. Eisen* 76(7):402.
42. Nicklin, T., and Brunner, E. 1961. *Inst. Gas Engrs. (British)* Pub. *593*.
43. Nicklin, T., and Holland, B.H. 1963. *Removal of Hydrogen Sulfide from Coke Oven Gas by the Stretford Process.* Paper presented at European Symposium on Cleaning of Coke Oven Gas, Saarbrucken (March).
44. Thompson, R.J.S., and Nicklin, T. 1964. *Le Procede Stretford.* Paper presented at Congress of Association Technique de l'Industrie du Gaz en France.
45. Ellwood, P. 1964. *Chem. Engr.* 71(July 20):128.
46. Anon. 1967. *The Stretford Process.* North Western Gas Board.
47. Ludberg, J.E. 1971. *Removal of Hydrogen Sulfide from Coke Oven Gas by the Stretford Process.* Paper presented at 64th Annual Meeting of Air Pollution Control Association (June).
48. Hasebe, N. 1970. *Chemical Economy & Engineering Review (Japan)* 2(Mar.):27.
49. Townsend, F.M. 1965. U.S. Patent No. 3,170,766.
50. Reid, L.S., and Townsend, F.M. 1958. *Oil Gas J.* 56(Oct. 13):120.
51. Renault, P. 1969. U.S. Patent No. 3,441,379.
52. Barthel, Y.; Bistri, Y.; Deschanps, A.; Renault, P.; Simandoux, J.C.; and Dutriau, R. 1971. *Hydroc. Process.* 50(May):89-91.
53. Wiewiorowski, T.K. 1969. U.S. Patent No. 3,447,903.
54. Anon. 1970. *Ch. Eng. News* 48(Apr. 27):68-69.
55. Anon. 1966. *Oil Gas J.* 64(Jun.):58.
56. Meuly, W.C., and Ruff, C.D. 1972. *Paper Trade Journal* (May 22).
57. Meuly, W.C. 1973. "Cataban Process for the Removal of Hydrogen Sulfide from Gaseous and Liquid Streams," Paper presented at Twelfth Annual Purdue Air Quality Conference, Purdue University, (Nov. 7-8).
58. Kasai, T. 1975. *Hydro. Process.* 54(2):93-95.
59. Anonym. 1977. *Sulphur* No. 129 (March/April):42-44.
60. Nicklin, T.; Riesenfeld, F.C.; and Vaell, R.P. 1973. "The Application of the Stretford Process to the Purification of Natural Gas," Paper presented at 12th World Gas Conference, Nice, France (June).
61. Moyes, A.J., and Wilkinson, J.S. 1974. *The Chemical Engineer* (Brit.), (February):84-90.
62. Vasan, S. 1978. *Oil Gas J.* 76(1):78-80.
63. Massey, M.J., and Dunlap, R.W. 1975. *J. of Air Pollution Control Asso.* 25(10):1019-1027.
64. Penderleith, Y. 1977. "Stretford Plant Operating Experience at DOFASCO," Paper presented at McMaster Symposium on Treatment of Coke-Oven Gas, McMaster University, Hamilton, Canada (May).
65. Carter, W.A.P.; Rodgers, P.; and Morris, L. 1977. "Gas Desulphurisation by the Stretford Process and the Development of a Process to Treat Stretford Effluent," ibid.
66. Mitachi, K. 1973. *Chem. Eng.* 80(24):78-79.
67. Kozumi, T.; Idzutsu, W.; Swain, C.D.; Tsuruok, H.; and Tsuchiya, T. 1977. "Coke Oven Gas Desulfurization by the Takahax Process," ibid.
68. Andrews, J.W.; Bonnifay, P.; Cha, B.Y.; Barthel, Y.; Deschamps, A.; Frankowiak, S.; and Renault, P. 1976. *J. of Air Pollution Control Asso.* 26(7):664-667.

69. Macák, J., Prosek, K., and Kustka, M. 1982. "Gasification of Lignite with High Sulfur Content and the Crude Gas Treatment," presented at the 15th World Gas Conference, Lausanne.

70. Mackinger, H., Rossati, F., and Schmidt, G. 1982. *Hydrocarbon Processing,* March, p. 169.

71. Anon, "Better Ways to Remove H₂S from Sour Gas, *Chemical Week,* August 3, 1983.

72. Wilson, B. M., and Newell, R. D. 1984. "H₂S Removal by the Stretford Process—Further Development by the British Gas Corporation," paper presented at National AIChE Meeting, Atlanta, Georgia, March 13.

73. Donovan, J. J., and Laroche, R. J. 1981. "Two Approaches to Effluent Treatment of a Coke Oven Gas Desulfurization Process," paper presented at 64th Chemical Conference and Exhibition, Chemical Institute of Canada, May 31 to June 3.

74. Fenton, D. M., and Gowdy, H. W. 1981. U.S. Patent 4,283,379.

75. Hass, R. H., Fenton, D. M., Gowdy, H. W., and Bingham, F. E. 1982. "Selectox and Unisulf: New Technologies for Sulfur Recovery," paper presented at International Sulfur '82 Conference, London, England, Nov. 14–17.

76. Hamamura, K., Sumitomo Metals Industries, Ltd., Japan, publication "The Operating Progress of Fumaks Desulfurization Facilities."

77. Brown, F. C. and Dyer, W. H. 1980. Geothermal Resources Council, TRANSACTIONS, Vol. 4 (September), p. 667.

78. Harvey, W. W., Makrides, A. C. 1980. U.S. Patent 4,192,854 (March 11).

79. Pacific Gas and Electric Company 1980. "Demonstration of EIC's Copper Sulfate Process for Removing Hydrogen Sulfide and Other Trace Contaminants from Geothermal Steam at Turbine Inlet Temperature and Pressures," DOE/RA/27181-01 Final Report (May).

80. Dagani, R. 1979. *Chemical and Engineering News 75* (December 3), p. 29.

10

Removal of Basic Nitrogen Compounds from Gas Streams

Nitrogen compounds occur in most gases resulting from carbonization or cracking operations—typically coal carbonization and cracking of petroleum or shale oil. By far the preponderant nitrogen-containing compound present in such gas streams is ammonia, and its removal is considered a necessity in practically all cases. In addition to ammonia, pyridine and its homologues (usually referred to as "pyridine bases") and certain acidic nitrogenous compounds are also present in carbonization and cracked gases. This chapter covers the removal of basic nitrogen compounds from gas streams either by solution in water or by reaction with strong acids, or a combination of both. Processes for the removal of acidic nitrogen compounds, such as hydrogen cyanide and oxides of nitrogen, are covered elsewhere in this text. Although the processes described in this chapter are primarily designed to remove basic materials, simultaneous removal of at least a portion of certain acidic compounds, if present in the gas, is inevitable in those processes using water as the absorbent.

Because of the great industrial importance and long history of the removal of basic nitrogen compounds from coal gases, particular emphasis is given to this subject although sufficient material of a general nature is presented to permit application to purification problems involving other gas streams.

During the carbonization or gasification of coal and other carbonaceous fuels, a portion of the nitrogenous material contained in the fuel is converted to volatile compounds which appear in the gaseous carbonization product. The principal nitrogen compounds which have been identified in such gases are ammonia, cyanogen, hydrogen cyanide, pyridine and its homologues, nitric oxide, and free nitrogen. Although the nitrogen present in the fuel is the primary source of these compounds, small amounts of atmospheric nitrogen

551

Table 10-1. Distribution of Nitrogen in Carbonization Products

Products	Distribution
In the coke	30-50 percent
As ammonia in the gas	10-15 percent
As cyanide compounds in the gas	1-2 percent
In the tar	1-3 percent
As free nitrogen in the gas	Balance

Table 10-2. Typical Concentrations of Nitrogen Compounds in Coal Gases

Compound	Volume %
Free nitrogen	0.5–1.5
Ammonia	1.1
Hydrogen cyanide	0.10–0.25
Pyridine bases	0.004
Nitric oxide	0.0001

(which enter the carbonization apparatus through leakage) also contribute to the presence of nitrogenous materials in the gas stream.

The distribution and concentration of nitrogen compounds in the products of coal carbonization vary, for a given coal, over a wide range, depending on the operating conditions of retorts and coke ovens. The principal operating variables influencing the distribution of nitrogen compounds are carbonization temperature, speed of carbonization, and the quantity of steam used, with the carbonization temperature having the most pronounced effect. For example, the fraction of the nitrogen contained in coal which is converted to ammonia varies from about 2 percent at a carbonization temperature of 400°C (750°F), which is typical for low-temperature carbonization processes, to 10 to 15 percent at 900°C (1650°F) or higher, the range of typical high-temperature carbonization installations. Under normal high-temperature carbonization conditions, the approximate distribution of nitrogen in the carbonization products is given in Table 10-1.

Rather detailed discussions of the effect of carbonization conditions on nitrogen distribution, with many data from European and American sources, are presented by Kirner (1) and Hill (2).

Typical concentrations of nitrogen compounds in coal gases are shown in Table 10-2.

Because of their corrosive and toxic nature, most of these compounds must be removed from coal gases prior to industrial or domestic use. In addition, ammonia and the pyridine bases are relatively valuable chemicals, and their recovery as by-products may be of considerable economic significance. In some instances it may be considered economical to remove hydrogen cyanide: as the pure chemical; or "blue" in iron oxide boxes (see Ch. 8); or in the form of potassium ferrocyanide for use in certain processes for removing hydrogen sulfide from gases (see Ch. 9). However, the most important by-product is ammonia, and ammonia-recovery plants are integral parts of most gasworks and steel mills.

Before the advent of synthetic ammonia processes, by-product ammonia from coal carbonization and gasification constituted the most important source of fixed nitrogen. At present, by-product ammonia, although accounting for only about 10 percent of the total ammonia produced in the United States, is still an important industrial chemical, especially in the form of ammonium sulfate. Approximately 80 to 90 percent of the ammonia recovered from coal carbonization is converted to ammonium sulfate, which is one of the principal sources of nitrogen in synthetic fertilizers. Although production of ammonium sulfate from synthetic ammonia has been rising rapidly since 1930, by-product ammonium sulfate still accounts for about 50 percent of the total ammonium sulfate produced in the United States. Figure 10-1 shows the production of synthetic and by-product ammonium sulfate in the United States between 1930 and 1958 (3).

The competition of ammonium sulfate made from synthetic ammonia and recent changes in the pattern of nitrogenous-fertilizer utilization initiated the development of processes capable of converting by-product ammonia to more valuable products. Possible alternate products which have been considered are diammonium phosphate, anhydrous ammonia, and, by utilization of the carbon dioxide present in the gas, urea (4). At present, several steel mills in the United States are manufacturing diammonium phosphate from by-product ammonia and furnace phosphoric acid (5). Recovery of anhydrous ammonia and production of urea is being investigated on a pilot-plant scale in Europe.

Since coal-gas purification has been practiced for many years, it is not surprising that the literature covering the removal of ammonia and the pyridine bases is quite abundant and repetitious. No attempt is made in this chapter to discuss all processes which have been developed. On the contrary, only the most important commercially used processes are described, and, in order to avoid further repetition, the discussion of well-known processes which are adequately treated in standard texts is confined to basic process features. Excellent reviews of many processes with very extensive bibliographies are presented by Hill (2) and Wilson (6).

The material presented in this chapter is further limited to the description of the gas-purification operation proper, specifically the absorption of ammonia and the pyridine bases in liquid media.

Detailed discussion of processes for converting the ammonia removed from gas streams to salable products, such as ammoniacal liquors, ammonium sul-

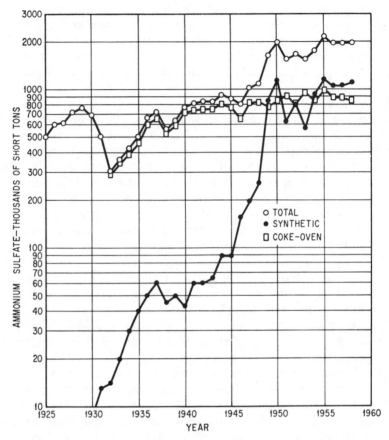

Figure 10-1. Production of ammonium sulfate in the United States between 1930 and 1958. *Stanford Research Institute*

fate, diammonium phosphate, and other ammonium salts, is considered to be beyond the scope of this text. The subject is treated quite extensively in the literature on industrial chemistry and coal-gas technology. The reader is especially referred to the comprehensive presentations of Hill (2), Wilson (6), Wilson and Wells (7) Bell (8) and Key (9).

Removal of Ammonia

Separation of ammonia from coal-carbonization gases is by far the most important industrial application of ammonia-removal processes. However, the basic principles underlying processes used for coal-gas purification are directly applicable to ammonia removal from other gas streams such as refinery gases and gases produced in shale retorting.

The technology of ammonia separation from coal gases has changed very little during the past 30 years; in fact, a modern installation is hardly distinguishable from a plant constructed many years ago. This is understandable if the simplicity of the physical and chemical phenomena involved is considered. Practically all processes in commercial use are based on washing the gas stream either with water or with a strong acid and removing the ammonia either in the form of aqueous solutions of ammonium hydroxide and ammonium salts of the acidic constituents also present in the gas or as the salt of the acid used as the absorbent.

Coal gases contain varying amounts of volatile acidic compounds, e.g., hydrogen chloride, hydrogen sulfide, hydrogen cyanide, carbon dioxide, and organic acids, which combine with ammonia during cooling of the gas and, because of the solubility of the resulting salts in water, are partially removed in processes using water as the absorbent. Ammonium salts of strong acids (principally ammonium chloride) are usually referred to as "fixed salts" while easily dissociable salts of weak acids such as the carbonate, bicarbonate, sulfide, hydrosulfide, and others are known as "free salts." Methods for the recovery of the ammonia from its various salts formed in the gas-purification processes are described briefly in later portions of this chapter.

As pointed out elsewhere in this text (Ch. 9), many attempts have been made to develop processes for the simultaneous removal of hydrogen sulfide and ammonia and the recovery of both compounds in the form of ammonium sulfate and elemental sulfur. Although some of these processes have been operated commercially with limited success, no generally accepted process was ever evolved. Another approach to the utilization of the ammonia contained in coal gases for the removal of H_2S is the recent development of the so-called selective H_2S-removal processes. These processes, which are based on the fact that under certain operating conditions H_2S reacts more rapidly than CO_2 with ammonia in aqueous solution, are discussed in Ch. 4.

Description of Processes

The three most commonly used commercial processes for removing ammonia from coal gases are the indirect, the direct, and the semidirect processes. Several modifications of the basic processes and a few other methods unrelated to these processes have been proposed. However, only the three above-mentioned processes are used on a large industrial scale. The direct process is fraught with many operational difficulties and, therefore, used only in a few installations. The indirect process, which is the oldest of the three, is most popular in Europe, while the semidirect process is used in most American installations. The indirect process is capable of yielding a variety of products, such as weak and strong ammoniacal liquors, anhydrous ammonia, and ammonium salts of weak and strong acids. Crude ammonium sulfate is the only product obtained in the direct process. In the semidirect process, which is a combination of the direct and indirect processes, either all of the ammonia is converted to the ammonium salt of a strong acid or only a portion is processed in this way while the rest may be treated in the same manner as in the indirect process.

In all three processes, the hot crude gas leaving the retorts or coke ovens is precooled by direct contact with a large amount of a mixture of cool tar and weakly ammoniacal aqueous solution which is sprayed directy into the collecting main. During this operation the gases are cooled to about 75° to 100°C (167° to 212°F) and most of the fixed ammonium salts (typically about 30 percent of the NH_3 originally present in the gas) as well as a major quantity of the tar are removed. The liquid, which is known as the "flushing liquor," is recycled to the collecting main after most of the tar has been decanted. A portion of the flushing liquor is continuously withdrawn from the cycle, combined with other liquid streams containing relatively low concentrations of ammonia (the so-called "weak ammoniacal liquor"), and further processed for recovery of the ammonia. The liquid removed from the cycle is replaced by condensate from direct or indirect coolers through which the pre-cooled gas passes after leaving the collecting main. From this point on the three processes take different routes.

The Indirect Process

This process has been in commercial operation for many years and is described in detail in many textbooks (10 to 13) and technical articles. A diagrammatic flow sheet is shown in Figure 10-2. The gas, which after partial cooling by direct contact with flushing liquor is saturated with water vapor, passes through the primary coolers where the temperature is lowered to about 30°C (86°F). Two types of primary coolers may be used, the indirect, shown in Figure 10-2, and the direct type. Indirect coolers, which are more commonly used, are shell-and-tube heat exchangers with the water flowing through the tubes. Direct coolers consist of packed towers, usually filled with wood hurdles, in which the gas is cooled by direct contact with a countercurrent stream of a weakly ammoniacal solution. The hot solution leaving the bottom of each tower is passed through water-cooled coils and recycled to the top of the towers. A portion of the circulating liquid is continuously withdrawn and added to the flushing-liquor cycle. The condensate from the indirect coolers, which contains some light tar and about 40 percent of the ammonia originally present in the coal gas, mostly as free salts, is collected in a decant tank. The tar is separated and the aqueous phase is added to the circulating flushing liquor.

The cooled gas flows to an exhauster and is pumped to the ammonia scrubbers where it is washed in several countercurrent stages operating in series, first with weak ammonia solution and finally with water. Electrostatic

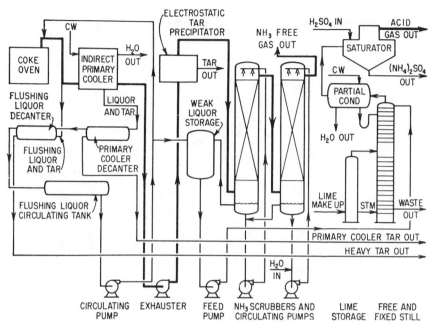

Figure 10-2. Diagrammatic flow sheet of indirect ammonia-removal process.

precipitators are usually installed ahead of the ammonia scrubbers in order to remove tar mist entrained in the gas. The essentially ammonia-free gas leaving the last ammonia scrubber passes to the light-oil recovery and desulfurization sections of the gas-purification system. The minute amount of ammonia remaining in the gas after passing the scrubbers is removed in the iron oxide boxes in the form of various ammonium salts. This ammonia aids in maintaining the proper pH of the iron oxide, and, in some cases, incomplete ammonia removal is practiced deliberately in order to obtain maximum efficiency of the dry-box purifiers (see Ch. 8). Although packed towers are commonly used as ammonia scrubbers in the United States, any type of device, cocurrent or countercurrent, static or dynamic, which ensures efficient contact between gas and liquid is satisfactory for ammonia absorption.

The effluent weak liquor from the bottom of the first ammonia scrubber is collected, together with a portion of the flushing liquor, in a storage tank which serves as a feed reservoir for an ammonia-distillation column. Typical weak liquors obtained from the treatment of coke-oven gas contain approximately 10 g/liter of free and 6 g/liter of fixed ammonium salts.

The weak liquor is pumped from the storage tank to the top of an ammonia distillation column, usually a bubble-plate column, which is heated at the bottom by direct steam. The weakly acidic ammonium salts are dissociated in the so-called "free still" while the fixed ammonium salts are decomposed,

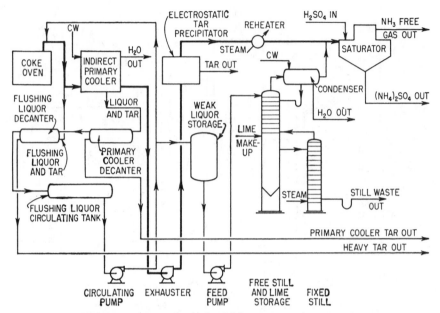

Figure 10-3. Diagrammatic flow sheet of semidirect ammonia-removal process.

after addition of a lime slurry, in the "fixed still." The free and fixed stills may be located in the same vessel—the free still occupying the upper section—and a separate vessel is used as the reservoir for the lime slurry. The two stills may also consist of two separate vessels arranged in such a manner that the bottom of the free still is at the same elevation as the top of the fixed still. In this case the lime-slurry reservoir is placed in the bottom section of the free still. The two different types of construction are shown in Figure 10-2 and 10-3 respectively.

The overhead vapors from the column contain ammonia, acid gases, and water vapor and are either condensed and processed to crude concentrated ammoniacal liquor or passed through strong acid in which the ammonia is absorbed and converted to the ammonium salt, usually ammonium sulfate. Pure concentrated ammoniacal liquor can be obtained if the acid gases are separated from the ammonia before condensation of the vapors. The separation is effected by conventional fractionation methods. Before contacting the overhead vapors with the acid, part of the water vapor is removed from the gas stream by condensation in a partial condenser. The acid-absorption step, which is essentially the same in all three methods, is described with the semidirect process. Practically ammonia-free water, containing calcium salts, leaves the bottom of the fixed still and is discarded.

The principal advantages of the indirect process are its flexibility in yielding a variety of products, its relative lack of interdependence of the various process steps, and the complete removal of tar. Its main disadvantages are the necessity of handling very large volumes of liquid and the attendant high steam consumption and, consequently, high operating costs. In addition, the capital investment is high and the ground-space required is appreciably larger than that for the other two processes. Finally, some ammonia is lost in the successive steps of the process and in storage.

The Direct Process

This process, which was proposed by Brunck (14), is designed to eliminate the necessity of recovering ammonia from aqueous solutions before its conversion to ammonium sulfate. The hot gas leaving the retorts or coke ovens is kept at a temperature above its water dew point and passed directly through concentrated sulfuric acid. During absorption of the ammonia, an appreciable portion of the tar is also removed from the gas; this results not only in serious contamination of the ammonium sulfate, but also in acid degradation and contamination of tar. In addition, ammonium chloride contained in the gas is decomposed by the concentrated sulfuric acid, and the hydrogen chloride evolved causes extremely severe corrosion of the equipment. Some of the difficulties of the process can be alleviated by installation of very elaborate tar-separation systems; others, however, have proved to be almost insurmountable, and, as a result, the process has only been used in a few installations. The relative merits of the process as compared with the indirect and semidirect methods are discussed in some detail by Ohnesorge (15, 16).

The Semidirect Process

This process, which was developed by Koopers Company, Inc. (17), is a combination of the direct and indirect processes. A schematic flow diagram is shown in Figure 10-3. The precooled gas coming from the collecting main is cooled further to about 30°C (86°F) in the same manner as in the indirect process. The condensate from the primary coolers, which is essentially of the same composition as that obtained in the indirect process, is freed of tar, returned to the flushing-liquor cycle, and eventually collected as weak ammoniacal liquor. This relatively small quantity of liquid is distilled in an ammonia column and the overhead vapors are either condensed and treated as in the indirect process or recombined with the main gas stream at a point located ahead of the acid scrubbing vessel.

The effluent gas from the primary cooler flows to an exhauster whence it is pumped through an electrostatic tar precipitator to a reheater where its temperature is raised above its dew point, usually to 60°C (140°F). The

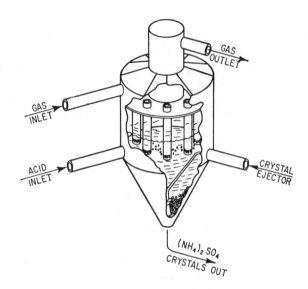

Figure 10-4. Saturator for production of ammonium sulfate.

reheated gas (which may contain the overhead vapors from the ammonia still) flows to the "saturator" where the ammonia is removed by reaction with a strong acid, in most cases sulfuric acid. The ammonia-free gas, which still contains most of the originally present weakly acidic gases, passes to the light-oil recovery and desulfurization operations.

Absorption of ammonia in acid (typically sulfuric acid) is carried out in a vessel called the "saturator." A diagrammatic view of a typical unit is shown in Figure 10-4. A discussion of different saturator designs is presented by Otto (18). Saturators are usually lead- or stainless steel-lined, although other materials such as Monel or acid-resistant refractories may also be used. A concentration of approximately 5 to 7 percent free acid is maintained in the saturator liquid by continuous addition of 60° Bé (77.6 percent) sulfuric acid. Periodically, usually every 24 hr, the liquid level in the saturator is raised and the free-acid content is increased to 10 to 12 percent. This is done to dissolve crystals of ammonium sulfate which have accumulated on the vessel walls and in the distributor pipes. The salt crystals formed are continuously removed from the bottom of the saturator by means of a compressed-air ejector. Liquid adhering to the crystals is first removed in a settling tank, then in a centrifuge, and is recycled to the saturator.

In the semidirect process the gas enters the saturator at a temperature of 50° to 60°C (122° to 140°F) which, together with the heat liberated by the reaction, is sufficient to maintain the saturator-bath temperature at approximately 60°C (140°F). Since the gas entering the saturator is unsaturated with water vapor, a considerable amount of water is evaporated from the aqueous acid solution, and the saturator functions in effect as an evaporator. Much higher temperatures—about 100°C (212°F)—are required in saturators operating in conjunction with installations, using the indirect process, in which the effluent gases from the ammonia still are saturated with water at 75° to 80°C (167° to 176°F). Thermal equilibria in saturators operating in indirect, direct, and semidirect processes are discussed in detail by Terres and Patscheke (19). A comprehensive review of ammonium sulfate production from by-product ammonia is given by Hill (2).

Diammonium phosphate may be obtained by using furnace phosphoric acid in conventional saturators. However, temperature control is somewhat more critical, and Monel is not suitable as a construction material. Production of many other ammonium salts, such as nitrate, bicarbonate, monophosphate, and others, has been proposed, but none of these are being manufactured on a large commercial scale.

The main advantage of the semidirect process over the indirect process is the fact that less than half the volume of aqueous ammonia solution is produced, and, therefore, the steam requirements for the ammonia distillation and operating costs are appreciably lower. In addition, the process has some flexibility because part of the ammonia can be converted to products other than the ammonium salts of strong acids. Furthermore, first cost and ground-space requirements are less than for installations using the indirect process, and, finally, smaller ammonia losses are incurred. Among the three major processes the semidirect method seems to be most advantageous from the standpoint of economics and freedom from operational difficulties.

Ammoniacal liquors produced in the indirect and semidirect processes contain, among other acidic constituents, varying amounts of so-called tar acids. These acids are a mixture of phenolic compounds, the compositions of which depends on the gasification process used. For example, liquors from coke-oven gases contain primarily phenol and only small quantities of higher molecular-weight phenolic compounds. In some cases recovery of tar acids from the ammoniacal liquors is economical. The processes most commonly used employ solvent extraction of the phenolic compounds with benzene or a phenol-free tar-oil fraction. The solvent is removed by distillation and recovered in the overhead vapors, while the tar acids remain as the residue. Another method involves neutralization of the rich solvent with aqueous sodium hydroxide, followed by separation of the solvent layer from the aqueous sodium phenolate layer. The tar acids are subsequently recovered from the aqueous phase by addition of sulfuric acid.

Typically, the extraction is carried out by countercurrently contacting the ammoniacal liquor with the solvent in a packed tower at a temperature of about 60°C (140°F) with the solvent entering at the bottom of the tower. About 80 to 95 percent of the phenol and some of the higher tar acids are normally- extracted from the liquor.

The USS Phosam Process

A process capable of yielding pure anhydrous ammonia from coke-oven gas has been developed by USS Engineers and Consultants, Inc., a subsidiary of United States Steel Corporation. The process, which has been named the USS Phosam Process (20) is covered by U.S. Patents 3,024,090 and 3,186,-795, as well as by patents in Canada, Germany and Great Britain. Although primarily developed for ammonia removal from coke-oven gases, the process is claimed to be suitable for selective absorption of ammonia from any gas stream containing acidic compounds such as hydrogen sulfide, hydrogen cyanide, and carbon dioxide. The selectivity for ammonia is very high, and the anhydrous ammonia product is reported to contain impurities in amounts ranging from undetectable to a few parts per milion (20).

The process flow scheme is a conventional absorption-regeneration cycle using an aqueous solution of ammonium phosphate as the absorbent which is stripped of ammonia by direct addition of steam. The absorbent liquid contains ammonia and phosphoric acid of the general formula $(NH_4)_n H_{3-n} PO_4$. When n is lower than 1.5, the solution is considered lean in ammonia, or regenerated, while values of n above 1.5 are characteristic for rich solutions.

In a typical coke oven gas application, the cooled gas which is free of entrained solids, water, and tar, enters the bottom of the absorber where it is contacted countercurrently with the absorbent and 98 to 99 percent of the ammonia is removed. The rich solution leaving the bottom of the absorber is heated by heat exchange with lean solution and subsequently with regenerator overhead vapors, and then enters the regenerator where the absorbed ammonia is stripped by direct addition of steam. The regenerator is usually operated under pressure, and the overhead vapors are condensed as aqueous ammonia. The aqueous ammonia may be further processed in an ammonia fractionator to obtain pure anhydrous ammonia.

At present one large unit using the Phosam process is operating in the Unites States, and five additional units are in operation or under construction in coke plants in Canada and Japan.

Miscellaneous Processes

Among the many other methods proposed for ammonia removal from gases, only absorption of ammonia in water at elevated pressures, condensation by refrigeration, and simultaneous absorption of ammonia, H_2S, and HCN in spent pickle liquor appear worth mentioning.

Still (21) proposed a process consisting of washing tar-free coal gas at pressures ranging from 100 to 200 atm by injection of water. The ammonia is collected as a concentrated solution of ammonium bicarbonate and ammonium sulfide. Solid salts are recovered from the solution after cooling, and the mother liquor is reused for further ammonia absorption.

A process developed in Germany (22) operates at a pressure of 4 atms and a temperature about 140°C (248°F). Dilute ammonia solution is injected into the gas stream, and the ammonia is recovered as a 10 percent liquor. Naphthalene is removed as a liquid by subsequent cooling of the gas to about 50°C (122°F). The last traces of ammonia are eliminated by absorption in water at 15°C (59°F).

Ammonia removal by refrigeration of the gas—using absorption refrigeration—has been proposed by Lenze and Rettenmaier (23). When typical coal gas is cooled to $-10°$ to 0°C (14° to 32°F), an aqueous solution containing 20 to 30 percent ammonium carbonate can be obtained. However, in order to prevent formation of solids, the ammonium carbonate concentration is usually kept at about 6 percent. In the process of Lenze and Rettenmaier, the crude gas from the primary coolers is contacted with a spray of refrigerated ammonia liquor and subsequently washed with water for removal of the last traces of ammonia. The advantages claimed for this process are primarily the production of a very small volume of liquid (less than in the semidirect process) and, consequently, a low steam requirement in the distillation step. In addition, almost complete tar removal and elimination of large quantities of naphthalene and water from the gas are obtained simultaneously with ammonia removal. Finally, the process is capable of yielding the same variety of end products as the indirect process. A number of plants utilizing this method have been operating in Europe for several years.

A process utilizing waste pickle liquor for the simultaneous removal of ammonia, hydrogen sulfide, and hydrogen cyanide from coke-oven gas has been described by Dixon (24). This process, which involves a rather complicated flow scheme, is known as the F-S process and was developed in Germany by F.J. Collin, A.G. The coke-oven gas is first contacted in a countercurrent six-stage spray tower with a weakly acidic ferrous sulfate solution. Ammonia reacts in the upper section of the tower with free acid and ferrous sulfate, forming ammonium sulfate and ferrous hydroxide. Some of the ferrous hydroxide may be removed from the solution by filtering a side stream withdrawn from the bottom of the second upper stage. The clear liquid is subsequently returned to the same stage. In the lower portion of the tower, the remaining ferrous hydroxide reacts with hydrogen sulfide and hydrogen cyanide to form insoluble ferrous sulfide and ammonium-ferrocyanide complexes. The solids are separated from the solution leaving the bottom of the absorption tower, and a portion of the liquid is recycled to the middle of the tower. The remainder of the solution, which contains the equivalent of the ammonia present in the inlet gas, is treated with hydrogen sulfide to precipitate dissolved iron, filtered and freed of H_2S by contact with a small stream of purified coke-oven gas. The purified solution is then evaporated, and ammonium sulfate crystals are recovered.

The solid sludge is treated with waste pickle liquor which results in the liberation of H_2S and the formation of ferrous sulfate solution. The H_2S is converted to sulfuric acid or elemental sulfur, and the solution is recycled to the top of the absorption tower, usually after adjustment of pH. Undissolved ammonium-ferrocyanide complexes and ferrous hydroxide obtained previously in the process are returned to the blast furnaces. Advantages claimed for the process are simultaneous removal of ammonia, hydrogen sulfide and hydrogen cyanide in one operation and utilization of pickel liquor, which is normally wasted, for the manufacture of saleable products. Drawbacks of the process are its complexity and the corrosiveness of the solution, which necessitates the use of special corrosion-resistant construction materials.

Process Design

Basic Data

The specific gravity of aqueous ammonia solutions as given by Ferguson (25) is shown in Figure 10-5.

Vapor pressures of ammonia over aqueous ammonia solutions and vapor-liquid equilibria of such solutions at the atmospheric boiling points are presented in Figures 10-6 and 10-7. Acidic compounds, e.g., H_2S, CO_2, and HCN, present in some industrial gases, especially coal-carbonization gases, materially affect the vapor pressure of ammonia over aqueous solutions obtained in the treating process. Equilibrium vapor pressure of ammonia, hydrogen sulfide, and carbon dioxide over aqueous solutions containing either two or all three of these components are shown in Ch. 4 (Figures 4-1 to 4-4), and calculation methods are given to extend the range of the data presented in the figures. Vapor/liquid equilibria of typical coke-oven gases

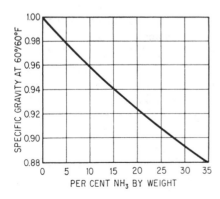

Figure 10-5. Specific gravity of aqueous ammonia solutions. *Data of Ferguson (25)*

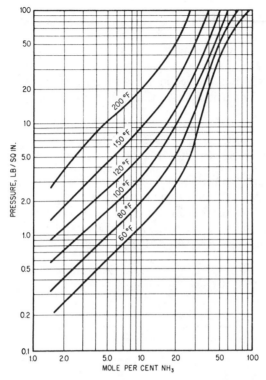

Figure 10-6. Equilibrium vapor pressure of ammonia over aqueous solutions. *Data from Perry*, Chemical Engineers' Handbook (27)

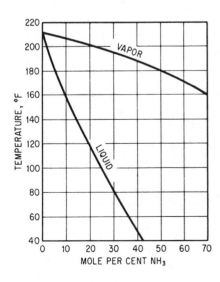

Figure 10-7. Boiling point diagram for aqueous ammonia solutions at 1 atm total pressure. *Data from Perry*, Chemical Engineers' Handbook (27)

and liquids obtained in ammonia scrubbers (indirect process) are shown in Figure 10-8 (26).

Differential and integral heats of solution of liquid ammonia in water are given in Tables 10-3 and 10-4 (27).

The heat of reaction evolved in the saturator during absorption of gaseous ammonia in sulfuric acid, based on heat-of-formation data, is shown in Equation 10-1).

$$2NH_3(g.) + H_2SO_4(aq.) \rightarrow (NH_4)_2SO_4(aq.)$$
$$-\Delta H = 86,000 \text{ Btu/lb mole} \tag{10-1}$$

This is augmented by the heat of dilution of sulfuric acid from 60° Bé (77.6 percent) to 7 percent which amounts to 15,000 Btu/lb mole. The total heat effect is, therefore, 101,000 Btu/lb mole of ammonium sulfate.

The heat evolved in the formation of diammonium phosphate from ammonia and phosphoric acid may be estimated from Equation (10-2).

$$2NH_3(g.) + H_3PO_4(aq.) \rightarrow (NH_4)_2HPO_4(aq.)$$
$$-\Delta H = 75,000 \text{ Btu/lb mole} \tag{10-2}$$

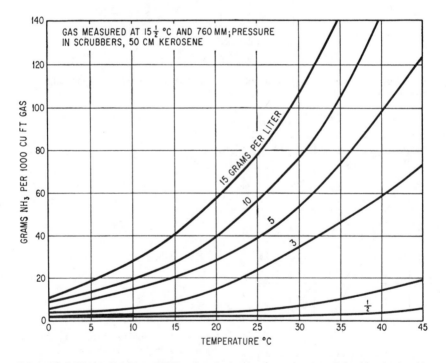

Figure 10-8. Vapor/liquid equilibria of typical coke-oven gases and liquids obtained in ammonia scrubbers (indirect process). *From* Gas Engineers' Handbook (26)

**Table 10-3. Differential Heats of Solution of Liquid Ammonia
(Btu per pound of ammonia dissolved)**

Concen-tration, wt %	Heat of solution	Concen-tration, wt %	Heat of solution	Concen-tration, wt %	Heat of solution	Concen-tration, wt %	Heat of solution	Concen-tration, wt %	Heat of solution
0	347.4	11	302.8	21	253.8	31	197.6	41	135.0
1	343.8	12	298.2	22	248.4	32	191.9	42	127.8
2	340.2	13	293.6	23	243.0	33	186.1	43	120.6
3	336.6	14	289.0	24	237.6	34	180.4	44	113.4
4	333.0	15	284.4	25	232.2	35	174.6	45	106.2
5	329.4	16	279.4	26	226.4	36	168.1	46	99.0
6	325.0	17	274.3	27	220.7	37	161.6	47	91.8
7	320.6	18	269.2	28	214.9	38	155.2	48	84.6
8	316.2	19	264.2	29	209.2	39	148.7	49	77.4
9	311.8	20	259.2	30	203.4	40	142.2	50	70.2
10	307.4								

Table 10-4. Integral Heats of Solution of Liquid Ammonia

Concentration, wt %	Btu/lb mixture	Btu/lb NH₃
0	0	358.0
10	34.4	343.8
20	65.7	328.5
30	92.5	308.2
40	108.2	270.0
50	109.4	218.8
60	101.9	169.7
70	84.8	121.1
80	60.7	75.8
90	31.9	35.5
100	0	0

The density, viscosity, and vapor pressure of aqueous ammonium sulfate solutions and the solubility of ammonium sulfate in water can be estimated from the nomogram presented in Figure 10-9 (28). To obtain the density, viscosity, and vapor pressure, the temperature on the t scale is selected and aligned with the concentration on the C scale. The intersects of the line of the p, γ, μ, and d scales give the values for vapor pressure, viscosity, and density of the solution. The solubility of ammonium sulfate in water is obtained by aligning the temperature (on the t scale) with the saturation point S and reading the solubility on the C scale. The values obtained by the nomogram agree very closely with published data.

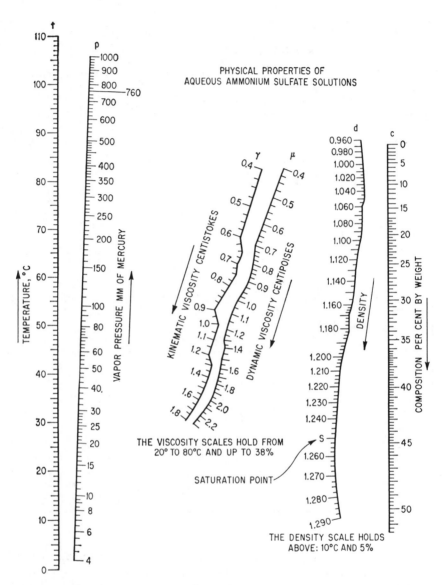

Figure 10-9. Nomogram for estimating density, viscosity, and vapor pressure of aqueous (NH₄)₂SO₄ solutions and the solubility of ammonium sulfate in water. *Data of Tans* (28)

Absorption

The rate of absorption of ammonia in different liquid media has been studied quite extensively, and data are available for a variety of absorption equipment. A list of the most important studies is given in Table 10-5. There

Table 10-5. Investigations on the Rate of Absorption of Ammonia in Towers

Authors	Column diameter in.	Packed height in.	Packing material	Gas	Solvent	Flow rate, lb/(hr) (sq ft) Liquid	Gas	Temp., °F	Remarks
Kowalke, Hougen, and Watson(30)	16	46	Spray tower	Air	Water	30–750	6–150	60	5 "Vermorel" nozzles
	16	41	Wood grids	Air	Water	21–670	19–240	68–110	
	16	41	No. 1 stoneware*	Air	Water	30–780	15–240	60–110	
	16	41	No. 2 stoneware†	Air	Water	30–780	15–240	60–110	
	10	19–31	Rings, 1-in.	Air	Water	570–830	55–530	77	Data of Borden and Squires
Sherwood and Holloway(31)	10	19–31	Rings, 1-in.	Air	Water	660–710	65–700	54	Data of Doherty and Johnson
	10	19–31	Rings, 1-in.	Air	0.5–4.5 N H₂SO₄	1,520–1,850	210	77	
	10	19–31	Rings and berl saddles, ½-in., 1-in.	Air	3.5–4.5 N H₂SO₄	75–480	3,300		Data of Withers
Dwyer and Dodge(32)	12	48	Rings, ½-in., 1-in., 1½-in.	Air	Water	100–1,000	100–1,000	74–88	
Fellinger(33)	20	9, 17, 25½, 25¾, 23¾ 20½, 22, 25	Rings, ⅜-in., ½-in., 1-in., 1½-in., 2-in.	Air	Water	500–4,500	200–1,000		
	20		Berl, saddles, ½-in., 1-in., 1½-in.	Air	Water	500–4,500	200–1,000		
	20	26	Triple spiral tile, 3-in.		Water	500–4,500	200–1,000		
Parsly, Molstad, Cress, and Bauer(34)			No. 6295 drip-point grid	Air	Water	1,900–15,000	100–1,000		
Pigford and Pyle(35)	31.5	26, 52	Spray tower	Air	Water	285–900	230–800		6 Sprayco 5-B nozzles

is agreement among all investigators that the rate of absorption of ammonia in water is primarily affected by the gas mass-velocity. Some authors found that, in wetted-wall columns and towers packed with certain materials, the effect of the liquid mass-velocity on the overall transfer coefficient is negligible and concluded that the rate of absorption of ammonia is entirely controlled by the gas-film resistance. However, other workers found definite effects of the liquid-flow rate on the overall transfer coefficient and therefore concluded that the absorption rate is controlled by both the gas- and liquid-film resistances. An excellent discussion of the theoretical aspects of ammonia absorption in both water and dilute acids is presented by Sherwood and Pigford (29).

Overall absorption coefficients (K_Ga) of ammonia in water, as a function of the gas mass-velocity, based on experimental data obtained by Kowalke, Hougen, and Watson (30) in a spray tower, with No. 1 and No. 2 stoneware and paraffined wood-grid packing, are shown in Figure 10-10. Similar data obtained by Parsly, Molstad, Cress, and Bauer (34) with No. 6295 drip-point grid packing are also included.

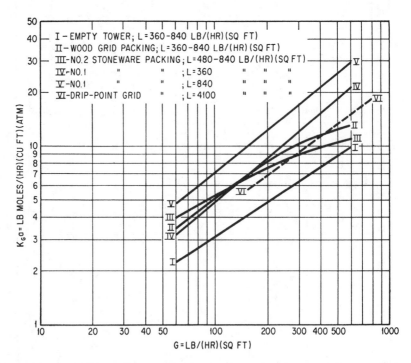

Figure 10-10. Over-all coefficients (K_Ga) for the absorption of ammonia in water. *Data of Kowalke et al. (30) and Parsly et al. (34)*

The experimental data of Kowalke, Hougen, and Watson show that, in general, K_Ga increases with liquid loading until a point is reached after which the liquid rate has very little effect. This point, which is a function of the packing, occurs at a liquid rate of approximately 400 lb/(hr)(sq ft) for wood grids, 500 for No. 2 stoneware, and higher than 840—the highest rate used in the experiments—for No. 1 stoneware. The experimental data also show that the gas-film resistance is about 50 times that of the liquid film, even at ammonia concentrations of 40 percent, and these authors conclude, therefore, that the transfer coefficients obtained represent essentially the transfer coefficients for the gas film alone. Parsly and coworkers (34) observed very small effects of the liquid-mass velocity on K_Ga at much higher rates [L = 1,900 to 15,000 lb/(hr)(sq ft)], which is consistent with the findings of Kowalke, Hougen, and Watson.

The most complete data on the rate of absorption of ammonia in water in packed towers have been presented by Fellinger (33) in terms of heights of overall transfer units based on the gas driving force (H_{OG}). Values obtained by Fellinger for nine different packing materials are presented in figures 10-11 to 10-19.

Fellinger's data indicate that H_{OG} is dependent on both the liquid and gas rates, and it is concluded that ammonia absorption in water is not controlled entirely by the gas-film resistance. This is contrary to the opinion of Kowalke and coworkers but confirms conclusions presented by Sherwood and Holloway (31) and Dwyer and Dodge (32). Values of H_{OG} decrease with increasing liquid rates, a phenomenon which is probably due to better wetting of the packing and greater liquid-film turbulence. There appears to be no simple relation between H_{OG} and packing size. Although H_{OG} increases slowly with increasing packing size, which is to be expected, this increase is not proportional to the reduction of packing-surface area as a result of larger packings.

Rates of absorption of ammonia in water in countercurrent spray towers have been determined by Kowalke, Hougen, and Watson (30) and by Pigford and Pyle (35). The data of the former authors are included in Figure 10-8. The data obtained by Pigord and Pyle for two different tower heights are shown in figures 10-20 and 10-21. The rate of absorption is expressed in terms of the number of overall transfer units (N_{OG}) for the entire tower height. Since performance of spray towers is almost entirely determined by the size and velocity of droplets, comparison between these data is difficult and no good explanation can be offered for the fact that Kowalke and coworkers find much higher transfer rates than Pigford and Pyle. Comparison of the values obtained by Pigford and Pyle with two different tower heights indicates that the number of transfer units does not change proportionally with the height of the tower and, therefore, that absorption is probably most rapid in the direct vicinity of the spray nozzles.

(text continued on page 576)

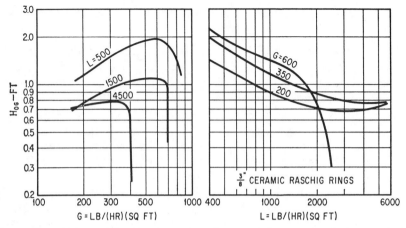

Figure 10-11. Values of H_{OG} for absorption of ammonia in water (packed tower, ⅜-in. ceramic raschig rings). *Data of Fellinger* (33)

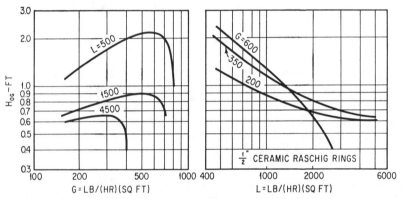

Figure 10-12. Values of H_{OG} for absorption of ammonia in water (packed tower, ½-in. ceramic raschig rings). *Data of Fellinger* (33)

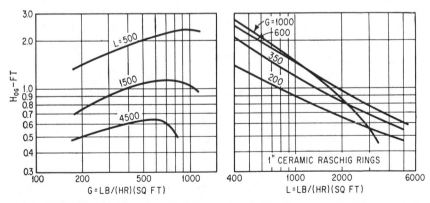

Figure 10-13. Values of H_{OG} for absorption of ammonia in water (packed tower, 1-in. ceramic raschig rings). *Data of Fellinger* (33)

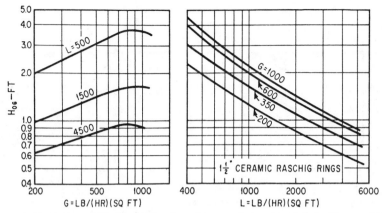

Figure 10-14. Values of H_{OG} for absorption of ammonia in water (packed tower, 1½-in. ceramic raschig rings). *Data of Fellinger* (33)

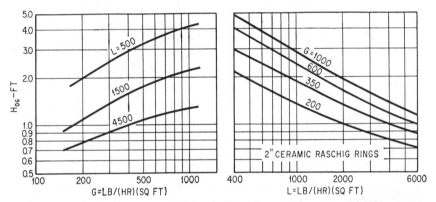

Figure 10-15. Values of H_{OG} for absorption of ammonia in water (packed tower, 2-in. ceramic raschig rings). *Data of Fellinger* (33)

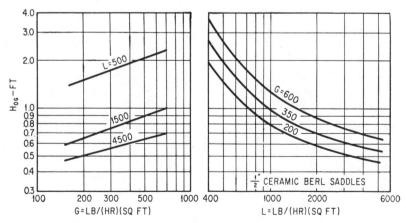

Figure 10-16. Values of H_{OG} for absorption of ammonia in water (packed tower, ½-in. ceramic berl saddles). *Data of Fellinger* (33)

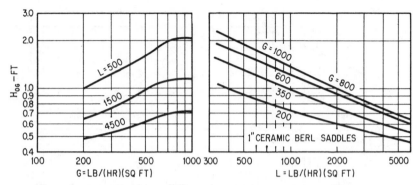

Figure 10-17. Values of H_{OG} for absorption of ammonia in water (packed tower, 1-in. ceramic berl saddles). *Data of Fellinger* (33)

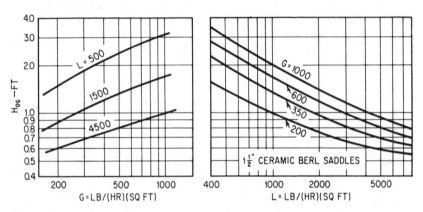

Figure 10-18. Values of H_{OG} for absorption of ammonia in water (packed tower, 1½-in. ceramic berl saddles). *Data of Fellinger* (33)

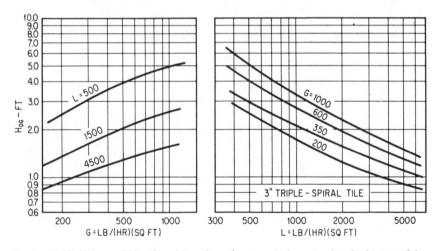

Figure 10-19. Values of H_{OG} for absorption of ammonia in water (packed tower, 3-in. triple-spiral tile). *Data of Fellinger* (33)

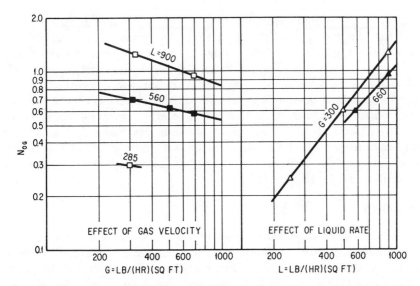

Figure 10-20. Number of transfer units (N_{OG}) for absorption of ammonia in water in a spray tower, 26 in. high. *Data of Pigford and Pyle* (35)

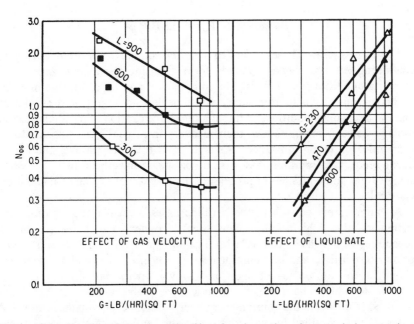

Figure 10-21. Number of transfer units (N_{OG}) for absorption of ammonia in water in a spray tower, 52 in. high. *Data of Pigford and Pyle* (35)

Conventional ammonia scrubbers packed with wood hurdles, such as those used for coal-gas purification, are normally designed by using the data of Kowalke, Hougen, and Watson. Design procedures for towers packed with different materials are discussed in detail by Sherwood and Pigford (29). Although packed towers and, in some instances, spray towers are customarily used for ammonia absorption in the United States, a variety of other contacting devices, employing mechanical means to effect optimum contact, have been proposed and are in use in European countries. Some of these devices are described by Hill (2). European practice for the design of multistage ammonia washers for coal-gas purification (indirect process) is discussed by Silver (36) and by Hopton (37).

Desorption

Ammonia-distillation columns used in most coke-oven and gas plants consist of flanged, cast-iron, bubble-cap sections and are heated at the bottom by admission of direct steam. The design of ammonia-distillation columns is usually based on practical experiences and certain empirical rules. Various designs are discussed by Gluud and Jacobson (11) and by Parrish (12). The latter recommends vapor velocities of 5 to 10 ft/sec. Although most stills operate at atmospheric pressure, vacuum operation has also been proposed and is practiced in some European installations.

Removal of Pyridine Bases

The term pyridine bases refers to a rather complex mixture of nitrogen-containing heterocyclic compounds formed during the carbonization and cracking of carbonaceous fuels. The composition and concentration of bases in the gaseous products from such operations depend largely on the fuel used and the operating conditions. Although the concentration of pyridine bases in most gas streams is quite low, their recovery from coal gases appears to be economical in certain instances. As a result of this, the removal and subsequent recovery of pyridine bases from coal gases have been investigated to some extent and a number of processes are presently practiced on a commercial scale.

An appreciable portion of the pyridine bases present in coal gases, consisting mainly of high-boiling constituents, condenses with the tar during precooling of the gas by contact with flushing liquor. The lighter compounds, primarily pyridine and its homologues, remain in the gas and are removed either in the ammonia- or light-oil-removal operations. In the indirect process for ammonia recovery, practically all of the pyridine bases are removed from gas in the ammonia scrubbers. Relatively complete removal of bases can be obtained in the saturator in the semi-direct process if proper operating conditions are maintained. Pyridine bases remaining in the gas after passing through the ammonia scrubbers or the saturator are washed out in the light-oil-recovery section of the gas-purification system.

Concentrations of pyridine bases in tar-free gas reported in the literature vary from 5 (38) to 30 (39) grains/100 cu ft. An analysis of bases recovered from coke-oven gas as reported by Klempt and Röber (38) is shown in Table 10-6.

Since, as pointed out above, pyridine bases are eliminated in the normal course of coal-gas purification, regardless of whether recovery is attempted or not, the main effort in the development of recovery processes has been directed toward absorption of the largest possible quantity of bases, in the operation of the purification system which would lend itself best for subsequent recovery. Because of the alkaline nature of the bases, absorption in sulfuric acid followed by separation upon neutralization of the absorption liquid appears to be the most logical method of recovery; in fact, all commercial processes are based on this principle. The bases are absorbed either simultaneously with the ammonia in one saturator or in a separate vessel located downstream from the ammonium sulfate saturator. Sulfates of pyridine bases are unstable at elevated temperatures; therefore, the efficiency of removal by absorption in sulfuric acid is determined by the equilibria between the pyridine bases contained in the gas and those in the absorption liquid. Equilibrium data for specific gas and liquid compositions at different temperatures were determined experimentally by Klempt and Röber (38), Dodge and Rhodes (40), Wald (42), and Meredith (39). Klempt and Röber (38) conclude from their experiments that a sulfuric acid concentration in the absorption liquid corresponding to about 200 percent of the stoichiometric quantity is required if essentially complete pyridine removal is to be obtained at operating temperatures around 100°C (212°F). Since operating conditions vary over a wide range in different installations no general conclusions regarding optimum operation can be drawn. In most cases it is necessary to determine vapor-liquid equilibria experimentally for the specific conditions involved before a recovery plant is designed.

Description of Processes

The first industrial process for recovery of pyridine bases from coal gases was developed by Dodge and Rhodes (40, 41). In this process, which is applicable to both the indirect and semidirect ammonia recovery processes, pyridine bases are absorbed simultaneously with the ammonia in the saturator until pyridine sulfate crystallizes with the ammonium sulfate. The saturator liquor is then transferred to a lead-lined still and neutralized with ammonia obtained from the overhead vapors of the ammonia-distillation column. The pyridine bases and some water are vaporized by the heat evolved during neutralization and subsequently condensed. After addition of 1 lb ammonium sulfate (solid) per gallon of condensate, the liquid separates into two

Table 10-6. Analysis of Pyridine Bases from Tar-free Coke-oven Gas	
Compound	Wt. %
Pyridine	69.0
α-Picoline	3.0
β-Picoline ⎱ 2-6,Lutidine ⎰	3.5
γ-Picoline	1.0
γ-Collidine	0.6
Aniline	4.0
Quinoline bases	12.0

phases, the upper of which consists of almost dry pyridine bases. The ammonium sulfate solution remaining in the still is returned to the saturator after removal of iron sulfide and other insoluble materials by filtration. This process was not generally accepted, mainly because its batch-type operation required a standby saturator.

A continuous process based on the method of Dodge and Rhodes was developed by Wald (42). In this process the pyridine content of the saturator liquor is maintained at a constant value of 10 g/liter by continuous removal of a side stream of 4 to 5 gpm. This liquor is neutralized with ammonia in a cooled tank whereupon it separates into two layers, one consisting of crude pyridine bases and the other of ammonium sulfate solution which is continuously recycled to the saturator. It is claimed that because of the low pyridine concentration maintained in the saturator liquor, almost complete removal of bases from the gas is achieved in this process.

A semicontinuous process was disclosed by Schutt (43), and its operation in conjunction with a semidirect ammonia-removal plant is described in some detail by Meredith (39). A diagrammatic flow sheet of this process is shown in Figure 10-22. Liquor is removed intermittently from the saturator and transferred to a small, lead-lined tank from which it is pumped, continuously at the rate of 1.5 gpm, to a neutralizing still. Overhead vapors from the ammonia-distillation column are introduced at the bottom of the neutralizing still, and the pyridine bases are steam-distilled. The flow of ammonia is controlled at such a rate that the effluent from the bottom of the neutralizing still is maintained at a pH of 6 to 7. Close pH control is quite critical because iron sulfide precipitates if the liquid is too alkaline and removal of pyridine bases is incomplete if it is too acid. The pyridine-free liquid is continuously returned to the saturator.

The overhead vapors from the neutralizing still are cooled in a tubular condenser to 92° to 96°C (198° to 205°F), and the condensate is collected in a separator, in which it splits into two layers. The upper layer, which contains all the pyridine bases, is continuously removed to storage while the lower layer, consisting of an aqueous solution of ammonium carbonate, is discharg-

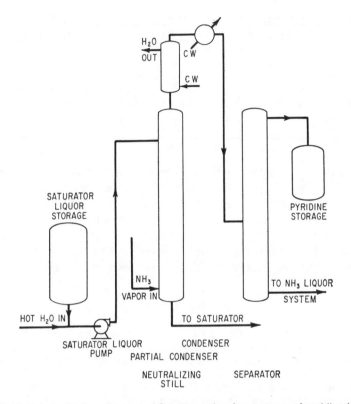

Figure 10-22. Schematic flow diagram of Schutt process for recovery of pyridine bases.

ed to the ammonia-liquor system. However, proper separation of pyridine bases from the condensate is possible only if the vapors leaving the condenser contain sufficient CO_2 to produce an aqueous phase of high ammonium carbonate content. Operating results reported by Meredith are shown in Table 10-7.

A process designed for the recovery of pyridine bases from coke-oven gas in an indirect ammonia process is described by Klempt and Röber (38). These authors found that under normal operating conditions about 82 percent of the pyridine bases originally contained in the tar-free gas are present in the vent gases from the saturator. The recovery process consists of passing the hot, saturator-vent gases through another saturator containing 50 percent sulfuric acid which is maintained at a temperature about 10°C (18°F) above that of the gas stream. In this manner the proper amount of steam is condensed to dilute the continuously added concentrated sulfuric acid to 50 percent. The pyridine sulfate solution is discharged into another vessel either continuously or intermittently, and neutralized with ammonia. By proper adjustment of the rate of ammonia addition, weak and strong bases may be obtained

Table 10-7. Operating Data-Schutt Process (39)
Gas to saturator . . . . 20–30 grains/100 cu ft (pyridine equivalent)
Gas from saturator . . . . 10–15 grains/100 cu ft (pyridine equivalent)
Saturator liquor . . . . 40 g/liter (pyridine equivalent)
Ammonium sulfate . . . . 0.07–0.18% (pyridine equivalent)

separately by fractional separation. The process is claimed to be capable of recovering about 90 percent of the pyridine bases present in the saturator-vent gases, provided a large excess of sulfuric acid (about 200 percent) is maintained in the absorption liquid. Typical saturated solutions contain about 250 to 300 g of pyridine bases/kg.

A few attempts have been made to recover pyridine bases from the light oil, but no commercial process of this type has been developed.

References

1. Kirner, W.R. 1945. *Chemistry of Coal Utilization.* Edited by H.H. Lowry. New York: John Wiley & Sons, Inc., Chap. 13.

2. Hill, W.H. 1945. *Chemistry of Coal Utilization.* Edited by H.H. Lowry. New York: John Wiley & Sons, Inc., Chap. 13.

3. Stanford Research Institute. 1958. *Chemical Economics Handbook,* vol. 7. Stanford University, Calif.: Inorganic Chemicals.

4. Terres, E. 1953. *Gas-u. Wasserfach* 94:260-265, 311-315.

5. Anon. 1958. *Chem. Week* 82(Feb. 1):65.

6. Wilson, P.J. 1945. *Chemistry of Coal Utilization.* Edited by H.H. Lowry. New York: John Wiley & Sons, Inc. Chap. 32.

7. Wilson, P.J., and Wells, J.H. 1948. *Blast Furnace Steel Plant* 36:806 and 961.

8. Bell, J. 1950. *Coke and Gas* 12(June):206-209 and 214.

9. Key, A. 1956. *Gas Works Effluents and Ammonia,* 2nd ed. London: The Institution of Gas Engineers.

10. Lunge, G., and Kohler, H. 1912. *Die Industrie des Steinkohlenteers und des Ammoniak,* 5th ed. Brunswick, Germany: F. Vieweg & Sons.

11. Gluud, W., and Jacobson, D.L. 1932. *International Handbook for the Byproduct Coke Industry.* New York: Reinhold Publishing Corporation.

12. Parrish, P. 1924. *Design and Working of Ammonia Stills.* London: Ernest Benn, Ltd.

13. Porter, H.C. 1924. *Coal Carbonization.* New York: Reinhold Publishing Corporation.

14. 1903. British Patent 8,287.

15. Ohnesorge, O. 1910. *Stahl. u. Eisen* 30:113-116.

16. Ohnesorge, O. 1923. *Brennstoff-Chem.* 4:118-122.

17. Koppers, H. 1907. U.S. Patents 846,035 and 862,976 (Re. 12,971 in 1909).

18. Otto, H. 1949. *Am. Inst. Min. Met. Engs., Proc. Blast Furnace, Coke Oven and Raw Materials Conference* 8:50-60.

19. Terres, E., and Patscheke, G. 1931. *Gas-u. Wasserfach* 74:761-764, 792-799, 810-814, 837-841.
20. USS Engineers and Consultants, Inc. 1971. *Bulletin No. 2-01* (Nov.).
21. Still, C. 1920. British Patent 147,787.
22. 1926. British Patent 281,288.
23. Lenze, F., and Rettenmaier, A. 1926. *Gas-u. Wasserfach* 69:689-691.
24. Dixon, T.E. 1955. *Iron Age* 175:91-93.
25. Ferguson, W.C. 1956. *Lange's Handbook of Chemistry,* 9th ed. New York: McGraw-Hill Book Company, Inc., p. 112.
26. Pacific Coast Gas Association. 1934. *Gas Engineers' Handbook.* New York: McGraw-Hill Book Company, Inc., p. 443.
27. Perry, J.H. (ed.). 1941. *Gas Engineers' Handbook.* New York: McGraw-Hill Book Company, Inc., pp. 404, 2542, 2544.
28. Tans, A.M.P. 1958. *Ind. Eng. Chem.* 50(6):971-972.
29. Sherwood, T.K., and Pigford, R.L. 1952. *Absorption and Extraction,* 2nd ed. New York: McGraw-Hill Book Company, Inc., p. 262.
30. Kowalke, O.L.; Hougen, O.A.; and Watson, K.M. 1925. *Chem. & Met. Eng.* 32(10):443-446 and (11):506-510.
31. Sherwood, T.K., and Holloway, F.A.L. 1940. *Trans. Am. Inst. Chem. Engrs.* 36:21-36.
32. Dwyer, O.E., and Dodge, B.F. 1941. *Ind. Eng. Chem.* 33:485-492.
33. Fellinger, L. 1941. Sc. D. Thesis in Chemical Engineering, Massachusetts Institute of Technology.
34. Parsly, L.F.; Molstad, M.C.; Cress, H.; and Bauer, L.G. 1950. *Chem. Eng. Progr.* 46:17-19.
35. Pigford, R.L., and Pyle, C. 1951. *Ind. Eng. Chem.* 43:1649-1662.
36. Silver, L. 1934. *Trans. Inst. Chem. Engrs.* 12:64.
37. Hopton, G.U. 1953. *The Cooling, Washing and Purification of Coal Gas.* London: North Thames Gas Board.
38. Klempt, W., and Röber, R. 1940. *Chem. Fabrik* 13:65-68.
39. Meredith, H.J. 1941. *Am. Gas Assoc., Proc.* 23:656-667.
40. Dodge, F.E., and Rhodes, F.H. 1920. *Chem. & Met. Eng.* 22:274-275.
41. Dodge, F.E., and Rhodes, F.H. 1919. *Trans. Am. Inst. Chem. Engrs.* 12:239-244.
42. Wald, M.D. 1940. *Iron Steel Engr.* 17:55-58.
43. Schutt, J.W. 1942. U.S. Patent 2,311,134.

11

Absorption of Water Vapor by Dehydrating Solutions

Water vapor is probably the most common undesirable impurity in gas streams. Usually, it is not the water vapor itself that is objectionable but rather the liquid or solid phase which may precipitate from the gas when it is compressed or cooled. Liquid water almost always accelerates corrosion, and ice (or solid hydrates) can plug valves, fittings, and even gas lines. To prevent such difficulties, essentially all fuel gas which is transported in transmission lines must be at least partially dehydrated. Compressed air used to operate automatic valves and instruments in refineries and chemical plants must also be thoroughly dry. There are occasionally other reasons why gas streams must be dehydrated as, for example, in catalytic processes where water constitutes a catalyst poison or source of undesirable side reaction and in the air-conditioning field, where dehumidification is frequently a requirement.

The quantity of water in saturated natural gas at various pressures and temperatures can be estimated from Figure 11-1, which is based on the correlation of McCarthy, Boyd, and Reid (1). This chart also shows a hydrate-formation line for 0.6 specific gravity gas, based on the data of Katz (2). To the left of this line, solid hydrates will form when saturated gas is cooled. It will be noted, for example, that if 2,000-psia gas (of 0.6 specific gravity) is cooled below about 69°F, hydrates will form if the gas contains more than 15 lb water/ MMscf. At pressures below about 150 psia, on the other hand, cooling to 32°F is necessary to precipitate a solid phase, and in this case, ordinary ice will form. The hydrates form more readily (i.e., at a higher temperature or lower pressure) with gases of greater density and less readily with very light gases. For example, at a pressure of 1,000 psia, hydrates form at about 62°F in natural gas of 0.60 specific gravity, while they form at 67° and 71°F, respectively, in gases of 0.75 and 1.00 specific gravity (3). The water vapor content of saturated natural gas is also affected by gas composition. The presence of substantial concentrations of CO_2 or H_2S, for example, increases the equilibrium concentration of water vapor, particularly, at pressures above 1,000 psi. This effect has been correlated by Robinson, et al (44), for gas pressures from 1,000 to 10,000 psi and H_2S concentrations up to 40 percent (in dry methane). Typical data from this correlation are given in Table 11-1. The points were selected to illustrate the effects of acid gas concentration on water content of the gas as a function of both temperature and pressure. It will be noted that the water concentration increase caused by the presence of acid gases is greatest at high pressures and low temperatures. For purposes of interpolation to other gas compositions, it can be assumed that CO_2 alone has the same effect as 0.75 times as much H_2S (e.g., 10 percent H_2S + 10 percent CO_2 in the gas is equivalent in effect to 17.5 percent H_2S).

The water content of saturated air at pressures from 1 to 1,000 atm is given in Figure 11-2, from an article by Landsbaum, Dodds, and Stutzman (4). The water content of air at atmospheric pressure and various degrees of saturation is most conveniently estimated from psychrometric charts, which are reproduced in most standard texts on air conditioning and chemical engineering.

Figures 11-1 and 11-2 follow the common practice of stating the water content of high-pressure natural gas in terms of lb/MMscf and that of air (high or low-pressure) in terms of lb water/lb dry air. Another useful method of indicating the water content of any gas is in terms of the dew point. This is defined as the temperature to which a gas must be cooled (at constant water content) in order for it to become saturated with respect to water vapor (i.e., attain equilibrium with liquid water). Since dehydration is frequently practiced to prevent the precipitation of water from gases when they are cooled, the dew point is a more direct indication of the dehydration effectiveness than the absolute water content. If a dew point of 40°F is desired, for example, a natural-gas stream would require dehydration to 62 lb H_2O/MM scf at 100 psia or 9 lb $H_2O/MMscf$ at 1,000 psia. Since the water-vapor pressure over dehydrating solutions normally varies with temperature in approximately the same manner as the vapor pressure of pure water, the effectiveness of a given solution can be evaluated in terms of the difference between the dew point of the dehydrated gas and the contact temperature. This difference is known as the "dew-point depression" and is roughly constant for a given dehydration system (i.e., solution strength and contact efficiency) over a fairly wide range of temperatures and pressures.

Commercial processes for removing water vapor from gas streams can be classified as follows:

1. Absorption by hygroscopic liquids (or by reactive solids)
2. Adsorption on activated solid desiccants
3. Condensation by compression and/or cooling

Table 11-1. Effect of H_2S and CO_2 on Water Vapor Content of Saturated Natural Gas [Data of Robinson et al. (44)]

Pressure psia	Temperature °F	H_2S Vol %	CO_2 Vol %	Water Concentration lb/MMSCF
1,000	100	0	0	58.9
1,000	100	10	10	63.9
1,000	100	20	20	71.9
1,000	200	0	0	630
1,000	200	20	20	733
6,000	100	0	0	23.1
6,000	100	10	10	38.5
6,000	100	20	20	73.6
6,000	200	0	0	197
6,000	200	20	20	397
10,000	100	0	0	19.9
10,000	100	10	10	36.1
10,000	100	20	20	71.8
10,000	200	0	0	159
10,000	200	20	20	378

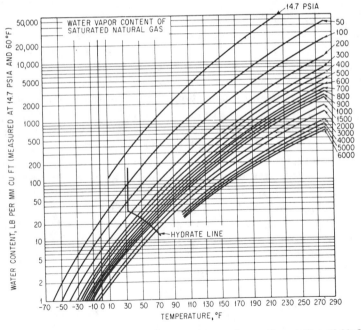

Figure 11-1. Water-vapor content of saturated natural gas. *Correlation of McCarthy, Boyd, and Reid (1). Hydrate line based on data of Katz (2)*

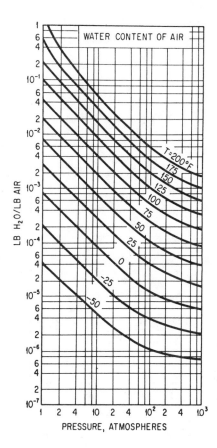

Figure 11-2. Water-vapor content of saturated air. *After Landsbaum, Dodds, and Stutzman* (4)

Only the first two methods are considered in detail (in this and the following chapter), as the third is primarily mechanical and is normally economical only under special circumstances.

Glycerin was one of the first liquids used for drying fuel gas, and the design of a plant utilizing it for city gas was described in 1929 by Tupholme (5). Calcium chloride solution was reportedly the first liquid used for natural gas; this was employed during the early 1930s (6). Diethylene glycol (DEG) was first used to dehydrate natural gas in the fall of 1936. This material and its close relative triethylene glycol (TEG) proved to be outstandingly effective, and, in September 1957, it was estimated that there were at least 5,000 glycol-type natural-gas dehydration plants in the United States and Canada (7). By 1966, triethylene glycol had become the industry standard for natural gas dehydration (45), however, other glycols are employed where they are able to meet the requirements at lower cost. In addition to DEG and TEG, tetraethylene glycol (T_4EG) and glycol blends have found application for the

Table 11-2. Properties of Glycols (9, 45)

Property	Ethylene Glycol (EG)	Diethylene Glycol (DEG)	Triethylene Glycol (TEG)	Tetraethylene Glycol (T_4EG)
Formula	CH_2OH / CH_2OH	CH_2CH_2OH / O CH_2CH_2OH	$CH_2\text{-}O\text{-}CH_2CH_2OH$ / $CH_2\text{-}O\text{-}CH_2CH_2OH$	$CH_2CH_2\text{-}O\text{-}CH_2CH_2OH$ / $CH_2CH_2\text{-}O\text{-}CH_2CH_2OH$
Molecular Weight	62.1	106.1	150.2	194.2
Boiling Point @ 760 mm Hg	197.6°C (387.7°F)	245.8°C (474.4°F)	288.0°C (550.4°F)	314.0°C (597.2°F)
Initial Decomposition Temp, °F (10)	329	328	404	—
Density @ 77°F (25°C), g/ml	1.110	1.113	1.119	1.120
Freezing Point	-12.7°C (9.1°F)	-7.8°C (17.6°F)	-7.2°C (19.04°F)	-5.6°C (22°F)
Viscosity, abs, cp @ 77°F (25°C)	16.5	28.2	37.3	39.9
@ 140°F (60°C)	5.08	7.6	9.6	10.2
Surface Tension @ 25°C, dyne/cm	47	44	45	45
Specific Heat @ 77°F (25°C), Btu/lb°F	0.58	0.55	0.53	0.52
Heat of Vaporization (760 mm Hg), Btu/lb	364	232	174	—
Heat of Solution of Water in Infinite Amount of Glycol (approx. 80°F) Btu/lb	—	58	86	—
Flash Point, °F (C.O.C.)	240	280	320	365

dehydration of natural gas. The blends represent impure products of the manufacturing operation and are, therefore, available at a lower cost than highly purified DEG, TEG, or T_4EG. However, they are significantly less efficient than the pure compounds (46). For air dehumidification, triethylene glycol and lithium chloride solutions are the only two liquid systems in common use (8).

Sulfuric acid is an excellent dehydrating agent, but because of its extreme corrosiveness, it is now used only for special applications, such as the drying of gas streams in sulfuric acid plants. Many other liquids possess dehydrating properties—including solutions of sodium or potassium hydroxide and the halides of several metals. However, these materials are not in wide use.

Glycol Dehydration Processes

Data on the physical properties of four glycols are given in Table 11-2. Diethylene glycol and triethylene glycol are the principal ones used for gas dehydration with triethylene glycol applications predominating. The factors which have led to the widespread use of glycols for gas dehydration are their unusual hygroscopicity, their excellent stability to heat and chemical decomposition, their low vapor pressures, and their ready availability at moderate cost. Photographs of typical glycol dehydration plants are presented in Figures 11-3 and 11-4.

A simplified flow diagram of a plant designed to use either of the glycols for the dehydration of natural gas is shown in Figure 11-5. The design shown includes vacuum regeneration and is typical of a large installation where maximum dehydration is required. As can be seen, the process is quite simple. The glycol stream containing from about 1 to 5 percent water contacts the gas in a short, countercurrent column. The water which is absorbed dilutes the glycol somewhat, and the dilute solution must be reconcentrated before it can be reused in the absorber. The reconcentration is accomplished by distilling water out of the solution in a regenerator. Because of the extreme difference in boiling points of water and glycol, a very sharp separation can be accomplished with a relatively short column. Some water reflux must be provided at the top of this column to effect rectification of the vapors and minimize glycol losses. In the flow arrangement shown, essentially all of the regenerator-vapor stream must be condensed to minimize the load on the vacuum pump (or steam jet), and a portion of this condensate is returned to the column as reflux. With atmospheric-pressure regenerators, it is common practice to condense only the amount of water required for reflux.

The degree of dehydration which can be attained with a glycol solution is primarily dependent upon the extent to which water is removed from the solution in the regenerator. In order to reduce the water content of solutions to the absolute minimum without using excessively high temperatures, the

Figure 11-3. Photograph of large natural-gas dehydration plant designed to operate at 1,000-psig absorption pressure. The flow arrangement of Fig. 11-6 is employed in this installation. High-pressure contactors are in the left foreground and the rectifier is in the right background behind glycol heat-exchanger and preheater units. *Southern Counties Gas Company of California*

process illustrated in Figure 11-5 makes use of vacuum regeneration. A second alternative is to employ an inert stripping vapor in conjunction with heating to regenerate the solution. Air is commonly used as a stripping vapor in air dehumidification systems, while natural gas stripping has been widely employed for natural gas dehydration plants. In the natural gas units, a separate gas stripping column is normally added downstream of the kettle reboiler. The reboiler is operated to produce approximately 99 percent TEG. This solution is fed to the column where it is contacted countercurrently by the dry inert stripping gas to further concentrate the glycol to 99.9 percent TEG or higher. It has been proven possible to achieve dew point depressions of approximately 220°F by this technique (47).

A related approach employs a volatile hydrocarbon to produce an inert stripping vapor which can then be condensed and recycled to the regenerator. The "Drizo" and "Super Drizo" processes developed by the Dow Chemical Company are of this type (48, 49, 50). They employ a hydrocarbon liquid which will vaporize at the temperature and pressure conditions of the regenerator to form a heterogeneous azeotrope with water. The hydrocarbon is then separated from the water as an immiscible phase after condensation of

Figure 11-4. Photograph of small "packaged" triethylene glycol-type natural-gas dehydrator. Unit shown is used for wellhead service and can handle 4 million SCF/day of gas at 1,000 psig. It is mounted on skids for ease of handling. *Black, Sivalls, & Bryson, Inc.*

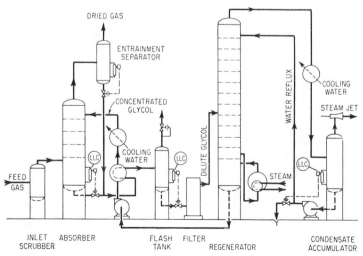

Figure 11-5. Flow diagram of typical glycol dehydration plant. Design shown would be employed where a large volume of high-pressure gas must be dehydrated to very low water-vapor levels.

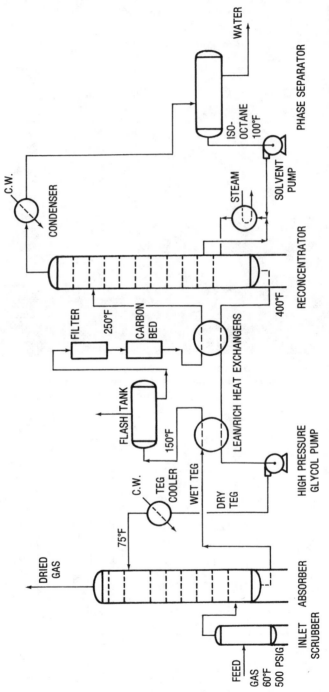

Figure 11-6. Flow diagram of Dow Chemical Company "Super-Drizo" Process as proposed for drying a "cracked" gas stream (48).

the stripper outlet vapors. A flow diagram of the Super Drizo process is shown in Figure 11-6 as proposed for the dehydration of an ethylene-containing "cracked" gas stream (48).

The flow pattern may also be modified to minimize corrosion. In a plant described by Senatoroff (11), the glycol from the absorber is first heated by heat exchange with the regenerated glycol and then passed through a steam-to-glycol preheater before it enters the regenerator column. The use of the preheater instead of a conventional reboiler is aimed at reducing the residence time at elevated temperature to as short a period as possible in order to minimize glycol breakdown and subsequent corrosion. A flow diagram of a plant of this type is shown in Figure 11-7. With this type of operation, the glycol solution is actually concentrated by boiling in the preheater, and the column which follows serves primarily to rectify the vapors which are released. In the particular arrangement shown in Figure 11-7, reflux is provided by adding a small stream of water to the top tray of the rectifier so that no condenser is required for the overhead vapors.

In low-pressure dehydration plants such as are used in air conditioning, the pressure drop through the absorber becomes a major design factor, and it is common practice to use spray nozzles in conjunction with a minimum of low-pressure-drop packing in the absorption zone. A schematic diagram of a unit of this type is shown in Figure 11-8. In this design, cooling coils are installed in the absorber to remove the latent heat of condensation of water. These also serve as packing. Cooling is required when low-pressure gas or air is dehydrated, because the relatively large amount of water in such gas streams otherwise causes an appreciable temperature increase. The increased temperature can reduce the dehydration efficiency, and increase the loss of glycol by vaporization. In the design of Figure 11-8, the regenerator is also a spray column, and air is used as a stripping vapor in conjunction with the heating coils. Reflux is provided by condensing a portion of the water from the regenerator air stream on cooling coils located above the glycol-feed point. Glycol from the absorber is used as coolant in the coils before it enters the regenerator. The solution-flow arrangement of this unit is such that a portion of the liquid pumped from the basin is passed through the regenerator as a slip stream operating in parallel to the absorber instead of in series, as is the practice in high-pressure gas-dehydration plants.

Glycol Injection Systems

Another method of utilizing glycol which is particularly useful for natural-gas gathering systems is illustrated in Figure 11-9. In this system, glycol is injected into the gas stream at the well head in order to protect the gathering lines from hydrate difficulties. The glycol flows with the gas as its temperature is reduced by heat exchange or expansion and continuously ab-

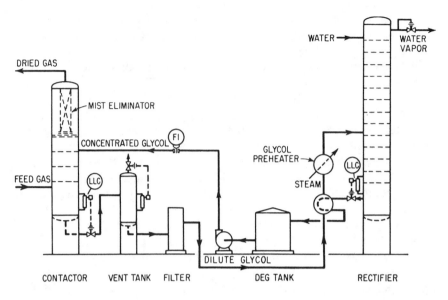

Figure 11-7. Flow sheet of glycol dehydration plant designed for minimum corrosion. *Data of Senatoroff* (11)

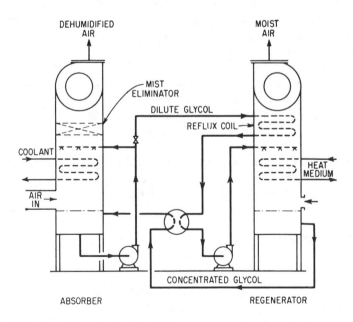

Figure 11-8. Diagram of glycol dehydration unit used in air-conditioning service.

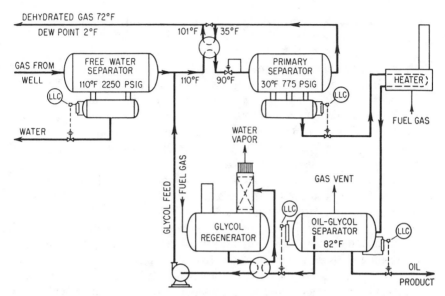

Figure 11-9. Flow diagram of process employing glycol injection and cooling to dehydrate natural gas. *Based on data of Sullivan* (12)

sorbs water, preventing the attainment of conditions under which solid hydrocarbon-hydrates can form. Typically, cooling of the gas stream also causes liquid hydrocarbons to precipitate so a hydrocarbon-glycol separator is required. A heater may be required in conjunction with the separator as the liquid mixture tends to form emulsions which are quite stable at low temperatures but separate more rapidly as the temperature is increased.

Ethylene glycol is generally preferred to diethylene or triethylene glycol for this type of operation because it is less soluble in liquid hydrocarbons. It is also more effective on a weight basis for hydrate inhibition and has a lower viscosity. On the other hand, the two higher glycols have lower vapor pressures. In order to minimize the possibility of solid-phase formation in the glycol, solutions near the eutectic composition (i.e., 60 to 80 percent) are commonly employed, compared to 95 percent or higher concentrations used in more conventional dehydrator designs. The dilute solutions are also preferred from the standpoint of solubility in hydrocarbon. The glycol provides some dehydration, but its primary function in this type of operation is to act as an "antifreeze" agent in suppressing the formation of solid hydrates. The high degree of dehydration attained in the case illustrated in Figure 11-9 may be attributed more to the cooling which occurs than to the effect of the glycol solution. There are many other process-flow schemes used in connection with glycol-injection systems. In its simplest form, the process is almost identical to conventional glycol dehydration with a section of the gas line serving as the gas-liquid contactor. Details of the design of glycol injection systems have been described by Arnold and Pearce (3, 13) and by Robirds and Martin (14).

Methanol has also been used in an injection process for hydrate control. Although methanol is not a glycol, the process is sufficiently similar to glycol injection to be included in this discussion. The use of methanol, which has been described in detail by Herrin and Armstrong (53) and Nielsen and Bucklin (54), is particularly applicable to natural gas turboexpander plants. In such plants, the feed gas is cooled in a gas-to-gas heat exchanger, and methanol can be conveniently sprayed on the exchanger tubesheets. The methanol inhibits hydrate formation by forming a water solution. This solution is separated from the gas stream and sent to a methanol still for methanol recovery. A significant amount of methanol dissolves in the liquid hydrocarbons produced in the plant; however, this can be recovered by using water produced in the methanol still to wash the liquid hydrocarbon.

Nielsen and Bucklin compared the cost of a methanol injection plant for 600-psig natural gas to the costs of solid-bed dehydration plants. They concluded that the methanol injection plant would have lower capital and operating costs than an activated alumina system for low-water content gas (4 lb H_2O/MM scf) and a lower capital cost but about the same operating cost compared to a molecular sieve system for high water content feed gas (23 lb H_2O/MM scf).

Basic Design Data

Figures 11-10 and 11-11, which are based on the data of Polderman (7) and Worley (15), present data on the dew points of gases in equilibrium with diethylene and triethylene glycol solutions at various temperatures. The Worley data included in Figure 11-11 agree closely with values computed by F.R. Scauzillo (16) but generally indicate appreciably lower equilibrium dew-point values for triethylene glycol solutions of 95 percent or higher concentration than the earlier publications of Polderman (7) and others.

The use of dew-point charts, rather than equilibrium data for the design of dehydration columns, is quite convenient because these charts can be used directly to estimate the solution composition and temperature required to provide a given degree of gas dehydration. Although some early investigations indicated that the dew-point depression obtainable with glycols drops off appreciably at pressures above 500 psia (17), more recent studies have indicated that the dew-point depression is relatively independent of pressure up to pressures of at least 2,000 lb (6, 7, 18).

Figures 11-12 and 11-13 are useful for the design of regenerators. The vertical lines drawn at 340°F for diethylene glycol solutions and 375°F for triethylene glycol solutions represent the maximum temperatures normally

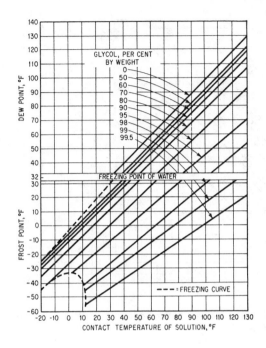

Figure 11-10. Equilibrium dew points of gases in contact with diethylene glycol solutions. *Data of Polderman* (7)

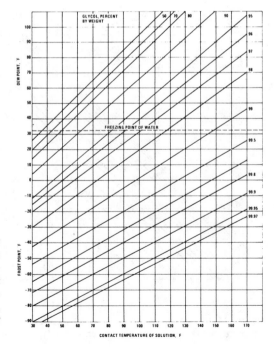

Figure 11-11. Equilibrium dew points of gases in contact with triethylene glycol solutions. *50-80% solution data from Polderman* (7), *90 percent—99.97 percent data from Worley* (15), *99.99 percent data from Dow Chemical Co.* (48)

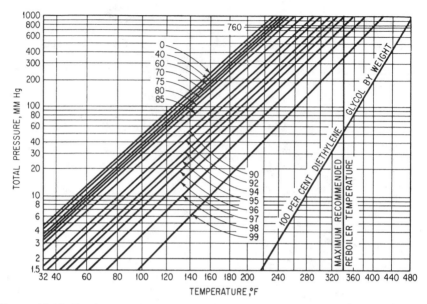

Figure 11-12. Total vapor pressure of various diethylene glycol solutions versus temperature. *Dow Chemical Company data* (19)

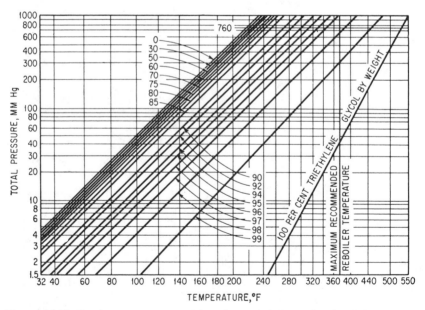

Figure 11-13. Total vapor pressure of various triethylene glycol solutions versus temperature. *Dow Chemical Company data* (19)

recommended for these materials. The intersection of these lines with the various boiling-point curves marks the recommended regenerator pressure for each solution concentration. Where maximum dehydration is required, triethylene glycol regenerators may be operated with a reboiler temperature as high as 400°F. In this case, the regeneration column is operated at as low a pressure as possible. With atmospheric pressure operation and no stripping vapor in the regenerator, 400°F operation results in the lean TEG concentration of about 98.6 percent.

Important physical-property data for diethylene glycol and triethylene glycol are presented in Figures 11-14 through 11-20. Figure 11-14 shows the effect of temperature and glycol concentration on the specific gravity of triethylene and diethylene glycol solutions. To estimate the specific gravity of glycol solutions at a temperature other than 60°F, it is safe to assume that the specific gravity-temperature curves will be approximately parallel to those for 100 percent glycols as long as the solutions are relatively concentrated.

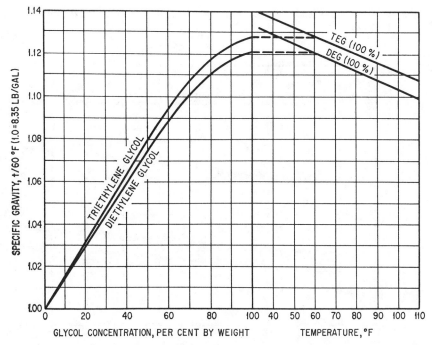

Figure 11-14. Specific gravity of diethylene and triethylene glycol solutions at 60°F and effect of temperature on specific gravity of the pure glycols. *Data of Union Carbide Corp.* (9)

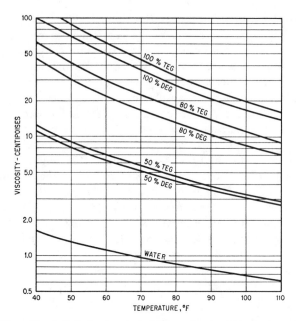

Figure 11-15. Viscosity of diethylene and triethylene glycol solutions. *Data of Union Carbide Corp.* (9)

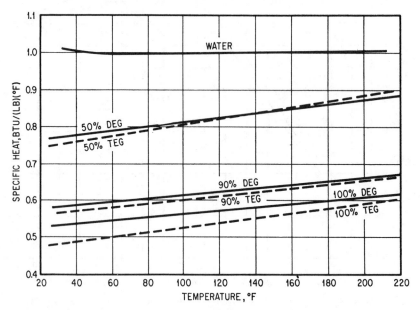

Figure 11-16. Specific heat diethylene and triethylene glycol solutions. *Data of Union Carbide Corp.* (9)

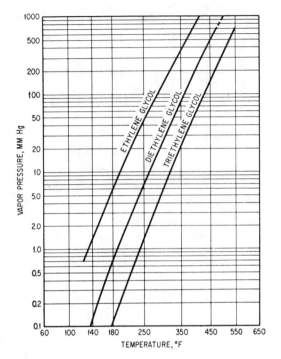

Figure 11-17. Vapor pressure of pure glycols. *Data of Union Carbide Corp.* (9)

The vapor-pressure chart, Figure 11-17, may be used as a basis for vapor-loss estimates by assuming Raoult's law holds for glycol in the concentrated solution employed. The vapor/liquid composition diagrams, Figures 11-19 and 11-20, are applicable to the design of vacuum regeneration columns.

Design and Operation

Absorption Column

In the design of glycol absorption columns, Figures 11-10 and 11-11 can be used to predict the minimum glycol concentration required for a given dehydration problem. The charts represent equilibrium conditions, which, however, cannot be attained in actual practice because the glycol becomes diluted as it passes through the column and the gas is not in contact with liquid of lean solution-concentration on a sufficient number of actual trays to permit equilibrium to be closely approached.

Fortunately, sufficient operating data are available on glycol-type dehydration plants to enable new units to be designed on the basis of rule-of-

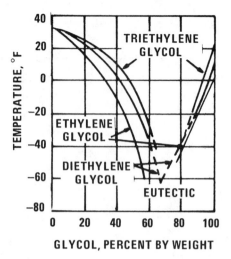

Figure 11-18. Freezing points of ethylene, diethylene, and triethylene glycol solutions. *Data of Union Carbide Corp.* (9)

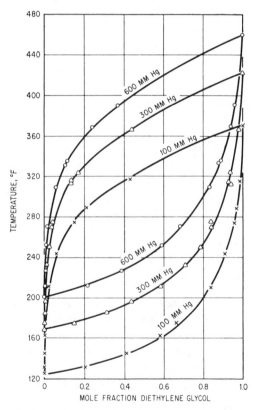

Figure 11-19. Vapor/liquid composition diagrams for diethylene glycol and water at 100, 300, and 600 mm Hg total pressure. *Data of Dow Chemical Company* (19)

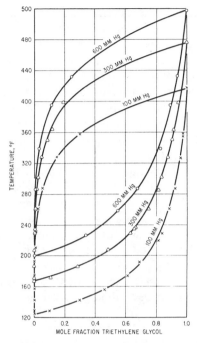

Figure 11-20. Vapor/liquid composition diagrams for triethylene glycol and water at 100, 300, and 600 mm Hg total pressure. *Data of Dow Chemical Company* (19)

thumb or experience factors. Two of these which are used in the design of high-pressure natural-gas plants follow.

1. At least 3 gal glycol should be circulated per pound of water absorbed.
2. The absorber should have the equivalent of at least four actual trays.

A comparison of dew-point depressions obtained by commercial units operating at the minimum of the above requirements, with the theoretical depression attainable if equilibrium with the lean solution were reached, is shown in Figure 11-21. Although an appreciable margin is noted between actual performance and the theoretical maximum, the dew-point depressions obtained are all over 60°F, a figure adequate for most pipeline transmission service.

Swerdloff (21) has calculated the effect of circulation rate for triethylene glycol solutions, assuming one equilibrium tray or about four actual trays in the absorber. The results of his calculations are shown in Figure 11-22. An interesting point brought out by this chart is that a high circulation rate is of more importance with very concentrated glycol streams than with the more dilute (96 percent) solution, if it is desired to obtain about the maximum dew-point depression possible with the given solution.

Many dehydration-plant absorbers are built with four contacting trays. It is estimated that these trays are usually about 25 to 40 percent efficient so that the column is about equivalent to one equilibrium contact. Increasing the number of trays has an effect similar to that of increasing the circulation rate, in that it makes possible a closer approach to equilibrium with the inlet solution. Because of this, it is sometimes desirable to use more than four trays in the absorber, particularly where maximum dehydration is required. As many as 10 trays have been used.

A clearer picture of gas and liquid composition changes on the trays of the absorber can be gained from Figure 11-23, which represents the tray diagram for a design of a typical natural-gas dehydration plant. This type of diagram is particularly useful for unusual dehydration cases, in which the use of rule-of-thumb procedures is risky, and for large plants, in which an appreciable economic saving can be realized by improving the plant design. For a given dehydration problem, the diagram can be used to estimate the required number of trays, solution rate, or solution concentration. Several combinations which will give the desired dehydration can be worked out and the most economical selected.

For the case shown in Figure 11-23, it is assumed that a natural-gas stream is saturated with water at 500 psia and 90°F and that it is desirable to dehydrate this gas to a water content of 10 lb/MMscf (dew point 28°F). Us-

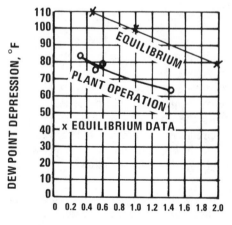

PERCENT WATER IN LEAN TRIETHYLENE GLYCOL

Figure 11-21. Effect of glycol concentration on dew-point depression attainable with commercial absorbers. *Plant data from Campbell* (20)

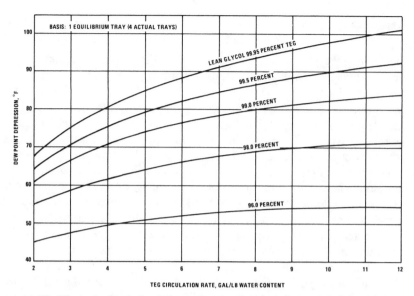

Figure 11-22. Effect of glycol circulation rate on calculated dew point depression attainable. *Data of Swerdloff* (21)

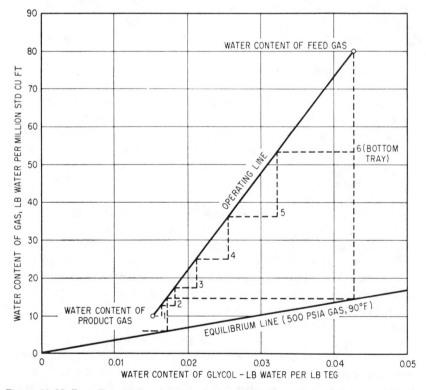

Figure 11-23. Tray diagram for absorber design. High-pressure natural-gas dehydration with triethylene glycol solution.

ing triethylene glycol, a concentration of 98.5 percent is about the maximum which can be obtained without requiring a vacuum regenerator. The dew-point chart, Figure 11-11, indicates that, at 90°F, a solution of this concentration is theoretically capable of producing the desired degree of dehydration. If 4 gal glycol/lb water absorbed are circulated, the solution will be diluted from 98.5 percent to about 95.9 percent. These two liquid compositions in combination with the inlet- and outlet-gas compositions (as estimated from Figure 11-1) are used to establish an operating line on the diagram. The equilibrium line is obtained by converting dew-point data from Figure 11-11 to water content of the gas for the specific temperature and pressure considered (using Figure 11-1). To simplify the analysis, it is assumed that temperature is constant over the length of the column. With the operating and equilibrium lines established, it can be seen that about 1½ theoretical trays would be required. If it is assumed that the Murphree vapor efficiency of the actual trays used will be about 40 percent, the number of actual trays required can be estimated by using vertical steps on the tray diagram which extend 40 percent of the distance from the operating line to the equilibrium line at each tray point. The use of this procedure would indicate that at least six trays should be used in the absorber of this plant. Examination of the figure will reveal that the glycol solution can be permitted to become much more dilute in passing through the column without approaching equilibrium with the feed gas. If an attempt is made to take advantage of this, however, more trays will be required. The optimum design must therefore take into account the costs of additional column height versus the cost of pumping solution.

In the foregoing case, it is assumed that the temperature does not change through the absorption column. For a more exact analysis, the effect of the heat of absorption of water should be taken into account. The heat liberated is equal to the latent heat of vaporization plus the heat of solution of liquid water in glycol. (See Table 11-1). In general, the gas stream has a considerably higher total heat capacity than the liquid stream so that the liquid leaving the bottom of the absorber will be at approximately the temperature of the entering gas. The exit-gas temperature can, therefore, be calculated by a heat balance around the absorber, taking into account the temperature of the entering solution and the heat liberated by the absorption of water. In the case of high-pressure gas dehydration, the net effect is usually a very slight increase in gas temperature on passing through the column (1° or 2°F), and this is normally of little significance.

In low-pressure gas dehydration and air dehumidification, on the other hand, the heat of absorption becomes of considerable importance and, in such cases, it is common practice to use cooling coils in the absorption zone to remove the heat which is liberated. In units used for air-conditioning dehumidification, the cooling coils may also serve as the packing in the air-glycol contact zone as illustrated in Figure 11-8.

The diameter of glycol absorbers can be calculated by conventional column-sizing techniques; however, a conservative gas velocity should be used because of the tendency for glycol solutions to foam under some circumstances. Swerdloff (6) presents a nomograph for calculating the constant in the Brown-Souders vapor-velocity correlation which takes into account the tray spacing and the acceptable triethylene glycol loss. For the typical case of a 24-in. tray spacing and an acceptable glycol loss of 1 lb/million cu ft, the correlation gives a value for the constant C of 400, to be used in the equation

$$W = C[d_2(d_1 - d_2)]^{1/2} \qquad (11\text{-}1)$$

where W = maximum allowable mass velocity of vapor, lb/(hr)(sq ft) of tower cross-sectional area
$\quad d_1$ = density of fluid, lb/cu ft
$\quad d_2$ = density of vapor, lb/cu ft

The carry-over rate can be reduced in practice by the installation of a suitable mist eliminator at the top of the absorber.

Charts for estimating the required internal diameter of natural-gas dehydration columns using triethylene glycol and equipped with valve type trays have been presented by Camerinelli (22). One of these charts, for the case of 24-in. tray spacing, is reproduced in Figure 11-24. The chart is based on a constant gas inlet temperature of 120°F; however, for pressures up to the maximum shown, 1,200 psia, gas temperature was found to have very little effect. The column size obtained from the chart is based on a design gas capacity of about 70 percent of flooding.

Regenerator Design

The regeneration of diethylene or triethylene glycol generally requires only the simple distillation of a binary mixture, the two components of which have widely different boiling points and do not form azeotropes. About the only difficulty in this otherwise straightforward engineering-design problem is that excessive decomposition of the glycols may occur if the temperature reaches too high a level. Recommended temperatures are about 340°F for DEG and 375°F for TEG; satisfactory operation has been reported in some instances, however, with appreciably higher reboiler temperatures. TEG reboilers, for example, have been reported to operate successfully with a glycol outlet temperature of 400°F. To overcome the temperature limitation when very concentrated solutions are required, the distillation process may be modified by the use of vacuum, an inert stripping gas, or a liquid hydrocarbon azeotrope former.

Because of the wide difference in boiling points, the separation of water from diethylene or triethylene glycol is a very easy operation and only a short column is required. The number of theoretical trays required for the separa-

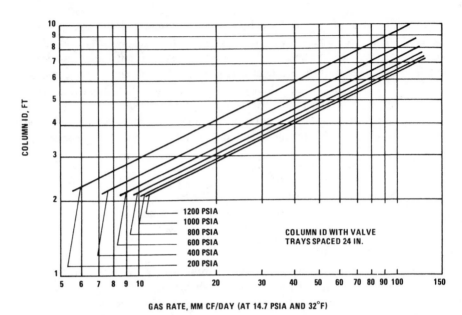

Figure 11-24. Chart for estimating the internal diameter of triethylene glycol absorption columns equipped with valve type trays. *Data of Camerinelli* (22)

tion can readily be calculated by means of a McCabe-Thiele diagram. Such an analysis usually shows that the job can be done by two or three theoretical trays, one of which is the reboiler. In view of this, the regenerator-column height is more commonly established by practical considerations, and, since the vapor and liquid quantities are usually small, sizing can be quite generous.

For relatively small plants, the regenerator column is frequently installed directly on top of the reboiler and packed with rasching rings. With 1-in. rings, packed heights from 6 to 15 ft are common. For larger plants, where a regenerator-column diameter of 24 in. or more is justifiable, bubble-cap columns are often used. The number of actual trays used in commercial plants ranges from 10 to 20 with the solution inlet located somewhat below the mid-point. The apparent large excess of trays is employed to minimize the loss of glycol with the overhead vapor. Because of the very low liquid-loading on trays above the feed, care should be taken that the trays are well sealed and weep holes are sufficiently small to prevent draining during operation.

Reflux to the top of the regenerator column may be supplied by several alternate systems. The simplest is to install an uninsulated or finned section at the top of the column to provide cooling and condense some water which flows back as reflux. This system is used in a large number of small dehydrators but is difficult to control under adverse weather conditions. A tubular, water-cooled condenser may be used either in the top of the tower for gravity return of reflux or in a separate vessel, in which case a reflux pump must be provided. This system provides accurate control for large plants but for many installations is unnecessarily expensive. Some first-cost savings can be realized by utilizing steam condensate or fresh water which is introduced directly into the top plate of the regenerator. Principal difficulties of this system result from the possible introduction of salts if plant water is used and the need for accurate control because of the small quantities of reflux required.

With regenerators employing kettle-type reboilers, equilibrium can be assumed to be attained in the reboiler, and the temperature-pressure-composition relationships can be obtained from Figure 11-12 or 11-13. The vapor-pressure values given on these charts represent the total pressure of water and glycol, which is equal to the total reboiler pressure if no inert stripping gas is added. In the latter case, the partial pressure of the inert gas added must be subtracted from the total reboiler pressure to give the solution-vapor-pressure to be used in reading the charts.

Triethylene glycol is generally the preferred absorbent where a maximum dew-point depression is required. The principal reason for this is its greater stability at the high temperatures required for adequate regeneration. Triethylene glycol also has the advantage of providing a slightly greater dew-point depression than diethylene glycol at equivalent concentration (by weight). The principal advantage of diethylene glycol is its lower cost.

As indicated by Figure 11-13, a 98.0 percent triethylene glycol solution is about the maximum that is considered good practice with an atmospheric-pressure regenerator. This can be increased to almost 99 percent by very careful design of the reboiler system to limit retention time and metal wall temperatures. If more concentrated solutions are required, the regeneration system must be designed to use vacuum, stripping gas, or azeotropic distillation. Any of the glycols may be used statisfactorily in conjunction with one of the high efficiency regeneration techniques; however, triethylene glycol is generally preferred.

Vapor/liquid composition diagrams for the glycol-water systems at 100, 300, and 600 mm Hg total pressure are presented in Figures 11-19 and 11-20. These can be used to construct McCabe-Thiele diagrams for detailed design of vacuum regenerators.

The effectiveness of stripping gas in reducing the water content of triethylene glycol is shown in Figure 11-25 as presented by Worley (15). The results indicate that a regenerated solution containing over 99.9 percent (wt) triethylene golycol can be obtained by using a 400° reboiler temperature and further stripping the hot glycol with 4 scf of gas per gallon in a separate countercurrent contactor. The water dew point for gas in equilibrium with such a solution (99.9 percent TEG) is indicated to be about $-44°F$ for a contact temperature of 100°F.

Equivalent results can be accomplished by the use of a liquid hydrocarbon which vaporizes in the regenerator to provide stripping vapor. The effects of feeding iso-octane into the regenerator of a triethylene glycol unit are indicated by the data of Table 11-3. These data are from a Dow Chemical Company paper on their "Super Drizo" process (48). The iso-octane reportedly forms a heterogeneous azeotrope system with water. This function is indicated by an overhead condenser temperature lower than the boiling points of either water (100°C) or iso-octane (99.2°C). For a plant designed to use a glycol circulation rate of 1.2 gal/lb H_2O and an iso-octane injection rate of about 0.15 gal/gal of TEG (to give a glycol concentration of 99.95 percent), the overhead condenser will operate at about 90°C (48).

The reboiler heat load values given in Table 11-3 are based on the simplified equation proposed by Sivalls (51) for conventional glycol dehydration units,

$$Q_t = 2,000 \, L \tag{11-2}$$

where Q_t is the total heat duty of the reboiler in Btu/hr and L is the glycol flow rate in gal/hr.

The equation assumes typical values for glycol sensible heat, latent heat of vaporization of water, reflux duty, and heat losses. For cases where iso-octane is used as a stripping aid, an additional amount of heat equivalent to approximately 1,000 Btu/gal of hydrocarbon used is required to provide the sensible and latent heat requirements of the additive.

Heat for the regeneration step is normally provided by direct combustion of natural gas in tubes in the reboiler or by steam. Hot oil and waste heat sources have also been proposed. The use of exhaust gases from engines driving gas compressors has been described by Carmichael (23). In the example he cites, the exhaust gases leave the engine at an average temperature of 1,260°F. They lose about 60°F enroute to the glycol regenerator and leave that unit at a temperature between 400° and 450°F. The advantages compared to a direct fired unit burning part of the natural gas are claimed to be lower cost, reduced maintenance because of lower temperatures in the heater tubes, and improved safety as a result of eliminating the open-flame system. When steam is used as the source of reboiler heat, it is often necessary to reduce the steam pressure upstream of the reboiler steam coil to limit the metal wall temperature and thereby minimize thermal degradation of the glycol. A maximum skin temperature of 435°F has been recommended (47) which implies a maximum steam-side temperature on the order of 463°F (465 psig steam). The same result can be accomplished by limiting the heat flux in the reboiler to values below about 7,000 Btu/hr ft.

Table 11-3. Effect of Iso-Octane Addition on Regeneration of Triethylene Glycol [Dow Super-Drizo Process (48)]

Iso-Octane/TEG Flow Ratio, gal/gal	Maximum Glycol Conc. Attainable* %	Reboiler Heat Load Btu/gallon of Glycol**
0	98.60	2000
0.10	99.90	2100
0.15	99.95	2150
0.20	99.98	2200
0.25	99.99	2250

*Assumes 400°F isothermal operation and maximum stripping efficiency.
**Based on simplified equation of Sivalls for reboiler duty of conventional plant (51).

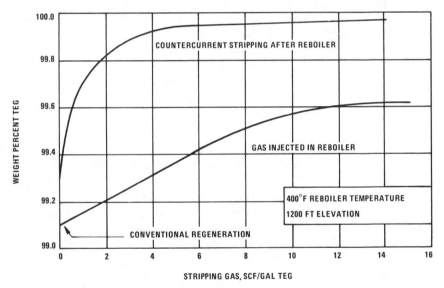

Figure 11-25. Effect of stripping gas quantity and method of injection on regeneration of triethylene glycol solutions. *Data of Worley* (15)

Glycol Injection Process Design

The design of glycol injection process systems is determined to a large extent by the overall gas handling system design, which establishes the gas quantities, pressures, temperatures and compositions at the glycol injection and separation points. The minimum required quantity of injected glycol can be determined by a material balance between the water removed from the gas and the amount which can be absorbed by the glycol without it becoming so dilute that hydrate formation can occur. The minimum glycol concentration required to provide protection can be estimated by the Hammerschmidt equation (24):

$$d = 2335W/(100\ M - MW) \tag{11-3}$$

where d = amount by which the hydrate formation point is lowered, °F

M = molecular weight of the glycol
W = weight percent glycol in the solution

If d is equal to the difference between the actual gas temperature and its normal hydrate formation temperature, the calculated value of W represents the minimum concentration of glycol required in the final solution.The quantity of feed glycol can then be estimated by assuming a feed concentration (e.g., 70 percent) and noting the total amount of water to be absorbed (from Figure 11-1). The above procedure provides an estimate of the minimum quantity of injection glycol (and maximum glycol dilution) that can be used. According to Robirds and Martin (14) a practical design target of 60 percent glycol returning to the regenerator is desirable with regeneration back to about 70 percent. As indicated in Figure 11-18, the 60 to 70 percent concentration range is near the eutectic point for all three glycols.

Plant Operation

Operating data for six typical glycol dehydration plants are presented in Table 11-4. As will be noted, the plants described cover a very wide range of conditions and dew-point depressions from 40° to 77°F. These values represent typical, but not maximum obtainable dew-point depressions. Polderman (7) has presented data on a plant which produces dew-point depressions of 85° to 100°F. This plant, which uses DEG, operates with a vacuum regenerator (175 mm Hg), an eight-tray absorber, and a glycol circulation rate amounting to 5 gal/lb of water removed. The reboiler temperatures of plants presented in Table 11-4 are well below the recommended maximum; however, higher temperature levels are not uncommon. Swerdloff (6) presents limited operating data for six TEG gas-dehydration plants, four of which have reboiler temperatures of 350°F or above, and one a temperature of 387°F. These units produced dew-point depressions from 57° to 73°F without requiring vacuum regeneration.

Worley (15) reported (in 1967) that up to 10 years of experience with several hundred triethylene glycol units utilizing reboiler temperature of 400°F had failed to indicate any evidence of measurable losses by degradation. Dew-point depressions in excess of 100°F are reported for these units.

Table 11-4. Glycol Dehydration Plant Operating Data

Plant variables	Plant					
	A (25, 26)	B (27)	C (28)	D (7)	E (7)	F (7)
Gas rate, MMSCF/day..	50*	60	10.8	70	73.5	130
Absorber pressure, psig..	1,000	750	385	375	720	450
Type of gas............	Nat.	Nat.	Nat.	Nat.	Nat.	H_2
Solution rate, gpm.......	5–10	6	2.27	11.2	15.6	11.7
Glycol used............	DEG	DEG	TEG	DEG	DEG	DEG
Lean Sol. Conc., % Glycol	95	95	98.25	97.6	97.8	97.8
Rich Sol. Conc., % Glycol	90	90	96.95	95.6	96.5	
Absorber:						
Diameter, in.........	50	36	36			
Height..............	33 ft 7 in.	28 ft	12 ft 6 in.			
No. of trays.........	4	4	4	6	7	5
Regenerator:						
Diameter, in.........	26	18	12¾			
Height..............	29 ft 10 in.	35 ft	6 ft 6 in.			
No. of trays.........	20	15	Rings			
Feed tray...	15	....	Rings			
	(from top)					
Reboiler:						
Temp., °F............	....	310	352	273	290	310
Pressure, psia........	Atm.	20	Atm.	4.9	7.4	7.7
Temp. of feed gas, °F....	78	60–68	55	64	84	82
Temp. of product gas, °F.	....	....	....	70	86	82
Dew point of product gas, °F..................	+38	+10, +15	−4	+2	+9	
Dew point depression†, °F..................	40	50, 53	59	68	77	
Glycol loss, lb/MMSCF..	....	....	....	0.21	0.28	0.48

*Rate to each of three absorbers.
†Depression based on feed-gas temperature if product-gas temperature not given.

Glycol loss constitutes one of the most important operating problems of dehydration units. Most of this loss occurs as carry-over of solution with the product gas, although a small amount of glycol is lost by vaporization into the gas stream; an additional small amount is always lost through mechanical leakage, and some may be lost with the vapors leaving the regenerator. By careful plant operation, total glycol losses can be maintained below ½ lb/MMscf of gas treated; however, a loss of 1 lb/MMscf is sometimes considered acceptable.

Since the major glycol loss is by entrainment, any operating practices which reduce this item can result in a considerable improvement in plant economics. Excessive amounts of entrainment can usually be traced to foaming of the glycol in the absorber. It has been found that foaming can result from contamination of the glycol by hydrocarbons, finely divided solids, or salt-water brought in with the feed gas. It is important, therefore, that the gas be passed through an efficient separator before it is admitted to the glycol absorber. When foaming does occur, it can usually be reduced by the addition of a foam inhibitor. The use of trioctyl phosphate for this purpose has been described (6). In this installation, the inhibitor was used in the concentration of 500 ppm and reduced the loss from as high as 15 lb glycol/MMscf to less than 0.5 lb/MMscf. An efficient mist eliminator should be used after the absorber to recover as much of the carry-over as possible. Metal-mesh materials are reported to perform well in this service. The major causes of glycol loss experienced with 1,200 package-type skid-mounted glycol units operated in the San Juan Basin are reported to be (a) overloading of the glycol absorber when shut-in wells are placed on the line and (b) failure of separator dump-valves to operate (29). An improvement developed for these units and intended for minimizing inlet-separator malfunctioning is the use of pipe coils inside the separator shell for the circulation of hot glycol. This prevents freezing of free water collected in the separator during cold weather and helps to break the heavy, viscous oil emulsion produced by some wells.

Corrosion can also be a serious problem in the operation of glycol dehydration plants. Since the pure glycol solutions are themselves essentially noncorrosive to carbon steel, it is generally believed that the corrosion is accelerated by the presence of other compounds which may come from the oxidation or thermal decomposition of the glycol, or enter the system with the gas stream. The rate of corrosion will, of course, be influenced by the temperature of the solution, velocity of the fluid, and other factors. In general, the principles which have been employed in combating corrosion are

1. The use of corrosion-resistant alloys
2. The use of corrosion inhibitors
3. The prevention of solution contamination
4. The use of process-design modifications to minimize temperatures and velocity

The principal chemical factors involved in glycol-plant corrosion are believed to be the oxidation of glycols to form organic acids and the absorption of acidic compounds, principally H_2S and CO_2, from the gas stream. The oxidation of diethylene glycol has been studied by Lloyd (30) with regard to oxidation product and rate-governing factors. He found that oxidation of diethylene glycol resulted in the formation of an organic peroxide as an intermediate product and formic acid and formaldehyde in copious quantities.

The oxidation rate was found to increase with increased oxygen partial-pressure and increased temperature and to be accelerated by the presence of acid.

Lloyd and Taylor (31) investigated the effect of glycol deterioration products and of various added chemicals on the rate of corrosion by diethylene glycol solutions. The corrosion tests were conducted by heating the glycol solution in a flask, with samples immersed in the solution and suspended in the vapor space. The vapor-phase samples, which were wetted with condensate, showed by far the most serious corrosion, and conclusions with regard to the corrosiveness of the various solutions were based on these samples. These conclusions were

1. Glycol solutions which had been made acid, either by autooxidation or by the addition of acetic acid, were consistently more corrosive than the neutral samples.
2. Low concentrations of neutral salts did not affect corrosion rates.
3. The glycol solutions which were made alkaline showed fairly low corrosion rates.

It is postulated that corrosion in this system is caused by the presence of a volatile acid which vaporizes and condenses with water on the mild-steel corrosion coupons. Alkaline buffers, such as potassium phosphate, combine with the free organic acids and reduce their vapor pressure to a negligible value. Organic alkaline materials, such as monoethanolamine, may also act by vaporizing with the organic acids and neutralizing them when condensation occurs.

Ballard (32, 52) suggests checking the glycol pH periodically and keeping it in the range of about 7.3 to 8-8.5 by addition of borax, ethanolamine, or other alkaline chemical. Too high a pH is undesirable because it can increase the tendency of the solution to foam and emulsify with hydrocarbons.

A number of corrosion inhibitors have been successfully applied in commercial plants; these include monoethanolamine (33) and sodium mercaptobenzothiazole. The use of the latter material in a plant which had previously encountered very severe corrosion has been described by Swerdloff and Duggan (34). The plant operated on a gas containing 58 grains H_2S/100 scf, 54 grains mercaptans/100 scf, 1.36 percent CO_2, and, at times, traces of oxygen. After 2 years' operation, the contactor trays were found to be severely corroded, and, after 3 years, the dried-gas line blew out, about 80 ft downstream from the contactor. When the corrosion was first noticed, the pH of the solution was found to range from 4.1 to 5.0. Steps taken to remedy the situation included the installation of stainless-steel lining and trays in the contactor, the addition of sodium mercaptobenzothiazole to the glycol, cooling of the gas stream from 100° to 80°F prior to its contact with the solution, and installation of a system to decrease entry of oxygen into the gas stream. The net result of all of these changes was a very great reduction in the corrosion rate of any iron in contact with the glycol, as measured by coupons. The inhibitor was used as a 45 percent solution of sodium mercaptobenzothiazole in water and added directly to drums of glycol used as makeup. Concentration in the dehydration solution amounted to about 1 percent.

Two additional steps were taken to minimize corrosion downstream to the glycol dehydrator unit. One was the use of a product-gas scrubber to minimize glycol entrainment in the gas stream, and the other was the use of a second inhibitor which was added to the dried-gas stream. This material was of the polyethanolrosinamine-type consisting of 70 percent solution in isopropyl alcohol of a mixture of 90 percent ethoxylated rosinamine (11 moles ethylene oxide with each mole of rosinamine) and 10 percent free rosinamine. This inhibitor was injected at the rate of 0.1 gal/day for a gas volume of 60 MMscf/day. As a result of the above measures, the corrosion rate of coupons suspended in the dried-gas line decreased from almost 30 mils/year to as low as 0.2 mil/year during about 4 years of testing.

In addition to causing corrosion, contamination of the glycol solution can result in fouling of heat-exchanger surfaces and loss in operating efficiency. The solution may become contaminated with oxidation products as described above, by corrosion products (usually iron oxide or iron sulfide), and by solid or liquid particles brought in with the gas stream. Solid contaminants are objectionable in that they settle out in tanks, contactor and still trays, heat exchangers, and other vessels. They may also be a factor in accelerating corrosion (or erosion). The use of some means for removing suspended particles is therefore usually justified. Filters of the common waste-pack or cartridge-type have proved quite successful and are usually located in the line carrying the solution from the contactor. Ballard (52) recommends cloth fabric filter elements in favor of paper or fiberglass and suggests that the solids content in the glycol be held below 0.01 wt percent (100 ppm).

Activated carbon is also employed to remove impurities from glycol solutions. It is particularly effective for removing nonfilterable heavy hydrocarbons. An activated carbon bed is shown in Figure 11-6 which depicts the flow diagram of a plant processing a "cracked" gas stream. Such gases are more apt to carry heavy hydrocarbons and polymer formers into the glycol than natural gas, so an activated carbon filter is particularly important for this type of application. The activated carbon bed should be located downstream of the filter to avoid plugging with solids. Coal-based carbons are reportedly more widely used than wood-based. Carbon particle sizes typically range from 8 by 30 to 4 by 10 mesh (52).

The carryover of carbon particles can also be a problem. Simmons (55) cites one case in which the top two trays of a glycol contactor were found to be plugged with a mixture of heavy oils and carbon fines. He recommends the use of a solids filter downstream of the carbon unit to prevent carbon fines from entering the system. According to Pearce (56), activated carbon filters usually operate on a 10 percent stream and are sized for a glycol flow of ~ 1 gal/min/ft^2 of cross-sectional area.

In high-pressure natural-gas service, an appreciable amount of the hydrocarbon gas is dissolved by the circulating glycol solution. The solubility of a typical natural-gas stream in two glycol solutions is shown in Figure 11-26. Although the hydrocarbon gas which is absorbed is not in itself corrosive, it can accelerate the corrosion caused by other components by flashing from the solution when its pressure is released and causing high-velocity turbulent flow of the two-phase mixture. To minimize this effect, and also to remove the maximum quantity of the acid gases which may be present in the glycol and which are known corrosive agents, it is common practice to provide a flash tank in the rich-solution line ahead of the regenerator. From a corrosion-control standpoint, the vent tank is most effective when located ahead of the heat exchanger, as in Figure 11-7, although a greater volume of gases will be vented if it is located downstream of the heat exchanger as shown in Figure 11-5.

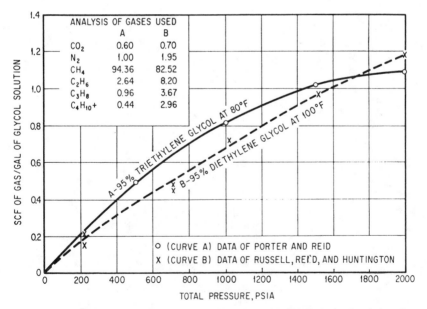

Figure 11-26. Solubility of natural gas in glycol solutions. *Triethylene glycol data from Porter et al. (17); diethylene glycol data from Russell et al. (35)*

When high-pressure gas streams containing very high concentrations of CO_2 are dehydrated, a simple flash tank may not be adequate for removing dissolved gas due to the high solubility of CO_2 in glycols at high partial pressure. It is desirable to remove dissolved CO_2 from the solution entering the regeneration system to minimize corrosion in the still and reboiler, reduce the heat load, and limit vapor traffic in the still. This problem has been studied by Glaves et al. who developed the design for a plant to dehydrate 550 MM scfd of 1080-psia gas containing 71.7 percent CO_2 (57). The original design was reevaluated on the basis of more recently published test data (58) (Figure 11-27). It was concluded that an intermediate pressure stripper was needed to remove CO_2 from the rich glycol before it enters the regeneration column. Calculations based on the new data showed that operation of the intermediate pressure stripping column at 450 psia with 1.5 MM scfd of stripping gas (containing 19 percent CO_2) would result in a reduction of CO_2 from 1.60 lb CO_2 per lb of glycol in the liquid leaving the absorber to 0.29 lb/lb in the liquid leaving the stripper. Although greatly reduced by the stripper, the

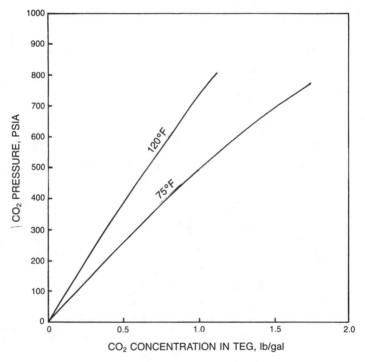

Figure 11-27. Solubility of Carbon Dioxide in Pure Triethylene Glycol. Data of Takahashi and Kobayashi (58).

amount of CO_2 in the glycol feed to the regenerator in this case would still be greater than the amount of water, significantly increasing the size of the regeneration column required. One positive aspect of the dissolved CO_2 is its action as a stripping vapor to assist in removing water from the glycol.

In addition to a filter and vent tank, the design shown in Figure 11-7 incorporates a regenerator preheater, instead of a reboiler, and stainless-steel regenerator-column construction—features which are also aimed at minimizing corrosion. Plants using this flow arrangement are designed for minimum liquid-flow velocities throughout and minimum temperatures consistent with the required dew-point depression; when they are in operation, inhibitors are added to control pH on the alkaline side.

A further precaution which is used at many plants to minimize the entry of oxygen into the solution is the use of a gas-blanketed glycol-storage tank. Since some gas streams contain free oxygen, this precaution will not always prevent oxidation of the glycol from occurring. In air dehydration, of course, there is no hope of excluding oxygen from the system, and the use of inhibitors is desirable.

Proper maintenance of the glycol solution is critical for trouble-free operation of the dehydration process. Chemical analysis of the solution can often prevent or identify the cause of operating problems. Analyses are typically made for water, pH, lower glycols, hydrocarbons, foaming tendency, and inorganic salts (particularly chlorides). Methods for conducting analysis of glycol are described in detail by Grosso et al. (59).

Dehydration with Saline Brines

The use of calcium chloride brine for water absorption is quite old. However, for most applications, it is gradually being replaced by the glycols and by more efficient salts, such as lithium bromide and lithium chloride. The lithium salts are used to a growing extent in air-conditioning installations for dehumidification.

Calcium Chloride for Gas Dehydration

The equilibrium dew point of gases in contact with aqueous solutions of calcium chloride is shown in Figure 11-28, which is based on the data of Brockschmidt (36). This author presents a comparison of operating data for a plant using a 35 percent calcium chloride solution and essentially the same unit employing a 95 percent solution of diethylene glycol. In order to permit glycol to be used in the plant, it was necessary to replace the calcium chloride-solution reboiler with a 13-plate regeneration column and to add heat exchangers and solution preheater. The comparison shows that the glycol gave a dew-point depression averaging about 45°F as compared with a dew-point depression averaging only 19°F for the calcium chloride solution. During comparison periods of about 7 months' operation with each liquid, the glycol was found to have removed about twice the quantity of water as the calcium chloride solution. In view of such poor performance, coupled with operating problems and corrosion, it is no wonder that conventional dehydration units utilizing calcium chloride solutions have been almost entirely replaced by glycol systems for natural-gas dehydration.

More recently, however, a novel application of calcium chloride to natural-gas dehydration has been introduced and a number of small units installed. A schematic diagram of this type of dehydration unit is shown in Figure 11-29, which is based on a description by Fowler (29). The unit contains a bed of 3/8- to 3/4-in. calcium chloride pellets and five specially designed brine-circulating trays. Gas enters the bottom of the column, passes up through the brine trays where it contacts progressively more concentrated calcium chloride so-

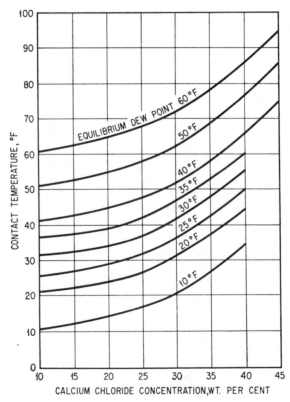

Figure 11-28. Equilibrium dew point of gases in contact with calcium chloride solutions. *Data of Brockschmidt (36)*

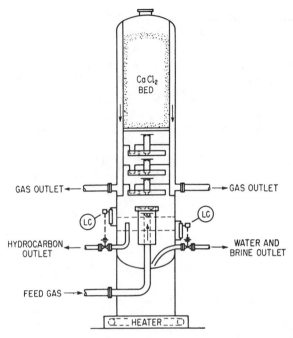

Figure 11-29. Schematic diagram of natural-gas dehydration unit employing calcium chloride pellets. *Data of Fowler* (29)

lutions, then continues upward through the bed itself, where additional water is absorbed on the surface of the pellets, forming concentrated brine which drips down onto the trays continuously. The design of the trays is such that the gas aspirates liquid upward to provide circulation and thereby maintains sufficient liquid on each tray without the need for a pump. The concentrated calcium chloride brine dripping from the bed of pellets has a specific gravity of 1.40, and this is reduced to approximately 1.15 to 1.20 by the time it reaches the bottom of the column. This brine is considered expendable and is dumped into a pit along with any free water produced by the well. The units are recharged with calcium chloride pellets periodically, bringing the calcium chloride bed to a depth of 8 ft. As the chemical is used up, the bed settles; however, the efficiency of the unit is not appreciably reduced as long as the level is above 24 in. It is reported that units of this type can give a dew point of as low as 7°F with a bed-depth as low as 2 ft and a gas temperature of 127°F.

Since the frequency of recharging and the cost of chemicals is proportional to the water content of the gas, it is desirable that the water content of feed gas be maintained as low as possible by operating at a temperature close to that of the hydrate freezing point. A plot of chemical costs, recharging days, and chemical consumption versus the flowing temperature of the gas, assuming 500 lb pressure, is presented in Figure 11-30. The principal operating difficulties of this unit have been caused by the freezing of brine on the trays.

The results of field tests on 250 units over an 18-month period have been summarized by Fowler (37). Calcium chloride has also been proposed for hydrate prevention in natural-gas gathering lines. In this application an aqueous solution is injected near the well head and collected at the downstream point after gas cooling has occurred. In tests reported in the Russian literature (38), the process was found to be very effective. However, purging of the solutions with natural gas prior to injection was found necessary to reduce their corrosive action.

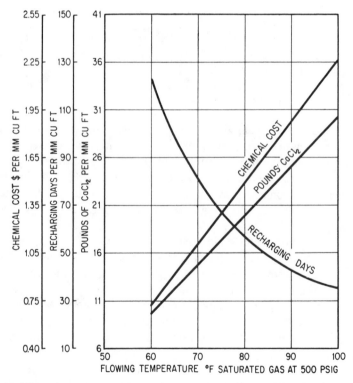

Figure 11-30. Effect of gas temperature on operation of calcium chloride pellet dehydration unit. *Data of Fowler* (29)

Lithium Halides for Air Dehydration

Data on the two lithium salts which are useful for air dehydration are presented in Table 11-5 and Figures 11-31 and 11-32. As shown in the figures, lithium bromide is considerably more soluble in water, and the saturated solutions of this salt have a lower vapor pressure than lithium chloride solutions at the same temperature and thus can provide a greater degree of dehydration. For most operations, however, the degree of dehydration provided by lithium chloride is adequate and, because of its somewhat lower cost, it is, therefore, the preferred compound. Dehydration units designed to employ lithium-halide solutions are essentially the same as those which use triethylene glycol. The principal difference between the halide solutions used for dehydration and the glycols is that the active component in the halide solution has essentially zero vapor pressure, and therefore, no rectification section is necessary in the regenerator.

The dew-point depression theoretically attainable with three lithium chloride solutions is presented in Figure 11-33, which is plotted on a psychrometric chart to illustrate the application of the data to air-conditioning problems. The solution of a typical air-conditioning dehumidification problem is shown in Figure 11-34 as presented by Gifford (8). In this problem, it is assumed that air is available at 95°F, 75°F wet bulb (99 grains water per pound dry air), and it is desired to determine the degree of dehydration attainable with 44 percent lithium chloride solution. Two cases are considered. In one (Case A) it is assumed that the absorbent solution can be cooled to 80°F by the available cooling medium. In Case B, it is assumed that a solution temperature of 60°F can be maintained. Further assumptions are that dilution effects are negligible and that, in both cases, the equipment design is such that a 90 percent approach to equilibrium can be attained. By

Table 11-5. Properties of Lithium Salts Used for Dehydration

Property	Lithium chloride	Lithium bromide
Formula......................	LiCl	LiBr
Molecular weight..............	42.40	86.86
Melting point, °C.............	614	547
Solubility in water, g/100g..... {	63.7 at 0°C 130.0 at 95°C	145 at 4°C 254 at 90°C
pH of 1% Sol.................	6.4	6.8
Heat of fusion, Cal/mole........	4	5

Physical-property data from *Chemical and Physical Properties of Lithium Compounds*, publication of the Foote Mineral Company, Philadelphia, Pennsylvania (39).

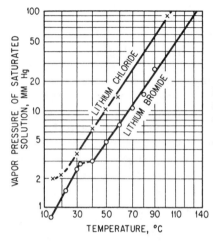

Figure 11-31. Vapor pressure of saturated solutions of lithium chloride and bromide. *Data of Foote Minerals Company* (39)

moving 90 percent of the distance along the line from the point representing the inlet-air condition to the point on the solution equilibrium curve corresponding to the solution temperature, it is seen that, for Case A, the air temperature can be reduced to 81°F (dry bulb) and 56½°F wet bulb. This corresponds to a 35°F dew point or a water content of 30 grains/lb dry air. In Case B, with a 60°F solution, the air can be dried and cooled to 64°F dry bulb and 46¼°F wet bulb, corresponding to a 24°F dew point or a water content of 19 grains/lb dry air.

Absorption and heat-transfer data for the dehumidification of air with lithium chloride solutions in a short column packed with 2-in. clay raschig rings (stacked) are given in Figure 11-35 as presented by Bichowsky and Kelley (40). The data are based on commercial-type work and are believed to be dependable within about 5 percent. Solution concentrations are not given; however, the authors report that, in the range of concentrations generally used, the coefficients are not found to be sharp functions of the concentration. Data for one of the experiments on which the curves of Figure 11-35 are based are given in Table 11-6.

The absence of a concentration effect was also observed by Tohata et al. (41, 42) for wetted wall and perforated plate columns. In the wetted wall column study lithium chloride solutions in the range of 18.7 to 28.4 percent were employed. The results showed the gas-phase resistance to be controlling. Very high-stage efficiencies (up to 90 percent) were observed in the perforated plate column study, also indicative of a gas-phase resistance controlled absorption.

Figure 11-36 is a photograph of a commercial Kathabar air-dehydration unit manufactured by the Ross Engineering Division of Midland Ross Corp.

Figure 11-32. Solubility of lithium chloride and lithium bromide in water. *Data of Foote Minerals Company* (39), *and International Critical Tables.*

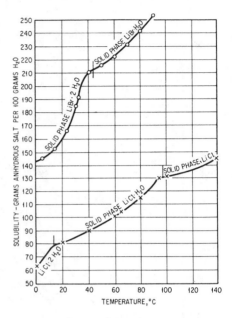

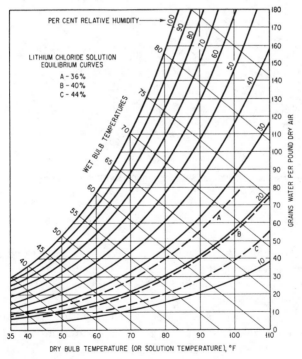

Figure 11-33. Equilibrium water content of air in contact with lithium chloride solutions. *Data of Gifford (8)*

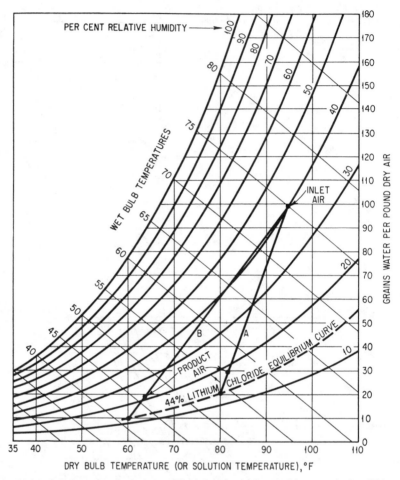

Figure 11-34. Solution of typical air-conditioning dehydration problem employing lithium chloride solution. *Data of Gifford* (8)

This is a 2,500-cfm system composed of the air-conditioner unit on the right, the regeneration unit on the left, and the interconnecting pump unit in between. The Kathene solution used in this unit is a solution of lithium chloride with appropriate additives.

A flow diagram of a duplex air-dehydration installation utilizing a 44 to 45 percent aqueous solution of lithium chloride is shown in Figure 11-37. This particular unit is installed in a penicillin-processing plant. Its job is to remove 30 gal/hr of water from 3,500 cfm of air, reducing the water content to 9 grains/lb in order that the humidity level in the plant can be maintained at 16

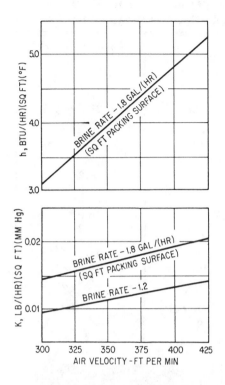

Figure 11-35. Effect of air velocity on heat transfer and absorption coefficients for air dehumidification with lithium chloride solutions. Column equipped with rotating distributor and 2-in. clay raschig-ring packing. Packed height 1.33 ft, cross-sectional area 2.4 sq ft, packing surface area 27 sq ft/cu ft. For operating conditions, see Table 11-6.

grains/lb to prevent moisture from damaging the hygroscopic penicillin (43). As shown in the diagram, outside air is drawn through unit A; this is cooled by circulating water at 85°F, and its moisture content is reduced from 122 grains to about 36 grains/lb. This partially dehydrated fresh air is combined with 2,850 cfm of recirculated air, and the mixture is passed through the second dehydration unit which utilizes Freon at 38°F as coolant. In this unit the moisture content is reduced to 9 grains/lb. In both absorbers, the principal contact surface is the outside of fin-type heat exchangers through which the coolant is circulated. About 90 percent of the lithium chloride solution from the basin is recycled in the absorbers, and the remainder is bypassed to a regenerator. The regenerator is heated by low-pressure steam to about 230°F, which is well below the boiling point of the solution. Regeneration is accomplished at this temperature by the use of air as stripping vapor to sweep evaporated water out of the regenerator. The regenerated solution then flows to the sump of the first absorber; here it is cooled by dilution and the sensible heat is ultimately removed by the circulating water in the cooling coil.

Although they are appreciably less corrosive than calcium chloride solution, lithium chloride and bromide brines are somewhat corrosive, particularly in the presence of impurities (especially copper) and it is desirable to utilize

Table 11-6. Sample Data for Experiment Using LiCl Brine and Absorber Packed with Raschig Rings (40)

Variable	Numerical Value
Air rate, cu ft/min	655
Tower cross section, sq ft	2.4
Packing height, ft	1.33
Packing surface, sq ft/cu ft	27
LiCl solution rate, gpm	2.75
LiCl solution, specific gravity	1.20
Temperatures, °F:	
Solution in	70
Solution out	78.2
Air in, dry bulb	82.8
Air in, wet bulb	66.0
Air out, dry bulb	79.0
Air out, wet bulb	61.2
Pressures, mm Hg:	
Partial pressure of water in inlet air	11.4
Partial pressure of water in outlet air	8.6
Vapor pressure of inlet brine	5.4
Vapor pressure of outlet brine	7.4
Absorption coefficient K, lb/(hr)(sq ft)(mmHg)	0.0224
Heat transfer coefficient h, Btu/(hr)(sq ft)(°F)	4.6

Figure 11-36. Commercial lithium chloride air-dehydration unit (Kathabar system). *Ross Engineering Div. Midland Ross Corp.*

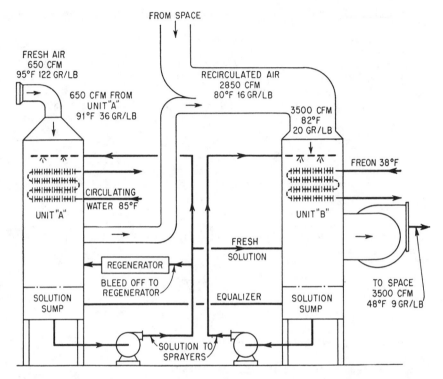

Figure 11-37. Flow arrangement and design specifications for air-dehydration plant employing lithium chloride solution. *From Chemical and Engineering News* (43)

inhibitors. One such inhibitor is lithium chromate, which is particularly useful in these solutions as it does not introduce a foreign cation.

An incidental benefit obtained with lithium chloride units is the appreciable degree of sterilization of the air treated. Research at the University of Toledo on lithium chloride solutions reportedly indicated 97 percent removal of air-borne microorganisms from the air processed (8). This feature is of particular importance in the evaluation of air-conditioning systems for hospitals and food-processing plants.

References

1. McCarthy, E.L.; Boyd, W.L.; and Reid, L.S. 1950. *J. Petrol. Tech.*, 189:241.
2. Katz, D.L. 1945. *Petrol. Engr.* 16(10):233-238.
3. Arnold, J.L., and Pearce, R.L. 1961. *Oil Gas J.* (June 19):92-95.
4. Landsbaum, E.M.; Dodds, W.S.; and Stutzman, L.F. 1955. *Ind. Eng. Chem.* 47(Jan.):101.
5. Tupholme. 1929. *Gas Age-Record* 63:311-313.
6. Swerdloff, W. 1957. *Oil Gas J.* 55(Apr. 29):122-129.
7. Polderman, L.D. 1957. *Oil Gas J.* 55(Sept. 23):107-112.

8. Gifford, E.W. 1957. *Heating, Piping Air Conditioning, J. Sect.* 29(Apr.):156-159.
9. Union Carbide Corp. 1971. *Glycols.*
10. Gallaugher, A.F., and Hibbert, H. 1937. *J. Am. Chem. Soc.* 59:2524.
11. Senatoroff, N.K. 1953. *Oil Gas J.* 51(Apr. 27):154.
12. Sullivan, J.H. 1952. *Oil Gas J.* 50(Mar. 3):70.
13. Arnold, J.L., and Pearce, R.L. 1961. *Oil Gas J.* (July 3):125-129.
14. Robirds, K.D., and Martin, J.C. III. 1962. *Oil Gas J.* (Apr. 30):85-89.
15. Worley, M.S. 1967. "Super-dehydration with glycols." *Proc. Gas Conditioning Conf.* Norman, Oklahoma.
16. Scauzillo, F.R. 1961. *J. Petrol. Tech.* (July).
17. Porter, J.A., and Reid, L.S. 1950. *J. Petrol. Tech.* 189:235.
18. Townsend, F.M. 1953. "Vapor liquid equilibrium data for diethylene glycol water and triethylene glycol water in natural gas systems." *Proc. Gas Hydrate Control Conf.*, University of Oklahoma (May 5-6).
19. Dow Chemical Company. 1956. *Properties and Uses of Glycols.* p. 1671.
20. Campbell, J.M. 1952. *Chem. Eng. Progr.* 48(Sept.):440-448.
21. Swerdloff, W. 1967. "Dehydration of natural gas." *Proc. Gas Conditioning Conf.*, Norman, Okla.
22. Camerinelli, I. 1970. *Hydroc. Process.* (Feb.):103-104.
23. Carmichael, C.J. 1964. *Oil Gas J.* (Nov. 2):72-74.
24. Hammerschmidt, E.G. 1934. *Ind. Eng. Chem.* 26:851-855.
25. Hull, R.H. 1945. *Calif. Oil World* 38(Aug.):4-9.
26. Senatoroff, H.K. 1945. *Oil Gas J.* 44(Dec.):98-108.
27. Love, F.H. 1942. *Petrol. Engrs.* 13(13):46.
28. Peahl, L.H. 1950. *Oil Gas J.* 49(July 13):92.
29. Fowler, O.W. 1957. *Oil Gas J.* 55(Apr.29):188.
30. Lloyd, W.G. 1956. *J. Am. Chem. Soc.* 78:72.
31. Lloyd, W.G., and Taylor, F.C. Jr. 1954. *Ind. Eng. Chem.* 46(Nov.):2407-2416.
32. Ballard, D. 1966. *Hydroc. Process.* 45(June):171-180.
33. Kruger, H.O., and Mazelli, J.R. 1952. *Proc. Pacific Coast Gas Assoc.* 43:179.
34. Swerdloff, W., and Duggan, M. 1955. *Petrol. Refiner* 34(Mar.):208.
35. Russell, George F.; Reid, L.S.; and Huntington, R.L. 1945. *Petrol. Refiner* 24(Dec.):137.
36. Brockschmidt, C.L. 1942. *Gas (Los Angeles)* 28(Apr.):28.
37. Fowler, O.W. 1964. *Oil Gas J.* 62(31):123-124.
38. Andryushchenko, F.K., and Vasilchenko, U.P. 1963. *Neft i Gas Prom. Nauchn.-Tekhn. Sb.* (4).
39. Foote Minerals Company. 1956. *Chemical and Physical Properties of Lithium Compounds.*
40. Bichowsky, F.R., and Kelley, G.A. 1935. *Ind. Eng. Chem.* 27(Aug.):879-882.
41. Tohata, H.; Yamada, T.; Nakada, T.; and Sasu, A. 1964. *Kagaku Kogaku* 28(2):155-158.
42. Tohata, H.; Yamada, T.; Nakada, T.; and Itorgaki, H. 1964. *Kagaku Kogaku* 28(10):832-836.
43. Anon. 1951. *Chem. Eng. News* 29(Feb. 26):819.
44. Robinson, J.N.; Wichert, E.; and Moore, R.G. 1978. *Oil Gas J.* (February 6):76-78.
45. Worley, M.S. 1966, "Twenty Years of Progress with TEG Dehydration," presented at CNGPA meeting, Calgary, Alberta, Canada (December 2).
46. Grosso, S. 1978. *Oil Gas J.* (February 13):106-110.

47. Valerius, M. 1974. "Dehydration of SNG," presented at the American Chemical Society meeting, Los Angeles, California (April 1).
48. Dow Chemical USA. 1975. "Super-Drizo; The Dow Dehydration Process," Technical Report presented at the Gas Conditioning Conference, Norman, Oklahoma (March 3-5).
49. Arnold, J.L.; Pearce, R.L.; and Schoelten, H.G. 1976. US Patent 3,349,544 (October 31).
50. Wall, J., ed. 1975. "Gas Processing Handbook," *Hydro. Process.* 54, No. 4(April):81.
51. Sivalls, C.R. 1974. "Glycol Dehydration Design Manual," presented at the Gas Conditioning Conference, Norman, Oklahoma.
52. Ballard, D. 1977. *Hydro. Process.* 56, No. 4(April):111-118.
53. Herrin, J. P. and Armstrong, R. A. 1972. "Methanol Injection and Recovery in a Turboexpander Plant," presented at the Gas Conditioning Conference, Norman, Oklahoma.
54. Nielsen, R. B. and Bucklin, R. W. 1983. *Hydrocarbon Processing 62* 4 (April) :71.
55. Simmons, Jr., C. V. 1981. *Oil & Gas Journal* (September 28) :313.
56. Pearce, R. L. 1982. "Fundamentals of Gas Dehydration with Glycol Solutions," presented at the Gas Conditioning Conference, Norman, Oklahoma.
57. Glaves, P. S., McKee, R. L., Kensell, W. W., and Kobayashi, R. 1983. *Hydrocarbon Processing 62* 11 (Nov) :213.
58. Takahashi, S. and Kobayashi, R. 1982. GPA TP-9, Gas Processors Association, Tulsa, Oklahoma (Dec).
59. Grosso, S. Pearce, R. L. and Hall, P. D. 1979. *Oil & Gas Journal,* Part 1 (September 24) :176 Part 2 (October 1) :56.

12

Gas Dehydration and Purification by Adsorption

The unit operation of adsorption is of increasing importance in gas purification and forms the basis for commercial processes which remove water vapor, organic solvents, odors, and other vapor-phase impurities from gas streams. In adsorption, materials are concentrated on the surface of a solid as a result of forces existing at this surface. Since the quantity of material adsorbed is directly related to the area of surface available for adsorption, commercial adsorbents are generally materials which have been prepared to have a very large surface area per unit weight. For gas purification, the adsorbent particles may be irregular granules or preformed shapes, such as tablets or spheres, and the gas to be purified is passed through a bed of the material. The gas-phase impurity is selectively concentrated on the internal surfaces of the adsorbent while the purified gas passes through the bed.

The nature of the forces which hold certain molecules at the solid surface is not thoroughly understood and numerous theories have been proposed to explain the phenomenon. The most familiar theory is that of Langmuir (1) who proposed that the forces acting in adsorption are similar in nature to those involved in chemical combination. Sites of residual valency are assumed to exist on the surface of solid crystals. When an adsorbable molecule from the gaseous phase strikes a suitable unoccupied site, the molecule will remain instead of rebounding into the gas. As in the evaporation of liquids, the adsorbed molecule may leave the surface when suitably activated; however, other molecules will continually adhere. When adsorption is first started, a large number of active sites exist, and the number of molecules adhering exceeds the number of those leaving the surface. As the surface becomes covered, the probability of a molecule in the gas finding an unoccupied space is decreased, until finally the rate of condensation equals the rate of evaporation which represents the condition of equilibrium. In accordance with the Langmuir theory, the adsorbed material is held onto the surface in a layer only one molecule deep, although it is recognized that these adsorbed molecules may have their force fields shifted in such a manner that they can attract a second layer of molecules which could have some attraction for a third layer, and so on. The adsorption mechanism postulated by Langmuir would require that the equilibrium quantity of a compound adsorbed from a gas increase with increased gas pressure but at a constantly decreasing rate. Equilibrium isotherms of this shape are referred to as Langmuir isotherms.

The forces holding adsorbed molecules to the surface may be quite weak, resembling those which cause molecules to coalesce and form the liquid phase, or so strong that the adsorbed material cannot be removed without a chemical change taking place. The weaker—physical or van der Waals—forces are apparently responsible for most adsorption phenomena; this explains why, in general, compounds which have low vapor pressures are adsorbed in greater quantity than relatively noncondensable gases. The chemical type of adsorption, which has been given the name "chemisorption," is of less importance in industrial adsorption processes. An example of chemisorption is the adsorption of oxygen on charcoal at temperatures above 0°C. When an attempt is made to desorb the oxygen by elevating the temperature, it is released as an oxide of carbon.

When a vapor- (or liquid-) phase component concentrates on a solid by adhering to the solid surfaces, even though the surfaces may consist of the interior of submicroscopic pores, the phenomenon is known as "adsorption." If, on the other hand, penetration of the solid or semisolid structure occurs and produces a solid solution, the phenomenon is termed "absorption." The general term "sorption" has been proposed to cover both cases.

Although adsorption can be practiced with many solid compositions, the great majority of gas-purification and dehydration adsorbents are based on some form of silica, alumina (including bauxite), carbon, or certain silicates, the so-called molecular sieves. The silica and alumina-base adsorbents are primarily used for dehydration, while activated carbon has the specific ability of adsorbing organic vapors and is very important for this purpose. The molecular sieves have very unusual properties with regard to both dehydration and the selective adsorption of other compounds.

Whether the process involves the removal of water vapor or of some other gas-phase impurity, the basic concepts involved in the design of the installation are similar. The gas must be passed through a bed of the adsorbent material at a velocity consistent with pressure drop and other requirements and under conditions which will allow the required material transfer to occur. The bed will eventually become loaded with the impurity and must then either be discarded, removed for reclaiming, or regenerated in place. When regeneration in place is practiced, this is almost always accomplished by the use of heat and a stripping vapor.

Most adsorption operations use fixed beds; however, processes have also been developed which employ fluidized or moving bed concepts. These are of particular interest for very large, low-pressure plants. Another recent process innovation makes use of system pressure reduction to remove adsorbed material from the bed during the regeneration cycle. The process, called pressure swing adsorption, is applicable only to operations in which the adsorption step takes place at an elevated pressure.

Water-Vapor Adsorption

A large number of solid materials will take up water vapor from gases; some by actual chemical reaction, others through formation of loose hydrated compounds, and a third group by adsorption as described above. Desiccants in the third group are of primary importance for commercial gas-dehydration processes. The adsorbents most commonly used for dehydration are

1. Silica gel
2. Silica-base beads
3. Activated alumina
4. Alumina-gel balls
5. Activated bauxite
6. Molecular sieves

With the possible exception of the molecular sieves which normally require somewhat higher regeneration temperatures, the equipment and process-flow arrangements for all of the adsorbents are essentially identical and need be described only once. In many cases, the adsorbents themselves are interchangeable and equipment designed for one can be operated quite effectively with another.

In its simplest form, a plant for removing water vapor from gases by adsorption will consist of two vessels filled with granular desiccant together with sufficient auxiliary equipment so that one bed of desiccant can be regenerated while the other is being used for dehydration. Regeneration is accomplished by passing hot gas through the bed. When the first bed is spent and the second completely regenerated, their effective positions in the flow pattern are reversed by suitable valving. The complete cycle is repeated periodically so that the process is in effect continuous with regard to gas dehydration. The principal difference between various adsorption-dehydration processes is the means of providing heat for regeneration. In order to make the process truly continuous, some work has been done on units in which the beds are made to move from the regeneration zone to the adsorption zone rather than switching the gas-feed point by valving. However, the moving-bed units have been industrially important only for low pressure applications such as air-conditioning dehydration.

Figure 12-1. Photograph of a typical natural-gas dehydration plant employing dry desiccant in vertical vessels. *United Gas Pipe Line Company*

Figure 12-2. Photograph of a large field installation for drying high-pressure natural gas with a solid desiccant. These adsorbers are of the horizontal compartmental type with four compartments per vessel. The gas is introduced through manifolds inside the units to provide even distribution. Regeneration gas is heated in a salt-bath indirect-fired heater. *Black, Sivalls & Bryson, Inc.*

Photographs of large field installations for solid-desiccant dehydration of high-pressure natural gas are shown in Figures 12-1 and 12-2, and a small package-type unit for drying instrument air is shown in Figure 12-3.

Many gas-dehydration problems can be solved by using either a solid desiccant or a liquid system. However, the principal areas of application of the dry-desiccant processes are

Figure 12-3. Package-type solid-desiccant gas dryer. Unit shown has a capacity of 1,000 SCF/min of air at 100-psig operating pressure. It is designed for an 8-hr tower-reversal cycle with 250-psig steam reactivation and a product air with —40°F dew point. *C.M. Kemp Manufacturing Company*

1. Cases where essentially complete water removal is desired
2. Installations (usually small) in which the operating simplicity of the granular-desiccant system makes it attractive.

In the dehydration of relatively large volumes of high-pressure natural gas, liquid dehydrating systems (diethylene glycol or triethylene glycol) are usually more economical if dew-point depressions of 40° to 50°F are required. If higher dew-point depressions, up to about 80°F, are necessary, either type may be selected on the basis of intangible factors. If dew-point depressions consistently higher than about 80°F are required, solid-desiccant dehydration is generally specified.

In general, where a simple triethylene glycol unit (atmospheric-pressure regeneration) is applicable, it is more economical from both an initial and operating-cost standpoint than a typical dry-desiccant system. A comparison

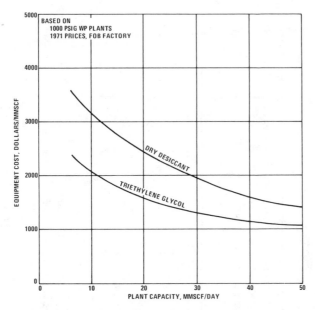

Figure 12-4. Comparative initial equipment costs for typical dry-desiccant and glycol dehydration processes in high-pressure natural-gas service.

of approximate equipment costs for natural-gas dehydration by dry-desiccant and glycol processes is presented in Figure 12-4. Operating costs for dry-desiccant systems are typically 20 to 30 percent higher than simple glycol dehydration units.

In comparison to liquid systems, solid-desiccant dehydration plants offer the following advantages:

1. Ability to provide extremely low dew points
2. Insensitivity to moderate changes in gas temperature, flow rate, pressure, etc.
3. Simplicity of operation and design of units
4. Relative freedom from problems of corrosion, foaming, etc.
5. Adaptability to dehydration of very small quantities of gas at low cost

The process has the following disadvantages:

1. High initial cost
2. Generally higher pressure drop
3. Susceptibility to poisoning or breakup
4. Relatively high heat requirement

Table 12-1. Important Physical Properties of Typical Desiccant Materials

Physical properties	Type of desiccant (typical commercial products)					
	1. Silica gel (2) (Davison 03)	2. Silica-base beads (3, 108) (Sorbead R)	3. Activated alumina (4) (Alcoa grade F-1)	4. Alumina-gel balls (4) (Alcoa grade H-151)	5. Activated bauxite (5) (Florite)	6. Molecular sieve (6) (Linde 4A & 5A)
True specific gravity	2.1-2.2	—	3.3	3.1-3.3	3.40	—
Bulk density, lb/cu ft (4-8 mesh)	45	49	52-55	51-53	50-52	40-45
Apparent specific gravity	1.2	—	1.6	—	1.6-2.0	1.1
Average porosity, %	50-65	—	51	65	35	
Specific heat, Btu/(lb)(°F)	0.22	0.25	0.24	—	0.24	0.2
Thermal conductivity, Btu/(sq ft)(hr)(°F)(in.)	1.0	1.37	1.0(100°F); 1.45(200°F)	—	1.09(360°F; 4-8mesh)	
Water content (regenerated),%	4.5-7	4-6	6.5	6.0	4-6	varies
Reactivation temperature, °F	250-450	300-450	350-600	350-850+	350+	300-600
Particle shape	granular	spheroidal	granular	spheroidal	granular	cylindrical pellets
Surface area, sq meter/g	720-760	650	210	390	—	—
Static sorption at 60% RH, %	29	33.3	14-16	22-25	10	22

Desiccant Materials

Important physical properties of typical desiccant materials are listed in Table 12-1. Equilibrium water-capacity data are presented in Figure 12-5.

Silica Gel

Silica gel is commercially available as a powdered, granular and spherical bead material of various size-ranges. The individual particles have a hard, glassy appearance resembling quartz. The material may be represented by the formula $SiO_2 \cdot nH_2O$. It is produced by reacting sodium silicate with sulfuric acid, coagulating the mixture into a hydrogel, washing to remove sodium sulfate, and drying the hydrogel to produce the commercial adsorbent. The end product is highly porous, with pores estimated to average 4×10^{-7} cm in diameter (2).

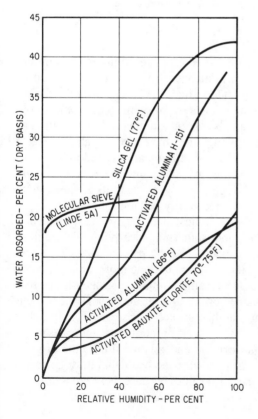

Figure 12-5. Equilibrium capacity of solid adsorbents versus relative humidity. *Silica-gel, data from Davison Chemical* (2); *Activated Alumina, data from Aluminum Company of America* (4); *activated bauxite, data from Amero et al.* (7); *molecular sieve, data from Union Carbide Corp.* (6)

A typical chemical analysis of commercial silica gel is given in Table 12-2 (2). The analysis is on a dry basis. The loss on ignition at 1750°F is reportedly 6.0 percent maximum.

The equilibrium partial pressure of water vapor over silica gel containing various concentrations of adsorbed water is shown in Figure 12-6. This figure is based upon the data of Taylor (8) and Hubard (9), extrapolated to cover the very low water range by application of the Freundlich relationship

$$W = KC^n \qquad (12\text{-}1)$$

where W = concentration of water in the silica gel
C = concentration (or partial pressure) of water in the gas phase
K and n = constants

In this figure, the residual-water-content of silica gel is included in the weight of the desiccant. This water, which normally amounts to about 6 percent of the activated weight, can be removed by heating to 1,750°F for 30 min but it is not removed at conventional regeneration temperatures.

Silica-Base Beads (Sorbead)

Sorbead is a chemically inert, solid, siliceous material in the form of uniform beads averaging about 0.14 in. in diameter (3). It is manufactured by Kali-Chemie AG of West Germany under a license from the Mobil Oil Corporation and distributed in the United States by Weskem, Inc. (108). It consists of ~97 percent silica (SiO_2) and 3 percent alumina (Al_2O_3) (109). The particles are hard and translucent although they have a very high percentage of internal porosity. The adsorption surface of Sorbead is stated to be over 3 million sq ft/lb. Its adsorption capacity on a per pound basis is essentially identical to that of conventional silica gel; however, due to its greater weight per cubic foot, it has somewhat greater capacity per unit of volume.

The water capacity of Sorbead R is indicated by Figure 12-7, which is based on atmospheric pressure, air, and isothermal adsorption. The equilibrium capacity shown on the chart was obtained by passing air at 77°F and various known humidities through the desiccant until no more moisture could be absorbed and the air emerged at the same humidity at which it entered. The break-point capacity was obtained by passing the air through the bed until the humidity of the air leaving the bed started to rise sharply. In either case, the capacity value was then determined by noting the increase in the weight of the bed due to water adsorption and relating this to the original weight of the dried material. The gas velocity and bed depth must be specified when reporting break-point capacity values because these are a function of operating conditions. The effects of adiabatic versus isothermal operation

Table 12-2. Typical Chemical Analysis of Commercial Silica Gel

Silica (SiO₂)	99.71 per cent
Iron as Fe₂O₃	0.03
Aluminum as Al₂O₃	0.10
Titanium as TiO₂	0.09
Sodium as Na₂O	0.02
Calcium as CaO	0.01
Zirconium as ZrO₂	0.01
Trace elements	0.03

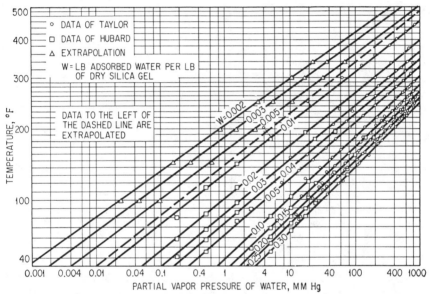

Figure 12-6. Equilibrium partial pressure of water vapor over silica gel containing various amounts of adsorbed water. Residual water which cannot be removed by conventional regeneration is included in the weight of the desiccant. *Based on data of Taylor* (8) *and Hubard* (9)

and other variables on break-point capacity are discussed in a later section.

Because of its spheroidal shape and mechanical strength, Sorbead R is more resistant to attrition by agitation than conventional granular silica gel. However, it is susceptible to damage if contacted with liquid water; in fact, regenerated beads shatter if dropped into water. To avoid deterioration of beds by the action of water droplets which may be entrained in the feed gas, a buffer desiccant is recommended for the portion of the bed nearest the feed-gas entry. This may be any solid desiccant which is not susceptible to damage by liquid water, including a more rugged bead-type desiccant which was developed for this purpose (Sorbead W).

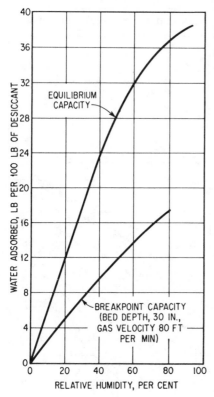

Figure 12-7. Effect of relative humidity on capacity of Sorbead R desiccant. Data obtained with air at atmospheric pressure and under essentially isothermal conditions (77°F). *Data of Mobil Oil Corp.* (3)

Activated Alumina

Activated Alumina is essentially a partially hydrated, porous, amorphous form of aluminum oxide containing very small quantities of other materials. It is made by a process which renders aluminum trihydrate highly porous and adsorptive. A typical analysis of Activated Alumina, grade F-1, manufactured by the Aluminum Company of America, is given in Table 12-3 (4). In the size ranges normally used for dehydration, Activated Alumina particles have a bulk density of about 52 lb/cu ft (equivalent to a specific gravity of about 0.8). Since the packed material contains about 50 percent voids, this indicates that the specific gravity of each particle, based on its overall volume including the pores, is about 1.6. The true specific gravity of solid material in the activated alumina particles is approximately 3.3 so the pores must occupy about 51 percent of the volume of each particle. More specific volume/weight relationship data for Activated Alumina, grade F-1, are presented in Table 12-4.

Table 12-3. Typical Analysis of Activated Alumina

Alumina (Al_2O_3) .... 92.00 percent
Loss on ignition (H_2O) 6.50
Soda (Na_2O)... 0.90
Ferric oxide (Fe_2O_3)... 0.08

Table 12-4. Volume-weight Relations for Activated Alumina

Particle size	Wt. of packed material, lb/cu ft,	Voids, %	Pores, %	Total air space, %
2–1 in......................	50	50	25	75
1–½-in.....................	50	50	25	75
½–¼ in....................	51	49	26	75
¼-in.–8 mesh	51	49	26	75
8–14 mesh.................	51	49	26	75
14–28 mesh................	54	46	28	74
28–48 mesh................	54	46	28	74
48–100 mesh...............	59	41	30	71
Minus 100 mesh............	64	36	33	69

The capacity of Activated Alumina for water is indicated by Figure 12-5, which includes a curve showing the water content of Activated Alumina F-1 in equilibrium with air at 86°F and various relative humidities.

Alumina-Gel Balls (Alcoa Activated Alumina H-151)

This material was developed by the Aluminum Company of America in order to provide increased water capacity and generally improved performance with an alumina-base desiccant. The manufacture of a gel-type alumina involves the mixing of sodium aluminate solution with sodium bicarbonate solution and the filtration, washing, drying, and activation of the gelatinuous mass thus formed. The initial version of the new material showed the desired high water-capacity as compared to conventional Activated Alumina; however, it did not prove satisfactory in granular form because of excessive particle breakdown. In order to overcome this problem and generally improve the appearance and performance of the product, a method of forming the gel-type alumina into balls was developed (10). The material is now offered as nominal ⅛- and ¼-in. diameter balls. Water-capacity data for Activated Alumina H-151 are presented in Figure 12-5 together with data for several other adsorbents. As can be seen, the equilibrium capacity of the alumina gel is intermediate between that of conventional Activated Alumina and silica gel—approaching the latter at 100 percent relative humidity.

A typical analysis of Alcoa H-151 Activated Alumina is given in Table 12-5.

A somewhat similar alumina product is offered by Kaiser Chemicals under the trademark KA-201 Active Alumina. This material is sold in the form of balls, typically ¼-in. diameter, and has a static equilibrium water absorption curve almost identical to that shown for the Alcoa H-151 desiccant in Figure 12-5. A typical chemical analysis of Kaiser Chemical KA-201 is given in Table 12-6 (11). The material has a porosity of 65 percent and a surface area of 380 sq m/g. It is claimed to possess an alumina structure significantly different from other desiccant grade aluminas (11).

Activated Bauxite

Activated bauxite used for gas dehydration usually appears as hard, brownish-red granules. The material is made from naturally occurring bauxite by heating it under controlled conditions to vaporize water from the hydrated alumina. The major component of the bauxite ore is usually alumina trihydrate. After activation, a typical activated bauxite (Florite) has the composition given in Table 12-7.

The principal advantage of activated bauxite is its low cost as compared to that of synthetic desiccants. It has the additional advantage of resisting breakup in the presence of liquid water, and, in spite of its low cost, it is capable of providing extremely low dew points. Its principal disadvantage is a somewhat smaller capacity for water than that of the other desiccants. The total capacity of Florite activated bauxite as a function of relative humidity is shown in Figure 12-5, which is based on air at atmospheric pressure. The break-point capacity for air, at room temperature and a 55°F dew point, flowing at a rate of 0.141 cu ft/min through 4- to 8-mesh Florite in a column 1.3 in. in diameter and 36 in. deep, is indicated to be about 6½ percent by weight according to the data of Amero et al (7). These authors recommend the use of 5 percent capacity for gases at temperatures less than 100°F, having a relatively high humidity, and containing no constituents that will cause undue fouling.

Molecular Sieves

Although naturally-occurring molecular-sieve adsorbents have been known for many years, this type of material did not become of commercial importance until the introduction of synthetic molecular sieves by Union Carbide Corporation's Linde Division in 1954. The molecular sieves differ from conventional adsorbents primarily in their ability to adsorb small molecules while excluding large ones, so that separations can be made based on

Table 12-5. Typical Analysis of Alcoa H-151 Activated Alumina

Al_2O_3 .... 90.0 percent
Na_2O 1.4
Fe_2O_3 .... 0.1
SiO_2 .... 1.1
Loss on ignition 6.0

Table 12-6. Typical Analysis of Kaiser Chemical KA-201

Al_2O_3 .... 93.60 percent
Na_2O 0.30
Fe_2O_3 .... 0.02
SiO_2 .... 0.02
Loss on Ignition 6.00

Table 12-7. Composition of Activated Bauxite

Al_2O_3 .... 70-75 percent
Fe_2O_3 3-4
SiO_2 11-12
TiO_2 3-4
Volatile (water) 4-6

molecular-size differences. They have the additional property of relatively high adsorption capacity at low concentrations of the material being adsorbed and have an unusually high affinity for unsaturated and polar-type compounds. The commercially available molecular sieves are crystalline sodium, potassium and calcium aluminosilicates which have been activated by heating to drive off water of crystallization. The crystals have a robust cubic structure which does not collapse on heating so that activation results in a geometric network of cavities connected by pores. The pores are of molecular dimension, in the range of 12 to 50 billionth of an inch in diameter, and cause the sieving action of these materials.

Properties of five types of molecular sieves offered by Union Carbide Corporation are listed in Table 12-8. All of the sieves are excellent desiccants for dehydrating gas; however, as indicated in the table, their special properties make certain types preferable for specific conditions. In addition to those listed, two special types, AW-300 and AW-500, are available for dehydration applications requiring resistance to acid components in the fluid being treated

Table 12-8. Basic Types of Union Carbide Corp. Linde Molecular Sieves

Basic type	Nominal pore diameter angstroms	Bulk density of pellets lb/cu ft	H_2O capacity (%/wt)*	Molecules adsorbed (typical)†	Molecules excluded	Typical applications
3A	3	47	20	H_2O, NH_3	Ethane and larger	Dehydration of unsaturated hydrocarbons
4A	4	45	22	H_2S, CO_2, SO_2, C_2H_4, C_2H_6, C_3H_6	Propane and larger	Static desiccant in refrigeration systems, etc. Drying saturated hydrocarbons
5A	5	43	21.5	$n\text{-}C_4H_9OH$,	Iso compounds, 4 carbon rings and larger	Separates n-paraffins from branched and cyclic
10X	8	36	28	Isoparaffins and olefins	Di-n-butylamine and larger	Aromatic hydrocarbon separations
13X	10	38	28.5	Di-n-propylamine	$(C_4F_9)_3N$ and larger	Coadsorption of H_2O, H_2S and CO_2

*Pounds H_2O/100 lbs activated adsorbent at 17.5 mm Hg partial pressure and 25°C, adsorbent in pellet form.
†Each type adsorbs listed compounds plus those of all preceding types.

(12). Molecular sieves are also offered by the Davison Chemical Division of W.R. Grace and Company and by the Norton Company. Norton offers several acid-resistant grades, designated Zeolon Series 100, 200, and 900 (95).

A major application of molecular sieves is for gas drying ahead of cryogenic processing, where extremely low dew points are required. Molecular sieve units are widely used prior to the cryogenic extraction of helium from natural gas, cryogenic air separation, liquefied natural-gas production, and deep ethane recovery from natural gas using the cryogenic turboexpander process. Although the molecular sieves are somewhat more expensive than other adsorbents, they offer the following advantages.

1. They provide good capacity with gases of low relative humidity.
2. They are applicable to gases at elevated temperatures.
3. They can be used to adsorb water selectively.
4. They can be used to remove other selected impurities together with water.
5. They can be used for adiabatic drying.
6. They provide extremely low dew points.
7. They are not damaged by liquid water.

The equilibrium capacity of molecular sieve, type 5A, for water vapor at 25°C is presented in Figure 12-5 with curves for other adsorbents which are included for purposes of comparison. The effect of temperature on the equilibrium adsorption-capacity of molecular sieve, Type 5A, Activated Alumina, and silica gel is shown in Figure 12-8 for a water-vapor partial pressure of 10 mm. These data show that at 200°F, for example, the molecular-sieve capacity is 15 percent by weight, while the other adsorbents have an almost negligible capacity at this temperature. The extremely high drying efficiency of molecular sieves makes them useful for "trimmer" beds in conjunction with silica-gel or alumina desiccants. In this type of installation, a small bed of molecular sieves following the conventional adsorbent removes the last traces of moisture from the gas stream, permitting the primary bed to be loaded to a higher capacity and producing a lower product-gas dew point.

Dehydrator Design

When a gas containing water is passed through a bed of freshly regenerated adsorbent, the water is adsorbed first near the inlet portion of the bed, and the dehydrated gas passes through the rest of the bed with only a small amount of additional drying taking place. As the adsorbent nearest the inlet becomes saturated with water at the condition of the feed gas, the zone of rapid water adsorption moves inward and ultimately progresses through the entire bed.

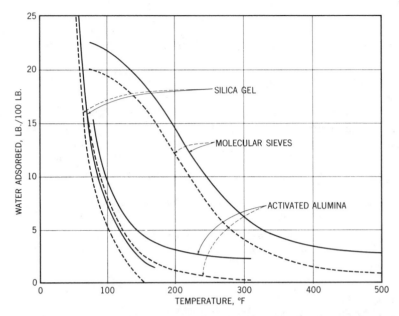

Figure 12-8. Comparison of the effects of temperature on the capacity of molecular-sieve type 5-A, silica gel, and activated alumina in equilibrium with water vapor at 10-mm partial pressure. The dotted lines show the effect of 2 percent residual water at the start of adsorption. *Data of Union Carbide Corp.* (6)

When this "adsorptive wave" reaches the outlet end, the water content of the product gas is observed to rise rapidly, signifying the "break point" for the particular operating conditions.

The adsorption of water results in the evolution of heat in the active adsorption zone. With high-pressure gas (above about 500 psig), the large weight of gas associated with each pound of water picks up the heat released with only a small temperature rise, on the order of 2° to 4°F. With low-pressure gas or air, on the other hand, there are fewer pounds of gas per pound of water vapor so that a much greater temperature rise is possible. The heat is actually liberated inside the adsorbent particles, as a result of water condensation and adsorbent wetting. If no cooling coils are provided in the bed, the heat is transferred to the gas stream in the active adsorption zone. On leaving this zone, however, the hot gas encounters cool (and dry) adsorbent, and heat transfer occurs in the reverse direction, warming the bed downstream to the adsorption zone and recooling the gas stream, which emerges from the bed only slightly warmer than it was upon entering. The high-temperature wave thus progresses through the bed somewhat in advance of the adsorptive wave, so that the exit-gas temperature starts to rise well before the dehydration break point occurs.

Table 12-9. Temperature Changes Within an Adiabatic Dehydration Bed						
Distance from air inlet, in.	Temperature, °F, for operation times given					
	0 hr	0.5 hr	2.5 hr	4.5 hr	7.0 hr	11 hr
0 (inlet)	75	75	75	75	75	75
3	75	165	105	95	90	85
12	75	95	245	140	115	100
21	75	92	145	235	160	120
30 (outlet)	75	90	93	185	213	170

Source: Data of Derr for air dehydration with activated alumina (13)

This effect is clearly shown by the data in Table 12-9, based on information presented by Derr (13) for air dehydration, at atmospheric pressure, with activated alumina. The table presents temperatures within an uncooled bed of desiccant, 32 in. high and 12 in. in diameter, drying air which is at 75°F dry-bulb temperature, contains 9 grains moisture/cu ft, and flows at a rate of 5.2 cu ft/hr per pound of alumina. The break point was first detected after 7 hr of operation at which time the exit-air temperature was 213°F. However, 1 hr later (after 8 hr of operation), the water content of the outlet gas was still only 0.34 grains/cu ft, indicating a water-removal efficiency of 96 percent.

The heat generated in adiabatic adsorbers not only raises the temperature of the bed and the gas but also decreases the operating capacity as a result of the effect of temperature on equilibrium. Cooling coils are sometimes placed within adsorbent beds to remove this heat, making the operation substantially isothermal, with an appreciable increase in capacity. However, the added expense of this type of construction is seldom justified, and it is more common practice to design for a larger bed and adiabatic operation. The extent of the capacity reduction as a result of adiabatic operation is not easily calculated because of such complicating factors as the cooling of the inlet portion of the bed by fresh gas with a subsequent increase in its capacity and readsorption of stripped water downstream to the adsorptive wave. The magnitude of the effect was investigated by Grayson (14) who compared adiabatic with isothermal operation for the case of Sorbead desiccant dehydrating air at atmospheric pressure. He found the adiabatic break-point capacity to be considerably below the isothermal break-point capacity and, under some conditions, to decrease with increased water content of the feed gas. This effect is illustrated in Table 12-10 by examples of Grayson's data which were obtained with air, at atmospheric pressure and 80°F dry-bulb temperature, passed through a bed of Sorbead 36 in. deep.

Table 12-10. Adiabatic and Isothermal Break-point Capacities for Sorbead Desiccant Dehydrating Air at Atmospheric Pressure (14)

Break-point capacity	Air velocity			
	30 ft/min		60 ft/min	
	Air wet-bulb temp.*		Air wet-bulb temp.*	
	35°F	60°F	35°F	60°F
Adiabatic, % by wt..........	7.4	5.6	5.8	4.2
Isothermal, % by wt.........	11.0	20.0	8.0	16.8

*Inlet air.

The effect of gas velocity on break-point capacity is also indicated by the data of Table 12-10. As can be seen, increasing from 30 to 60 ft/min decreases both the adiabatic and isothermal break-point capacities with a given quantity of adsorbent. Grayson also investigated the effects of humidity and bed depth. As indicated in the table, increased humidity generally (although not always) decreased the capacity at break point. Increasing the bed depth was found to increase the break-point capacity (on a unit weight basis) appreciably, as a result of the cooling effect of the inlet gas on the first portion of the bed contacted.

Equations describing nonisothermal adsorption in large fixed beds have been developed by Leavitt (15). His analysis indicates that for the case of a feed gas containing one adsorbable and one nonadsorbable component in a large diameter bed, two separate transfer zones tend to form. The equations permit calculation of concentrations, loadings, and temperature in both zones and in the interzone region as well as the speeds and lengths of the zones as they move through the adsorber.

As pointed out above, an increase in gas velocity decreases the break-point capacity for a given bed of desiccant. However, if the bed is made deeper to compensate for the increased gas velocity and provide the same cycle time as with lower gas rate, the average capacity of the bed at the break point is increased. This is due primarily to an increase in the mass-transfer coefficient as a result of the high gas velocity.

As in other mass-transfer operations, the rate of transfer of water vapor from the gaseous phase is a function of the gas-flow rate, the size and shape of the desiccant particles, and the properties of the gaseous and adsorbate phases. If the mass-transfer coefficient is very high, the adsorptive zone will be quite abrupt; i.e., complete dehydration will be obtained until the break point is reached; at this time the water content of the product gas will rise very rapidly. If the mass-transfer coefficient is quite low, on the other hand (or if the bed is very shallow), some water vapor may pass through with the gas from the start, and the water content of the exit gas will increase slowly as the entire bed becomes saturated. Most commercial installations fall between these two extremes in that there is first a period of maximum dehydration, then after a definite break point, the water content of the product gas is observed to rise at a moderate rate. Break-point capacity curves of this type for a typical desiccant (Activated Alumina H-151) in high-pressure natural-gas service are shown in Figure 12-9. The data illustrated were obtained in a twin-tower dehydration unit, each tower 16 ft high and 3 ft in diameter. Natural gas at approximately 850 psig was dehydrated using downflow during dehydration and upflow (with 360°F gas) during regeneration. One tower required 3,900 lb H-151 Activated Alumina Gel Balls, ¼ in. to 6 mesh, and handled 16 to 19 MMscf/day of gas.

In the dehydration of air at atmospheric pressure for air conditioning, pressure through the desiccant must be kept to an absolute minimum, and very shallow beds are employed. Because of this, the initial break point occurs very quickly, and the dew point of the product gas climbs steadily during most of the adsorption cycle. This type of operation has been analyzed quite thoroughly by Ross and McLaughlin (16) for silica-gel adsorption. The results of typical adsorption and desorption tests by these authors are shown in Figure 12-10.

Design Capacity of Desiccants

Although the equilibrium capacity is of interest in comparing desiccants, the design capacity is always lower for the following reasons.

1. As shown in Figure 12-9, which illustrates a typical drying period, water vapor appears in the product gas well before the average moisture content of the bed reaches the value which it would have if equilibrium were attained with the inlet gas.
2. Since the break point represents the limit of capacity with maximum dehydration, it is common design-practice to provide a factor of safety and end the dehydration period before the break occurs. (Where a higher dew point is allowable, units are sometimes designed to exceed the break-point capacity.)
3. An appreciable decrease in capacity normally occurs with extended operation. It is, therefore, necessary to design for the minimum capacity expected before desiccant replacement is required.

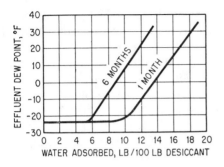

Figure 12-9. Capacity-test curves for Activated Alumina H-151 after 1 month and 6 months of service, dehydrating natural gas at approximately 850 psig and 80°F; gas-flow rate, 16 to 18 million SCF/day; tower size, 16 ft high, 3 ft diameter. *Data of Getty et al.* (10)

	NOMINAL	ADSORPTION	DESORPTION
BED THICKNESS (IN.)	4		
ROOM DB (°F)	70	70.0	70.1
ROOM RH (%)	35	34.8	34.6
AIR FLOW (CFM)	110-112.5	110.6	113.2
FACE VEL (FPM)	55-56.25	55.3	56.6
WATTS	4500		4531
CFM/WATT	0.025		0.025

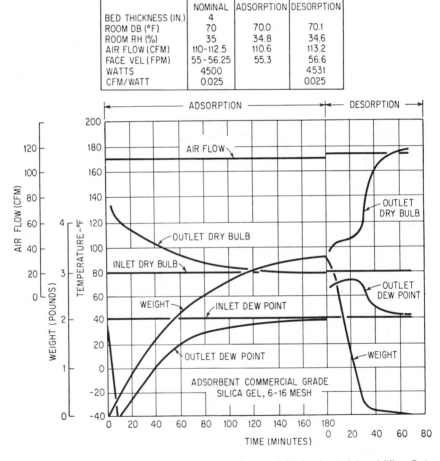

Figure 12-10. Characteristic operation curves for a solid adsorbent dehumidifier. *Data from Ross and McLaughlin* (16)

Figure 12-11 illustrates the decrease in adsorbent capacity with time in service for several adsorbents. All of the data in this figure are for units dehydrating natural gas; however, the operating conditions and gas analyses vary considerably so that curves for different adsorbents are not strictly comparable. The curves are typical, however, and indicate the basis for establishing design capacities. Recommended design capacities for adsorbents in high-pressure natural-gas service, assuming the gas to be clean and essentially saturated with water, are presented in Table 12-11. For comparison, the drying capacities of the adsorbents when new are included. The latter data are based on air of 70 percent humidity (at 80°F), dehydrated at a rate of 15 cu ft/hr per pound of adsorbent, to a dew point of −50°F.

The phenomenon of decreased capacity with service is not fully understood; however, some of the factors which contribute to this effect are known. Relatively nonvolatile compounds, such as absorption oil or compressor oil are known to reduce the adsorptive capacity. It has also been shown that, even without contamination, the adsorption capacity is gradually reduced by the effect of water and heat during regeneration (22).

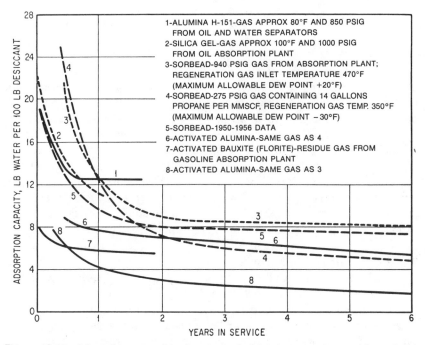

Figure 12-11. Adsorption capacity of various desiccants versus years of service in dehydrating high-pressure natural gas. *Curve 1, data of Getty et al* (10); *curve 2, data of Herrmann* (17); *curves 3, 4, 6, and 8, data of Swerdloff* (18); *curve 5, data of Hammerschmidt* (19); *and curve 7, data of Cappel et al* (20)

Table 12-11. Typical Design Capacities of Adsorbents		
Adsorbent	Typical design capacity, %	New capacity, %
Activated bauxite-type desiccants	3-5	12-13
Activated Alumina	4-6	13-15
Silica-gel and silica-base beads	5-8	18-21
Alumina-gel balls (Alumina H-151)	5-12	20-24
Molecular Sieve Type A	7-14	14-18

Table 12-12. Capacity of Bauxite after 2-Years Service	
Position in Adsorber	Capacity, percent
Tray 1 (gas inlet)	3.2
Tray 2	3.9
Tray 3	4.6
Tray 4	5.7
Tray 5 (gas outlet)	5.7

It has been observed that the desiccant at the top of the bed (gas inlet) decreases in capacity most rapidly. This is not unexpected inasmuch as the top portion of the bed contacts the wet inlet gas during the entire adsorption period and, in addition, is more apt to adsorb small quantities of heavy hydrocarbons or other desiccant poisons which may be present. During regeneration, which is normally conducted with upflow of heated gas, all of the steam generated in the lower portion of the bed must pass upward through the top section. Hydrocarbons adsorbed near the top are thus carried out without contacting the lower portion of the bed. Data relative to the decline in capacity of bauxite desiccant as a function of position in the adsorber have been presented by Hammerschmidt (19). These data show bauxite, with an original capacity of 8 percent water, to have the capacity shown in Table 12-12 after 2½ years of service.

As indicated by the equilibrium-capacity data for the various desiccants, the quantity adsorbed is a function of the vapor pressure of water in the gas stream (and thus of the relative humidity of the feed-gas stream). Capacity values used for design must also be adjusted downward if the gas being dehydrated is less than 100 percent saturated. Data on two desiccants used to dehydrate natural gas containing a maximum of 7 lb water/million cu ft at 72°F and 570 psig are presented by Harrell (23). Very complete gas dehydration is required at this plant, since the gas stream is chilled to −105°F, and it is necessary to prevent ice formation on the chillers. Several desiccants were

Table 12-13. Capacity of Desiccants (Used in dehydrating gas containing a maximum of 7 lb H$_2$O/MM cu ft at 72°F and 570 psig) (23)

Age of desiccant, months	Capacity, per cent		No. of reactivations	
	Silica-gel beads	Alumina pellets	Silica-gel beads	Alumina pellets
0	7.0	7.0	1	1
8	5.0	3.9	145	145
18	2.7	2.5	272	283
27	2.0	2.2	425	436
35	2.2	2.1	587	598

tested, and silica-gel beads and alumina-base pellets were selected as most favorable. Both of these showed capacities of 7 percent immediately after installation and decreased in capacity as shown in Table 12-13.

The drying of cracked-gas streams poses a particular problem because of polymerization of the olefins on the desiccant. Molecular sieves are affected less than silica or alumina in this type of service because of their capability to exclude large molecules. A comparison of four desiccants for drying a cracked gas stream was conducted at Union Carbide Corporation's Texas City olefins plant (24). For both high- and low-density alumina, the capacity was found to decrease from about 5 lb H$_2$O/100 lb desiccant after 25 cycles to about 1 lb/100 lb after 200 cycles. Silica gel's capacity dropped from about 10 to 2 lb/100 lb in the same period, while the capacity of molecular sieve 3A decreased by only about 25 percent from 13.7 to 10.2 lb H$_2$O/100 lb desiccant, in the period between 25 and 200 drying cycles. As a result, all of the adsorption units were recharged with Type 3A molecular sieve (replacing alumina). Because of its greater capacity for water, a depth of about 45 percent of the original bed depth was used.

At the Phillips Petroleum Company, Sweeny, Texas, plant, considerable cost savings were effected by converting two dehydration units treating cracked gas streams from activated alumina to Type 3A molecular sieves. In addition to replacing the bed material, the system was modified from a lead-trim mode to a parallel mode, and a compound bed consisting of both 1/8-in. and 1/16-in. pellets was used. In the lead-trim mode of operation, two beds are used for dehydration in series while the third is being regenerated. This is advisable with activated alumina to prevent leakage of moisture as the material loses capacity due to accelerated aging. In the parallel mode, two beds are used for dehydration in parallel while the third is regenerated. This reduces pressure drop and therefore compression energy in the main gas stream and permits the use of smaller pellets. The overall effect of the changes was an estimated savings of $300,000 per year (110).

According to Silbernagel (25), butadiene and acetylene (including methyl acetylene and propadiene) cannot usually be dried at all with conventional desiccants because of rapid fouling of the bed by adsorbed hydrocarbons. With Type 3A molecular sieve, on the other hand, design water loadings of 7 to 14 lb H_2O/100 lb desiccant can be used with a 2- to 4-year molecular sieve life. A 380° to 425°F purge-gas temperature is normally used for regeneration of the adsorbent in this service. Although the Linde 3A molecular sieve is able to exclude all molecules larger than methane and thereby resists fouling due to polymerization of adsorbed hydrocarbons, it is apparently quite sensitive to attack by acids including H_2S and CO_2 when present in high concentrations. This problem has been resolved by the development of acid-resistant molecular sieves. In one plant employing an acid resistant Type A sieve to dehydrate natural gas containing 26 percent H_2S and 5 percent CO_2 at 2,200 psig, a normal service life of 2 years per desiccant charge has been achieved (26).

Another problem relating to the use of molecular sieves for processing gas containing H_2S and CO_2 is the formation of COS. The phenomenon has been identified as the simultaneous H_2S adsorption and rate-limited catalytic reaction of H_2S and CO_2 to form carbonyl sulphide (COS) and water (96). An investigation of the problem as it relates to the removal of low concentrations of H_2S (less than 160 ppm) from natural gas has been reported by Cines, et al (97). They found sieves of the 5A type to be least catalytic, the 4A type to be of intermediate activity, and the 13A sieve to be the most catalytic. The COS formed is not strongly adsorbed and therefore appears in the treated gas long before H_2S appears. The problem can also be encountered in molecular sieve beds performing primarily as dehydrators. McAllister and Westerveld report that a large cryogenic plant in the Middle East produced a raw gas liquid containing 800 ppm COS as a result of the H_2S reaction proceeding in the molecular sieve dehydrator (98). They recommend removal of acid gas prior to dehydration to avoid this problem.

Desiccant Regeneration

Regeneration (or reactivation) of the desiccants is accomplished by taking advantage of the fact that the capacities of all of the desiccants decrease with increased temperature. Usually a stripping gas is also employed to flush the released water vapor from the bed and reduce the partial pressure of water vapor in the gas to the lowest point possible during regeneration. Sufficient heat is required to provide the latent heat of vaporization of the adsorbed water and to raise the temperature of the bed and associated equipment to the final regeneration temperature. Theoretically, the heat of wetting must also be provided; however, this is quite small relative to the heat of vaporization and is usually neglected. The required heat is generally supplied by passing a stream of preheated gas through the bed as pointed out above, the preferred flow direction for the regeneration gas being the reverse of that taken by the gas during dehydration. A regeneration-gas temperature of about 350° to 400°F is typical for all of the adsorbents, except molecular sieves. These materials require regeneration at a temperature in the range of 500° to 700°F if they are to provide maximum capacity and minimum dew point ($-120°$ to $-150°F$). However, they reportedly will still provide one and one-half to two times the capacity, at dew points 15° to 40°F lower than conventional desiccants, when regenerated in the same manner as the conventional desiccants (27). The regeneration pressure for all adsorbents may be the same as that used for dehydration; however, somewhat lower pressures are sometimes used to improve the stripping of adsorbed hydrocarbons.

A reactivation period for a natural-gas dehydration unit is shown in Figure 12-12. In this unit, gas at 300°F is used for regeneration, and it will be noted that the temperature of the regeneration gas leaving the unit (bottom of bed)

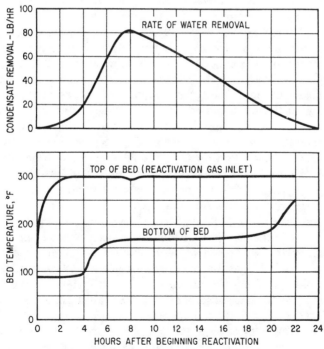

Figure 12-12. Adsorber reactivation period. Pressure, 150 psia; natural-gas specific gravity, 0.68 (dry). *Data of Cappell et al.* (20)

rises rapidly to about 170°F and then remains relatively constant while evaporation of the bulk of the adsorbed water takes place. When the outlet-gas temperature rises rapidly to within about 50° of the inlet-gas temperature (in this case 300°F), reactivation is considered to be complete.

The temperature of the gas leaving the desiccant bed during the period when the major portion of the water is being removed can be estimated by a trial-and-error, heat-balance calculation (20). This calculation can be made by assuming a value for the outlet-gas temperature and adjusting it until the sensible heat loss of the regeneration gas is equal to the heat required to vaporize water and raise the temperature of the adsorbent and equipment. This heat quantity is calculated on the basis of the latent heat of vaporization of the water required to saturate a unit quantity of the gas at the assumed temperature plus the sensible heat necessary to raise the temperature of adsorbent, regenerated by this unit gas quantity and its surrounding vessel, to the inlet-gas temperature. In natural-gas dehydration, the water is usually evolved at a temperature between 150° and 250°F, regardless of regeneration pressure.

Although the regeneration period illustrated in Figure 12-12 is of interest in emphasizing the constant-temperature water-removal period and indicating the rate of water removal during the operation, a more typical period would probably be completed in less time and with gas of a higher temperature. Where the dehydrator vessels are cycled every 8 hr, for example, the regeneration-gas quantity and temperature are usually set to provide complete regeneration in about 4 hr, with cooling in about 2 to 3 hr, leaving some margin of safety before the dehydrator is brought on stream. Cooling is most commonly accomplished at the same gas-flow rate as regeneration, with the regeneration heater bypassed. It is sometimes satisfactory and much simpler to pass the cooling gas through the bed in the same direction taken during regeneration. Where a consistently high degree of dehydration is required, however, the gas-flow direction should be reversed, from upflow during heating to downflow during cooling, in order that any water deposited by the cooling gas will be at the top of the bed and, therefore, not picked up by the product gas during the adsorption cycle.

In a process offered by Maloney Steel of Calgary, Canada, the customary cooling step is eliminated (99). The Maloney process typically employs three beds of molecular sieve adsorbent. Regeneration is accomplished by a continuous stream of inlet gas that is heated first in a gas-to-gas heat exchanger by regeneration gas from the adsorber tower and finally in a gas-fired salt bath heater to 500°F. Each adsorber is in service two-thirds of the time and in regeneration one-third of the time. Since no time is required for cooling, the adsorbent is actively drying gas for a greater fraction of the overall cycle period than in conventional designs and, as a result, the process requires a smaller volume of adsorbent.

The principal drawback of the process is the possibility of producing poorly dried gas for a brief period when a hot, regenerated bed is first put on stream for dehydration service. However, operating experience of more than a year with a commercial unit did not show evidence of such occurrences (possibly due to instrumentation limitations) (99). In theory, the heat transfer zone moves through the bed several hundred times faster than the mass transfer zone. As a result, except for a brief period at the beginning of the cycle, the heat transfer zone moves rapidly ahead of the mass transfer zone, providing an adequate supply of cool adsorbent to pick up water contained in the feed gas.

Drying-Tower Dimensions

The approximate volume of disiccant required can be estimated from consideration of its assumed design capacity and bulk density as well as the desired cycle time. A more rigorous approach to the problem has been developed by Hougen and Marshall (28) for the case of isothermal operation. This method employs equations relating the fraction of water remaining in the gas (or fractional approach to saturation of the bed at any point) to the height of the gas-film mass-transfer unit, the slope of the equilibrium line, the gas velocity, and the bed-depth. An expression is also presented which relates the height of an overall mass-transfer unit for water-vapor adsorption by silica gel to the Reynolds number and surface area. This approach and more recent attempts to develop rigorous mathematical models for fixed-bed adsorber behavior are described in standard chemical engineering texts (29, 30).

The zone of active adsorption within a fixed bed is sometimes called the mass-transfer zone or MTZ. The length of this zone is a measure of how rapidly adsorption is occurring. A short MTZ is desirable in that it enables a large fraction of the total bed to approach equilibrium with the gas before breakthrough occurs. An equation for calculating MTZ length is given by Barry (31). A simplifying approach which is claimed to be standard design practice for the sizing of fixed-bed molecular sieve systems views the total bed as the sum of two sections, the equilibrium section and a section representing the equivalent length of unused bed (LUB) (32). The LUB is the quantity of adsorbent required to compensate for the presence of a mass transfer zone during dynamic adsorption. When breakthrough occurs, it is defined by the following equation:

$$LUB = L_o - L_s \qquad (12\text{-}2)$$

where LUB = length equivalent to unused bed, ft
L_o = total bed length, ft
L_s = position of stoichiometric front in bed, ft

The position of the hypothetical stoichiometric front at any time is given by the equation:

$$L_s = 100G/\rho_b \times (\Delta Y/\Delta X)\theta \qquad (12\text{-}3)$$

where G = the gas-feed rate, lb mole/hr ft²

ρ_b = adsorbent bulk density, lb/ft³

$\Delta X = Xe - Xo$ where Xe and Xo are the adsorbate loadings in equilibrium with the feed gas and on the regenerated adsorbent respectively, lb mole/100 lb activated adsorbent

$\Delta Y = Ye - Yo$ where Ye and Yo are the concentrations of adsorbable component in the gas feed and in the gas in equilibrium with regenerated adsorbent respectively, lb mole/lb mole

θ = time from the start of adsorption, hrs

The equations are best used in conjunction with actual test data. L_s is calculated from Equation 12-3 by substituting θ_B, the time at which breakthrough occurs, for θ and using adsorption isotherms to establish ΔX and ΔY. LUB is then determined by Equation 12-2. This LUB value can then be applied to the design of a plant with an entirely different length of equilibrium section, provided factors affecting the mass transfer zone are the same. A detailed review of adsorption column design using the mass-transfer-zone concept has been presented by Lukchis (100).

In general, the dimensions of the tower needed to contain the required volume of adsorbent are established on the basis of construction costs and permissible gas-pressure drop. According to Amero et al. (7), this usually results in a height/diameter ratio between 2:1 and 5:1 and a gas velocity between 20 and 60 ft/min, based on the empty vessel. Where deep beds are indicated, it is common practice to install intermediate support-trays at intervals of 4 or 5 ft in order to minimize the load on particles at the bottom of the bed and aid in gas distribution. Even in high-pressure gas streams, pressure drop is of some importance, and several design modifications have been proposed to minimize it, including the use of horizontal rather than vertical vessels and the use of radial gas flow from a central core to an outer annulus in vertical vessels. In low-pressure gas, and atmospheric-pressure—air service, pressure drop is of extreme importance, and large-diameter shallow beds with a height/diameter ratio of 1:1 (or less) are frequently employed.

For a given total gas flow and desiccant-bed volume, a deep bed is more effective than a shallow bed, in that it permits the desiccant to attain a higher average loading and provides a higher degree of dehydration of the gas. These advantages are gained, however, at the expense of pressure drop since the deep bed must be operated at a higher gas velocity.

Allowable gas velocities during both adsorption and regeneration are limited by considerations of particle entrainment and bed agitation as well as pressure drop. This problem has been analyzed by Ledoux for the case of up-flow of gas, and the following semiempirical equation has been proposed (33):

$$G^2/d_g d_a Dg = 0.0167 \tag{12-4}$$

where G = mass velocity of gas, lb/(sec)(sq ft)
$\quad d_g$ = gas density (flow conditions), lb/cu ft
$\quad d_a$ = adsorbent bed density, lb/cu ft
$\quad D$ = average particle diameter, ft
$\quad g$ = acceleration due to gravity, ft/(sec)^2

As the equation is dimensionless, any other consistent set of units could be used. Actual design velocities should be somewhat lower than might be indicated by the equation because of uncertainties inherent in the use of an average particle size and bed density for commercial adsorbents.

An alternative approach mentioned by Wunder (34), for establishing the maximum permissible downflow gas velocity in beds of granular alumina makes use of a momentum concept:

$$N_M = V \times M \times P \le 30,000 \tag{12-5}$$

where N_M = a design parameter proportional to the gas momentum
$\quad V$ = superficial gas velocity, ft/min
$\quad M$ = molecular weight of gas
$\quad P$ = system pressure, atm

It is stated that desiccant attrition should not be a problem if N_M is equal to or less than 30,000. This value is based on granular alumina but is also recommended for silica gel.

The following gas velocities are given by Barrow as typical design values which will give an acceptable pressure drop for ⅛-in. molecular sieve in gas dehydration service (111).

Pressure (psia)	Velocity (fpm)
200	55–71
600	32–48
1000	25–40

It is recommended that the velocity of the gas being dehydrated be limited to a value which will result in a pressure drop no higher than 7 psi. During regeneration, the gas velocity must be high enough to create a pressure drop of 0.01 psi/ft. This leads to velocities in the range of 5–8 fpm for $\frac{1}{8}$-in. molecular sieve and 2–3 fpm for $\frac{1}{16}$-in. material (111).

Pressure Drop

In the design of solid-desiccant dehydration plants, it is important that the pressure drop through the bed be estimated as accurately as possible, because the work required to overcome this pressure drop can represent a major operating cost.

Generalized correlations for estimating pressure drop of gas flowing through a bed of granular particles have been proposed by Rose (35), Brownell and Katz (36), Leva (37), and Ergun (38). A modified form of the Ergun equation is suggested by Union Carbide Corporation for use with molecular sieve beds (39). The equation, which appears to be quite suitable for general use with all fixed bed adsorbents, is given below:

$$\Delta P/L = (f_t\, C_t\, G^2/\rho D_P)\, 10^{-10} \tag{12-6}$$

where C_t = pressure-drop coefficient $(ft)(hr^2)(in.^2)$
   D_P = effective particle diameter, ft
   f_t = friction factor
   G = superficial mass velocity, $lb/(hr)(ft^2)$
   L = distance from bed entrance, ft
   ΔP = pressure drop, psi
   ρ = fluid density, lb/ft^3
   $\Delta P/L$ = the pressure drop per unit length of bed, psi/ft

The value of D_P for cylindrical pellets is given by Equation 12-7.

$$D_P = D_c/[\tfrac{2}{3} + \tfrac{1}{3}(D_c/L_c)] \tag{12-7}$$

where D_c = particle diameter, ft
   L_c = particle length, ft

The friction factor, f_t, and the pressure-drop coefficient, C_t, are determined from Figure 12-13. The friction factor is plotted against a modified Reynolds number in which

   μ = fluid viscosity, $lb/(hr)(ft)$

and the other factors are as defined above. The pressure-drop coefficient, which takes into account the packing density and also includes required con-

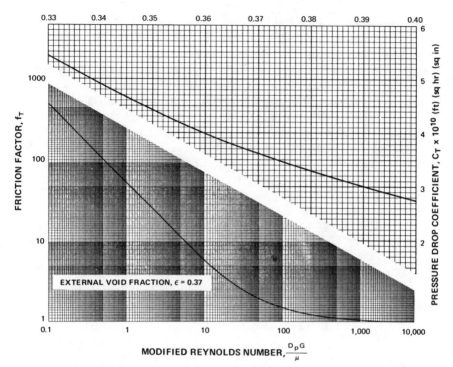

Figure 12-13. Correlation for estimating friction factor and pressure drop coefficient for use in calculation of pressure drop of gas flowing through beds of granular adsorbents. *Data of Union Carbinde Corp.*

version factors and constants, is plotted against the external void fraction of the bed. Suggested values for D_P, the effective particle diameter, and ϵ, the external void fraction, are given in Table 12-14 for a number of desiccants based on the data of Fair (40) and manufacturers bulletins.

Since the above equations and charts are somewhat cumbersome to use, Figures 12-14 and 12-15 are presented to permit quick pressure-drop estimation for the two conditions most frequently encountered: i.e., high-pressure natural gas and air at atmospheric pressure. Figure 12-15, which is based on the data of Allen (41), has been extended to include two grades of silica gel. Available data for pressure drop through beds of Sorbead R and Kaiser Chemicals KA-201 Active Alumina ($\frac{1}{4}$ in. × 8 mesh) indicate that both of these ball-shaped desiccants would be represented quite closely by the 4–8 mesh curve of Figure 12-15.

Since some bed-settling and particle-breakage may occur, it is not unusual for the pressure drop through desiccant beds to increase appreciably with time. The pressure-drop record of a silica-gel dehydration unit is shown in Figure 12-16 (17). The data for this figure were obtained in a twin-tower unit, each bed 15 ft high and 38 in. in diameter. The unit handled 15 MMscf/24 hr of natural gas at about 970 psig and 95°F inlet temperature. Data presented by Harrell (23) also show increased pressure drop versus time for both silica-gel beads and alumina pellets. His data are summarized in Table 12-15 for the case of 3.0 MMscf/hr of natural gas at 570 psig and 72°F. The dehydrator beds are 6 ft 6 in. in diameter and 20 ft deep.

Process Flow System

A typical flow diagram for a high-pressure natural-gas dehydration plant is shown in Figure 12-17. In this arrangement, the regeneration gas is taken from the main wet-gas stream ahead of a pressure-reducing valve, which maintains sufficient pressure drop to enable the regeneration gas to flow through a heater, dehydrator, cooler, and separator and back into the wet-gas feed stream. In an alternate arrangement which has been proposed, the cool-

Table 12-14. Desiccant Properties for Pressure Drop Calculation

Desiccant	Particle form	Mesh size	Bulk density lb/ft³	Effective diameter D_P ft	External void fraction, ϵ
Silica Gel	Granules	3 x 8	45	0.0127	0.35
	"	6 x 16	45	0.0062	0.35
	Spheres	4 x 8	50	0.0130	0.36
Alumina	Granules	4 x 8	52	0.0130	0.25
	"	8 x 14	52	0.0058	0.25
	"	14 x 28	54	0.0027	0.25
	Spheres	¼ in.	52	0.0208	0.30
	"	⅛ in.	54	0.0104	0.30
Molecular sieves	Granules	14 x 28	30	0.0027	0.25
	Pellets	⅛ in.	45	0.0122	0.37
	"	1/16 in.	45	0.0061	0.37
	Spheres	4 x 8	45	0.0109	0.37
	"	8 x 12	45	0.0067	0.37

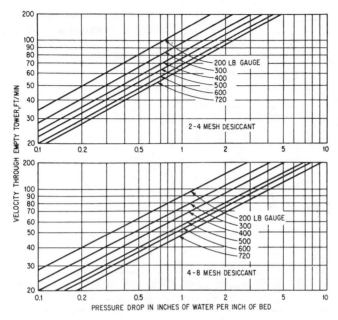

Figure 12-14. Pressure drop of gas through adsorbent beds at elevated pressures. Specific gravity of gas, 0.677. *Data of Cappell et al.* (20)

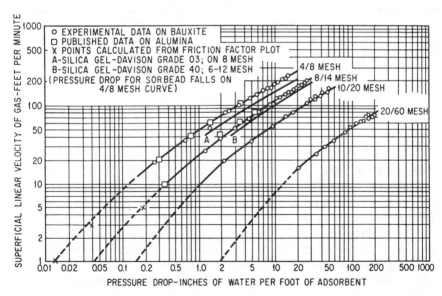

Figure 12-15. Pressure drop for air at atmospheric pressure and ambient temperature through beds of granular adsorbents. *Data of Allen* (41); *silica gel data from Davison Chemical Company* (2)

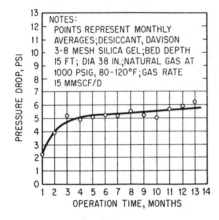

PRESSURE DROP, PSI

NOTES:
POINTS REPRESENT MONTHLY
AVERAGES;DESICCANT, DAVISON
3-8 MESH SILICA GEL;BED DEPTH
15 FT; DIA 38 IN.;NATURAL GAS AT
1000 PSIG, 80-120°F;GAS RATE
15 MMSCF/D

OPERATION TIME, MONTHS

Figure 12-16. Effect of time in service on pressure drop through silica-gel desiccant handling high-pressure natural gas. *Data of Herrmann* (17)

Table 12-15. Pressure-drop Changes in Desiccant Beds with Length of Service

Length of service, months	Pressure drop, psi	
	Silica-gel beads	Alumina pellets
8	8.5	1.8
18	9.0	2.6
35	10.5	2.9

ed regeneration gas reenters the system at a mid-point of the tower in dehydrating service. This uses the pressure drop across the first half of the bed instead of the pressure drop across a valve to provide the driving pressure for the regeneration gas and has the effect of reducing the required pressure drop across the entire dehydration plant. In the unit shown, steam at 386°F is used to heat the regeneration gas to 360°F. When regeneration of the bed is completed, as evidenced by a rise in the exit-gas temperature to approximately that of the inlet gas, the steam to the heater is shut off (or the heater may be bypassed), and the slip stream of gas is used to cool the regenerated bed. When the bed-temperature approximates that of the dehydration-plant feed, it is ready for dehydration service. Switching the two dehydrators from dehydration service to regeneration heating and then to regeneration cooling

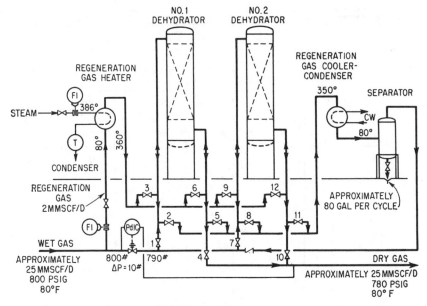

Figure 12-17. Process flow diagram of typical natural-gas dehydration plant.

is accomplished by the use of 12 valves which are numbered on the drawing. The valve positions for the various operations are given in Table 12-16.

In the arrangement shown, the gas flow is downward during dehydration, when the gas velocity is highest, in order to avoid disturbing the bed. It is upward during regeneration in order that the bottom of the bed will be thoroughly regenerated, as it is the last point of contact with the gas being dehydrated; and it is downward during cooling in order that any water deposited from the cooling gas will be at the top of the bed where it cannot be revaporized into the gas stream during dehydration.

The length of the dehydration period may be varied within wide limits. Shorter periods are more economical with regard to equipment size and desiccant charge; however, operating costs may be higher because of the more frequent valve operations. Obviously, sufficient time must be allowed for complete regeneration and cooling of one tower while the other is in service; however, the regeneration-time requirement may also be varied by adjusting the quantity and temperature of the regeneration gas stream. In general, the length of the drying cycle is established on the basis of operating-labor schedules, resulting in the usual specification of 8-, 12-, 16-, and 24-hr

	Bed No. 1 on dehydrating service		Bed No. 2 on dehydrating service	
Valve No.	Bed No. 2, regenerating	Bed No. 2, cooling	Bed No. 1, regenerating	Bed No. 1, cooling
1	O	O	C	C
2	C	C	O	C
3	C	C	C	O
4	O	O	C	C
5	C	C	C	O
6	C	C	O	C
7	C	C	O	O
8	O	C	C	C
9	C	O	C	C
10	C	C	O	O
11	C	O	C	C
12	O	C	C	C

Table 12-16. Positions of Valves in Dehydration Plant of Figure 12-17 During Operating Cycles*

*C = closed;   O = open.

cycles. An 8-hr cycle is probably most typical. It is common practice to switch beds on the basis of the established time schedule rather than to take advantage of the full capacity of the desiccant. The desiccant-bed capacity decreases with age, however, and ultimately a time is reached when the bed just barely accomplishes the desired dehydration at the end of the drying cycle and must be replaced.

A process flow system using three towers packed with silica gel at the top and molecular sieve at the bottom is often selected to provide a combination of natural gas dehydration, control of hydrocarbon dewpoint temperature, and recovery of hydrocarbon liquids (112). At any one time, two towers are in adsorption service while the third is being regenerated or cooled. The feed gas is first cooled and passed through a liquid separator before it flows to the top of the adsorber towers. The silica gel in the top portion of the bed adsorbs both water and heavy hydrocarbons and also serves as a guard for the molecular sieve which does the final dehydration. Natural gas heated to 500–600°F is used for regeneration. Adsorbed water and heavy hydrocarbons are recovered from the regeneration gas by passing it through a cooler/condenser and separating the products. A portion of the dried product gas is used for cooling before the adsorber is placed back in adsorption service. Gas used for regeneration and cooling is recycled to the active adsorber.

Operating Practices

As pointed out above, the capacity of solid desiccants decreases with time, and ultimately the beds must be replaced. In order to predict the replacement time well in advance, it is good practice to determine periodically the maximum capacity of the beds. This is accomplished by allowing the beds to remain on stream past the normal cycle time and then monitoring the dew point of the product gas to detect the break point and the point at which it reaches the maximum allowable value.

Since the decrease in capacity is believed to be due to a considerable extent to heavy hydrocarbons which are deposited on the bed, all possible precautions should be followed to prevent such materials from entering the unit. An efficient scrubber, ahead of the dehydration plant, is recommended for removing liquid hydrocarbons, water, and other particulate impurities which may be present in the feed-gas stream.

Liquid water is particularly objectionable with high-capacity gel-type adsorbents because it can cause particle breakage. To minimize this effect, it is common practice to guard susceptible desiccants, such as Sorbead R, with a layer of liquid-water-resistant desiccant, such as Sorbead W or Activated Alumina. Another precaution is to insulate the tower surfaces and connecting pipelines so as to prevent the condensation of water on the cooled metal surfaces and avoid the possibility of such water reaching the desiccant.

Although regeneration can be accomplished at 300°F for all desiccants except molecular sieves, higher temperatures (up to 400°F) provide lower dew points and may actually increase the desiccant's useful life by providing more complete removal of adsorbed hydrocarbons. Periodic regeneration at a higher than normal temperature is also occasionally of value in reducing the rate of capacity decline. Laboratory tests have indicated that the regeneration of Activated Alumina H-151 may be possible at temperatures as high as 850° to 1,250°F to burn off carbonaceous deposits (42).

Hydrogen sulfide in the feed-gas to solid-bed dehydrators can cause difficulties, particularly if the gas also contains a trace of oxygen. In the latter case, the desiccant acts as a catalyst for the oxidation of hydrogen sulfide to elemental sulfur which deposits on the particles. Some of this is vaporized during regeneration and can cause plugging of cooler condensers. Wilkinson and Sterk (43) present data on the dehydration of a very sour gas, containing 1,800 to 2,000 grains H_2S/100 scf (and presumably no oxygen), with Sorbead and Activated Alumina. They found the Activated Alumina to be inac-

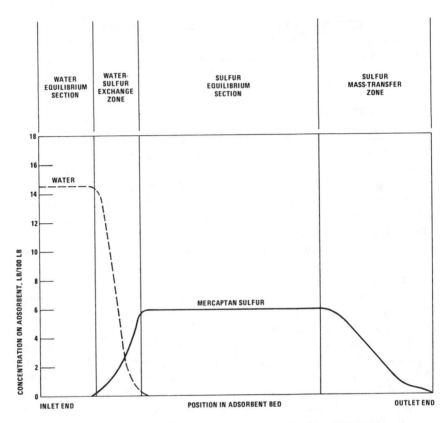

Figure 12-18. Adsorption zones in molecular-sieve bed, adsorbing both water and mercaptans from natural gas. *Data of Conviser*

tivated rapidly while the Sorbead was relatively unaffected by the H_2S. The special acid-resistant molecular sieves are also recommended for drying gas streams containing high concentrations of H_2S and carbon dioxide (12, 26).

The presence of H_2S in the gas to be dehydrated is particularly serious when bauxite which contains iron oxide is used as the desiccant. The iron oxide reacts with the hydrogen sulfide to form iron sulfide, which changes the characteristics of the bauxite; this results in deactivation and disintegration of the particles.

An operating precaution which may seem rather obvious but which nevertheless is frequently overlooked is the prevention of sudden pressure surges. If the dehydration vessel is vented rapidly, for example, very high, localized gas velocity may occur in the bed, causing bed motion, attrition, and even entrainment of the desiccant particles in the gas stream.

Use of Molecular Sieves for Gas Purification

In addition to their use for dehydration as described in a previous section of this chapter, the molecular-sieve adsorbents are finding application for many other gas-purification problems. Most of these take advantage of the fact that the molecular sieves show a high adsorptive selectivity for polar and unsaturated compounds. Polar compounds, such as water, carbon dioxide, hydrogen sulfide, sulfur dioxide, ammonia, carbonyl sulfide and mercaptans, are very strongly adsorbed and can readily be removed from such nonpolar systems as natural gas or hydrogen.

In most gas-purification cases water vapor is also present in the impure gas and is removed by the molecular-sieve adsorbent along with the other impurities. Since water is adsorbed more strongly than any of the other components, it concentrates initally at the inlet portion of the bed where it displaces the other impurities which had previously been adsorbed. These desorbed impurities are then readsorbed farther down the column, and this impurity adsorption zone moves through the bed in advance of the water adsorption zone. This phenomenon is illustrated graphically in Figure 12-18 for the case of mercaptan adsorption from natural gas (44). As indicated by the illustration, a much smaller bed would be required, for the same adsorption period, if dehydration alone were desired. In that case a small amount of mercaptan sulfur would be removed with the water, but, at the water breakthrough point, the sulfur content of the product gas would be the same as that of the feed. In a similar manner, the presence of any more strongly adsorbed component will affect capacity and breakthrough point for a given impurity.

Equilibrium capacity data for molecular sieves and the gases CO_2, H_2S, SO_2, and NH_3 are presented in Figures 12-19, 12-20, 12-21, and 12-22, respectively. Data for silica gel are included on the NH_3 chart for comparison purposes. As pointed out above, the capacity attainable in plant operations involving multicomponent gas streams cannot be estimated directly from equilibrium data obtained with a single component. However, such data are useful as an indication of maximum possible capacities and as a base for more complex design analyses.

One approach to the correlation and prediction of multicomponent adsorption equilibria for molecular sieves has been proposed by Yon and Turnock (45). This approach uses a Loading Ratio Correlation (LRC) which makes use of constants which are derived from pure-component behavior. Accord-

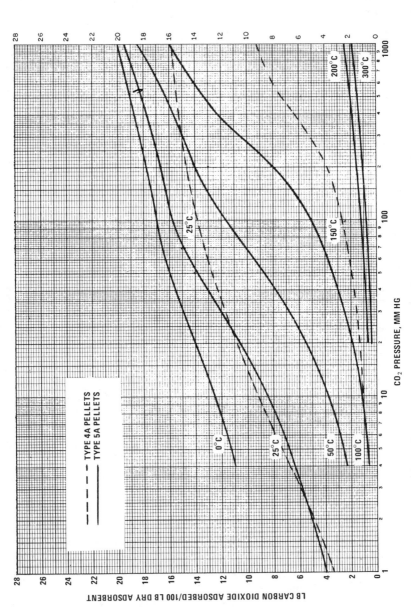

Figure 12-19. Equilibrium isotherms for carbon dioxide adsorption on molecular sieves, Types 4A and 5A. *Data of Union Carbide Corp.*

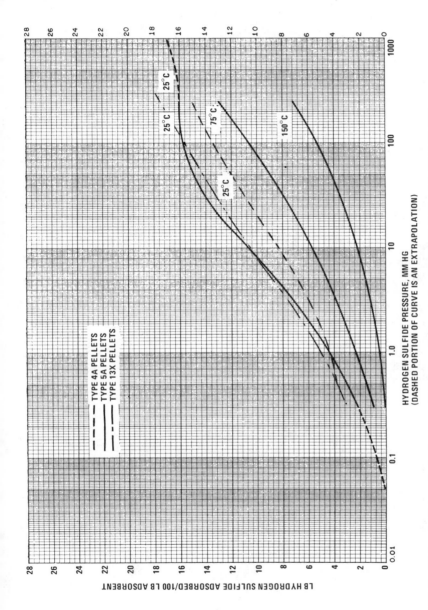

Figure 12-20. Equilibrium isotherms for hydrogen sulfide adsorption on molecular sieves, Types 4A, 5A, and 13X. *Data of Union Carbide Corp.*

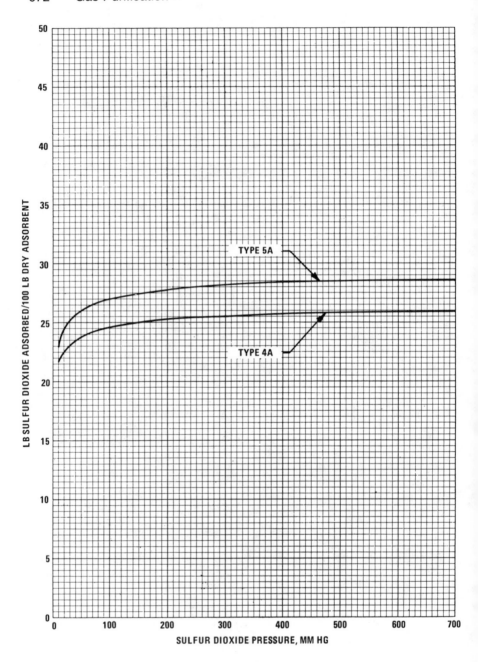

Figure 12-21. Equilibrium isotherms for sulfur dioxide adsorption at 25°C (77°F) on molecular sieves, Types 4A and 5A. *Data of Union Carbide Corp.*

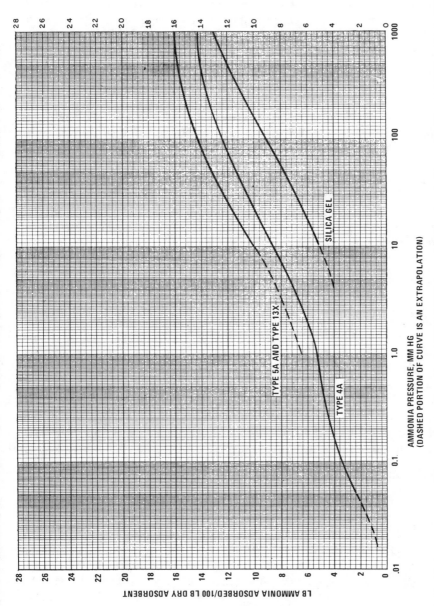

Figure 12-22. Equilibrium isotherms for ammonia adsorption at 25°C (77°F) on molecular sieves, Types 4A, 5A, and 13X and on silica gel. *Data of Union Carbide Corp.*

ing to the authors, a single set of constants can characterize adsorption equilibria over broad ranges of temperature, pressure, and concentration. Although the technique is not excessively cumbersome to use, a detailed description is beyond the scope of this text, and the reader is referred to the original paper for additional information (45). For the specific case of natural gas purification (H_2S and water removal) using a 5A molecular sieve, a detailed design calculation procedure has been presented by Chi and Lee (101). The proposed design method has been compared with experimental data over a wide range of conditions and found to adequately predict hydrogen sulfide break time. It is based on the use of a dynamic H_2S saturated capacity, which is correlated with H_2S and CO_2 partial pressures, and a "Lost Bed Height" (similar to LUD) which correlates with H_2S flux (lb/ft^2 hr) and the ratio of CO_2 to H_2S partial pressures.

Process cycles used for gas purification are generally similar to those used for dehydration. Fixed beds of adsorbent are employed, with gas typically flowing downward during the adsorption cycle, upward during regeneration and downward during cooling. Once the bed capacity has been established from pilot-plant tests or fundamental data, the general process design techniques described for dehydration in the initial section of this chapter can be applied. An exception is pressure swing adsorption which employs pressure reduction instead of a hot purge gas for bed regeneration. Another nontypical molecular sieve process, which makes use of both the catalytic and adsorptive properties of molecular sieves, is the Pura Siv-N process developed by Union Carbide Corporation for the removal of NO_x from nitric acid plant tail gas (102). The process utilizes two beds of a special acid-resistant molecular sieve. When tail gas from the acid plant is passed through one of the beds, the NO is catalytically converted to NO_2 which is immediately adsorbed as an equilibrium mixture of NO_2 and N_2O_4. The operation is cycled to the other bed prior to NO_2 breakthrough, and the NO_2 in the spent bed is desorbed by heating to about 600°F. Additional specific applications of molecular sieves for gas purification are described in the following sections.

Inert-Gas Purification

Inert gas, for use in annealing and other operations requiring a nonreactive atmosphere, is customarily made by removing carbon dioxide and water vapor from a combustion gas made under carefully controlled conditions. A flow diagram of the purification process employing molecular sieves is shown in Figure 12-23. Natural gas is burned with air in near stoichiometric proportions to give a product gas containing approximately 89 percent nitrogen and 11 percent carbon dioxide plus the water originally present in the air and that formed during the combustion. The combustion gas is cooled by heat exchange with regeneration air and then by passage through a water-cooled exchanger. The cooled gas is then passed through one of the three molecular-

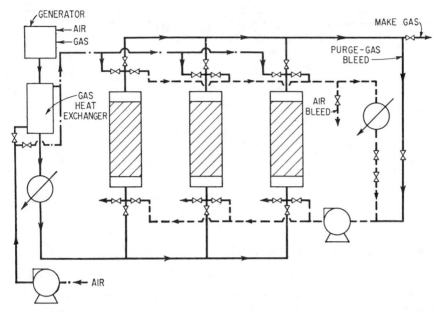

Figure 12-23. Molecular-sieve process for purification of inert gas used in annealing operations. *From Clark (27)*

sieve beds for water and CO_2 removal. During this time, the second molecular-sieve bed is regenerated, and the third is cooled. At the end of a 1-hr period, the valves are automatically switched so that the bed previously used for adsorption is regenerated, the regenerated bed is cooled, and the cooled bed is used for purification. Operating data for a relatively small commercial unit are presented in Table 12-17.

During the regeneration period, the bed is brought up to a temperature of 450°F by the heated air stream. Cooling is accomplished in two parts—for 5 min, product gas is passed through the bed at a rate of 200 scf/hr and vented; and for 55 min, gas is recirculated through the bed and a cooler at the rate of 12,500 scf/hr. However, even during the 55-min period, fresh product gas is added and recirculated gas vented at the rate of 200 scf/hr. At the end of the 1-hr cooling and purging period, the oxygen content in the recirculation gas has been reduced to less than 50 ppm.

The CO_2 content of the combustion gas is reduced from about 11 percent to less than 0.01 percent, and the dew point is lowered to below −75°F by the molecular-sieve adsorption process. Oxides of nitrogen which may be present in trace quantities in the combustion gas are completely removed.

Table 12-17. Operating Data for Inert-gas Generator with Molecular-Sieve Gas Purification (27)

Adsorbent-bed diameter, ft............................	2.25
Adsorbent-bed height, ft.............................	3.5
Weight of adsorbent per bed, lb.......................	600
Air required for regeneration, SCF/hr..................	12,500
Purified gas from adsorption bed, SCF/hr..............	2,200
Purified gas used as purge during cooling, SCF/hr	200
Net inert-gas production rate, SCF/hr.................	2,000

Carbon Dioxide and Water Removal from Ethylene

Molecular sieves have been employed to coadsorb CO_2 and H_2O from ethylene gas which is used for the production of polyethylene plastic. A very low CO_2 content is required in the feed gas to the polymerization plants, and liquid systems normally require more than one stage of absorption. In the molecular-sieve process, a standard two-bed system is used with regeneration accomplished by heating the bed with 600-psig steam and purging with heated methane. Pertinent design data for one installation are presented in Table 12-18.

It is reported that the initial investment for this type of carbon dioxide-water removal plant is lower than that of a more conventional installation employing liquid processes for CO_2 removal followed by adsorptive dehydration and that operating costs are competitive (27).

Carbon Dioxide Removal from Cryogenic Plant Feed Gas

Molecular sieves are used in many installations for the simultaneous removal of carbon dioxide and water from gas streams which undergo low-temperature liquefaction in a subsequent operation. Even relatively low concentrations of carbon dioxide and water can cause the formation of frost, or "rime," on the surfaces of cryogenic heat exchanger fins. One of the earliest CO_2 removal applications of molecular sieves was for purifying the air feed to cryogenic air separation plants. In this service continuous product gas purities of less than 1 ppm CO_2 and 1 ppm H_2O are readily attained.

A growing field of application is front-end feed purification for natural-gas liquefaction plants. In mid-1971 it was reported that over three-fourths of the approximately 30 peak shaving LNG plants in operation or under construction utilized molecular-sieve purification units (46). Figure 12-24 is a

**Table 12-18. Design Data for Molecular-Sieve Ethylene
Gas-purification Plant (27)**

Ethylene-gas feed rate, MMSCF/day	1.75
Feed-gas pressure, psig	430
Adsorber diameter (ID), in	60
Adsorber height, ft	32
Adsorbent charge (per vessel), lb	15,000
	(Linde type 5-A molecular sieve)
Steam pressure, psig	600
Steam rate, lb/day	11,400
Cooling-water rate, gal/day	57,200
Ethylene in purged gas (recycled to furnace) MMSCF/day	0.05
Methane in purged gas, MMSCF/day	0.06
Inlet-gas CO_2 content, ppm	3,000
Product-gas CO_2 content, ppm	<1
Product-gas dew point, °F	$< -100°F$

simplified flow diagram of a three-tower molecular-sieve adsorption plant, and Figure 12-25 is a photograph of a plant of this type. This arrangement has one tower drying and purifying gas, another tower being cooled, and a third tower being regenerated by a hot purge at the same time. The towers are manifolded to common inlet and outlet headers and are automatically switched from adsorption to heating to cooling in sequence. The same gas is used for cooling one bed and heating another so purge gas requirements are about one-half that of a dual bed system. In the flow diagram shown, the purge gas is withdrawn from the purified gas stream, and after use it is cooled to knock out most of the water and is then returned to the pipeline. Boil off gas from the liquefaction plant is also frequently used for purging.

Removal of Sulfur Compounds

In natural-gas processing the mercaptans are generally the most strongly adsorbed impurities (other than water), followed by H_2S with CO_2 being the most weakly adsorbed compound in this series. However, all mercaptans are not adsorbable by Type 4A or 5A sieves because pore size limitations prevent adsorption of any but the lightest members of the family. As a consequence, 13X is the preferred adsorbent for complete sulfur removal from natural-gas streams. Existing commercial sulfur removal units treat natural-gas streams at flow rates ranging from 2 to 200 MMscfd. Data on one of the largest plants is given in Table 12-19. The presence of a trace of glycol, glycol degradation products, and absorber oil in the feed gas is noted as these heavy

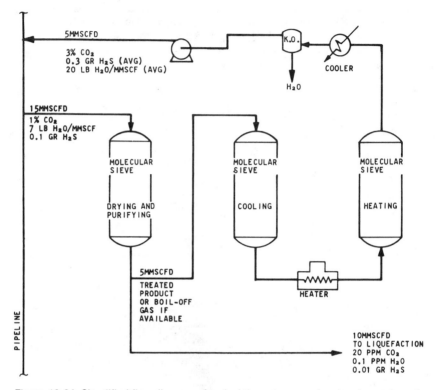

Figure 12-24. Simplified flow diagram of typical three-tower molecular-sieve adsorption system for removing carbon dioxide and water from natural gas prior to liquefaction. *Data of Thomas and Clark* (47)

molecules affect both capacity (by coadsorption) and bed life (due to coke formation). It is claimed that a properly designed molecular sieve, sulfur removal system will perform for 3 to 5 years before new adsorbent is required (44).

Molecular sieves are very effective for H_2S removal and can produce gas containing extremely low levels of this impurity. Since H_2S is adsorbed more strongly than CO_2, the molecular sieves also offer the capability for selectively removing H_2S from gas streams which contain both impurities. However, the process has not gained general acceptance because of the problem of disposing of the sulfur-rich regeneration gas. The problem has been bypassed in one process scheme which is applicable to cases where it is desired to leave part of the CO_2 in the gas stream while removing essentially all of the H_2S and water. This flow arrangement is shown diagrammatically in Figure 12-26 for a typical case. The gas feed to the dry bed and liquid absorption units is ratioed in such a manner that the final mixed product meets

Figure 12-25. Molecular-sieve gas-purification plant. Automated three-tower system removes carbon dioxide and water from natural gas prior to liquefaction. *Memphis Gas, Light and Water Division*

pipeline CO_2 specifications. The increased gas volume which results from leaving 3 percent CO_2 in the product gas provides a credit for gas sold on a volume basis. Use of the dry-bed adsorption process on a portion of the gas has the further advantage of permitting the liquid absorber system to operate closer to maximum capacity, as a product gas which is slightly off specification with regard to H_2S, and water can be tolerated for dilution with the very pure adsorber product. It is reported that 10 plants of this type having the capability of treating over 1 billion scfd were in operation in mid-1971 (46).

Molecular sieves have also found application for desulfurization of natural-gas feed to ammonia plants. Removal of all types of sulfur compounds ahead of these plants is desirable because sulfur acts as a temporary poison to steam-hydrocarbon reforming catalysts and a permanent poison to expensive low-temperature shift conversion catalysts. An installation employing a standard dual bed adsorption system has been described by Lee and Collins (48). The authors also describe comparative tests of a molecular sieve and a commercial grade of impregnated activated carbon in a dual-bed mobile pilot unit. The test results indicated that the molecular sieve could treat 2 to 4 times as much gas per unit volume of adsorbent as the carbon.

Table 12-19. Design and Operating Data for Large Mercaptan Removal Plant (44)

Natural gas feed rate, MMscf/day	200
Feed gas pressure, psig	750
Adsorber description	(two horizontal vessels)
Diameter, ft	6
Length, ft	36
Bed depth, ft	3
Adsorbent charge, (per vessel) lb	25,000 (Type 13X)
Inlet gas composition	
Mercaptan and heavy sulfur, grains/100 SCF as H_2S	2
Water, lbs/MSCF	<7
Glycols, glycol degradation products, and absorber oil	trace
H_2S and CO_2	nil
C_4 hydrocarbons, % vol.	0.03
C_1, C_2, C_3, and N_2	balance
Product gas, total sulfur, grains/100 SCF as H_2S	<0.06
Operating Cycle:	

Step	Flow Direction	Time
1. Purify	Down	12 hours
2. Hot purge (600°F)	Up	8 hours
3. Cool purge	Down	4 hours

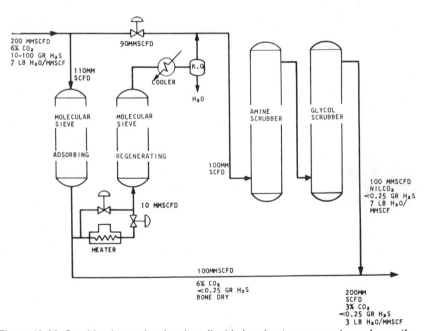

Figure 12-26. Combination molecular sieve-liquid absorbent process scheme for purifying CO_2-rich natural gas. *Data of Thomas and Clark* (47)

The commercial plant consistently provided gas to the primary reformer containing less than 0.3 ppm (vol) peak total sulfur from a feed gas averaging about 0.6 ppm H_2S, 0.7 ppm mercaptans, 1.6 ppm sulfides, and 1.2 ppm residual sulfides. In this application the high-sulfur purge gas can be fed into the plant fuel system so that the sulfur is finally vented to the atmosphere as dilute sulfur dioxide. Although this disposal method is preferable to venting the original sulfur compounds, it is not broadly acceptable and constitutes a significant drawback to the process.

Hydrogen Purification by Pressure Swing Adsorption

Although the pressure swing cycle is not limited to either hydrogen purification or molecular-sieve adsorbents, this combination appears to represent the principal current application. The cycle operates between two pressures, adsorbing impurities at the higher pressure and desorbing them at the lower pressure with no temperature change except for that caused by the heat of adsorption and desorption. The absence of a heat requirement leads to a simple installation compared to thermal cycle systems. This advantage is counterbalanced, however, by a greater gas loss resulting from the venting and low-pressure purge operation. According to Stewart and Heck (49), the process can produce an ultrapure hydrogen from any of the sources listed in Table 12-20.

The performance of a plant designed to produce high purity hydrogen from demethanizer off-gas by pressure swing adsorption is given in Table 12-21. The data indicate that a very high purity hydrogen is obtained, but less than 70 percent of the hydrogen in the feed is recovered as product. The waste gas can be recompressed and used as fuel, however.

A recent development in pressure swing adsorption for hydrogen purification is the Polybed PSA process which is offered by the Union Carbide Corporation. This process utilizes a large number of relatively small beds in a complex valving, sequencing, and pressure reduction cycle to provide improved separation efficiency (103). A comparison of the Polybed PSA process with a "conventional" process for producing pure hydrogen by naphtha reforming has been presented by Heck and Johanson (104). The

Table 12-20. Sources for Ultrapure Hydrogen	
Hydrogen Source	Impurities
Steam reformer hydrogen	$CO_2, H_2O, CH_4, CO, N_2$
Demethanizer overhead	$CH_4, CO, C_2H_4,$
	$C_2H_6, C_2H_2, CO_2, N_2$
Electrolytic off-gas	N_2, CO_2, H_2O
Dissociated ammonia	N_2, NH_3
Ammonia reactor loop purge	CH_4, A, N_2, NH_3
Methanol reactor loop purge	CO, CO_2, CH_4, N_2

results of this comparison are summarized in Table 12-22. The indicated production costs are a strong function of the assumptions used in the evaluation and would be much less favorable for pressure swing adsorption if, for example, the cost of fuel was assumed to be significantly less than the cost of naphtha feedstock, and if less credit were given for export steam. Of perhaps greater significance is the extremely high purity of the hydrogen produced by the adsorption process. A 40 MMscf/d Polybed PSA unit making 99.999 percent pure hydrogen recently went on stream at a refinery in West Germany. The unit will supply hydrogen to a variety of refinery processes, including a hydrocracking unit (104).

The first U.S. Polybed PSA Unit was started in July 1978 and gave an on-stream factor of 0.999 during the first 14 months of operation (113). Many additional units have been built since that time. Because of the very large number of valves which must be opened and closed in a precise sequence, the efficient operation and control of valves is a key requirement of Polybed PSA systems. The early PSA units were controlled by pneumatic systems, but, with the increased complexity of the polybed configurations, the use of microprocessors and computer control systems has become necessary (114).

Air and Gas Purification with Silica Gel

A number of serious mishaps have occurred in low-temperature air-separation plants; these have been attributed to the accumulation of acetylene (and possibly other unstable or oxidizable compounds) in the liquid oxygen. As a result, several operations designed to increase safety have been developed for these plants, including the use of silica gel to remove adsorbable impurities from the air feed and liquid-air streams.

According to Kerry (50), the air feed to a low-temperature oxygen plant should meet the specifications in Table 12-23 with regard to contaminants.

Even with air meeting these specifications, the use of silica-gel adsorbers in the line carrying the liquid from the bottom of the high-pressure column to the middle of the low-pressure column is common practice to prevent any accumulation of impurities in the system. In industrial areas, where appreciable amounts of atmospheric contaminants are present, a preliminary purification of the feed air stream is also required. This may be accomplished by the use of silica-gel "prefilters" which are placed between the warm exchanger system and the liquefier. These units purify the air while it is still in the gaseous phase and operate in a typical adsorption-regeneration cycle. Regeneration is accomplished by the use of dry oil-free air or nitrogen heated to 250° to 350°F.

Silica gel is also used for recovering hydrocarbons from natural gas. This is more of a product-recovery operation than a gas-purification step, so only a brief description is presented. The technique is particularly applicable to lean

Table 12-21. Performance of Pressure Swing Adsorption Process (49)

	Feed	Product	Waste
Flow, MMSCFD	1.2	0.5	0.7
Pressure, psig	200	198	2.0
Composition, mol %			
H_2	61.1	99.999+	33.3
CH_4	36.7	1 ppm	63.0
C_2H_4	0.6	1 ppm	1.0
C_2H_6	0.1	1 ppm	0.1
CO	1.1	1 ppm	1.9
N_2	0.4	1 ppm	0.7

Table 12-22. Comparison of Polybed PSA (Pressure Swing Adsorption) Hydrogen Plant with "Conventional" Design (104)

	Polybed PSA	Conventional*
Costs:		
Total Installed Cost (1977 basis) $ Million	16.5	16.0
Production Cost**, $/1000 scf H_2	1.30	1.40
Design Basis:		
Capacity, MMSCFD contained H_2	50	50
Product Pressure, psig	350	350
Hydrogen Purity, %	97	99.999
Feedstock	Naphtha	Naphtha
Feed and Utilities:		
Feedstock, MM Btu/h (low heat value)	796	533
Fuel, MM Btu/h (low heat value)	50	306
Boiler Feedwater, lb/h	157,500	63,800
Cooling Water, gpm (25°F rise)	4,120	7,480
Electric Power (kW)	730	920
Export Steam, lb/h	70,400	11,700

*"Conventional" plant uses a liquid process for CO_2 removal followed by methanation.

**Based on feed and fuel, $2.50/MM Btu (low heat value); boiler feedwater, $0.20/1000 lb; cooling water, $0.05/1000 gal; electric power, $0.03/kWh; superheated HP steam, $3.50/1000 lb; capital charges, 32%/year of investment; and maintenance, 2%/year of investment for PSA plant and 3% for conventional plant.

Table 12-23. Specifications for Air Feed to Low-temperature Oxygen Plant	
Acetylene	<1 ppm
NO + NO$_2$	<1 ppm
Ozone	<0.1 ppm
CO$_2$	<5 ppm
Oil vapor	Not detectable

gas streams containing relatively low concentrations of hydrocarbons heavier than propane. Dehydration is also accomplished during the process because of the greater affinity of silica gel for water compared to hydrocarbons. However, the hydrocarbons are normally present in higher concentrations than water; therefore, the silica gel bed adsorbs considerably more hydrocarbon liquids than water during each cycle.

Since the hydrocarbons in natural gas are present in a continuous series with only slightly different volatilities between adjacent members of the series, no clear-cut breakthrough point is observed. As a result, somewhat more complex design procedures are required to optimize the bed size and cycle time than with single component adsorption units. Empirical design methods for both the adsorption and condensation steps have been developed by Humphries (51). His correlations are based on a considerable amount of data obtained with a small three-tower test unit processing gas from the Lolita Field, Jackson County, Texas. Most of the test data were obtained with Sorbead H adsorbent. This adsorbent is a spherical silica gel base material very similar to Sorbead R (see dehydration) but with considerably larger pores to provide a greater capacity for liquid hydrocarbons.

Silica gel adsorption plants are frequently operated with extremely short cycle times (as short as 20 minutes) in order to provide maximum hydrocarbon recover. The effect of cycle time on recovery is illustrated by the data in Table 12-24 for a plant processing a gas containing 0.192 gal/MMscf of pentanes and higher hydrocarbons (3).

The operation of short-cycle units has been described in detail by Ballard in several papers (52)(53)(54). Many of the operating precautions which he mentions are applicable to solid-bed adsorption plants generally.

1. An adequate inlet scrubber should be provided to remove all solids and free liquids from the incoming gas stream.
2. Inlet gas nozzles and/or baffle plates should be designed to prevent bed agitation at the top of the tower.
3. A buffer layer equal to about 5 percent of the total bed volume should be used on top of adsorbents which fracture in the presence of free liquids. The buffer particle density should be similar to that of the main bed.

4. Gas flow through the bed should be held to about 50 percent of the design capacity for the first 10 or 12 cycles to let the adsorbent cure properly.

5. Rapid pressure changes should be avoided. Pressure reduction in particular can cause adsorbent breakage.

Organic-Vapor Adsorption on Active Carbon

The removal of organic vapors from air and other gases by active carbon is probably second in importance only to dehydration as an industrial application of adsorption. Active carbon is the preferred adsorbent in this application because of its selectivity for organic compounds.

The process is of particular importance in the removal and recovery of volatile solvents, the removal of obnoxious odors or other trace impurities from air, and the recovery of liquid hydrocarbons from gases. These applications are closely related and may employ the same adsorbent; however, they are discussed in separate sections because of important differences in equipment design, particularly with regard to bed-thickness and regeneration procedure.

The use of active carbons to control emissions of volatile organic compounds for purposes of air pollution abatement is of increasing importance. Over 700 plant locations employing adsorption systems are listed in a 1983 EPA report (115). The report covers full-scale activated carbon vapor-phase adsorption applications and gives specific flow rates, chemicals adsorbed, and sources of emissions. A comprehensive study of the costs of controlling emissions from 102 plants by carbon adsorption or catalytic incineration was conducted by Du Pont (105). Only sources emitting more than 3 lb/hr were reported, and carbon adsorption was the method selected for 60 percent of the sources in the small, 3 to 8 lb/hr, and large, over 15 lb/hr, categories and for 90 percent of the intermediate size sources. The study showed that the cost of abatement increased very rapidly (per lb of organic material recovered) as the size of the source decreased below about 100 lb/hr. It was con-

Table 12-24. Effect of Cycle Time on Recovery	
Cycle time, hr.	Hydrocarbon Recovery, gal/MMscf
32	10
16	20
8	30
4	40

cluded that about 75 percent of the inventoried emissions came from sources that emit more than 500 lb/hr, and these could be controlled (80 percent removal) for only about 10 percent of the cost required to control all of the inventoried sources.

The term "active carbon" covers a multitude of carbon-based materials that possess adsorptive power. Many manufacturers of these materials refer to their products as "activated carbon," and the two terms are considered synonymous and used interchangeably. The term "activated charcoal" is generally reserved for those carbons which are derived from woody materials.

The gas-adsorptive properties of wood charcoal were recognized as early as 1773 by Scheele (55); however, it was not until the middle of the nineteenth century that the property was utilized commercially. This application consisted of the use of charcoal air-filters in the ventilation and disinfection of sewers (56). The next major step in the development of active carbon for gas adsorption occurred during World War I, when the use of poison gas by the Germans made it necessary to develop a hard, granular carbon for gas-mask use.

The principal operation in the manufacture of active carbon is the heating of carbon-containing material so that volatile components, either originally present or formed during the heating operation, are distilled off, leaving a highly porous structure. Source materials that have been used for the production of active carbon include bones, coal, coconut shells, coffee grounds, fish, fruit pits, kelp, molasses, nutshells, peat, petroleum coke, rice hulls, sawdust, and wood of all types. The amount of activity developed by simple heating is dependent upon the composition and properties of the raw materials. With many substances, additional steps are required to obtain a highly active carbon. Two such methods which have been used extensively are the incorporation of chemical additives (particularly metallic chlorides) into the pulverized carbonaceous material before heating and the controlled oxidation of the char by using suitable oxidizing gases at elevated temperatures.

Properties of Gas-Adsorption Carbons

In addition to a high degree of activity, special requirements must be met by gas-adsorption carbons. They must possess sufficient strength to withstand abrasion, they must be available in granular form to provide low-pressure-drop beds, and they must be as dense as possible to minimize adsorber-space requirement.

Of the many carbonaceous materials which form active charcoal, only relatively few—coconut shells, fruit pits, and cohune and babassu nutshells—readily yield chars with all of the properties desired for gas-adsorbent use. Because of the limited supply of these materials, special preparatory treatments have been developed to enable other base materials to be used for gas-adsorbent carbon. In its most common form, the pretreatment consists of pulverizing carbonaceous material, incorporating a suitable binder, and pelleting or extruding to form a dense, compressed material. The pellets or spaghetti-like extrusions are then carbonized at temperatures from 700° to 900°C. Various types of wood and coal have been found to be suitable base materials, and materials such as sugar, tar, and lignin can be used as binders.

As can be realized from the above discussion, active carbon is not a single, standardized adsorbent. Not only do carbons from different sources differ in appearance and adsorptive capacity, but also in selectivity for various gases. In general, however, for compounds of similar molecular structure, the quantity retained on any active carbon will increase with increased molecular weight or critical temperature. These effects are illustrated in Figures 12-27 and 12-28, which present adsorption isotherms for several hydrocarbons on a coconut-shell carbon (Columbia G grade activated carbon) and on a coal-base material (57). The figures show the coconut-shell carbon to have a higher capacity, although somewhat less selectivity, for propane over ethane. Several exceptions to the molecular-weight, critical-temperature rule are noted. Propylene is adsorbed more strongly than propane and 1-butene more strongly than isobutane, both cases being in reverse order to molecular

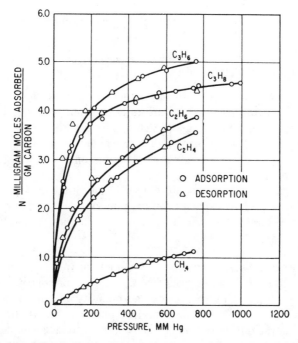

Figure 12-27. Adsorption isotherms for hydrocarbons on activated coconut-shell carbon at 25°C. *Data of Lewis et al.* (57)

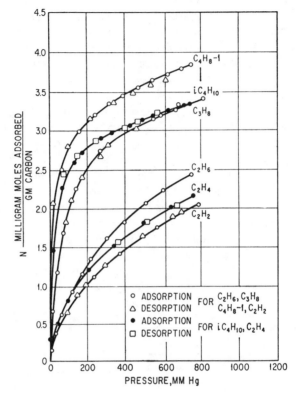

Figure 12-28. Adsorption isotherms for hydrocarbons on coal-base activated carbon at 25°C. *Data of Lewis et al. (57)*

weight. Acetylene is in the correct order relative to molecular weight although it is out of place with regard to critical temperature. These exceptions illustrate an additional rule (which applies even more strongly to silica gel and some other adsorbents), that more unsaturated compounds are generally adsorbed more strongly than saturated ones with the same number of carbon atoms.

Standardized tests have been developed to aid in the evaluation of active carbons with regard to capacity for organic vapors, retentivity of the adsorbed vapors, service life, hardness, and other factors. The following tests are of particular value.

Carbon Tetrachloride Activity. The numerical value obtained from this test indicates the adsorptive capacity of the carbon for concentrated organic vapors. It is obtained by measuring the quantity of carbon tetrachloride vapor adsorbed, at 25°C and 760 mm Hg, from air which has been saturated with carbon tetrachloride vapors at 0°C, and is expressed as a percentage of the original charcoal weight.

Carbon Tetrachloride Retentivity. This factor represents the ability of a carbon to retain a previously adsorbed vapor. Its numerical value indicates the percent by weight of carbon tetrachloride remaining in the carbon (based on the weight of activated carbon) after blowing dry air at 25°C for 6 hr through the bed saturated from the activity test.

Minute Service. This is a measure of the length of time during which a specified thin bed of active carbon will completely adsorb an organic vapor, preventing any breakthrough of the vapor through the carbon. The service life is measured with chloropicrin vapors under standardized test conditions (58).

Hardness. (C.W.S. ball-abrasion method). This is a measure of the resistance of activated carbon to mechanical breakage during handling and use. The value represents the percentage of 6- to 8-mesh carbon remaining on a 14-mesh screen after shaking with steel balls under specified conditions for 30 min.

Typical test values and other properties of several grades of commercial activated carbon are presented in Table 12-25.

Solvent Recovery With Activated Carbon

In many industrial processes, relatively volatile organic solvents are used as carrier liquids and dissolving agents. During certain processing steps, these solvents are vaporized into the air. In many cases, the removal of vaporized

Table 12-25. Typical Properties of Commercial Activated Carbons			
	PCB*	SXWC†	JXC†
Total surface area (N_2, BET method), m²/g	1150-1250	—	—
Packed apparent density, g/cc	0.45	0.45	0.48
Carbon tetrachloride activity, % min	60	65	60
Hardness (ball abrasion) %	92	95	95
Moisture, % (max)	4.0	2.0	2.0
Total ash, % (max)	4.0	3.0	2.0
Specific heat at 100°C	0.25	—	—
Kindling point, °C (min. in oxygen)	—	350	—

*Pittsburg activated carbon, coconut-shell type designed for vapor phase applications.
†Union Carbide Corp., "Columbia" activated carbon pellets, petroleum base type, developed especially for solvent adsorption.

Figure 12-29. Automatically operated package-type activated-carbon unit for gas purification or solvent recovery. *Barnebey-Cheney Company*

solvent from the air and its recovery for reuse is an economic necessity. In other cases, its removal is desirable to prevent air pollution. Typical commercial units are illustrated in Figures 12-29 and 12-30. Solvent-recovery processes are important in plants manufacturing cellulose acetate rayon, plastic-coated paper or cloth, plastic films, rubber products, and smokeless powder. They are also used in connection with such operations as solvent extraction, high-speed printing, painting and varnishing, and degreasing of metal parts.

Some form of activated carbon is used in these processes rather than silica- and alumina-base adsorbents, because of carbon's selectivity for organic vapors in the presence of water. Typical solvents, which can be recovered from air streams by activated carbon, include: hydrocarbons such as naphtha or petroleum ether; methyl, ethyl, isopropyl, butyl, and other alcohols; chlorinated hydrocarbons such as carbon tetrachloride, ethylene dichloride, and propylene dichloride; esters such as methyl, ethyl, isopropyl, butyl, and amyl acetate; acetone and other ketones; ethers; aromatic hydrocarbons such as benzene, toluene, and xylene; carbon disulfide and many other compounds.

In general, the solvents should have a boiling point less than about 350°F, in order that they may be removed readily from the adsorbent by low-pressure steam. In addition, the process will not operate satisfactorily with solvents which are very easily oxidized, e.g., the nitroparaffins and cyclohexanones.

Process Description

A flow diagram of a typical solvent-recovery plant is shown in Figure 12-31. As in dehydration, two adsorbers are normally used, with one adsorbing while the other is being regenerated. Regeneration is accomplished by passing low-pressure steam upward through the bed. The steam raises the temperature of the bed, reducing its equilibrium capacity for the adsorbed vapors; provides latent heat of vaporization for the solvent; and acts as a stripping vapor to reduce the partial pressure of solvent in the vapor phase. The steam passes out of the vessel with the desorbed solvent, and both are then condensed to permit solvent recovery.

In the flow diagram shown, steam and solvent vapors discharged from the adsorber being regenerated are condensed in a water-cooled heat exchanger;

Figure 12-30. Large adsorption unit installed to recover acetone from acetate-yarn manufacturing operations. Outdoor installation was possible in this case because of the mild climate. This unit was preassembled in New Jersey, then dismantled and shipped in sections for reassembly at the site (Ocotlán, Mexico). *Union Carbide Corp.*

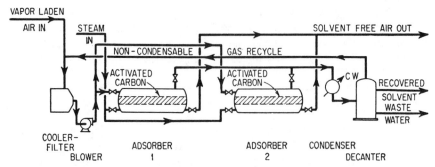

Figure 12-31. Flow diagram of typical solvent-recovery plant.

the condensate is collected in a separate vessel; and the solvent, which is insoluble in water, is decanted and returned to the process. When the solvent is partially or completely soluble in water, a more elaborate separation step is required. This step, which usually includes distillation, must, of course, be designed for the specific separation involved.

Some steam is condensed and adsorbed in the bed during the stripping operation, and this quantity of water must be removed from the bed before the next regeneration cycle in order to prevent a buildup of water in the adsorption bed. With carbons which have a high, selective adsorption-capacity for the solvent in the presence of moisture, this excess water can be removed during the adsorbing period by vaporization into the air which passes through the bed. If the solvent is not adsorbed strongly in the presence of water, it is necessary to dry the bed between regeneration and adsorption cycles. This may be accomplished by passing dry air through the bed for a short period of time.

The concentration of solvent vapors, in the air entering most solvent-recovery units, ranges from about ¼ to 2 lb/1,000 cu ft. With flammable solvents, fire-hazard considerations set the maximum allowable solvent concentration; this should be less than about 50 percent of the lower explosive-limit.

A compilation of fire-hazard properties of a large number of organic solvents and other solutions, including the reported lower and upper explosive-limits in air, is presented in a report on Lighting of Chemical Plants (59). Data are also available from Underwriters' Laboratories, Inc. (60). Values of the lower explosive-limit for common solvents in air at 75°C (167°F) range from 2 lb/1,000 cu ft (measured at 100°F) for carbon disulfide to over 5 lb/1,000 cu ft for methyl acetate. The vapor-laden air is usually filtered at the inlet of the recovery plant to prevent dust or other particulate matter from entering the activated carbon bed.

According to Browning (61), the overall recovery efficiency of such systems is usually between 80 and 95 percent, depending primarily upon the effectiveness of the design of the hoods and vapor-collection systems.

However, 99 percent (or more) of the solvent which enters the adsorption plant is recovered.

Where a very high efficiency of solvent recovery is desired, series operation of the adsorbers is recommended. In this arrangement, four adsorbers are typically employed. At any time in the cycle, one of the four adsorbers will be undergoing regeneration while the other three are in air-purification service. Two of these are used to purify separate portions of the air stream in parallel, and the third adsorber is used in series behind the most nearly spent adsorber. This permits each adsorber to be operated until it actually passes some solvent and thus acquires a higher solvent charge than would be possible if it were taken off stream before any solvent appeared in the product air. This type of series operation reportedly results in a solvent-recovery efficiency of 99.7 to 99.8 percent of the solvent in the entering air (61).

Two adsorption process modifications have been proposed by Mattia (62) for application to contaminated air streams in which the organic vapor concentrations are too low to be recovered economically by conventional means. Both techniques involve passing the contaminated air through a bed of active carbon as the primary gas-purification step. They differ from conventional systems, however, in the carbon regeneration step. In one version, called the Zorbcin Process (63), regeneration is accomplished by blowing hot air through the bed. A blowdown stream of the regeneration air is passed to an incinerator (either thermal or catalytic) where the stripped organic vapor is burned and a hot flue gas is produced. A regulated amount of this hot flue gas is returned to the regeneration cycle to provide heat for regeneration. A small amount of natural gas is also burned in the incinerator to provide the initial heat requirements.

In the second process modification (64), solvents contained in the stripping gas from the primary adsorbers are recovered by passing the gas through a secondary adsorber. In this case heat is provided for regeneration of the primary adsorbers by passing a portion of the recirculating gas stream through a fired heater. The secondary adsorber is regenerated by passing steam through the bed, and the solvents are recovered by condensing the steam-solvent mixture and separating the solvent by decantation or distillation.

A third process modification, based on the use of vacuum to aid regeneration, has also been described (116, 117). Vacuum regeneration can be cost effective when conventional steam injection causes a corrosion problem or when the absorbate may undergo polymerization or reaction due to the high temperature or presence of water during steam regeneration. Vacuum-stripped carbon adsorption systems have been used for a wide variety of industrial emission sources, including household products, magnetic tape, pharmaceutical, and polymer manufacture. In one system, controlling emissions from a pill-coating operation, the bed is heated to ~65°C by a combination of convective heating using a preheated gas stream and conductive heating from elements embedded in the carbon, and regeneration is accomplished by subjecting the heated bed to a 1 mm Hg vacuum. The desorbed organics are removed from the adsorber by a closed-loop purge gas stream which is passed through a refrigerated condenser and returned to the adsorber. The condensed organic liquid is collected in a condensate recovery tank.

In many solvent recovery systems, adsorption represents only one step in a complex series of chemical engineering operations. The design of a complete system for recovering methylene chloride and methanol from air emitted from a dryer in a resin processing plant has been described by Drew (106). The overall solvent recovery system includes a water scrubber to remove resins and cool the air to 100-110°F; a standard 2-bed carbon adsorber unit designed for 95 percent solvent removal efficiency; a condenser and decanter to handle the vapors which are stripped from the carbon by steam; an extraction column in which water is used to remove the water soluble methanol from the methylene chloride phase; a stripping column to remove dissolved methanol and methylene chloride from the waste water; and a drying column to remove water from the recovered methylene chloride. These items of equipment and operations are representative of those required for complete solvent recovery systems, however, each system must, of course, be tailored to the properties of the specific solvent involved.

Process Design

Adsorption Vessels. The adsorbers and carbon beds are normally sized on the basis of a desired cycle-time and pressure drop. A further limitation is the air velocity, which must be kept sufficiently low to minimize movement of the carbon and consequent attrition. Since large volumes of air at atmospheric pressure are usually handled, pressure drop is an important consideration, and relatively shallow beds, 1 to 3 ft in depth, are employed. The beds are contained in horizontal cylindrical vessels for large plants and in vertical vessels for smaller installations. Downflow is employed to minimize movement of the adsorbent (which is relatively light), and superficial air velocities in the range of 30 to 100 ft/min are used (50 is a typical value). The active carbon is ordinarily supported by structural members. The screen size recommended by one manufacturer for support of granular carbon is presented in Table 12-26.

The quantity of solvent adsorbed per unit weight of carbon is dependent on the carbon/solvent equilibrium relationship and the operating conditions. The relative capacity of activated carbon for several solvents is indicated by the data in Table 12-27. The values given, which are based on the data of Lamb and Coolidge (65), do not represent complete equilibrium but are very close to it for the conditions 0°C (32°F) and 10 mm Hg pressure of solvent vapor over the carbon. Equilibrium data for a specific solvent (benzene) and one particular carbon (a type of activated coconut-shell charcoal) are

Table 12-26. Screens Recommended for Support of Granular Activated Carbon

Carbon mesh-size	Screen specifications		
	Mesh	Wire diameter, in.	Opening, in.
4 × 8	10	0.035	0.065
6 × 10	12	0.028	0.055
8 × 12	14	0.025	0.046
10 × 20	24	0.014	0.0277
12 × 30	35	0.012	0.0166

Courtesy Barnebey-Cheney Company.

Table 12-27. Carbon Capacity and Integral Heats of Adsorption for Organic Compounds on Activated Coconut-shell Charcoal

Compound	Formula	Solvent adsorbed (0°C and 10 mm Hg vapor pressure)		Integral heat of adsorption (1 g-mole of solvent on 500 g of carbon at 0°C†), cal
		ml gas/g*	lb/100 lb carbon	
Ethyl chloride......	C_2H_5Cl	109	31	12,000
Carbon disulfide....	CS_2	125	43	12,500
Methyl alcohol......	CH_3OH	110	16	13,100
Ethyl bromide......	C_2H_5Br	92	45	13,900
Ethyl iodide........	C_2H_5I	106	74	14,000
Chloroform.........	$CHCl_3$	92	49	14,500
Ethyl formate......	$HCOOC_2H_5$	97	32	14,500
Benzene...........	C_6H_6	97	34	14,700
Ethyl alcohol.......	C_2H_5OH	...	...	15,000
Carbon tetrachloride......	CCl_4	78	54	15,300
Diethyl ether.......	$(C_2H_5)_2O$	84	28	15,500

*Gas volume measured at standard condition (0°C, 760 mm Hg).
†Equivalent to a solvent concentration of 44.6 ml gas/g carbon.
Source: Based on data of Lamb and Coolidge. (65)

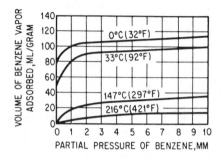

Figure 12-32. Equilibrium capacity of activated coconut-shell charcoal for benzene. Gas volume measured in milliliters at standard conditions (0°C, 760 mm Hg); charcoal outgassed at 550°C before tests. *Data of Coolidge* (66)

presented in Figure 12-32 to illustrate the typical effects of the partial pressure of the solvent and temperature on the quantity adsorbed. The two upper curves represent adsorber operating conditions, and the two lower curves represent conditions during regeneration. Adsorption isotherms for other solvents are generally similar in that the quantity adsorbed rises very sharply with pressure at low solvent-pressures and then levels off as the partial pressure of solvent in the gaseous phase is further increased. The net capacity of activated carbon for solvents in commercial units is also dependent upon the efficiency of regeneration and operating factors, e.g., air velocity, bed depth, and cycle timing. The effective capacity of the adsorbent ranges from about 6 lb of solvent per 100 lb carbon for simple hydrocarbons such as hexane and heptane, to as high as 20 lb per 100 lb for more strongly adsorbed molecules such as perchloroethylene. For preliminary estimation of airflow (or bed size), it can be assumed that under average bed conditions, 100 lb of carbon can efficiently treat 200 ft^3/min of solvent-laden air (107).

In most installations, the beds are switched at the end of each cycle by automatically operated valves. The cycle time may be based on a fixed period, which is known to be adequate to prevent solvent carry-through with the air stream, or may be varied to accommodate different conditions. In one arrangement, a vapor detector is installed to monitor air discharged from one portion of the carbon bed so that a cycle change can be made when the bed is fully charged with solvent (67).

Air Velocity and Pressure Drop. The cost of power to move air through the system constitutes one of the major operating expenses of active-carbon, solvent-recovery installations. In view of this, it is imperative that pressure drop through the beds be kept to a minimum, and adsorbent depths greater than about 3 ft are seldom used. The pressure drop can be estimated by the use of generalized correlations (see Figure 12-13); however, since these units are normally operated with the same gas (air), at approximately the same temperature and pressure (atmospheric), specific pressure-drop correlations can be conveniently used. The chart of Figure 12-33 presents pressure-drop data for activated carbon of various particle-size grades (based on Tylerscreen cuts) with atmospheric air at 70°F flowing downward. The maximum allowable velocity to prevent attrition can be estimated from the correlation of Ledoux (33) (Equation 12-4) for the case of upflow gases. Although it is not strictly applicable, this equation can also be used as a guide for the downflow case.

Heat Effects. As described in the preceding section for the case of dehydration, heat is evolved when a vapor is taken up by an adsorbent. For the case of organic-vapor adsorption on active carbon, this heat of adsorption cannot be assumed to be equal to the heat of condensation from vapor to liquid phase, although in general, compounds with high heats of condensation (or vaporization) also have relatively high heats of adsorption. The initial portion of a vapor adsorbed on carbon has the highest heat of adsorption, and thereafter the heat of adsorption normally drops somewhat for each increment as additional quantities are adsorbed. The heat of adsorption for a small increment of vapor at a constant concentration in the carbon is called the differential or instantaneous heat of adsorption, and the heat quantity evolved between two adsorbate concentrations is known as the integral heat of adsorption. This latter value is of more interest for engineering calculations and is most commonly measured with an initial adsorbate concentration of zero. Typical integral heat-of-adsorption values for organic compounds are presented in Table 12-27, which is based on the data of Lamb and Coolidge (65).

As solvent is adsorbed, the heat of adsorption causes the temperature of the carbon to rise in the active adsorption zone. This heat is transferred to the

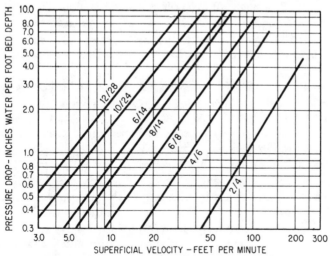

Figure 12-33. Pressure drop through dry-packed activated-carbon beds for air flowing downward at 1 atm pressure and 70°F. *Union Carbide Corp.*

gas, which carries it further into the bed, raising the temperature of the carbon in advance of the adsorptive wave, and partially cooling the air stream. The hot zone thus proceeds through the adsorption bed, ultimately causing an appreciable increase in the temperature of the exit gas and a corresponding decrease in capacity over that which would be obtainable under isothermal conditions. Loss in capacity due to this effect can be minimized by evaporating water from the carbon during the adsorption cycle. Since steam is normally used for regeneration, the regenerated carbon is saturated with water. When adsorbing solvents which readily displace water from carbon, a sufficient cooling effect is often obtained to make a precooling period unnecessary. During the regeneration cycle, low-pressure steam is passed through the carbon bed for a period of 15 to 60 min. The initial portion of steam serves primarily to raise the temperature of the carbon bed to the regeneration temperature and is condensed in the process. Some additional steam is condensed to provide heat of desorption for the adsorbed solvent, and the remainder of the steam serves as stripping vapor to lower the partial pressure of the solvent in the vapor phase and to purge the solvent vapors from the system. According to Ray (68), the quantity of steam required for regeneration in a well-designed system ranges from 3 to 5 lb/lb solvent recovered. Approximately the same range is considered typical for the Acticarbone process (see Table 12-28).

Operating Data

Typical operating requirements for activated carbon systems designed to control a wide range of volatile organic compound emissions are: electricity—2.9 kWh per 1,000 cfm-hr; carbon replacement—$6.66 per lb organic per hr per year (3-year life); steam consumption—5 lb steam/lb organic for carbon regeneration, 1 lb steam per lb of organic for the separation step (105).

The steam consumption and other utility requirements of a commercial acetone-recovery installation are given by Ray (68). In this plant, 4 lb of steam are used per pound of acetone recovered, and an acetone-water condensate is obtained which averages 20 to 33 percent acetone. A portion of the steam is required to distill this mixture. This plant also uses 8.75 gal of cooling water (at 55°F) and 0.082 kwhr of electricity per pound of solvent. Moffett (69) reports that 3.5 to 4.3 lb of 5-psig steam are required, per pound of recovered solvents, in a plant recovering solvents from printing operations at the New York Daily News Rotogravure printing plant. This steam consumption apparently also includes the amount necessary to recover the solvent dissolved in the water layer discharged from the decanter. The plant recovers over 300,000 gal/year of ink solvent. In addition to the steam requirement, the plant uses approximately 10 gal of cooling water (at 70°F) and 0.13 to 0.17 kwhr of electricity per pound of recovered solvent.

At the Ford Motor Company's spark plug plant in Fostoria, Ohio, trichloroethylene vapor in the air from degreaser units is recovered by adsorption on activated carbon. Two 72-in. diameter adsorbers are used, each containing 1,500 lb of carbon pellets. The system is reported to be capable of recovering 400 to 450 gal of liquid trichloroethylene per day with a collection efficiency of over 90 percent. Operating costs (in late 1969) were only about 3 percent of the value of the recovered solvent (70).

Data on European practice for solvent recovery in the printing industry are presented by Benson and Courouleau (71), who discuss the Acticarbone process of the Parisian concern Carbonisation et Charbons Actifs. A flow diagram for a typical application of this process, employed to recover a mixed solvent containing 40 percent toluene, 14 percent butyl acetate, and 46 percent ethanol from printing-plant exhaust air is shown in Figure 12-34. Operating data for the installation and typical design ranges are given in Table 12-28.

The major operating problems encountered in activated-carbon solvent-recovery processes are contamination of the activated carbon and corrosion of equipment. In improperly designed systems, attrition and plugging of the bed may also constitute problems. Contamination can occur as a result of resinous or polymerizable substances in the air stream which remain on the carbon during the regeneration cycle and reduce its activity. Certain types of contaminants which cannot normally be removed and recovered satisfactorily with active carbon can be tolerated in very small amounts, as they ac-

---

Table 12-28. Typical Operating Data and Ranges of Process-design Factors for Acticarbone Process Solvent-adsorption Plants (71)

Operating data:

Air rate to adsorbers (2 in parallel), SCF/min	11,000
Solvent content of air feed, lb/MSCF	0.43
Air temperature at adsorber inlet, °F	90
Steam rate to adsorber being regenerated, lb/hr	1,800
Steam/solvent ratio	3.5:1
Time cycle for each adsorber:	
Adsorption, hr	2
Steaming, min	30
Drying [with hot (220°F] air), min	15
Cooling (with cold air), min	15
Typical range of design factors:	
Solvent adsorbed by activated carbon per cycle, % by wt	10 to 14
Solvent concentration in air stream, lb/MSCF	0.3 to 0.5
Power required, kwhr/lb of solvent recovered	0.10 to 0.15
Cooling water, gal/lb of solvent recovered	7 to 10
Steam, lb/lb of solvent recovered	3.5 to 5
Activated carbon, lb/ton of solvent recovered	0.5 to 1

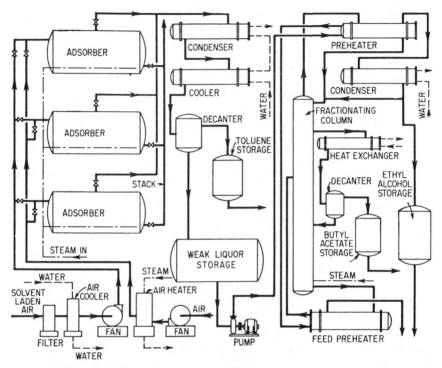

Figure 12-34. Flow diagram for Acticarbone process of Carbonisation et Charbons Actifs (a Parisian concern). This installation recovers a mixed solvent containing toluene, butyl acetate, and ethanol from a printing-plant air stream (71).

cumulate in the top portion of the bed which is first contacted with the air and partially removed during the regenerative cycle. Since the major portion of the bed remains in good condition, a reasonable service life can be realized, and the carbon may occasionally be returned to the manufacturer for reactivation. Some contaminants must be removed from the vapor-laden air stream prior to its entry into the adsorbent bed. Phenolic materials, for example, which are encountered in operations designed to recover alcohol from certain resin-impregnating operations, can be removed by scrubbing with caustic solution in a packed or spray-type scrubber. Traces of polymerizable or very heavy compounds can also be removed in a guard chamber which is placed in the gas stream ahead of the main adsorbers. The carbon or other adsorbent in the guard chamber is permitted to become contaminated in order to protect the main carbon-charge.

Although most dry solvents are not particularly corrosive, corrosive conditions may occur in the adsorption equipment as a result of the steaming operation. Many solvents hydrolyze in the presence of liquid water or steam at elevated temperatures, and activated carbon may act as a catalyst for this and other decomposition reactions. Esters such as ethyl acetate are particularly corrosive because hydrolysis results in the formation of organic acids. When electrolytes are present or are formed during the process, corrosion of iron and steel can be further accelerated by galvanic action between the carbon and the metal screen or vessel. Solvents such as hydrocarbons which do not decompose can usually be handled in plain steel equipment. Special construction materials, e.g., copper, Everdur, Monel, and stainless steel, may be required for other solvents. Special adsorption-vessel designs may also be employed to minimize corrosion by preventing the wet steam-solvent mixture from contacting the vessel shell.

Oxidizing gases such as oxygen and chlorine may react with carbon under some conditions, and contact with these gases at high temperatures should be avoided. It is reported that activated carbon can be used with high-velocity air at temperatures up to about 150°C; with low air velocity, somewhat higher temperatures are permissible (72).

Attrition of activated carbon can be minimized by proper design of the adsorption vessels. The air-flow rate should be below 100 ft/min, preferably below 60 ft/min, and good air distribution to the top of the bed should be provided. Plugging of activated carbon beds can be prevented by eliminating any sources of carbon attrition and providing an adequate screen or filter ahead of the adsorption vessel.

Fluidized Bed Active Carbon Adsorption Processes

In the viscose (rayon) process a major fraction of the carbon disulfide consumed leaves the plant in the ventilating air. In many plants no attempt is made to recover the CS_2 because new solvent can be purchased for less than the recovery cost. It is expected, however, that tightening air pollution laws will make recovery plants more common, so the first-of-a-kind large commercial fluidized bed-adsorption plant installed by Courtaulds, Ltd. at its Holzwell, Wales, viscose plant is of particular interest.

A detailed description of the process and its development is provided by Rowson (73). A 250,000-cfm air stream containing 1,000 ppm (vol) CS_2 and 20 to 30 ppm H_2S required treatment. Preliminary process studies indicated that a fixed-bed plant would be uneconomical. A large cross-sectional area of bed would by required (about 6,000 ft²) to provide a reasonable air velocity. Because of the low concentration of CS_2, fixed-bed adsorption would be inefficient and the break point would occur early. The decision was therefore made to develop a fluidized bed process. Such a process offers the potential advantages of fully continuous operation, countercurrent contacting with attendant high and uniform solvent loadings on the carbon, and high gas velocities leading to small vessel diameters.

Commercial operation of the Courtaulds, Ltd. fluidized bed plant commenced in 1960, about 8 years after initiation of development work on the project. It is reported that the plant—code named Landmark—cost $1.7

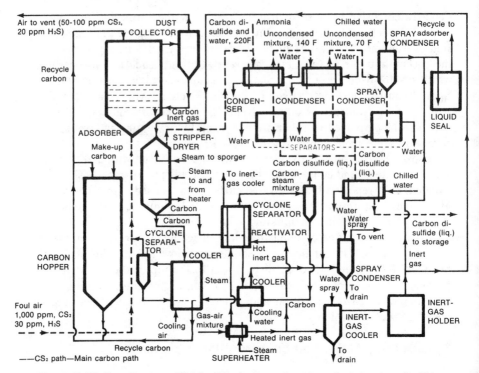

Figure 12-35. Flow diagram of fluidized bed process for adsorption of carbon disulfide on active carbon.

million (74). A flow diagram of the process is shown in Figure 12-35. Design data and typical operating conditions are listed in Table 12-29.

Air from the viscose plant is first contacted by an alkaline ferric oxide suspension in a spray scrubber (not shown in the flow diagram) to remove the bulk of the H₂S. Removal is necessary because hydrogen sulfide is catalytically oxidized by air, in the presence of active carbon, to elemental sulfur which is extremely difficult to strip from the carbon.

The air is forced through the fluidized bed adsorber and a dust collection system by means of a 1,000-hp fan. Carbon is conveyed to the top of the adsorber column and on to the top tray. Carbon level on the tray is maintained by a weir which it flows over into a downcomer leading to the tray beneath. After passing over a total of five trays, the carbon flows downward through seal legs into the stripper vessel. The seal legs are purged with inert gas to prevent steam and CS₂ from entering the adsorber.

The adsorbent is stripped of CS₂ by countercurrent contact with 300°F steam while passing downward in plug flow through an 18-ft diameter vessel. The bottom half of the vessel contains a number of vertical, externally finned tubes. High-pressure steam inside the tubes heats and dries the carbon which flows downward on the outside. Steam liberated from the carbon flows upward and augments the stripping steam. The hot dry carbon is split into two streams; 10 percent is reactivated and the remainder is cooled to 160°F by contact with fresh air. This is accomplished in a single 7-ft diameter fluidized bed.

Reactivation of a continuous bypass stream of carbon serves to remove elemental sulfur, sulfuric acid, and other low volatility compounds which cannot be removed by normal stripping. The operation is conducted by passing low-pressure, highly superheated steam through the carbon in a single fluidized bed.

The carbon disulfide is recovered from the exit gases leaving the stripper by a four-stage condensation train. About 2 lb/hr of ammonia is added to the exit gas before it enters the condensers to protect the aluminum tubes from corrosion. The first two stages serve to condense the bulk of the water (shown as a single unit in the flow diagram). The third stage condenses most of the CS_2, and final cleanup is obtained in the fourth stage—a spray tower using 35°F water. CS_2 and water are virtually immiscible so the heavier CS_2 can be separated from the water readily by decantation.

Operating experience with the Courtaulds plant has shown the recovery efficiency to be at least as high as for a fixed-bed plant and steam consumption a little lower. Carbon attrition has proven to be somewhat of a problem as the accumulation of fines in the bed increases the residence time which in turn

---

Table 12-29. Design and Operating Data for Fluidized Bed Carbon Disulfide Adsorption Plant (73, 74)

Air flow rate	250,000 cfm
Inlet concentration	1000 ppm CS_2, 30 ppm H_2S
Outlet concentration	50-100 ppm CS_2, 20 ppm H_2S
CS_2 recovery rate	1.2 tons/hr
CS_2 adsorption efficiency	85-95%
Adsorbent	6-14 mesh coconut charcoal
Adsorbent circulation rate	50,000 lb/hr
CS_2 loading on adsorbent	
lean	0.5%
rich	7.0%
Stripping steam required	12,000 lb/hr
Adsorber design	
size	38 ft dia x 45 ft high
number of beds	5
bed depth	2-3 inches (each)
pressure drop	3 psi

leads to a greater adsorption of water at the expense of CS_2. This problem can be partially resolved by drawing off fines collected from the exit gas instead of returning them to the bottom adsorber tray as shown in the flow diagram. However, this solution leads to an increased requirement for makeup carbon.

The Purasiv HR process is an improved fluidized bed activated carbon process based on the use of a spherical beaded adsorbent developed by Kureha Chemical of Japan. The process is marketed in the United States and Canada by Union Carbide Corporation who introduced it in 1978 (118). The unique form of the carbon adsorbent represents the key feature of the process. The beads, which are about 0.7 mm in diameter, create a homogeneous fluidized bed in the adsorption section and a free-flowing dense bed in the desorption section while providing a much higher resistance to attrition than conventional granular or pelletized material. The beads are produced by a proprietary process which involves shaping molten petroleum pitch into spherical particles which are subsequently carbonized and activated under controlled conditions.

A simplified flow diagram of the Purasiv HR system is given in Figure 12-36. The solvent-laden air is fed into the bottom of the adsorption section and passes upward countercurrent to the carbon beads which are fluidized on a series of perforated trays. The beads flow from each tray to the one beneath it by an overflow weir arrangement and finally flow from the bottom tray into a seal pot where they form a dense-packed bed. The carbon then flows downward as a moving dense bed through a secondary adsorber section where it removes hydrocarbons from the nitrogen used for regeneration; a desorption section where it is heated while being stripped of hydrocarbons by a rising stream of nitrogen; and finally a cooler where the carbon is cooled to ambient temperature. The regenerated and cooled carbon is then conveyed to the top tray of the adsorber column. High temperature solvent-laden nitrogen from the desorption section passes through a water-cooled condenser and a separator which removes liquid solvent. Nitrogen from the separator, which is cool but still saturated with solvent, is passed through the secondary adsorber section and is then recycled to the bottom of the desorption section.

This description and the flow diagram refer to the nitrogen-stripping technique, designation Type N, which is particularly suitable for the recovery of water-soluble solvent. The process can also be used with a more conventional steam-stripping technique, designated Type S, which is useful for the recovery of chlorinated hydrocarbons and other water-insoluble solvents. The Type S system is similar to the Type N except that the secondary adsorber and provisions to recycle the stripping vapor are not required. Commercial Purasiv HR systems recovering ketones, tetrahydrofuran, toluene, isopropyl alcohol, acetone, kerosene, methylene chloride, and various solvent mixtures are reported to be in operation in the United States (118).

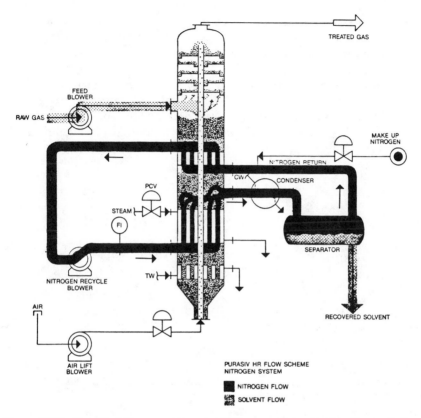

Figure 12-36. Simplified Diagram of Purasiv HR Process (Union Carbide Corporation).

Odor and Trace-Impurity Removal by Adsorption

The problem of odor removal differs from that of solvent recovery primarily in that the impurities are present in much lower quantities and that usually no attempt is made to recover the adsorbed compounds. Large volumes of atmospheric-pressure air must be handled in air-conditioning types of equipment, and a very low pressure drop is mandatory. The pressure-drop requirement makes the use of very thin beds of activated carbon desirable. Fortunately, most odorous vapors are compounds of relatively high molecular weight, which are adsorbed readily, and can therefore be removed in shallow beds. Commercial equipment for the removal of odors from air streams is designed to give maximum face-area for the passage of air in a minimum space. The equipment is unitized for ease of replacement of spent carbon, and the units are commonly in the form of cylindrical canisters or larger unit cells containing flat or corrugated beds.

Principal requirements of carbon for use in odor-removal equipment are a high capacity and low pressure drop. The material must also be dustless. Specifications for activated carbon used in air purification are presented in Table 12-30. Typical commercial equipment is illustrated in Figures 12-37, 12-38, and 12-39.

Table 12-30. Specifications for Activated Carbon Used in Air Purification

Property	Specification, Minimum
Activity for carbon tetrachloride	50 per cent
Retentivity for carbon tetrachloride	30 per cent
Apparent density	0.40 g/ml
Hardness (ball abrasion)	80 per cent
Size distribution	6 to 14 mesh (Tyler-sieve series)

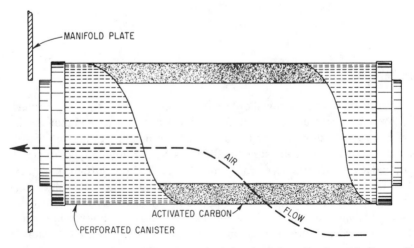

Figure 12-37. Diagram of canister-type activated-carbon air-purification unit. *Connor Engineering Corporation*

Process Design

According to Ray (75), an air velocity of 60 ft/min can be used with a ½-in. bed of suitable activated carbon. In general, the canisters or cells of activated carbon are removed from the equipment when saturated and returned to the manufacturer for regeneration (which consists of stripping with superheated steam at 1,000° to 1,300°F). The units may last for as long as 4 or 5 years before replacement or regeneration is required, depending, of course, on the concentration of odorous vapors in the air stream. Extremely long cycles are possible because even relatively severe odor problems are usually caused by only very small concentration of odorous compounds, and a good grade of carbon can pick up 0.1 to 0.2 lb of odorant per pound of carbon before regeneration is required. The upper limit of odorant concentration for the practical use of replaceable cartridge-type activated-carbon adsorption equipment is stated to be about 2 to 5 ppm (76). Above this level, the service life of the cartridges is greatly diminished, necessitating frequent reactivation.

Figure 12-38. Photograph of air-purification unit employing cylindrical canisters, showing typical method of installation. *Connor Engineering Corporation*

Figure 12-39. Corrugated bed-type activated charcoal air-purification cell. This unit contains 35 lbs of 50-min activated coconut-shell charcoal and handles 1,000 cfm of air with a pressure drop of 0.25 in. of water. *Barnebey-Cheney Company*

Turk (77) has proposed the following formula to predict the service life of an adsorption system between regenerations:

$$t = \frac{6.43(10^6)SW}{eQ_rMC_v}$$

(12-8)

where t = service life, hr
S = ultimate proportionate saturation of carbon (retentivity, fractional)
W = weight of carbon, lb
e = adsorption efficiency (fractional)
Q_r = quantity of air processed by adsorption equipment, cfm
M = average molecular weight of contaminants
C_v = concentration of contaminants, ppm

For a commercial cell, containing 45 lb carbon, and rated at 700 cfm air at an operating efficiency of about 95 percent, the service life in hours (t) is calculated to be $4.35 \times 10^5 S/MC_v$. Constants of some important atmospheric

Table 12-31. Characteristics of Some Atmospheric Odorants (76)

Substance	Molecular weight, M	Retentivity, S	Odor-threshold concentration, ppm	Service life for odor-threshold concentrations, hr*
Acrolein (heated fats).....	56	0.15	1.8	650
Hexane (hydrocarbon)....	86	0.16	almost odorless	
Phenol (carbolic acid)....	94	0.30	0.28	5,000
Valeric acid (body odor)..	102	0.35	0.00062	2.4×10^6

*Calculated value.

odorants for use in this equation are given in Table 12-31. Recent data obtained on the adsorption of contaminants from air streams in which they are present in low ppm concentrations indicate that generally reported retentivities represent maximum values (78). Appreciably lower capacities can be expected when contaminants are present in concentrations near the odor threshold.

The relative capacity (or retentivity, S, of Table 12-31) of a good grade of activated coconut-shell charcoal (50 min by chloropicrin test), for a number of vapors (including some which may be classed as odors) has been presented by Barnebey (79), and a portion of his data is reproduced in Table 12-32. Most materials encountered in odor-removal problems fall in classes 3 or 4, which means that the weight of carbon required will be three to six times the weight of odorant present in the air purified during the period of operation. If specific chemical odorants are not identified in the air to be purified, Barnebey suggests that the required quantity of carbon be estimated on the basis of

1. Possible odor sources (e.g., number of occupants of a room)
2. Type of occupancy (e.g., laboratories, toilets, offices, etc.)
3. Consideration of the amount of outside air which normal air-conditioning-design practice would call for to eliminate odors and stuffiness
4. Use of a preliminary test with a small adsorber
5. Assumption of an average odor composition which, for example, will be just perceptible at a concentration of 0.01 lb of odorant per million cubic feet of air
6. Rules of thumb and experience

Table 12-32. Capacity of 50-min Activated Coconut-shell Charcoal for Vapors (79)					
Vapor	Capacity Index No.	Vapor	Capacity Index No.	Vapor	Capacity Index No.
Acetaldehyde	2	Carbon dioxide	1	Hydrogen sulfide	3
Acetic acid	4	Carbon monoxide	1	Isopropyl alcohol	4
Acetone	3	Carbon tetrachloride	4	Masking agents	4
Acrolein	3	Chlorine	3	Mercaptans	4
Alcohol	4	Chloropicrin	4	Ozone	4
Amines	2	Cigarette smoke	4	Perfumes, cosmetics	4
Ammonia	2	Cresol	4	Perspiration	4
Animal odors	3	Diesel fumes	3	Phenol	4
Anesthetics	3	Disinfectants	4	Propane	2
Benzene	4	Ethyl acetate	4	Pyridine	4
Body odors	4	Ethylene	1	Ripening fruits	4
Butane	2	Essential oils	4	Smog	4
Butyl alcohol	4	Formaldehyde	2	Solvents	3
Butyric acid	4	Gasoline	4	Stuffiness	4
Cancer odor	4	Hospital odors	4	Toluene	4
Caprylic acid	4	Household smells	4	Turpentine	4

The capacity index numbers have the following meanings:

4: High capacity for all materials in this category. One pound takes up about 20 to 50 per cent of its own weight—averaging about ⅓ (33⅓ per cent). This category includes most of the odor-causing substances.

3: Satisfactory capacity for all items in this category. These constitute good applications but the capacity is not as high as for category 4. Adsorbs about 10 to 25 per cent of its weight—averaging about ⅙ (16.7 per cent).

2: Includes substances which are not highly adsorbed but which might be taken up sufficiently to give good service under the particular conditions of operation. These require individual checking.

1: Adsorption capacity is low for these materials. Activated charcoal cannot be satisfactorily used to remove them under ordinary circumstances.

Typical yearly requirements in pounds of 50-min activated coconut-shell charcoal for various occupancies have been given as follows (80):

Residences or offices—total per person	
Smokers	2 lbs
NonSmokers	1
Hotels, per person average occupancy	3
Bar or tavern, per occupant	5

A detailed discussion of air-conditioning systems incorporating active carbon for air purification is beyond the scope of this book. Additional details are presented by Barnebey (79), Mantell (81), and others. It is interesting to note that in air-conditioning applications, the use of adsorptive air purification can frequently show economies over the use of fresh outside air to reduce odors and stuffiness because of the high cost of heating or cooling the outside air to the desired interior temperature.

Operating Data

Activated carbon filters have been used successfully for over 20 years at the District of Columbia's refuse transfer station to remove garbage odors from air and prevent contamination of the surrounding atmosphere (82). The filters operate only about 2 hours per day during periods when the collection trucks return to the station to transfer their loads. A fan capacity of 85,300 cfm is provided on the basis of seven air changes per hour. The air is first passed through cyclone type collectors to remove large dust particles, then through glass fiber filters to remove the remaining dust and finally through 3420 canister type carbon filters for odor removal. The canisters are each 10½ in. long and 4⅜ in. outside diameter with a ¾-in. thick annular bed of carbon. Pressure drop through the canisters is 0.15 in. of water at the rated throughput of 25 cfm each. The canisters have an initial odor removal efficiency of 95 percent and require replacement every 4 years.

An example of the commercial application of active-carbon adsorption in the control of air pollution by removing odorous materials from the exhaust air of a chemical plant has been described by Lorentz (83). The odors in this case arise during steps in the synthesis of Vitamin B-1. These steps involve the production and handling of sodium dithioformate, and the vapors have a tenacious garlic-sulfide stench. The odor-containing air is drawn from exhaust vents of the chemical-process equipment and adjacent work area at a rate of 12,000 cfm at a suction pressure of ⅜ in. of water. Air from the fan is first passed through air-conditioning-type dust filters with a total cross-sectional area of about 33 sq ft and then through a bank of 503 Dorex Type 42 carbon canisters. Pressure drop through the canisters is 0.14 in. of water at an air-flow rate of 24 cfm per unit. Approximately 800 lb of carbon are contained in the bank of canisters and between 300 and 400 lb of air-borne material is collected in the carbon by the end of the 6-month cycle. In addition to the odorous compounds, the carbon picks up some water and traces of organic solvents from the air stream, which leaves the unit entirely free from objectionable odor. The spent carbon is returned to the suppliers for reactivation, and about ¾ of the original carbon is reclaimed and returned. The carbon bed in the canister is about 1 in. thick, and a total cross-sectional area of about 1,000 sq ft of carbon surface is exposed to the flowing air stream. Total pressure drop through the system is about ¾ in. of water.

An interesting application which combines the features of odor-removal and solvent-recovery plants is described by Larkin and Davis (84). This installation removes solvent vapors from air from a silk-screening operation. The principal objective is air purification rather than solvent recovery as the vapors were found to be extremely irritating to workers. The unit utilizes a single large canister, 7½ ft long by 5-ft OD, containing a 6-in. thick cylindrical bed of carbon. The charge consists of 1,100 lb of 50-min. activated coconut-shell carbon. Regeneration is accomplished *in situ* by passing low-pressure steam through the unit, for a short period of time, once a month. Steam and solvent vapors are condensed; however, the removed mixed solvents have little value. The carbon has been found to last for approximately 5 years in this installation.

A rather specialized but important use of activated carbon is for the trapping of airborne radioactive iodine which may result from nuclear reactor operations. Problems encountered in this application include weathering by humid air, the presence of a portion of the iodine in compounds such as methyl iodide which are not as readily adsorbed, and the effects of radiation. At the Savannah River Plant (SRP) activated carbon beds must be replaced after 3 to 5 years to ensure that iodine adsorption efficiency meets the required design level (99.85 percent removal) (85). In this plant Barneby-Cheney Type 416 carbon is used to purify a flow of 100,000 cu ft/min of air. The effect of radiation on the adsorption of iodine and methyl iodide is described by Jones (86).

Adsorption of Hydrocarbons from Gas Streams with Active Carbon

Active carbon is used in commercial installations to remove benzol (a benzene-rich light oil) and other impurities from manufactured and coke-oven gas. The principal reason for removing benzol from such gas streams is undoubtedly to recover it as a valuable by-product. A second reason is to improve the properties of the gas as a household fuel. Benzol gives the gas a tendency to burn with a sooty flame.

A typical benzol-removal installation has been described by Howell (87) and Walker et al (88). The plant is located at the Manchester Gas Department in England and consists of four adsorbers, each holding approximately 1 ton of 12- to 20-mesh active charcoal. The operating cycle consists of adsorption, about 60 min; steaming, 30 min; and drying, 30 min. Drying is accomplished by recirculating stripped coal gas through an indirect steam heater and then through the adsorber. Regeneration is accomplished with steam introduced countercurrent to the normal gas flow. The unit adsorbs practically all of the benzene in the gas stream and a portion of the sulfur content ranging from 64 percent with fresh charcoal to 40 percent after about 7 months. The light oil recovered from the charcoal during the regeneration period contains between 1.2 and 1.4 percent sulfur, about three-fourths of which is CS_2, the balance being thiophene and other compounds.

According to Deringer (89), the use of an inert gas, rather than steam, in this process results in a substantial saving of steam. Any essentially air-free inert gas, such as recycled coal gas or internal-combustion-engine exhaust gas, can be used as stripping vapor. Both the adsorber and the stripping gas are heated during the regeneration. Steam requirements are stated to average 2.2 lb/lb light oil, which is about one-half of that consumed in the direct steam process.

Propane and heavier hydrocarbons are removed from natural gas by active carbon adsorption processes which are quite similar to those used for benzol recovery. They also closely resemble processes which employ silica gel for hydrocarbon recovery from natural gas and which are described in the preceding section. The principal difference is that the active carbon adsorbents do not provide simultaneous dehydration and hydrocarbon recovery. However, they do provide a higher capacity for light hydrocarbons than silica gel.

A comprehensive set of equilibrium data for hydrocarbons and natural-gas sulfur compounds on activated carbon (Pittsburg BPL) has been presented by Grant, et al (90,91). These data and some additional experimental data for isobutane and carbon dioxide adsorption on Columbia NXC, 8 × 10 carbon have been used by Hasz and Barrere (92) as a basis for predicting isotherms for all of the common constituents of natural gas using the Polyani potential theory of adsorption (93).

Field test data showing the effects of natural gas composition and flow rate are given by Enneking (94). The natural-gas processing systems are not described in detail herein because of their similarity to the other processes mentioned and the fact that they are primarily recovery rather than purification operations.

References

1. Langmuir, I. 1916. *J. Am. Chem. Soc.* 38:2207. *Also:* 1917. 39:1883. *Also:* 1918. 40:1361.

2. Davison Chemical Division, W. R. Grace & Company, Bulletins 201, PC/ADS 2-777, 1C-15-780, 1C-5-1082.

3. Mobil Corporation. 1971. *Mobil Sorbead Desiccants.*

4. Aluminum Company of America. 1969. *Activated and Catalytic Aluminas.* (July 14).

5. Floridin Company. *Floridin Activated Bauxite Products.*

6. Union Carbide Corporation. *Linde Molecular Sieves.*

7. Amero, R.C.; Moore, J.W.; and Capell, R.G. 1949. *Chem. Eng. Progr.* 43(July):349.

8. Taylor, R.K. 1945. *Ind. Eng. Chem.* 37(July):649.

9. Hubard, S.S. 1954. *Ind. Eng. Chem.* 46(Feb.):356.

10. Getty, R.J.; Lamb, C.E.; and Montgomery, W.C. 1953. *Petrol. Eng.* 25(Aug.):B-7.

11. Kaiser Chemicals. 1967. *Gas Drying with KA-201TM Active Alumina* (July).

12. Collins, J.J. *A Report on Acid-Resistant Molecular Sieve Types AW-300 and AW-500.* Union Carbide Corporation, Linde Molecular Sieve Bulletin.

13. Derr, R.B. 1938. *Ind. Eng. Chem.* 30(Apr.):384.

14. Grayson, H.G. 1955. *Ind. Eng. Chem.* 47(Jan.):41.

15. Leavitt, F.W. 1962. *Chem. Eng. Progr.* 58(Aug.):54-59.

16. Ross, W.L., and McLaughlin, E.R. 1951. *Refrig. Eng.* 59(Feb.):167.

17. Herrmann, R.H. 1955. *Oil Gas J.* 53(Apr. 18):144-147. (Also presented at Gas Conditioning Conference, March 29 to 30, 1955.)

18. Swerdloff, W. 1967. "Dehydration of natural gas." *Proc. Gas Conditioning Conf.,* Norman, Okla.

19. Hammerschmidt, E.G. 1957. *CEP-57-7.* Presented at Production Conference, Operating Section, American Gas Association, Bal Harbor, Fla. (May 20-22).

20. Cappell, R.G.; Hammerschmidt, E.G.; and Derschner, W.W. 1944. *Ind. Eng. Chem.* 36(Sept.):779.

21. Heinemann, F., and Heinemann, H. 1954. *Petrol. Refiner* 33(June):159.

22. Hammerschmidt, E.G. 1945. *Gas (Los Angeles)* 21(Jan.):32,33.

23. Harrell, A.G. 1957. *Oil Gas J.* 55(Oct. 28):121.

24. Pierce, J.E., and Stieghan, D.L. 1966. *Hydroc. Process. Petrol. Refiner* 45(Mar.):170-172.

25. Silbernagel, D.R. 1967. *Chem. Eng. Progr.* 63(4):99-102.

26. Kraychy, P.N., and Masuda, A. 1966. *Oil Gas J.* 64(Aug. 8):66.

27. Clark, E.L. 1958. *Proc. Gas Conditioning Conf.* Norman, Okla.

28. Hougen, O.A., and Marshall, W.R. Jr. 1947. *Chem. Eng. Progr.* 43(4):197.

29. Perry, R.H., and Chilton, C.H. ed. 1973. *Chemical Engineers' Handbook,* 5th ed. New York: McGraw-Hill Book Company, Inc.

30. Hougen, O.A., and Watson, K.M. 1947. *Chemical Process Principles,* Part 3. New York: John Wiley & Sons, Inc., p. 1080.

31. Barry, H.M. 1960. *Chem. Engr.* (Feb. 8):105-120.

32. Collins, J.J. *Chem. Eng. Progr.* (Symposium Series No. 74).

33. Ledoux, E. 1948. *Chem. Engr.* 55(Mar.):118.

34. Wunder, J.W.J. 1962. *Oil Gas J.* (Aug. 6):137-148.

35. Rose, H.E. 1945. *Inst. Mech. Engrs. (London)* 153:141-68.

36. Brownell, L.E., and Katz, D.E. 1947. *Chem. Eng. Progr.* 43:537-549.

37. Levas, M. 1947. *Chem. Eng. Progr.* 43:549-554.

38. Ergun, S. 1952. *Chem. Eng. Progr.* 48:89-94.

39. Union Carbide Corp. "Linde molecular sieve fixed bed pressure drop calculations." *Linde Molecular Sieve Adsorbent Bulletin F-34.*

40. Fair, J.R. 1969. *Chem. Engr.,* (July 14):90-102.

41. Allen, H.V., Jr. 1944. *Petrol. Refiner* 23(July):93-98.

42. Aluminum Company of America. 1956. *Activated and Catalytic Aluminas* (June 29).

43. Wilkinson, E.P., and Sterk, B.J. 1950. *Petrol. Engr., Reference Annual,* p. D-30.

44. Conviser, S.A. 1965. *Oil Gas J.* 63(Dec. 6):130-135.

45. Yon, C.M., and Turnock, P.H. 1971. *Multicomponent Adsorption Equilibrium on Molecular Sieves.* Paper presented at the AIChE 68th National Meeting (March 1). (Reprinted by Union Carbide Corporation as *Linde Molecular Sieves, Adsorbent Bulletin F-59.)*

46. Bancroft, W.G. (Union Carbide Corp.). 1971. Personal communication (Sept. 14).

47. Thomas, T.L., and Clark, E.L. 1967. *Oil Gas J.* 65(12):112-115.

48. Lee, M.N.Y., and Collins, J.J. 1968. *Ammonia Plant Feed Desulfurization with Molecular Sieves.* Paper presented at the Tripartite AIChE Meeting, Montreal, Canada (Sept. 25).
49. Stewart, H.A., and Heck, J.L. 1967. *Hydrogen Purification by Pressure Swing Adsorption.* Paper presented at the AIChE 64th National Meeting, New Orleans, La. (March 16-20).
50. Kerry, F.G. 1956. *Chem. Eng. Progr.* 52(Nov.):441.
51. Humphries, C.L. 1966. *Hydroc. Process.* 45(Dec.):88-96.
52. Ballard, D. 1965. *Oil Gas J.* (July 12):101-110.
53. Ballard, D. 1965. *Hydroc. Process.* 44(Apr.):131-136.
54. Ballard, D. 1963. *A Discussion of Short-Cycles, Solid Desiccant Hydrocarbon Recovery Units.* Paper presented at the Natural Gas Processor's Association Meeting, Lafayette, La. (Jan. 8).
55. Deitz, V.R. 1944. *Bibliography of Solid Adsorbents* Washington, D.C.: U.S. Cane Sugar Refiners and Bone Char Manufacturers and National Bureau of Standards, p. ix.
56. Stenhouse, J. 1961. *Chem. News* 3:78.
57. Lewis, W.K.; Gilliland, E.R.; Chertow, B.; and Cadogan, W.P. 1950. *Ind. Eng. Chem.* 42:1326-1332.
58. Fieldner, A.C., et al. 1919. *Ind. Eng. Chem.* 11:524.
59. Anon. 1940. *Chem. & Met. Eng.* 47(Jan.):31.
60. Anon. 1950. *Underwriters' Labs. Bull. Research No. 43* (Jan.).
61. Browning, F.M. 1952. *Chem. Engr.* 59(Oct.):158.
62. Mattia, M.M. 1970. *Chem. Eng. Progr.* 66(Dec.):74-79.
63. Mattia, M.M. 1969. U.S. Patent 3,455,089.
64. Mattia, M.M. 1970. U.S. Patent 3,534,529.
65. Lamb, A.B., and Coolidge, A.S. 1920. *J. Am. Chem. Soc.* 42:1146.
66. Coolidge, A.S. 1924. *J. Am. Chem. Soc.* 46(Mar.):596.
67. Ray, A.B., and Logan, L.A. U.S. Patent 2,211,162.
68. Ray, A.B. 1940. *Chem. & Met. Eng.* 47(May):329-332.
69. Moffett, T.F. 1943. *Heating and Ventilating* 40(Apr.):33-36.
70. Anon. 1969. *Filtration Engineering* (Dec.).
71. Benson, R.E., and Courouleau, P.H. 1948. *Chem. Eng. Progr.* 44(June):459.
72. Barnebey-Cheney Company. *Process Development and Industrial Use of Adsorbite* (11):5.
73. Rowson, H.M. 1963. *Brit. Chem. Eng.* (Mar.).
74. Anon. 1963. *Chem. Engr.* (Apr. 15):92-93.
75. Ray, A.B. 1936. U.S. Patent 2,055,774.
76. Turk, A. 1958. *Ind. Wastes* 3(Jan./Feb.):9-13.
77. Turk, A. 1955. *Ind. Eng. Chem.* 47:966.
78. Stankovich, A.J. 1969. "The capacity of activated charcoals under dynamic conditions for selected atmospheric contaminants in the low parts per million range." *ASHRAE Symposium Bulletin, Odors and Odorants: An Engineering View.*
79. Barnebey, H.L. 1958. *Heating, Piping Air Conditioning, J. Sec.* 153(Mar.).
80. 1970. *ASHRAE Guide and Data Book,* p. 463.
81. Mantell, C.L. 1951. *Adsorption.* New York: McGraw-Hill Book Company, Inc.
82. Enneking, J.C., and Todd, H.H. Jr. 1969. *ASHRAE J.* (July).
83. Lorentz, F. 1950. *Chem. Eng. Progr.* 46(Aug.):377.
84. Larkin, S.C., and Davis, W. 1957. *Industry Power* (May).
85. Milham, R.C. 1968. *Proc. 10th AEC Air Cleaning Conf.* (CONF-680821), pp.167-169.

86. Jones, L.R. 1968. *Proc. 10th AEC Air Cleaning Conf.* (CONF-680821), pp. 204-215.

87. Howell, A.K. 1943. *Gas J.* 242(Sept. 15):337.

88. Walker, C.R.; Applebee, H.C.; and Howell, A.K. 1944. *Gas J.* 243:310-346, 379.

89. Deringer, H. 1947. *Trans. Fuel Conf., World Power Conf.*, Sec. A4, Paper No. 4.

90. Grant, R.J.; Manes, M.; and Smith, S.B. 1962. *AIChE J.* 8(3):403-406.

91. Grant, R.J. and Manes, M. 1964. *Ind. Eng. Chem. Fundamentals* 3(3):221-224.

92. Hasz, J.W., and Barrere, C.A. Jr. 1964. *Chem. Eng. Progr. Symposium Series* 96(65):48-56.

93. Polyani, M. 1914. *Verh. Dtsch. Phys. Ges.* 16:1012.

94. Enneking, J.C. 1966. *Hydroc. Process.* 45(Oct.):189-192.

95. Norton Company, "Zeolon Acid Resistant Molecular Sieves," Bulletin Z-51-R1.

96. Turnock, P.H., and Gustafson, K.J. 1972. "Advances in Molecular Sieve Technology for Natural Gas Sweetening," Pro. of the Gas Conditioning Conference, University of Oklahoma, Norman, Oklahoma.

97. Cines, M.R.; Haskell, D.M.; and Houser, C.G. 1976. *Chem. Eng. Progr.* 72,8(August):89-93.

98. McAllister, W.S., and Westerveld, W.W. 1978. *Oil Gas J.* (January 16):66-77.

99. Palmer, G.H. 1977. *Hydroc. Process.* (April):103-106.

100. Lukchis, G.M. 1973. *Chem. Engr.* 80,13(June 11):111-116.

101. Chi, Chang W., and Lee, H. 1973. "Gas Purification by Adsorption," AIChE Symposium Series No. 134, Vol. 69:96-101.

102. Fornoff, L.L. 1971. "The Pura Siv-N Process," Presented at the AIChE 64th Annual Meeting, San Francisco, CA. (November 28).

103. Fuderer, A., and Rudelstorfer, E. 1976. U.S. Patent No. 3,986,849 (October 19).

104. Heck, J.L., and Johanson, T. 1978. Abstract of Paper Presented at Western Gas Processors and Oil Refiners Association Meeting, Long Beach, CA. Published in *Oil Gas J.* (January 9):91-92.

105. Kittleman, T.A., and Akell, R.B. 1978. *Chem. Eng. Progr.* 74,4(April):87-91.

106. Drew, J.W. 1975. *Chem. Eng. Progr.* 71,2(February):92-99.

107. Manzone, R.R., and Oakes, D.W. 1977. *Pollution Engineering* 5,10 (October):23-24.

108. Weskem, Inc. 1982. *Sorbead R/H/W,* Product Data Sheet.

109. Kali-Chemie. 1979. KC-Trockenperlen, P2000 12.79.

110. Ezell, E. L., Gelo, G. F. 1982. *Hydrocarbon Processing,* May, p. 191.

111. Barrow, J. A. 1983. *Hydrocarbon Processing,* January, p. 117.

112. Badger Engineers, Inc. 1984. *Hydrocarbon Processing,* April, p. 55.

113. Heck, J. L. 1980. *Oil and Gas Journal,* February 11, p. 122.

114. Watson, A. M. 1983. *Hydrocarbon Processing,* p. 91.

115. Troxler, W. L., Parmele, C. S., Barton, C. A., and Hobbs, F. D. 1983. "Survey of Industrial Applications of Vapor-Phase Activated Carbon Adsorption for Control of Pollutant Compounds from Manufacture of Organic Compounds," EPA-600/2-83-035, August (PB83-200618).

116. Parmele, C. S., O'Connell, W. L., and Basdekis, H. S. 1979. *Chemical Engineering,* December 31, p. 59.

117. Kenson, R. D. 1979. *Pollution Engineering,* July, p. 38.

118. Union Carbide Corporation. 1983. "Purasiv HR" Brochures F-48668A5M and F-48668A15M.

13

Catalytic Conversion of Gas Impurities

Catalytic processes used in gas purification differ from the techniques discussed so far in this text in that the objectionable impurities are not physically removed from the gas stream but, rather, converted to compounds which are either not objectionable and, therefore, may remain in the gas stream or which can subsequently be removed with greater ease than the compounds originally present. While, in the first case, the catalytic conversion step constitutes the entire purification operation, additional steps such as absorption or adsorption are required in the alternative case.

The field of chemical catalysis is extremely complex; and any attempt to discuss its theoretical aspects, even in the most elementary form, would go far beyond the scope of this book. Excellent textbooks have been written on the subject which should be consulted by readers who wish to acquire a more extensive knowledge of catalysis (1,2,3). As is well known, catalysis is based on the fact that the rate of certain chemical reactions is influenced by the presence of substances—called catalysts—which appear unchanged in the reaction products. Catalysts may increase (positive catalysis) or decrease the rate of reaction (negative catalysis), or direct the reaction along a specific path. Depending on the phase relationship between the catalyst and the initial reactants, catalytic processes may involve homogeneous and heterogeneous catalysis. Homogeneous catalysis is characterized by the fact that the catalyst and the reactants are in the same phase, while in heterogeneous catalysis the catalyst and the reactants are in different phases. Catalytic gas-purification processes utilize positive catalysis and, since the catalysts are generally solids, involve heterogeneous catalysis.

In heterogeneous catalysis the contact between the catalyst and the reactants at the phase boundary is of utmost importance. This contact is established by adsorption of the reactants on the catalyst surface, resulting in increased concentration of the reactants. Adsorption takes place on the catalyst surface in localized areas of especially high attractive power, called "active centers." The mechanism of adsorption of gases on solids is discussed in some detail in Chapter 12. The overall mechanism of heterogeneous catalysis involves (a) mass transfer of the reactants from the fluid to the catalytic surface and of reaction products from the catalytic surface to the fluid; (b) diffusion of reactants and reaction products into and out of the pores of the catalyst; (c) activated adsorption of reactants and desorption of reaction products at the interphase; and (d) surface reactions of adsorbed reactants to form chemically adsorbed products. Considering that each of these steps is quite complex in itself, the highly complicated nature of the overall mechanism can be appreciated.

Catalysts used in most gas-purification processes are metal salts or metals, usually supported on an inert carrier of large surface area. However, unsupported catalysts are also used. Typical carriers are alumina, bauxite, asbestos, china clay, activated carbon, and metal wires. In cases where very highly active catalysts are required, the catalytic surface is "activated" by special procedures. It is not uncommon to use two or more catalytic materials in one preparation because the activity of one catalytic material may frequently be increased beyond that expected by the simple additive effect of the additional components. Additives of this type are known as promoters.

Since, as pointed out above, the catalysts are not changed by the reaction, they are, ideally, usable for indefinite periods of time. In practice, however, most catalysts deteriorate or are gradually deactivated during operation and have to be regenerated or replaced periodically. Catalyst deterioration may be physical or chemical. Physical deterioration is caused by mechanical attrition or overheating and sintering of the catalysts. Attrition causes excessive entrainment losses and high pressure drop through the catalyst beds, while sintering results in change of the catalyst surface and rapid loss of activity. Chemical deterioration is primarily the result of chemical reactions between the catalytic materials and substances present in the feed stream with formation of stable reaction products. Another cause is the accumulation of heavy compounds which are either present in the gas stream or produced by side reactions during the process. In both cases, the number of active centers on the catalytic surface is diminished and the activity of the catalyst reduced. Deactivation by reaction with impurities present in the gas stream is usually called catalyst poisoning. Gradual catalyst deactivation by accumulation of deposits on the surface and, consequently, the need for periodic regeneration, is normal in processes where one or more reaction products are nonvolatile. For example, carbon is formed in the conversion of carbon disulfide to hydrogen sulfide by hydrogenation, and periodic removal of carbon from the catalyst is required.

Figure 13-1. Typical commercial catalysts. *United Catalysts Inc.*

Commercial catalysts not only must have the required activity and resistance to catalyst poisons but also must be rugged, especially if used in continuous-flow processes. Furthermore, it is important to prepare the catalysts in such shapes and sizes that minimum pressure drop through the bed is incurred. Typical commercially used catalysts are shown in Figure 13-1.

A typical catalytic gas-purification installation consists of a contacting or reaction vessel, often called the converter, in which the catalyst is arranged in single or multiple fixed beds, in tubes or in specially constructed containers. The size of the converter is governed primarily by the space-velocity required for a given reaction. Space velocity is an indirect measure of the contact time between the gas and the catalyst. In general it is defined as the volume of gas at standard conditions passing through a unit volume of catalyst per unit of time. However space velocity may also be expressed in terms of actual flowing gas volume per volume of catalyst per unit of time. Since the objectionable impurities to be removed from gas streams are usually present in low concentrations, the temperature rise in the converter caused by exothermic reactions is rather small and installation of internal cooling coils is normally not necessary. The heat required for endothermic reactions is generally supplied by preheating the gas before it enters the converter. In many cases, preheating of the gas to the temperature required for the reaction to occur is necessary. The volume of the catalyst bed through which the gas passes is determined by the required space-velocity. Attrition of the catalyst and bed-stability conditions must be considered when selecting the linear gas velocity through the bed, which in turn determines the bed-depth and cross-sectional area. Larger areas, with attendant lower gas velocities, are indicated when pressure drop through the catalyst bed is an important consideration. Correlations for the calculation of pressure drop through granular materials are given in Chapter 12.

A mechanical design-feature which may require consideration is the ease of quick removal of spent catalyst and its replacement with fresh material. This is of importance in installations in which regeneration is not carried out in the reaction vessel or in which the catalyst is not regenerated at all. Since most catalytic gas-purification processes are operated at high temperatures, the plants also contain heating, cooling, and heat-exchange equipment. Equipment for the removal of reaction products is also provided in many installations. Finally, some processes require separate catalyst-regeneration facilities.

The most logical classification of catalytic-conversion processes would be on the basis of the chemical reaction involved, i.e., oxidation, hydrogenation, and hydrolysis. However, such a clear division is not always possible because in some processes several types of reactions occur simultaneously, and, at times, it is difficult to decide which reaction is predominant. It has, therefore, become customary to identify processes either by the compounds to be removed or by the nature of the chemical reaction, and this (admittedly unsystematic) method is used in the following presentation. The most important catalytic gas-purification processes used commercially are (a) conversion of organic and inorganic (i.e. sulfur dioxide) sulfur compounds present in flue, refinery, synthesis, and Claus unit tail gases to hydrogen sulfide or oxides of sulfur; (b) removal of carbon monoxide from synthesis, fuel, or inert gases by conversion to carbon dioxide or methane; (c) conversion of acetylene contained in olefinic-gas streams to ethylene by selective hydrogenation; and, finally, (d) oxidation and reduction of a multitude of objectionable organic and inorganic compounds present in industrial exhaust gases. Processes involving catalytic oxidation of sulfur compounds, both hydrogen sulfide and organic derivatives, are discussed in Chapter 8 of this text because these processes are closely related to the iron oxide process and, therefore, fall more logically into the category of dry oxidative processes for sulfur removal.

Catalytic conversion of sulfur dioxide as applied to processes for the removal of sulfur dioxide from power-plant and other industrial exhaust gases is discussed in detail in Chapter 7.

Catalytic gas-purification processes are becoming increasingly more important as more stringent regulations for air pollution prevention are enacted. Because of their relative simplicity and suitability for handling large volumes of gas containing low concentrations of pollutants at high temperatures and low pressures, such processes are, in many cases, considerably more economical than absorption or adsorption techniques for removing undesirable compounds from exhaust gas streams.

The following presentation covers the description of the most important industrially used processes without going into a detailed discussion of the characteristics of the many commercially available catalysts. Such information is readily available from catalyst manufacturers.

Conversion of Organic Sulfur Compounds to Hydrogen Sulfide

The principal organic sulfur compounds present in industrial gas streams are carbonyl sulfide, carbon disulfide, mercaptans of low molecular weight, and thiophene. Natural and refinery gases contain primarily mercaptans and, in some instances, traces of carbonyl sulfide and thiophene; but carbonyl sulfide, carbon disulfide, and thiophene are the principal organic sulfur compounds found in coal and synthesis gases. The concentrations of these compounds vary over a considerable range and, in some cases, depend on prior processing-steps to which the gas has been subjected. For example, the mercaptan content of natural gas may vary from less than 1 grain/100 cu ft to 5 to 10 grains/100 cu ft, depending on the extent of ethane, propane, and butane removal in the preceding natural-gasoline-recovery process. Total organic sulfur concentrations in coal and synthesis gases range typically from 20 to 50 grains/100 cu ft with carbonyl sulfide and carbon disulfide as the major constituents.

The organic sulfur compounds are much less chemically reactive than hydrogen sulfide and, therefore, are either not at all or incompletely removed in conventional H_2S-removal processes. Some absorption, adsorption, and oxidation processes used for H_2S removal are capable of partially eliminating organic sulfur (see Chapters 5, 8, 9, and 14) but, in general, catalytic conversion at high temperatures is used to eliminate organic sulfur compounds from most gas streams. Most catalytic organic-sulfur-removal processes require an inlet gas essentially free of hydrogen sulfide. However, certain catalysts are able to tolerate appreciable quantities of H_2S and other sulfur compounds in the gas to be treated without loss of activity. Such catalysts are of economic significance in the purification of synthesis gases where removal of H_2S by conventional processes prior to organic sulfur conversion necessitates cooling and reheating of the synthesis gas.

Another application which requires catalysts capable of functioning in the presence of hydrogen sulfide and sulfur dioxide is the purification of Claus unit tail gases which is discussed later in this chapter.

In most countries no legal limitations for the organic sulfur content of domestic and industrial fuel gases are in existence, and, consequently, little effort is made to remove these compounds. However, certain applications do require a lower total sulfur content of fuel gases than is obtained by removing only hydrogen sulfide. In these instances, at least partial removal of organic sulfur compounds is necessary. Synthesis gases used in catalytic processes usually require almost complete absence of all sulfur compounds because many synthesis catalysts are extremely sensitive to sulfur poisoning. Consequently, two types of organic-sulfur-removal processes have been developed, depending on the ultimate use of the purified gas. In the first case, the organic sulfur content of fuel gases (mostly coal gases) is reduced sufficiently to make them suitable for use in such applications as flueless heating appliances, direct-fired pottery kilns, heat-treating processes of nonferrous metals, and related processes. When fuel gases containing relatively high concentrations of sulfur are used in such service, the resulting high concentrations of sulfur dioxide in the flue gas cannot be tolerated because (a) irritation of the eyes and the respiratory organs of the workers is caused, and (b) undesirable properties are imparted to the materials treated in the processes. However, a total sulfur content of 3 to 10 grains/100 cu ft is satisfactory for practically all such applications. A number of catalytic-purification processes of this type have been in commercial operation for several years. The efficiency of total organic sulfur removal achieved in these processes depends primarily on the sulfur compounds present in the feed-gas stream. Carbonyl sulfide, carbon disulfide, and mercaptans are relatively easily converted to hydrogen sulfide in the presence of certain catalysts, but thiophene is very resistant to destructive hydrogenation under the operating conditions of processes of this type.

The second type of process is intended for essentially complete removal of organic sulfur compounds from industrial gases, especially synthesis gases. Because of the extremely low total-sulfur concentrations required in gases used for many catalytic syntheses—sometimes less than 0.01 grain/100 cu ft—these processes should, ideally, be capable of completely converting not only carbonyl sulfide, carbon disulfide, and mercaptans, but also thiophene.

Basic Chemistry

The principal chemical reactions taking place in the catalytic conversion of organic sulfur compounds to hydrogen sulfide can be expressed by the following equations:

Hydrogenation reactions

$$CS_2 + 2H_2 \rightleftharpoons C + 2H_2S \tag{13-1}$$

$$COS + H_2 \rightleftharpoons CO + H_2S \tag{13-2}$$

$$RCH_2SH + H_2 \rightleftharpoons RCH_3 + H_2S \tag{13-3}$$

$$C_4H_4S + 4H_2 \rightleftharpoons C_4H_{10} + H_2S \tag{13-4}$$

Hydrolysis (in the presence of water vapor)

$$CS_2 + 2H_2O \rightleftharpoons CO_2 + 2H_2S \tag{13-5}$$

$$COS + H_2O \rightleftharpoons CO_2 + H_2S \tag{13-6}$$

Besides the reactions shown in Equations 13-1 to 13-6, a number of other reactions, involving oxygen, hydrogen, hydrogen sulfide, hydrocarbons, nitrogen compounds, and carbon monoxide may also take place, depending on the feed-gas composition and the operating conditions. For example, hydrogenation of carbon disulfide may be represented by the following mechanism:

$$CS_2 + 4H_2 \rightleftharpoons 2H_2S + CH_4 \tag{13-7}$$

$$CS_2 + H_2O + H_2 \rightleftharpoons CO + 2H_2S \tag{13-8}$$

Side reactions occurring in the treatment of coal gas with different catalysts have been investigated by Key and Eastwood (4). These authors found that significant amounts of mercaptans may be formed from hydrogen sulfide and unsaturated hydrocarbons, depending on the temperature and catalysts used. Wedgwood (5) presents a detailed study of side reactions occurring in a plant using the Holmes-Maxted process which is described later in this chapter. Side reactions resulting in the formation of polymeric materials, which deactivate the catalyst, are especially undesirable.

Equilibrium constants for the hydrolysis of carbon disulfide according to Equation 13-5 have been determined experimentally by Terres and Wesemann (6). Since the reaction takes place in two steps, the equilibrium constants for the two intermediate reactions—shown in Equations 13-9 and 13-10—were determined, and the constant for the overall reaction was obtained by simple multiplication.

$$H_2S + CO_2 \rightleftharpoons H_2O + COS \tag{13-9}$$

$$COS + H_2S \rightleftharpoons CS_2 + H_2O \tag{13-10}$$

As can be seen, the reaction in Equation 13-9 is the same as that in Equation 13-6, which represents the hydrolysis of carbonyl sulfide. The first reaction, Equation 13-9, was studied within the temperature range of 350° to 600°C (662° to 1,112°F) and the second reaction, Equation 13-10, at 700° to 900°C (1,292° to 1,652°F). The experimental data were extrapolated over the range of 20° to 1,000°C (68° to 1,832°F) by conventional thermodynamic calculations. The equilibrium constants thus obtained for the reactions in Equations 13-5 and 13-6 and the opposite of the reaction in Equation 13-10 are shown in Table 13-1.

Key and Eastwood (4) determined equilibrium constants for the hydrogenation of carbonyl sulfide (Equation 13-2) at 300°C (572°F) and 600°C (1,112°F). A value of 15.0 was found for the equilibrium constant at both temperatures. Although these authors do not claim high accuracy for their experiments, calculation of the constants for the hydrolysis reaction (Equation 13-6), from the experimental values and accepted values for the equilibrium constants of the water-gas reaction, gives good agreement with the experimental data of Terres and Wesemann.

The equilibrium constant for other reactions can be calculated with fairly good accuracy from thermodynamic data such as those presented by Kelley (7). Calculated equilibrium constants for the reactions in Equations 13-1 and 13-2 are shown in Table 13-2.

The Carpenter-Evans Process

The first catalytic process used commercially for the removal of organic sulfur compounds from manufactured gas was developed in England by Carpenter and Evans (8,9,10). The catalyst used in this process is a sulfide of nickel, shown by Evans and Stanier (11) to be nickel subsulfide of the formula Ni_3S_2. This catalyst is prepared by soaking broken firebrick in a solution of nickel chloride and then heating the firebrick to 930°F. The nickel oxide thus obtained is subsequently sulfided with the hydrogen sulfide contained in crude coal gas. The catalyst is most active at 800° to 850°F and loses its effectiveness quite rapidly at lower temperatures. The basic reaction occurring in the process is hydrogenation of the organic sulfur compounds, primarily carbon disulfide, to hydrogen sulfide. Since hydrogenation of carbon disulfide yields carbon, which is deposited on the catalyst surface, periodic regeneration of the catalyst is required. The presence of appreciable quantities of hydrogen sulfide in the inlet gas decreases the efficiency of organic sulfur conversion (10).

The process operates at essentially atmospheric pressure and temperatures of 790° to 840°F. The operation begins with preheating the gas, first by heat exchange with product gas and then in a coke-fired heater, almost to reaction temperature before admission to the catalyst chamber. After passage through the catalyst, which is contained in steel tubes, the gas is heat-exchanged with the inlet gas and cooled, and the hydrogen sulfide formed in the converter is

Table 13-1. Equilibrium Constants for the Hydrolysis of Carbonyl Sulfide and Carbon Disulfide

Temperature, °C	$K_1 = \dfrac{(H_2S)(CO_2)}{(COS)(H_2O)}$	$K_2 = \dfrac{(COS)(H_2S)}{(CS_2)(H_2O)}$	$K_3 = \dfrac{(H_2S)^2(CO_2)}{(CS_2)(H_2O)^2}$
20	7.25×10^5	7.15×10^9	5.21×10^{15}
100	3.16×10^4	3.84×10^7	1.21×10^{12}
200	2.75×10^3	7.15×10^5	1.91×10^9
300	5.55×10^2	5.28×10^4	2.9×10^7
400	1.85×10^2	8.33×10^3	1.5×10^6
500	81.0	2.22×10^3	1.8×10^5
600	43.1	7.94×10^2	3.4×10^4
700	26.1	3.45×10^2	9.2×10^3
800	17.3	1.74×10^2	3.0×10^3
900	12.3	100.0	1.2×10^3
1000	4.5	63.0	2.8×10^2

Source: Data of Terres and Wesemann (6)

Table 13-2. Equilibrium Constants for the Hydrogenation of Carbonyl Sulfide and Carbon Disulfide

Temperature		$K = \dfrac{(H_2S)(CO)}{(COS)(H_2)}$	$K = \dfrac{(H_2S)^2(C)}{(CS_2)(H_2)^2}$
°K	°C		
500	227	7.55	6.99×10^{11}
600	327	10.9	1.03×10^9
700	427	13.8	1.27×10^7
800	527	16.0	2.72×10^5
900	627	17.8	1.68×10^4
1000	727	19.0	1.81×10^3

Source: Calculated from data of Kelly (7)

removed in iron oxide purifiers. Regeneration of the catalyst is accomplished by contact with air at a temperature somewhat below that used during conversion. Operating data given by Evans (10) indicate a typical organic-sulfur-removal efficiency of about 80 percent (from 35 to 7.5 grains/100 cu ft) and a catalyst life, between regenerations, of 30 to 35 days.

Three commercial plants with daily capacities of 2, 10, and 15 million cu ft of gas, respectively, have been operated for a considerable length of time.

However, the process has not been accepted widely, primarily because of its high fuel and maintenance costs.

A modification of the Carpenter-Evans process, used for the removal of small amounts of organic sulfur compounds (8 grains/100 cu ft) from coal gas, is described by Trutnovski (12). In this process nickel turnings are used as the catalyst. Approximately 90 percent conversion of carbon disulfide to hydrogen sulfide is claimed at operating temperatures of about 860°F and a space-velocity of 2,000 volumes per volume per hour. The catalyst is regenerated periodically by air blowing.

The Peoples Gas Company Process

This process was developed by the Peoples Gas Company of Glassboro, New Jersey, for removing a sufficient amount of organic sulfur compounds from city gas to make it suitable for use in a glass-annealing furnace at a New Jersey glass factory. The process, which is described by Menerey (13) and Darlington (14), employs a catalyst consisting of magnesium sulfate and zinc oxide and is stated to be capable of reducing the organic sulfur content of city gas from 15 grains to less than 3 grains/100 cu ft.

In the installation described, the gas to be purified is first preheated to about 800°F in a preheater which is located in the annealing furnace. It then passes downward through four layers of catalyst, each 1 ft thick. The catalyst vessel consists of brick-lined, 24-in. standard steel pipe insulated on the outside. The gas, which emerges from the bottom of the catalyst vessel at about 650°F, is cooled to 100° to 150°F in an air-cooled heat exchanger and finally freed of H_2S in an iron oxide box. The catalyst is regenerated weekly by circulating air through the vessel for 12 hr at the rate of 8 cfm. Replacement of catalyst is usually required after six months of operation. The plant operates at a pressure of 10 psig and has an average capacity of 2,500 cu ft gas/hr, with a maximum of about 4,000 cu ft/hr. The entire plant is constructed of carbon steel, with the exception of stainless-steel catalyst trays and supports.

The Holmes-Maxted Process

The most extensively used commercial process for removing organic sulfur compounds from coal-derived fuel gases was developed by W.C. Holmes & Company, Ltd., of Huddersfield, England, and is known as the Holmes-Maxted process (15). This process, which has been described in some detail (5, 16-21), employs a catalyst consisting of a metal thiomolybdate which promotes conversion of organic sulfur compounds to hydrogen sulfide by hydrogenation. The process has been used in more than 50 installations ranging in capacity from 5,000 to 4 million cu ft gas/day (22).

Process Description

The catalyst used in the Holmes-Maxted process consists of a normal thiomolybdate: typically copper, iron, zinc, cobalt, or nickel thiomolybdate. Its normal composition in the case of a bivalent metal, M, can be represented by the formula $MMoS_4$. The composition of this compound, especially its sulfur content, is subject to variation in the presence of gaseous sulfur compounds and other constituents of coal gas at high temperatures, and its ability to establish a kinetic sulfur balance under these conditions is presumed to be a major factor in determining its catalytic activity (16).

A commercial catalyst, e.g., copper thiomolybdate, is prepared by dissolving molybdenum trioxide in a 25 percent aqueous ammonia solution to which copper sulfate has been added. The pale-blue precipitate of copper molybdate is redissolved by addition of a slight excess of ammonia, and this solution is then used for the impregnation of the catalyst support. Granular bauxite of ⅛- to ¼-in. diameter is used as the carrier material. After impregnation, the bauxite is heated to approximately 750°F in order to drive off excess ammonia and to decompose ammonium sulfate. The catalyst is subsequently converted to copper thiomolybdate by treatment with hydrogen sulfide—containing gas at temperatures between 570° and 750°F (16).

A diagrammatic flow sheet of the Holmes-Maxted process, as practiced in a large industrial installation (5), is shown in Figure 13-2.

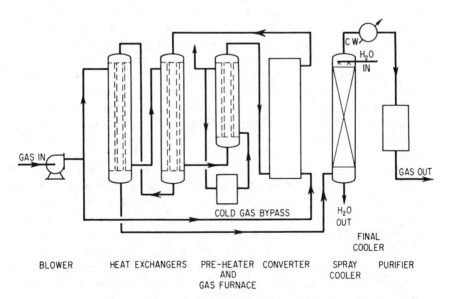

Figure 13-2. Diagrammatic flow sheet of Holmes-Maxted process. *W.C. Holmes & Co., Ltd.*

Hydrogen sulfide-free coal gas coming from iron oxide purifiers flows to a blower and from there to shell-and-tube heat exchangers where its temperature is raised to approximately 500°F by exchange with the treated gas. From the heat exchanger the gas passes to a preheater where it is further heated (to about 570° to 645°F) and then enters the bottom of the converter which contains the catalyst in an annular bed, 12 in. thick, with a central gas passage. The preheater is a shell-and-tube heat exchanger where the coal gas flows through the tubes and combustion gases from a gas-fired furnace flow through the shell. Heat is released in the converter by the hydrogenation of organic sulfur compounds and by the reaction of oxygen with hydrogen. In order to avoid overheating the catalyst, in cases where the gas contains substantial amounts of oxygen, a portion of the gas may be bypassed around the heat exchangers and preheater and fed directly into the converter. After passing twice through the catalyst the gas leaves the top of the converter and flows through the heat exchangers, a spray cooler where it is cooled by direct contact with an aqueous solution of sodium carbonate, a water-cooled condenser for final cooling and, finally, through iron oxide boxes for removal of the hydrogen sulfide formed in the process. A diagram of the converter is shown in Figure 13-3.

Catalyst is removed semicontinuously from the bottom of the converter through a gas lock—at the rate of 1 cu ft/MM cu ft of gas—and fresh catalyst is added to the top of the vessel through a feed hopper. The spent catalyst, which contains about 10 percent carbon, is regenerated in a separate vessel with a mixture of air and flue gas, containing a maximum of 5 percent oxygen, at 480° to 660°F. Temperature control during regeneration is necessary because the catalyst is permanently damaged if the temperature exceeds 1,100°F. The regenerator as described by Priestley (19) consists of a cylindrical vessel with a perforated conical bottom, containing the catalyst in a continuous bed. Spent catalyst is added to the top of the vessel, and the hot air-flue gas mixture is blown through the perforations of the cone at a rate of about 10 cu ft/lb of catalyst. Regenerated catalyst is withdrawn periodically from the bottom of the cone (through a valve) without interruption of the operation.

Different arrangements of process equipment may be used, mainly for reasons of plant compactness. The heat exchanger, preheater, and reactor may be placed in one vessel, and in very small installations, the heat exchanger may be completely omitted. An installation using a combined heat exchanger, preheater, and converter is described by Priestley (19). The rather complicated flow scheme—designed for maximum heat recovery—can be followed on the diagram shown in Figure 13-4. The external heat required for startup and, in some cases, during operation is supplied by burners located in a combustion chamber surrounding the converter.

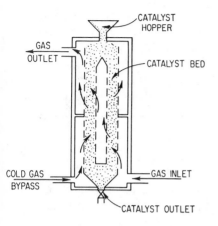

Figure 13-3. Diagram of converter, Holmes-Maxted process. *W.C. Holmes & Co., Ltd.*

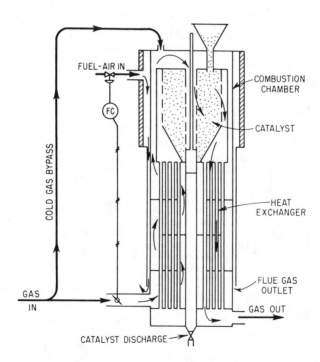

Figure 13-4. Diagram of combination converter—heat-exchanger—preheater, Holmes-Maxted process. *W.C. Holmes & Co., Ltd.*

Since all the oxygen contained in the inlet gas is consumed in the converter, addition of air to the treated gas ahead of the iron oxide beds is usually necessary in order to obtain continuous regeneration *in situ*. This, however, can be avoided by providing two iron oxide beds, one before and the other after the converter, as shown in Figure 13-5. By alternating the pattern of gas flow, one oxide bed is regenerated by oxygen in the feed gas while the other is removing H_2S from the treated gas. Figures 13-6 and 13-7 show photographs of commercial plants.

Process Design and Operation

The conversion efficiency of the catalyst is unaffected by fairly large variations in space-velocity. Maxted and Priestley (16) investigated space velocities ranging from 1,000 to 20,000 volumes per volume per hour with copper, tin, zinc, and cobalt thiomolybdate catalysts. The organic sulfur content of the treated gas was found to be the same during initial operation for all space velocities and to increase after passage of a given volume of gas. This indicates that the activity of the catalyst within this range of space velocities is entirely a function of the amount of carbon deposited on the surface. Space velocities of 2,000 volumes per volume per hour are practical for large plants and up to 5,000 volumes per volume per hour for small installations.

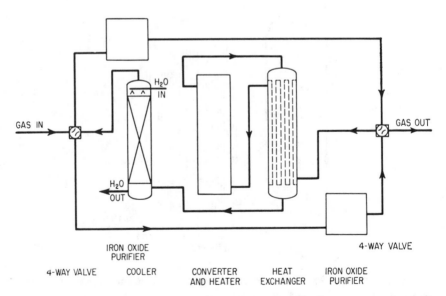

Figure 13-5. Flow diagram of Holmes-Maxted process with alternate operation and regeneration of iron oxide box. *W.C. Holmes & Co., Ltd.*

Figure 13-6. Medium-sized Holmes-Maxted plant. *W.C. Holmes & Co., Ltd.*

Figure 13-7. Large Holmes-Maxted plant. *W.C. Holmes & Co, Ltd.*

The process operates at essentially atmospheric pressure (1 to 2 psig) and within a temperature range of 580° to 720°F, with an optimum around 650°F. No benefit is obtained from operating at higher pressures. The reaction temperature is maintained either by an external heat supply or, if the gas contains sufficient oxygen, by the heat of reaction liberated during conversion. As stated earlier, the converter temperature may be controlled by bypassing the heat exchangers with a portion of the inlet-gas stream.

The operation of the process is unaffected by the presence of water vapor (up to 50 percent of the gas volume) and ammonia (up to 700 grains/100 cu ft) in the inlet gas. Hydrogen sulfide in concentrations up to 200 grains/100 cu ft reportedly has no detrimental effect on the efficiency of the catalyst. However, at higher H_2S concentrations and at temperatures above 700°F, carbonyl sulfide is synthesized from carbon monoxide and H_2S, resulting in less complete organic-sulfur conversion (15).

Results from the continuous operation (for 1 year) of a plant treating about 500,000 cu ft/day of coal gas with a copper thiomolybdate catalyst were reported by Priestley (19) and are shown in Table 13-3. Removal efficiencies for individual sulfur compounds in the same installation are given in Table 13-4. These data show that the total organic-sulfur-removal efficiency of the process depends primarily on the type of sulfur compounds present in the gas. The plant has been operated with gas containing as much as 100 grains H_2S/100 cu ft without noticeable effect on the organic-sulfur-removal efficiency.

Wedgwood (5) reports some operating data from a plant treating 4 million cu ft/day of coal gas in two parallel units of equal capacity. In this installation, either the gas was heated to the reaction temperature of 590° to 680°F by the use of external heat in a preheater, or sufficient air or oxygen was added to the gas to maintain the proper converter temperature. On the basis of data obtained during a short test-run, Wedgwood concludes that temperature control by adding oxygen to the gas is preferable to either external heating or air addition, provided a source of low-cost oxygen is available.

The chemistry of the process was studied extensively by Wedgwood, who found that as many as 14 different reactions might occur during the catalytic conversion operation. An interesting result of this work is the discovery that significant amounts of ammonia are formed during conversion, presumably by direct synthesis from nitrogen and hydrogen as well as by reduction of nitric oxide. The ammonia present in the converter-outlet gas was found to promote growth of hydrogen sulfide-generating bacteria in the iron oxide boxes and thus led to slippage of hydrogen sulfide through the beds. Removal of ammonia from the gas stream ahead of the iron oxide boxes resulted in proper functioning of these units. This experience is contrary to normal operating practices with iron oxide boxes (see Chapter 8) where a small amount of ammonia is often deliberately maintained in the feed gas to ensure proper pH of the iron oxide.

Table 13-3. Typical Operating Results of Holmes-Maxted Process (19) (One year's operation)

Operating variables	
Total gas volume treated, cu ft	198,105,000
Average daily gas volume treated, cu ft	542,000
Fuel gas used for heating converter, cu ft	755,000
Fuel gas used for catalyst regeneration, cu ft	126,000
Catalyst discharged from converter, lb	12,310
Catalyst regenerated, lb	10,800
Catalyst temperature, °F	680–725
Space-velocity, vol/(vol)(hr)	1,300–2,000
Pressure drop through catalyst, in. H_2O	2.8
Organic sulfur, grains/100 SCF:	
Inlet	23.1 (average)
Outlet	6.7 (average)
Oxygen, %:	
Inlet	1.1 (average)
Outlet	0
Carbon on spent catalyst, %	7.5

Table 13-4. Typical Removal of Individual Organic Sulfur Compounds (19) (One year's operation)

Compound	Conc., grains/100 cu ft		Removal, %
	Inlet	Outlet	
Thiophene	4.5	3.6	20
Carbon disulfide	14.5	1.3	91
Carbonyl sulfide	4.1	1.8	56
Total organic sulfur	23.1	6.7	71

Organic sulfur removal obtained during Wedgwood's test-run is shown in Table 13-5. It is interesting to note that appreciably better removal was obtained than in the installation described by Priestley (see Table 13-4).

No expensive construction materials are required in a Holmes-Maxted plant. Carbon steel is satisfactory for all components with the exception of the catalyst baskets; these are constructed from either cast iron or aluminized carbon steel.

British Gas Council* Processes

Key and Eastwood (4) describe two catalysts for organic sulfur removal from water and coal gases which were developed in a joint study of the British

*Now British Gas Corporation

Table 13-5. Removal of Individual Organic Sulfur Compounds (5)
(Short plant test)

| Compound | Concentration, grains/100 cu ft | | | | Removal, % | |
| | Inlet | | Outlet | | | |
	Maximum	Minimum	Maximum	Minimum	Maximum	Minimum
Thiophene.............	3.2	2.8	3.1	1.6	43	3
Carbon disulfide.........	16.3	10.0	0.8	0.6	96	95
Carbonyl sulfide.........	18.8	12.7	1.8	0.6	95	91
Mercaptans.............	1.3	0.7	0.2	0	100	85
Total organic sulfur......	39.8	26.2	5.9	2.8	89	85

Gas Council and the University of Leeds. A composition consisting of a mixture of copper sulfide and chromium oxide, supported on active carbon, was found to be quite effective for practically complete removal of organic sulfur compounds (carbonyl sulfide and carbon disulfide) from water gas. This catalyst, which promotes primarily the hydrolysis of carbonyl sulfide, is reported to be capable of removing 98 percent of the organic sulfur—with the concentration dropping from 15 to 0.3 grain/100 cu ft—at a temperature of 250°C (482°F) and space velocities up to 6,000 volumes per volume per hour. The activity of the catalyst appears to be unimpaired by the presence of relatively large amounts of hydrogen sulfide in the inlet gas (250 grains/100 cu ft) and is definitely increased by the addition of steam. Removal of about 95 percent of organic sulfur compounds present in gas streams containing appreciable amounts of H_2S and 30 percent (by volume) of steam is reported when operating with a space velocity of 2,000 volumes per volume per hour. Multistage treatment of water gas, with hydrogen sulfide removal after each stage, results in much more complete removal of organic sulfur compounds. It is claimed that a three-stage process yields a purified gas containing less than 0.002 grain sulfur/100 cu ft (4).

The copper sulfide-chromium oxide catalyst is unfortunately poisoned by acetylenic compounds and, therefore, is not usable for treatment of coal gas, which always contains small amounts of such compounds. A process modification, applicable to coal-gas treatment, consists of pre-treatment of the gas with a nickel sulfide or molybdenum sulfide hydrogenation catalyst, supported on Activated Alumina, for conversion of acetylenic compounds, followed by treatment with the copper sulfide-chromium oxide preparation (4). This double treatment is claimed to be effective for quite complete conversion of organic sulfur compounds present in coal gas, with the exception of thiophene, which is hardly affected. The molybdenum sulfide hydrogenation catalyst alone is reportedly capable of promoting 85 to 90 percent conversion of organic sulfur compounds, besides hydrogenation of acetylene, at a temperature of 300°C (572°F), provided a space velocity of 2,-000 volumes per volume per hour is not exceeded (4). Although somewhat lower final sulfur contents in the treated gas are obtainable with the two-stage treatment, the ineffectiveness of both catalysts in removing thiophene makes the economic justification of the two-stage process doubtful.

Removal of carbonyl sulfide and carbon disulfide from Claus unit tail gases by catalytic conversion to carbon dioxide and elemental sulphur, according to Equations 13-11 and 13-12, has been proposed by Nicklin (23) of the British Gas Council's North Western Gas Board.

$$2COS + SO_2 \rightleftharpoons 2CO_2 + 3S \qquad (13\text{-}11)$$

$$CS_2 + SO_2 \rightleftharpoons CO_2 + 3S \qquad (13\text{-}12)$$

The catalysts used, containing thorium-uranium or molybdenum-uranium disposed as the oxides on active alumina, reportedly are specific for promoting Reactions 13-11 and 13-12 (23). Laboratory experiments show better than 90 percent conversion of carbonyl sulfide and carbon disulfide contained in a synthetic Claus unit tail gas at a temperature of 330°C and a space velocity of 3,340 cu ft gas/(cu ft catalyst)/(hr) (23). Kinetic data presented by Nicklin (23) indicate that close approach to equilibrium can be reached at temperatures as low as 200°C and space velocities of 2,000 and 1,000 for carbon disulfide and carbonyl sulfide, respectively.

Organic Sulfur Removal from Synthesis Gases

The principal organic sulfur compound present in synthesis gases obtained by partial oxidation of sulfur-bearing hydrocarbons is carbonyl sulfide which is readily converted to hydrogen sulfide, either by hydrogenation or by hydrolysis, in the presence of certain catalysts. Iron oxide catalysts are effective for simultaneous shift conversion and conversion of carbonyl sulfide to hydrogen sulfide while chromia-alumina and copper-chromia-alumina catalysts are used for selective hydrolysis of carbonyl sulfide in the presence of large amounts of carbon monoxide. In addition, catalysts containing copper, chromium, and vanadium oxides have been developed for removal of hydrogen sulfide and organic sulfur compounds from synthesis gases.

Iron Oxide Catalysts

Typical catalysts used in the purification of ammonia-synthesis gases, for simultaneous shift conversion and conversion of carbonyl sulfide to hydrogen sulfide, contain 5 to 15 percent chromic oxide, besides iron oxide and a small percent of inert binder material. These catalysts promote both hydrogenation and hydrolysis of carbonyl sulfide. The extent to which carbonyl sulfide is either hydrogenated or hydrolyzed is largely determined by the simultaneously occurring water-gas reaction. If the water-gas reaction is not at equilibrium, hydrogenation is predominant, while both reactions take place to about an equal extent if the water-gas reaction is at equilibrium.

Under typical operating conditions, involving temperatures of 650° to 950°F, pressures from near atmospheric to 450 psig, and space velocities from 400 to 4,000 volumes per volume per hour, a shift-converter outlet gas containing only a few ppm of carbonyl sulfide is obtained (24). Moderate quantities of H_2S present in the inlet gas have no detrimental effect on the activity of the catalyst.

Use of an iron oxide catalyst containing 6 to 7 percent chromic oxide, for the purification of essentially H_2S-free ammonia-synthesis gas in a German installation, has been reported by Holroyd (25). Although no analyses are given, it is stated that the treated gas was suitable for ammonia synthesis, indicating a residual sulfur content of only a few ppm. Sands, Wainwright, and Egleson of the U.S. Bureau of Mines (26) investigated the suitability of an iron oxide catalyst for the removal of organic sulfur from gases containing large amounts of carbon monoxide and hydrogen sulfide. The results from this study were quite disappointing, as it was found that the catalyst promotes the synthesis of organic sulfur compounds from CO and H_2S and that the organic sulfur content of the outlet gas is primarily determined by the CO content of the inlet gas.

Chromia-Alumina and Copper-Chromia-Alumina Catalysts

The chromia-alumina catalyst is used for the removal of carbonyl sulfide and carbon disulfide from synthesis gases containing large amounts of carbon monoxide. The catalyst promotes selective hydrolysis of the sulfur compounds with essentially no conversion of carbon monoxide and is unaffected by the presence of hydrogen sulfide in the inlet gas. Practically complete conversion of organic sulfur compounds is obtained at temperatures of 600° to 800°F, elevated pressures, and space velocities of 250 to 1,000 volumes per volume per hour (27). Representative operating conditions for carbonyl sulfide removal from a gas stream consisting of a mixture of hydrocarbons and carbon monoxide are shown in Table 13-6 (28).

A catalyst containing copper-chromia-alumina is useful for the same applications as the chromia-alumina catalyst. However, its activity is

**Table 13-6. Typical Operating Conditions for
Carbonyl Sulfide Conversion in Hydrocarbon-carbon
Monoxide Stream with Chromia-alumina Catalyst**

Temperature, °F...................	600
Pressure, psig.....................	117
Space-velocity, vol/(vol)(hr)........	1,000
Steam/gas ratio...................	1.0
Carbonyl sulfide, inlet, ppm..........	90.5
Carbonyl sulfide, outlet, ppm........	1.5

Source: Courtesy of United Catalysts Inc. (28)

**Table 13-7. Typical Operating Conditions for
Carbonyl Sulfide Conversion in Methanol-synthesis Gases
with Copper-chromia-alumina Catalyst**

Inlet temperature, °F..............	500–525
Pressure, psig....................	380
Space-velocity, vol/(vol)(hr).......	2,000
Carbonyl sulfide, inlet, ppm........	8,000
Carbonyl sulfide, outlet, ppm.......	10

Source: Courtesy of United Catalysts Inc. (24)

reported to be much higher, permitting operation at higher space velocities and lower temperatures (24). Typical operating conditions for carbonyl sulfide conversion in raw methanol-synthesis-gas streams using a copper-chromia-alumina catalyst are presented in Table 13-7.

Both the chromia-alumina and copper-chromia-alumina catalysts are also effective for removing organic sulfur compounds, mainly carbonyl sulfide, from carbon dioxide-gas streams used for dry-ice manufacture or urea synthesis. Typical operating conditions for a chromia-alumina catalyst in this service are shown in Table 13-8 (28).

Copper-Chromium-Vanadium Oxides (Huff Catalysts)

These catalysts, which were developed by Huff and Logan (29), promote conversion of organic sulfur compounds to hydrogen sulfide which is retained on the catalyst as metal sulfide, together with any hydrogen sulfide originally present in the gas. Thus, conversion and actual removal of sulfur compounds from the gas are effected in one operation. The catalyst has to be regenerated with air periodically to convert the metal sulfide back to the oxides.

Table 13-8. Typical Operating Conditions for Carbonyl Sulfide Removal from Carbon Dioxide with Chromia-alumina Catalyst
Temperature, °F.   600
Pressure, psig.   2
Space-velocity, vol/(vol)(hr).   1,500
Water vapor. . . . .   Saturated at ambient temperature
Carbonyl sulfide, inlet, ppm.   4
Carbonyl sulfide, outlet, ppm 0.5
Source: Courtesy of United Catalysts Inc. (28)

Removal of several organic sulfur compounds from synthesis gas with Huff catalysts has been investigated by Sands and coworkers (26). The catalysts consist of mixtures of copper, chromium, and vanadium oxides; in some cases, the vanadium is replaced by uranium. The catalysts are of four types: fused, precipitated, precipitated on a carrier, and simple mixed oxides. A typical composition, based on the pure-metal content, consists of 80 parts copper, 10 parts chromium, and 10 parts vanadium. The preparation of the various catalyst forms is described by Sands, Wainwright, and Egleson(26).

These catalysts appear to be very effective for the removal of organic sulfur compounds, especially carbon disulfide and mercaptans, from synthesis gas in the presence of hydrogen sulfide. However, thiophene is not decomposed and, if present in appreciable amounts, causes catalyst poisoning. Removal of 13 to 35 grains of organic sulfur (present as carbon disulfide) and 150 to 300 grains of hydrogen sulfide/100 cu ft to yield a concentration less than 0.1 grain of total sulfur/100 cu ft is reported at two operating temperatures (570° and 840°F) and a space velocity of 2,000 volumes per volume per hour (26). Under these conditions the capacity of the catalyst for sulfur is about 6 to 14 percent of its weight, depending on the degree of regeneration.

The catalyst is regenerated by contact with air at approximately 850°F and subsequent reduction with raw synthesis gas at a space velocity of 2,000 volumes per volume per hour.

A commercial process using the Huff catalysts would have the obvious advantage of simultaneous hydrogen sulfide and organic sulfur removal in one step. However, the cyclic operation and the consumption of synthesis gas required for reduction of the regenerated catalyst are definite disadvantages. In addition, the catalyst is ineffective for the removal of thiophene.

Organic Sulfur Removal from Hydrocarbon-gas Streams

Organic sulfur compounds such as carbonyl sulfide and alkyl mercaptans may be removed from hydrocarbon streams by catalytic hydrogenation according to Equations 13-2 and 13-3. Depending on the catalyst used, simultaneous hydrogenation of the unsaturated hydrocarbons also present in the gas stream or selective conversion of organic sulfur compounds may be obtained. Although several catalysts have been proposed for such applications, two types, a cobalt-molybdate and the copper-chromia-alumina catalyst mentioned above, are used most commonly in commercial operations.

Cobalt-Molybdate Catalyst

This catalyst, which contains about 12 to 13 percent of cobalt and molybdenum oxides, is primarily used for organic sulfur removal from hydrocarbon streams prior to steam-reforming of the hydrocarbons. It is effective for simultaneous hydrogenation of the organic sulfur compounds and unsaturated hydrocarbons at temperatures of 500° to 750°F, pressures above 100 psi and space velocities of 300 to 1,000 volumes per volume per hour. Typical commercial operating conditions are shown in Table 13-9.

Copper-Chromia-Alumina Catalyst

This catalyst is suitable for the selective hydrogenation of the carbonyl sulfide contained in olefinic hydrocarbon streams. A typical application is the removal of carbonyl sulfide from ethylene and propylene, used for the manufacture of synthetic alcohols, where hydrogenation of the olefins during the organic-sulfur-removal operation is undesirable. The reaction proceeds at temperatures of 400° to 600°F, pressures ranging from near atmospheric to several hundred psig and space velocities of 500 to 2,000 volumes per volume per hour (24). Another application of the catalyst is the selective conversion of carbonyl sulfide in refinery-gas streams. Typical operating conditions are given in Table 13-10.

Platinum Catalyst

This catalyst, which is composed of platinum supported on an active catalytic alumina base, is reported to be capable of promoting the hydrolysis of carbonyl sulfide at low temperatures (30). Its suggested applications include carbonyl sulfide removal from natural, refinery and partial oxidation synthesis gases. Laboratory studies indicate that this catalyst may be suitable for conversion of carbonyl sulfide and carbon disulfide to hydrogen sulfide in Claus unit tail gases (31). Typical operating conditions are given in Table 13-11.

The Beavon Sulfur Removal (BSR) Process

This process, which was developed jointly by Ralph M. Parsons Company and Union Oil Company of California, is specifically designed for essentially

Table 13-9. Typical Operating Conditions for Simultaneous Hydrogenation of Unsaturates and Carbonyl Sulfide with Cobalt-molybdate Catalyst

Inlet temperature, °F	550
Outlet temperature, °F	750
Pressure, psig	150
Space-velocity, vol/(vol)(hr)	300
Hydrogen added, vol/vol propylene	0.5
Inlet gas:	
Propane, mole %	88
Propylene, mole %	12
Carbonyl sulfide, ppm	20
Outlet gas:	
Propane, mole %	99.9
Propylene, mole %	0.1
Carbonyl sulfide, ppm	1

Source: Courtesy of United Catalysts Inc. (28)

Table 13-10. Typical Operating Conditions for Selective Conversion of Carbonyl Sulfide in Refinery-gas Stream

Inlet temperature, °F	600
Pressure, psig	350
Space-velocity, vol/(vol)(hr)	1,000
Moisture	saturated at 100°F
Inlet gas:	
H_2, mole %	21.5
CO, mole %	0.1
N_2, mole %	3.6
CO_2, mole %	0.1
O_2, mole %	0.8
CH_4, mole %	39.6
C_2H_4, mole %	9.3
C_2H_6, mole %	18.8
C_3H_6, mole %	4.2
C_3H_8, mole %	1.5
C_4H_8 and heavier, mole %	0.4
H_2S, ppm	700
· COS, ppm	70
Outlet gas:	
COS, ppm	5
Olefin hydrogenation	nil

Source: Courtesy of United Catalysts Inc. (24)

**Table 13-11. Typical Operating Conditions for Hydrolysis
of Carbonyl Sulfide with Platinum Catalyst**

Inlet temperature, °F	100-450 (normally 250-350)
Pressure, psig	25-500
Space velocity, vol/(vol) (hr)	2,000-4,000
Water requirement	Saturation at 150-200°F
Inlet gas, COS, ppm	1-50
Conversion, %	85-99

Source: United Catalysts Inc.   c.

complete removal of all sulfur compounds remaining in Claus unit tail gases. Although a well-designed Claus plant is capable of converting about 96 to 97 percent of the hydrogen sulfide contained in the feed to elemental sulfur (see Chapter 8), the tail gases from such units contain sufficiently high concentrations of sulfur compounds, such as hydrogen sulfide, sulfur dioxide, carbonyl sulfide, carbon disulfide, and sulfur vapor to require further purification in order to satisfy air pollution control regulations. The Beavon process is reportedly capable of reducing the total sulfur content of Claus unit tail gases to less than 250 ppm by volume (calculated as sulfur dioxide) and thus to attain an overall conversion of more than 99.9 percent of the hydrogen sulfide fed to the Claus unit (32, 33). The residual sulfur compounds consist almost entirely of carbonyl sulfide, with only traces of carbon disulfide and hydrogen sulfide. The effluent gas is practically odorless and can be vented directly to the atmosphere, obviating the need for incineration and the attendant consumption of fuel. Since the introduction of the process in 1973, over 50 units have been installed or are in various stages of design and construction. These units remove about 575 long tons of sulfur per day from tail gases of Claus plants processing more than 11,000 daily long tons of sulfur. Process improvements and operating experience have been reported by Andrews et al. (75), Fenton et al. (76), Beavon and Brocoff (77), and Kouzel et al. (78).

Process Description

The Beavon Sulfur Removal Process provides a means to overcome the limitations of the Claus process due to the thermodynamic equilibrium. Even in a four converter Claus unit the maximum conversion of hydrogen sulfide to sulfur attainable is only about 98 percent, leaving too much unconverted sulfur compounds in the exhaust gas for compliance with air pollution control requirements. Conversion could be increased if water, a reaction product, could be removed between catalytic stages, thus disturbing the equilibrium and driving the reaction to the right. However, all attempts to do so have so far failed because the condensate obtained is extremely corrosive and contains solid sulfur which deposits in lines, valves, and vessels.

In the Beavon Sulfur Removal Process, all sulfur compounds (other than hydrogen sulfide) contained in the tail gas are catalytically converted to hydrogen sulfide, which is subsequently removed by any convenient method. If complete removal of hydrogen sulfide is required, chemical absorption or oxidation in the liquid phase may be used. However, if partial removal is adequate, the effluent from the catalytic hydrogenation reactor may be treated, after cooling and condensation of the bulk of the water, either in a final Claus reactor or in a Selectox reactor, which may be followed by a Claus reactor (see following section). The overall conversions attainable by these two versions of the process range from 98 to more than 99 percent.

The conversion step which is carried out at elevated temperatures over a cobalt-molybdate catalyst involves hydrogenation and hydrolysis of sulfur compound according to Equations 13-1, 13-2, 13-5, 13-6, 13-13, and 13-14.

$$SO_2 + 3H_2 \rightleftharpoons H_2S + 2H_2O \qquad (13\text{-}13)$$

$$S_2 + 2H_2 \rightleftharpoons 2H_2S \qquad (13\text{-}14)$$

Although it is probable that carbonyl sulfide and carbon disulfide are converted primarily by hydrolysis, especially since the tail gas contains about 30 percent water vapor, it is conceivable that hydrogenation also takes place, although to a minor extent.

A schematic flow diagram of the process is presented in Figure 13-8. In this version of the process, the hydrogen sulfide formed in the catalytic step is removed by the Stretford or Unisulf process (see Chapter 9), both of which have been demonstrated to be very effective for reducing the hydrogen sulfide content of the hydrogenated gas to less than 10 ppm. Use of an aqueous methyldiethanolamine (MDEA) solution as a selective solvent (see Chapter 2) for hydrogen sulfide removal has been described by Meissner (84). With this solvent an overall conversion, including the Claus unit, of 99.9 percent of the H_2S fed to the Claus unit is attainable. However, the treated gas contains some residual H_2S which requires incineration before discharge to the atmosphere.

A photograph of a BSR process plant incorporating a Stretford unit for H_2S removal is shown in Figure 13-9.

By following the flow diagram, it is seen that the Claus plant tail gas is first heated to the temperature required for the catalytic reaction by adding a hot stream of gas resulting from partial combustion of hydrocarbon gas in a line burner. This gas not only supplies the necessary heat but also enough hydrogen to satisfy the hydrogen demand for the hydrogenation reactions.

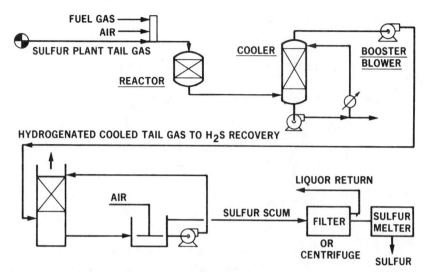

Figure 13-8. Typical process flow diagram, Beavon sulfur-removal process using Stretford process for H₂S removal.

After passing through the reactor, the gas is cooled to ambient temperature by direct contact with water. For better heat economy, especially in larger installations, a steam generator may be placed ahead of the direct contact water cooler. The cooled gas which contains primarily nitrogen, carbon dioxide, hydrogen sulfide, and traces of carbonyl sulfide is then treated for hydrogen sulfide removal. The water condensed from the gas in the direct contact cooler is stripped of hydrogen sulfide in a sour water stripper and then discarded. The stripped hydrogen sulfide is either fed to the hydrogen sulfide removal section or, when permissible, incinerated.

Typical results from plant operations and utilities' costs for a plant processing tail gas from a 100 long tons per day Claus Unit are shown in Table 13-12.

The Selectox Process

This process, which was also developed by the Union Oil Company of California and The Ralph M. Parsons Company, utilizes a proprietary catalyst—Selectox 33—for the oxidation of relatively low concentrations of hydrogen sulfide to elemental sulfur in a one-step operation. Three applications of the process are shown in Figure 13-10.

In the BSR/Selectox version, the process is used for hydrogen sulfide removal from Claus tail gas after hydrogenation in a BSR process hydrogenation section. About 99.5 percent overall sulfur recovery, including the Claus

Figure 13-9. Beavon Sulfur Removal Process Plant (The Ralph M. Parsons Company).

unit, is attainable. Even higher recovery can be achieved if the effluent from the Selectox reactor is treated in a final Claus stage.

The Once Through Selectox process is suitable for gas streams, for example geothermal off-gas, containing up to 5 percent hydrogen sulfide, and the Two Stage Selectox process with recycle can be used for gas streams containing more than 5 percent hydrogen sulfide. About 80 percent conversion of hydrogen sulfide to sulfur is reported for these two versions of the process (86).

The principal advantage of the Selectox process, if used in conjunction with the BSR process, is that substantial capital cost savings can be realized by replacing a Stretford system with a Selectox reactor. In addition, problems with liquid effluents are eliminated. The advantage of the other two versions of the process is that they are suitable for the treatment of very di-

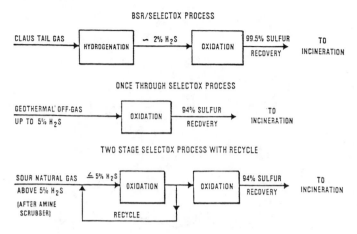

Figure 13-10. Selectox Applications (86).

Table 13-12. Beavon Sulfur Removal Process—Typical Plant Operating Results and Utilities Costs (78)

Component	Claus Unit Tail Gas	Gas From Stretford Absorber
H_2S, ppmv	3,000-6,000	1
SO_2, ppmv	1,500-3,000	Not Detectable
COS, ppmv	200-3,300	30-100
CS_2, ppmv	200-3,300	9-20

Commodity	Unit Cost	Consumption	Daily Cost ($)
Power	1.3¢/KW-Hr.	300 KW	94
Fuel Gas	$2.25/MMBTU	125 MMBTU/day	281
Soft Water	20¢/Gal.	10M Gal/day	2
Chemicals	—	—	75
Catalysts	—	—	8
Total Costs			460
Less Credit for 50 psig Steam	$2.50/M lb.	2,500 lb/hr.	(150)
Less Credit for Recovered Sulfur	$50/LT	4.9 LT/day	(245)
Net Cost			65
Net Cost/Ton of Sulfur Recovered from Tail Gas			13

Note: Utility costs for unit treating tail gas from 100 LT/day Claus unit operating at 95 percent conversion.

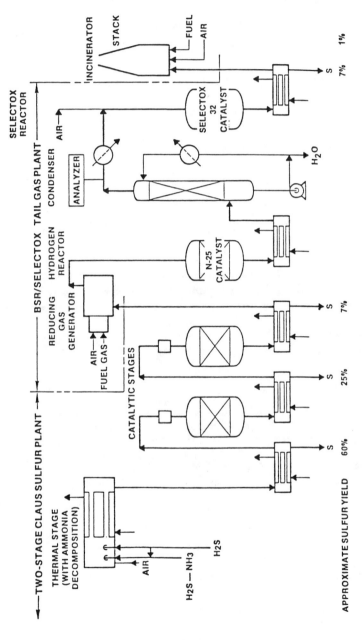

Figure 13-11. BSR/Selectox Process for Claus Plant Emission Control (86).

lute H_2S containing gas streams which cannot be processed in conventional Claus units. The operation is reported to be simple and reliable and the conversion remarkably high (86, 89).

Process Description

The process and its first commercial applications have been described by Beavon, Hass and Muke (85), Beavon et al. (86), Hass et al. (87, 88), Goar (89), and Warner (90). In the BSR/Selectox process, which is shown schematically in Figure 13-11, the Claus tail gas passes first through the hydrogenation reactor and a two-stage cooling step, where a substantial portion of the water vapor contained in the gas is condensed. After addition of a carefully controlled amount of air to the cooled gas, it enters the Selectox reactor where hydrogen sulfide is catalytically oxidized according to reaction 13-15:

$$H_2S + 1/2\ O_2 \rightleftarrows S + H_2O \tag{13-15}$$

The effluent is cooled and sulfur is condensed. When required, the cooled gas may be further processed in a final Claus stage. The gas leaving the Selectox (or final Claus) reactor is incinerated either thermally or catalytically, using Selectox catalyst, before discharge to the atmosphere. Typical gas composition changes in the course of the process are shown in Table 13-13.

Inspection of Table 13-13 shows that the Claus tail gas contains typically 30 to 35 percent water vapor. Since the Claus reaction is reversible, the high water vapor content of the Claus tail gas limits conversion of H_2S to sulfur. The cooled hydrogenated gas flowing to the Selectox reactor contains only about 1.5 to 4 percent water vapor, allowing the reaction to proceed much closer to completion in the Selectox reactor. About 80 to 90 percent of the H_2S entering the reactor is converted to sulfur, the conversion being limited only by the increase in temperature due to the heat of reaction (see Chapter 8).

The Selectox 33 catalyst is reported to be highly selective for the oxidation of hydrogen sulfide to sulfur, without formation of SO_3 and without oxidation of either hydrogen or of saturated hydrocarbons. It is claimed to be highly active and stable and to retain its activity over long periods of time, without regeneration, when operating at temperatures similar to those encountered in Claus reactors (85, 86).

The first commercial BSR/Selectox plant started operations in 1978 and reportedly is attaining consistent overall recovery efficiencies of 98.5 to 99.5 percent even though the preceding Claus unit recovery efficiency varies between 93 to 96 percent.

The Once Through Selectox process consists of the Selectox reactor and sulfur condenser shown in Figure 13-11. The allowable concentration of hy-

Table 13-13. Gas Composition Changes In BSR/Selectox Process (85)

Component	Unit	Claus tail gas	After hydrogenation	After cooling	After BSR/Selectox	After final Claus	After incineration
H_2S	ppmv	4,000–10,000	10,000–15,000	12,000–20,000	2,000–3,000	400–600	*1
SO_2	ppmv	3,000–6,000	*0	*0	1,000–1,500	200–300	1,000–1,500
COS	ppmv	300–5,000	10–30	15–40	15–40	15–40	*1
CS_2	ppmv	300–5,000	*0	*0	*0	*0	0
S†	ppmv	700–1,000	*0	*0	700–800	700–800	*1
H_2	vol %	1–3	2–3	3–4	2–3	2–3	0
CO	vol %	0.5–1	*0	*0	*0	*0	0
CO_2	vol %	1–15	1–15	1–20	1–20	1–20	1–15
H_2O	vol %	30–35	30–35	1.5–4	3–6	3–6	8–12
N_2	vol %	60–70	80–90	80–90	80–90	80–90	80–90
Cumulative percent of Claus feed recovered		93–96	93–96	93–96	98.5–99	99.4–99.6	99.4–99.6

*Approximate. †As S_1.

Reprinted with permission from *Oil & Gas Journal*, March 12, 1979, Copyright Pennwell Publishing Company drogen sulfide in the gas stream to be treated is limited by the fact that the entire heat of reaction is liberated in the reactor, resulting in excessively high temperature levels if the H_2S concentration exceeds about 5 percent. At the 5 percent level the reactor outlet temperature is about 700°F which is considered acceptable with respect to equilibrium and corrosion of carbon steel equipment (86).

The Two Stage Selectox Process with Recycle, which is reportedly suitable for the treatment of gas streams containing up to 40 percent H_2S (86), is shown schematically in Figure 13-12. In order to overcome the temperature effect due to heat of reaction, a portion of the effluent from the first sulfur condenser is recycled and mixed with the feed gas to the Selectox reactor. In this manner the H_2S concentration in the feed is adjusted to about 5 percent. The remaining portion of the gas leaving the first sulfur condenser flows to a Claus stage using either alumina or Selectox catalyst. About 82 percent of the hydrogen sulfide in the feed is recovered as sulfur after the Selectox reactor and an additional 12 percent is recovered after the Claus stage.

A commercial Recycle Selectox Process plant has been described by Goar (89). This unit is reported to recover about 20 LTPD of sulfur from an acid gas stream containing 13 percent H_2S (balance mostly CO_2) with a recovery efficiency of better than 95.0 percent.

Further variations of Selectox process including packaged units and three stages with recycle have been reported by Hass et al. (87).

The Shell Claus Off-Gas Treating (SCOT) Process

This process, which has been developed by Shell International Petroleum Maatschappij, The Hague, Netherlands, is similar in basic principle to the Beavon Sulfur Removal Process. It also relies on catalytic conversion of sulfur compounds, other than hydrogen sulfide, contained in the Claus tail gas to hydrogen sulfide which is subsequently removed. However, while the preferred method of hydrogen sulfide removal for the Beavon process is the Stretford, or similar liquid oxidation processes, the SCOT process makes use of selective absorption of hydrogen sulfide with only partial absorption of carbon dioxide in alkanolamine solutions. The acid gas is stripped from the amine solution and recycled to the Claus unit. Although it is reported that the process is capable of producing a purified gas stream containing less than 300 ppm of total sulfur (measured as sulfur dioxide after incineration) there is enough hydrogen sulfide left in the effluent from the amine absorber to require incineration before venting to the atmosphere. The process has been described, and operating experience has been reported by Groenendaal and Van Meurs (79), Naber et al (80), and Harvey and Verloop (81).

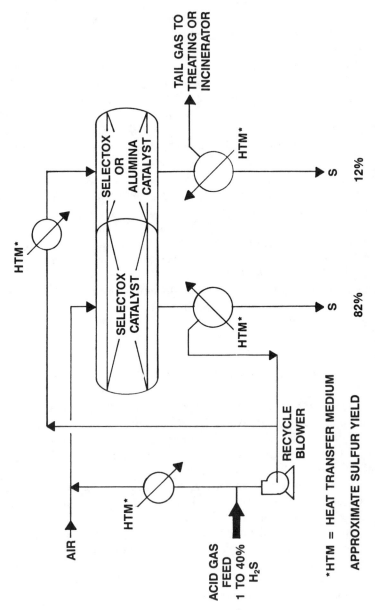

Figure 13-12. Two Stage Selectox Process with Recycle (86).

Process Description

A schematic flow diagram of the process is shown in Figure 13-13. The flow of gas and liquids in the process is quite similar to that of the Beavon Sulfur Removal Process. The Claus tail gas is first heated in a line gas heater which may have the double purpose of supplying reducing gas required in the subsequent catalytic step. However, reducing gas may also be furnished from an outside source. The hot gas then flows to the reactor where reduction of sulfur compounds occurs over a cobalt-molybdenum catalyst at about 300°C (572°F). The effluent from the reactor is subsequently cooled in two steps, first in a waste heat boiler where low pressure steam is produced, and then by water washing in a cooling tower. In this step, most of the water vapor contained in the gas is condensed. Excess water from the cooling tower, which contains a small amount of hydrogen sulfide, is treated in a sour water stripper where hydrogen sulfide is expelled and returned to the Claus unit.

After cooling, the gas enters the amine absorber where essentially all of the hydrogen sulfide is removed, but only a portion of the carbon dioxide is co-absorbed. The rich amine solution is stripped of acid gas in the regenerator by application of indirect heat supplied by a steam heated reboiler, and the acid gas is returned to the Claus unit. This portion of the process is quite similar to the conventional ethanolamine processes discussed in Chapters 2 and 3, with the exception that amines are used which are selective for hydrogen sulfide in the presence of carbon dioxide.

Selective absorption is based on the fact that the rate of absorption of hydrogen sulfide in alkanolamines is substantially more rapid than that of carbon dioxide. This phenomenon is more pronounced with secondary and tertiary than with primary amines (see Chapter 2). Consequently, appreciable selectivity may be attainable by proper selection of the amine and by design of the absorber providing for short contact time between the gas and the amine solution. Depending on gas composition, choice of amine and absorber design, co-absorption of carbon dioxide can be limited to about 10 to 40 percent of the carbon dioxide contained in the feed gas to the absorber (80). In most applications, especially in petroleum refineries where the Claus tail gas contains relatively small amounts of carbon dioxide, diisopropanolamine (DIPA) is the preferred solvent. However, for Claus tail gases containing higher concentrations of carbon dioxide, methyldiethanolamine (MDEA) is recommended (81).

The SCOT process is reported to be quite flexible and not very sensitive to upsets in the preceding Claus unit. Relatively wide variations in the H_2S/SO_2 ratio in the Claus plant as well as fluctuations in the feed gas volume can be tolerated in a properly designed SCOT unit.

A SCOT unit can be integrated with the desulfurization unit supplying the acid gas to the Claus plant by using the same amine, for example DIPA, in

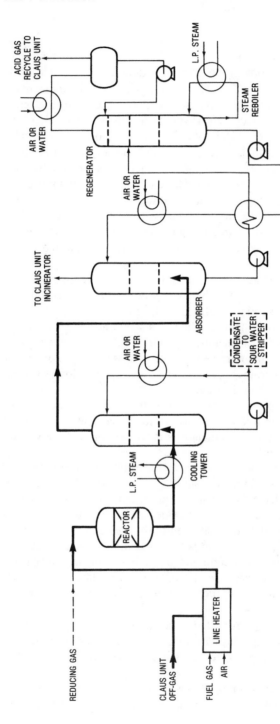

Figure 13-13. Schematic flow diagram, SCOT Process.

both the desulfurization and the SCOT unit, and stripping the combined rich solutions from the two absorbers in a single regenerator. This scheme is best applied in petroleum refineries where the ADIP process, employing DIPA as the active agent, is used for refinery gas desulfurization.

One rather important advantage of the SCOT process is the absence of liquid effluent which could cause pollution problems. Possible disadvantages are the use of appreciable amounts of fuel for amine regeneration and incineration of the purified gas, and recycle of hydrogen sulfide which results in an increase of about seven to eight percent in the load of the Claus unit (80).

Capital costs for SCOT units as reported by Harvey and Verloop (81) in 1976 are about the same as those for the preceding Claus unit if the SCOT unit is added to an existing Claus unit. The costs are lower if a common amine regenerator is used. Based on 1976 utilities' cost, operating costs reported by the same authors are about $16 per long ton of sulfur produced in the SCOT plant. In 1976, 14 SCOT units were in operation and 20 were being designed or under construction (81).

Design and operation of a rather large SCOT unit has been described by Herfkens (91). This unit treats tail gas from two Claus units with a combined capacity of 1,225 tons of sulfur per day and a required overall recovery of 98.5 percent. Aqueous MDEA is used for H_2S removal after hydrogenation and the acid gas is recycled to the Claus units. The MDEA unit produces a treated tail gas containing less than 500 ppmv of H_2S with only about 10 percent of the CO_2 being coabsorbed.

Complete Desulfurization of Hydrocarbon and Synthesis Gas Steams

Certain catalysts used for steam reforming of hydrocarbon feed stocks and shift conversion of reformer effluents are extremely sensitive to poisoning by sulfur compounds. Feed-gas streams for such operations must, therefore, be completely desulfurized, i.e., the concentration of total sulfur compounds has to be reduced to less than 0.1 part per million by volume.

The techniques available for complete sulfur removal include adsorption of sulfur compounds on activated or metal impregnated carbon at ambient temperature and reaction with zinc oxide at elevated temperatures. Various combinations of these processes as applied to the pretreatment of hydrocarbon feed stocks for steam-hydrocarbon reformers have been discussed by Livingston (35). Experience with desulfurization of ammonia synthesis gases produced from coal, and a comparison of cost and performance of activated carbon and zinc oxide has been reported by Allen (92). Additional data on typical applications, operating conditions, and adsorbent properties are available from catalyst manufacturers (36, 37, 93).

Activated or metal-impregnated carbons are used for removal of small quantities—about 20 ppmv—of sulfur compounds, primarily hydrogen sulfide. If carbonyl sulfide, organic sulfides and disulfides are present in relatively high concentrations, carbon is not recommended because, at the required operating temperatures (40 to 120°F), the organic sulfur compounds do not hydrolyze to hydrogen sulfide which is necessary for maximum removal (92). Adsorption is carried out at temperatures ranging from 60° to 120°F and at pressures from atmospheric to several hundred psig. Depending on the total sulfur content of the feed gas and the type of carbon used, carbon-bed capacities of 100,000 cu ft of gas per cu ft of carbon (for unimpregnated carbon) to more than 200,000 cu ft of gas per cu ft of carbon (for copper-impregnated carbon) can be attained (35). After the bed is saturated, it is regenerated with superheated steam or hot desulfurized gas at temperatures between 325° and 750°F. Space velocities are set by minimum cycle requirement of the carbon bed and range from 800 to 1,600 (vol)/(vol) (hr). Details of the design of activated carbon adsorption systems are presented in Chapter 12.

In some cases it is advantageous to follow the treatment with activated carbon by adsorption of sulfur compounds on zinc oxide at elevated temperatures. Since activated carbon has a limited capacity for adsorbing hydrogen sulfide, some aftertreatment is necessary if the gas contains appreciable quantities of hydrogen sulfide.

Since sulfur removal with zinc oxide involves not only the chemical reaction of the zinc oxide with hydrogen sulfide but also catalytic conversion of other sulfur compounds to hydrogen sulfide, which subsequently reacts with the zinc oxide, this process can be used without prior treatment of the gas with activated carbon.

Typical adsorption temperatures are within the range of 400° to 800°F with an optimum between 700° and 750°F. Zinc oxide catalyst is capable of adsorbing about 15 to 20 weight percent sulfur. Regeneration is not considered to be economical. Generally space velocities will range from 400 to 2,000 vol/(vol) (hr).

Catalytic Removal of Carbon Oxides from Synthesis Gases

Removal of carbon monoxide and dioxide from synthesis-gas streams is required when such gases are used in catalytic processes which are adversely affected by the presence of these compounds. For example, the presence of CO, CO_2, and oxygen cannot be tolerated in the hydrogen used for ammonia synthesis because even small quantities of these compounds act as very powerful catalyst poisons. The bulk of the carbon dioxide present in synthesis gases is always removed by means of regenerative liquid absorption. Following such processes the concentration of CO_2 remaining in the gas may be further reduced either by nonregenerative absorption in sodium hydroxide solutions or by catalytic conversion to methane. Carbon monoxide may also be absorbed in regenerable liquids; however, catalytic conversion to carbon dioxide, which is subsequently absorbed in a liquid system, or to methane is normally employed if the gas to be treated contains relatively small quantities of carbon monoxide. Since catalytic conversion of oxides of carbon to methane involves hydrogenation, any oxygen present in the gas is removed concomitantly.

Conversion of Carbon Monoxide to Carbon Dioxide (Shift Conversion)

Formation of hydrogen and carbon dioxide by the reaction of carbon monoxide with water vapor in the presence of iron oxide catalysts is one of the earliest industrial applications of catalysis. The technology of this process, which is known as shift conversion, has been highly developed since its early inception, and, at present, it is used very extensively for the production and purification of hydrogen. The process is applicable to the purification of hydrogen produced in water-gas generators, steam-hydrocarbon reformers, partial oxidation, and steam-iron processes. It is also suitable for adjustment of the H_2/CO ratio in synthesis gases and the purification of gases produced in controlled-atmosphere generators. Since the process has been adequately covered in the literature, the following discussion is confined to its fundamental features and primarily stresses its application to gas-purification problems.

Catalysts

Two types of catalysts are used for shift conversion, depending on the temperature at which the reaction is carried out. The high-temperature catalyst is chromium-promoted iron oxide which is active within the temperature range of 650° to 1100°F. Typical preparations contain 70 to 80 percent ferric oxide and 5 to 15 percent chromic oxide and are available in ¼- by ¼-in. or ⅜- by ³⁄₁₆-in. tablets. This catalyst is relatively insensitive to sulfur compounds and is not permanently affected by limited exposure to liquid water.

Low-temperature catalysts (copper-zinc) are effective over a temperature range of 350° to 700°F and are used advantageously when very low carbon monoxide contents are required. These catalysts are extremely sensitive to poisoning by sulfur compounds, and the feed gas has to be thoroughly desulfurized before contacting the catalyst. Since the equilibrium for the shift reaction is more favorable at lower temperatures, almost complete conversion of carbon monoxide—on the order of 99 percent—is possible with low-temperature catalysts, even in a single stage operation. Further advantages of the low-temperature catalysts are significantly lower steam requirements, smaller equipment, and a less complex plant, as interstage carbon dioxide removal and reheating of the process gas is eliminated.

The low-temperature catalyst is substantially more expensive than the high-temperature catalyst, and the most economical installation usually consists of a combination arrangement in which both the high- and low-temperature catalysts are used in one conversion stage, with intercooling of the process gas by addition of steam condensate. About 90 to 95 percent conversion is obtained in the high-temperature section, and essentially all of the remaining carbon monoxide is removed in the low-temperature section of the reactor.

A discussion of the development and commercial operation of a low-temperature catalyst has been presented by Habermehl and Atwood (38), and complete information on physical properties of the various catalysts as well as design methods for shift converters employing both high- and low-temperature catalysts are available from several catalyst manufacturers.

Process Description

The process is illustrated in the schematic flow diagram shown in Figure 13-14, using the purification of hydrogen obtained by the reforming of natural-gas hydrocarbons with steam as an example. Also for the purpose of this example, use of a conventional high-temperature catalyst is assumed. The gas mixture emerging from the reforming furnace, which contains mainly hydrogen, carbon monoxide, and carbon dioxide, is cooled to about 700°F by addition of steam or condensate and is then passed over the shift-conversion catalyst in the first conversion stage. About 90 to 95 percent of the carbon monoxide is converted to carbon dioxide, and a quantity of hydrogen, equivalent to that of the carbon monoxide reacted, is produced. The first conversion stage is primarily intended for the production of additional hydrogen and therefore is not a gas-purification operation in the true sense of the word. The hot gas leaving the converter is cooled to about 100°F, and the carbon dioxide is removed in a conventional regenerative liquid system (ethanolamine or hot potassium carbonate). The carbon dioxide-free gas is reheated in a direct-fired heater and, after addition of steam, passed through the second conversion stage, which is followed by another carbon dioxide scrubber. If hydrogen of very high purity is required, a third conversion and carbon dioxide-removal stage may be added. A typical product gas from a three-stage shift-conversion plant contains 0.02 percent (by volume) CO, 0.01 percent CO_2, 0.27 percent CH_4, and 99.7 percent hydrogen.

The flow scheme may be modified by (a) the use of heat exchange between the outlet gas from one converter stage and the inlet gas to the following stage

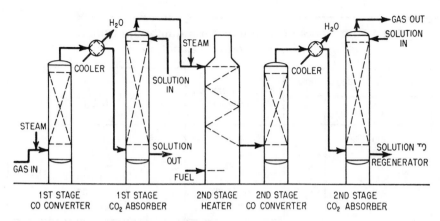

Figure 13-14. Schematic flow diagram of shift-conversion plant.

and (b) the addition of the methanator for conversion of the last remaining traces of carbon monoxide and carbon dioxide. The process may be operated over a considerable range of pressures and temperatures. Pressures of 30, 125, and 350 psig are typical for operation in conjunction with low-pressure or high-pressure steam hydrocarbon-reforming and partial oxidation of hydrocarbons, respectively. Typical operating temperatures range from 600° to 900°F.

Basic Chemistry

Carbon monoxide reacts exothermally with steam at elevated temperatures according to Equation (13-16).

$$CO + H_2O = CO_2 + H_2 \tag{13-16}$$

The heats of reaction and equilibrium constants, within the range of 500° to 1,000°K (439° to 1,341°F), are shown in Table 13-14 (39).

Design and Operation

The important design variables for a shift converter are temperature, pressure, space velocity, steam/gas ratio, and the carbon monoxide content of the inlet gas. Since these variables interact, some working in opposite directions, it is not possible to define generally valid optimum operating conditions, and, therefore, each case has to be analyzed individually. However,

Table 13-14. Heats of Reaction and Equilibrium Constants for Shift-Conversion Reaction

Temperature		Heat of Reaction ΔH, Cal/g mole	$K = \dfrac{(CO_2)(H_2)}{(CO)(H_2O)}$
°K	°C		
500	227	−9.520	126.0
600	327	−9.294	27.08
700	427	−9.051	7.017
800	527	−8.802	4.038
900	627	−8.553	2.204
1000	727	−8.311	1.374

Source: Data of Wagman et al. (39)

certain generalizations with respect to individual variables can be stated which must, of course, be evaluated in the proper relationship to overall operating conditions.

The activity of high-temperature shift-conversion catalysts increases markedly with temperature and is sufficiently high at 650°F for commercial operation at pressures above 100 psig. Operation at near-atmospheric pressure requires temperatures of 750°F or higher. The low-temperature catalyst is, as stated earlier, active at about 400°F. Although high operating temperatures are desirable from the standpoint of catalyst activity or rate of reaction, they are undesirable from the standpoint of equilibrium, which becomes unfavorable with increasing temperature (see Table 13-14).

The conversion efficiency is increased appreciably for a given temperature, by an increase in operating pressure from atmospheric to about 300 psig; above 300 psig, the effect of pressures is practically negligible.

The space velocity used in commercial installations depends to a considerable extent on other operating conditions. Typical space-velocities used in the first conversion stage operating with high-temperature catalyst are 500, 800, and 1,000 volumes per volume per hour for operating pressures of 40, 125, and 350 psig, respectively, and for temperatures between 700° and 800°F. Somewhat higher space velocities may be used in second- and third-stage converters where the carbon monoxide concentration in the inlet gas is quite low.

The steam/gas ratio required for optimum conversion varies with both temperature and pressure. For a given set of operating conditions, with respect to pressure and temperature, conversion first increases with an increasing steam/gas ratio and, after reaching an optimum, decreases upon further addition of steam. This behavior is due to the favorable effect on the equilibrium of high concentrations of steam and the unfavorable effect of decreased contact time. Steam/dry gas ratios ranging from 1:1 to 5:1 are typical for first stage conversion, and 0.5:1 to 1:1 are normally used in the second and third stages. Somewhat higher space velocities and appreciably lower steam/dry gas ratios are used for operation with low-temperature catalyst.

The carbon monoxide content of the inlet gas is not an important design variable. Concentrations ranging from a few percent to more than 50 percent have very little effect on the percentage of conversion obtained in a properly designed plant.

Conversion of Oxides of Carbon to Methane (Methanation)

Catalytic hydrogenation of carbon monoxide and carbon dioxide to methane is normally used to eliminate small quantities of these compounds remaining in gas streams after bulk removal by other techniques. A typical application of the process is the removal of small amounts of residual CO and CO_2 from hydrogen, following shift conversion and absorption of CO_2 in a liquid system. The process is suitable for the treatment of gas streams containing the oxides of carbon at a maximum concentration of about 2 mole percent and is particularly advantageous if the presence of methane in the treated gas is not objectionable in processes following the purification step. Under proper operating conditions, the reaction goes almost to completion, and exit gases containing only a few ppm of the oxides of carbon are obtained.

Removal of oxides of carbon by methanation is required for the protection of certain hydrogenation and ammonia synthesis catalysts against rapid deactivation. Furthermore, methanation is an essential step in the reaction systems associated with the Fisher-Tropsch synthesis and with the production of substitute natural gas from liquid hydrocarbons and coal. An extensive discussion of methanation in connection with the Fischer-Tropsch process is presented by Grayson (40). The methanation process used in the production of substitute natural gas and liquid hydrocarbons is substantially different from the one described in this section. It involves methanation of large amounts of oxides of carbon and is conducted at high temperatures. The heat evolved during the reaction is recovered by generation of high pressure steam. One such substitute natural gas process in which methanation is carried out in several stages is the RMProcess, described by White, Roszkowski, and Stanbridge of The Ralph M. Parsons Company (82).

Process Description

Although many types of catalysts have been investigated for hydrogenation of carbon monoxide and carbon dioxide to methane (40), catalysts of high nickel content are commonly used for gas purification. Typical commercial preparations contain 25 to 30 percent nickel and are used in ⅜- by ⅜-in. or ¼- by ¼-in. cylindrical tablets. The catalyst does not require regeneration and has an expected service life of several years.

Methanation catalysts are easily poisoned by sulfur, and, therefore, sulfur compounds must be removed from the gas streams before they enter the methanator. The flow scheme of the process is quite simple. A diagrammatic flow sheet typical of final purification of the hydrogen used for ammonia synthesis is shown in Figure 13-15. The effluent gas from the last carbon dioxide scrubber is first passed through a heat exchanger, where it is preheated by exchange with the converter exit-gas and then through a heater where its temperature is raised to 500° to 700°F. The hot gas flows downward through the converter and from there through the heat exchanger to a final cooler. Commercial converters normally contain a single bed of catalyst with a minimum depth/diameter ratio of 1:1 (28).

Basic Chemistry

The reactions involved in methanation of carbon monoxide and carbon dioxide can be represented by Equations 13-17 and 13-18.

$$CO + 3H_2 = CH_4 + H_2O \tag{13-17}$$

$$CO_2 + 4H_2 = CH_4 + 2H_2O \tag{13-18}$$

The heats of reaction and equilibrium constants within the range of 400° to 1,000°K (261° to 1341°F) are shown in Table 13-15.

Process Design and Operation

The process can be operated over a considerable range of temperatures and pressures. Inlet-gas temperatures may vary from 350° to 750°F; however, because of the unfavorable equilibrium at elevated temperatures the exit-gas temperature should not exceed 825°F. The strongly exothermic nature of the reaction results in a temperature rise of 134°F for each 1 percent of carbon monoxide in the gas during conversion, and, therefore, the proper temperature must be selected to obtain the desired operating range. The large amount of heat generated by the reaction is also a primary reason that the process is considered unsuitable for the treatment of gases containing more than 2.5 mole percent of the oxides of carbon and oxygen.

The pressure range over which the process can be operated extends from essentially atmospheric pressure to 12,000 psig. Space velocities of 1,000 to

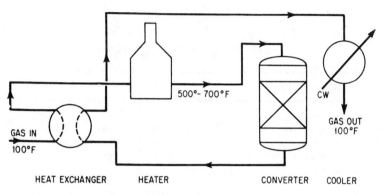

Figure 13-15. Schematic flow diagram of methanator.

Table 13-15. Heats of Reaction and Equilibrium Constants for Methanation of Carbon Monoxide and Carbon Dioxide

Temperature		Heat of Reaction ΔH Cal/g mole		$K = \dfrac{(CH_4)(H_2O)}{(CO)(H_2)^3}$	$K = \dfrac{(CH_4)(H_2O)^2}{(CO_2)(H_2)^4}$
°K	°C	Eq. (13-10)	Eq. (13-11)		
400	127	−50.353	−40.643	4.009×10^{15}	2.709×10^{12}
500	227	−51.283	−41.763	1.148×10^{10}	8.712×10^{7}
600	327	−52.061	−42.768	1.980×10^{6}	7.310×10^{4}
700	427	−52.703	−43.652	3.726×10^{3}	4.130×10^{2}
800	527	−53.214	−44.412	27.97	6.032
900	627	−53.610	−45.058	0.131	0.2888
1000	727	−53.903	−45.592	0.0265	0.0365

Source: Data of Wagman et al. (39)

2,000 volumes per volume per hour are typical for operation at atmospheric pressure. Much higher space velocities—up to 20,000 volumes per volume per hour—can be used at elevated operating pressures. Operation at very high pressures usually results in carbon-oxide concentrations below detectable limits. Typical operating conditions of commercial installations are shown in Table 13-16 (28).

Table 13-16. Typical Operating Conditions for Methanation of Oxides of Carbon

Design Variables	Plant A	Plant B	Plant C
Inlet temperature, °F.............	520	500	500–600
Outlet temperature, °F............	550	550	500–600
Pressure, psig....................	10	750	3,600
Space-velocity, vol/(vol)(hr).......	1,000	7,500	10,000
Inlet gas, ppm:			
CO......................	2,300	6,000	1,000
CO$_2$........................	500	200	50
O$_2$........................	.....	.....	1,000
Outlet gas, ppm:			
CO........................	0.7	10	1
CO$_2$........................	1.5	10	1
O$_2$........................	.....	.....	5

Source: Courtesy of United Catalysts Inc. (28)

Catalytic Removal of Acetylenic Compounds from Olefins by Selective Hydrogenation

Most olefinic gas streams obtained from refinery off-gases or from the cracking of saturated hydrocarbons contain acetylenic compounds, in concentrations ranging from a few tenths of 1 percent to about 2 percent. These compounds have to be removed if the olefins are to be used for the production of certain petrochemicals. Although liquid-purification processes employing selective solvents are available, selective catalytic hydrogenation is usually more economical, especially if the acetylenic compounds are present in relatively small quantities and gas of very high purity is required. Acetylene concentrations of less than 10 ppm can be obtained in the final product gas by catalytic treatment of either the cracked gas or the purified-olefin stream. Removal of acetylene may be effected at several points during the olefin-purification process. The selection of the most economical operation depends on a large number of factors and has to be evaluated for each individual case. Basically, the choice is between treating the cracked gas or the purified-olefin stream. In most commercial installations presently in operation, the acetylenic compounds are removed from the cracked gas, usually after removal of aromatics and acid gases (41). In some installations, selective catalytic hydrogenation is used to eliminate small quantities of acetylene and its homologues from purified-olefin streams. A detailed discussion of the advantages and disadvantages of acetylene removal at different points of the olefin-recovery and -purification system is presented by Reitmeier and Fleming (41).

Process Description

Catalysts suitable for selective hydrogenation of acetylenic compounds in cracked-gas streams contain elements of groups VI and VIII of the periodic system. Use of molybdenum sulfide supported on Activated Alumina is reported by Key and Eastwood (4). The catalysts most commonly used in the United States for the treatment of cracked gases containing an excess of hydrogen are of the cobalt molybdate (42), nickel-base (43), and nickel-cobalt-chromium types (44). The conversion efficiency and selectivity of each of these catalysts is affected primarily by the inlet-gas composition and the contact time. Efficiency increases, in general, with increased partial pressure of hydrogen and contact time and decreases with increased partial pressure of methane and heavier hydrocarbons and sulfur content. The selectivity increases with decreasing partial pressure of hydrogen and ethylene, shortening of the contact time, and increasing sulfur content of the gas (41).

Palladium and promoted palladium catalysts are used for the hydrogenation of small amounts of acetylene present in purified-olefin streams. These catalysts are sensitive to sulfur poisoning, and, therefore, the feed gas has to be freed of sulfur compounds prior to hydrogenation. Several noble metal catalysts used for the removal of acetylene from straight ethylene and ethylene-ethane streams, and, also, from high hydrogen streams low in unsaturates are available from Engelhard Industries (94).

Commercial plants for selective acetylene hydrogenation consist usually of two or three reactors containing the catalyst in fixed beds. One or two beds (operating in parallel) are in operation while the third is being regenerated. The gas usually flows downward through the catalyst and is not cooled during the reaction. Regeneration, which is necessary to remove polymeric materials from the catalyst, is accomplished by application of superheated steam which flows upward through the catalyst bed. Provisions should be made for addition of steam to the inlet gas to control selectivity and for removal of particulate matter.

Basic Chemistry

Hydrogenation of acetylene and methylacetylene can be represented by the following equations:

$$C_2H_2 + H_2 = C_2H_4 \tag{13-19}$$

$$C_2H_2 + 2H_2 = C_2H_6 \tag{13-20}$$

Table 13-17. Heats of Reaction and Equilibrium Constants for Hydrogenation of Acetylene to Ethylene and Ethane

Temperature		Heat of Reaction ΔH, Cal/g mole		$K = \dfrac{(C_2H_4)}{(C_2H_2)(H_2)}$	$K = \dfrac{(C_2H_6)}{(C_2H_2)(H_2)^2}$
°K	°C	Eq. (13-12)	Eq. (13-13)		
300	27	−41.711	−74.451	3.37×10^{24}	1.19×10^{42}
400	127	−42.368	−75.553	7.63×10^{16}	2.65×10^{28}
500	227	−42.911	−76.485	1.65×10^{12}	1.31×10^{20}
600	327	−43.311	−77.211	1.19×10^{9}	3.31×10^{14}
700	427	−43.645	−77.767	6.50×10^{6}	3.10×10^{10}
800	527	−43.676	−78.157	1.28×10^{5}	2.82×10^{7}
900	627	−44.014	−78.432	5.88×10^{3}	1.17×10^{5}
1000	727	−44.099	−78.584	2.23×10^{2}	1.46×10^{3}

Source: Data of Wagman et al. (46), Kilpatrick et al. (47), and Prosen et al. (48)

$$C_3H_4 + H_2 = C_3H_6 \qquad\qquad (13\text{-}21)$$

$$C_3H_4 + 2H_2 = C_3H_8 \qquad\qquad (13\text{-}22)$$

The heats of reaction and equilibrium constants for Equations 13-19 and 13-20 within the range of 300° to 1000°K (80° to 1341°F) are shown in Table 13-17. The mechanism and kinetics of the reactions are discussed in detail by Bond (49).

Design and Operation

Reactors most commonly used in the process consist of cylindrical vessels containing the catalyst in a fixed bed of a maximum depth of 10 ft. The bed-depth/diameter ratio is normally not greater than 1:1 (28). In some instances tubular reactors have been used.

Process operating conditions depend on the type of gas and catalyst used. Acetylene removal from unpurified cracked-gas streams by selective hydrogenation over a special nickel-type catalyst may be carried out at pressures above 50 psig and temperatures ranging from 260° to 660°F and in the presence of sulfur compounds (43). Under these conditions the acetylene concentration can be reduced to less than 100 ppm while only a relatively small amount of ethylene is hydrogenated simultaneously. The catalyst requires periodic regeneration with steam, and resulfiding of the catalyst is necessary after several cycles.

The cobalt molybdate-type catalyst is used for removal of acetylenic compounds from cracked gases, usually after elimination of aromatics and acid gases. Typical commercial operating conditions involve pressures of 75 psig to more than 225 psig, temperatures of 350° to 600°F, and space velocities of about 500 to 1,000 volumes per volume per hour (41). Steam may be added to the inlet gas to improve selectivity and to reduce the rate of polymer formation during the reaction. The activity of the catalyst decreases gradually as polymeric material is deposited, and eventually the catalyst has to be regenerated. The loss of activity can be compensated for by operating at increasingly higher temperatures. The catalyst is somewhat sensitive to the presence of small amounts of sulfur compounds in the gas. This adverse effect can also be counteracted by increasing the operating temperature. The fouled catalyst is regenerated, usually after 4 to 6 weeks of operation, by treatment with steam or a mixture of steam and air and subsequent reduction with hydrogen at 750° to 850°F (41). Concentrations of acetylenic compounds may be reduced from 1 to 2 percent to less than 10 ppm with small loss of olefins.

The nickel-cobalt-chromium-type catalyst which is also used for the treatment of cracked gases is stated to be different in several respects from the cobalt molybdate-type catalyst (41). The selectivity of the catalyst appears to be less affected by wide variations in gas composition, partial pressures of hydrogen and ethylene, and contact time. Catalyst activity reportedly is not lowered to an appreciable extent by the presence of sulfur compounds in concentrations of 25 to 50 grains/100 cu ft. Both selectivity and activity can be kept at high levels by relatively small variations in the operating temperature. The rate of polymer formation is stated to be considerably lower, thus permitting longer operating periods between regenerations (3 to 6 months) than is possible with the cobalt molybdate catalyst.

The catalyst may be used over a considerable range of pressures, temperatures, and space velocities, depending on the composition of the gas stream to be treated. Pressures above 50 psig, inlet temperatures of 250° to 400°F and space velocities ranging from 1,000 to 3,000 volumes per volume per hour are representative for commercial installations (41).

Typical industrial operating conditions for acetylene removal from cracked-gas streams with a catalyst containing elements from groups VI and VIII of the periodic system (exact composition undisclosed) and with nickel-cobalt-chromium catalyst are shown in Tables 13-18 (28) and 13-19 (24), respectively.

As pointed out earlier, palladium and promoted palladium catalysts are used for the removal of relatively small amounts of acetylenic compounds from purified-olefin streams. The primary advantage of removing acetylene from the purified stream is the smaller size of the installation, which, however, is partially offset by the considerably higher cost of the catalyst. In

Table 13-18. Typical Commercial Operating Conditions for Selective Hydrogenation of Acetylene in Cracked-gas Streams

Design Variables	Plant A	Plant B	Plant C
Inlet temperature, °F	350–500	350–500	320–500
Outlet temperature, °F	350–500	350–500	320–500
Pressure, psig	65–80	150	250
Space-velocity, vol/(vol)(hr)	750	600	460
Olefins in inlet gas, mole %	36.0	21.2	41.5
Hydrogen in inlet gas, mole %	24.0	21.2	11.3
Acetylene in inlet gas, mole %	0.4	0.4	0.14
Acetylene in outlet gas, ppm	50˙	50	20

Source: Courtesy of United Catalysts Inc. (28)

Table 13-19. Typical Commercial Operating Conditions for Selective Hydrogenation of Acetylene in Cracked-gas Streams with Nickel-Cobalt-Chromium Catalyst

Operating Variable	Plant A	Plant B	Plant C
Inlet temperature, °F	285	355	375
Pressure, psig	170	150	210
Space-velocity, vol/(vol)(hr)	1,050	1,500	1,500
Inlet gas (key components):			
Ethylene, mole %	30	31	23
Propylene, mole %	12	1	11
Hydrogen, mole %	14	30	14
Acetylene, mole %	1.0*	0.4	0.3
Sulfur, ppm	16–32†	6–8	6–8
Outlet gas:			
Acetylene, ppm	10	10	10

Source: Courtesy of United Catalysts Inc. (24)
*Methyl acetylene generally comprises about two-thirds of total acetylenes in this stream.
†Plant has operated at concentrations up to 1,200 ppm for short periods of time.

addition, butadiene is not hydrogenated during the reaction and, therefore, can be recovered from the gas stream. The catalysts are effective at pressures above 100 psig, temperatures of about 150° to 350°F, and space velocities of 1,000 to 5,000 volumes per volume per hour. An excess of hydrogen, equivalent to two to four times the stoichiometric amount required to hydrogenate the acetylenic compounds, has to be added to the inlet gas (28, 41). Some olefin is lost in the process by reaction with the excess hydrogen. Typical operating conditions with this type of catalyst are shown in Table 13-20 (28).

A mathematical analysis of optimum operating conditions for the catalytic removal of acetylene from ethylene with a palladium catalyst has been presented by Huang (95). Parameters investigated include temperature, space velocity, and feed gas composition. The results of the analysis show that, for the case studied, temperature is the most sensitive parameter. Optimum operating conditions, yielding the lowest acetylene concentration (1 ppmv) and the lowest ethylene loss include a reactor temperature of 700°R (240°F) and a space velocity of 7,000 (vol)/(vol) (hr).

Purification of Gas Streams by Catalytic Oxidation and Reduction

Vapor-phase catalytic oxidation and reduction is used for the removal of a large variety of objectionable compounds from many types of gas streams. Typical applications of this technique in gas-purification operations are (a) removal of oxygen and carbon monoxide from hydrogen and synthesis gases, (b) removal of oxygen from nitrogen and inert-gas streams, (c) removal of organic compounds from air and industrial exhaust gases, (d) removal of nitrogen oxides from exhaust streams, and (e) elimination of hydrogen from oxygen streams. Catalytic oxidation is particularly suitable for removing small amounts of combustible contaminants from gas streams containing these compounds in concentrations below the flammable limit and, therefore, has found wide application in the field of air pollution and odor control (50, 51). Catalytic oxidation and reduction processes are used most advantageously when the impurities to be removed yield innocuous reaction products, such as water, nitrogen, and carbon dioxide. However, they are also useful for converting offensive and poisonous chemicals to less objectionable compounds, as for example, oxidation of hydrogen sulfide and organic sulfur compounds to sulfur dioxide. Another application is catalytic combustion of low heating value (LHV) gases with recovery of heat, even when the gases will not burn in a normal manner (96). Such gas streams include gases obtained by in situ or surface gasification of coal with air and a large number of industrial waste gases with heating values as low as 20 Btu/scf (96).

Basic Data

Oxidation and reduction catalysts most commonly used in gas-purification processes are precious metals, primarily platinum and palladium, supported on various carrier materials. In addition to precious metals, copper chromite, various metallic oxides, and Fischer-Tropsch catalysts have also been mentioned (52). These catalysts are effective over a considerable range of temperatures and pressures and are quite versatile with respect to the type and concentration of compounds to be removed.

The simplest form of an oxidation-reduction reaction, the formation of water from hydrogen and oxygen, is represented in Equation 13-23.

$$H_2 + 1/2 \, O_2 \;=\; H_2O \; (g) \tag{13-23}$$
$$- \, \Delta H \;=\; 57.8 \; Cal/g \; mole$$

This reaction proceeds readily at room temperature in the presence of certain catalysts and goes practically to completion. Other compounds, e.g., hydrocarbons, require appreciably higher temperatures to initiate catalytic reactions. Minimum temperatures for initiation of catalytic oxidation of aliphatic and aromatic hydrocarbons reported by Suter (50) are shown in Table 13-21. These data were obtained with a precious-metal catalyst under typical commercial operating conditions at hydrocarbon concentrations corresponding to approximately 10 percent of the lower limit of flammability. Although these values vary considerably with space velocity, catalyst activ-

Table 13-20. Typical Operating Conditions for Hydrogenation of Acetylenic Compounds in Purified-olefin Streams

Hydrocarbon	Initiation Temperature, °F
Methane	760
Ethane	680
Propane	650
n-Butane	570
n-Pentane	590
n-Hexane	630
n-Heptane	580
n-Octane	490
n-Decane	500
n-Dodecane	540
n-Tetradecar.e	550
Benzene	440
Toluene	460
o-Xylene	470

Source: Data of Suter (50)

ity, and hydrocarbon and oxygen concentrations, they are useful for estimating the gas temperature required at the catalytic-converter inlet.

The temperature change in the converter and the maximum degree of removal to be expected can be calculated from published thermodynamic and equilibrium data such as those presented by Rossini et al. (53, 54). Space velocities to be used in plant design depend on the concentration and type of gas impurities present, the pressure and temperature of the gas to be treated, the concentration of oxygen, and the activity of the catalyst. Depending on these variables, space velocities for conventional installations utilizing fixed catalyst beds may range from 2,000 to 20,000 volumes per volume per hour (24). In general, the space velocity required for a given set of conditions is determined on the basis of experience.

Description of Processes

The principal distinction between the various commercially practiced gas-purification processes based on catalytic oxidation and reduction lies in the type of catalyst used and the form of the catalytic element. The catalyst may be supported on a granular carrier and arranged in fixed beds in conventional converters. It may also be contained in specially constructed units which are installed in the converter in specific geometrical configurations. A rather detailed discussion of catalytic elements is given by Tucci (96). Various types of ceramic and metallic elements are described, with special emphasis on the catalyst support to provide a maximum active catalytic surface area. Also, several designs of tubular combusters are proposed.

Table 13-21. Minimum Temperatures for Initiation of Catalytic Oxidation of Hydrocarbons

Operating Variable	Gas Stream	
	Ethylene	Propylene
Inlet temperature, °F	130–350	130–350
Outlet temperature, °F	130–350	130–350
Pressure, psig	200	300
Space-velocity, vol/(vol)(hr)	2,000	5,000
Acetylene in inlet gas, ppm	3,000	
Acetylene in outlet gas, ppm	18	
Methylacetylene in inlet gas, ppm	....	4,000
Methylacetylene in outlet gas, ppm	....	50

Source: Courtesy of United Catalysts Inc. (28)

The basic gas-purification installation employing this type of process consists of a preheater, where the temperature of the inlet gas is raised to the level required for the reaction, an injection and mixing device for addition of either oxygen or a reducing gas, and the catalytic converter. In some applications the reaction proceeds at room temperature and the preheater may, therefore, be omitted. Heat savings may be realized by use of heat exchange between the inlet and outlet gas or by recirculation of a portion of the treated gas. Recirculation has the added advantage that the concentration of contaminants in the converter-inlet gas may be adjusted; this is of considerable importance in cases where the lower explosive limits of the gas mixture may be approached. Recirculation or heat exchange is always recommended when the concentration and nature of contaminants in the inlet gas is such that the heat generated by the reaction is sufficient to heat the gas to the required reaction temperature. When gases containing very large amounts of impurities are treated, the heat of reaction may be used to generate steam in a waste-heat boiler. Four different schemes are shown diagrammatically in Figure 13-16.

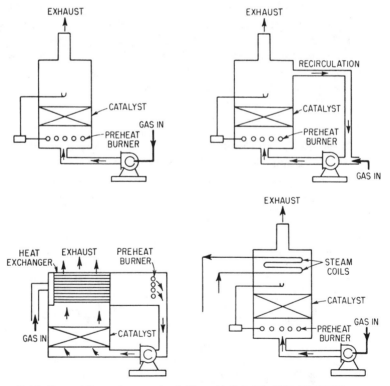

Figure 13-16. Several flow schemes of catalitic oxidation-reduction processes.

The Deoxo Processes

These processes, which were developed by Engelhard Industries, Inc., Chemical Division, utilize precious-metal catalysts supported on Activated Alumina (55,56). The catalytic converter consists of a conventional circular vessel containing the catalyst in a fixed bed. Purification units ranging in capacity from 5 to 500,000 cu ft/hr are in commercial operation. Typical applications of the processes are removal of oxygen or hydrogen from hydrogen or oxygen, respectively, or from noble, inert, and synthesis-gas streams and the removal of carbon monoxide from synthesis gases.

A catalyst consisting of palladium supported on pelleted Activated Alumina is used to remove oxygen from hydrogen, nitrogen, argon, neon, carbon dioxide, and saturated-hydrocarbon streams. Hydrogen has to be present in the gas stream in at least stoichiometric proportion to react with the oxygen. The catalyst is effective at room temperature, provided the gas does not contain chlorides, sulfur compounds, carbon monoxide, oil, or unsaturated hydrocarbons. The same catalyst may be used for oxygen removal from gases containing carbon monoxide and from ethylene streams if operating temperatures above 250° and 430°F, respectively, are maintained (57). In all cases the reaction is practically complete and the treated gas contains less than 1 ppm oxygen.

Oxidation of carbon monoxide to carbon dioxide is promoted by a supported platinum catalyst—containing 0.1 to 0.5 percent platinum—at temperatures of 212° to 390°F (55). The process is reportedly suitable for purifying gas streams of high hydrogen content provided a 200 percent excess of oxygen with respect to carbon monoxide is available in the gas and the temperature is controlled to range between 260° and 320°F (55).

Selective carbon monoxide oxidation as applied to the purification of impure hydrogen streams and ammonia synthesis gases has been studied by Brown, et al. (58) and Andersen, et al. (59). These authors report the effectiveness of supported noble metal catalysts, especially platinum, for the catalytic oxidation of up to 2.5 percent carbon monoxide at temperatures within the range of 250° to 320°F, pressures ranging from 12 to 200 psig and space velocities of up to 14,000 scf/(cf)(hr).

A photograph of a typical Deoxo installation is shown in Figure 13-17. Commercial units incorporating a catalytic deoxidizer, operating at room temperature, and a solid-bed dehydrator are available from several manufacturers. An installation of this type is shown in Figure 13-18.

The Oxycat Process

This process, which was developed by Houdry, employs a catalyst consisting of a combination of platinum and alumina supported on a porcelain carrier in a specially designed element (60). The Oxycat element, which is shown in Figure 13-19, consists of 71 porcelain rods held in place by two porcelain end plates. The catalytic agent is deposited on the surface of the rods. The unit is 5½ in. long, 3³/16 in. high, and 3 in. wide; it weighs 1.6 lb. The elements are arranged in several layers, usually three or more, in the combustion chamber, as shown in Figure 3-20. The catalyst elements are supported by suitable metal grates which may be constructed from carbon steel when the operating temperature does not exceed 750°F. However, special alloys and refractory materials are required for operation at higher temperatures. Accessory equipment, e.g., a preheater, a gas-mixing device, and a heat exchanger, is normally provided in a typical installation. The number of Oxycat elements required is estimated on the basis of a space velocity of 5 to 15 cu ft of gas/min per element, for operation at atmospheric pressure. Somewhat higher space velocities are permissible if the gas is available at elevated pressures. The number of catalyst layers and the shape and dimensions of the catalyst chamber are chosen on the basis of experience. The pressure drop through the catalyst beds normally does not exceed 1 in. of water (60).

The Oxycat process is primarily used for the removal of organic and other combustible vapors from air and industrial exhaust gases. Several industrial installations have been described by Houdry and Hayes (61), Uhleen and Hayes (62), and Resen (63). A list of typical commercial applications is given in Table 13-22 (60).

The Catalytic Combustion Corporation Process

The catalysts used in this process are precious metals (platinum group) supported on high-nickel-alloy ribbon media, which are contained in elements of all-metal construction. Several types of standard elements, which resemble metallic air-filter mats, are shown in Figure 13-21. The principal advantages claimed for this type of element are low pressure drop, high throughput capacity, high heat resistance, and resistance to thermal shock (64). The elements can be installed in a large variety of configurations permitting gas flow in practically any desired direction. A cartridge-type design, such as shown in Figure 13-22, is most commonly used for gas-purification installations, especially for operation at elevated pressures (64). The catalytic elements are reported to be suitable for operation at temperatures ranging from 100° to 1,700°F and pressures from atmospheric to as high as 2,000 psig (64).

The applications of the process are essentially the same as those discussed previously, i.e., removal of oxygen and hydrogen from various gas streams and elimination of combustible compounds from air and exhaust gases. Descriptions of the process and of typical industrial applications have been presented by Suter (50) and by Ruff (65, 66, 67).

Figure 13-17. Deoxo unit. *Engelhard Industries, Inc.*

Catalytic Removal of Nitrogen Oxides

In view of the role played by nitrogen oxides in the formation of atmospheric pollutants, normally referred to as "smog," a great deal of work has been done to develop processes for the removal of such compounds from exhaust gases of all types prior to venting to the atmosphere. The most serious offenders with respect to nitrogen oxide emission are power plant stack gases and exhaust gases from internal combustion engines. In stack gases, nitrogen oxide removal is often associated with the removal of sulfur dioxide. Processes of this type, which at present are either in commercial use or under development, are discussed in Chapter 7. A comprehensive study of the state of the art nitrogen oxide removal processes for utility application has been published by the Tennessee Valley Authority under a grant from the Environmental Protection Agency and the Electric Power Research Institute (83). More recently a review of methods used for the control of nitrogen oxides emission in flue gases has been presented by Siddigi and Tenini (97).

Figure 13-18. Combined deoxidizer and gas dehydrator. *C. M. Kemp Manufacturing Company*

Figure 13-19. The Oxycat element. *Oxy-Catalyst, Inc.*

Figure 13-20. Arrangement of Oxycat elements in the converter. Top layer of elements shown in lower portion of photograph. *Oxy-Catalyst, Inc.*

Table 13-22. Typical Commercial Applications of Oxycat Process

Industrial Process	Contaminating Agents in Waste Gases	Approximate Temperature, °F, Required for Catalytic Oxidation
Asphalt oxidizing..........	aldehydes, anthracenes, oil vapors, hydrocarbons	600–700
Carbon black mfg..........	H_2, CO, CH_4, carbon	1200–1800*
Catalytic cracking units....	CO, hydrocarbons	650–800
Coke ovens...............	Wax, oil vapors	600–700
Formaldehyde mfg.........	H_2, CH_4, CO, HCHO	650
HNO_3, mfg..............	NO, NO_2	500–1200†
Metal-lithography ovens....	solvents, resins	500–750
Octyl-phenol mfg..........	C_6H_5OH (phenol)	600–800
Phthalic anhydride mfg.....	maleic acid, phthalic acid, naph-thoquinones, carbon monoxide, formaldehyde	600–650
Polyethylene mfg..........	hydrocarbons	500–1200
Printing presses............	solvents	600
Varnish cooking..........	hydrocarbon vapors	600–700
Wire-coating and enameling ovens.................	solvents, varnish	600–700

Source: Courtesy of Oxy-Catalyst, Inc.
*Temperatures in excess of 1,200°F required to oxidize carbon.
†Reducing atmosphere required.

Figure 13-21. Typical elements used in Catalytic Combustion Corporation Process. *Catalytic Combustion Corporation*

Figure 13-22. Cartridge-type converter for high-pressure operation, Catalytic Combustion Corporation process. *Catalytic Combustion Corporation*

There are several approaches to the prevention of nitrogen oxides emission into the atmosphere from stationary sources, including modification of combustion, thermal decomposition, selective or nonselective catalytic conversion to non-objectionable compounds, wet absorption, adsorption on solids and finally, selection of fuels low in nitrogen compounds. All these methods are discussed by Siddigi and Tenini (97) and lists of available processes and operating commercial installations using them are given. Most of the technology was developed and is practiced in Japan where air pollution is a particularly serious problem. However, a number of processes has been developed in the United States, the most noteworthy being those offered by Exxon, Shell, The Ralph M. Parsons Company (BSR process), and Foster Wheeler (97). Among the processes used industrially for nitrogen oxides removal, nonselective and selective catalytic reduction are predominant (97, 98, 99, 100, 101).

In nonselective processes, the reducing agent, e.g., hydrogen, natural gas, or light naphtha, reacts with oxygen, and with nitrogen oxides—NO_2 and NO— yielding nitrogen and water plus CO_2 if the reducing agent is a hydrocarbon. Initiation temperatures for the reaction range from 300°F, if hydrogen is used, to 750°F or 840°F if naphtha or natural gas respectively are the reducing agents. Although this type of process is very effective in nitrogen oxide abatement, its drawbacks are that it requires sophisticated control, large amounts of reducing agent and, because of the high temperatures, expensive materials of construction (99).

The predominant technology used in flue gas treating processes is selective catalytic reduction with ammonia. Since oxygen is present in practically all flue gases, the chemical reactions can be expressed as follows:

$$4NH_3 + 4NO + O_2 \xrightarrow{\text{cat.}} 4N_2 + 6H_2O \qquad (13\text{-}24)$$

$$4NH_3 + 2NO_2 + O_2 \xrightarrow{\text{cat.}} 3N_2 + 6H_2O \qquad (13\text{-}25)$$

The NH_3:NO mol ratio commonly used to effect about 90 percent NO_x reduction and leave about 20 ppm of ammonia in the exhaust ranges from 1.0 to 1.1. Depending on the catalyst used, operating temperatures range from 570 to 850°F.

Catalysts used comprise base metals such as iron, vanadium, chromium, manganese, cobalt, nickel, copper, or barium deposited on alumina, titanium dioxide, and silica (97). Laboratory and pilot plant studies on the use of catalysts containing noble metals, copper-zinc-chrome, barium promoted copper chromite and chromium promoted iron oxide were reported by Andersen et al. (70) and Ayen and Amirnazmi (71).

The copper oxide flue gas treatment process, which has the capability to remove SO_2 and NO_x simultaneously, is described in Chapter 7. The process effectively removes NO_x by catalytically reacting it with NH_3; however, it has not yet been applied commercially. Several selective catalytic reduction processes have been commercialized in Japan where air pollution control standards for NO_x are very stringent. One of these, the Hitachi Zosen NO_x Flue Gas Treatment Process, has been extensively tested and evaluated in the United States (102, 103). The process uses a low pressure drop honeycomb configuration catalyst element with an expected life of about two years. The active components are compounds of vanadium and titanium. Ammonia is injected upstream of the catalyst bed, which is installed at a point in the flue gas system where the temperature is in the range of 300–450°C.

Catalytic conversion of nitrogen oxides to innocuous compounds which can be vented without causing air pollution has been in use for some time. One process, utilizing the Catalytic Combustion Corporation catalyst and reducing gases such as hydrogen, carbon monoxide, methane, and other hydrocarbon vapors has been described by Donahue (68). The process is claimed to be operable at atmospheric as well as elevated pressures and over a temperature range extending from ambient to 1,000°F. A typical atmospheric-pressure installation is shown diagrammatically in Figure 13-23. Cartridge-type reactors, such as the one shown in Figure 13-24, are used for systems operating at elevated pressures. In such installations the gas enters the reactor through a side-inlet nozzle and flows upward through an annular space surrounding the main shell. At the top of the vessel the gas flow turns downward into a second annulus and enters a hollow cylindrical catalyst bed. The gas then passes inwardly through the catalyst bed and is discharged from the bottom of the vessel. It is claimed that nitrogen oxide concentrations of less than 200 ppm in the treated gas can be obtained in a properly operated installation. Although long-term operating data are not available, a catalyst life exceeding 20,000 operating hours is expected on the basis of experimental data (68).

A photograph of a commercial installation used for the removal of nitrogen oxides from waste gases of a nitric acid plant is shown in Figure 13-25 (69). In this installation the entering waste gases are diluted and preheated by a recycle stream of hot, treated gas; they are further heated, if required, by a preheater burner and, after addition of propane, passed through the catalyst elements. The effluent gases, which are color- and odorless, are vented to the atmosphere.

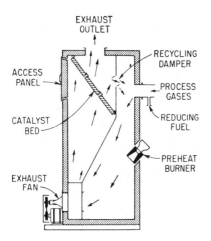

Figure 13-23. Typical flow diagram of low-pressure installation, Catalytic Combustion Corporation process. *Catalytic Combustion Corporation*

Figure 13-24. Reactor for catalytic removal of nitric oxide from exhaust gases. *Catalytic Combustion Corporation*

Laboratory and pilot-plant studies on catalytic reduction of nitrogen oxide by use of catalysts containing noble metals, copper-zinc-chrome, barium-promoted copper chromite, and chromium-promoted iron oxide have been reported by Andersen et al. (70) and Ayen and Amirnazmi (71).

Since most gases containing nitrogen oxides also contain appreciable amounts of oxygen, nitrogen oxide removal can be combined with power recovery by utilizaton of the substantial heats of reaction. For example, a typical tail gas from a nitric acid plant contains 1 to 3 percent oxygen and 0.2 to 0.5 percent oxides of nitrogen. Typical operating conditions, when using a commercial catalyst containing platinum and nickel on an alumina support as given by the catalyst manufacturer (72) are shown in Table 13-23. The oxidizing or reducing designations refer to the presence or absence of excess oxygen over that stoichiometrically required. The NO_x is reduced in both cases.

Figure 13-25. Commercial installation for catalytic removal of nitric oxide from the exhaust gases of a nitric acid plant. *Catalytic Combustion Corporation*

The Mine Safety Appliance Company Process

This process is primarily used for the complete removal of small amounts of acetylene (0.1 to 1.0 ppm) and other hydrocarbons from air streams fed to low-temperature separation plants. Complete removal of acetylene from such streams is of great importance because of its low solubility in liquid oxygen. Accumulation of solid acetylene on heat-exchanger surfaces may result in local concentrations exceeding the explosive limits of the mixture; in fact, many explosions in air-separation plants have been attributed to this phenomenon. The hydrocarbons are completely oxidized to carbon dioxide and water at relatively low temperatures in the presence of Hopcalite, which is a mixture of 60 percent manganese dioxide and 40 percent copper oxide. The catalyst is also effective for the oxidation of carbon monoxide to carbon dioxide and for the decomposition of ozone. Promoted Hopcalites containing relatively small amounts of silver salts have been found particularly effective for the treatment of moist air streams (73). Commercial Hopcalite promotes essentially complete combustion of acetylene at 305° to 315°F. However, higher temperatures, ranging up to 800°F, are required for the oxidation of

	Power Recovery Primary		Nitrogen Oxide Removal Primary	
	H_2 Fuel	CH_4 Fuel	H_2 Fuel	CH_4 Fuel
Ignition temp., °F	450-500	900	400	800
Outlet temp., °F	900-950	1,200	1,000-1,200	1,400-1,500
Space velocity*	40,000	40,000	20,000	20,000
Oxidizing or reducing	Oxid.	Oxid.	Red.	Red.
Outlet NO_x, ppm	300	300-500	100	100

Table 13-23. Operating Conditions for Nitrogen Oxide Removal; Platinum-nickel Catalyst

Source: United Catalysts, Inc. (72)
*SCF gas/(hr) (cu.ft. of catalyst @ 60° F and 1 atm.)

other hydrocarbons. The degree of conversion of several hydrocarbons in the presence of a commercial catalyst as a function of temperature is shown in Figure 13-26 (74).

The catalytic element used in the process is shown in Figure 13-27 (74). The element is about 20 in. in diameter and 20 in. high; it contains about 80 lb catalyst, which is confined in an annular space between two stainless-steel screens. The air flow radially through the catalyst, from the outside to the hollow center, and is discharged through an opening at the bottom of the unit. The catalytic element is designed for a capacity of 300 cfm of air measured at operating conditions. It is, therefore, advantageous to operate the converter at elevated pressures. A temperature-resistant mechanical filter-blanket is wrapped around the outside of the element to protect the catalyst from particulate matter. This filter is usually replaced when the pressure drop reaches twice the original value. If properly operated, the catalyst has an expected service life of several years, after which it is returned to the manufacturer for regeneration.

A typical catalytic converter consists of an insulated pressure vessel where the individual elements are mounted on one or several decks, as shown in Figure 13-28. The converter is usually located in the stream just after the air-compression stage in the separation plant, before the air is refrigerated and liquefied. The temperature of the air discharged from the compressor is normally at the level desired for the catalytic reaction. However, if higher temperatures are required, the converter may also be placed after one of the intermediate compression stages.

Special elements for the purification of cabin air in jet aircraft and a complete unit for the elimination of hydrogen from air have also been developed. (74).

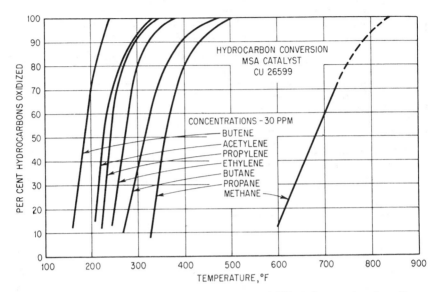

Figure 13-26. Oxidation of hydrocarbons over commercial Hopcalite as a function of temperature. (59) *Mine Safety Appliance Company*

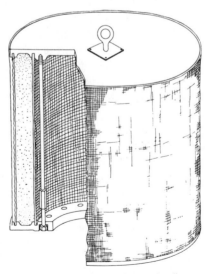

Figure 13-27. Catalytic element used in Mine Safety Appliance Company process. *Mine Safety Appliance Company*

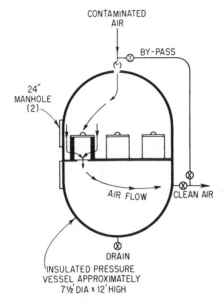

CONTAMINATED
AIR

BY-PASS

24"
MANHOLE
(2)

AIR FLOW   CLEAN AIR

DRAIN
INSULATED PRESSURE
VESSEL APPROXIMATELY
7½' DIA x 12' HIGH

Figure 13-28. Diagram of typical converter, Mine Safety Appliance Company process. *Mine Safety Appliance Company*

References

1. Berkman, S.; Morrell, J.C.; and Egloff, G. 1940. *Catalysis.* New York: Reinhold Publishing Corp.

2. Emmett, T.H. (ed.). 1956. *Catalysis.* New York: Reinhold Publishing Corp.

3. Frankenburg, W.G.; Komarewsky, V.I.; and Rideal, E.K. (eds.). 1948. *Advances in Catalysis.* New York: Academic Press, Inc.

4. Key, A., and Eastwood, A.H. 1946. Forty-fifth Report of the Joint Research Committee of the Gas Research Board and the University of Leeds. London: The Gas Research Board (presently The Gas Council), Publ. GRB 14/4.

5. Wedgwood, W. 1958. *Inst. Gas Engrs.,Publ. No. 525, London.*

6. Terres, E., and Wesemann, H. 1932. *Angew. Chem.* 45:795-801.

7. Kelley, K.K. 1937. *U.S. Bur. Mines, Bull., No. 406.*

8. Carpenter, C. 1913. *J. Gas Lighting* 122:1010.

9. Carpenter, C. 1913. *J. Gas Lighting* 123:30.

10. Evans, E.V. 1915. *J. Soc. Chem. Ind. (London)* 34:9-14.

11. Evans, E.V., and Stanier, H. 1924. *Proc. Roy. Soc. (London)* 105A:626.

12. Trutnovski, H. 1935. *Gas-u. Wasserfach* 78:462-465.

13. Menerey, E.J. 1931. *Am. Gas Assoc. Proc.* 13:1082-1085.

14. Darlington, F.H. 1947. *Am. Gas Assoc. Monthly* 29:147.

15. Maxted, E.B. 1937. British Patent 490,775.

16. Maxted, E.B., and Priestly, J.J. 1946. *Gas J.* 247:471, 515, 556, 593.

17. Maxted, E.B., and Marsen, A. 1946. *J. Soc. Chem. Ind. (London)* 65:51.

18. Priestly, J.J., and Marris, H.Q. 1947. *Gas Times* 50:396-399.

19. Priestly, J.J. 1957. *Gas Times* 91(Oct./Nov.):640.

20. Anon. 1958. *Gas J.* 294(Mar.):595-597.
21. Priestly, J.J. 1958. *Gas in Ind.* 1:21-28.
22. W.C. Holmes & Company, Ltd. 1958. Private communication.
23. Nicklin, T. 1971. *Some Aspects of Environmental Control.* Paper presented at Inst. of Gas Eng., Scottish Section, Annual General Meeting, Andrews, Scotland (Oct. 14).
24. Catalysts and Chemicals, Inc. 1958. Private communication.
25. Holroyd, R. 1946. *U.S. Bur. Mines, Inform. Circ. 7370.*
26. Sands, A.E.; Wainwright, H.W.; and Egleson, G.C. 1950. *U.S. Bur. Mines Rept. Invest. 4699.*
27. Chemetron Corporation. *Girdler Catalysts, Bulletin GC-1256.*
28. Chemetron Corporation. 1958. Private communication.
29. Huff, W.J., and Logan, L. 1936. *The Purification of Commercial Gases at Elevated Temperatures.* New York: American Gas Association, Inc.
30. Catalysts and Chemicals Inc. *Catalyst Data C53-2.*
31. Catalysts and Chemicals Inc. 1971. Private communication.
32. Beavon, D.K., and King, F.W. 1970. *Can. Gas J.* (Sept./Oct.):22-26.
33. Beavon, D.K., and Vaell, R.P. 1971. Paper presented at the AIChE Eighth Annual Technical Meeting, Southern California Section (Apr. 20).
34. The Ralph M. Parsons Company. *The Beavon Sulfur Removal Process.*
35. Livingston, J.Y. 1971. *Hydroc. Process.* 50(Jan.):126-132.
36. Catalysts and Chemicals Inc. *Technical Bulletins C8-048, C8-120, C7-081 and C7-086.*
37. Girdler Catalysts. 1970. *Hydrogen and Synthesis Gas Production.*
38. Habermehl, R., and Atwood, K. 1964. Paper presented before the Division of Fuel Chemistry, American Chemical Society, Philadelphia, Penn. (April).
39. Wagman, D.D.; Kilpatrick, J.E.; Taylor, W.J.; Pitzer, K.S.; and Rossini, F.D. 1945. *J. Research Nat. Bur. Standards* 34:143-161.
40. Grayson, M. 1956. *Catalysis,* vol. 4. Edited by P.H. Emmett. New York: Reinhold Publishing Corp., p. 473.
41. Reitmeier, R.E., and Fleming, H.W. 1958. *Chem. Eng. Progr.* 54(Dec.):48-51.
42. Giaro, J.A. 1956. U.S. Patent 2,735,897.
43. Barry, A.W. 1950. U.S. Patent 2,511,453.
44. Catalysts and Chemicals, Inc. *Brochure C-36.*
45. Andersen, C.A.; Haley, A.J.; and Egbert, W. 1960. *Ind. Eng. Chem.* 52(Nov.):901-904.
46. Wagman, D.D.; Kilpatrick, J.E.; Pitzer, K.S.; and Rossini, F.D. 1945. *J. Research Nat. Bur. Standards* 35:467-496.
47. Kilpatrick, J.E.; Prosen, E.J.; Pitzer, K.S.; and Rossini, F.D. 1946. *J. Research Nat. Bur. Standards* 36:599-612.
48. Prosen, E.J.; Pitzer, K.S.; and Rossini, F.D. 1945. *J. Research Nat. Bur. Standards* 34:403-411.
49. Bond, G.C. *Catalysis.* vol. 3. Edited by P.H. Emmett. New York: Reinhold Publishing Corp., p. 109.
50. Suter, H.R. 1955. *J. Air Pollution Control Assoc.* 5(3):173.
51. Turk, A. 1958. *Ind. Wastes* 3(1):9.
52. Turk, A. 1954. *Am. Ind. Hyg. Assoc. Quart.* 15(2):119-123.
53. Rossini, F.D.; Pitzer, K.S.; Arnett, R.L.; Braun, R.M.; and Pimentel, G.C. 1953. *Selected Values of Physical and Thermodynamic Properties of Hydrocarbons and Related Compounds.* American Petroleum Institute Research Project 44. Pittsburgh: Carnegie Press.

54. Rossini, F.D.; Wagman, D.D.; Evans, W.H.; Levine, S.; and Jaffe, J. 1952. *U.S. Nat. Bur. Standard Circ. No. 500.*

55. Engelhard Industries, Inc., Chemical Division. *Bulletin E1-4419A.*

56. U.S. Patents 2,475,155 (1949), 2,582,885 (1952), 2,577,220 (1951).

57. Baker & Company, Inc. *The Role of Platinum Group Metals as Catalysts in Industry.* Form 4683-56-1MX1.

58. Brown, M.L.; Green, A.W.; Cohn, G.; and Andersen, H.C. 1960. *Ind. Eng. Chem.* 52(10):841-844.

59. Andersen, H.C., and Green, W.J. 1961. *Ind. Eng. Chem.* 53(8):645-646.

60. Oxy-Catalyst Inc. 1957. *Technical Manual.*

61. Houdry, J.H., and Hayes, C.T. 1957. *J. Air Pollution Control Assoc.* 7(3):182-186.

62. Uhleen, A.E., and Hayes, C.T. 1955. *Instrumentation* 8(3).

63. Resen, L. 1958. *Oil Gas J.* 56(1):110.

64. Catalytic Combustion Corporation. 1958. Private communication.

65. Ruff, R.J. 1957. *Wire and Wire Products* 32(Jan.):62-64, 98.

66. Ruff, R.J. 1957. *Chem. Eng. Progr.* 53:377-380.

67. Ruff, R.J. 1958. *Am. Ind. Hyg. Assoc. Quart.* 14(Sept.):183-187.

68. Donahue, J.L. 1958. *J. Air Pollution Control Assoc.* 8(Nov.):209.

69. Anon. 1958. *Chem. Processing* 21(4):194-197.

70. Andersen, H.C.; Green, W.J.; and Steele, D.R. 1961. *Ind. Eng. Chem.* 53(3):199-204.

71. Ayen, R.J., and Amirnazmi, A. 1970. *Ind. Eng. Chem. Process Des. Develop.* 9(2):247-254.

72. Girdler Catalysts. *Technical Data Sheet, Girdler G-43 Catalyst.*

73. Rushton, J.H. 1954. *Advances in Catalysis,* vol. 2. Edited by W.G. Frankenburg, et al. New York, Academic Press, Inc., p. 107.

74. Mine Safety Appliance Company. *MSA Technical Products Release No. 1501.*

75. Andrews, E.J., and Kouzel, B. 1974. "Beavon Sulfur Removal Process Eliminates Sulfur in Claus Plant Tail Gas." Paper presented at 53rd GPA Annual Convention, Denver (March 25-27).

76. Fenton, D.M.; Woertz, B.B.; Brocoff, J.C.; and Jirus, E.J. 1975. "Tail Gas Clean-up With the Beavon Sulfur Removal Process." Paper presented at NPRA Annual Meeting, San Antonio (March 23-25).

77. Beavon, D.K., and Brocoff, J.C. 1976. "Recent Advances in the Beavon Sulfur Removal Process and the Stretford Process." Paper presented at the Second International Conference of the European Federation of Chemical Engineers, University of Salford, England (April 6-8).

78. Kouzel, B.; Fuller, R.H.; Jirus, E.J.; and Woertz, B.B. 1977. "Treat Low Sulfur Gases with Beavon Sulfur Removal Process and the Improved Stretford Process." Paper presented at 27th Annual Gas Conditioning Conference, University of Oklahoma, Norman, Oklahoma (March 7-8).

79. Groenendaal, W., and Van Meurs, H.C.A. 1972. *Petroleum and Petrochemical International* 12(9):54-58.

80. Naber, J.E.; Wesselingh, J.A.; and Groenendaal, W. 1973. *Chem. Eng. Progr.* 69(12):29-34.

81. Harvey, C.G., and Verloop, J. 1976. "Experience Confirms Adaptability of the SCOT Process." Paper presented at Second International Conference of the European Federation of Chemical Engineers, University of Salford, England (April 6-8).

82. White, G.A.; Roszkowski, T.R.; and Stanbridge, D.W. 1974. "The RM Process," Proceedings, 168th National Meeting, American Chemical Society, Division of Fuel Chemistry, Atlantic City, New Jersey (September 8-13).

83. Technical Assessment of NO_x Removal Processes for Utility Application, EPRI AF-568, API-600/7-77-127 (November, 1977).

84. Meissner, R. E. 1983. "Claus Tail Gas Treating with MDEA," paper presented at Union Oil Company of California/The Ralph M. Parsons Company Third BSR and Selectox Users Conference, June 1–2.

85. Beavon, D. K., Hass, R. H., Muke, B. 1979. *The Oil and Gas Journal,* March 12, p. 76.

86. Beavon, D. K., Hass, R. H., Kouzel, B., and Ward, J. W. 1980. "Developments in Selectox Technology" paper presented at 7th Canadiam Symposium of Catalysis, Edmonton, Alberta, October 19–22.

87. Hass, R. H., Ingalls, M. N., Trinker, T. A., Goar, B. G., and Purgason, R. S. 1981. *Hydrocarbon Processing* 60:5, May p. 104.

88. Hass, R. H., Fenton, D. M., Gowdy, H. W., and Bingham, F. E. 1982. "Selectox and Unisulf: New Techniques for Sulfur Recovery," paper presented at International Sulphur '82 Conference, London, England, November 14–17.

89. Goar, B. G. 1982. "World's First Recycle Selectox Sulfur Recovery Unit," paper presented at 32nd Gas Conditioning Conference, University of Oklahoma, March 10.

90. Warner, R. E. 1982. "Save with Selectox," paper presented at Canadian Gas Processors Association Quarterly Meeting, Calgary, Alberta, Sept. 8.

91. Herfkens, A. H. 1982. *Hydrocarbon Processing* 61:11, November p. 199.

92. Allen, D. W., United Catalysts, Inc. Louisville, Kentucky.

93. Katalco 32-4 Desulfurization Catalyst, Katalco Corporation, Oak Brook, Illinois.

94. Engelhard Catalysts for Acetylene Removal from Gas Streams, Technical Bulletin.

95. Huang, W. 1979. *Hydrocarbon Processing* 58:10, October. p. 131.

96. Tucci, E. R. 1982. *Hydrocarbon Processing* 61:3, March. p. 159.

97. Siddigi, A. A., and Tenini, J. W. 1981. *Hydrocarbon Processing,* 60:10, October. p. 115.

98. Oka, H., Eiichi, I., and Tatsuo, S. 1974. *Hydrocarbon Processing* 53:10, October. p. 113.

99. Marzo, L. and Fernandez, L. 1980. *Hydrocarbon Processing,* 59:2, February. p. 87.

100. Hill, H. L. 1981. *Hydrocarbon Processing* 60:2, February. p. 141.

101. Gillespie, G. R., Boyum, A. A., and Collins, M. F. 1972. *Chemical Engineering Progress,* 68:4, April. p. 72.

102. Tanaka, S., and Wiener, R. 1983. "Hitachi Zosen NO_x Flue Gas Treatment Process: Vol. 1. Pilot Plant Evaluation," EPA-600/7-82-057a, March.

103. Burke, J. M. 1983. "Hitachi Zosen NO_x Flue Gas Treatment Process: Vol. 2, Independent Evaluation," EPA-700/7-82-057b, March.

14

Miscellaneous
Gas-Purification Techniques

Absorption with Complex Formation

CO Removal by Absorption in Copper-ammonium-Salt Solutions

Carbon monoxide removal is an important step in the purification of ammonia-synthesis gas produced by the partial oxidation of hydrocarbons, the water-gas reaction, or steam hydrocarbon reforming. A major portion of the carbon monoxide in the raw synthesis-gas mixture is first reacted catalytically with water to form carbon dioxide and hydrogen by the shift-conversion reaction. Carbon dioxide can be removed readily by absorption in water or alkaline solvents as discussed in previous chapters; however, the resulting gas may still contain 2 to 4 percent carbon monoxide which must be removed completely to prevent poisoning of the ammonia conversion catalyst. Although other processes, such as methanation and liquid-nitrogen wash, have been developed to dispose of small quantities of carbon monoxide, the copper-ammonium-salt process has retained a position of importance for many years.

In this process, which is illustrated diagrammatically in Figure 14-1, carbon monoxide at high pressure is absorbed by an aqueous solution of a copper-ammonium salt in a countercurrent contactor with the formation of a cuprous-ammonium-carbon monoxide complex. The preferred compounds for this service are copper-ammonium formate, carbonate, and acetate although other salts of weak acids have been proposed. The solution is regenerated by the application of heat which destroys the complex to liberate almost all of the absorbed carbon monoxide. Because of its mildly alkaline nature, the solution also absorbs carbon dioxide, which is also liberated dur-

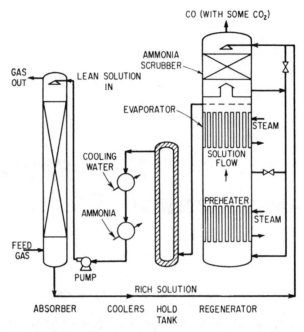

Figure 14-1. Simplified flow diagram of copper-ammonium—salt process for carbon monoxide removal from gases.

ing the regeneration. The process is complicated somewhat by side reactions such as the reduction of cupric to cuprous ions by carbon monoxide, and the autooxidation of cuprous to cupric ions with the precipitation of elemental copper. In order to prevent the latter reaction, it is necessary to provide oxygen to the system to maintain a sufficiently high concentration of cupric ions.

The ability of copper salts to dissolve carbon monoxide was first discovered by Leblanc in 1850 (1). Early interest in the phenomenon centered around its use in gas analysis for the determination of carbon monoxide. The early solutions were generally acid and unsuited for use in conventional equipment because they caused severe corrosion. This problem was ultimately solved by the use of salts of weak acids, such as carbonic and formic, rather than of hydrochloric or sulfuric acids as previously used. The use of cuprous-ammonium carbonate and formate solutions was first described in a German patent in 1914 (2). Since that time the process has been applied in a great many plants, and the chemistry of the system has been studied extensively. The majority of the studies have been concerned with formate and carbonate systems although the complex acetate solution appears to be gaining favor. A view of the CO absorbers and copper-solution regeneration towers of an ammonia plant is shown in Figure 14-2.

Figure 14-2. View of gas-purification unit utilizing copper-ammonium—salt process as part of ammonia plant. *Left foreground:* secondary copper-solution scrubbers (one for each train) and behind them the primary copper-solution scrubbers (one for each train); *right center:* copper-solution rectifying towers (one active, one idle); *right:* water scrubber for ammonia in gas leaving rectifying tower. *Tennessee Valley Authority*

Basic Data

Chemical Reactions. There is no general agreement as to the exact nature of the reaction occurring during the absorption of carbon monoxide by copper-ammonium-salt solutions, although it is generally accepted that the quantity of CO absorbed approaches one mole of CO per mole of cuprous ion, indicating that the compound formed contains these two components in a 1:1 relationship. The cupric ion is inactive in the absorption of carbon monoxide but is a necessary component of the solution because it prevents copper precipitation. Considerable evidence exists that cuprous salts occur primarily in the form $Cu(NH_3)_2^+$ in the presence of excess ammonia. There is some evidence that this complex ion adds at least one additional molecule of ammonia (or possibly water) when CO is added. On this basis, the following carbon monoxide absorption-desorption reaction, which has been proposed by van Krevelen and Baans (3), appears to be reasonable:

$$Cu(NH_3)_2^+ + CO + NH_3 = Cu(NH_3)_3(CO)^+ \qquad (14\text{-}1)$$

Secondary reactions—occurring in the absorber and reversed to some extent in the regenerator—which result in the absorption of carbon dioxide are

$$2NH_4OH + CO_2 = (NH_4)_2CO_3 \qquad (14\text{-}2)$$

$$(NH_4)_2CO_3 + CO_2 + H_2O = 2NH_4HCO_3 \qquad (14\text{-}3)$$

In addition to the above absorption-desorption reactions, certain oxidation-reduction reactions can occur in this system. One of the most important of these is the reduction of cupric to cuprous ion by dissolved carbon monoxide as follows:

$$2Cu^{++} + CO + 4OH^- = 2Cu^+ + CO_3^= + 2H_2O \qquad (14\text{-}4)$$

The reaction in Equation 14-4 depletes the supply of cupric ions in the solution, shifting the equilibrium of the following reaction (Equation 14-5) to the right:

$$2Cu^+ = Cu + Cu^{++} \qquad (14\text{-}5)$$

To avoid the precipitation of elemental copper by the foregoing reaction, it is necessary to maintain about $\frac{1}{5}$ of the copper in the cupric state. This is accomplished by injecting air into the system to oxidize cuprous ion as follows:

$$4Cu^+ + O_2 + 2H_2O = 4Cu^{++} + 4OH^- \qquad (14\text{-}6)$$

Note that in Equations 14-4 to 14-6, Cu^+ and Cu^{++} actually represent complex copper-ammonium ions; however, the ammonia and acid groups have been left out of the equations as they do not play an important part in the reactions.

Vapor-Liquid Equilibria. The vapor-liquid equilibrium relationships for copper-ammonium-salt solutions containing dissolved carbon monoxide have been studied by a number of investigators. Hainsworth and Titus (4) measured the vapor pressure of carbon monoxide over copper-ammonium carbonate solutions. Zhavoronkov and Reshchikov (6) studied solutions of chlorides, formates, lactates, and acetates. Zhavoronkov and Chagunava (7) made a detailed study of formate-carbonate mixtures, including the solution from an operating plant. More recently van Krevelen and Baans (3) derived theoretical equations to correlate the vapor-pressure data of previous investigators and conducted a limited experimental investigation of chloride and formate solutions.

Copper-ammonium carbonate, formate, and carbonate-formate mixtures are of considerable importance industrially, and vapor-pressure data on these systems selected from the above-mentioned investigations are presented in Figures 14-3 to 14-5. Figure 14-3, which is based on the data of Hainsworth and Titus, illustrates the basis for the generally accepted conclusion that carbon monoxide is absorbed by formation of an unstable complex containing one mole of CO per mole of cuprous ion. Figure 14-4 shows the effect of the partial pressure of carbon monoxide on the quantity of CO absorbed by solutions of various compositions. This chart represents data for a single temperature (20°C); however, data for other temperatures can be estimated by reference to Figure 14-5, which shows the effect of temperature on CO vapor pressure for several CO concentrations in a typical solution.

A method of calculating carbon monoxide vapor pressures over copper-ammonium-salt solutions, under conditions outside the range of experimental

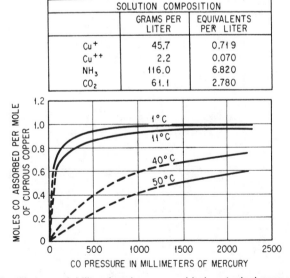

SOLUTION COMPOSITION		
	GRAMS PER LITER	EQUIVALENTS PER LITER
Cu^+	45.7	0.719
Cu^{++}	2.2	0.070
NH_3	116.0	6.820
CO_2	61.1	2.780

Figure 14-3. Equilibrium solubility of carbon monoxide in a typical copper-ammonium carbonate solution as a function of carbon monoxide pressure for several temperatures. Solubility data are presented as moles CO absorbed per mole of cuprous ion, to show approach to 1:1 relationship as pressure increases. *Data of Hainsworth and Titus* (4)

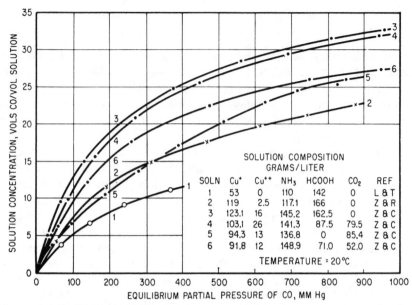

Figure 14-4. Equilibrium solubility of carbon monoxide in various copper-ammonium formate and carbonate solutions as a function of the partial pressure of CO; temperature, 20°C. *Data of Larson and Teitsworth (5), Zhavoronkov and Reshchikov (6), and Zhavoronkov and Chagunava (7)*

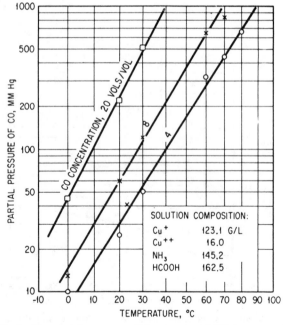

Figure 14-5. Effect of temperature on partial pressure of CO for several CO concentrations in a typical copper-ammonium formate solution. *Data of Zhavoronkov and Chagunava (7)*

data, has been proposed by van Krevelen and Baans (3). These authors concluded that the basic chemical reaction involved in the absorption of carbon monoxide by ammoniacal solutions is expressed in Equation 14-4. By considering equilibrium relationships in this reaction, they derived an equation relating the carbon monoxide partial pressure to the solution composition as follows:

$$\frac{KH}{\gamma_t} = \frac{m}{(m-1)(mB-A)p_{\text{CO}}} = C_{eq} \tag{14-7}$$

where K = equilibrium constant for the reaction in Equation 14-1
   H = Henry's law constant in the equation $(\text{CO}) = Hp_{co}$
   γ_t = the overall activity coefficient
   m = moles of monovalent copper ion per mole of carbon monoxide absorbed
   A = total cuprous salt concentration, g moles/liter
   B = total free ammonia concentration before carbon monoxide absorption, g mole/liter
   p_{co} = partial pressure of carbon monoxide, atm
   C_{eq} = apparent equilibrium constant

Equation 14-7 can be used for calculating CO vapor pressures providing that C_{eq} is known. Van Krevelen and Baans have developed satisfactory equations relating C_{eq} to temperature and ionic strength of the solution for copper-ammonium chloride solutions (based on their own experiments) and copper-ammonium formate solutions (based on the data of Larson and Teitsworth). These equations follow.
For copper-ammonium chloride.

$$\log_{10} C_{eq} = \frac{11,900}{2.3RT} - 0.040I - 8.790 \tag{14-8}$$

and for copper-ammonium formate,

$$\log_{10} C_{eq} = \frac{13,500}{2.3RT} - 0.040I - 9.830 \tag{14-9}$$

where T = the absolute temperature, °K
   I = ionic strength of the solution
   $= \frac{1}{2}\Sigma c_i z_i^2$
where c_i = concentration of the ion i, g moles/liter
   z_i = valence of the ion i

As an example of the application of Equations 14-8 and 14-9, consider the case of a solution with the following analysis:

Cu^+.............. 0.94 g atoms/liter
Cu^{++}.............. 0.02 g atoms/liter
$HCOO^-$.......... 1.72 g equiv/liter
NH_3 (total)....... 7.01 g moles/liter

If it is assumed that each cuprous ion is associated with two molecules of ammonia and one formate ion, while each cupric ion is associated with four molecules of ammonia and two formate ions, the following concentrations can be deduced:

$$NH_3 \text{ (as uncomplexed ammonium formate)} = 1.72 - 0.94 - 2(0.02)$$
$$= 0.74$$

$$NH_3 \text{ (free + complex)} \quad 7.01 - 0.74 = 6.27$$

$$NH_3 \text{ (free)} \quad 6.27 - 2(0.94) - 4(0.02) = 4.31$$

For this system the ionic strength I is equal to

$$\tfrac{1}{2}[0.94 + (0.02)(4) + 1.72 + 0.74] = 1.74$$

For the case of T = 295°K (22°C, 71.6°F) and a carbon monoxide concentration in the solution equivalent to 0.777 moles CO/mole Cu^+ ($m = 1.29$), C_{eq} is calculated to be 1.25 by Equation 14-9 and p_{CO} is calculated to be 0.78 atm by Equation 14-7. This may be compared with an experimental value of about 0.84 atm obtained by van Krevelen and Baans. They show that the equations also fit the data of Larson and Teitsworth remarkably well.

A much simpler, though not as rigorous, relation between carbon monoxide pressure and solution composition is presented by Zhavoronkov and Chagunava (7), and a nomograph for quickly determining the approximate quantity of carbon monoxide absorbed in various copper-ammonium carbonate solutions at 25°C and 1 atm carbon monoxide pressure has been presented by Egalon, Vanhille, and Willemyns (8).

In addition to the carbon monoxide vapor pressure, that of other solution components is of interest in assisting the calculation of possible losses and heat requirements. Copper-ammonium carbonate solutions have an appreciably higher vapor pressure than solutions of the formate or stronger acids because of the relatively high decomposition pressure of ammonium carbonate. The total vapor pressures of a number of solutions have been measured by Zhavoronkov and Reshchikov (6) and Zhavoronkov and Chagunava (7), and some of their data are reproduced in Table 14-1.

Table 14-1. Total Vapor Pressures of CO-free Solutions

Solution	Acid*	Solution vapor pressure, mm HG at				Ref
		0°C	20°C	60°C	80°C	
1	formic	25.8	71.1	347.0	657.0	6
2	formic	16.0	50.0	284.0	591.0	6
3	lactic	27.5	64.5	212.2	453.0	6
4	lactic	31.0	60.0	216.0	454.0	6
5	formic	29.0	69.0	403.0	790.0	7
6	formic-carbonic	45.0	109.0	563.0	1148.0	7
7	formic-carbonic	26.0	64.0	399.0	829.0	7
8	formic-carbonic	12.0	32.0	327.0	......	7
9	carbonic	33.0	94.0	378.0	881.0	7
10	formic-carbonic	19.0	42.0	286.0	670.0	7
11	formic-carbonic	33.0	82.0	447.0	939.0	7

*Detailed compositions of the solutions are given in Table 14-3 in the section entitled "Heat effects."

The vapor pressures of the individual components of typical plant solutions have been determined by Zhavoronkov (9). One set of his data is presented in Figure 14-6 as a plot of pressure versus temperature on a log p versus $1/T$ scale.

Heat Effects. The molar heats of vaporization of ammonia, carbon dioxide, and water from a typical copper-ammonium-salt solution (mixed formate and carbonate) have been calculated by Zhavoronkov (9) using the Clausius-Clapeyron equation and the slopes of the log p versus $1/T$ lines as plotted in Figure 14-6. For the solution illustrated in this figure, he obtained the results in Table 14-2.

As indicated in Table 14-2, heat is required to release any of the volatile components from the solution, and conversely heat is liberated if they are absorbed. The absorption of carbon dioxide can, in fact, account for a considerable portion of the heat generated in the absorber.

The absorption of carbon monoxide is also exothermic, and its heat of absorption can be estimated by a similar analysis of available vapor-pressure data. Zhavoronkov and Reshchikov (6) and Zhavoronkov and Chagunava (7) have done this for the solutions which they studied and some of their results are presented in Table 14-3.

The calculated values for the differential heat of solution do not appear to differ significantly for formate, carbonate, or lactate solutions. A value of about 11,000 cal/g mole of CO absorbed (19,800 Btu/lb mole) is believed to

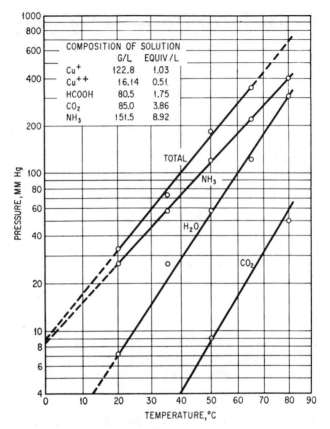

Figure 14-6. Vapor pressures of the individual components of a typical copper-ammonium formate solution. *Data of Zhavoronkov* (9)

Table 14-2. Approximate Heats of Vaporization of Copper-solution Components

Component	Calculated heat of vaporization	
	cal/g mole	Btu/lb mole
Ammonia.............	9,280	16,700
Carbon dioxide.........	15,100	27,200
Water................	12,610	22,700
Mixed vapors..........	10,500	18,900

Source: Mixed formate and carbonate solution, data of Zhavoronkov (9)

Table 14-3. Average Values of Differential Heats of Solution for Carbon Monoxide in Copper-ammonium—Salt Solutions*

Solution	Solution composition, g/liter						Diff. heat of sol. cal/g mole	Ref
	Cu+	Cu++	NH$_3$	Acid				
				Formic	CO$_2$	Lactic		
1	119.0	2.5	117.1	166.0	....	.....	12,000	6
2	104.1	20.0	123.9	166.0	....	.....	9,600	6
3	106.2	2.5	103.1	.....	....	373.0	11,600	6
4	88.7	20.0	106.4	.....	....	373.0	10,600	6
5	123.1	16.0	145.2	162.5	....	.....	12,200	7
6	88.2	15.0	163.4	128.0	40.7	.....	9,600	7
7	103.1	26.0	141.3	87.5	79.5	.....	10,500	7
8	101.3	40.0	156.8	37.5	47.0	.....	9,200	7
9	94.3	13.0	136.8	0.0	85.4	.....	13,100	7
10	131.4	28.0	145.9	158.7	83.0	.....	9,400	7
11	91.8	12.0	148.9	71.0	52.0	.....	13,200	7

*Temperature, 0-80°C.

be a reasonable engineering approximation. Similar calculations based on the data of Hainsworth and Titus (4) for a copper-ammonium carbonate solution indicate a heat absorption of the same magnitude.

Calculation of heat effects also requires a knowledge of the heat capacity of the solution. For engineering estimates, this can be taken to be 0.8 Btu/(lb)(°F).

Design and Operation

Solution Composition. The optimum solution composition is determined by a delicate balance of many factors including capacity for CO, stability, and cost. Since the primary active ingredient is the cuprous ion, it is desirable to have the solution as concentrated in respect to this component as possible. Ammonia increases copper solubility and also increases the effectiveness of the cuprous ion for absorbing carbon monoxide; however, the allowable ammonia concentration is limited by ammonia vapor-pressure considerations. The acid ion is necessary to maintain the copper in solution. The most inexpensive acid is carbonic acid, but carbonate solutions have higher ammonia and CO$_2$ vapor pressures and cannot hold as much copper as do solutions of organic acids such as formic, acetic, or lactic. The organic acids are subject to decomposition and loss; however, their advantages make their use justifiable in many cases. The cupric ion is also a necessary component of the solution. However, too high a concentration is uneconomical as this ion is inactive as a carbon monoxide absorbent.

Table 14-4. Recommended Composition for High Copper Concentration Solution (10)

Component	Concentration, g/liter
Total Cu	170–175
Cu^{++}	25–40
HCOOH	110–120
CO_2	60–70
NH_3	140–150

Table 14-5. Recommended Composition for Low Copper Concentration Solution (10)

Component	Concentration, g/liter
Total Cu	135–140
Cu^{++}	15–20
HCOOH	90–100
CO_2	60–70
NH_3	110–115

A study of copper solubility and the stability of carbonate and formate solutions was made by Pavlov and Lopatin (10). These authors found the maximum possible concentration of copper to be 160 g/liter in carbonate solutions and about 210 g/liter in formate solutions. These values require uneconomically high ammonia concentrations, however. The authors recommend the composition in Table 14-4 for solutions of high copper concentration.

The composition in Table 14-5 is suggested as an alternate for solutions of lower copper content.

The vapor pressure of ammonia over such solutions is appreciably lower than over the more concentrated solutions and ammonia losses are, therefore, reduced.

The analyses of formate-complex solutions from operating installations have been presented by several investigators, and six of these are given in Table 14-6. Data on an acetate solution are presented with other operating data in Table 14-7.

The plant solutions are normally prepared by dissolving metallic copper in a mixture of ammonia, acid, and water. The use of distilled water is desirable since chlorides or sulfates which may be introduced with less-pure water sources can result in corrosion. It is necessary to blow air into the dissolver to oxidize the copper to a soluble form. Oxygen from the air also oxidizes dissolved copper to the cupric state and the cupric copper which is formed is capable of dissolving additional elemental copper by oxidizing it to the solu-

Table 14-6. Analyses of Copper-ammonium—Salt Solutions from Operating Plants

Components	Solution					
	1	2	3	4	5	6
Cu$^+$:						
g/liter................	98.6	90.8	123	135	104	120
g atoms/liter..........	1.55	1.44	1.93	2.12	1.64	1.89
Cu^{++}:						
g/liter................	15	12	16	15	13	16
g atoms/liter..........	0.23	0.19	0.25	0.24	0.21	0.25
HCOOH:						
g/liter................	70.5	71.0	80.8	57.9	55.1	87.7
moles/liter...........	1.53	1.54	1.76	1.26	1.20	1.91
CO$_2$:						
g/liter................	0	52.0	84.9	188	180	180
moles/liter...........	0	1.08	1.93	4.26	4.09	4.09
NH$_3$:						
g/liter................	129	149	151	144	125	145
moles/liter...........	7.60	8.75	8.92	8.47	7.36	8.53
Reference.............	14	7	9	9	15	15

Table 14-7. Typical Plant-Operating Data for Absorption of CO and CO$_2$ in Copper-ammonium—Salt Solution (Courtesy Tennessee Valley Authority) (12)

Operating variables	July	December
Temperatures, °F:		
Gas entering primary scrubber.........................	106	87
Gas leaving secondary scrubber........................	46	36
Solution entering primary scrubber.....................	32	30
Solution leaving primary scrubber.....................	69	65
Pressures:		
Gas entering primary scrubber, psig....................	1635	1750
Gas leaving secondary scrubber, psig..................	1535	1710
Flows:		
Gas entering primary scrubber, cfm*...................	8,350	9,800
Gas leaving secondary scrubber, cfm*.................	8,050	9,425
Gas dissolved in solution (calculated), cfm†.............	300	375
Solution entering primary scrubber, gpm...............	117	119
Solution entering secondary scrubber, gpm.............	7	7
Ammonia addn. to copper solution, lb/min..............	0.8	0.7

Table 14-7 continued

Table 14-7 continued

Operating variables	July	December
Ammonia production in synthesis section, lb/min	163	189
Compositions:		
Gas‡ entering primary scrubber:		
CO_2, %§	0.81	0.72
CO, %§	2.70	2.87
Gas entering secondary scrubber, CO, ppm§	<5	<5
Gas leaving secondary scrubber:		
CO_2, ppm§	<5	<5
CO_2 + CO, ppm§	<5	<5
H_2, %¶	74.6	74.7
CH_4, %¶	0.2	0.2
N_2, %¶	25.2	25.1
Solution entering primary scrubbers:		
Temp. of sample when sp. gr. was measured, °F	88	82
Specific gravity	1.153	1.157
Cu^{++}, g/liter	22.2	21.2
Cu^+, g/liter	84.4	90.3
Total Cu, g/liter	106.6	111.5
Cu^+/Cu^{++}, g/liter	3.80	4.26
NH_3, g/liter	150.1	147.8
HCOOH, g/liter	8.5	10.9
CO_2, g/liter	100.2	92.7
CH_3COOH, g/liter	51.1	55.0
H_2O, g/liter	736.2	739.1
Free NH_3, eq/liter	3.24	3.33
Free NH_3/total Cu, eq/liter	1.60	1.59
Ammonia activity§	3.0	3.0
Solution entering secondary scrubber:		
Temp. of sample when sp. gr. was measured, °F	77	72
Specific gravity	1.091	1.092
Cu^{++}, g/liter	45.7	28.0
Cu^+, g/liter	22.9	63.0
Total Cu, g/liter	68.6	91.0
Total NH_3, g/liter	212.3	208.1
CO_2, g/liter	96.3	76.9
Free NH_3, mole %	17.5	18.0

* Gas volumes are reduced to 1 atmosphere and 60°F (calculated flows).

† This gas contains some H_2, CH_4, and N_2.

‡ Gas entering the primary scrubber contains approximately 0.1 per cent oxygen. Dissolved oxygen in water used in CO_2 scrubber is the source of this oxygen.

§ These analyses were made according to the procedures described by Brown and coworkers.[13]

¶ This analysis was made with an Orsat gas analyzer.

ble cuprous form. This reaction is also of value as a means of controlling the cupric/cuprous ionic ratio in the solution as it is made. Careful control is required during the operation of manufacturing the copper solution to keep sufficient quantities of both ammonia and acid in the system. An excess of acid relative to the ammonia content can cause the solution to become corrosive, while insufficient acid or ammonia can result in precipitation of copper compounds from the solution.

Absorber. A discussion of CO absorption-tower design is presented by Egalon et al. (8). However, the design method which they propose requires data on a commercial column of similar characteristics and, in effect, assumes that the new installation will have the same $K_L a$. In an illustrative example, they present the commercial-column data presented in Figure 14-7 and use these to estimate the size of a new column by assuming the required tower volume to be directly proportional to the quantity of CO absorbed and inversely proportional to the driving force. The driving forces are determined by graphical integration adjusting the equilibrium curves for temperature changes in the columns.

A theoretical treatment of the absorption phenomenon has been presented by van Krevelen and Baans (3). They point out that the physical solubility of CO is very low and that in the presence of excess ammonia the reaction is very rapid. Therefore, in this case, it is probable that the reaction zone in the liquid coincides with the liquid-gas interface. The rate of absorption, which is very rapid, is determined entirely by the mass transfer of reactants and reaction products to and from the interface.

In neutral or weakly ammonical solutions, on the other hand, the rate of absorption is much lower and apparently influenced by the rate of reaction. For the case of a solution containing substantial quantities of carbon dioxide, for example, the chemical-reaction velocity is apparently the rate-controlling factor. Van Krevelen and Baans (3) conducted tests with the solution given in Table 14-8.

They observed an absorption coefficient K_G for CO of 2.1×10^{-6} kg moles/(atm)(sec)(sq m) [1.3×10^{-3} lb moles/(hr)(sq ft)(atm)]. This value was obtained in a glass absorber 30-mm ID by 40 cm high, packed with 5.5-mm glass raschig rings. The absorption experiment was performed with mixtures of N_2 and CO at atmospheric pressure and about 20°C (68°F).

According to Yeandle and Klein (11) a typical CO-absorption unit operating at 1,800 psig will reduce the CO content of the gas stream from about 3.5 percent to less than 25 ppm, using about 75 gpm of copper-ammonium formate solution for 4,000 scf/min of feed gas. They claim that in practical plant experience with a solution inlet-temperature of 0° to 5°C (32° to 41°F), about 70 percent of the theoretical CO/Cu+ ratio of 1:1 can be attained.

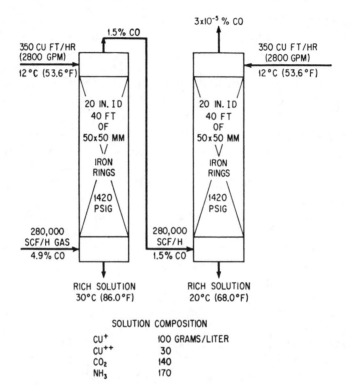

Figure 14-7. Absorption-tower operating data for removal of CO with copper-ammonium carbonate solution. *Data of Egalon et al.* (8)

Table 14-8. Composition of Solution Used in CO Absorption Tests by van Krevelen and Baans (3)

Component	Concentration	
	g moles/liter	g/liter
Cu+	1.67	106.1
Cu++	0.16	10.2
HCOO-	2.35	82.7
Fixed CO₂	1.84	81.0
Total NH₃	8.19	139.0

Detailed operating data on the absorption stages of a plant utilizing a solution containing acetic acid (as well as CO_2 and some formic acid) are presented in Table 14-7. The absorber in this plant is 3 ft in diameter by 61 ft high and is packed with two 21-ft sections of 2 in. steel raschig rings. It is followed by a secondary scrubber, 2 ft in diameter by 47 ft high, which is packed with two 17-ft sections of 1-in. steel raschig rings. The secondary scrubber employs a solution fortified with excess ammonia to remove the last traces of CO_2. The solution from this column is recirculated until it begins to lose CO_2-absorption capacity; at that time it is drained to the regeneration system and replaced with regenerated solution (12).

Solution Regeneration. Regeneration of the copper-ammonium-salt solution is accomplished primarily by pressure reduction and the application of heat. Unfortunately, these operations have other effects, e.g., the vaporization of ammonia and the production of side reactions, which result in the requirement for a somewhat complex regenerator design. The regeneration temperature should be below 180°F in order to minimize the vaporization of ammonia and occurrence of side reactions. Zhavoronkov and Reshchikov (6) recommend the range of 158° to 176°F, while Egalon et al. (8) specify the very narrow range of 174° to 176°F (for copper-ammonium carbonate solution). The pressure of regeneration should be as low as economically feasible. Operation at 1 atm is probably most common, although several patents specify regeneration at subatmospheric pressures (16, 17).

A number of schemes have been proposed to minimize ammonia losses from the regeneration system. The use of a water wash to absorb vaporized ammonia has been proposed by Christensen (18). In this process, the resulting aqueous ammonia solution is distilled to drive the ammonia back into the copper-ammonium-salt solution. A system which is simpler from the equipment standpoint has been described in a patent to Dely (19) which involves the use of the cool copper solution itself to absorb ammonia from the gases escaping from the regeneration zone. A third alternative is to pass the regenerator off-gases to a by-product plant where the ammonia content is recovered by the production of marketable ammonium salts. It is believed that most commercial installations make use of the second method; i.e., washing with cold solution.

The temperature limitation of regeneration (below 180°F) satisfactorily prevents undue losses of ammonia and solution decomposition but, unfortunately, also limits the degree of carbon monoxide removal that is attainable by simple evaporation. The last traces of carbon monoxide can be removed from the solution, however, by permitting the reaction in Equation 14-4 to proceed. By this reaction, carbon monoxide is oxidized to carbonate by the cupric ion, which is itself reduced to the cuprous state. This reaction is very slow at room temperature but quite rapid at 170° to 180°F, so that it can be made to proceed most satisfactorily while the solution is still at regeneration temperature. The rate of cupric-ion reduction by dissolved CO has been investigated in some detail by Dontsova (20), who determined concentration changes in solution by measuring the emf of a concentration cell. The reaction was found to be very slow in the absence of cuprous ion but to accelerate as some cupric ion was reduced. It was concluded that the reaction proceeds with complexed CO and that the cuprous ion increases the rate by bringing more CO into solution.

According to Egalon et al. (8), a solution hold-time of 15 to 20 minutes is sufficient to ensure that the last traces of carbon monoxide will be oxidized. A recent French regeneration-system design, described by these authors, is shown in the flow-sheet, Figure 14-1. In this design, the temperature at the top of the scrubbing section is 40°C (104°F). The solution flows out of the bottom of this section at 65°C (149°F) and into the bottom of the lower section of the column. The solution flows upward through the tubes of the first heat-exchanger section which serves as a preheater, then through the tubes of the second steam-heated exchanger which serves as an evaporator. Vapors from the evaporator pass through a chimney tray and directly up through the scrubber zone while the solution overflows at a temperature of 72°C (161.6°F) into an insulated holding tank, where removal of the last traces of carbon monoxide is achieved by oxidation with cupric ions. This reaction is slightly exothermic, and the solution leaves the holding tank at approximately 77°C (170.6°F). No provision is made in this design for the addition of air to adjust the cupric/cuprous ion-ratio because, in the plant for which the system was designed, sufficient air was present in the gas feed to the absorber to reoxidize the necessary amount of cuprous ion. In other designs, air is added at the base of the regenerator column or in a separate vessel if dilution of the carbon-monoxide stream is undesirable.

The regeneration section of a CO-removal installation in an ammonia plant is shown in Figure 14-8. In this installation, the regenerator surge tanks are horizontal vessels, 9 ft-ID by 73½ ft long, which are visible in the foreground of the illustration. The regeneration towers are each 61 ft high, with a 6-ft ID by 23-ft high reflux section packed with 10½ ft of raschig rings at the top and a 7½-ft ID by 38-ft high preheater section containing two steam baskets in the lower portion. There are two complete regeneration systems in the plant; however, only one is used in the process, the other being a spare. Detailed operating data for this installation are presented in Table 14-9.

Operating Problems. Principal operating problems of the copper-ammonium salt CO-removal process may be listed as follows:

1. Loss of active solution components
2. Control of cuprous/cupric ion-ratio

Figure 14-8. View of regeneration section of CO-removal plant using copper-ammonium—salt solution. Regeneration towers in background; regenerator surge vessels in foreground; solution-preparation tank partially visible at right. *Tennessee Valley Authority*

3. Recovery of pure carbon monoxide or a carbon monoxide-hydrogen mixture
4. Formation of precipitates
5. Corrosion

These problems are, of course, closely interrelated. Improper control of the cuprous/cupric ion-ratio, for example, can result in the precipitation of elemental copper and a consequent loss of one of the major solution components. In the interest of clarity, however, the problems are discussed separately below.

Loss of ammonia with the purified gas from the absorber can be minimized by operating at a sufficiently high pressure and with a low-temperature solution. The loss which is incurred in this operation can be estimated from

Table 14-9. Typical Operating Data for Regeneration System of Copper-ammonium—Salt Process for CO Removal (Courtesy Tennessee Valley Authority) (12)

Operating Variables	July	December
Temperature of solution, °F:		
Entering reflux.	69	65
Leaving reflux.	143–150	142–146
Between preheaters.	173–177	179–181
Leaving preheaters.	174–177	177–179
Center of regenerator (storage).	175–177	176–180
Leaving regenerator (storage).	175–177	176–179
Leaving water cooler.	96–99	75–76
Leaving refrigerated cooler.	28–29	32–33
Entering primary scrubbers.	32	30
Pressure of gas over solution in reflux, preheaters, and regeneration, psig*.	0–1.5	0–1.5
Flows:		
Solution entering regeneration system, gpm†.	248	251
Air to sol. below preheaters, cfm.	0–3.5	0–3.5
Gas (CO_2, CO, H_2, CH_4, N_2) leaving reflux, cfm‡.	600	750
Ammonia make-up, lb/min§.	0.8	0.7
Acetic acid make-up, lb/month.	1070	1070
Formic acid make-up, lb/month.	480	480
Total volume of sol. in gas scrubbers and regeneration system, gal.	25,250	26,050
Typical composition¶ of gas leaving reflux, %:		
CO_2.	17.7	15.1
CO.	70.2	77.0
H_2.	3.6	2.6
CH_4.	0.4	0.6
N_2.	2.6	2.0
NH_3.	5.5	3.1

* This pressure is varied by means of a damper installed in the gas vent of the reflux tower.

† Total from primary scrubbers.

‡ This volume does not include NH_3 or H_2O.

§ Liquid anhydrous ammonia is added to the copper solution near its inlet to the refrigerated cooler.

¶ Composition of the gas from which NH_3 had been removed was determined by means of an Orsat analyzer. The NH_3 in the gas was determined by means of a compensometer.

partial-pressure data such as those presented in Figure 14-6. Ammonia losses can be more serious in the regeneration step because of the higher temperatures and lower pressures involved, and, as mentioned in the preceding section, several process schemes have been developed to minimize losses from this operation. When ammonia is recovered by washing the exhaust gases with the rich copper solution, it is important that this solution be as cool as possible. Since a temperature rise occurs in the absorber, this in turn means that the lean solution fed to the absorber should be at the lowest possible temperature. Plants generally operate with the lean-solution temperature in the range of 32° to 68°F with perhaps 40°F a preferred value.

When copper-ammonium carbonate is employed in the solution, the carbonate content equilibrates at a satisfactory operating level as a result of the absorption of CO_2 from the gas being purified and the stripping of an equal quantity in the regeneration system. With copper-ammonium formate, on the other hand, a gradual loss of formic acid occurs. It is conceivable that some of this loss can result from vaporization of formic acid in the absorber and regenerator; however, the vapor pressure of formic acid over solutions of this type is extremely low, and it is believed that this loss is negligible. Of considerably more importance is the decomposition of formic acid which is probably due to oxidation to carbon dioxide. In the presence of cupric ion, it would be expected that the reaction would proceed as follows:

$$HCOOH + 2Cu^{++} + 2(OH)^- = CO_2 + 2H_2O + 2Cu^+ \qquad (14\text{-}10)$$

If formic acid is lost more rapidly than ammonia or if it is not replaced at the rate it is lost, the resulting excess of ammonia reacts with carbon dioxide so that the solution gradually approaches the copper-ammonium carbonate composition. This problem can be minimized by the use of more stable organic acids such as acetic or lactic. Cuprous lactate has also been reported to have the advantage of a high stability to reduction by CO (21).

The cuprous/cupric ion-ratio can generally be controlled at the desired value (about 5:1 on a weight basis) by adding the proper amount of oxygen with the feed gas or as air to the regenerator. In some cases where excess oxygen is present, control of the ratio by reduction of Cu^{++} with CO is required. The problem associated with this operation is in determining the amount of oxidation or reduction required. This can only be accomplished by regular or continuous analysis of the solution. An automatic system for accomplishing this has been developed by Brown et al. (13).

In a conventional regeneration system, the CO produced is contaminated with CO_2 and also with nitrogen (if air is added to the regenerator). If pure CO or a CO-H_2 mixture is desired for recycle to the shift converter, special steps are required. The extent to which these are warranted depends on the relative value of the recovered CO. Some increase in purity can be realized by adding air in a separate vessel (similar to the hold tank in Figure 14-1), so that the gas from the primary regenerator is at least nitrogen-free. A relatively CO_2-free stream of $CO-H_2$ can be obtained by stripping in stages. Extra wash operations are required, however, to prevent losses of ammonia with the gas vented from each stripping stage or air-oxidation step.

In addition to copper precipitation, which has been mentioned previously, it is possible for other solids such as basic copper carbonate to precipitate from the solution under some conditions. Such difficulties are most frequently caused by an insufficient quantity of ammonia or of acid ions in the solution. These will show up in solution analyses and can readily be corrected by addition of the proper solution components.

The copper-ammonium-salt solutions are generally not corrosive to mild steel; however, the gases evolved during regeneration can be quite corrosive due to the presence of carbon dioxide. To prevent corrosion in the vapor zones, it is good practice to use stainless steel (vessels or liners) in the exhaust-gas scrubber section and above the liquid in the evaporator section. Rubber-lined equipment has also been proposed (22).

Cosorb Process

The Cosorb process, developed by Tenneco Chemicals, Inc. in the early 1970s, was initially aimed primarily at the recovery of high-purity carbon monoxide from nonconventional gas-stream sources (92). However, the process can also serve as a gas purification technique when, for example, CO is removed from synthesis gas to yield purified streams of both hydrogen and carbon monoxide.

The Cosorb process resembles the copper-ammonium-salt process described in the preceding section in that it also uses a copper compound which forms a complex with absorbed CO. It is significantly different, however, in that the absorbent is nonaqueous. The active component is cuprous aluminum chloride ($CuAlCl_4$) dissolved in an aromatic base (toluene) (93).

The absorbent is inert to gases such as hydrogen, carbon dioxide, methane, and nitrogen normally found in synthesis gas, although these compounds exhibit some degree of physical solubility in the toluene base. Water in the feed gas reacts quantitatively with the active ingredient in the absorbent to form HCl gas and a waste product which is soluble in the solvent. Because of this reaction, it is necessary to predry the feed gas to a Cosorb unit, preferably to less than 1 ppm H_2O. Other gaseous impurities, which may be present in the feed gas such as hydrogen sulfide, sulfur dioxide, and ammonia, can react with the absorbent and should also be reduced to low levels before the gas enters the Cosorb process absorber.

The Cosorb process consists of four sections:

1. Feed gas preparation
2. The complexor-decomplexor circuit
3. Aromatic recovery
4. Gas compression

The feed gas preparation section typically includes refrigeration and molecular sieve adsorption for water removal and activated carbon adsorption for removal of sulfur compounds to acceptable levels.

The complexor-decomplexor section is a standard absorber-stripper loop using tray or packed absorption and stripping columns, a steam-heated reboiler, a conventional rich-lean solution heat exchanger, and a lean solution cooler which uses plant cooling water. The solvent is heated to about 135°C in the stripper. Optionally, a third column or flash tank may be used to remove dissolved inert gases from the rich solution before it enters the stripper.

Because of the high volatility of the aromatic solvent used in the process, it is necessary to include aromatic recovery operations on all product gas streams. This is accomplished for the absorber and stripper off-gas streams by a combination of compression, refrigeration, and treatment with activated carbon. Chilling is generally adequate to recover aromatics from the small amount of gas produced by the inerts removal column.

It is reported that mild steel is generally applicable as the material of construction for Cosorb units; however, some specifically selected stainless steels, copper nickel alloys, and brass are also acceptable (92).

Low-Temperature Gas-Purification Processes

Separation of gaseous mixtures by fractional condensation and distillation at low temperature has been practiced extensively since Carl von Linde disclosed his air-liquefaction process at the beginning of the twentieth century. In general, low-temperature processes are not aimed at the removal of small amounts of impurities from gas streams but rather at the separation and recovery of pure components, e.g., oxygen, nitrogen, helium, carbon monoxide, hydrogen, and various hydrocarbons and, therefore, cannot be called true gas-purification processes. However, low-temperature techniques are used in such processes as the purification of the hydrogen used for ammonia synthesis, the recently proposed CNG acid gas removal process and the Rectisol process. In these processes the crude gas is first precooled and some of the impurities are removed by condensation. Final purification is attained by absorption of the remaining impurities in liquid nitrogen, liquid carbon dioxide and methanol.

One process which, in a broad sense, may be classified as a gas-purification process, involves low-temperature separation of methane and nitrogen to increase the heating value of natural gas containing substantial amounts of nitrogen. A typical version of the process employing a rather complicated flow scheme has been described by Streich (23). A commercial installation operating in France is reported to reduce the nitrogen content of natural gas originating in the Groningen field in Holland, from about 14 percent to about 2.5 percent.

The nitrogen is separated by a two-stage distillation, with practically all of the separation work, energy losses, and refrigeration demands being provided by pressure reduction of the gas. The product gas leaves the plant in two streams, one containing about 70 percent of the methane feed at 355 psig and the other, containing the balance of the methane, at 16 psig. Both streams contain approximately 2 percent nitrogen. The waste nitrogen is essentially free of hydrocarbon, containing less than 0.5 percent of methane (23). A discussion of nitrogen removal from natural gas with several flow schemes for processing gases of different nitrogen concentrations has been presented by Harris (94).

Removal of Gas Impurities by Condensation and Absorption in Liquid Nitrogen

Separation of hydrogen from coke-oven gas by liquefaction of the higher-boiling components was first practiced on an industrial scale in France and Germany shortly after the end of World War I. Since then this technique has been applied successfully to the separation of pure hydrogen from a variety of industrial-gas streams, such as petroleum refinery off-gases and the product streams from steam hydrocarbon-reforming and the partial oxidation of natural gas. As stated earlier, the process consists of successive steps of gas cooling with fractional condensation of the impurities followed by absorption of the carbon monoxide and methane still present in the gas in a stream of liquid nitrogen. Hydrogen containing about 2 percent carbon monoxide can be obtained from coke-oven gas without nitrogen wash (24). However, most modern installations include a liquid-nitrogen wash column, especially if the product gas is used for ammonia synthesis.

Process Description

Low-temperature gas-purification plants consist essentially of complex refrigeration and heat-exchange systems. The various designs used in industrial installations differ principally in the method of refrigeration and the arrangement of heat-exchange equipment. Either an external coolant such as nitrogen is used in the main refrigeration cycle or the purified gas itself serves as the refrigerant. The use of nitrogen in the refrigeration cycle has the advantage of permitting the purified gas to leave the plant under pressure. Various process-design modifications have been described by Guillaumeron (24), Ruhemann (25), Bardin and Beery (26), Baker (27), and Jester (28).

A flow diagram of a typical low-temperature gas-purification process operated at elevated pressure for the production of pure hydrogen for ammonia synthesis is shown in Figure 14-9. This flow scheme represents the process of the Gesellschaft für Linde's Eismachinen A.G. of Munich, Germany (German Linde Company), in which conventional refrigeration (ammonia) is used for precooling the feed gas and nitrogen, and a high-pressure nitrogen cycle is used for the primary refrigeration requirements. Similar refrigeration methods are used in other processes, such as those developed by L'Air Liquide (24) and Air Products (27), when it is desired to obtain the purified-hydrogen stream under pressure.

In the process shown in Figure 14-9 the hydrogen-rich feed gas, which is usually at a pressure of 150 to 300 psig, is first precooled by returning gases

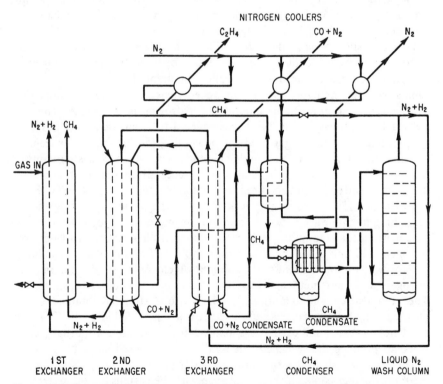

Figure 14-9. Schematic flow diagram of low-temperature gas-purification process using high-pressure nitrogen-refrigeration cycle and liquid-nitrogen wash column. *From Ruhemann (25)*

before it enters the low-temperature apparatus. The gas is then dehydrated and further cooled (to −50°F) by use of a conventional ammonia cycle. High-purity nitrogen obtained in an air-separation plant is compressed to approximately 3,000 psig and cooled to −50°F together with the feed gas. By following the flow of the different streams in Figure 14-9, it can be seen that the precooled gas passes first through three heat exchangers in which it is cooled by outgoing products, i.e., evaporating methane, carbon monoxide, and nitrogen from the bottom of the liquid-nitrogen wash column, and hydrogen-nitrogen product from the top of the column. In the first exchanger, where the gas temperature is lowered to about −150°F, small amounts of liquid hydrocarbons are condensed and discharged periodically. The gas temperature is reduced to about −230°F in the second exchanger; this results in the condensation of the so-called ethylene fraction, which contains most of the ethylene contained in the gas, the remaining heavier hydrocarbons, and a small amount of methane. The ethylene fraction is vaporized and used to cool a portion of the incoming nitrogen. In the third exchanger the gas is cooled to approximately −290°F by evaporating methane and a mixture of carbon monoxide and nitrogen. Additional methane and ethylene are condensed in this operation.

After passing through the three exchangers, the gas enters the methane condenser where it is cooled to about −300°F by liquid nitrogen boiling at atmospheric pressure. Here the so-called methane fraction, consisting of almost all the methane, the remaining ethylene, some carbon monoxide, nitrogen, and an appreciable quantity of hydrogen, is condensed. The condensed liquid flows to the third exchanger after a portion of it has been evaporated in an exchanger used for precooling the incoming nitrogen.

The effluent gas from the methane condenser, which consists mainly of hydrogen, carbon monoxide, and small traces of methane, is contacted countercurrently in a plate column with liquid nitrogen at about −300°F and a pressure of 150 to 300 psig. Practically all of the carbon monoxide and the remaining methane are removed from the gas in this operation. The overhead stream from the column contains 85 to 95 percent hydrogen, 5 to 15 percent nitrogen, and only a few ppm of carbon monoxide and methane. The liquid withdrawn from the bottom of the column, which contains carbon monoxide, nitrogen, and some methane, is vaporized and used to cool the incoming gas. The composition and quantities of the various streams obtained in the purification of a typical coke-oven gas are given in Table 14-10 (25). The ethylene, methane, and carbon monoxide-nitrogen streams are usually combined to produce the so-called "rich gas" which is used as fuel.

The precooled, high-pressure nitrogen entering the low-temperature unit at −50°F is further cooled by heat exchange with the various outgoing streams, as shown in Figure 14-9, and is used for three different purposes. One part of the cooled nitrogen is expanded to atmospheric pressure, liquefied, and used as refrigerant in the methane condenser. The second part is flashed to the operating pressure of the nitrogen-wash column, liquefied in coils located in the methane condenser, and fed into the top of the wash column. The third part is reduced in pressure and added to the overhead stream from the wash column to establish the required hydrogen/nitrogen ratio in the final product stream.

The basic process scheme is amenable to modification and can be adapted to the treatment of a variety of gas streams. Several modifications have been described by Bardin and Beery (26).

A process using the purified hydrogen as the main refrigerant was developed by Claude and is known as the L'Air Liquide process. A flow diagram of this process is shown in Figure 14-10. The feed gas is first compressed to 175 to 220 psig and flows through one of two alternate heat exchangers (the so-called warm exchanger) where it is cooled by returning streams, to $-85°F$. Two exchangers are usually provided to permit periodic removal of accumulated solids without interruption of the gas flow. The gas then enters the second (cold) exchanger where it is cooled to about $-150°F$ and the ethylene fraction is condensed. After passage through an additional heat exchanger, the gas flows through the tubes of the methane condenser in which the methane fraction is liquefied. The liquid from the bottom of the condenser is expanded and introduced into the shell-side of the vessel, thus providing refrigeration. The effluent gas, which still contains most of the carbon monoxide and some methane, flows to the final condenser in which most of these impurities are condensed. Refrigeration is supplied partly by evaporation of condensed liquid and partly—in the upper section—by the product gas which has been cooled by pressure reduction (to 1 atm) in an expansion engine. Because of the negative Joule-Thomson coefficient of hydrogen, external work in an expansion engine is required to obtain the desired lowering of the gas temperature. The expanded gas, which is at a temperature of about $-330°F$, is used as refrigerant in the final condenser and in the two exchangers as shown in the diagram.

The purified gas from such an installation contains about 2 to 3 percent carbon monoxide if a typical coke-oven gas is processed. In cases where gas of higher purity is required, the remaining carbon monoxide is removed in a liquid-nitrogen wash column which operates in essentially the same manner as described above. The nitrogen, which usually enters the system at a pressure of 300 to 450 psig, is cooled and liquefied by heat exchange with a portion of the vaporized liquids from the two condensers and the bottoms from the wash column and is then expanded and fed into the top of the wash column. The overhead and bottom products from the column are heat exchanged with the feed gas and, as stated above, with the incoming nitrogen.

The L'Air Liquide process has been described in some detail by Ruhemann (25), and Guillaumeron (24). The process has some advantages, especially

Table 14-10. Composition of Gas Streams in Low-Temperature Purification of Coke-oven Gas

Component	Feed gas, per cent by vol	Ethylene fraction, per cent by vol	Methane fraction, per cent by vol	Carbon monoxide fraction, per cent by vol	Product, per cent by vol
H_2	49.3	. . . . .	4.2	. . . . .	75.0
CH_4	26.6	30.6	74.4	6.9	
N_2	13.7	. . . . .	8.7	73.1	25.0
CO	6.6	2.0	9.6	18.0	
C_nH_m*	1.8	36.7	1.9		
C_2H_6	1.0	30.7			
O_2	1.0	. . . . .	1.2	2.0	
Total.....................	100.0	100.0	100.0	100.0	100.0
Relative quantity, SCF/min..	100.0	3.2	32.6	18.9	64.0

*Mostly ethylene with some propylene

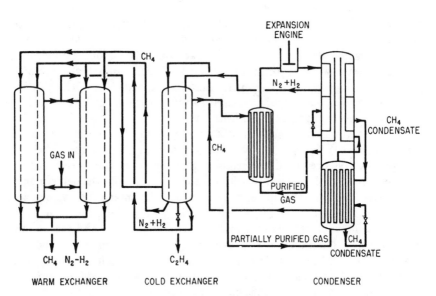

Figure 14-10. Schematic flow diagram of low-temperature gas-purification process using the purified gas as refrigerant. *From Ruhemann* (25)

elimination of the high-pressure nitrogen cycle, when purified gas at atmospheric pressure is satisfactory. However, in cases where high-pressure gas is required, refrigeration by means of the high-pressure nitrogen cycle appears to be more economical (24).

Process Design

The fact that hydrogen boils at a considerably lower temperature than all the other compounds present in the gas streams treated in low-temperature processes permits essentially complete removal of most impurities by fractional condensation. However, in order to obtain complete separation, it is necessary to cool the gas mixture to a temperature considerably below the boiling point of the constituent to be removed. This temperature can be estimated from the vapor pressures of the pure constituents, shown in Figure 14-11 (29), assuming ideal-gas behavior.

Calculations of condensation curves for binary gas mixtures have been presented by Ruhemann (25) and Guillaumeron (24). The example given by Ruhemann shows that a temperature of $-190°C$ ($-310°F$) is required for essentially complete condensation of methane from a mixture of 70 percent hydrogen and 30 percent methane at a pressure of 8 atm. In the case given by Guillaumeron where carbon monoxide is removed from a mixture containing 60 percent hydrogen and 40 percent carbon monoxide, it is shown that at a pressure of 20 atmospheres, the temperature has to be lowered to $-210°C$ ($-346°F$) in order to remove the carbon monoxide, to a point where the final concentration is 0.3 percent.

The refrigeration required in low-temperature processes can be estimated from the specific heats and the latent heats of vaporization (or condensation) of the constituents of the gas mixture. Values for the most frequently encountered gases are given in Tables 14-11 and 14-12.

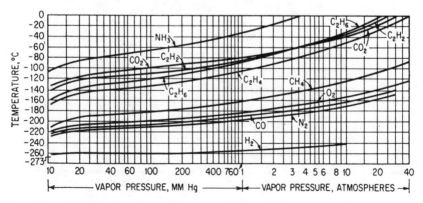

Figure 14-11. Vapor pressures of various gases at low temperatures; left half in mm Hg right half in atm. *Data of Stull* (29)

While the sensible heat of the incoming gas can be removed by heat exchange with cold product-streams, the latent heat required for condensing the impurities is much greater and can most conveniently be supplied by a boiling liquid. Since the entropy of the system increases with the difference between the condensation temperature of the impurity and the boiling temperature of the refrigerant, it is advantageous to use liquids boiling close to the condensation temperatures desired. Typical liquid refrigerants used in the process are methane, carbon monoxide-nitrogen mixtures, and nitrogen. A detailed discussion of the calculation method as applied to the separation of coke-oven gas by partial condensation is presented by Ruhemann (25).

The design of the liquid-nitrogen wash column for CO removal has been investigated by Ruhemann (25) and Ruhemann and Zinn (36). Since the boiling points of nitrogen and carbon monoxide are only 6°C (10.8°F) apart, the temperature gradient from the top to the bottom of the column is quite small. By neglecting this small temperature effect and by assuming equal volumes of liquid and vapor on each tray (because of the almost identical latent heats of

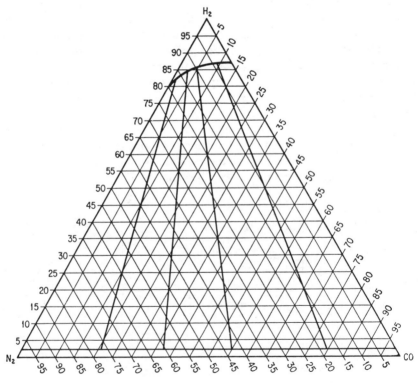

Figure 14-12. Equilibrium diagram of the system CO—H₂—N₂ at −190°C (−310°F) and 12 atm. *Data of Ruhemann and Zinn (36)*

Table 14-11. Boiling Points and Latent Heats of Vaporization of Gases

Gas	Boiling Point, °C	Latent heat of vaporization at normal boiling point and 1 atm, calories/g mole
Hydrogen (30)	−249.4	206
Oxygen (31)	−183.0	1630
Nitrogen (32)	−195.8	1333
Carbon monoxide (33)..	−191.5	1443
Methane (31)	−161.5	1955
Ethane (31)	−88.6	3517
Ethylene (35)	−103.8	3237

Table 14-12. Specific Heats of Gases at Constant Pressure* (cal/g mole)

Gas	Temperature, °C				
	−50	−80	−100	−140	−200
Hydrogen (30)	6.654	6.472	6.309	5.864	5.120
Oxygen (30)	6.990	6.994	7.004	7.051	
Nitrogen (30)	6.978	6.974	6.994	7.034	
Carbon monoxide (33)	6.99	7.01	7.05	7.19	
Methane (34)	.....	7.97 (−74°C)	7.21 (−115°C)	7.24 (−131°C)	

*One atmosphere.

vaporization of nitrogen and carbon monoxide), the minimum amount of liquid nitrogen required to remove carbon monoxide can be determined from the ternary equilibrium diagram CO-H_2-N_2. A diagram representing typical operating conditions of commercial columns is shown in Figure 14-12 (36). Additional diagrams covering temperatures of −183 and −195°C (−298 and −319°F) and pressures of 20, 26, 35, and 50 atmospheres are given by Ruhemann and Zinn (36). Data reported by Guillaumeron (24) and by Bardin and Beery (26) indicate that the quantity of nitrogen used in commercial treatment of coke-oven and refinery gases amounts to about 10 and 20 percent of the inlet-gas volumes, respectively.

The number of theoretical plates required for a specified gas purity may be determined graphically from the equilibrium diagram by a method also developed by Ruhemann (25). In practice, the number of actual plates required is considerably higher than would appear theoretically necessary, indicating very low plate-efficiency.

The design of liquid-nitrogen wash columns is always based on carbon monoxide removal, even when the gas to be treated contains methane. If the column is designed properly, all the methane will also be condensed and leave the column with the liquid withdrawn from the bottom.

A modification of the nitrogen wash, as part of the purification system of ammonia synthesis gases obtained by steam-hydrocarbon reforming, has been described by Grotz (37). In this process a low-temperature purifier is provided after methanation and drying of the synthesis gas, and methane is removed almost completely by washing with condensed nitrogen. The nitrogen required is obtained by using more than the stoichiometric amount of air in the secondary reformer.

The advantages claimed for this process include substantial savings in the overall process fuel requirements and capital savings in the ammonia synthesis loop, as purge requirements are minimized. The refrigeration requirements are supplied by passing the gas through a turboexpander which, however, results in a pressure loss of only about 30 to 50 psi.

The CNG Process

This process which was described by Hise et al. (95) is a joint development of CNG Research Corporation, the U.S. Department of Energy, and Helipump Corporation. The process, which is still in the developmental stage, is proposed for acid gas removal from coal-derived gases available at pressures above 300 psig and containing high concentrations of carbon dioxide, typically in excess of 25 mole percent. The carbon dioxide is condensed at a temperature near its triple point and subsequently used as the solvent for the removal of H_2S, CO_2 and other impurities such as COS, CS_2, HCN, and mercaptans.

A simplified flow diagram of the process is shown in Figure 14-13. The crude gas (containing 32.45 mole percent CO_2 and 1.13 mole percent H_2S) is first cooled in two heat exchangers by exchange with the purified gas, to $-56°C$ ($-69°F$), near the carbon dioxide triple point. Water is removed in the first exchanger and in a subsequent dehydration system. About 67 percent of the CO_2 contained in the crude gas is condensed in the second heat exchanger. The next step is complete isothermal absorption of H_2S and all other trace impurities from the main gas stream in Absorber C-2 by a stream of pure liquid CO_2. The CO_2 remaining in the gas stream—about 14 percent—is subsequently contacted in Absorber C-1 with a refrigerant-absorb-

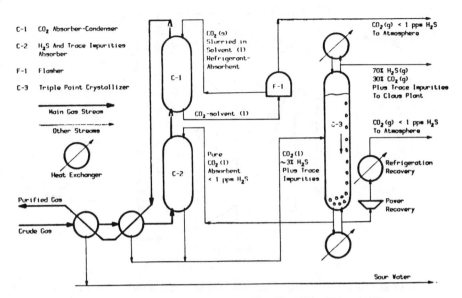

C-1	CO_2 Absorber-Condenser
C-2	H_2S And Trace Impurities Absorber
F-1	Flasher
C-3	Triple Point Crystallizer

Figure 14-13. CNG Acid Gas Removal Process Simplified Flow Diagram (95).

ent consisting of CO_2 crystals slurried in a solution of CO_2 in an organic liquid. The heat of absorption of CO_2 melts the CO_2 crystals which thus provide refrigeration. The cold purified gas exiting Absorber C-1 supplies refrigeration for the incoming crude gas.

The contaminated liquid CO_2 streams from the second heat exchanger and Absorber C-2 are combined and treated in a triple point crystallizer which produces extremely pure CO_2 (less than 1 ppm impurities) and a stream of high H_2S content gas (70 percent H_2S, 30 percent CO_2) which is suitable for processing in a Claus unit.

Adiabatic flashing at the top of the crystallizer cools the liquid and causes CO_2 crystals to form. The crystals fall to the bottom of the crystallizer where they are melted by vaporized liquid withdrawn from the bottom of the crystallizer. Both liquid and vapor are withdrawn from the bottom of the crystallizer, with pure CO_2 vapor rejected to the atmosphere, after recovery of refrigeration and power, and pure CO_2 liquid recycled as solvent to the H_2S absorber.

The organic liquid leaving the bottom of the CO_2 Absorber C-1 is regenerated by adiabatically flashing pure CO_2 to the atmosphere to form the slurry absorbent, and then recycled to the CO_2 absorber.

The advantages claimed for the process are complete removal of all sulfur compounds and trace impurities, production of a gas stream rich in H_2S, production of pure CO_2, low solvent flow rates, and low energy consumption.

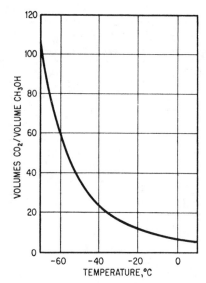

Figure 14-14. Solubility of carbon dioxide in methanol; partial pressure of carbon dioxide = 1 atm. *Data of Herbert* (38)

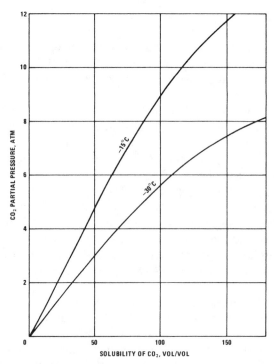

Figure 14-15. Effect of partial pressure on solubility of carbon dioxide in methanol. *Data of Hochgesand* (39)

The Rectisol Process

In this process, which was developed in Germany by Lurgi Gesellschaft für Wärmetechnik, removal of carbon dioxide, hydrogen sulfide, organic sulfur compounds, hydrogen cyanide, benzene, and gum-forming hydrocarbons from synthesis gases is accomplished by physical absorption in methanol at relatively low temperatures. Advantage is taken of the fact that these compounds, especially carbon dioxide and hydrogen sulfide, are highly soluble in methanol at low temperatures and elevated pressures and readily released from the solvent when the pressure is reduced. The solubility of carbon dioxide in methanol as a function of temperature at a partial pressure of 1 atm is shown in a Figure 14-14 (38). The effect of partial pressure on the equilibrium solubility at two temperatures is shown in Figure 14-15 (39). The heat requirements of the process are very low because the solution is cooled by pressure reduction in the regeneration step, and the inlet gas is refrigerated by efficient heat exchange with the outgoing purified and acid-gas stream.

The principal advantages claimed for the process are (a) considerably lower energy consumption than is required in conventional processes for acid-gas removal (such as absorption in water or in ethanolamine solutions), (b) satisfactory removal of all undesirable impurities in a single process, and (c) production of a product gas of very low water content (38, 40).

An additional advantage of the process is the considerably higher solubility of hydrogen sulfide, over that of carbon dioxide, in methanol, permitting selective removal of hydrogen sulfide from gas streams containing both acid gases (42). Equilibrium solubilities of hydrogen sulfide and carbon dioxide in methanol at a partial pressure of 1 atm and at two different temperatures are shown in Table 14-13. The main disadvantages of the process are (a) its complex flow scheme and (b) relatively high vaporization losses of the solvent, caused by the appreciable vapor pressure of methanol even at low temperatures. These losses may be estimated from Figure 14-16 (41). The process appears to be best suited for the treatment of gases containing large amounts of impurities—primarily carbon dioxide and hydrogen sulfide—at pressures above 150 psig.

Process Description and Operation

The basic flow scheme of the Rectisol process as described by Herbert (38) for the treatment of synthesis gas produced by coal gasification is shown diagrammatically in Figure 14-17. The inlet gas enters at the bottom of a two-stage contactor which operates at elevated pressure, typically 300 psig. In the first stage where the bulk of the CO_2, practically all of the H_2S and hydrocarbons, and an appreciable amount of organic sulfur compounds are

Table 14-13. Equilibrium Solubilities of H₂S and CO₂ in Methanol

Temperature °C	Solubility, vol/vol H₂S	CO₂	Selectivity H₂S/CO₂
−10	41	8	5.1
−30	92	15	6.1

Source: Data of Hochgesand (43)

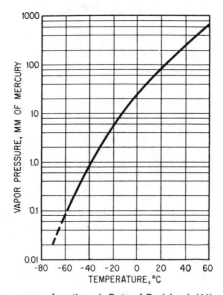

Figure 14-16. Vapor pressure of methanol. *Data of Dreisbach (41)*

removed, the gas is washed countercurrently with methanol, which is fed into the middle of the column at a temperature of approximately −100°F. The temperature of the solvent increases due to the heat of absorption, until it reaches −4°F at the contactor outlet. The solvent is regenerated by two successive pressure reductions and flashing of the dissolved gases. In the first step in which the pressure is reduced to one atmosphere, the solvent is cooled to about −30°F, and in the second step in which the pressure is lowered to 3 psia, the temperature of the methanol is reduced to −100°F. The cooled regeneration solvent which still contains some CO_2 is recycled into the middle of the absorption column.

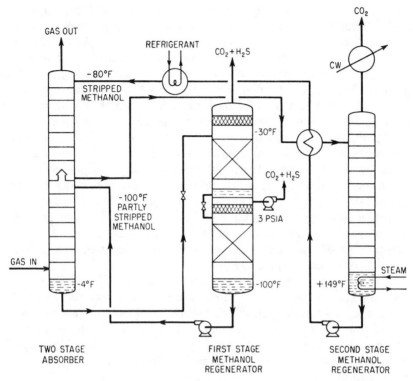

Figure 14-17. Schematic flow diagram of basic Rectisol process.

The partially purified gas leaving the lower section of the contactor flows to the second treating stage, located in the upper portion of the column, where it is washed countercurrently with a small stream of thoroughly stripped methanol which enters the column at about −80°F. In this operation most of the remaining CO_2 and practically all of the residual organic sulfur compounds are removed from the gas.

The rich solvent is withdrawn at the bottom of the second contacting stage, stripped of acid gas by heating with indirect steam in a conventional stripping column, cooled, and recycled into the top of the contactor.

The basic flow scheme may be modified in various ways: precooling of the feed gas by heat exchange with the purified gas, successive flashing at three different pressure levels with recycle of the gases disengaged in the first stage to the contactor inlet, regeneration by using inert stripping gas, and regeneration at elevated temperature.

Hochgesand (43) describes two flow schemes suitable for the purification of hydrogen or ammonia synthesis gas obtained by partial oxidation of hydrocarbon liquids. In the first version, shown in Figure 14-18, the crude gas, which after quenching and scrubbing with water for carbon removal is available at ambient temperature and a pressure ranging from 500 to 800 psig, is purified in two stages, both employing the Rectisol process. The first stage which precedes shift conversion, serves to reduce the sulfur content of the gas to a sufficiently low level to permit the use of low-temperature shift conversion catalyst. The second Rectisol stage, following shift conversion, reduces the carbon dioxide content of the gas stream to about 0.1 percent.

Following the flow in Figure 14-18, it is seen that the feed gas to the first Rectisol stage, which contains typically 5 to 6 percent carbon dioxide and 1 percent hydrogen sulfide and carbonyl sulfide, is precooled by heat exchange with the outlet gas and by ammonia refrigeration. It is then contacted with a stream of partially regenerated methanol withdrawn from the second flash of the carbon dioxide removal section of the plant. The sulfur-free gas passes through the shift converter, is again cooled and is then treated in the second Rectisol stage where the carbon dioxide content is reduced from about 35 percent to less than 0.1 percent.

The methanol leaving the first stage absorber is flashed at an intermediate pressure level, and the disengaged gases are recompressed and recycled to the absorber inlet. Hydrogen sulfide and carbonyl sulfide are subsequently stripped from the methanol in a reboiled regeneration column. Residual carbon dioxide is removed from the partially regenerated methanol by stripping nitrogen, and the completely regenerated methanol is fed to the top of the second absorber.

The methanol leaving the second (carbon dioxide removal) Rectisol stage is regenerated by flashing at two pressure levels and subsequently by stripping with nitrogen. One portion of the partially regenerated methanol flows to the first stage Rectisol absorber, and the other is fed to the middle of the second stage absorber.

The second version, which is suitable for acid-gas removal from high-pressure partial oxidation gases (on the order of 1,200 psig), differs from the first in that the gas is passed through the shift converter prior to acid-gas removal. The flow scheme which in some respects is similar to that described above is shown in Figure 14-19. This process also involves two absorption stages, one for selective removal of sulfur compounds and the second for removal of carbon dioxide.

A large industrial Rectisol plant used for the purification of gas obtained by coal gasification in Lurgi gasifiers at the Fischer-Tropsch synthesis plant of South African Oil, Coal, and Gas Corporation (Sasol) has been described by Hoogendoorn and Solomon (40) and by Ranke (96). This installation consists of three identical purification units with a total capacity of 164 MMscf/day of gas and a common regeneration section.

A schematic flow diagram showing one purification unit is presented in Figure 14-20. The inlet gas to each unit, which is at a pressure of 350 psig, is

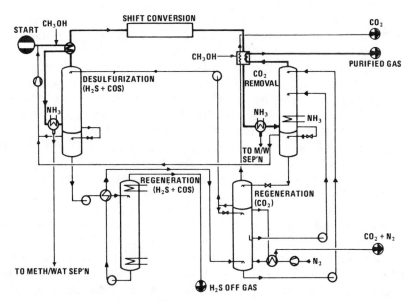

Figure 14-18. Flow diagram of Rectisol process for selective hydrogen sulfide removal, followed by carbon dioxide removal.

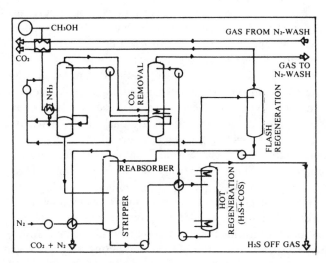

Figure 14-19. Flow diagram of Rectisol process for hydrogen sulfide and carbon dioxide removal from partial oxidation effluent gas.

split into three streams: one stream is cooled by evaporation of high-pressure ammonia, the second by heat exchange with cold, acid gas flashed from the rich methanol, and the third by exchange with the purified gas. The three streams, which have been cooled from about 77° to 40°F, are recombined and condensed water and hydrocarbons are drained off. The gas stream is subsequently split again into three portions which are cooled further (to about −30°F): one by refrigeration, one by heat exchange with flashed acid gases, and the third by heat exchange with the purified gas. In order to prevent freezing of the water vapor present in the gas, methanol is injected ahead of the second gas-cooling stage. Condensed methanol, aromatics, organic sulfur compounds, and some water are removed from the gas stream before it enters the contactor.

The gas is contacted with methanol in three consecutive stages, two of which are located in one column, with a separate column being used for the final stage. In the first or prewash stage, the methanol enters the column at −70°F and is discharged at −20°F. The last traces of aromatics, some CO_2 and some H_2S, and most of the organic sulfur compounds are removed in this operation. The rich methanol is combined with the condensate from the second gas-cooling stage and regenerated as described later.

The bulk of the CO_2 and practically all of the H_2S are removed in the second absorption stage, where the gas is again contacted with methanol entering the column at −70°F and leaving at −20°F. Refrigeration coils are installed in the column to remove the heat generated by absorption of acid gas and to keep the temperature of the solvent at the desired level. The effluent gas from the second stage is subjected to a final treatment in a separate column by contact with completely stripped methanol at a temperature of −50°F. The purified gas contains about 1 percent CO_2 and no detectable amounts of H_2S, benzene, or organic sulfur compounds. Its water content is equivalent to a dew point of less than −70°F. It is claimed that further reduction of the CO_2 content of the gas is possible (40).

Regeneration of the rich methanol from the prewash stage and the methanol condensed from the second gas-cooling stage is accomplished in the following manner. The two streams are collected in a tank operating at atmospheric pressure and dissolved gases are flashed off. Water is added continuously to the tank and two liquid phases are formed: an aqueous methanol phase and a hydrocarbon phase which contains the aromatics and organic sulfur compounds. The hydrocarbon phase is further processed for the recovery of aromatics and the aqueous methanol is fed to a distillation column. The overhead vapors from the column, which now consist of practically pure methanol, are condensed and pumped to a specially designed tower used for regenerating the rich solvent from the second absorption stage.

The rich methanol from the second contacting stage is regenerated by pressure reduction in six stages to a final pressure of about 4 psia. The flashed gas from the four superatmospheric stages, which contains about 98.5 percent

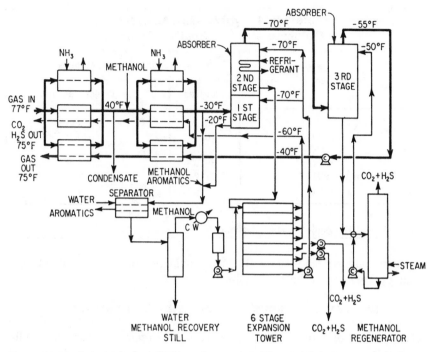

Figure 14-20. Schematic flow diagram of commercial Rectisol installation. *South African Oil, Coal, and Gas Corporation* (40)

CO_2 and 1.5 percent H_2S, is heat-exchanged with the plant-inlet gas before being discharged to the atmosphere. The gas from the two subatmospheric stages is compressed and vented to the atmosphere.

The rich methanol from the final purification stage is partly regenerated by flashing to atmospheric pressure and is then completely stripped of acid gas in a conventional distillation column. Typical analyses of inlet and outlet gases obtained at the Sasol plant are given in Table 14-14 (40).

Selective H_2S removal with methanol from synthesis gas, obtained by partial oxidation of a sulfur-rich residual oil, and from gas produced by coal gasification is discussed by Ranke (96). An evaluation of methanol as the solvent for acid gas removal in coal gasification processes has been presented by Rousseau et al. (97) and Ferrell et al. (98). This work, which is based on pilot scale tests, demonstrates that acid gases can be separated from sour gas streams produced by coal gasification and that there is no unusual accumulation of carbonyl sulfide in the acid gas removal system. The results indicate that gasification of devolatilized coal produces a sour gas that does not lead to accumulation of high molecular weight hydrocarbons or sulfur compounds in methanol, while gas produced by gasification of New Mexico

Table 14-14. Typical Gas Analyses
Obtained at Sasol Rectisol Plant

Constituent	Quantity Present, Mole %	
	Inlet Gas	Outlet Gas
$CO_2 + H_2S$	30.2	0.9*
O_2	0.0	0.0
CO	19.8	28.5
H_2	38.6	55.6
CH_4	11.0	14.2
H_2	0.4	0.8

*No detectable H_2S
Source: Data of Hoogendoorn and Solomon (40)

subbituminous coal leads to significant accumulation of such compounds, including thiophene, in the methanol solvent.

Removal of Hydrocarbon Vapors by Oil Absorption

The removal of hydrocarbon vapors from gas streams by absorption in liquid oils is an important part of a great many industrial processes. In some of these, such as natural-gasoline recovery, the absorption step is but a portion of the refining process which produces a number of extremely valuable industrial commodities, and the gas-purification aspects of the operation are of little or no importance. In other cases, however, such as the removal of naphthalene, and to some extent benzol, from coal gas, the absorption operation is conducted to improve the value of the gas stream.

In general, the design of hydrocarbon-absorption systems is relatively straightforward. No chemical reactions or side reactions are involved in the primary process and vapor/liquid equilibrium correlations based on the laws of ideal solutions are frequently applicable. Since mass transfer is not complicated by the occurrence of chemical reactions in the liquid phase, conventional absorption-coefficient and theoretical-plate concepts may be employed for design calculations. The principal complicating factor in the design of hydrocarbon-absorption plants is frequently the presence of numerous components. This not only makes the calculations cumbersome but results in the necessity for knowledge of extensive equilibrium data. Equilibrium data are readily available for the relatively simple paraffinic hydrocarbons encountered in natural-gas absorption operations, and the procedures for the design of such installations are well established and adequately presented in standard texts (44, 45, 46). Coke-oven gas, on the other hand, contains numerous cyclic hydrocarbons as well as nitrogen, sulfur, and oxygen com-

**Table 14-15. Typical Composition and Yield of
Crude Light Oil from Coke-oven Gas Operation (49)**

Constituent	Original Conc., gal/ton coal	Conc. in Crude light oil, %
Benzene...	1.85	57.8
Toluene...	0.45	14.1
Xylenes and light solvent naphtha..............	0.30	9.4
Unsaturated hydrocarbons, etc.................	0.16	5.0
Naphthalene and other heavy hydrocarbons.......	0.24	7.5
Wash oil...	0.20	6.2
Total crude light oil........................	3.20	100.0

pounds so that design procedures for hydrocarbon removal from such gas systems are somewhat less precise. Since naphthalene removal is an important purification step for the coal gas used as a household fuel (because of naphthalene's tendency to deposit as a solid which plugs lines and equipment), and light-oil removal has some significance as a purification step, brief descriptions of these processes are presented in the subsequent paragraphs. For more detailed descriptions of these and other coal-gas purification steps, the reader is referred to the comprehensive work of the Committee on Chemical Utilization of Coal, Division of Chemisty and Chemical Technology of the National Research Council (47, 48).

Light-Oil Removal from Coal Gas

The principal sources of coal gas in the United States are coke ovens operated to produce coke for iron- and steel-producing operations. However, in other parts of the world, a considerable amount of coal gas is manufactured in retorts operated for the primary purpose of producing fuel gas. The light-oil content of the gas depends upon both the manner of carbonization and the nature of the coal. Typically, it accounts for about 1 percent (by volume) of the raw-gas stream, and the recovered liquid has a composition as indicated by Table 14-15.

As indicated by the analysis, benzene is by far the most abundant compound in the light oil. For this reason the total liquid recovered is sometimes referred to as benzol (also spelled benzole), particularly by British writers. The marketable benzene products of specified purity are also referred to as benzol (e.g., 1° benzol, industrial benzol, etc.) and the word "toluol" has the same significance with regard to toluene.* Carbon disulfide can also be

---

*To avoid confusion, in this text the word "benzol" will be used to refer to the crude light oil or impure benzene; the word "benzene" will refer to the pure chemical.

removed from the gas stream during the light-oil recovery operation, and in some installations, which are specially designed for this purpose, sulfur removal is of equal importance to benzol recovery.

A flow sheet of a typical light-oil recovery plant is shown in Figure 14-21, which is based on a plant description presented by Hopton (50). As can be seen, the operation is quite simple. The gas stream is contacted with lean oil in a countercurrent absorption tower. The rich oil from the bottom of the absorber is pumped through heat exchangers in which it is heated first by the vapors leaving the top of the still, then by the lean solution leaving the bottom of the still, and finally by steam before it is fed into the still or the stripping column near the top. In the process illustrated, the still is operated under vacuum, although atmospheric pressure operation is more conventional. Steam is admitted into the bottom of the still to provide stripping vapor for the light oils and thereby to minimize the temperature required for adequate stripping. The lean oil is pumped through the heat exchanger, then through a cooler and back to the top of the absorber. Vapors from the still are cooled with partial condensation by indirect heat exchange with the rich oil. The condensate is passed to a separator which removes water, and the liquid hydrocarbons are returned to the still as reflux. The partially cooled vapors are then passed to the benzol condenser where relatively complete condensation is obtained by the use of cooling water. In this particular plant, the condensate and uncondensed vapors are then passed through a vacuum pump to a separator. From this noncondensed gases are returned to the fuel-gas main, water condensate is removed for disposal, and the hydrocarbon layer is transferred to the crude-benzol storage tanks.

Practically all light-oil plants employ a flow scheme basically similar to that of Figure 14-21, the commercial processes differing from each other primarily in the selection of wash oil, the mechanical design of the absorber, and the conditions of stripping.

Wash Oils

Although a large number of hydrocarbon liquids have been proposed for absorbing the light-oil fraction of coal gas, the most commonly used materials are petroleum liquids (gas oil or straw oil) and coal-tar fractions (creosote oil). The coal-tar oils have the obvious advantage of being available at any coke plant. However, they have several disadvantages, chief among these being their tendency to become more viscous with use. Because of this factor, and frequently because of cost considerations, petroleum-base wash oils appear to be growing in favor. The properties of several typical wash oils of the two types are presented in Table 14-16.

As will be noted from the table, the coal-tar oils have appreciably higher capacities for benzene than do the petroleum oils. However, this advantage can be lost as a result of the "thickening" phenomenon noted above.

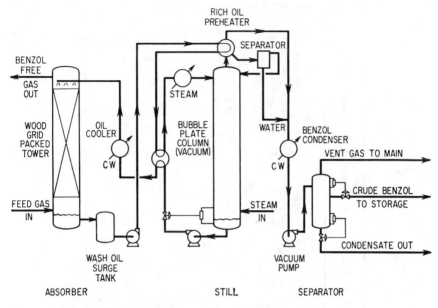

Figure 14-21. Simplified flow diagram of light-oil absorption plant.

The solubilities of important light-oil constituents in a typical hydrocarbon oil have been determined by Silver and Hopton (52), and their data are presented in Table 14-17.

The use of partition coefficients (as given in Table 14-17) rather than equilibrium constants is quite common among British authors, who have done a considerable amount of work on light-oil-absorber design. If Raoult's law is assumed to hold, i.e., $p = Px$ for each component, the partition factor, k, can be defined as follows:

$$k = \frac{c_L}{c_G} = \frac{62,300\,sT}{MP} \tag{14-11}$$

where c_L and c_G = equilibrium concentrations of solute in the liquid and gas phases respectively, expressed in weight per unit volume units (e.g., lb/cu ft)

s = specific gravity of solvent

T = absolute temperature, °K

M = molecular weight of solvent

P = vapor pressure of the pure component at temperature T, mm Hg

p = partial pressure of the component in the gaseous phase, mm Hg

x = mole fraction of the component in the liquid

Table 14-16. Properties of Typical Petroleum and Coal-tar Oils Used for Benzol Recovery (51)

Wash Oil (fresh)	Distillation Range, °C	Specific Gravity (20/20°C)	Viscosity, 20°C, cp	Volatility (loss to gas) at 20°C		Absorptive capacity, concentration of benzene in liquid in equilibrium with benzene vapor at 6.5 mm Hg pressure	
				g/cu m	gal/million cu ft	g/100 g solution	vol./100 vol. absorbent
Creosote oils:							
Light fraction (free from tar acids)	205–265	0.9535	25	2.8	18.3	4.55	5.15
Light (tar acids 17.5% by volume)	200–300 (95%)	1.0135	35	1.9	11.7	3.95	4.75
Medium (tar acids 20% by volume)	200–350 (90%)	1.031	51	2.0	12.1	3.6	4.35
Petroleum-gas oils:							
Light fraction, (a)	214–285 (95%)	0.8295	36	1.7	13.0	3.2	3.1
Fraction (b)	210–400 (95%)	0.8635	70	1.75	12.5	2.9	2.95
Fraction (c)	260–365 (95%)	0.849	92	0.9	6.7	2.8	2.8

Table 14-17. Solubilities of Light-oil Constituents in Typical Petroleum Oil*

Solute	Solubility (Partition coefficient, *k*) lb per cu ft in liquid/ lb per cu ft in gas			Apparent molecular wt. of oil		
	25°C (77°F)	80°C (176°F)	130°C (266°F)	25°C (77°F)	80°C (176°F)	130°C (266°F)
Carbon disulfide.......	235	56	22	187	161	143
Benzene.........	650	114	36	260	215	199
Toluene.........	2030	280	79	272	228	193
Xylene.........	7570	716	170	248	222	200
Naphthalene.....	171,000	8,620	1,320	321	276	251

*Designated "Benzol Absorption Oil," specific gravity at 20°C (68°F), 0.871; viscosity at 20°C (68°F), 12.4 cp; boiling point: first drop 270°C (518°F), 50 per cent 331°C (627.8°F), 75 per cent 380°C (716°F).
Source: Data of Silver and Hopton (52)

Measured partition coefficients can be used to calculate apparent molecular-weight values for the solvent by application of Equation 14-11; this has been done for the data of Table 14-17. Since the apparent molecular-weight values do not vary over a very wide range, they can be interpolated readily to provide a reasonably accurate basis for estimating partition coefficients at intermediate temperatures or for alternate components (of a similar nature).

In addition to capacity, the following factors are of importance in the selection of suitable wash oil:

1. The specific gravity should be far enough removed from that of water to permit satisfactory settling without the formation of an emulsion.
2. The viscosity should be as low as possible.
3. The initial boiling point of the oil should be as high as possible to minimize vaporization into the purified gas and to permit separation of the light oil and wash oil to be accomplished readily.

Absorber Design

In view of the long history of light-oil recovery from coal gas (dating back to before 1880), it is not surprising that a great number of absorber designs have been developed. In general, the absorbers can be classified as countercurrent towers, horizontal multiple-chamber scrubbers, and mechanical washers. The tower scrubbers are more or less conventional packed or tray columns. Wood-grid packing is frequently used because of its low pressure drop. Bubble-cap or perforated trays are occasionally specified for installations which operate at a pressure of several atmospheres—where pressure drop is not an important consideration. Horizontal multiple-chamber scrubbers have some utility in cases where the height requirement of towers is objectionable or where very low oil rates are required. A counter-current stagewise action can be obtained by pumping oil from one chamber to the next in the direction opposite to that of the gas flow. Mechanical washers are usually provided with rotating elements to disperse the oil and provide intimate contact with the gas. Both horizontal and vertical types have been developed, incorporating a number of ingenious ideas. However, it is believed that their use is diminishing in favor of simple countercurrent towers.

Since the equilibrium curves for light-oil components are reasonably close to linear and the operating line is relatively straight in typical light-oil absorption operations, simplified design equations such as Equation 1-8, based on $K_G a$; Equation 1-14, based on transfer units; and Equation 1-15, based on theoretical plates, can be employed. Modified forms of these equations based upon weight/volume partition factors (see Table 14-17) rather than mole fractions have been used by a number of investigators (50, 52, 53, 54). In these units, the parameter L_M/mG_M of Equations 1-14 and 1-15 becomes Lk/G, where the gas and liquid rates are based on volume and k is the partition coefficient c_L/c_G. As with the parameter L_M/mG_m, an Lk/G value of 1.0 represents the minimum liquid/gas ratio which can result in complete removal of the component in a tower of infinite height. In actual practice, the above-mentioned investigators report Lk/G values on the order of 1.3 as being typical for light-oil absorption. The required flow ratio of oil to gas for obtaining essentially complete removal of carbon disulfide, benzene, toluene, xylene, or naphthalene with the typical petroleum fraction used to obtain the data in Table 14-17 can be calculated directly by dividing the assumed Lk/G value (e.g., 1.3) by the partition coefficients given in the table. Such calculations show that an oil rate of approximately 35 gal/1,000 cu ft of gas is required to provide substantially complete removal of carbon disulfide (at 77°F) as compared to 13 gal/1,000 cu ft for benzene removal and only 4 gal/1000 cu ft for toluene removal.

The magnitude of the absorption coefficient obtainable for benzene absorption in petroleum oil using wood-grid packing can be estimated from data presented by Silver and Hopton (52). One relatively complete set of absorption-column data obtained by these investigators is presented in Table 14-18.

It will be noted that Lk/G in the above instance is equal to 1.41. Silver and Hopton report that other tests gave K_G values ranging from 24 to 47 for benzene at gas velocities of 9,000 to 10,000 ft/hr and about 22 for carbon disulfide at gas velocities of 6,000 to 7,000 ft/hr. They recommend a design value of 34 lb/(hr)(sq ft)(lb/cu ft) for benzene which corresponds to a $K_G a$

Table 14-18. Benzene Absorption-tower Test Data

Tower diameter, ft..	9
Packing:	
No. of sections..	6
No. of layers (of boards) per section.........................	22
Dimensions of boards used, in.............................	5¾ by ⅜
Spacing (horizontal) between boards, in......................	½
Total board surface, A, sq ft..............................	110,000
Oil-flow rate, gal/hr......................................	5,000
L, cu ft/hr..	802
Gas-flow rate:	
G, cu ft/hr..	418,000
Gas velocity based on free cross-sectional area (36.4 sq ft) available	
for gas flow, ft/hr......................................	11,500
Gas density, lb/cu ft......................................	0.037
L/G (802/418,000).......................................	0.00192
Temperature:	
°C...	23
°F...	73.4
Partition coefficient, k (for benzene in petroleum oil at 23°C).......	733
Benzene absorption, observed efficiency, %......................	98.6
$K_G a$, lb benzene/(hr)(cu ft)(lb/cu ft).......................	1,100
K_G, lb benzene/(hr)(sq ft)(lb/cu ft), assuming all of A is available for	
absorption...	40.3

Source: Data of Silver and Hopton (52)

value of about 930 lb/(hr)(cu ft)(lb/cu ft) [or about 2.4 lb moles/(hr)(cu ft)(atm)] for wood-grid packed towers of the type described in Table 14-18.

Stripper Design

The rich "benzolized" wash oil is stripped of its light-oil content by fractional distillation—most commonly in a bubble-plate column operated at substantially atmospheric pressure with direct-steam addition to the bottom. The number of theoretical trays required for the separation can be calculated by conventional techniques and the actual requirement estimated on the basis of an assumed tray-efficiency. Overall efficiencies of 45 to 70 percent have been reported as typical for light-oil stripper stills (52, 55), and 10 to 15 actual trays are commonly employed below the feed.

Raschig-ring packed stripping stills are also used to some extent and data on several such columns have been presented by Silver and Hopton (52). For purposes of design, these authors recommend a $K_L a$ for benzene of between 5 and 8 lb/(hr)(cu ft)(lb/cu ft) [or lb moles/(hr)(cu ft)(lb mole/cu ft)] for 1½-in. rings at oil rates from 100 to 150 gal/(hr)(sq ft) and superficial steam velocities of from 0.2 to 0.3 fps. Both liquid- and steam-flow rates are based on the empty column, and the steam-velocity values assume that no other vapors are present.

The quantity of steam required is dependent on the overall objectives of the operation. The minimum quantity for complete removal of each component can be estimated by setting Gm/L (or G/Lk) equal to unity. This type of analysis indicates that a much lower steam/oil ratio is required for complete carbon disulfide removal from the liquid than for removal of benzene or the other light-oil hydrocarbons. However, since the quantity of oil required for efficient carbon disulfide removal is higher than that for light-oil removal alone, the net steam requirement of plants designed for the two types of service may not be appreciably different. This relationship is brought out in Table 14-19, which is based on design calculations presented by Hopton (50).

Operating Data

Operating data from four plants have been presented by Silver and Hopton (52). A summary of data on two of these which illustrate the performance of commercial installations designed primarily for benzol removal and sulfur removal, respectively, is presented in Table 14-20.

The principal operating difficulties in light-oil recovery plants are wash-oil deterioration, as evidenced by an increase in viscosity or the formation of sludge, and corrosion—which has been shown to be due primarily to the presence of ammonium thiocyanate in the oil (56). The formation of sludge is believed due to oxidation and/or polymerization of certain coal-gas constituents such as indene or styrene. The sludge deposits interfere with plant operation by coating heat-transfer surfaces and tower packings. The sludge problem can be minimized by prior treatment of the gas with an electrostatic precipitator to remove sludge-forming components which are present in particulate form; however, a more practical approach is to regenerate the wash oil. One successful method of accomplishing this involves the use of steam distillation of a small slip stream of the wash oil which is bypassed into a separate vessel. A relatively large amount of steam must be passed through the oil to vaporize the wash-oil components. However, the steam and wash-oil vapors are passed directly into the main stripper so that their heat content is not lost. Nonvolatile sludge residue is periodically discarded from the regenerator vessel.

Naphthalene Removal from Coal Gas

As pointed out earlier, the removal of naphthalene from coal gases is a gas-purification operation of considerable importance because, unlike other hydrocarbons present in the coal gases, naphthalene condenses as a solid when the temperature of the gas is lowered, resulting in the plugging of processing equipment and distribution lines. Naphthalene is formed in vary-

Table 14-19. Comparison of Theoretical Designs for Plants to Remove Benzol and Carbon Disulfide

Design Variables	Benzol-Removal Plant	Sulfur-Removal Plant
Wash-oil circulation rate (petroleum oil) gal/ 1,000 cu ft gas at 20°C	10.2	28.7
Still pressure	1 psig	10 in. Hg (abs)
Steam-flow rate, lb/gal oil	0.295	0.157
Steam temperature	130°C (266°F)	80°C (176°F)
Removal efficiency, %:		
Carbon disulfide	45	88
Benzene	88	85
Toluene	85	80
Xylene	86	83
Naphthalene	93	89

Source: Data of Hopton (50)

Table 14-20. Light-oil Recovery Plant Operating Data (52)

Plant Variables	Plant A *	Plant B†
Gas-flow rate, SCF/hr	45,700	264,500
Oil circulation rate:		
gal/hr	510	9,000
gal/1,000 cu ft	11.2	34.1
Mean wash temperature, °F	68.9	63.5
Steam-flow rate:		
lb/hr	201	1,080
lb/gal of oil	0.39	0.12
Mean still temperature, °F	239.9	176
Still pressure, psia	15.4	4.5
Light-oil recovery:		
gal/10,000 cu ft	2.23	2.60
per cent	88	87
Carbon disulfide in feed, grains/100 cu ft	18.7	22.8
Carbon disulfide recovery, %	52	91

*Designed primarily for benzol recovery.
†Designed primarily for carbon disulfide recovery.

ing quantities during coal carbonization, depending on the carbonization temperature. In typical high-temperature carbonization operations the gas may contain as much as 250 grains of naphthalene per 100 cu ft, while less than 100 grains/100 cu ft may be formed under low-temperature carbonization conditions (50). Because of its high boiling point (218°C), most of the naphthalene condenses with the tar in the primary gas coolers. However, appreciable amounts, typically 15 to 50 grains/100 cu ft, remain in the gas after the primary cooling step. In order to protect distribution lines against deposition of solid naphthalene, it is customary to remove the naphthalene so that the final concentration is about 2 grains/100 cu ft (50).

The principal method used industrially for removing naphthalene from coal gases consists of absorption in oil in regenerative or nonregenerative systems. Other processes, such as condensation of all the naphthalene by refrigeration of the gas and adsorption of naphthalene on solids, have been proposed but have not been applied commercially to a significant extent (47).

The types of oils used for light-oil absorption are also suitable for naphthalene removal. Petroleum oils which are free of naphthalene are preferable from the standpoint of maximum naphthalene removal. However, coal-tar oils have a somewhat higher absorptive capacity.

Naphthalene is removed satisfactorily in the oil-washing operation for benzol recovery, and a separate napthalene-removal plant is not required if benzol is recovered. However, in many cases it may be advantageous to remove naphthalene separately at the beginning of the gas-purification train, even if benzol is removed from the gas before it enters the distribution system. In general, the preferred location for the naphthalene-removal plant is at the inlet of the secondary gas cooler (50). Removal at this point prevents accumulation of naphthalene in the ammonia washers and the iron oxide purifiers and results in more efficient operation of that equipment.

A schematic flow diagram of a naphthalene-removal installation is shown in Figure 14-22. The gas is washed with the oil in an absorber of special design, permitting efficient contact between a large volume of gas and a small quantity of liquid. Because of the similarity of the operation, the same absorber designs are used for naphthalene removal as for benzol absorption. Countercurrent towers containing two stages of wood-grid packing are normally used in the United States. In the lower section of the tower, where the bulk of the naphthalene is removed, the gas is washed with partially rich oil which is recirculated at a high rate. The partially purified gas enters the upper section of the tower where the naphthalene content is reduced to the desired level by contact with a small stream of fresh oil. The fresh oil is required at such a low rate that continuous countercurrent addition would result in poor contact between the gas and the liquid, and the fresh oil is, therefore, often injected intermittently at a sufficiently high rate to wet the packing of the upper tower section. The oil leaving the upper section of the absorber

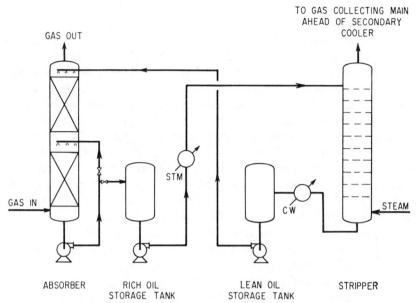

Figure 14-22. Simplified flow diagram of naphthalene-removal installation.

joins the recirculating oil stream in the lower section, thereby providing fresh oil for bulk naphthalene absorption. Rich oil is continuously withdrawn from the system at the same rate at which fresh oil is added; it is then stripped of naphthalene by steam distillation and reused, as shown in Figure 14-22. A convenient way of disposing of the stripped naphthalene consists of returning the stripper overhead vapors to the gas-collecting main ahead of the primary cooler in order to condense the naphthalene with the tar in the primary cooler. However, the rich oil is not always regenerated and may be processed in a variety of ways, depending on the overall economics of the installation. In gas works producing carburetted water gas, for example, gas oil may be used for absorption of naphthalene and the rich oil added to the oil used for carburetting the water gas. The naphthalene content of the gas is not increased significantly because the rich oil constitutes a small fraction of the total oil used for carburetting, and most of the naphthalene is removed in the gas-cooling system.

An installation using a venturi-type scrubber for absorption of naphthalene in tar oil has been described recently (57). Naphthalene removal to a dew point of less than 75°F is reported for an absorption temperature of 95°F.

Process Design

Absorption columns can be designed by the use of simplified equations as described in the preceding section for light-oil absorption. Because of the relatively low vapor pressure of naphthalene and its high solubility in various oils (see Table 14-17), very low oil rates are required to obtain a high removal efficiency. The vapor pressure of naphthalene as a function of temperature is shown in Figure 14-23 and vapor/liquid equilibria for naphthalene and three different petroleum oils are presented in Figure 14-24 (58). In general, oil-circulation rates of about 0.01 to 0.1 gal/1000 cu ft of gas are required to reduce the naphthalene content of typical coal gases to less than 2 grains/100 cu ft. An oil-circulation rate of 0.035 gal/1,000 cu ft of gas has been calculated by Hopton (50) for an Lk/G value of 1.3 by using partition factors for petroleum oil given in Table 14-17.

Stripping columns are also designed in the same manner as those used in light-oil recovery plants. Hopton (50) recommends a steam rate of 16.5 lb/gal of oil for operation at 248°F and 1 psig pressure. This value is based on G/Lk value of 1.3 and the use of petroleum oil.

Because of the very low oil-circulation rate, naphthalene is absorbed selectively, without removal of significant amounts of other hydrocarbons. This is of importance in cases where the hydrocarbons are left in the gas in order to obtain a higher calorific value. In general, the calorific value of coal gas is decreased by less than 1 Btu as a result of naphthalene removal.

Acid-Gas Removal by Physical Absorption in Organic Solvents

When the acid-gas impurities make up an appreciable fraction of the total gas stream, the cost of removing them by heat regenerable reactive solvents may be out of proportion to the value of the treated gas. This is particularly true when the acid-gas fraction consists solely or primarily of carbon dioxide, and the entire heat requirements for solvent regeneration have to be supplied by burning gas which, in many operations, is the only saleable product.

To overcome the economic disadvantages of heat regenerable processes, several processes have been developed which are based on the use of essentially anhydrous organic solvents which physically dissolve the acid gases and can be stripped by reducing the acid-gas partial pressure without the application of heat.

Such processes, of course, require a high partial pressure of the acid gases in the feed gas to be purified, but gases containing high concentrations of CO_2 or H_2S are frequently available at a high pressure. One such process (the Rectisol proceess described in a preceding section of this chapter) uses

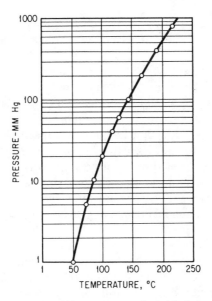

Figure 14-23. Vapor pressure of naphthalene.

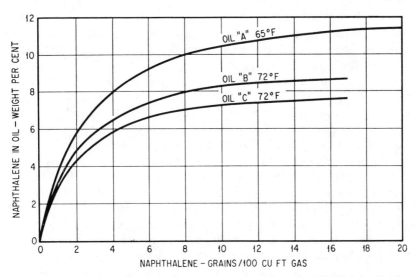

Figure 14-24. Vapor liquid equilibria of naphthalene in various oils (58). Oil A: crude oil saturated with benzene. Oil B: residuum gas oil. Oil C: crude oil.

methanol as the solvent. This process, however, requires the use of very low temperatures to minimize loss of solvent by vaporization into the purified gas and acid-gas streams.

Although many organic solvents appear to be suitable for this type of process, their actual number is limited by certain criteria which must be fulfilled to make them acceptable for economic operation. In order to be practical, the solvents must have an equilibrium capacity for acid gases several times that of water, coupled with a low capacity for the primary constituents of the gas stream, e.g., hydrocarbons and hydrogen. In addition, they must have an extremely low vapor pressure, permitting operation at essentially ambient temperatures without excessive vaporization losses; they must have low viscosity and low or moderate hygroscopicity. They must also be noncorrosive to common metals as well as nonreactive with all components in the gas. Finally, they must be available commercially at a reasonable cost.

Physical solvent processes are being used commercially for acid-gas removal from high-pressure natural-gas streams and for carbon dioxide removal from crude hydrogen and ammonia synthesis gases produced both by partial oxidation and steam-hydrocarbon reforming. Since solvent processes are most efficient when operated at the highest possible pressure carbon dioxide removal from reformer effluents is best carried out after compression of the gas to the ultimate pressure required for such processes as ammonia synthesis or hydrocracking. A very important advantage of this type of operation is the fact that by leaving the carbon dioxide in the gas mixture before compression, the molecular weight of the gas is sufficiently high to permit use of relatively inexpensive centrifugal compressors and still reach the required discharge pressure.

Most organic solvents have an appreciably higher solubility for hydrogen sulfide than for carbon dioxide, and a certain degree of selective hydrogen sulfide removal can be attained. This feature is of significance when the ratio of carbon dioxide to hydrogen sulfide in the crude gas is so high that the acid gases cannot be processed in a standard Claus unit. By removing essentially all of the hydrogen sulfide and only a portion of the carbon dioxide, this ratio can be lowered sufficiently to permit normal processing in a Claus plant.

Minor gas impurities such as carbonyl sulfide, carbon disulfide, and mercaptans are quite soluble in most organic solvents, and these compounds are removed to a large extent, together with the acid gases. The solubility of hydrocarbons in organic solvents increases with molecular weight. Consequently, hydrocarbons above butane are also largely removed with the acid-gas fraction. Although special designs for the recovery of these compounds have been proposed, physical solvent processes are generally not economical for the treatment of "wet" hydrocarbon streams.

Processes based on physical organic solvents are most economical for bulk removal of acid-gas constituents. When high-purity treated gas is required, the remaining acid gas is usually removed with chemical solvents, e.g., ethanolamine or hot potassium carbonate solutions. However, essentially complete acid-gas removal, for example, reduction of the carbon dioxide content of synthesis gases to less than 0.1 percent, is possible, if solvent regeneration methods beyond simple flashing are employed.

Several processes using organic physical solvents either alone or in combination with chemical solvents (Sulfinol) are in commercial operation and will be described in this section. Although many of these processes are quite successful, the search for more economical solvents continues. A rather extensive study to optimize physical solvent processes for the purification of gases at high pressure has been presented by Zawacki et al. (99). In this work a large number of physical solvents were screened and, after selection of two solvents, (the dimethyl ether of tetraethylene glycol and N-formyl morpholine) process schemes were proposed for a variety of applications. With the exception of the Sulfinol process, which uses a combination of a physical and a chemical solvent, all processes are quite similar, differing only in the selection of the solvent.

Process Description

In their simplest form, physical solvent processes require little more than an absorber, atmospheric flash vessel, and recycle pump. No steam or other heat source is required. When the absorbed gases are desorbed from the solution by flashing at atmospheric pressure, the lean solution contains acid gas in an amount corresponding to equilibrium at 1 atm acid-gas partial pressure; and this, therefore, represents the theoretical minimum partial pressure of acid gas in the purified-gas stream. To obtain a higher degree of purification, vacuum or inert gas stripping or heating of the solvent may be employed. Other process modifications are being used to minimize loss of valuable gas components, provide a relatively low temperature of operation, and otherwise improve process economics.

The operation of a typical solvent process is illustrated in the schematic flow diagram shown in Figure 14-25. This example incorporates three possible modes of solvent regeneration, i.e., simple flashing, inert gas stripping, and heat regeneration. In addition, a split-stream cycle is shown (broken lines) which can be used for bulk removal of acid gases with partially stripped solvent, followed by final purification with completely regenerated solvent.

The gas enters the bottom of the absorber, containing either packing or trays, and is washed by a descending stream of regenerated solvent. The rich solvent leaving the bottom of the absorber is regenerated by pressure reduction, usually at two or three different pressure levels, the last of which may be atmospheric or subatmospheric. The gases flashed at the highest pressure level, which contain most of the dissolved nonacidic gases, are usually recom-

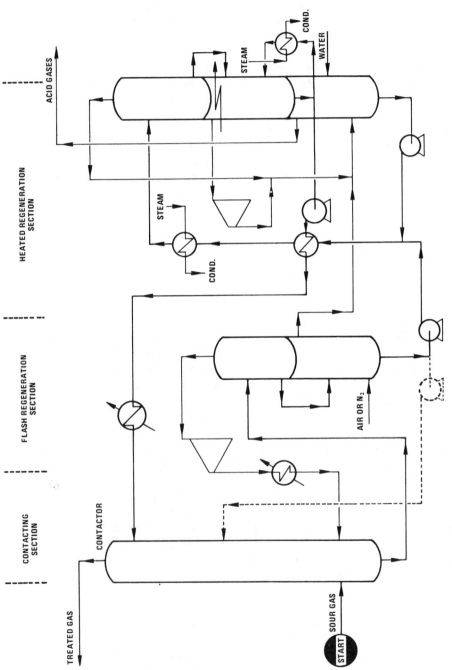

Figure 14-25. Typical physical solvent process flow diagram.

pressed and returned to the absorber inlet, in order to avoid losses of product gas. Quite frequently, solvent regeneration by pressure reduction is adequate for attaining the required gas purity, and the solvent leaving the lowest pressure flash is directly recycled to the top of the absorber. If product gas of higher purity is required, the residual acid-gas content of the solvent is further reduced by stripping with air or an inert gas. Complete removal of the last remaining acid gas is effected by heat regeneration and reboiling of the solvent.

Although not shown on the flow diagram, the rich solvent and the flashed acid gases are frequently expanded through power recovery turbines. This has the double purpose of reducing energy requirements and cooling the regenerated solvent before reuse. Since the absorption capacity of the solvents increases as the temperature is lowered, it is advantageous to operate at the lowest possible temperature. Generally sufficient autorefrigeration is available in the system to make outside refrigeration unnecessary.

Solvent losses are minimized by water washing all effluent gas streams. Most commercially used solvents are very stable, and reclaiming is not required. However, water removal by distillation is sometimes necessary to maintain the system in water balance.

The Fluor Solvent Process

The Fluor Solvent process, which is licensed by the Fluor Corporation of Los Angeles, was first described by Kohl and Buckingham in 1960 (59). Although several solvents have been covered by U.S. patents (60), only propylene carbonate has so far been used commercially. In 1975, the process was reported to be in use in ten commercial installations, seven processing natural gas, one ammonia synthesis gas, and two hydrogen (61).

Basic Data

Selected physical properties of propylene carbonate are tabulated in Table 14-21 and shown graphically in Figures 14-26 to 14-29. Data on the solubility of various gases in propylene carbonate have been reported by Kohl, et al. (59), Dow Chemical Company (62), Schmack and Bittrich (63), and Makranczy, et al. (64). The equilibrium solubilities of hydrogen sulfide and carbon dioxide as a function of pressure are shown in Figure 14-30. Although there is some scattering of points, it is evident that the solubilities of both acid gases follow Henry's law up to a pressure of about 20 atm. The effect of temperature is shown, for carbon dioxide and hydrogen, in Figure 14-31. It is interesting to note that the solubility of hydrogen increases with increasing temperature.

846

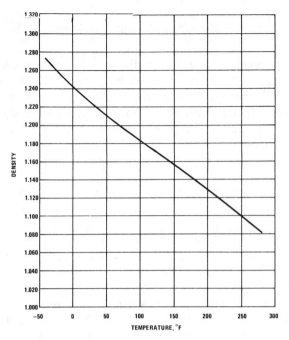

Figure 14-26. Density versus temperature of propylene carbonate. *Data of Dow Chemical of Canada, Ltd.* (62)

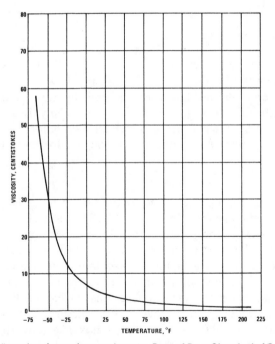

Figure 14-27. Viscosity of propylene carbonate. *Data of Dow Chemical of Canada, Ltd.* (62) (62)

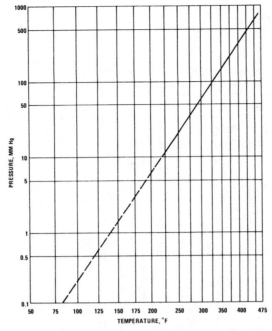

Figure 14-28. Vapor pressure of propylene carbonate. *Data of Dow Chemical of Canada, Ltd.* (62)

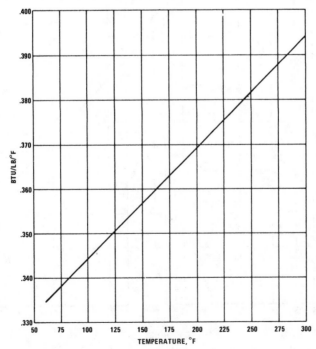

Figure 14-29. Specific heat of propylene carbonate. *Data of Dow Chemical of Canada, Ltd.* (62)

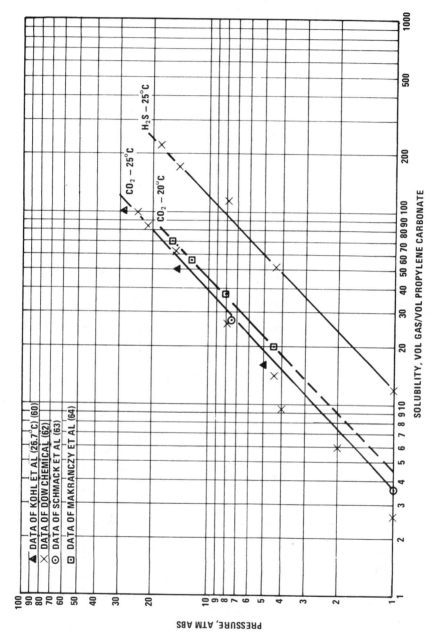

Figure 14-30. Effect of pressure on the solubility of carbon dioxide and hydrogen sulfide in propylene carbonate (gas volumes at 0°C and 760 mm Hg).

Table 14-21. Physical Properties of Propylene Carbonate	
Empirical formula	$C_4H_6O_3$
Molecular weight	102.09
Boiling point, 760 mm Hg	464°F
Freezing point	−55.2°F
Thermal conductivity	0.12 Btu/(hr)(ft^2)(°F)(ft)
Heat of vaporization, 464°F	208 Btu/lb.
Heat of solution-saturation	
with CO_2 at 80°F	0.70 Btu/lb.

Source: Dow Chemical of Canada, Limited (62)

No data are presented in the literature to indicate the effect of dissolved carbon dioxide, hydrogen sulfide, and hydrogen on the solubility of the individual compounds. However, the sketchy information available on gas mixtures containing carbon dioxide and methane indicates that at high partial pressures the solubility of carbon dioxide is somewhat reduced while the solubility of methane is appreciably increased by the presence of the other compound (62, 64). In view of the low solubility of hydrogen in propylene carbonate, it is reasonable to assume that its presence has no significant effect on the solubility of carbon dioxide or hydrogen sulfide. Solubilities of several gases, in terms of Bunsen coefficients (volume of gas at 0°C and 760 mm Hg per volume of propylene carbonate), are shown in Table 14-22.

Process Design and Operation

The basic design of the Fluor Solvent process follows essentially that shown in Figure 14-25. It involves several flashes of the rich solvent at decreasing pressure levels with recycling of the gas evolved in the high pressure flash. A modification of the process resulting in an appreciable reduction of hydrocarbon gas loss has been described by Freireich and Tennyson (79). In this scheme, the gas from the second intermediate pressure flash is washed with solvent in a small absorber. The overhead from this absorber is recompressed and recycled to the plant inlet. It is claimed that the cost of the recovered gas amply compensates for the cost of the additional equipment required, and that a payout of as low as three months is realized with a fuel cost of $2 per MMBTU (79).

The operation of the Fluor Solvent process has been described by Buckingham (65), and performance data on four plants treating natural gas have been reported by the same author (66). These data are given in Table 14-23.

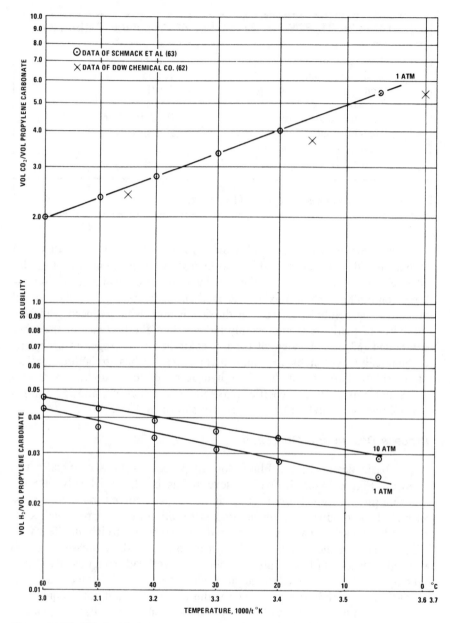

Figure 14-31. Effect of temperature on the solubility of hydrogen and carbon dioxide in propylene carbonate (gas volumes at 0°C and 760 mm Hg).

An economic study comparing the Fluor Solvent process with the activated hot potassium carbonate process for the removal of carbon dioxide from synthesis gas for the production of ammonia and urea was presented by Cook and Tennyson (67). The authors concluded that the process is more economical than activated hot potassium carbonate in all cases studied, which included production of ammonia and urea by steam reforming of natural gas and naphtha and by partial oxidation.

A photograph of a Fluor Solvent plant processing natural gas is shown in Figure 14-32.

The Purisol Process

This process which was developed by Lurgi Gesellschaft für Wärme und Chemotechnik m.b.H. of Frankfurt, West Germany, is presently being used in four commercial installations, two treating natural gas and two in hydrogen plant service (61). Discussions of the basic features of the process and of its application to the purification of natural gas, hydrogen, and synthesis gases have been presented by Hochgesand (39, 43), Stein (68), Kapp (69), and Beavon and Roszkowski (70). The solvent used in the Purisol process is N-methyl-2-pyrrolidone (NMP), a high boiling liquid, which has an exceptionally high solubility for hydrogen sulfide; thus, it is particularly suitable for selective hydrogen sulfide absorption in the presence of carbon dioxide. The process is reportedly capable of yeilding gas streams containing less than 0.1 percent carbon dioxide and a few parts per million of hydrogen sulfide (72).

Basic Data

Physical properties of NMP are given in Table 14-24 and in Figure 14-33. Equilibrium solubility data for carbon dioxide, hydrogen sulfide, methane, and propane reported by Stein (68), Hochgesand (39, 43), and Boston and

Table 14-22. Solubility of Gases in Propylene Carbonate	
Solute	Bunsen coefficient
C_2H_2	8.6 (62)
CO	0.5 (62)
H_2	0.03 (63)
CH_4	0.3 (62)
N_2	0.0 (62)
COS	5.0 (62)
Temp. = 25°C, press. = 760 mm Hg	

Table 14-23. Operating Data of Fluor Solvent Process Plants (66)

Plant	Design feed gas rate MMSCFD	Feed gas composition CO₂	Feed gas composition H₂S	Absorption pressure, psig	Acid gas partial pressure in feed, psi	Sales gas specifications CO₂	Sales gas specifications H₂S
A	220	53%	3 gr/100 scf	850	CO₂ - 432	2%	0.25 gr/100 scf
B	10	17%	—	450	CO₂ - 75	5%	—
C	20	22.8%	—	800	CO₂ - 180	1%	—
D	28	10-13%	5-15%	1,000	CO₂+H₂S - 250-300	0.1%	0.8 gr/100 scf

Figure 14-32. Fluor solvent process plant processing high-pressure natural gas. *The Fluor Corporation and El Paso Natural Gas Company*

Table 14-24. Physical Properties of N-Methyl-2-Pyrrolidone

Molecular weight	99
Specific gravity	1.027
Viscosity (25°C)	1.65 cp
Boiling Point	202°C (395°F)
Freezing point	−24.4°C
Specific heat (at 20°C)	0.4 Btu/(lb)(°F)

Source: General Aniline and Film Corporation

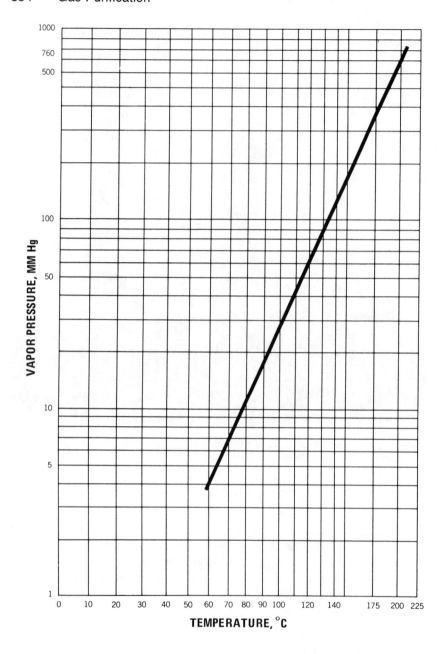

Figure 14-33. Vapor pressure of N-methyl-2-pyrrolidone. *Data of General Aniline and Film Corp.*

**Table 14-25. Equilibrium Solubility of Gases in
N-Methyl-2-Pyrrolidone**

Gas	Temperature, °C	Partial pressure atm. abs.	Solubility vol/vol*
CO₂	20	1.0	3.95 (68)
CO₂	23.5	0.67	2.0 (71)
CO₂	35	1.0	3.0 (43)
CO₂	35	10.0	32.0 (43)
H₂S	20	1.0	48.8 (68)
H₂S	23.5	0.41	14.3 (71)
H₂S	35	1.0	25.0 (43)
CH₄	20	1.0	0.28 (68)
CH₄	23.5	56	12.2 (71)
C₃H₈	23.5	0.59	1.9 (71)

*Gas volume at 0°C and 760 mm Hg.

Schneider (71) are presented in Table 14-25. The data of Boston and Schneider were obtained with a gas mixture containing 97.10 percent methane, 1.02 percent propane, 1.17 percent carbon dioxide, and 0.71 percent hydrogen sulfide at a total pressure of 850 psia and a temperature of 74°F.

Process Operation

Operation of a Purisol plant processing about 40 MMscf per day of natural gas containing hydrogen sulfide and carbon dioxide has been described by Stein (68). The principal objective in this installation is to remove a maximum of hydrogen sulfide with only partial removal of carbon dioxide. Absorption of the acid gases is carried out in two columns, operating in series, the first provided with trays and the second with packing. The solvent passes from the second to the first absorber and is then regenerated by flashing at three different pressure levels, augmented by stripping with inert gas at elevated temperature in the last flash. Cooling of the solvent is required between the two absorption columns and, of course, after regeneration. The gas released in the first flash is recompressed and returned to the inlet of the first absorber. The acid gases are further processed in a Claus type sulfur recovery unit.

Typical operating data from this installation are shown in Table 14-26. Examples of process conditions proposed by Lurgi (72) for three cases involving (a) essentially complete carbon dioxide removal from high-pressure gases with high carbon dioxide content, (b) bulk removal of hydrogen sulfide

Table 14-26. Operating Data of Purisol Plant Treating Natural Gas

Gas vol. MMscf/d	Inlet Gas	Outlet Gas
H₂S, vol %	1-10	0.02-0.2
CO₂, vol %	8-26	6-20
N₂, vol %	4-5	4-5
CH₄, vol %	70-80	75-90
Temperature, °F	32-59	77
Pressure, psig	720	570

Temperature, °F:	
Solvent to first absorber	70
Solvent from first absorber	86
Solvent to second absorber	75-82
Solvent from second absorber	104-111
Solvent to flash	266
Pressure, psig:	
First flash	500
Seccond flash	210
Third flash	7.5

from natural gas, and (c) selective removal of hydrogen sulfide from natural gas, are presented in Table 14-27. The first case involves solvent regeneration by flashing and inert-gas stripping. In the second case the solvent is simply flashed at three pressure levels. In the third case, requiring complete solvent regeneration, flashing and high temperature regeneration with reboiling are employed (see Figure 14-25).

The Selexol Process

This process also uses a physical solvent and, as a result, is quite similar to the other processes discussed in this section. The process, which was developed by Allied Chemical Corporation and, since 1982, is licensed by the Norton Company, has been described quite extensively in the literature (73, 74, 80 to 86, 100–102). It is used widely for bulk removal of CO_2 from natural and synthesis gases and for selective sulfur removal from a variety of gas streams. In 1984, about 40 plants were either in operation or under construction worldwide.

Basic Data

Physical properties of the SELEXOL solvent, the dimethyl ether of polyethylene glycol, and solubilities of several gases in the solvent are shown in

Table 14-27. Typical Process Conditions for Acid-Gas Removal with Purisol Process

Case:		1	2	3
Feed gas:				
Volume, MMscf/day		100	100	100
Pressure, psig		1,070	510	1,070
Temperature, °F		110	80	80
H_2	vol. %	64.53	—	—
CO_2	"	33.15	1.0	15.0
H_2S	"	—	34.0	6.0
CO	"	1.50	—	—
CH_4	"	0.44	63.7	75.0
C_2H_6-C_4H_{10}	"	—	1.1	—
C_6H_{12+}	"	—	0.2	—
N_2	"	0.38	—	4.0
Treated gas:				
H_2,	vol. %	96.44	—	—
CO_2	"	0.10	1.2	13.6
H_2S	"	—	2.0	2 ppm
CO	"	2.24	—	—
CH_4	"	0.59	95.4	82.0
C_2H_8-C_4H_{10}	"	—	1.4	—
C_6H_{12+}	"	—	—	—
N_2	"	0.63	—	4.4
Utilities:				
Power, KW*		2,100	1,600	1,100
Steam, (45 psig), lb/hr		3,750	1,500	13,000
				(60 psig)
Cooling water (75°F) gpm		1,300	750	820
Condensate, lb/hr		2,850	2,000	2,200
NMP loss, lb/hr		6.5	11	9

*Without power recovery.

Table 14-28 and Figure 14-34, respectively. As shown in Figure 14-34, hydrogen sulfide is about nine times as soluble in the SELEXOL solvent as is carbon dioxide, which makes the process suitable for selective removal of hydrogen sulfide. Furthermore, methyl mercaptan, a common constituent of natural gas, and carbonyl sulfide which is present in some natural gases and in the product gases from coal gasification processes are also quite soluble in the SELEXOL solvent. This enables the process to essentially achieve complete removal of all sulfur compounds from a variety of gas streams.

The solvent is reported to be very stable and not susceptible to heat or oxidative degradation. It is non-reactive with gas constituents, and its extremely

Table 14-28. Physical Properties of Selexol Solvent (74)

Freezing point, °F	−8 to −20
Flash point, °F	304
Vapor pressure (77°F), mm Hg	<0.01
Specific heat (77°F) Btu/(lb)(°F)	0.49
Density (77°F), lb/gal	8.60
Viscosity (77°F), cp	5.8
Thermal conductivity, Btu/(hr)(ft²)(°F/ft)	0.11

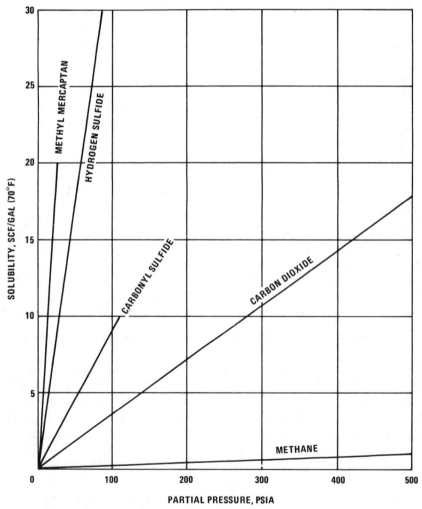

Figure 14-34. Solubility of gases in selexol solvent. *Data of Sweny and Valentine* (74)

low vapor pressure results in negligible vaporization losses under normal operating conditions. Other important properties are its low viscosity, non-toxicity, and low specific heat.

Process Design and Operation

The basic flow scheme of the process is very simple, requiring only an absorption stage and regeneration by flashing at successively decreasing pressure levels (see Figure 14-25). If the objective is bulk removal of carbon dioxide from a gas stream, this is all that is required for regenerating the solution. More extensive regeneration can be achieved by vacuum flashing, stripping with air or an inert gas or stripping by application of heat. In the case of hydrogen sulfide removal, stripping with inert gas or heat is the normal procedure.

If selective removal of sulfur compounds is required, the SELEXOL plant consists of an absorber, a flash, and a steam-heated stripping column. The flashed gas is compressed and recycled to the absorber inlet, and the regenerator effluent is treated for production of elemental sulfur.

In cases where selective sulfur removal and complete carbon dioxide removal are required, such as with coal-derived substitute natural gases, two successive independent absorption-regeneration cycles are used as shown in Figure 14-35. The different flow schemes are discussed in some detail by Van Deraerschot and Valentine (86) and by Sweny (84, 102). These authors claim that with proper plant design removal of sulfur compounds to concentrations as low as a few parts per million can be achieved with relatively low co-absorption of carbon dioxide (84, 86).

Operating results from three plants treating natural gas containing various amounts of hydrogen sulfide and carbon dioxide (73) are summarized in Table 14-29. In Plant A the bulk of the solvent is regenerated by flashing at three pressure levels (400 psig, 200 psig, and atmospheric), and a side stream is further regenerated by an additional atmospheric flash at elevated temperature. A split-flow circuit is used, with the semistripped solvent and the completely stripped solvent being fed to the absorber at different points. Extensive power recovery by hydraulic turbines and gas expanders supplies the total pumping energy required. A single stream circuit is employed in Plant B. The solvent is regenerated first by flashing at 300 psig, 175 psig, and atmospheric pressure and then by stripping with air at elevated temperature. The flow scheme of Plant C is similar to that of Plant B, except that a split-flow cycle for semistripped and completely stripped solvent is used. Solvent flashing is carried out at 250 psig, 190 psig, and atmospheric pressure. Power recovery is also practiced in Plants B and C.

A typical material balance for acid gas removal from a product gas from coal gasification as reported by Sweny (84) is shown in Table 14-30. It should be noted that the low hydrogen sulfide content of the product and the high

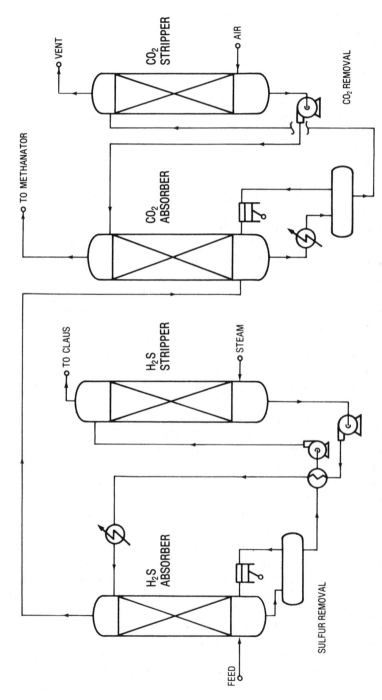

Figure 14-35. Flow diagram of SELEXOL Process for acid gas removal from coal-derived synthesis gas.

Table 14-29. Operating Data of Selexol Plants Processing Natural Gas (73)

Plant	A	B	C
Feed gas:			
Volume, MMscf/day	275	250	130
Pressure, psig	1,000	1,000	1,000
CO_2 %	43	3.5	18
H_2S, gr/100 scf	1	8	8
Treated gas:			
CO_2 %	3.5	3.0	2.5
H_2S, gr/100 scf	0.25	0.25	0.25

Table 14-30. Selexol Process for Purification of Coal Gas, Material Balance (84)

	Feed Gas	Claus Feed	Product Gas	Vent Gas*
CO_2, Mole%	31	68	0.5	98
H_2S, ppm	7,000	32%	<1	5
C_1, Mole%	8	—	11	—
H_2, Mole%	46	—	67	2
CO, Mole %	15	—	22	—
*Air Free				

hydrogen sulfide content of the acid gas indicates excellent selectivity. The ratio of carbon dioxide to hydrogen sulfide in the feed, about 44:1, is reduced to about 2:1 in the acid gas making it a suitable feed to a Claus unit.

A comparison of the relative economics of the Selexol process, the aqueous MEA and the hot potassium carbonate processes is shown in Table 14-31. The cost data presented are expressed as per cent of the MEA plant cost.

The Estasolvan Process

Disclosed jointly by Institut Francais du Petrole of France and Friedrich Uhde, G.m.b.H. of West Germany (75), this process is again similar in its characteristics to the solvent processes described previously in this chapter. However, one proposed modification of the process incorporates simultaneous absorption of acid gases and liquefied petroleum gases (LPG) in the solvent, followed by separation of the absorbed components by fractional distillation. The process was demonstrated in two pilot plants, but so far no commercial application has been reported. The solvent used in the

	20% MEA	Hot Potasium Carbonate	Selexol
Table 14-31. Comparative Economics of Selexol Process (74)			
Plant capacity, MMscf/day	100	100	100
Feed gas pressure psig	1,000	1,000	1,000
Feed gas:			
H_2S gr/100 scf	20	20	20
CO_2, %	30	30	30
Treated gas:			
H_2S gr/100 scf	0.25	1	0.25
CO_2, %	25 ppm	2	2
Grassroots plant cost, %	100	80	70
Direct operating costs,* %	100	75	40
Steam, %	100	70	10
Electricity, %	100	95	20
Cooling and process water, %	100	95	25
Indirect operating cost, %	100	80	75
Total operating costs, %	100	75	50

*Includes operating labor, maintenance, and chemical losses.

Estasolvan process, tributylphosphate, is reported to have high capacity for acid gases and good selectivity for hydrogen sulfide with respect to carbon dioxide (75).

Basic Data

Physical properties of tributylphosphate and data on the solubility of hydrogen sulfide, carbon dioxide, and methane in the solvent are given in Table 14-32 and Figure 14-36, respectively.

Process Operation

When used for acid-gas removal only, the flow scheme of the Estasolvan process is no different from that of other processes using physical solvents. However, an appreciably more complex flow arrangement is required for simultaneous removal of acid gases and LPG, involving the following steps:

1. Absorption at ambient temperature in a reboiled absorber-stripper
2. Solvent regeneration in two stages, including stripping of all dissolved gases with propane at medium pressure and slightly elevated temperature followed by recovery of propane by flashing

Table 14-32. Physical Properties of Tributylphosphate (75)

Formula	$(C_4H_8)_3PO_4$
Molecular weight	266.32
Specific gravity (25°C)	0.973 gram/ml
Melting point	−80°C
Boiling point (30 mm Hg)	180°C
Viscosity 20°C	3.19 cp
40°C	2.15 cp
100°C	1 cp
Vapor pressure 20°C	0.0037 mm Hg
40°C	0.018 mm Hg
100°C	1 mm Hg
Solubility	
TBP in water (25°C)	0.42 gram/liter
Water in TBP (25°C)	65 grams/liter

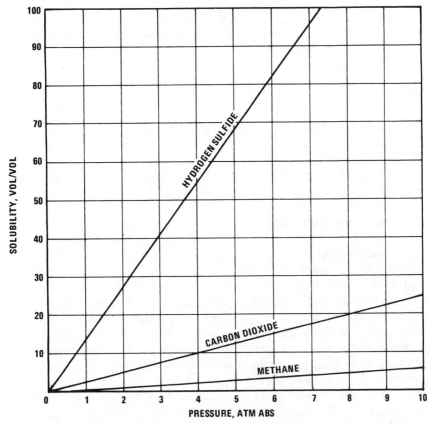

Figure 14-36. Solubility of H_2S, CO_2, and CH_4 in tributylphosphate at 25°C (gas volumes at 0°C and 760 mm Hg).

Table 14-33. Design Data for Estasolvan Plant; Removal of H₂S from Natural Gas (75)			
Plant capacity		115 MMscf/day	
Absorber pressure		1,000 psig	
Solvent circulation		1,118 gpm	
	Feed Gas	Treated Gas	Acid Gas
H₂S, %	10.0	3 ppm	85.75
COS, g/100 scf	21.0	< 0.25	—
RSH, g/100 scf	62.0	< 2.0	—
CO₂, %	7.0	6.4	11.40
N₂, %	7.5	8.0	—
CH₄, %	75.5	85.6	2.20
C₂₊, %	Trace		
Utilities:			
Electric power		2,100 KWH/hr	
Steam		11,000 lbs/hr	
Cooling water		1,200 gpm	

3. Separation of acid gases and hydrocarbons by distillation in two consecutive operations

Design data for a commercial plant designed to remove hydrogen sulfide from a very dry natural gas are given in Table 14-33 (75).

The Methylcyanoacetate Process

The Methylcyanoacetate process, disclosed by Woertz of the Union Oil Company of California (87), has been tested in a pilot plant but has not been used commercially. The solvent, methylcyanoacetate (MCA), is reported to be stable and have high capacity for acid gases. An interesting feature of the process is that the solvent is appreciably more selective for acid gases contained in hydrocarbon gas streams than the solvents used in commercial processes of this type. For example, it is reported (87) that under comparable operating conditions, propylene carbonate removes about 50 percent of the propane and 100 percent of the butanes present in a hydrocarbon feed gas, while MCA only removes about 30 percent of the propane and 75 percent of the butanes. Vapor-liquid equilibrium data of a number of gases in methylcyanoacetate have been reported by Woertz (88).

The Sepasolv MPE Process

This process, which was developed by BASF of West Germany, is quite similar to the Selexol process, both with respect to the solvent used and its mode of operation (103, 104, 105). It was initially developed primarily for the selective removal of H_2S from natural gas, but reportedly is also suitable for CO_2 removal from synthesis gases.

The Sepasolv MPE solvent is described as a mixture of polyethylene glycol methyl isopropyl ethers with a mean molecular weight of 316. Physical properties of the solvent are shown in Table 14-34 and the solubility of acid gases is given in Figure 14-37. A comparison of vapor pressures of Sepasolv MPE with Selexol, NMP (Purisol process) and methanol is shown in Figure 14-38.

Table 14-34. Physical Properties of Sepasolv MPE (104)	
Molecular Weight	316
Density, 20°C, kg/m³	1.002
Specific Heat, 0°C, KJ/kgK	1.94
Specific Heat, 100°C, KJ/kgK	2.18
Viscosity, 20°C, in Pa. S	7.2
Viscosity, 0°C, in Pa. S	15.0
Freezing Point, °C	−25

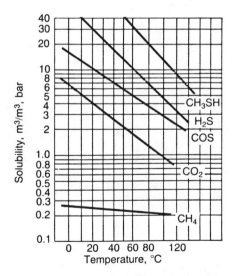

Figure 14-37. Acid Gas Solubility in Sepasolv MPE (104). Courtesy *Hydrocarbon Processing,* November 1982

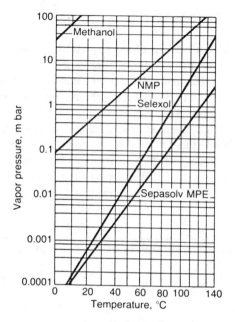

Figure 14-38. Vapor Pressure of Various Physical Solvents (104). Courtesy *Hydrocarbon Processing,* November 1982

Table 14-35. Plant Operating Data Sepasolv MPE Process*

	Plant II			Plant III		
	Crude Gas	Pure Gas	Claus Gas	Crude Gas	Pure Gas	Claus Gas
Gas Rate, m^3/h	50,000	42,250	4,897	50,000	45,600	4,536
Pressure, bar (gauge)	70	69	0.6	70	69	0.6
Temperature, °C	10	3	32	10	3	32
Analysis, He, Vol %	0.05	0.05	—	0.05	0.05	—
N_2 Vol %	3.9	4.24	0.64	3.9	4.25	0.26
CO_2 Vol %	8.73	7.08	23.71	8.73	7.54	20.43
CH_4 Vol %	80.81	88.45	7.79	80.81	87.96	6.51
C_2H_6 Vol %	0.19	0.19	0.18	0.19	0.19	0.18
H_2S Vol %	6.32	0.0003	64.53	6.32	0.0003	69.66
H_2O Vol %	—	—	3.0	—	—	3.0
COS, mg/m^3	110	25	—	110	25	961
RSH, mgS/m^3	91	<1	—	91	<1	993
Solvent Rate, kg/h		86,000			71,000	
Lean Solvent, °C		−1			−1	
Heating Steam, kg/h		5,600			4,600	

* Courtesy *Hydrocarbon Processing,* November 1982

The process scheme used is essentially the same as that used for the other physical absorption processes described in this section, incorporating absorption at relatively low temperature and regeneration of the rich solution by flashing at successively lower pressures. When high purity of the treated gas is required, a portion of the solvent may be stripped with air or inert gas, or by steam heating.

Operating data for two plants used for selective absorption of H_2S from natural gas reported by Wolfer are shown in Table 14-35 (103, 104). Tests conducted in a semi-industrial plant removing CO_2 from synthesis gas were reported by Volkamer et al. (105).

The Sulfinol Process

In contrast to the other solvent processes described in this section, the Sulfinol process employs a mixture of a chemical and physical solvent as the absorption medium and is, in many respects, comparable to the Ethanolamine process described in Chapters 2 and 3. However, the presence of the physical solvent enhances the solution capacity appreciably over that of a conventional ethanolamine solution, especially when the gas stream to be treated is available at high pressure and the acidic components are present in high concentrations. Licensed by Shell International Research Mij. N.V., The Hague, The Netherlands, and Shell Development Co., Houston, the process has found wide application in the treatment of natural, refinery, and synthesis gases. In 1975, over 100 commercial units were reported to be in operation or under construction (61).

Basic Data

The Sulfinol solvent consists of an ethanolamine, usually diisopropanolamine (DIPA), sulfolane (tetrahydrothiophene dioxide), and water. The equilibrium solubility of hydrogen sulfide in sulfolane and in the Sulfinol solvent, as a function of partial pressure, is shown in Figure 14-39 (76). For comparison, the solubility of hydrogen sulfide in water and in 20 percent aqueous monoethanolamine solution is also shown. It should be noted that, from the standpoint of solvent capacity, the Sulfinol solvent is inferior to aqueous monoethanolamine at low hydrogen sulfide partial pressures. However, as the partial pressure of hydrogen sulfide becomes higher, the capacity of the Sulfinol solvent continues to increase while that of the monoethanolamine solution remains essentially the same after the stoichiometric proportion of the chemical reaction between hydrogen sulfide and the amine has been reached.

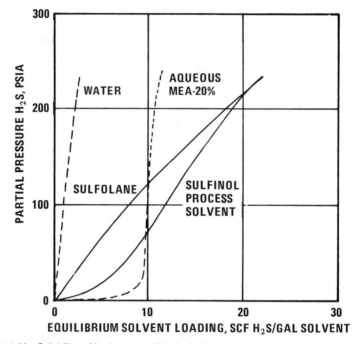

Figure 14-39. Solubility of hydrogen sulfide in Sulfinol solvent (76).

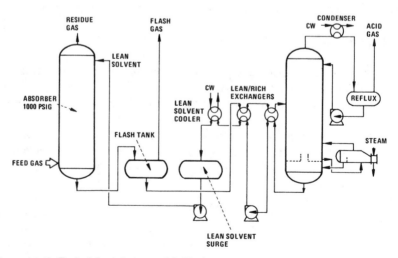

Figure 14-40. Typical flow diagram of Sulfinol process.

Process Description

A typical flow diagram of a Sulfinol unit is shown in Figure 14-40. The scheme is identical with that shown in Figure 2-9, with the exception of the flash tank, which is optional in aqueous ethanolamine systems but almost a necessity in a Sulfinol unit, especially when high-pressure gases are processed. Because of the relatively high solubility of hydrocarbons in the solvents, as in most organic solvents, omission of the flash tank would lead to high concentrations of hydrocarbons in the acid gas and to operational difficulties in Claus units where the hydrogen sulfide is converted to elemental sulfur. The flash gas can either be recompressed and recycled to the absorber inlet or, as usually practiced in natural gas-treating plants, used as plant fuel. In some instances it may be necessary to make special provisions for the removal of heavy hydrocarbons, e.g., pentanes and aromatics from the acid-gas stream, as such compounds are definitely troublesome in a Claus unit.

Process Operation

Operating data from pilot-plant tests and from commercial units have been reported by Dunn, et al. (76, 77), Frazier (78), and Klein and Verloop (89). Pilot-plant data obtained with an East Texas natural gas are shown in Table 14-36 (76). For comparison, performance of the pilot plant with an aqueous 20 percent monethanolamine solution is also presented. Additional pilot-plant data obtained in processing four different Canadian natural-gas streams are given in Table 14-37 (76).

Studies of hydrocarbon solubility in the Sulfinol solvent conducted during the pilot-plant tests indicate that aliphatic hydrocarbons up to pentane are largely rejected by the solvent. However, aromatics are absorbed quite efficiently (76). The Sulfinol solvent proved to be very stable, resulting in losses due to degradation of less than 5 lbs per MMscf of raw gas. Corrosion is, in general, no problem in Sulfinol units, and carbon steel is a satisfactory material of construction (76). However, isolated cases of absorber corrosion, particularly in the lower section of the vessel, have been reported by Schmeal, MacNab, and Rhodes (90). Blisters and "ring pits" were observed on trays, downcomers, and vessel walls. The corrosion was ascribed to boiling and flashing of carbon dioxide. It was especially pronounced in units processing gases where the acid gases are present in low ratios of carbon dioxide to hydrogen sulfide. Lowering of temperatures in the absorber bottom and protection of that area with stainless steel liners were found to be effective remedies.

In addition to its effectiveness for acid-gas removal, the Sulfinol solvent showed excellent capability to remove carbonyl sulfide and mercaptans. In one test about 96 percent removal of methyl mercaptan was reported (76).

Table 14-36. Pilot-Plant Data for Sulfinol Process and Aqueous MEA (76) (East Texas Natural Gas)

	Sulfinol	20% MEA
Absorber pressure, psig	1,000	1,000
Feed gas rate, scf/min	40.7	23.6
Solvent rate, gpm	1.0	1.0
Solvent loading, scf/gal.	8.5	4.9
Total steam, lb/lb acid gas*	1.42	3.80
Feed gas:		
H_2S %	15.00	
CO_2 %	6.00	
COS, ppm	60	
N_2, %	7.50	
CH_4, %	57.69	
C_2H_6, %	6.24	
$C_3H_8 + $ %	7.57	
Treated gas:		
H_2S, gr/100 scf	<1	<1

*Combined duty of preheater and reboiler

Table 14-37. Pilot-Plant Data for Sulfinol Process (76) (Canadian Natural Gas)

Gas stream	A	B	C	D
Pressure, psia	995	935	935	715
Feed gas:				
H_2S, %	26.40	16.20	15.60	3.40
CO_2, %	5.20	5.40	5.50	1.30
COS, %	0.03	0.01	0.01	—
RSH, gr/100 scf	0.75	0.14	0.14	—
Treated gas:				
H_2S, gr/100 scf	0.4	0.13	0.1	0.15
CO_2, gr/100 scf	<1	<1	<1	<1
RSH, gr/100 scf	<0.01	—	—	—
COS, ppm	2.0	1.0	1.0	—
Acid gas:				
Hydrocarbon, %	1.9	0.8	0.7	3.6
Steam, lb/lb acid gas	0.9	1.0	1.0	1.3

Table 14-38. Commercial Operating Data of Sulfinol Process (77, 78)				
Plant	A(78)	B(77)	C(77)	D(77)
Feed gas:				
Volume, MMscf/D	32	150	50	150
Pressure, psig	1,000	1,000	1,000	1,000
H₂S, gr/100 scf	1.60(%)	6.35	27.00	19.00
CO₂, %	6.90	3.30	9.16	6.81
COS, ppm	7	—	—	—
RSH, gr/100 scf	19(ppm)	0.75	0.20	—
Total hydrocarbon, %	91.00	96.33	89.88	91.40
Treated gas:				
CO₂, %	—	0.30	0.30	0.30
H₂S, gr/100 scf	<0.1-0.6	0.25	0.25	0.25
RSH, gr/100 scf	—	0.25	—	—
Solvent rate, gpm	315-335	—	—	—
Solvent loading, SCF/gal	6.0	—	—	—
Reboiler duty, MMBtu/MMscf gas	10.1	—	—	—

The commercial operating data reported by Dunn, et al. (77) and Frazier (78) are summarized in Table 14-38.

The Amisol Process

The Amisol process disclosed by Bratzler and Doerges (91) is similar to the Sulfinol process in that it uses a combination of a physical and chemical solvent for acid gas removal. The solvent consists of methanol with an undisclosed additive and mono- or diethanolamine. The process is reportedly capable of complete removal of all sulfur compounds—down to residual sulfur levels of less than 0.1 ppm—from gases of relatively low hydrogen sulfide and carbon dioxide content. Carbon dioxide can be removed to a residual of less than 5 ppm.

The flow scheme of the process is quite conventional. However, since absorption is carried out at about 35°C (95°F) and regeneration at 80°C (176°F), no solution heat exchanger is required. Because of the volatility of methanol, both the absorber and regenerator effluent gas streams have to be washed with water to recover vaporized methanol. The methanol is recovered from the water in a small distillation column.

Semi-commercial plant tests with a gas produced by pressure gasification of residual oil with oxygen and steam showed removal of sulfur compounds to less than 1 ppm and carbon dioxide removal to 10 ppm. The observed solvent capacity was 26.6 N m³ of $H_2S + CO_2$ per m³ of solution (91).

Membrane Permeation Processes

The use of membranes to separate and purify gases is a relatively new development. Although the principles of selective permeation have been known for many years, the process did not become commercially feasible until the development of high-flux cellulose-acetate membranes for gas separation in 1970 (106). The original membranes were in the form of flat sheets made up of a very thin "active" layer on the order of 0.1 μm in thickness, supported on thick, open-pore material, typically about 100 μm thick. This development was followed by the introduction of hollow fibers incorporating a similar asymmetric membrance structure (107).

Commercial configurations have been developed based on both the sheet and hollow-fiber membrane forms. The sheet material is typically formed into spiral-wound elements which are inserted into cylindrical metal tubes (108, 109); while the hollow fibers are arranged in the form of bundles which also fit conveniently into tubular containers (110). Two polymers have attained commercial importance, cellulose acetate, and polysulfone. These base materials may be coated on the active face with a thin layer of another polymer to seal surface pores or modify separation properties.

The operation of gas permeation separators is very simple; high pressure feed gas is supplied to one side of the membrane (the outside of hollow fibers): gases which pass through the membrane form a permeate gas stream on the low-pressure side, while residue product remains on the high-pressure side. The permeability of gases through polymer membranes is believed to involve both solubilization and diffusion. Selectivity is determined by the relative rates of passage of gas-stream components through the membrane. The principles of membrane gas separation and basic design equations are discussed by MacLean et al. (111), Hogsett and Mazur (112), and Schell and Hoernschemeyer (113).

It has been predicted that three applications of membrane technology have the greatest economic potential when compared to more established processes (114). These are (1) recovery of hydrogen from ammonia purge streams, (2) recovery of hydrogen from hydrotreater offgas, and (3) separation of carbon dioxide from hydrocarbons for enhanced oil recovery. All of these are considered to be more properly classified as separation processes than gas purification processes and are therefore not described in this text. However, several gas purification applications have also been noted. Although they have not yet achieved significant commercial usage, it is believed that the technology is sufficiently promising to warrant further discussion.

Pilot-scale studies of CO_2 removal from hydrocarbon gas streams are reported by Schell and Houston (108). The results of one test are shown in Table 14-39. In this case, high pressure, wellhead hydrocarbon gas was de-

**Table 14-39. Pilot Plant Data on CO₂ Removal
by Membrane Permeation (from Schell and Houston (108)**

	Feed Gas	Residual Gas	Permeate Gas
Pressure, psig	800	772	10
Temperature, °F	87	87	87
Gas Composition, Vol. %			
CO_2	26.7	1.7	69.0
CH_4	70.5	93.2	31.0
C_2H_6	1.5	2.7	trace
$C_3{}^+$	0.1	0.5	trace
N_2	1.2	1.9	trace

hydrated, then fed to a membrane separator system consisting of twenty-four 2-in.-diameter spiral-wound cellulose-acetate elements operating in series. The results indicate that the membrane system is capable of reducing the carbon dioxide level to a value suitable for pipeline transmission with very low pressure drop.

Membrane processes have also been proposed for the removal of H_2S and water vapor from gas streams. Test data on the dehydration of methane at 250 and 350 psig indicate that greater than 97 percent water removal can be attained (115). The process becomes more economical for dehydration as the feed gas pressure is raised due to both the increased permeation rate and the reduced amount of water in the feed gas.

Three specific examples of membrane separation for natural gas purification are described by Grey and Mazur (116). These are:

1. Reducing the CO_2 concentration from 8–9 percent to 5 percent in order to meet the sales gas specification.
2. Removing H_2S from a gas containing 6000 ppm to produce a gas containing <50 ppm for use on onsite fuel.
3. Reducing the CO_2 concentration from about 50 percent to about 40 percent to meet the fuel heating valve requirements of a gas turbine.

Commercial spiral-wound membrane elements were tested for the three cases. The test data were used to project the performance of the commercial systems as given in Table 14-40.

The use of a combination membrane/amine system to purify high-pressure gas streams is described by Miller et al. (117). The combination process is shown to be more economical than amine alone for the case of 1050-psig natural gas containing a relatively high concentration of CO_2 (8.3 percent) and some H_2S (0.01 percent). The gas stream is passed through the mem-

Table 14-40. Typical Membrane Applications for Natural Gas Purification*

	Case 1			Case 2			Case 3		
	Feed	Permeate	Residue	Feed	Permeate	Residue	Feed	Permeate	Residue
Flow, MMSCFD	1.13	0.13	1.00	7.41	3.71	3.70	3.00	0.66	2.34
Pressure, psig	190	10	188	745	18.2	740	215	30	210
Temp, °F	90	85	85	90	85	85	105	90	90
Composition									
N_2, %	0.8	0.6	0.8	—	—	—	0.8	0.1	0.9
CO_2, %	8.2	32.3	5.0	3.5	6.8	0.1	53.6	94.7	42.0
HC, %	91.0	67.1	94.2	95.9	92.0	99.9	45.7	5.2	57.1
H_2S ppm	—	—	—	6,000	12,000	30	—	—	—

*Data of Grey and Mazur.

brane unit first where the more permeable gases (CO_2 and H_2S) are separated from the less permeable gases (CH_4, C_2H_6, N_2). The partially purified high-pressure product gas then flows through amine and glycol absorbers in series to provide final purification and dehydration. The permeate stream, which has a heating value of 350–450 Btu/scf is used as fuel for the amine and glycol reboilers.

As the membrane area is increased, the amount of acid gas which must be removed by the amine plant decreases, causing a reduction in capital and operating costs of this unit. However, when the membrane area is quite high, the permeate flow exceeds fuel gas requirements, adversely affecting overall economics. As a result of these compensating factors, a minimum cost combination system can be defined for gas purification cases where this process scheme is applicable.

References

1. Leblanc, F. 1850. *Compt. Rend.* 30:483.
2. Badische Anilin und Soda Fabrik. 1914. German Patent 289,694.
3. Van Krevelen, D.W., and Baans, C.M.E. 1950. *J. Phys. & Colloid Chem.* 54:370-390.
4. Hainsworth, W.R., and Titus, T.Y. 1921. *J. Am. Chem. Soc.* 43:1-11.
5. Larson, A.T., and Teitsworth, C.S. 1922. *J. Am. Chem. Soc.* 44:2878.
6. Zhavoronkov, N.M., and Reshchikov, P.M. 1933. *J. Chem. Ind. (U.S.S.R.)* 10(8):41-49.
7. Zhavoronkov, N.M., and Chagunava, V.T. 1940. *J. Chem. Ind. (U.S.S.R.)* 17(2):25-29.
8. Egalon, R.; Vanhille, R.; and Willemyns, M. 1955. *Ind. Eng. Chem.* 47(May):887-899.
9. Zhavoronkov, N.M. 1939. *J. Chem. Ind. (U.S.S.R.)* 16(10):36-37.
10. Pavlov, K.F., and Lopatin, K.I. 1947. *J. Appl. Chem. (U.S.S.R.)* 20(12):1223-1234.
11. Yeandle, W.W., and Klein, G.F. 1952. *Chem. Eng. Progr.* 48(July):349-352.
12. Walthall, J.H. (Tennessee Valley Authority) 1958. Personal communication. (Oct. 29).
13. Brown, E.H.; Cline, J.E.; Felger, M.M.; and Howard, R.B. Jr. 1945. *Ind. Eng. Chem., Anal. Ed.* 17:280-282.
14. Markevich, K.Y. 1939. *J. Chem. Ind. (U.S.S.R.)* 16(10):31-35.
15. Levitskaya, E.P. 1937. *J. Chem. Ind. (U.S.S.R.)* 14:342-346.
16. Badische Anilin und Soda Fabrik. 1914. British Patent 8030 (March 30).
17. I.G. Farbenindustrie, A.G. 1938. British Patent 478,973.
18. Christensen, A. 1936. U.S. Patent 2,029,411.
19. Dely, J.G. 1936. U.S. Patent 2,047,550.
20. Dontsova, E.I. 1942. *J. Appl. Chem. (U.S.S.R.)* 15(6):447-452.
21. Gump, W., and Ernst, I. 1940. *Ind. Eng. Chem.* 22(Apr.):382-384.
22. Lombard-Gerin, L. 1939. French Patent 848,704.
23. Streich, M. 1970. *Hydroc. Process.* 49(Apr.):86-88.
24. Guillaumeron, P. 1949. *Chem. Engr.* 56:105.

25. Ruhemann, M. 1949. *The Separation of Gases*, 2nd ed. New York: Oxford University Press, Chap. 10.

26. Bardin, J.S., and Beery, D.W. 1953. *Petrol. Refiner* 32(2):99-102.

27. Baker, D.F. 1955. *Chem. Eng. Progr.* 51(9):399-402.

28. Jester, M.R. 1958. *Trans. Inst. Chem. Engrs. (London)* 36:133-136.

29. Stull, D.R. 1947. *Ind. Eng. Chem.* 39:517-550.

30. *U.S. Nat. Bur. Standards Circ. No. 564* (1955).

31. *U.S. Nat. Bur. Standards Circ. No. 500* (1952).

32. Kirk, R.E., and Othmer, D.F. (eds.). 1952. *Encyclopedia of Chemical Technology*, vol. 9. New York: Interscience Publishers, Inc., 404.

33. Din, F. (ed.). 1956. *Thermodynamic Functions of Gases*, vol. 1. London: Butterworths Scientific Publications, p. 164.

34. Millar, R.W. 1923. *J. Am. Chem. Soc.* 45:874.

35. Egan, C.J., and Kempt, J.D. 1937. *J. Am. Chem. Soc.* 59:1264-1268.

36. Ruhemann, F., and Zinn, N. 1937. *Physik. Z. Sowjetunion* 12:389-403.

37. Grotz, B.J. 1967. *Hydroc. Process.* 46(Apr.):197-202.

38. Herbert, W. 1956. *Erdöl u. Kohle* 9(2):77-81.

39. Hochgesand, G. 1968. *Chemie-Ing.-Techn.* 40(9/10):432-440.

40. Hoogendoorn, J.C., and Solomon, J.M. 1957. *Brit. Chem. Eng.* 2(May):238-244.

41. Dreisbach, R.R. 1952. *Pressure-Volume-Temperature Relationships of Organic Compounds*, 3rd ed. New York: McGraw-Hill Book Company, Inc., Chart 20.

42. Lurgi Gesellschaft für Wärme-und Chemotechnik. *Rectisol for Gas Treating.*

43. Hochgesand, G. 1970. *Ind. Eng. Chem.* 62(7):37-43.

44. Sherwood, T.K., and Pigford, R.L. 1952. *Absorption and Extraction*, 2nd ed. New York: McGraw-Hill Book Company, Inc.

45. Perry, J.H. (ed.). 1963. *Chemical Engineers' Handbook* 3rd ed. New York: McGraw-Hill Book Company, Inc.

46. Huntington, R.L. 1950. *Natural Gas and Natural Gasoline*. New York: McGraw-Hill Book Company, Inc.

47. Glowacki, W.M. 1945. *Chemistry of Coal Utilization*, vol. 2. Edited by H.H. Lowry. New York: John Wiley & Sons, Inc., pp. 1136-1231.

48. Powell, J.R. 1945. *Chemistry of Coal Utilization*, vol. 2. Edited by H.H. Lowry. New York: John Wiley & Sons, Inc., pp. 1232-1251.

49. Porter, H.C. 1924. *Coal Carbonization*. New York: Reinhold Publishing Corporation.

50. Hopton, G.U. 1953. *The Cooling, Washing and Purification of Coal Gas*. London: North Thames Gas Board.

51. Hoffert, W.H., and Claxton, G. 1938. *Motor Benzole: Its Production and Use*. 2nd ed. London: The National Benzole Association.

52. Silver, L., and Hopton, G.U. 1942. *J. Soc. Chem. Ind. (London)* 61(Mar.):37.

53. Hutchison, W.K. 1937. *A Plant for the Removal of Sulfur Compounds by Oil Washing*. Paper presented at the 9th Autumn Meeting, The Institution of Gas Engineers (November 2).

54. Hollings, H., and Silver, L. 1934. *Trans. Inst. Chem. Engrs. (London)* 12:49.

55. Glowacki, W.M. 1945. *Chemistry of Coal Utilization*, vol. 2. Edited by H.H. Lowry. New York: John Wiley & Sons, Inc., p. 1195.

56. Cawley, C.M., and Newall, H.E. 1945. *J. Soc. Chem. Ind. (London)* 64:285.

57. Anon. 1958. *Chem. Engr.* 65(May):68-70.

58. Speece, J.E. 1925. *Proc. Pacific Coast Gas Assoc.* 16:345-385.

59. Kohl, A.L., and Buckingham, P.A. 1960. *Petr. Refiner* 39(May):193-196.

60. U.S. Patents 2,926,751, 2,926,752 and 2,926,753 (March, 1960).

61. Anon. 1975. *Hydro. Process.—Gas Processing Handbook* Issue 54 (Apr):79-138.

62. Dow Chemical of Canada, Ltd. 1962. *Gas Conditioning Fact Book.*

63. Schmack, P., and Bittrich, H.J. 1966. *Wissenschaftl. Zeitschrift,* 8(2/3):182-186.

64. Makranczy, J.; Szeness, M.M. and Rusz, L. 1965. *Veszpremi Vegyipari Egyetem Kozlemenyei* 9:95-105 (University of Chemical Industries, Institute of General and Inorganic Chemistry, Veszprem, Hungary).

65. Buckingham, P.A. 1961. Paper presented at NGAA Regional Meeting, Odessa, Tex. (May).

66. Buckingham, P.A. 1964. *Hydroc. Process.* 43(Apr.):113-116.

67. Cook, T.P., and Tennyson, R.N. 1969. *Chem. Eng. Progr.,* 65(11):61-64.

68. Stein, W.H. 1969. *Erdöl-Erdgas-Zeitschrift* 85:467-470.

69. Kapp, E. 1970. *Erdöl and Kohle-Erdgas-Petrochemie Vereinigt mit Brennstoff Chemie* 23(9):566-571.

70. Beavon, D.K., and Roszkowski, T.R. 1969. *Oil Gas J.* 67(Apr. 14):138-142.

71. Boston, F.C., and Schneider, M.L. 1971. Paper presented at Gas Conditioning Conference of the University of Oklahoma.

72. Lurgi Gesellschaft für Wärme und Chemotechnik m.b.H. *Purisol for Gas Treating.*

73. Hegwer, A.M., and Harris, R.A. 1970. *Hydroc. Process.* 49(Apr.):103-104.

74. Sweny, J.W., and Valentine, J.P. 1970. *Chem. Engr.* 49(Sept.7):54-56.

75. Franckowiak, S., and Nitschke, E. 1970. *Hydroc. Process.* 49(May):145-148.

76. Dunn, C.L.; Freitas, E.R.; Goodenbour, J.W.; Henderson, H.T.; and Papadopoulos, M.N. 1964. *Hydroc. Process* 43(Mar.):150-154.

77. Dunn, C.L.; Freitas, E.R.; and Hill, E.S. 1965. *Hydroc. Process.* 44(Apr.):137-140.

78. Frazier, J. 1970. *Hydroc. Process.* 49(Apr.):101-102.

79. Freireich, E., and Tennyson, R.N. 1977. "Increased Natural Gas Recovery from Physical Solvent Gas Treating Systems." Paper presented at the Annual Gas Conditioning Conference, University of Oklahoma, Norman, Oklahoma (March 7-8).

80. Sweny, J.W. 1973. "Synthetic Fuel Gas Purification by the SELEXOL Process." Paper presented at 165th National Meeting of the American Chemical Society, Division of Fuel Chemistry, Dallas, Texas (April 8-12).

81. Valentine, J.P. 1974. *Oil Gas J.* 72(46):60-62.

82. Clare, R.T., and Valentine, J.P. 1975. "Acid Gas Removal using the SELEXOL Process." Paper presented at Second Quarterly Meeting of the Canadian Gas Processors Association, Edmonton, Alberta (June 5).

83. Valentine, J.P. 1975. "Economics of the SELEXOL Solvent Gas Purification Process." Paper presented at the 79th National Meeting of the American Institute of Chemical Engineers, Houston, Texas (March 19).

84. Sweny, J.W. 1976. "The SELEXOL Process in Fuel Gas Treating." Paper presented at 81st National Meeting of the American Institute of Chemical Engineers, Kansas City, Missouri (April 11-14).

85. Raney, D.R. 1976. *Hydro. Process.* 55(4):73-75.

86. Van Deraerschot, R., and Valentine, J.P. 1976. "The SELEXOL Solvent Process for Selective Removal of Sulfur Compounds." Paper presented at 2nd International Conference on the Control of Gaseous Sulphur and Nitrogen Emission, Salford University, England (April 6-8).

87. Woertz, B.B. 1971. *J. of Pet. Tech.* (April):483-490.

88. Woertz, B.B. 1975. *Soc. Pet. Eng. J.* (February):7-12.

89. Klein, J.P., and Verloop, J. 1975. "The Purification of Natural Gas Prior to Liquefaction." Paper presented at Second Iranian Congress of Chemical Engineering, Tehr'an (May 11-14).

90. Schmeal, W.R.; MacNab, A.J.; and Rhodes, P.R. 1978. *Chem. Eng. Progr.* 74(3):37-42.

91. Bratzler, K., and Doerges, A. 1974. *Hydro. Process.* 53(4):78-80.

92. Haase, D. J., Duke, P. M., and Cates, J. W. 1982. *Hydrocarbon Processing,* March, p. 103.

93. Haase, D. J. 1975. *Chemical Engineering* 82, 16, August 4, p. 52.

94. Harris, R. A. 1980. *Hydrocarbon Processing,* 59(8), August p. 111.

95. Hise, R. E., Massey, L. G., Adler, R. J., Brosilow, C. B., Gardner, N. C., Brown, W. R., Cook, W. J., and Petrik, M. 1982. "The CNG Process, A New Approach to Physical Absorption Acid Gas Removal," paper presented at AIChE Annual Meeting, November 14–18, 1982.

96. Ranke, G. 1973. Linde Reports on Science and Technology 18/1973.

97. Rousseau, R. W., Kelley, R. M., and Ferrell, J. K. 1981, "Evaluation of Methanol as a Solvent for Acid Gas Removal In Coal Gasification Processes," paper presented at AIChE Spring Meeting, Houston, Texas, April 5–9, 1981.

98. Ferrell, J. K., Felder, R. M., Rousseau, R. W., Ganesan, S., Kelley, R. M., McCue, J. C., and Purdy, M. J. 1982. Project Summary, Coal Gasification/Gas Cleanup Test Facility: Volume II. Environmental Assessment of Operation with Devolatilized Bituminous Coal and Chilled Methanol. EPA-600/S7-82-023. August.

99. Zawacki, T. S., Duncan, D. A., and Macriss, R. A. 1981. *Hydrocarbon Processing,* 59(4), April p. 143.

100. Judd, D. K. 1978. *Hydrocarbon Processing* 57(4), April p. 122.

101. Swanson, C. G. 1978. "Carbon Dioxide Removal in Ammonia Synthesis Gas by Selexol," paper presented at 71st AIChE Meeting, Miami Beach, Florida, November 12–16.

102. Sweny, J. W. 1980. "High CO_2-High H_2S Removal with Selexol Solvent," paper presented at 59th Annual GPA Convention, Houston, Texas, March 17–19, 1980.

103. Wolfer, W., Schwartz, E., Wodrazka, W., and Volkamer, K. 1980. *Oil and Gas Journal,* January 21, p. 66.

104. Wolfer, W. 1982. *Hydrocarbon Processing,* 61(11), November, p. 193.

105. Volkamer, K., Wagner, E., and Schubert, F. 1982. *Plant Operations Progress* 1(2), April.

106. Gantzel, P. K. and Merten, U. 1970. *Ind. Eng. Chem. Process Des, Dev.,* Vol. 9, No. 2, p. 331.

107. Gardner, R. J., Crane, R. A., and Hannon, J. F. 1977. *Chem. Eng. Prog.* October, p. 76.

108. Schell, W. J. and Houston, C. D. 1982. *Chem. Eng. Prog.* October, p. 33.

109. Russell, F. G. 1983. *Hydrocarbon Processing,* August, p. 55.

110. Rosenzweig, M. D. 1981. *Chemical Engineering,* November 30, p. 61.

111. MacLean, D. L., Stookey, D. J., and Metzger, T. R. 1983. *Hydrocarbon Processing,* August, p. 47.

112. Hogsett, J. E. and Mazur, W. H. 1983. *Hydrocarbon Processing,* August, p. 52.

113. Schell, W. J., and Hoernschemeyer, D. L. 1982. "Principles of Membrane Gas Separation," presented at AIChE Symposium, June 7–10, Anaheim, California.

114. Schendel, R. L., Mariz, C. L., and Mak, J. Y. 1983. *Hydrocarbon Processing,* August, p. 58.

115. Schell, W.J. and Houston, C. D. 1982. *Hydrocarbon Processing,* September, p. 149.

116. Grey, N. R. and Mazur, W. H. 1984. "Membrane Separation of CO_2 and H_2S from Natural Gas—Field Experience," presented at the 1984 Spring National Meeting of the AIChE, Anaheim, California, May 20-23.

117. Miller, B. D., Richards, R., and Schott, M. E. 1984. "Separex™ System Makes Hydrocarbon Recovery Feasible," presented at the 1984 Spring National Meeting of the AIChE, Anaheim, California, May 20-23.

Appendix

Units and Conversion Factors

Quantity	SI Unit	Symbol
Length	metre	m
Mass	kilogram	kg
Time	second	s
Temperature	kelvin	K
Amount of substance	mole	mol
Force	newton	$N = kg\ m/s^2$
Work or energy	joule	$J = Nm$
Power	watt	$W = J/s$
Pressure	pascal	$Pa = N/m^2$
Dynamic viscosity		$PA\ s$ or $N\ s/m^2$
Kinematic viscosity		m^2/s
Diffusivity		m^2/s
Surface tension		J/m^s or N/m
Enthalpy		J/kg
Entropy		$J/kg\ K$
Heat capacity		$J/kg\ K$
Thermal conductivity		$W/m\ K$
Heat transfer coefficient		$W/m^2\ K$
Mass transfer coefficient		m/s

Frequently Used Conversion Factors

Length	1 in.		= 2.54 cm
	1 ft		= 0.3048 m
Area	1 in^2		= 6.4516 cm^2
	1 ft^2		= 0.092903 m^2
Volume	1 in^3		= 16.387 cm^3
	1 ft^3	= 7.4805 gallons (U.S.)	= 0.02832 m^3
	1 gallon (U.S.)		= 3785.4 cm^3
	1 barrel (oil)	= 42 gallons (U.S.)	= 0.15987 m^3
Mass	1 lb	= 7000 grains	= 0.45359 kg
	1 ton (long)	= 2240 lb	= 1016.06 kg
Force	1 lbf		= 4.4482 N
	1 dyn		= 10^{-5} N
Temperature difference	1°F (R°)		= 5/9 °C (°K)
Energy (work, heat)	1 cal (I.T.)		= 4.1868 J
	1 erg		= 10^{-7} J
	1 Btu (I.T.)	= 252 cal (I.T.)	= 1055.056 J
Heating value (volumetric)	1 Btu/ft^3		= 37.259 kJ/m^3
Velocity	1 f/s		= 0.3048 m/s
Volumetric flow	1 ft^3/s		= 0.028316 m^3/s
	1 U.S. gal/min (gpm)		= 63.09 cm^3/s
Mass flow	1 lb/h		= 0.1260 g/s
Density	1 lb/in.^{3}		= 27.680 g/cm^3
	1 lb/ft^3		= 16.019 kg/m^3
Pressure	1 atm (std)	= 14.696 lbf/in.^{2}	= 101.3250 kPa
	1 lbf/in.^{2} (psi)		= 6.8948 kPa
	1 mm Hg (0°C)	= 1 torr	= 133.32 Pa
Power	1 hp (British)		= 745.70 W
	1 Btu/h		= 0.29307 W
	1 ton of refrigeration		= 3516.9 W
Viscosity, dynamic	1 P (poise)		= 0.1 N s/m^2
	1 lb/ft s		= 1.4882 N s/m^2
Viscosity, kinematic	1 St (stoke)		= 10^{-4} m^2/s
Surface tension	1 erg/cm^2	= 1 dyn/cm	= 10^{-3} N/m
Mass transfer coefficient	1 ft^{3}h/ ft^2	= 1 ft/h	= 0.084667 m/s
Heat transfer coefficient	1 Btu/h ft^2°F		= 5.6783 W/m^2 K
Enthalpy	1 Btu/lb		= 2.326 kJ/kg
Heat capacity	1 Btu/lb°F		= 4.1868 kJ/kg K

Abbreviations and Constants

For SI units m = milli, 10^{-3}; C = centi, 10^{-2}; k = kilo, 10^3; and M = mega, 10^6.
SCF (or scf) = standard cubic feet of gas (60°F, 1 atm)
MSCF = thousand standard cubic feet of gas
MMSCF = million standard cubic feet of gas
s.t.p. = standard temperature and pressure (273.15 K and 1.013 × 10^5 Pa)
1 grain per 100 SCF = 24.19 mg/m^3 (s.t.p.)
Volume of 1 lb mol of ideal gas at 60°F and 1 atm = 379 ft^3
Volume of 1 kmol of ideal gas at s.t.p. = 22.41 m^3
Gas constant, R = 1.986 Btu/lb mol°R = 8.314 J/mol k

Index

Absorbers (*see also* Columns; Towers)
Absorption (*see also* specific absorbents and
  absorbates)
  with chemical reaction, 14–16
  design procedures, 2–21
Absorption coefficients, for air dehydration
  with lithium chloride solutions, 625
  for ammonia in water, 568–575
  for benzene in petroleum oil, 831–835
  for carbon dioxide, in diethanolamine
    solutions, 86, 87
    in hot potassium carbonate solutions, 226,
    227
    in monoethanolamine solutions, 82–86
    in sodium carbonate-bicarbonate solutions,
    184, 185
    in sodium hydroxide-carbonate solutions,
    186, 187
    in water, 252–259
  for carbon monoxide in
    copper-ammonium-salt solutions, 801
  for chlorine in water, 290–296
  definition of, 3
  for fluorides in water, 273–282
  for hydrogen sulfide, in diethanolamine
    solutions, 89, 90
    in monoethanolamine solutions, 89, 90
    in thioarsenate solutions, 504
Absorption towers (*see* Columns; Towers)
Acetic acid in copper-ammonium-salt
  solutions, 787–808

Acetylene, catalytic conversion of, 762–767
  hydrogenation of, catalysts for, 765
    equilibrium constants for, 764
    heats of reaction in, 764
  vapor pressure of, 815
Acetylene removal by selective hydrogenation,
  762–767
Acticarbone process for solvent recovery, 699
Activated alumina, capacity for water of, 637,
  646
  gas dehydration with, 640
  physical properties of, 636, 641
  volume-weight relations, 641
Activated bauxite, 642
  capacity for water, 637
  dehydration with, 642
  physical properties of, 636, 643
Activated carbon process for hydrogen sulfide
  removal, 442–446
  organic sulfur removal in, 445, 446
  for sulfur dioxide removal, 405–409
Activated carbons, 685
  adsorption isotherms for hydrocarbons on,
    687, 688
  air purification with, 685–713
  benzol removal with, 712
  capacity of, for organic compounds, 687,
    688, 695, 710
  heat of adsorption of organic compounds on,
    695
  pressure drop through beds of, 697

properties of, 685
solvent recovery with, 689–701
(see also Active carbons)
Activated charcoal, 685
(see also Activated carbons)
Active alumina, 643
(see also Activated alumina)
Active carbons, adsorption of organic vapors
on, 689–701
tests for evaluation of, 688
(see also Activated carbons)
Adip process, 41, 117
Adip solutions, specific gravity of, 69
specific heat, 74
viscosity of, 72
(see also diisopropanolamine solutions)
Adipic acid, 319–321
Adsorbents, dehydration, 632–668
capacity for water of, 637, 639, 646
design capacity of, 649–654
pressure drop through beds of, 660–664
properties of, 636
regeneration of, 654–657
(see also specific adsorbents)
Adsorption, for air purification, 682–712
for benzol removal, 712
definition, 21
dehydration by (see Dehydration)
design methods, 21–24
fluidized bed, 701
heat effects during, 646
of hydrogen sulfide, ammonia, carbon
dioxide, and sulfur dioxide on
molecular sieves, 668–682
mechanism of, 630–632
of monoethanolamine vapor, 127–129
for odor removal, 705–712
of organic vapors on active carbon,
685–689
pressure swing, 681
for sulfur dioxide removal, 403–409
for trace-impurity removal, 705–712
of water vapor, 632–668
Adsorption vessels, design of, 645–662, 694
Air, water content of, in contact with lithium
chloride solutions, 623
water vapor content of saturated, 585
Air conditioning, dehydration with lithium
chloride solutions in, 621–627
glycol dehydration in, 589
Air dehydration, by Kathabar system, 626
Air pollution by sulfur dioxide, 299–303

Air purification, by adsorption, 682–712
by catalytic oxidation, 767–783
for low temperature separation, 685
process design for, 694–698
with silica gel, 682–685
Alkacid process, for hydrogen sulfide and
carbon dioxide removal, 203–211
solution "dik," capacity for hydrogen
sulfide, 204
solution M, 208
vapor pressure of hydrogen sulfide over
solutions, 210
Alkali carbonate solutions, acid gas removal
with, 182
Alkali metal compounds for sulfur dioxide
removal, 340–359
Alkali metal sulfite-bisulfite process for sulfur
dioxide removal, 351–356
Alkaline earth compounds, removal of sulfur
dioxide with, 307–340
Alkalized alumina process for sulfur dioxide
removal, 399
Alkanolamines, 29–109
(see also Ethanolamines)
Alumina, activated (see Activated alumina)
as catalyst for hydrogen sulfide-sulfur
dioxide reaction, 457
Alumina-gel balls, dehydration with, 641
physical properties of, 636, 643
Aluminum, corrosion of, in ethanolamine
solutions, 123
Aluminum sulfate solution, 371
Amine recovery, glycol wash for, 47
water wash for, 46, 47
Amines, aromatic, sulfur dioxide recovery
with, 378
(see also Dimethylaniline; Toluidine;
Xylidine)
Amisol process, 871
Ammonex process, 164–165
Ammonia, in coal gases, 552
distillation of, 557, 558, 576
heat of vaporization of, from
copper-ammonium-salt solutions, 796, 797
heats of reaction of, with hydrogen sulfide
and carbon dioxide, 150, 151
heats of solution of, 567
Henry coefficient for, in pure water, 152
losses of, from copper-ammonium-salt
solutions, 804–808
reactions with hydrogen sulfide and carbon
dioxide, 150, 151

recovery of, as ammonium salts, 554
removal of hydrogen sulfide with, by
    Katasulf process, 467–470
separation from carbon dioxide by
    Chemo-Trenn process, 164
vapor pressure of, 151–157
    over ammonia-sulfur dioxide-water
        solutions, 361
    over aqueous solutions, 565
Ammonia, separation from hydrogen sulfide
    by Ammonex process, 164
Ammonia absorption by water, in packed
    towers, 568
    in spray towers, 575
Ammonia removal, 554–576
    by direct process, 559
    by indirect process, 556–559
    with pickle liquor, 562
    under pressure, 562
    by refrigeration, 562
    by semidirect process, 559–561
    by USS Phosam process, 562
Ammonia solutions, boiling point diagram of,
    565
    carbon dioxide removal with, 177–180
    effect of hydrogen cyanide on, 175, 176
    equilibrium with coke oven gases, 566
    heat of absorption of sulfur dioxide in, 362,
        367
    hydrogen sulfide removal with, 159–177
    recovery of sulfur dioxide with, in
        heat-regenerative process, 363, 364
    regeneration of, 176, 177
    selective hydrogen sulfide removal with,
        159–177
    specific gravity of, 564
    sulfur dioxide recovery with, 359–370
    vapor pressure of ammonia, hydrogen
        sulfide, carbon dioxide, and water over,
        151–157
Ammonia-sulfur dioxide-water solutions, pH
    of, 363
Ammonium sulfate, production by
    Katasulf process, 467–470
    in United States, 557
    saturator for production of, 560
    solubility of, in water, 568
Ammonium sulfate solutions, physical
    properties of, 568
Ammonium sulfide, sulfur extraction with, 442
Ammonium sulfite solutions, removal of
    sulfur dioxide with, 359–369

AMOCO cold bed absorption process, 449
Appleby-Frodingham process for hydrogen
    sulfide and organic sulfur removal, 479
Aqueous Carbonate process for sulfur dioxide
    removal, 389–393
Aqueous Aluminum Sulfate Process, 371
Arsenic trioxide, use of, in
    Giammarco-Vetrocoke process, 238
ASARCO process for sulfur dioxide recovery,
    382–386
    materials of construction in, 386
Autopurification process for hydrogen sulfide
    removal, 512–515

BASF (MDEA) process for carbon dioxide
    removal, 42
Battersea process for sulfur dioxide removal,
    307
Bauxite, activiated (see Activated bauxite)
Beavon Sulfur Removal Process, 739–743
Benzene, capacity of activated carbon for, 696
    heat of adsorption on activated carbon, 695
    solubility of, in petroleum oil, 833
    (see also Light oil)
Benzol removal, by adsorption, 652
    from coal gas, 821–836
Binax process, 263
Bog ore, use of, in iron oxide process, 423
Boiling-point diagram for ammonia solutions,
    565
Boiling points of gases, 816
Break point of adsorbents, 646
British Gas Council processes for organic
    sulfur removal, 733
Burkheiser process for hydrogen sulfide
    removal, 488

Calcium chloride, dehydration with, 617–621
    dew point of gases in contact with solutions
        of, 618
    pellets, dehydrations with, 618–620
Carbon, activated (see Activated carbons;
    Active carbons)
Carbon dioxide, adsorption isotherms on
    molecular sieve, 671
    catalytic conversion of, 759–762
    (see also Methanation)
    corrosion of steel in ethanolamine process
        by, 110, 111
    heat of reaction, with ammonia, 151
        with ethanolamines, 79

with potassium carbonate, 231
heat of vaporization of, from
    copper-ammonium-salt solutions, 796
ionization constant of, 183
rate of absorption, with hydrogen sulfide in
    sodium carbonate solutions, 192
in potassium carbonate solutions, 226–230
in sodium carbonate, bicarbonate, and
    hydroxide, 183, 184
reactions with ammonia, 150, 151
separation from ammonia by Chemo-Trenn
    process, 164
solubility of, in methanol, 820
in triethylene glycol, 616
in water, 253, 254
vapor pressure of, 815
    over ammonia solutions, 151–157
    over diethanolamine solutions, 54, 55
    over ethanolamine solutions, 51–65
    over monoethanolamine solutions, 51–53
    over potassium carbonate solutions,
        217–225
    over sodium carbonate-bicarbonate
        solutions, 184
Carbon dioxide absorption, by diethanolamine
    solutions in packed columns, 86
by ethanolamine-solutions with hydrogen
    sulfide, 49–75, 89–91
by monoethanolamine solutions, in packed
    columns, 82–86
in plate columns, 86–89
by water, in packed columns, 249–257
in plate columns, 257–260
Carbon dioxide removal, with alkali carbonate
    solutions, 182–196
with ammonia solutions, 148–181
cost of, 178
by BASF (MDEA) process, 42
by Binax process, 263
by CNG process, 818
with copper-ammonium-salt solutions,
    799–800
economics of, 234, 235, 862
by Estasolvan process, 861–864
by ethanolamine process, 29
by Fluor Solvent process, 845–851
by Giammarco-Vertrocoke process, 238–244
by methanation, 759–767
by membraine permeation, 872
with molecular sieves, 669–677
with organic solvents, 840–871
with potassium carbonate solutions, 182, 211

by Purisol process, 851–856
by Rectisol process, 821–828
by Selexol process, 856–861
by Sepasolv MPE process, 865
with sodium carbonate solutions, 182–196
with sodium hydroxide solutions, 182–196
by Sulfinol process, 867–871
with water, 249–265
Carbon disulfide, capacity of activated carbon
    for, 695
effect on ethanolamines, 38, 39, 102
heat of adsorption on activated carbon, 695
hydrogenation of, 725
hydrolysis in Claus process, 456
oxidation of, 465
solubility of, in petroleum oil, 833
as sulfur solvent in iron oxide process, 441
(see also Organic sulfur)
Carbon disulfide removal, from coal gas, 829,
    837
Carbon monoxide, catalytic conversion of,
    755–762
(see also Methanation; Shift conversion for
    carbon monoxide removal)
equilibrium with hydrogen and nitrogen,
    817
heat of solution of, in
    copper-ammonium-salt solutions, 797
latent heat of, 816
solubility of, in copper-ammonium-salt
    solutions, 790–795
in water, 250
specific heat of, 816
vapor pressure of, 815
    over copper-ammonium-salt solutions,
        790–795
Carbon monoxide removal, by condensation,
    810–818
by Cosorb process, 808
with copper-ammonium-salt solutions,
    787–809
by L'Air Liquide process, 813–815
by Linde process, 811–813
by methanation, 759–762
by nitrogen wash process, 809–818
by shift conversion, 755–759
Carbon tetrachloride activity of active carbons,
    688
Carbon tetrachloride retentivity of active
    carbons, 688
Carbonyl sulfide, effect of, on ethanolamines,
    39–41, 100

formation of, by molecular sieve
  adsorbents, 654
hydrogenation of, 725
hydrolysis of, 456, 725
oxidation of, 465
(see also Organic sulfur)
Carbonyl sulfide removal, from carbon
  dioxide, 738
  with hot carbonate solutions, 230
  from hydrocarbon gases, 738
  from synthesis gases, 737
Carboxylic acids, 319–321
Carl Still process, 171
Carpenter-Evans process for organic sulfur
  removal, 724
Cataban process, 518
Catacarb process, 236–238
Catalysis in gas purification, 717–783
Catalysts, for acetylene hydrogenation,
  763–767
  for Catalytic Combustion Corporation
    process, 772
  for Deoxo processes, 771
  for gas purification by oxidation and
    reduction, 767
  for hydrogen sulfide oxidation, 469, 519
  for methanation, 759
  for Mine Safety Appliance Company
    process, 779
  for organic sulfur conversion, 721–755
  for organic sulfur oxidation, 469, 471, 475
  for Oxycat process, 771
  for shift conversion, 755
Catalytic Combustion Corporation process,
  772
Catalytic conversion of organic sulfur
  compounds, 721–755
Catalytic/IFP/CEC Ammonia scrubbing
  process, 369–371
Catalytic oxidation of hydrocarbons, minimum
  temperature for, 769
  process for, 767–773
Catalytic oxidation for sulfur dioxide removal,
  409–411
Catalytic reactors, design of, 24, 25
Catalytic reduction, process for, 767–773
Cat-Ox process for sulfur dioxide removal,
  409–411
Caustic wash for carbon dioxide removal,
  184, 185
Chemical conversion, 24
(see also Catalysis)

Chemo-Trenn process for separation of
  ammonia and carbon dioxide, 164
Chiyoda thoroughbred 101 process for sulfur
  dioxide removal, 374
Chiyoda Thoroughbred 121 System, 333
Chlorine, absorption of, in water, 291–296
  corrosion during, 296
  rate of, 294
  reaction of, with water, 292
Chlorine, solubility of, in water, 292
Chromium-aluminum oxide catalysts, 736
Citrate process for sulfur dioxide removal,
  356, 357
Claus process for hydrogen sulfide
  conversion, 451–464
Claus process, tail gas purification, 446–451,
  739–753
CNG process, 818
Coal gas, impurities in, 149
  light oil content of, 829
  naphthalene removal from, 736–740
  organic sulfur removal from, 724–735
  removal of light oil from, 829–836
Cobalt-molybdate catalysts, 739, 740
Coke oven gas, purification of, 148
Collin process for selective hydrogen sulfide
  removal, 167, 177
Columns, absorption, for fluorides, 273–282
  calculation of diameter of, 16–21
  height of, 3–16
  Kittel tray, for selective hydrogen sulfide
    removal, 173
  packed, absorption of fluorides by water in,
    273
    absorption of carbon dioxide, by
      ethanolamine solutions in, 82–89
    absorption of carbon dioxide by water in,
      249
    calculation of height, 3–16
    flooding correlations, 16–21
    for selective hydrogen sulfide removal, 171
  plate, absorption of carbon dioxide, by
    monoethanolamine solutions in, 86–89
    by water in, 257–260
  relative cost of, 6
  selection of, 4, 5
  selective hydrogen sulfide removal in, 171
  spray, for selective hydrogen sulfide
    removal, 173
  stripping, in ethanolamine plants, 91–99
  types of, 2–6
(see also Towers)

Cominco process for sulfur dioxide recovery, 364–368
  absorption step, 365
  stripping step, 368
Conversion, catalytic (*see* Catalytic conversions)
Copper, solubility of, in copper-ammonium-salt solutions, 798
Copper-ammonium-salt process for carbon monoxide removal, 787–808
  corrosion in, 808
Copper-ammonium-salt solutions, absor-coefficients for carbon monoxide in, 801
  components of, 791, 792, 794
  of copper in, 798
  heat capacity of, 797
  heat of solution of carbon monoxide in, 797
  heat of vaporization of components from, 796
  regeneration of, 803, 804
  solubility, of carbon monoxide in, 791
  vapor pressure, of carbon monoxide over, 792–796
Copper-chromium-aluminum oxide catalysts, 736, 738
Copper-chromium-vanadium oxide catalysts, 737
Copper oxide process for sulfur dioxide removal, 400
Copper sulfide-chromium oxide catalysts, 743
Corrosion, in absorption of chlorine by water, 296
  in ammonia solutions for hydrogen sulfide and carbon dioxide removal, 177, 180
  in copper-ammonium-salt process for carbon monoxide removal, 808
  in dehydration with lithium halide solutions, 625
  in ethanolamine process, 110–125
    control of, 116
  in Ferrox process, 492
  in fluoride removal, 285
  in glycol dehydration process, 612–614
  in hot potassium carbonate process, 233
  in hydrogen chloride absorption, 291
  in Thylox process, 503
  in water-wash process, for carbon dioxide removal, 263
    for hydrogen sulfide removal, 267
    inhibitors in amine systems, 118
Cosorb process, 808

Cost, of carbon dioxide removal, with ammonia solutions, 178
  of columns, 6
  by hot potassium carbonate process, 236
  by various processes, 862
  of ethanolamines, 37
  of hydrogen sulfide removal, by Ferrox process, 490
  by Thylox process, 499
  by vacuum carbonate process, 197, 198
  natural gas dehydration equipment, 635
  of sulfur dioxide removal, 308
Cuprous-ammonium-salt solutions (*see* Copper-ammonium-salt solutions)
Cyclic lime process for sulfur dioxide removal, 307

Dehumidification (*see* Dehydration)
Dehydration, by adsorption, 632–669
  adsorbent regeneration in, 654–656
  adsorbents for, 636
  dehydrator design for, 645
  flow systems in, 662
  operating practices in, 667
  pressure drop in, 660
  tower dimensions for, 657
  of air, by Kathabar system, 625
  with calcium chloride, 617–621
    pellets, 619
  with glycols, 587–617
    flow schemes for, 587
  with lithium halide solutions, 621–627
  of natural gas, cost of equipment for, 635
  by Rectisol process, 821
  with saline brines, 617–627
Density of ammonium sulfate solutions, 568
  (*see also* Specific gravity)
Deoxo processes, 771
Desiccants, solids (*see* Adsorbents; Dehydration)
Desorption, definition, 2
Dew point of gases in contact with calcium chloride solutions, 618
  in contact with diethylene glycol solutions, 595
  in contact with triethylene glycol solutions, 595
Dew-point depression, 583
  in commercial glycol dehydrators, 602
Diammonium phosphate, 561, 566
  heat of formation of, 566
DIAMOX process, 168

Dichromate solutions, hydrogen sulfide
  removal with, 545
Diethanolamine, 38–41
  degraduation of, 129–140
  irreversible reaction with carbon dioxide of,
    129–136
  reaction with carbonyl sulfide of, 133
  vapor pressure of, over aqueous solutions,
    128
Diethanolamine solutions, carbon dioxide
    absorption by, in packed columns, 86
  corrosion in, 110–125
  heat capacity of, 72
  rate of absorption, of carbon dioxide in,
    86–89
    of hydrogen sulfide in, 89–91
  specific gravity of, 68
  vapor pressure of carbon dioxide over, 54, 55
  vapor pressure of H₂S over, 60
  viscosity of, 70
Diethylene glycol, dehydration with, 587–617
  physical properties of, 586
  vapor pressure of, 596, 599
Diethylene glycol solutions, dew points of
    gases in contact with, 595
  freezing points of, 600
  solubility of natural gas in, 615
  specific gravity of, 597
  specific heat of, 598
  vapor/liquid equilibrium diagrams of, 600
  vapor pressure of, 599
  viscosity of, 598
Diglycolamine, 29, 32, 37, 40, 103
Diglycolamine solutions, density of, 68
  heat capacity of, 73
  heat of reaction with H₂S and CO₂, 79
  viscosity of, 71
Diisopropanolamine, 29–31, 41
  use of in Sulfinol process, 867
  (see also Adip)
Diisopropanolamine solutions, corrosion by,
    113, 117
  heat of reaction with H₂S and CO₂, 79
  vapor pressure of H₂S over, 62
Dimethylaniline, properties of, 379
  solubility of sulfur dioxide in, 380
  sulfur dioxide recovery with, 382–386
Double alkali process for sulfur dioxide
    removal, 341–350
Dowa Dual Alkali process, 371
Dry-box purification (see Iron oxide process
    for hydrogen sulfide removal)

El Copper Sulfate process (Cuprosol), 547
Econamine process, 40, 103
Economics of carbon dioxide removal
    processes, 235–237
Ejectors, absorption of fluorides in, 276
Electrolytic Regeneration process for sulfur
    dioxide removal, 357
Equilibrium, vapor/liquid, of
    monoethanolamine solutions, 138
Equilibrium constants, for hydrogenation, of
    acetylene, 764
  of carbon disulfide, 725
  of carbonyl sulfide, 725
  for hydrolysis of carbon disulfide, 725
  of carbonyl sulfide, 725
  for methanation, 761
  for shift conversion, 758
Equilibrium vapor pressure (see Vapor
    pressure)
Estasolvan process, 861–864
Ethane, boiling point of, 816
  latent heat of, 816
  vapor pressure of, 815
Ethanolamine process, 29–145
  absorber, diameter, 81
    height, 81–91
    operating data, 99–104
  absorption columns for, 49
  basic flow scheme of, 45
  chemical losses in, 127–140
  corrosion in, 110–125
  heat exchanger-absorber for, 50
  solution rate estimation in, 50, 61
  split stream cycle for, 48
  stripping column design for, 91–99
  temperatures, 75, 80
Ethanolamines, cost of, 37
  effect of gas impurities on, 35
  heats of reaction of hydrogen sulfide and
    carbon dioxide with, 79
  pH of, during neutralization, 34
  vapor pressure of, 37
  (see also Diethanolamine;
      Methyl-diethanolamine;
      Monoethanolamine; Triethanolamine;
      Diglycolamine; Diisopropanolamine;
      Adip)
Ethanolamine solutions, absorption of
    hydrogen sulfide and carbon dioxide in,
    81–91
  degradation of, 129–140
  foaming of, 125

purification of, 136–140
selection of, 35–45
solubility of methane in, 143–145
vapor pressure, of carbon dioxide over, 51–56
of hydrogen sulfide over, 57–65
Ethylene, boiling point of, 816
latent heat of, 816
vapor pressure of, 815
Ethylene glycol, dehydration with, 591–594
physical properties of, 586
vapor pressure of, 599
Ethylene purification with molecular sieve, 676

Feld process for hydrogen sulfide removal, 484
Ferrox process for hydrogen sulfide removal, 489
Film coefficients, 3–7
Filtration of ethanolamine solution, 136
Fire-hazard properties of organic solvents, 692
Fischer process for hydrogen sulfide removal, 511
Flexitray plates, 2
Flooding of packed towers, 16–21
Flue gas, sulfur dioxide concentration in, 303
Fluidized-bed active carbon adsorption process, 701
Fluidized-bed iron oxide process (Appleby-Frodingham process), 479
Fluor Econamine process, 40, 103
Fluorides, absorbers for, 273–282
absorption of, in ejectors, 276
Fluosilicic acid solutions, pH of, 273
in packed columns, 274
rate of, in water, 273–287
disposal of absorbed, 287
removal with water, 268–288
materials of construction for, 285
operating data for, 282–285
in spray towers, 273, 281
vapor pressure of silicon tetrafluoride over, 271
in Venturi scrubbers, 276, 277
in water, 273–287
Fluor Solvent carbon dioxide removal process, 845–851
Fluor Solvent hydrogen sulfide removal process, 845–851
Foaming of ethanolamine solutions, 125

Formic acid in copper-ammonium-salt solutions, 792
Freezing points of glycols, 600
F.S. process for purifying coke oven gas, 563
Fumaks process, 539

Gastechnik process for hydrogen sulfide removal, 429
Giammarco-Vetrocoke process, for carbon dioxide removal, 238–244
for hydrogen sulfide removal, 505–509
Girbotol process (see Ethanolamine process)
Gluud combination process for hydrogen sulfide removal, 485
Gluud process for hydrogen sulfide removal, 493
Glycol-amine process for carbon dioxide and hydrogen sulfide removal, 36
absorber operating data for, 100
corrosion in, 114–116
Glycol-amine solutions, purification of, 139, 142
solubility of methane in, 143
Glycol dehydration in air conditioning, 591
Glycol dehydration process, absorber design for, 599–605
corrosion in, 612–614
dew-point depression in, 601
flow schemes for, 588–591
foaming in, 614
glycol loss in, 614
plant operation in, 610–617
regenerator design for, 605–609
Glycol injection process, 591, 609
Glycol wash for amine recovery, 47

Heat of adsorption, of benzene on activated carbon, 695
of carbon disulfide on activated carbon, 695
of organic compounds on activated carbon, 695, 696
Heat capacity, of copper-ammonium-salt solutions, 797
of diethanolamine solutions, 72
of monoethanolamine solutions, 72
of potassium carbonate solutions, 621
(see also Specific heat)
Heat of formation of diammonium phosphate, 566
Heat of fusion of lithium halides, 621

Heat of reaction, of ammonia with hydrogen
    sulfide and carbon dioxide, 151
  of carbon dioxide with potassium carbonate,
    231
  of ethanolamines with carbon dioxide and
    hydrogen sulfide, 79
  in hydrogenation of acetylene, 764
  of hydrogen sulfide with iron oxide, 436
  of hydrogen sulfide with oxygen, 456
  in methanation, 761
  in shift conversion, 758
  of xylidine with sulfur dioxide, 379
Heat of solution, of ammonia, 567
  of carbon monoxide in
    copper-ammonium-salt solutions, 796
  of water in glycols, 586
Heat of vaporization of components from
    copper-ammonium-salt solution, 796
  of glycols, 586
  of hydrofluoric acid, 270
Heat-transfer coefficients for air dehydration
    with lithium chloride solutions, 625
Height of transfer units (see Transfer units)
Henry's law constants for chlorine in water,
    294
Hi-Pure process for hydrogen sulfide removal,
    214
Hitachi process for sulfur dioxide removal,
    405
Holmes-Maxted process for organic sulfur
    removal, 726-733
Hopcalite catalyst, 780
Hot activation process (see Vacuum carbonate
    process for hydrogen sulfide removal.)
Hot potassium carbonate process for carbon
    dioxide removal, 211-236
  corrosion in, 233, 234
  economics of, 236
  materials of construction in, 233
  solution regeneration in, 231, 232
  solution strength for, 214
  split-stream operation of, 213, 214
  (see also Potassium carbonate)
Huff catalysts, 737, 738
Hydrates, 582-584
Hydrocarbon gases, removal of organic sulfur
    from, 738, 753
Hydrocarbons, adsorption isotherms for, on
    activated carbon, 687, 688
  catalytic oxidation of, 767-775
Hydrochloric acid, removal of, with water,
    288-291

Hydrochloric acid solutions, heat of solution
    of hydrogen chloride in, 289
  vapor pressure of hydrogen chloride over,
    289
Hydrofluoric acid, heat of vaporization of,
    270
  ionization constant of, 270
  removal of, with water, 268-288
  vapor pressure of, over solutions, 270-272
Hydrogen, boiling point of, 816
  equilibrium with carbon monoxide and
    nitrogen, 817
  latent heat of, 816
  solubility of, in water, 250
  specific heat of, 816
  vapor pressure of, 815
Hydrogen chloride, heat of solution of, in
    hydrochloric acid solution, 289
  vapor pressure over hydrochloric acid
    solutions, 289
  (see also Hydrochloric acid)
Hydrogen chloride removal with water,
    288-291
  absorber design, 289-291
  materials of construction for, 291
Hydrogen cyanide, in coal gases, 552
  effect on ammonia solutions of, 173-175
Hydrogen cyanide removal, by iron oxide
    process, 441
  by iron oxide suspension, 487
  by Katasulf process, 467
  by Perox process, 520
  by Scaboard process, 189
  by Staatsmijnen-Otto process, 512-515
  by Stretford process, 525
  by Thylox process, 502
  by vacuum carbonate process, 189-192
Hydrogen fluoride (see Fluorides;
    Hydrofluoric acid)
Hydrogen sulfide, adsorption isotherms for, on
    molecular sieves, 671
  corrosion of steel in ethanolamine process
    by, 110-120
  heat of reaction, with ammonia, 150, 151
    with ethanolamines, 79
    with iron oxide, 436
    with oxygen, 456
  ionization constant of, 183
  oxidation of, 436, 506
  rate of absorption of, in diethanolamine
    solutions, 89-91
    in monoethanolamine solutions, 89

in potassium carbonate solutions, 192
in sodium carbonate solutions, 192
in thioarsenate solutions, 505
reactions of, with ammonia, 150, 151
solubility of, in water, 250
vapor pressure of, over Adip solutions, 62
    over Alkacid solutions, 206–209
    over ammonia solutions, 151–155
    over diethanolamine solutions, 60
    over monoethanolamine solutions, 56–58
    over potassium carbonate solutions, 222
    over tripotassium phosphate solutions, 201
Hydrogen sulfide absorption, in ethanolamine
    solutions, 29–145
in potassium carbonate solutions, 213–236
Hydrogen sulfide removal, by activated carbon
    process, 442–446
by Alkacid process, 203
by alkanolamines 29–145
by Amisol process, 871
with ammonia solutions, 148–181
by Appleby-Frodingham process, 479
by Autopurification process, 512
by Burkheiser process, 488
by Cataban process, 518
by CBA process, 449
by DIAMOX process, 168
with dichromate solutions, 545
by dry oxidation process, 420–479
by EIC copper sulfate process, 547
by Estasolvan process, 861
by ethanolamine process, 29–145
by Feld process, 484
by Ferrox process, 489–493
by Fischer process, 511–512
by F-S process, 563
by Fumaks process, 539
by Gastechnik process, 429
by Gas/Spec solvents, 30, 119
by Giammarco-Vetrocoke process, 505–509
by Gluud combination process, 485
by Gluud process, 493
by Hi-Pure process, 214
by IFP Process, 542–544
by iron complex solutions, 515–519
by iron cyanide complexes, 509–515
by iron oxide process, 421–442
with iron oxide suspension, 486
by Jefferson Lake Sulfur Company process,
    464
by Katasulf process, 467–470
by Konox process, 519

by Koppers C.A.S. process, 485
by Lacey-Keller process, 546
by liquid oxidation processes, 483–547
by Lo-Cat process, 521
by Manchester process, 494
by Methylcyanoacetate process, 864
with molecular sieves, 677–681
with organic catalyst, 519–539
with organic solvents, 840–871
by oxidation to oxides of sulfur, 465–479
by oxidation to sulfur, 420–465
with permanganate solutions, 545
by Perox process, 520
with polythionate solutions, 484–486
by Purisol process, 851–856
by Rectisol process, 821–827
by Seaboard process, 187–189
selective, with ammonia solutions, 159
    by Collin process, 167
    with methyldiethanolamine, 41
    by Selexol process, 856–861
    by Sepasolv MPE process, 865–867
    by tripotassium phosphate process,
        196–203
by Selectox process, 743–749
with sodium carbonate solutions, 187–196
by Sodium phenolate process, 203
by Staatsmijnen-Otto process, 512–515
with sterically hindered amines, 30, 43
by Stretford process, 521–536
by Sulfiban process, 104
by Sulfinol process, 867–871
by Sulfint process, 516, 517
by Sulfreen process, 446–449
by Takahax process, 538
with thioarsenate solutions, 497–509
by Thylox process, 497–505
in Thyssen-Lenze towers, 427
by Townsend process, 539
by tripotassium phosphate process, 196
by UCARSOL process, 30, 42, 119
by Unisulf process, 536
by vacuum carbonate process, 189–196
with water, 265–268
by Wierwiorowski process, 545
Hydrogenation, of acetylene, 762–767
    catalysts for, 763
    heats of reaction in, 764
    of carbon disulfide, 725
    of carbonyl sulfide, 725
Hydrogen sulfide separation from ammonia by
    Ammonex process, 164

Hydrolysis, of carbon disulfide, 725
    of carbonyl sulfide, 725
B.B′-Hydroxyaminoethyl ether, 29

IFP process for hydrogen sulfide removal,
    542–544
Indirect process for ammonia removal,
    556–559
Inert gas purification with molecular sieves,
    674
Intalox saddles, 2
Ionization constant, of carbon dioxide, 183
    of hydrofluoric acid, 270
    of hydrogen sulfide, 183
Iron complex solutions for hydrogen sulfide
    removal, 515–519
Iron-cyanide complexes for hydrogen sulfide
    removal, 509–515
Iron oxide, catalysts, 736
    heat of reaction of hydrogen sulfide with,
        436
    pressure drop through beds of, 436
    properties of, 424
Iron oxide boxes, 421–442
Iron oxide process for hydrogen sulfide
    removal, 421
    basic chemistry of, 422
    bed size design for, 432
    continuous, 388
    heat effects in, 436
    high-pressure operation of, 431
    hydrogen cyanide removal in, 441
    nitric oxide removal in, 441
    operation of, 438
    pressure drop in, 436
    purifier types, 425
    sulfur recovery in, 441
    tower purifies for, 427
    treating efficiency of, 435
Iron oxide suspensions for hydrogen sulfide
    removal, 486–488

Jefferson Lake Sulfur Company process for
    hydrogen sulfide removal, 464

Katasulf process for hydrogen sulfide removal,
    467–470
Kathabar air dehydration systems, 569
Kittel plates, 2
    carbon dioxide absorption in water with, 260

    selective hydrogen sulfide removal with, 173
Konox process for hydrogen sulfide removal,
    519
Koppers C.A.S. process for hydrogen sulfide
    removal, 485

Lacey-Keller process for hydrogen sulfide
    removal, 546
L'Air Liquide process for carbon monoxide
    and methane removal, 813
Latent heats of gases, 816
Light oil, composition of, 829
    solubility of constituents of, in wash oils,
        833
Light oil removal from coal gas, 829–834
    mass transfer in, 834
    partition coefficient for, 831–835
    properties of wash oils for, 832
Lime, use for sulfur dioxide absorption,
    307–332
Lime injection with wet scrubbing for sulfur
    dioxide removal, 307
Lime/limestone process for sulfur dioxide
    removal, 307–332
Lime slurry spray dryer process for sulfur
    dioxide removal, 393
Limestone injection process for sulfur dioxide
    removal, 397
Linde molecular sieves (see Molecular sieves)
Linde process for carbon monoxide and
    methane removal, 811
Lithium bromide, solubility of, in water, 623
Lithium bromide solutions, vapor pressure of,
    622
Lithium chloride, solubility of, in water, 623
Lithium chloride solutions, air-conditioning
    dehydration with, 621–627
    rate of absorption of water vapor in, 625
    vapor pressure of, 622
    water content of air in contact with, 623
Lithium halides, dehydration with 621–627
    physical properties of, 621
Lo-Cat process for hydrogen sulfide removal,
    521
Low-temperature processes for gas
    purification, 809–821

Magnesium Oxide Process for sulfur dioxide
    absorption, 334–340
Maloney process for gas dehydration, 656

Manchester process for hydrogen sulfide removal, 494
    absorber design for, 495
Manganese dioxide-copper oxide catalysts, 780
Mass transfer, in light oil removal, 834
    in solid bed dehydration, 657
Materials of construction, for absorption of chlorine in water, 276
    for ASARCO process for sulfur dioxide recovery, 386
    for carbon dioxide removal with ammonia solutions, 180
    for ethanolamine process, 110–125
    for Ferrox process, 492
    for fluorides removal, 285
    for hot potassium carbonate process, 233
    for hydrogen chloride absorption, 289
    for hydrogen sulfide removal with water, 267
    for hydrogen sulfide removal with ammonia solutions, 177
    for lime/limestone process, 332
    for Thylox process, 502
    for vacuum carbonate process, 196
    for water-wash process for carbon dioxide removal, 263
    for Wellman-Lord process, 356
MCRC Sulfur Recovery process, 451
Membrane permeation process, 872–875
Mercaptan removal from synthesis gases, 739
Mercaptans, removal by Sulfinol process, 869
Mercaptans (see Organic sulfur)
Metal impregnated carbon, 754
Methanation for removal of oxides of carbon, 759–762
    catalysts for, 760
    equilibrium constants for, 761
    heats of reaction in, 761
Methane, boiling point of, 816
    latent heat of, 816
    solubility of, in monoethanolamine solutions, 143
    in water, 250
    specific heat of, 816
    vapor pressure of, 815
Methane removal, by condensation, 810
    by L'Air Liquide process, 813
    by Linde process, 811
    by nitrogen wash process, 810–818
Methanol, acid gas removal with, 821–828
    (see also Rectisol process)

injection for hydrate control, 594
solubility of carbon dioxide in, 820
vapor pressure of, 822
Methylacetylene, removal of, 768
Methylcyanoacetate process for acid gas removal, 864
Methyldiethanolamine, 41, 75
    selective removal of hydrogen sulfide with, 41, 75
    solutions, corrosion in, 121
Methylene chloride, removal from air, 694
Mine Safety, Appliance Company process, 779–783
Minute service of active carbons, 689
Molecular sieves, adsorption isotherms of hydrogen sulfide, carbon dioxide on, 670, 671
    capacity for water of, 637, 646
    carbon dioxide and water removal with, 676
    dehydration with, 586–669
    ethylene purification with, 676
    gas purification with, 669–685
    hydrogen sulfide removal with, 677–681
    inert gas purification with, 674
    natural gas purification with, 677–681
    physical properties of, 636, 644
Molten Carbonate process for sulfur dioxide removal, 386–388
Molybdenum sulfide catalysts, 734
Monoethanolamine,
    with carbonyl sulfide, 133
    degradation of, 129–136
    irreversible reaction of, with carbon dioxide, 130
    reactions of, with carbon dioxide and hydrogen sulfide, 33
    recovery of vaporized, by adsorption, 127, 128
    vapor presure of, over aqueous solutions, 128
Monoethanolamine solutions,
    corrosion in, 110–125
    heat capacity of, 72
    purification of, 125, 136
    rate of absorption, of carbon dioxide in, 82–86
    solubility of methane in, 143
    specific gravity of, 67
    stripping of, 94–99
    viscosity of, 70
Monoethanolamine-water, vapor/liquid composition diagram, 138
Murphree plate efficiency, 13

Napththalene effect of on vacuum carbonate process, 195
  removal of, from coal gas, 836–841
  solubility of, in petroleum oil, 833
  vapor pressure of, 841
Natural gas, solubility of, in glycols, 615
  water vapor content of saturated, 584
Nickel alloys, corrosion of, in ethanolamine solutions, 110–125
Nickel-aluminum-oxide catalysts, 760
Nickel-cobalt-chromium catalysts, 765
Nickel subsulfide catalysts, 724
Nitric oxide in coal gases, 552
Nitric oxide removal, by catalytic reduction, 773
  by CEC process, 371
  by iron oxide process, 441
  by Pura Siv-N process, 674
Nitrogen, boiling point of, 816
  in carbonization products, 552
  equilibrium with hydrogen and carbon monoxide, 817
  latent heat of, 816
  solubility of, in water, 250
  specific heat of, 816
  vapor pressure of, 815
Nitrogen compounds in coal gases, 552
Nitrogen oxides, catalytic removal of, 773
Nitrogen wash process for carbon monoxide and methane removal, 810–818
Nittetsu process, 535
N-methyl-2 pyrrolidone (NMP), physical properties of, 853
  use of in Purisol process, 851–856
North Thames Gas Board process for organic sulfur removal, 470–475

Odorants, characteristics of, 709
Odor removal by adsorption, 705–712
Organic catalysts for hydrogen sulfide removal, 519–539
Organic compounds, capacity of activated carbon for, 695
  heat of adsorption on activated carbon, 695
Organic solvents, carbon dioxide and hydrogen sulfide removal with, 840–871
Organic sulfur, catalytic conversion of, 721–754
  content of gases, 721, 722
  (see also Carbon disulfide; Carbonyl sulfide; Thiophene)

Organic sulfur removal, by activated carbon, 445, 446
  by Appleby-Frodingham process, 479
  by British Gas Council processes, 733
  by Carpenter, Evans process, 724
  by catalytic conversion, 721–754
  from coal gases, 724–735
  by Holmes-Maxted process, 726–733
  from hydrocarbon gases, 738
  by North Thames Gas Board process, 470–475
  by Peoples Gas Company process, 726
  by Rectisol process, 821–827
  by soda-iron process, 475–479
  from synthesis gases, 735, 753
Organic vapor adsorption on active carbon, 685–713
Oxidation, catalytic (see Catalytic oxidation of hydrocarbons)
Oxycat process, 771
Oxygen, boiling point of, 816
  latent heat of, 816
  solubility of, in water, 250
  specific heat of, 816
  vapor pressure of, 816

Packed columns (see Columns)
Packed towers (see Towers)
Palladium catalysts, 771
Pall rings, 2
Partial pressure of water over silica gel, 639
  (see also Vapor pressure)
Particulate removal with wet scrubbers, 248
Partition coefficients for light oil removal, 831–833
Peoples Gas Company process for organic sulfur removal, 726
Perchloroethylene as sulfur solvent for iron oxide purifiers, 441
Permanganate solutions for hydrogen sulfide removal, 545
Perox process for hydrogen sulfide removal, 520
pH, of ammonia-sulfur dioxide-water solutions, 362
  of Cominco process solutions, 367, 368
  control in iron oxide purifiers, 437
  of ethanolamine solutions, 34
  of fluosilicic acid solutions, 273
Phenol removal from ammoniacal liquor, 561
Pickle liquor, ammonia, hydrogen sulfide and

hydrogen cyanide removal with, 563
Plate columns (*see* Columns)
Plate efficiency, 13–15
  for carbon dioxide absorption in
    monoethanolamine solutions, 86–89
Plates, design of, 11–21
  theoretical, 11
Platinum catalysts, 735, 771
Polybed PSA adsorption process, 681
Polyethylene glycol dimethyl ether, properties
    of, 858
  solubility of gases in, 858
  use of, in Selexol process, 856–861
Polythionate solutions for hydrogen sulfide
    removal, 484–486
Porosity of adsorbents for gas dehydration,
    636
Potassium bicarbonate, solubility of, in
    potassium carbonate-bicarbonate solutions,
    216
Potassium carbonate, heat of reaction of, with
    carbon dioxide, 231
  solubility of, in potassium
    carbonate-bicarbonate solutions, 216
Potassium carbonate solutions, carbon dioxide
    removal with, 182, 211–238
  heat capacity of, 221
  hydrogen sulfide absorption in, 218
  rate of absorption of carbon dioxide in, 226
  regeneration of, 231
  specific gravity of, 220
  vapor pressure, of carbon dioxide over, 222,
    223
  of hydrogen sulfide over, 222
  of water over, 219, 224
  (*see also* Hot potassium carbonate process)
Potassium N-dimethylglycine (*see* Alkacid
    process)
Potassium permanganate, hydrogen sulfide
    removal with, 545
Potassium sulfite-bisulfite cycle for sulfur
    dioxide removal, 351–356
Power recovery in carbon dioxide removal
    with water, 263
Pressure drop, through activated carbon beds,
    697
  through iron oxide beds, 436
  through solid desiccant beds, 661–663
Pressure swing adsorption, 681
Propylene carbonate, physical properties of,
    846–849
Purasiv HR process, 704

Pura-Siv-N process for NO_X removal, 674
Purisol process, 851–856
Pyridine bases, in coal gases, 576
  in coke-oven gas, 577
  removal of, 576–580
    by Schutt process, 578
Reactors, catalytic, design of, 24, 25
Rectisol process for acid gas removal,
    821–827
Reduction, catalytic (*see* Catalytic reduction)
Reinluft process for sulfur dioxide removal,
    407
Richards sulfur recovery process, 464

Saline brines, dehydration with, 617–627
Saturator for ammonium sulfate production,
    560
Schutt process for pyridine bases removal, 578
SCOT process for hydrogen sulfide removal,
    749
Seaboard process for hydrogen sulfide
    removal, 187
Selectox process, 743–749
Selexol process, 856–861
Semidirect process for ammonia removal,
    559–562
Sepasolv MPE process, 865–867
Shed trays, 2
Shift conversion for carbon monoxide
    removal, 755
  catalysts for, 755
  equilibrium constants for, 758
  heats of reaction for, 758
Silica-base beads, capacity for water of, 640,
    648
  gas dehydration with, 638
  physical properties of, 636
Silica gel, air purification with, 682
  capacity for water of, 636, 639
  chemical analysis of, 639
  equilibrium partial pressure of water over,
    639
  gas dehydration with, 637
  physical properties of, 636
Silicon tetrafluoride, vapor pressure of, over
    fluosilicic acid solutions, 270–272
  (*see also* Fluorides)
SNPA-DEA process, 39
Soda-iron process for organic sulfur removal,
    475–479
  hydrogen sulfide removal in, 478

tower design for, 476
Sodium alanine, 203
  (*see also* Alkacid process)
Sodium bicarbonate solutions, rate of
  absorption of carbon dioxide in, 185
Sodium carbonate, use of, in soda-iron
  process, 475
Sodium carbonate-bicarbonate solutions, vapor
  pressure of carbon dioxide over, 184
Sodium carbonate solutions, carbon dioxide
  removal with, 183–187
  hydrogen sulfide removal with, 187–196
  rate of absorption of carbon dioxide in, 185,
  186
Sodium dichromate, hydrogen sulfide removal
  with, 545
Sodium hydroxide solutions, carbon dioxide
  removal with, 183–187
  rate of absorption of carbon dioxide in, 185
Sodium phenolate process for hydrogen
  sulfide-removal, 203
Sodium sulfite-bisulfite cycle for sulfur
  dioxide removal, 351
Solid-desiccant dehydration process, 632–669
  (*see also* Adsorption; Dehydration)
Solid desiccants (*see* Adsorbents, dehydration)
Solubility, of ammonium sulfate in water, 568
  of benzene in petroleum oil, 833
  of carbon dioxide in methanol, 822
  in water, 253, 254
  of carbon disulfide in petroleum oil, 833
  of carbon monoxide in
    copper-ammonium-salt solutions,
    790–795
  of copper in copper-ammonium-salt
    solutions, 798
  of gases in N-methyl-2 pyrrolidone, 855
  of gases in water, 250, 253, 254, 267, 270
  (*see also* individual gases)
  of light oil constituents in wash oils, 833
  of lithium halides in water, 623
  of naphthalene in petroleum oil, 833
  of natural gas in glycols, 615
  of potassium carbonate-bicarbonate mixtures
    in water, 216
  of sulfur dioxide, in dimethylaniline, 380
    in water, 314
  of toluene in petroleum oil, 833
  of xylene in petroleum oil, 833
    in xylidine-water mixtures, 381
Solvent recovery, by Acticarbone process, 699
  with activated carbon, 689–705

by Purasiv HR process, 704
Solvents, organic, fire-hazard properties of,
  692
Sorbead, 638
  (*see also* silica-base beads)
Space velocity, 25
  for activated carbon hydrogen sulfide
    removal process, 446
  for iron oxide hydrogen sulfide removal
    process, 432
  for soda iron process, 477
Specific gravity, of adsorbents for gas
  dehydration, 636
  of Adip solutions, 69
  of ammonia solutions, 564
  of diethanolamine solutions, 68
  of Diglycolamine solutions, 68
  of glycol-amine solutions, 67
  of glycols, 586, 597
  of glycol solutions, 597
  of monoethanolamine solutions, 67
  of potassium carbonate solutions, 220
Specific heat, of adsorbents for gas
  dehydration, 636
  of gases, 816
  of glycols, 586
  of glycol solutions, 597
Sponge iron oxide, 423
Spray dryer process for sulfur dioxide
  removal, 388–396
Staatsmijnen-Otto process for hydrogen sulfide
  removal, 512–515
Steam distillation of monoethanolamine
  solutions, 137–139
Steels, corrosion of, in ethanolamine process,
  110–125
Sterically hindered amines, 30, 43
Stone & Webster Ionics process for sulfur
  dioxide removal, 357
Stretford process for hydrogen sulfide
  removal, 521–536
Stripping of ammonia solutions used for
  hydrogen sulfide removal, 176
Stripping columns, design of, for ethanolamine
  plants, 91–99
Sulfacid process for sulfur dioxide removal,
  405
Sulfiban process for hydrogen sulfide
  removal, 104
Sulfidine process, for sulfur dioxide recovery,
  380–382
Sulfinol process for acid gas removal, 867–871

Sulfint process, 516
Sulfolane, use of in Sulfinol process, 867-871
Sulfreen process, 446-449
Sulfur recovery of, by Claus process, 451
    by Cold Bed Adsorption (CBA) process, 449
    in iron oxide process, 441
    by MCRC process, 451
    by Sulfreen process, 446
Sulfur dioxide
    adsorption isotherms for, 404
    concentration in flue gases, 303
    heat of absorption of, in ammonia solutions,
        362
    heat of reaction of, with xylidine, 379
    oxidation of, in ammonia solutions, 362
    permissible concentration of, in atmosphere,
        300
    reactions of, with lime/limestone, 311-317
    solubility of, in dimethylaniline, 380
        in pure water, 314
        in xylidine-water mixtures, 381
    vapor pressure of, over ammonia-sulfur
        dioxide-water solutions, 362
Sulfur dioxide removal, 299
    by ABS process, 369
    by Activated Carbon process, 405-409
    by adsorption, 403-409
    with alkali metal compounds, 340-359
    by alkali metal sulfite-bisulfite process,
        351-356
    with alkaline earth compounds, 307-340
    by Alkalized Alumina process, 399
    with ammonia solutions, 359-369
    by Aqueous Aluminum Sulfate process, 371
    by Aqueous Carbonate process, 389-393
    with aromatic amines, 378-380
    by ASARCO process, 382-386
    by Battersea process, 307-310
    by Catalytic/IFP/CEC ammonia scrubbing
        process, 369
    by catalytic oxidation, 409-411
    by Cat-Ox process, 410
    by Chiyoda thououghbred 101 process,
        374-377
    by Chiyoda Thoroughbred 121 process, 333
    by Citrate process, 356
    by Cominco process, 364-368
    by copper oxide process, 400-403
    by Dowa process, 371
    by double alkali process, 341-351
    by dry sorption with chemical reaction, 396
    by Electrolytic Regeneration process, 357

    by Hitachi, activated carbon process, 405
    by lime injection with wet scrubbing, 307
    by lime/limestone process, 307-322
    by lime slurry spray dryer process, 393
    by limestone injection, 397
    by magnesium oxide process, 334-340
    by Molten Carbonate process, 386-388
    by Reinluft process, 407
    by spray dryer process, 388-396
    by Sulfacid process, 405
    by Sulphidine process, 380-382
    by Wellman Lord process, 351-356
    by zinc oxide process, 350
Sulfur dioxide removal process, categorization
    of, 304-306
    cost comparison, 305, 308, 399
Sulfuric acid, use of for sulfur dioxide
    stripping, 368
Sulfur trioxide formation, 302
Super Drizo process for gas dehydration, 588,
    590, 608
Superphosphate plant exhaust gases (see
    Fluorides, removal of)
Synthesis gases, removal of hydrogen sulfide
    and carbon dioxide from, 102
    removal of organic sulfur from, 735, 753
    removal of oxides of carbon from, 754-762,
        787

Takahax process for hydrogen sulfide removal,
    538
Tar acids, 561
Tetraethylene glycol, properties of, 586
Thermal conductivity of adsorbents for gas
    dehydration, 636
Thioarsenate solutions, hydrogen sulfide
    removal with, 497-509
    rate of absorption of hydrogen sulfide in, 505
Thiomolybdate catalysts, 726, 727
Thiophene, removal of, by North Thames Gas
    Board process, 473
    (see also Organic sulfur)
Thiosulfate formation, in iron oxide
    suspensions, 487
    in Stretford process, 527
    in Thylox process, 502
Thylox process for hydrogen sulfide removal,
    497-505
    cost of operation of, 498
    materials of construction for, 502
    modified, 500

Thyssen-Lenze towers for hydrogen sulfide removal, 427
Toluene, solubility of, in petroleum oil, 833
Toluidine, properties of, 379
  sulfur dioxide recovery with, 378
Towers, absorption, types of, 2
  (see also Columns)
  adsorption, dimensions of, for gas dehydration, 657
  bubble-cap, design of, for ethanolamine process, 81
  packed, ammonia absorption in, 568-576
  calculation of height of, 3-16
  in ethanolamine process, flooding of, 81
  flooding of, 16-18
  spray, absorption of fluorides in, 273
  ammonia absorption in, 571
  traytype, design of, 19-21
Townsend proces for hydrogen sulfide removal, 539
Trace impurity removal of adsorption, 705
Trail, sulfur dioxide recovery plants at, 364-368
Transfer units, height of, 10
  for ammonia absorption in packed towers, 571-575
  for carbon dioxide absorption in water, 253-257
  number of, 10, 11
  for ammonia absorption in water in spray towers, 575
Tray diagram for water absorption in glycol, 603
Tray efficiency (see Plate efficiency)
Trays, design of, 19
Tributyl phosphate, physical properties of, 863
  solubility of gases in, 863
  use of in Estasolvan process, 861-864
Triethanolamine, 29, 37, 86
  solutions, carbon dioxide absorption with, in packed columns, 86
Triethylene glycol, dehydration with, 591-617
  physical properties of, 586
  solubility of carbon dioxide in, 616
  solubility of natural gas in, 615
  vapor pressure of, 596, 599
Triethylene glycol solutions, dew point of gases in contact with, 595
  freezing points of, 600
  specific gravity of, 597
  specific heat of, 598
  vapor/liquid equilibrium diagrams for, 601

  viscosity of, 598
Tripotassium phosphate process for hydrogen sulfide removal, 196-203
  steam consumption in, 202
Tripotassium phosphate solutions, capacity of, 201
Tripotassium phoshate solutions, vapor pressure of hydrogen sulfide over, 201
Turbogrid plates, 2
Turbulent contact absorber, 3, 322, 327

UCARSOL Process, 30, 42, 119
Uniflux plates, 2
Unisulf process, 536
USS Phosam process for ammonia removal, 562

Vacuum carbonate process for hydrogen sulfide removal, 189-196
  cost of operation of, 197
  operating problems in, 195
  solution regeneration in, 194
Valve plates, 2
Vapor pressure, of ammonia
  over ammonia-sulfur dioxide-water solutions, 360
  over aqueous solutions, 565
  hydrogen sulfide, carbon dioxide, and water over ammonia solutions, 152-157
  of ammonium sulfate, 568
  of carbon dioxide, over diethanolamine solutions, 54-55
  over monoethanolamine solutions, 51-53
  over potasium carbonate, 222, 223
  over sodium carbonate-bicarbonate solutions, 184
  of carbon monoxide over copper-ammonium-salt solutions, 790-795
  of copper-ammonium-salt solutions, 790-795
  of diethanolamine over aqueous solutions, 128
  of ethanolamines, 37
  of gases at low temperature, 815
  of glycol solutions, 596
  of glycols, 599
  of hydrofluoric acid solutions, 270-272
  of hydrogen chloride over hydrochloric acid solutions, 289

of hydrogen sulfide, over Alkacid solutions,
206, 208
over diethanolamine solutions, 60, 61
over monoethanolamine solutions, 57–59
over potassium carbonate solutions, 222
over tripotassium phosphate solutions, 201
of lithium halide solutions, 622
of monoethanolamine over aqueous
solutions, 128
of naphthalene, 841
of silicon tetrafluoride over fluosilicic acid
solutions, 271, 272
of sulfur dioxide
over ammonia-sulfur dioxide-water
solutions, 360
over magnesium salt solution, 339
of water, over ammonia solutions, 152
over ammonia-sulfur dioxide-water
solutions, 362
over potassium carbonate solutions, 219,
224
Venturi scrubbers, absorption of fluorides in,
274, 276
Viscosity, of Adip solutions, 72
of ammonium sulfate solutions, 568
of DGA solutions, 70
of diethanolamine solutions, 70
of glycol solutions, 598
of glycols, 586
of monoethanolamine solutions, 70

Water absorption of carbon dioxide
in packed columns, 252–257
in plate columns, 257–260
absorption of chlorine in, 291–296
capacity of adsorbents for, 636, 639, 646
equilibrium partial pressure of, over silica
gel, 639
heat of vaporization of, from
copper-ammonium-salt solutions, 795,
796
rate of absorption, of chlorine in, 293
of fluorides in, 273–281

removal of carbon dioxide with, 249–265
plant corrosion in, 263
plant operation, 261
power recovery in, 263
Water, removal of fluorides with, 268–288
removal of hydrogen chloride with,
288–291
removal of hydrogen sulfide with, 265–268
plant corrosion in, 267
solubility, of carbon dioxide in, 253, 254
of gases in, 250, 253, 254, 267, 270
(see also individual gases)
of hydrogen sulfide in, 267
of sulfur dioxide in, 314
vapor pressure of, over ammonia solutions,
152
over ammonia-sulfur dioxide-water
solutions, 362
over potassium carbonate solutions, 219,
224
Water vapor, adsorption of, 632–669
in saturated air, 585
in saturated natural gas, 584
Water-vapor removal (see Dehydration)
Water wash for amine recovery, 46
Wellman Lord process for sulfur dioxide
removal, 351–356
Wet scrubbers, 248
Wiewiorowski process for hydrogen sulfide
removal, 545

Xylene, solubility of, in petroleum oil, 833
Xylidine, heat of reaction with sulfur dioxide,
379
properties of, 379
sulfur dioxide recovery with, 378–382
Xylidine-water mixtures, solubility of sulfur
dioxide in, 381

Zeolites, use of for hydrogen sulfide removal,
465
Zinc oxide process for sulfur dioxide removal,
350